Rapid Detection and Characterization of Foodborne Pathogens by Molecular Techniques

T0187480

Robert E. Levin

CRC Press
Taylor & Francis Group
Boca Raton London New York

CRC Press is an imprint of the
Taylor & Francis Group, an **informa** business

CRC Press
Taylor & Francis Group
6000 Broken Sound Parkway NW, Suite 300
Boca Raton, FL 33487-2742

First issued in paperback 2019

© 2010 by Taylor and Francis Group, LLC
CRC Press is an imprint of Taylor & Francis Group, an Informa business

No claim to original U.S. Government works

ISBN: 978-1-4200-9242-4 (hbk)
ISBN: 978-0-367-38502-6 (pbk)

Library of Congress Cataloging-in-Publication Data

Levin, Robert E., 1930-
  Rapid detection and characterization of foodborne pathogens by molecular techniques
/ Robert E. Levin.
    p. ; cm.
  Includes bibliographical references and index.
  ISBN 978-1-4200-9242-4 (hardcover : alk. paper)
  1. Food--Microbiology. 2. Food--Analysis. I. Title.
  [DNLM: 1. Food Microbiology. 2. Food Analysis--methods. 3. Food
Contamination--prevention & control. 4. Food Poisoning--prevention & control. QW 85
L665r 2010]

  QR115.L38 2010
  664.001'579--dc22   2009031428

Visit the Taylor & Francis Web site at
http://www.taylorandfrancis.com

and the CRC Press Web site at
http://www.crcpress.com

Dedication

This volume is dedicated to the memories of Max Levine, PhD, without whose early inspiration this volume would not have been possible, and Eugene Levine, MD, who departed this world too soon.

Contents

Preface

During the three decades following the development of the polymerase chain reaction (PCR), a significant array of associated techniques has been developed, along with the enormous expansion of Genbank. This has now made it possible to rapidly detect low numbers of all known pathogenic microorganisms in foods, water, and environmental samples without the traditional methods of cultivation and phenotypic characterization, which are labor intensive and require several days to complete. This volume has been written specifically with the intent that it be of significant utility to individuals in the field. Readers unfamiliar with the details of real-time PCR will find that Chapter 1 is comprehensive and presents the theoretical and operational aspects of this method in complete detail. Each of the following chapters deals with a different bacterial pathogen associated with foods.

The more recent development of intercollating dyes, such as ethidium bromide monoazide, has notably expanded the application of PCR to allow the selective amplification of DNA from only viable bacterial cells, which greatly enhances the public health utility of PCR in detecting only viable pathogens. The recent development of loop-mediated isothermal amplification of DNA has allowed the quantitative yield of amplicons to be increased by three orders of magnitude, has eliminated thermal cycling, and has reduced amplification time to 60 minutes.

The purpose of this volume is to serve as a comprehensive presentation of the literature and methodology pertaining to the use of these and other molecular techniques to detect and quantify foodborne pathogenic bacteria, and to characterize isolates of a given species below the species level. Each chapter presents a comprehensive list of DNA primers and probes with respect to conventional and real-time PCR for the convenience of the reader.

The Author

The author, Robert E. Levin, is professor of food microbiology in the Department of Food Science at the University of Massachusetts, Amherst. He obtained his BS degree in biology from Los Angeles State University, his MS degree in bacteriology from the University of Southern California, and his PhD in microbiology from the University of California, Davis. His early academic career involved the isolation and study of obligately psychrotrophic bacteriophages, characterization of seafood food spoilage enzymes of bacterial origin, and fermentative production of microbial gums. His present research program focuses on the conventional and real-time PCR quantification of viable bacterial pathogens in foods.

CHAPTER 1

Molecular Techniques for Detecting, Quantifying, and Subspecies Typing of Foodborne Pathogenic Bacteria

I. THE POLYMERASE CHAIN REACTION (PCR)

A. Introduction

PCR is one of the most powerful analytical techniques ever developed. It allows segments of minute amounts of double-stranded DNA to be amplified several millionfold in several hours. Its most notable application to foods is for the detection of low numbers of foodborne pathogenic and toxigenic bacteria in a wide variety of food products, in addition to confirming the identification of such organisms isolated from food.

B. Requirements for PCR

PCR uses repeated temperature cycling involving template denaturation, primer annealing, and the activity of DNA polymerase for extension of the annealed primers from the 3′-ends of both DNA strands (Figure 1.1). This results in exponential amplification of the specific target DNA sequence. The availability of the thermostable *Taq* DNA polymerase, from the extreme thermophile *Thermus aquaticus,* greatly facilitates repeated thermal cycling at ~95°C for template denaturation without having to repeatedly add a less thermally stabile DNA polymerase after each cycle. The notably high optimum temperature for *Taq* polymerase activity (75–80°C) allows high extension temperatures (72–75°C), which, when coupled with a high annealing temperature (50–65°C) and denaturation at 95°C, increase specificity, yield, and sensitivity of the PCR reaction (Innis and Gelfand, 1990). PCR reactions are usually performed in 0.5-ml or 0.2-ml thin-walled polyethylene PCR tubes containing 50 μl total reaction volume. The availability of second-generation thermal cyclers with heated lids has eliminated the previous need for overlaying the reaction volumes with

1

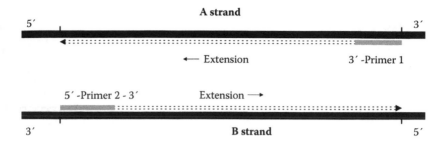

Figure 1.1   Amplification of a known target sequence with a set of two primers.

Table 1.1   Typical PCR Reaction Components

1. Template DNA	1–10 μl
2. Tris-HCl (pH 8.3 at 20°C)	20 mM
3. KCl	25 mM
4. Triton X-100	0.1%
5. dNTPs (dATP, dCTP, dGTP, dTTP)	50 mM each
6. *Taq* polymerase	1.0 unit
7. MgCl$_2$	1.5 mM
8. Primers	10 pmoles each

50 μl of mineral oil to prevent evaporation. The four deoxynucleotide triphosphates (Table 1.1) are presently available commercially premixed. Variables that require optimization include components 5–8 in Table 1.1. The concentration of MgCl$_2$ is particularly critical. Innis and Gelfand (1990) have discussed the optimization of PCRs in detail. Most thermal cycling of PCRs encompasses 35 cycles; rarely are more than 35 cycles of benefit.

A typical thermal cycling protocol is given in Table 1.2. After an initial denaturation step at 95°C, steps 2–4 are then sequentially performed for 35 cycles followed by a final extension (step 5) at 72°C to ensure that the final round of strand synthesis at high substrate concentration is completed. The sixth step involving reduction of the temperature to 4°C is used to terminate all reactions for convenient holding until agarose gel electrophoresis is performed. The time required to traverse from one temperature to another is referred to as the ramp time and usually contributes

Table 1.2   Typical PCR Thermal Cycling Protocol

Step	Process	Temperature	Time
1	Initial denaturation	94°C	3 min
2	Denaturation	94°C	1 min
3	Annealing	60°C	1 min
4	Extension	72°C	1 min
5	Final extension step	72°C	4–7 min
6	Terminate reactions and hold	4°C	—

significantly to the total thermal cycling time. Most 35-cycle amplification protocols for the PCR are completed within 3 to 4 hr. The total cycling time can often be significantly reduced to less than 90 min by lowering each holding time to 10–20 s.

C. Sample Preparation without Enrichment

A variety of components in foods and various tissues is capable of inhibiting the PCR. Sample preparation is therefore one of the more critical steps that can adversely influence the PCR. Meat and cheese products are particularly challenging. With ground beef, a 25-g sample is stomached with 225 ml of 0.01 M phosphate buffered saline (PBS, pH 6.0) in a Wirl-pak stomacher bag with a mesh insert and homogenized at normal speed for 60 s. A majority of the homogenate (~200 ml) is then centrifuged at low speed (160 g, or 1000 rpm) for 3 min at 4°C to pellet tissue debris. Most of the supernatant (~190 ml) is then centrifuged at high speed (16,000 g or 10,000 rpm) for 10 min at 4°C to pellet bacterial cells. The supernatant is discarded and the pellet is resuspended in a minimum volume of saline. Bacterial cells are then lysed and the DNA is purified.

D. Lysing of Cells and Isolation of Bacterial DNA

The author has found that most or all gram-negative bacterial cells are readily lysed using TZ lysing solution (Abolmaaty et al., 2000) at 2× concentration (5.0 mg/ml of sodium azide and 4.0% Triton X-100 in 0.2 M Tris-HCl buffer, pH 8.0. One ml of 2× TZ lysing solution is added to the resuspended pellet from above and heated at 100°C for 10 min to achieve cell lysis. The lysate is then centrifuged at 10,000 g for 10 min to remove cellular debris.

Gram-positive bacteria are notably more difficult to lyse than gram-negative organisms. The method of Wiedman, Barany, and Batt (1995) has been found to be effective for lysing *Listeria monocytogenes* and also *Staphylococcus aureus*. Cells are suspended in 100 µl of PCR buffer containing 2 mg/ml of lysozyme followed by incubation at room temperature for 15 min. Proteinase K (1 µl, 20 mg/ml) is then added followed by incubation for 1 hr at 55°C. The sample is then boiled for 10 min to inactivate the proteinase K and to lyse the cells.

Bacterial DNA is then isolated from the above lysate with the "Wizard" DNA clean-up system (Promega Co., Madison, Wisconsin) by mixing 0.45 ml of the lysate with 1.0 ml of the DNA clean-up resin. The mixture is then passed through a "Wizard" microcolumn, washed with 2.0 ml of 80% isopropanol, and dried by centrifugation (2 min at 10,000 g). DNA is then eluted from the column with 0.45 µl of sterile deionized water dH$_2$O. The eluate is then mixed with 1 µl of pellet paint (Novogen Co., Madison, Wisconsin), which is a visible fluorescent-dye-labeled carrier to facilitate visualization of the ultimate DNA pellet, followed by the addition of 5 µl of 3 M sodium acetate and 102 µl of ice cold 100% ethanol. The preparation is then stored at 4°C for 30 min and then centrifuged at 14,000 g at 4°C for 10 min. The supernatant is removed, the pellet air dried and dissolved in 20 µl of sterile deionized water, and the entire preparation incorporated into the PCR reaction.

E. PCR for Identification of Pure Cultures

Confirmation of the identity of pure cultures using PCR is most readily achieved by picking up a small visible amount of cells from an agar culture on the end of a needle and suspending the cells in 20 μl of dH$_2$O. Gene releaser (20 μl, Bioventure, Inc., Murfreesboro, Tennessee) is then added and the heating protocol of the manufacturer is used to lyse the cells. Then 1.0 μl of the resulting cell lysate is added directly to the PCR reaction mixture.

F. Quantitative PCR

The author has found that the methodology described below is ideally suited for quantitative assessment of target DNA incorporating an internal standard with the use of a conventional thermal cycler.

The operational assumption is that PCR yields an exponential rate of increase of the initial number of target DNA molecules. This is described by the following equation:

$$P = T\ (2^n)$$

where P is the number of PCR product molecules formed, T is the number of input target sequences, and n is the number of amplification cycles. This exponential equation, however, does not apply to the entire amplification process. During the first few cycles, when the number of initial target molecules is very low, the rate of amplification can be anticipated to be low (Diaco, 1995). During the last few cycles, when the ratio of PCR product molecules to unreacted primer and *Taq* polymerase molecules has greatly increased, the overall amplification rate can be expected to decrease significantly from an exponential rate. This is illustrated in Figure 1.2. It is only along the linear portion of the plot that a reliable quantitative relationship between the number of input molecules and final accumulated products can be derived. A direct approach to determine when amplification enters the plateau phase is to increase the number of input target molecules and measure the number of product molecules after cycling (Figure 1.2). It is important with quantitative PCR assays that the initial number of target DNA molecules be low, because the linear range of amplification is represented by no more than about a two log increase in DNA. This will often result in a working range of input target colony-forming units (CFU) of about 1.5×10^2 to 1.5×10^4 (Guan and Levin, 2002a). If an enrichment step (usually 4.5–6.0 hr) is used, then an appropriate dilution of the resulting CFU should be used for the PCR to ensure that the number of input target molecules falls within the linear PCR amplification range. There are several advantages to the use of nonselective enrichment cultivation of food samples:

1. Only viable CFU are increased.
2. Detection of the initial viable CFU is lowered to 0.5 to 1.0 CFU/g of sample (Guan and Levin, 2002b).
3. Dilution of PCR inhibitors in the food sample. It is important to keep in mind that a 10% reduction in the efficiency of amplification will result in a reduction of PCR product accumulation by more than 95%.

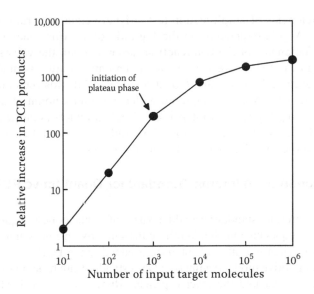

Figure 1.2 Dependence of PCR product accumulation on number of input targets from 35 amplification cycles.

4. The inclusion of an internal standard involving an identical set of primers and a similar target can be used to eliminate variations in amplification efficiency, allowing reliable PCR quantitation.

G. Use of an Internal Standard for Quantitative PCR

The application of an internal standard is based on the coamplification of the target sequence and an internal DNA standard, which is amplified with equal efficiency as the target sequence by using the same primer pair. The internal standard contains the same primer binding sites as the target, and the two DNAs compete for reaction reagents to produce PCR products of different sizes, which can be separated in an agarose gel. The log ratio of intensities of amplified target DNA to internal standard is determined by the following equation given by Zarchar, Thomas, and Goustin (1993):

$$-\mathrm{Log}\,(N_{n1}/N_{n2}) = \log\,(N_{01}/N_{02}) + n\log\,(\mathrm{eff}_1/\mathrm{eff}_2)$$

where

N_{n1} = number of resulting target molecules after amplification.

N_{n2} = number of resulting internal standard molecules after amplification.

N_{01} = number of initial target molecules.

N_{02} = number of initial internal standard molecules.

n = a given cycle.

eff_1 = efficiency of amplification of target molecules.

eff_2 = efficiency of amplification of internal standard molecules.

If the efficiencies of amplification (eff$_1$/eff$_2$) are equal, the ratio of amplified products (N_{n1}/N_{n2}) is dependent on the log ratio of starting reactants (N_{01}/N_{02}). Even if the efficiencies of the two reactions are not equal, the values for N_{n1}/N_{n2} still hold assuming that eff$_1$/eff$_2$ is constant and amplification is in the exponential phase (Zarchar, Thomas, and Goustin, 1993). With this technique, varying amounts of target DNA are coamplified with a constant amount of internal standard. The resulting log ratio of intensities of PCR products is plotted against the log of target CFU for the construction of a standard curve that is used for determining the number of CFU per gram of food product.

H. Construction of an Internal Standard for Quantitative PCR

The ideal internal standard should consist of a nucleotide sequence derived from the target sequence and should utilize the same two primers and primer binding sites so that when the internal standard is added to the PCR mix, amplification occurs in a competitive mode. An internal DNA standard can be synthesized by PCR using a hybrid primer and the reverse primer (Rupf, Merte, and Eschrich, 1997). The hybrid primer contains two components: a 5′ portion consisting of the original forward primer, and a 3′ portion consisting of a small DNA fragment derived from the target sequence. A 20-nucleotide sequence about 200 to 300 nucleotides in from the 5′-terminal of the target sequence is identified and a hybrid primer is synthesized (Figure 1.3) for use in the construction of an internal DNA standard. Complete details are given by Guan and Levin (2002b).

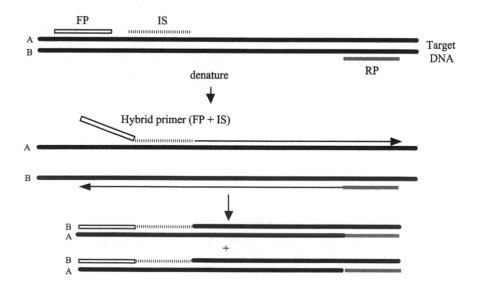

Figure 1.3   Construction of an internal DNA standard for quantitative PCR. FP, forward primer; RP, reverse primer; IS, internal sequence.

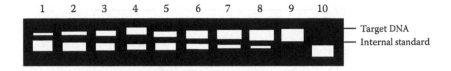

Figure 1.4 Idealized representation of competitive amplification of a constant amount of internal standard and varying amounts of target DNA. Lane 9, target DNA alone. Lane 10, internal standard alone.

I. Construction of a Calibration Curve for Quantitative PCR

Varying amounts of target DNA from a pure culture are coamplified with a constant amount of the internal standard (IS). The amplified co-PCR products are separated by electrophoresis in a 1.5% agarose gel (Figure 1.4). The gel images are then captured by a digital camera and analyzed with National Institutes of Health (NIH) Image 1.61 software. To correct differences in the intensity of fluorescence of ethidium bromide-stained PCR fragments of different sizes, the intensity of the internal standard is multiplied by the ratio of the number of nucleotides in the target sequence to that of the internal standard. The log of the ratio of fluorescence intensity of the amplified target sequence to that of the internal standard is then plotted against the log of CFU to establish a standard curve, using a constant amount of internal standard and varying the number of CFU (Figure 1.5). It is important to note that the calibration curve is operational over little more than a one log cycle of input target sequences (CFU).

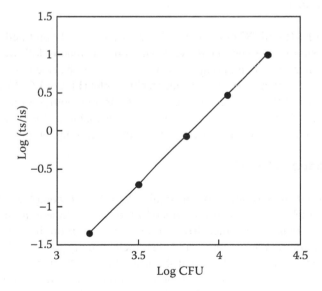

Figure 1.5 Standard curve for quantitative PCR.

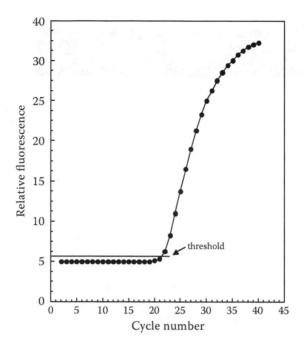

Figure 1.6   Real-time PCR screen display of DNA target amplification.

II. REAL-TIME PCR (RTI-PCR)

A. Introduction

Real-time PCR (Rti-PCR) refers to the detection of PCR-amplified target DNA (amplicons) usually after each PCR cycle. The signal is readily followed on a computer screen where each point is automatically plotted and the extent of amplification is followed as an ongoing continual direct graphical plot (Figure 1.6). Computer software handles all of the preprogrammed calculations and plotting of data. Table 1.2 lists the major Rti-PCR thermal cyclers presently available and the manufacturers. Table 1.3 lists a number of commercial sources of primers and fluorescent probes.

B. Advantages of Rti-PCR

Conventional thermocyclers often require 2 to 3 hr to complete 35–40 "thermal" cycles. Much of this time is consumed by the "ramp" time required to traverse from one temperature to another. Rti thermocyclers incorporating air heating and cooling, capillary sample systems, and thermoelectrically controlled blocks have greatly reduced ramp times. The use of shortened target DNA sequences (60–70 bp) in Rti-PCR results in more efficient amplification than standard PCR where amplicons are required to be at least 200 bp in length to allow detection by electrophoretic separation and also allows reduced extension times. Short PCR product yields are

Table 1.3 Real-Time PCR Instrumentation

Name of Instrument	Manufacturer	Heating/ Cooling	Cycling Time	Observation of Data	Number of Wells
IQ5	BioRad	Thermoelectric	~2 hr	Each PCR cycle	96
Smart Cycler	Cepheid	Ceramic heating plate	~40 min to 1 hr	Each PCR cycle	6–96
Gene Amp 5700 & Prism 7700	Applied Biosystems	Heating block	~2 hr	End-point	96–384
Rotor Gene	Corbett Research	Air	~50 min	Each PCR cycle	32
ICycler iQ	BioRad	Heating block	·2 hr	Each PCR cycle	60, 06, 384
MX 4000	Stratagene	Heating block	~90 min	Each PCR cycle	96
LightCycler	Roche Applied Biosystems	Air	~20 min to 1 hr	Each PCR cycle	32

significantly improved by lower denaturation temperatures (Yap and McGee, 1991) so that a denaturation temperature of 90°C instead of 95°C is preferred. This also reduces ramp time. In addition, conventional PCR requires visualization of amplified products after agarose gel electrophoresis, which usually involves an additional 30–60 min. Rti-PCR completely eliminates this step through the use of a fluorometer built into the Rti-PCR thermal cycler that measures the intensity of fluorescence after each amplification cycle, unless one wishes to confirm the identity of the products on the basis of molecular size. Agarose gel electrophoresis is now usually replaced with Rti-PCR systems by programmed generation of a thermal denaturation curve of the amplified product after the PCR that allows automatic calculation of the T_m value of the amplicon when SYBR Green is the fluorescent reporter molecule. This is most useful in confirming the identity of an amplicon.

The quantitative range with conventional PCR is no more than 1.5–2.0 log cycles, whereas with Rti-PCR an operational range of at least 5–6 log cycles is usually achieved. Conventional thermocyclers can presently be acquired for $2000 to $3000. In contrast, Rti-PCR systems presently range in price from $20,000 to $40,000. An inexpensive approach is the use of a fluorescence-activating microplate reader where final fluorescence is measured after amplification in a conventional thermal cycler. Rti-PCR can be performed with units furnishing 16 wells, with systems accommodating 96 or 384 well microplates, with chains of eight linked PCR tubes, or with individual PCR tubes. Because of the significantly increased cost of reagents, particularly the fluorescent probes and dyes compared to reagents used with conventional PCR, the reaction volume is usually reduced from 50–100 µl to 10–20 µl.

In addition to detection of amplicons, Rti-PCR units can quantitate amplified target DNA and differentiate alleles (determine point mutation or sequence variation). Allelic variation is assessed on the basis of T_m variations derived from analysis of melting curves of duplexes formed by fluorescent probes and amplicons. In addition, two or more different PCR reactions (multiplex PCR) amplifying different target sequences can be followed and quantifed in the same PCR tube or well. Cockerill and Uhl (2002) have extensively discussed the advantages of Rti-PCR.

The GeneAmp 5700 and PRISM 7700 (Applied Biosystems) and the LightCycler system (Roche Applied Science; Table 1.3) can be obtained with coupled automated nucleic acid extraction instruments resulting in completely automated DNA extraction, amplification, and detection. Rti-PCR has great potential for the meat-processing industry where massive recalls have occurred during the past few years due to the presence of *Escherichia coli* O157:H7 and *L. monocytogenes* in various meat products. More recently, costly recalls in vegetables due to *E. coli* O157:H7 and outbreaks of salmonellosis from tomatoes have occurred The present state of Rti-PCR technology should allow processors to detect product contamination from the production line in essentially near real-time to reduce such massive recalls by allowing detection prior to shipment of the product.

C. Mechanisms of Rti-PCR

Rti-PCR depends on the emission of a an ultraviolet (UV)-induced fluorescent signal that is proportional to the quantity of DNA that has been synthesized. Several fluorescent systems have been developed for this purpose and are discussed below.

1. SYBR Green

The simplest, least expensive, and most direct fluorescent system for Rti-PCR involves the incorporation of the dye SYBR green whose fluorescence under UV greatly increases when bound to the minor groove of double-helical DNA (Figure 1.7).

SYBR green lacks the specificity of fluorescent DNA probes but has the advantage of allowing a DNA melting curve to be generated and software calculation of

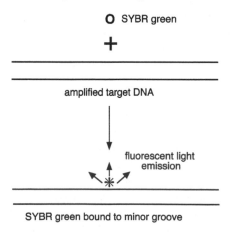

Figure 1.7 Mechanism of SYBR green fluorescence. SYBR green dye binds to the minor groove of the DNA double helix. The unbound dye emits little fluorescence, which is greatly enhanced when bound to DNA.

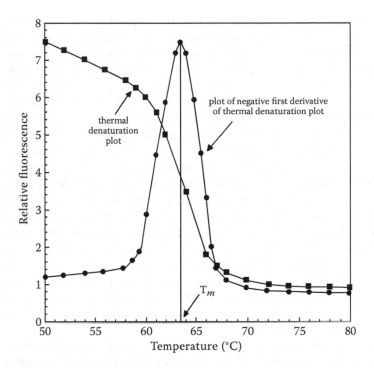

Figure 1.8 Thermal denaturation curve of double-stranded amplified target DNA with SYBR green I as reporter dye. As the double helix is denatured to the single-strand state, increasing amounts of SYBR green dissociate from the double-stranded DNA resulting in a linear decrease in fluorescence. Bell-shaped curve results from plotting the second derivative of the thermal denaturation plot with the apex coincident with the T_m value.

the T_m of the amplicon after the PCR (Figure 1.8). This allows identification of the amplified product and its differentiation from primer–dimers, which also result in a fluorescent signal with SYBR green but which usually have a lower T_m value. The fluorescent signal is measured immediately after the extension step of each cycle because thermal denaturation yielding single-stranded DNA eliminates fluorescence. A software plot of the negative first derivative of the thermal denaturation plot yields a bell-shaped symmetrical curve, the midpoint of which yields the T_m value for the amplified product (Figure 1.8). Interference of the amplicon's signal by the signal resulting from primer–dimer formation can be eliminated by raising the temperature to a critical point that is above the T_m of the primer–dimer formed (resulting in thermal denaturation of the primer–dimers) but below the T_m of the amplicons prior to measuring the intensity of fluorescence emission.

2. TaqMan™ Probes

The TaqMan probes are proprietary double-dye probes synthesized by Perkin Elmer Applied Biosystems. A variety of such double-dye probes is available from a number of commercial sources (Table 1.4). The TaqMan system makes use of the 5′ 3′-exonuclease activity of *Taq* polymerase to produce a fluorescent signal. A custom-synthesized TaqMan probe is incorporated into the PCR containing a sequence of nucleotides homologous to a specific nucleotide sequence of one strand of the amplicon internal to both primers. The probe harbors a fluorophore (reporter dye) such as 6-carboxyfluorescein (FAM) as the reporter dye at the 5′ end and 6-carboxytetramethyl-rhodamine (TAMRA) as the quenching dye at the 3′ end (Figure 1.9), which are close enough to prevent emitted fluorescence of the reporter. A phosphate molecule is usually attached to the terminal 3′-thymine residue to prevent extension of the bound probe during amplification. The fluorescent emission spectrum of FAM is 500–650 nm. The fluorescent intensity of the quenching dye TAMRA changes very little over the course of PCR amplification. The intensity of TAMRA dye emission therefore serves as an internal standard with which to normalize the reporter (FAM) emission variation. Following each thermal denaturation step, the temperature is lowered to allow annealing of the probe to single-stranded amplicons. Increasing amounts of the single-stranded amplicons will bind increasing amounts of the probe. During primer extension, *Taq* polymerase cleaves the probe from the 5′ to the 3′ direction releasing the reporter dye, which then emits fluorescence as a result of its increased distance from the quencher. Fluorescence is then measured following each extension stage of every cycle.

Eclipse probes are proprietary double-dye probes available from Epoch Biosciences and Fluoresentric (Table 1.4). They differ from TaqMan probes in that the fluorophore is bound to the 3′ end and the quencher to the 5′ end. Prior to binding, the fluorophore is in close proximity to the quencher. The presence of the quencher at the 5′ position prevents digestion by *Taq* polymerase. Fluorescence is measured during the annealing step. Black hole quenchers are molecular species that exhibit no inherent fluorescence themselves that are used in conjunction with fluorogenic reporter dyes with dual-labeled probes. Primer–dimers are not detected by any of these dual-labeled probes, which constitutes an additional advantage in their use.

3. Fluorescent Resonance Energy Transfer (FRET)

FRET involves the incorporation of two different custom-synthesized oligonucleotide probes into the PCR. One probe (light donor) harbors a fluorescein label at its 3′ end and the other probe (light acceptor) is labeled with LightCycler Red 640 at its 5′ end (Figure 1.10). The sequences of the two probes are selected so that they can hybridize to the same strand of an amplicon in a head-to-tail orientation internal to both primers resulting in the two fluorescent dyes coming into close proximity to each other. Under UV, the fluorescein emits a green light, which then excites the Red 640 dye because of their close proximity, which in turn emits a red light proportional to the amount of amplicon present. The red emission is measured at 640 nm. This energy transfer is referred to as FRET and occurs when no more than one to five nucleotides separate the two dyes. Fluorescence is measured after each annealing step of every cycle. After annealing, the temperature is raised and the hybridization probes are displaced by the *Taq* polymerase. Primer–dimers are not detected.

Table 1.4 Commercial Sources of Real-Time PCR Primers and Fluorescent Probes

Source	Double Dye[a]	FRET Probes	Molecular Beacons	Other
BioNexus Inc./ABP www.bionexus.net	+	+	+	
Biosearch Technologies www.biosearchtech.com	+	+	+	Black hole quenchers
Proligo (formerly Genset) www.proligo.com	+	+	+	
Epoch Biosciences www.epochbio.com	–	–	–	Eclipse probes
Qiagen Operon www.operon.com	+	+	+	Black hole quenchers
Eurogentec SA www.eurogentec.com	+	+	+	Scorpions primers
Molecular Research Laboratories www.molecula.com	+	+	+	
Gene Link Inc. www.genelink.com	+	+	+	
Integrated DNA Technologies www.idtdna.com	+	+	+	
Midland Certified Reagent Co. www.mcrc.com	+	+	+	
Sigma-Genosys www.sigma-genosys.com	+	+	+	
TriLink BioTechnologies www.trilinkbiotech.com	+	+	+	
Applied Biosystems www.appliedbiosystems.com	+	+	+	
Invitrogen www.invitrogen.com	+	+	–	Lux primers
Fluoresentric www.fluoresentric.com	+	+	+	Eclipse primers
Roche Applied Science www.Roche-Applied-Science.com	+	+	+	
Molecular Probes www.probes.com	+	+	+	
Fluorosentric www.fluoresentric.com	+	+	+	Eclipse probes

[a] TaqMan™ probes are proprietary double-dye probes synthesized by Perkin Elmer Applied Biosystems.

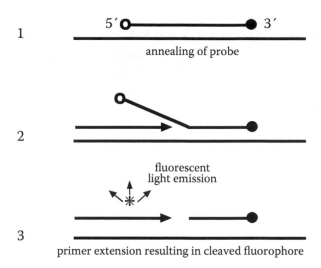

Figure 1.9 Mechanism of TaqMan fluorescence probes. The 5′ nuclease activity of *Taq* DNA polymerase is utilized to cleave a TaqMan probe during DNA amplification. (1) The TaqMan probe is annealed to the target DNA sequence. The probe contains a reporter dye (o) at the 5′ end of the probe and a quencher dye (●) at the 3′ end of the probe. (2) During PCR amplification, a complementary strand of DNA is synthesized and the 5′ exonuclease activity of *Taq* polymerase excises the reporter dye. (3) Fluorescence of the reporter dye (o) occurs when it is separated from the quencher dye and the intensity is measured.

An interesting variation of FRET involves the use of a single probe labeled at the 5′-end with a reporter dye LCRed64 or Cy5 and the addition of SYBR green I to the PCR mix (Loh and Yap, 2002). The labeled probe hybridizes to the homologous sequence of the amplified target strand and then SYBR green I binds to the resulting double-stranded DNA so as to excite the reporter dye. Primer–dimers are distinguished from the amplified target DNA on the basis of different T_m values obtained after the PCR.

4. Molecular Beacons™

Molecular Beacons™ is the proprietary trade name of custom-synthesized nucleotide probes available from Stratagene with GC-rich complementary terminal nucleotides that form a hairpin configuration with a hybrid stem (Figure 1.11).

A variety of such probes is available from a number of commercial sources (Table 1.4). The probes harbor a reporter dye at one end and a quencher dye at the other end. The reporter dye or fluorophore is close to the quencher dye and is nonfluorescent in the hairpin configuration. The energy taken up by the fluorophore is transferred to the quencher and released as heat rather than being emitted as light. When the probe hybridizes to its homologous nucleotide sequence of the amplicons, a rigid double helix is formed that separates the quencher from the reporter dye resulting in emission of UV-induced fluorescence. Fluorescence is then measured after the annealing step of each cycle.

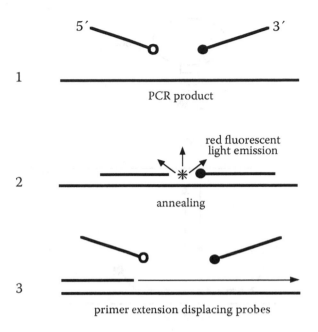

Figure 1.10   Mechanism of resonance fluorescence energy transfer (FRET). Two different oligonucleotide labeled probes are depicted. Probe 1 carries a fluorescein label at its 3′ end and probe 2 is labeled with Light Cycler Red 640 or another suitable fluorescent dye such as Cyanin 5 at its 5′ end (frame 1). The sequences of the two probes are chosen so that they can hybridize to the PCR product in a head-to-tail orientation, resulting in close proximity of the two fluorescent dyes (frame 2). The first dye (fluorescein) is excited by the cycler's light source and emits a green fluorescent light. Close proximity (within one to five nucleotides) of the two dyes results in the emitted green fluorescence exciting the Light Cycler Red 640 dye attached to the 5′ end of the second probe, which in turn emits a red fluorescent light that is measured. This energy transfer is known as FRET. Increasing amounts of DNA resulting from the PCR yields a proportional increase in red fluorescence. Because red fluorescence at 640 nm occurs only when both probes are hybridized, fluorescence is measured after annealing. During primer extension, the hybridized probes are displaced by the *Taq* polymerase and are then too distant from each other for FRET to occur (frame 3).

5. Unique Fluorogenic Primers

Nazarenko et al. (2002) reported on the development of unique fluorescent primers for use with multiplex quantitative Rti-PCR (QRti-PCR). Fluorogenic primers are labeled with a single fluorophore on a base close to the 3′ end with no quencher. A tail of 5–7 nucleotides is added to the 5′ end of the primer to form a blunt-end hairpin loop when the primer is not incorporated into a PCR product. This design provides a low initial fluorescence of the primers that increases up to eightfold upon formation of the PCR product. The hairpin oligonucleotide primers provide additional specificity to the PCR by preventing primer–dimer formation and mispriming. Invitrogen custom-synthesizes such single-fluorophore hairpin loop primers under the proprietary designation LUX™ primers. No probes are required.

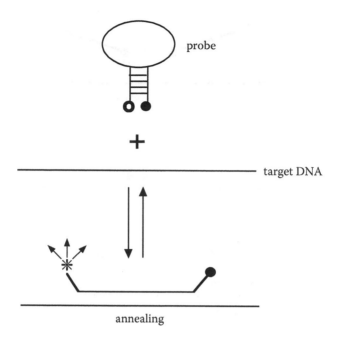

Figure 1.11 Mechanism of molecular beacon probes. The molecular beacon probe in its hairpin configuration is nonfluorescent because of the close proximity of the fluorescent dye such as fluorescein (6-FAM) (o) and quencher dye such as DABCYI (●) maintained by the hybrid stem. Hybridization of the probe sequence in the loop with its homologous target sequence results in physical separation of the two dyes allowing fluorescence to occur. Fluorescence is measured after thermal denaturation and annealing of the probe at a low temperature (~41°C), notably lower than the optimum temperature for primer annealing (~52°C).

Scorpions™ are PCR primers with a "stem-loop" tail containing a fluorophore and quencher (Figure 1.12). The "stem-loop" tail is separated from the PCR primer sequence by a "PCR blocker" that prevents the *Taq* polymerase from copying the stem-loop sequence. During PCR, Scorpion primers are extended to form PCR products. During the annealing phase, the probe sequence in the Scorpion's tail curls back to hybridize to the target sequence in the newly formed PCR product. Because the tail of the Scorpions and PCR product are now part of the same strand, the interaction is considered intramolecular and the incorporated fluorophore is at a distance from the quencher. The recommended length of the "loop" is 20 to 35 nucleotides.

D. Theory of Quantitative Real-Time PCR (QRti-PCR)

QRti-PCR requires the design and use of proper controls for quantitation of the initial target sequences. Heid et al. (1996) were the first to develop such QRti-PCR methodology and made use of the TaqMan reaction with FAM as the reporter dye and TAMRA as the quencher. The software calculates a value termed ΔRn or ΔRQ from the following: $\Delta Rn = (Rn^+) - (Rn^-)$, where $Rn^+ =$ emission intensity of reporter/

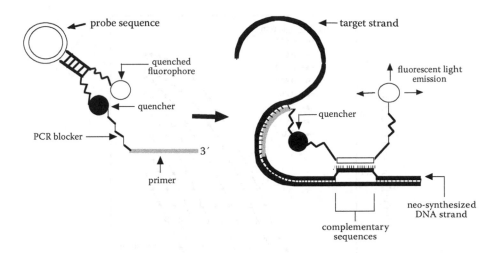

Figure 1.12 Mechanism of Scorpions™ labeled primers. During PCR, Scorpions primers are extended to form PCR products. During annealing the probe sequence in the Scorpion's tail curls back to hybridize to the target sequence in the newly formed PCR product so that the fluorophore (reporter dye) (o) is incorporated into the newly synthesized strand at a considerable distance from the quencher (●).

emission intensity of quencher at any given time in a reaction tube, and $Rn^- =$ emission intensity of reporter/emission intensity of quencher measured prior to PCR amplification in the same reaction tube. The ΔRn mean values are plotted on the y-axis, and the number of cycles is plotted on the x-axis. During the early cycles, the ΔRn remains at baseline.

When a sufficient amount of hybridization probe has been cleaved by the 5′-nuclease activity of *Taq* polymerase, the intensity of reporter fluorescence emission increases. A threshold level of emission above the baseline is selected, and the point at which the amplification plot crosses the threshold is defined as C_T (threshold cycle) and is reported as the number of cycles at which the log phase of product accumulation is initiated (Figure 1.6). The threshold is usually set at 10 times the standard deviation of the baseline. By setting up a series of wells containing a 4–5 log span of genomic DNA concentration (each concentration in triplicate wells) a series of amplification plots is generated by the software in real time (Figure 1.13). The amplification plots shift to the right as the quantity of input target DNA is reduced. Note that the flattened slopes and early plateaus do not influence the calculated C_T values. The C_T values decrease linearly with increasing target quantity. A plot of the resulting C_T values on the y-axis versus the log of the ng of input genomic DNA yields a straight line (Figure 1.14), which is then used as a standard curve for quantitation of samples with unknown levels of genomic DNA. This approach and the original nomenclature of Heid et al. (1996) have been universally adapted for quantitative PCR. However, this methodology does not address the issue that PCR inhibitors may be present in a DNA sample derived from a complex food product.

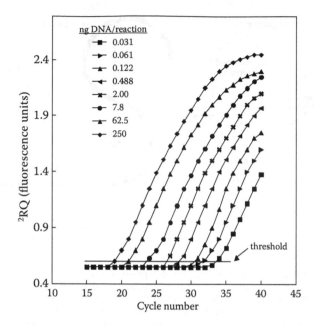

Figure 1.13 Appearance of real-time PCR screen display of thermal cycling plots of target DNA amplification. Decreasing the quantity of target DNA results in an increased number of cycles required to detect fluorescence above the threshold level (shift of the plot to the right).

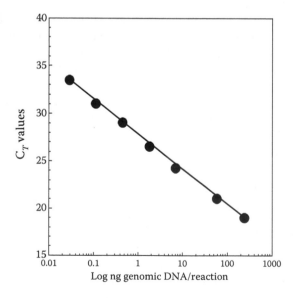

Figure 1.14 Standard curve for real-time PCR quantification of target DNA.

Detection of PCR inhibition and normalization of C_T values can be accomplished with the use of either of two methods: (1) splitting the sample into two parts, where one portion is subjected to QRti-PCR and the second portion is used to amplify an external control standard, or (2) quantitative comparative PCR using a normalization or "housekeeping" gene contained within the sample for QRti-PCR for correction of the observed C_T values. If equal amounts of nucleic acid are analyzed for each sample and if the amplification efficiency is identical for each sample, then the internal control (normalization gene or competitor) should give equal signals for all samples (Heid et al., 1996).

E. Problems and Limitations of QRti-PCR

Wilhelm, Hahn, and Pingoud (2000) found that with the LightCycler instrument unexpected variations in QRTi-PCR experiments occurred. This unit is a rapid thermal cycler that utilizes temperature-controlled airflow for heating and cooling samples. The PCRs were performed in glass capillaries placed along the perimeter of a circular carousel with 32 positions. The shape and amplitude of fluorescence curves as well as the calculated amounts of target DNA exhibited wide variations that were found to be dependent on the position of a particular capillary in the carousel. Interestingly, the presence of capillaries in the carousel was found to alter the temperature stability or airflow inside the carousel chamber. Revised software (version 5.32) ensuring the continued rotation of the sample carousel significantly reduced temperature variation in the carousel and eliminated the large variations in results.

The incorporation of an internal control in a multiplex format to detect inhibition of amplification is critical for the detection of false negative results. However, if QRti-PCR is to be successfully used for enumerating the initial number of gene targets, then a control is necessary to detect the extent to which positive amplification may be partially inhibited. Such a "quantitative" control can assume several formats. The simplest and most direct approach is to split the sample into two PCR tubes, where one tube is used to amplify the target sequence to be enumerated and the second tube is used to amplify a different control target sequence but with ideally identical primers (external control). Standard curves are prepared for both targets. The extent to which the external control deviates from its standard curve is then used to correct and normalize the C_T value for the target being quantified.

A second approach is to establish a multiplex PCR incorporating primers for the primary target sequence of interest and primers for a control sequence plus a known number of control targets to be simultaneously amplified. The use of TaqMan probes differing with respect to the fluorophores and their emission wavelengths will then allow the amplification of both target sequences to be followed. For example, if 100 control target sequences in a PCR that normally have a C_T of 30 cycles without inhibition yield a C_T of 35 cycles, then a corresponding correction factor can be applied to the C_T derived from the target being quantified. Alternatively, both TaqMan probes can be labeled identically and the multiplex PCR performed in two tubes: one tube containing the TaqMan probe for the primary target and the second tube containing the other TaqMan probe for the internal control sequence. Such multiplex reactions with an internal control are predicated on an abundance of *Taq* polymerase and dNTP precursors such that they do not become limiting as a result of competitive PCR.

The use of universal primers for QRti-PCR of Eubacteria using the highly conserved 16S or 23S rRNA gene sequences is ideally suited for samples containing large numbers of bacteria ($>10^4$/g). The detection or quantitation of low numbers of bacteria with nonspecies- or nongenus-specific primers is fraught with problems arising from the presence of low numbers of contaminating organisms and DNA from PCR reagents. It is well documented that *Taq* DNA polymerase preparations frequently contain contaminating DNA as a result of incomplete purification during manufacture (Böttger, 1990; Rand and Houck, 1990).

Taq DNA polymerase is commonly produced as a recombinant protein in *E. coli* or is obtained as a native enzyme from *T. aquaticus*. Corless et al. (2000) examined this problem in detail with Rti-PCR using a set of primers targeting conserved flanking sequences of the eubacterial 16S rRNA gene and a TaqMan probe yielding an 87-bp amplicon from *E. coli*. The study included a comparison of four *Taq* DNA polymerases from three commercial sources with respect to their ability to produce amplicons in Rti-PCR assays without added target DNA (negative controls). All four *Taq* DNA polymerases yielded false Rti-PCR positive results. More than one amplicon was identified by DNA sequencing, reflecting contamination by DNA from more than one organism. The authors concluded that the problem is exacerbated by the highly sensitive nature of the Rti-PCR process and the fact that the 16S rRNA gene is present in multiple copies (up to seven) in the genomes of many bacteria. The study indicated that low-DNA LD Ampli*Taq* polymerase (Applied Biosystems) yielded the highest C_T values in Rti-PCR assays, among the four *Taq* enzymes examined, without added DNA, reflecting the lowest amount of contaminating bacterial DNA.

III. NESTED PCR

A. Introduction

The use of a pair of "nested primers" flanking a DNA sequence internal to the sequence encompassed by the external pair of primers allows a greater level of amplification than is normally achieved with a single set of primers. A target sequence is initially amplified by an external pair of primers for 20–40 cycles. A small aliquot of this reaction is then amplified a second time for 20–40 cycles using the internal or nested primer set (Figure 1.15). The inner pair of primers anneals to complementary sequences internal to the initially amplified product. This has been shown to result in greater amplification than reamplifying with the same initial pair of primers (Albert and Fenyo, 1990). The number of nucleotides flanked by the internally nested primers determines the final product size. The nucleotide sequence of the nested product of the second stage of amplification is always shorter than the primary PCR product.

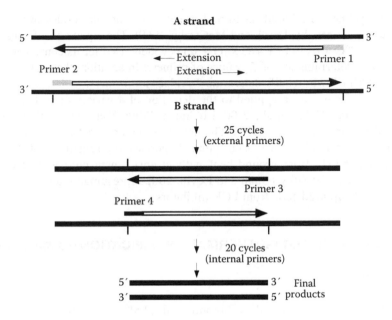

Figure 1.15 Two-tube nested PCR. Amplification with external primers for 25 cycles followed by amplification of an aliquot for 25 cycles using internal nested primers.

Nested PCR is of particular value with foods in the absence of enrichment cultivation where inhibitors present from the food may significantly reduce the efficiency of amplification. In addition, the nested primers serve as a control for the specificity of the amplified external sequence and can therefore improve both sensitivity and specificity of DNA amplification (Jackson, Hayden, and Quirke, 1991). A major advantage of nested PCR is that the second round of amplification can result in an increase in the sensitivity of detection of several orders of magnitude over that achieved with just primary amplification alone. An additional advantage is that of enhanced specificity in that it is unlikely that any nonspecific product of amplification in the primary PCR will contain sequences complementary to the primers of the secondary PCR.

Nested PCR, however, introduces an additional risk of amplification of contaminating DNA when used routinely. This problem can be circumvented by using a "one-tube nested PCR" involving the use of two primer pairs of different melting points (T_m values) in a single reaction mixture. Annealing of the nested primers during the first set of cycles is prevented by their lower T_m. The larger external fragment should ideally have a higher G+C content so that it will denature at a slightly higher temperature than the nested fragment. This difference in T_m values for the two amplified products has to be critically determined for both the larger and the nested fragment. A number of mathematical approaches have been developed for calculating the T_m based on either the G+C content of the fragments or slightly more accurate values based on nearest-neighbor thermodynamics (Breslauer et al., 1986; Freier et al., 1986; Rychlik, Spencer, and Rhoads, 1990). Other calculations of T_m

values for primers are based on formulas developed for nucleotides over 100 bp in length (McConaughy, Laird, and McCarthy, 1969). The equation of Suggs et al. (1981), $T_m = 2°C \times (A + T) + 4°C \times (G + C)$ is widely used for its convenience and approximate determination of T_m values of primers. In addition, the GCG software program "Prime" will readily determine T_m values for primers.

Nested PCR has been applied to the detection of a variety of pathogens in a number of foods (Ozbas et al., 2000; Lindquist, 1999; Waage et al., 1999; Gilgen et al., 1998; Kapperud et al., 1993). The use of immunomagnetic capture of target cells prior to nested PCR has been found to increase the sensitivity of detection of *E. coli* O157:H7 from ground beef without enrichment cultivation from 110 CFU/10 g to 24 CFU/10 g (Guan and Levin, 2002a) presumably due to magnetic separation of captured cells from PCR inhibitors.

IV. LOOP-MEDIATED ISOTHERMAL AMPLIFICATION (LAMP) ASSAY

A. Introduction

LAMP was first developed by Notomi et al. (2000) and utilizes a DNA polymerase isothermally at 60–65°C and a set of four primers that recognize a total of six distinct sequences on the target DNA. An inner primer containing sequences of the sense and anti-sense strands of the target DNA initiates LAMP. The resulting strand displacement DNA synthesis primed by an outer primer releases a single strand of DNA. This then serves as a template for DNA synthesis primed by the second inner and outer primers that hybridize to the other end of the target, which produces a stem-loop DNA structure. In subsequent LAMP cycling one inner primer hybridizes to the loop on the product and initiates displacement DNA synthesis, yielding the original stem-loop DNA and a new stem-loop DNA with a stem twice as long. The cycling reaction continues with an accumulation of 10^9 copies of target in less then 60 min. LAMP reactions usually result in about 10^3-fold or higher levels of amplification than conventional PCR. The final products are stem-loop DNAs.

With an additional one or two primers termed "loop" primers, the reaction can be accelerated. The method is capable of yielding an unusually large amount of DNA, more then 500 µg. At the completion of amplification a white precipitate forms that has been identified as magnesium pyrophosphate (Mori et al., 2001), which can be used to confirm amplification. The yield of synthesized DNA, can be quantified by measuring the intensity of fluorescence using ethidium bromide in agarose gels or in an Rti-PCR unit or alternatively by measuring turbidity. The level of turbidity has been found to correlate with the amount of synthesized DNA, which is related to the number of initial target sequences. The identity of the DNA polymerase was found to be a critical factor for efficient amplifcation. *Bst* polymerase (Takara) is preferred along with the presence of betaine or L-proline, which were found to stimulate the rate of amplification and to increase selectivity (Notomi et al., 2000). DNA amplification can be further greatly accelerated by the use of two additional primers termed loop primers resulting in detection of 600 templates in 13 min (Nagamine, Hase, and Notomi, 2002).

V. PULSED-FIELD GEL ELECTROPHORESIS (PFGE)

A. Introduction

PFGE is presently considered the method of choice and is frequently referred to as the "gold standard" for discriminating genetic differences and lineage among strains of the same bacterial species. The method is particularly useful in epidemiology and has also been applied to the detection of clonal strains of pathogens that have been found to persist in food manufacturing facilities (Johansson et al., 1999; Autio et al., 1999; Brett, Short, and McLauchlin, 1998).

B. Mechanism of PFGE

PFGE is based on the use of low-frequency restriction nucleases to generate a family of high molecular weight fragments derived from genomic DNA and their subsequent resolution based on size using changes in the direction of the electric field during agarose electrophoresis. Large fragments stretch out linearly in the direction of the electric field. When the direction of the field changes, the fragments undergo an initial relaxation in conformation and then form multiple undulations or kinks in the direction of the new field, followed by linearization (Gurrieri et al., 1990). At present the most widely adapted system is the contour-clamped homogeneous electric field (CHEF) apparatus, involving an hexagonal distribution of electrodes that undergo periodic alternate uniform electric fields with an angle of 120° (Figure 1.16). CHEF instruments are available from Bio-Rad, Pharmacia, and BRL. A longer pulse time increases the size of the fragment that can be separated but results in a decrease in resolution of fragments of similar size.

Progressive increases in pulse time (pulse time ramping) can significantly increase resolution. A ramp from 5 to 40 s will allow optimal separation of DNA fragments from 50 kb to 600 kb, and increasing the pulse time to 75 s will extend the separation

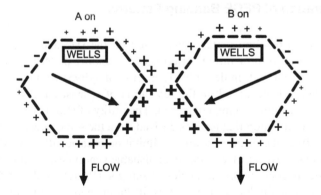

Figure 1.16 Voltage clamping by the CHEF-DR-II system. A: Relative electrode potentials when the +60 volt field vector is activated. B: Relative electrode potentials when the −60 volt field vector is activated.

to 1 Mb (Struelens, De Ryck, and Deplano, 2001). PFGE is usually performed at 12–15°C for enhanced band resolution and to prevent the development of temperature gradients in the gel. Agarose of high gel strength and low electro-endosmosis (EEO) is used. DNA fragments from 50 kb to 1 Mb are usually separated with 1.0% agarose. Reducing the agarose concentration will increase the pore size and allow separation of larger fragments. Higher concentrations of agarose (1.2 to 1.6%) reduce pore size and will increase resolution and sharpness of bands, but reduce separation of larger fragments and result in longer running times (Struelens, De Ryck, and Deplano, 2001). Birren et al. (1988) have reported on optimized conditions for PFGE and the effects of a variety of variables on the resolution of DNA fragments.

The restriction nucleases used with PFGE are selected on the basis of rarity of their recognition sequence in the target genome. Struelens, De Ryck, and Deplano (2001) and Tenover et al. (1995) have listed restriction nucleases useful for PFGE analysis of a number of bacterial genera and species.

The usual methodology for DNA extraction is unsuitable for PFGE because the large size of genomic bacterial DNA ($\sim 5 \times 10^9$ Da) results in rapid shearing during pipetting to $\sim 1 \times 10^7$ Da. To prevent such rupture of large DNA molecules, intact cells are embedded in an equal volume of 2.0% nuclease-free, low-melting-temperature agarose plugs. The cells are then lysed in situ, immersing the plugs first in lysozyme and then proteinase K, or other suitably lytic enzymes for the specific organism involved. To save time, lysozyme can be added to the agarose–cell mixture before solidification. The agarose plugs are then placed in wells for PFGE. Uniform cell densities of $1–5 \times 10^9$ cells/ml are critical for valid comparison of resolved bands. Adjusting cell suspensions to a uniform A_{600} value is reliably effective and convenient, provided each culture is of the same age. To ensure adequate alignment and normalization of banding patterns and accurate fragment size estimates, appropriate DNA ladders should be included in at least every fifth lane (Struelens, De Ryck, and Deplano, 2001).

C. Interpretation of PFGE Banding Patterns

Distinctions in banding patterns are based on the size and shape of resolved bands. The problem of how minor differences in banding patterns should be interpreted has been dealt with in detail by Tenover et al. (1995) with the establishment of well-reasoned criteria based on the resolution of at least ten distinct fragments derived from each culture being compared. The utility of strain typing using PFGE is based on the assumption that isolates derived from the same reference strain are of recent lineage from the original strain. A limitation occasionally arises when unrelated isolates may have similar or indistinguishable genotypes, especially if there is limited genetic diversity within a species or subtype (Barrett et al., 1995). It should also be assumed that random genetic events, including endpoint mutations and deletions and insertions of nucleotides, will alter PFGE patterns of DNA from progeny derived from an original reference strain. An isolate is interpreted to be closely related to a reference strain if its PFGE pattern differs from the reference strain by

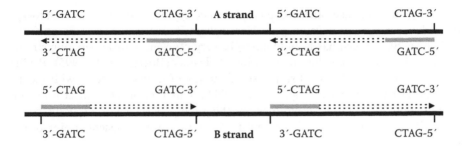

Figure 1.17 Random amplification of polymorphic sequences with a single random primer.

changes reflecting a single genetic event, that is, a point mutation or a frame shift mutation involving the insertion or deletion of one or more sequential nucleotides.

Such changes usually result in two to three band differences. For example, a single spontaneous mutation that creates a new genomic restriction site will split one restriction fragment into two smaller ones. The loss of the original large fragment plus the appearance of two new smaller fragments will result in a three-band difference between the reference strain and its direct progeny that has undergone such a single mutational event (Tenover et al., 1995). In addition, variations of two to three bands have been reported in strains repeatedly cultured or isolated multiple times from the same patient (Arbeit et al., 1993; Sader et al., 1993). Patterns that are distinctly different from an outbreak or reference strain by only two or three fragments should therefore be considered subtypes of the same clonal lineage. An isolate is considered to be possibly related to a reference strain if there are four to six band differences. An isolate should be considered unrelated to a reference strain if its PFGE pattern differs by seven or more bands reflecting three or more independent genetic events.

VI. RANDOM AMPLIFICATION OF POLYMORPHIC DNA (RAPD)

A. Introduction

The use of a single random primer with RAPD will yield significantly fewer DNA bands than PFGE with a single restriction nuclease and is therefore notably less discriminating then PFGE in terms of distinguishing various isolates of the same species. However, when the results derived from the use of three random primers independently are combined, the discriminatory power is usually equal to that of PFGE and is still less costly and time consuming.

RAPD, also referred to as arbitrarily primed PCR (AP-PCR) analysis, has been found to be a rapid and valuable technique for distinguishing different strains of the same species (Lawrence, Harvey, and Gilmour, 1993) with a high level of strain discrimination (Lawrence and Gilmour, 1995). It is a particularly useful technique for genetic typing of human pathogens from foods, processing plants, and food-borne outbreaks. In addition, it has been found to be a powerful method that can advantageously replace other more cumbersome typing methods such as serotyping, ribotyping, multilocus enzyme electrophoresis, restriction enzyme analysis, and phage typing, and has been found to be very efficient in differentiating strains while still allowing the clear recognition of clusters (Boerlin et al., 1995). RAPD has been found capable of distinguishing strains of a given species with identical 16s rDNA sequences (Czajka et al., 1993). This high level of discrimination should allow RAPD to be used in establishing the persistence of a specific strain in foods and in processing plants and its distinction from transient strains of the same species.

B. Mechanism of RAPD

In conventional PCR, a known DNA sequence is amplified by using two primers, one that anneals to the 3′ end of the sequence on the A strand and is extended inward with *Taq* polymerase from the 3′ end of the primer. The second primer anneals to the 3′ end of the B strand and is also extended inward from the 3′ end (Figure 1.1). After the first cycle and denaturation, four target sequences are then available for duplication. The sequence is usually highly specific for the target gene and occurs only once (monomorphic) in the genomic DNA of the organism.

With RAPD a single random primer of about ten nucleotides is used with no known target sequence being required, and the first round of amplification results in single strands having palindromatic termini (Figure 1.17). The single randomly chosen primer targets specific but unknown sites in the genomic DNA, which are polymorphic (repeating) with respect to the terminal sequences to which the single primer anneals. During subsequent cycles a number of different target sequences are amplified, many of which will be of differing base pair length so as to generate a variety of DNA agarose bands resulting in a specific DNA banding pattern for each culture (Figure 1.18).

The ratio of primer to template in the RAPD reaction is critical. Template DNA concentration should be carefully titered against a standard concentration of primer so as to reveal the most reproducible amplified products (del Tufo and Tingey, 1994). A hazy smear obscuring the amplified bands on the agarose gel is usually caused by failure to saturate the DNA template with primer. This is easily corrected by adjusting the ratio of primer to template DNA (del Tufo and Tingey, 1995). Meuner and Grimont (1993) expressed concern about the reproducibility of RAPD profiles but concluded that reproducibility was excellent with standardized methodology. They found that reproducibility of banding patterns was dependent on the make of the thermal cycler, with variations in patterns occurring with less rigid temperature control. Reproducibility of RAPD with whole cells is critically dependent on all strains being in the same stage of growth (Boerlin et al., 1995). Mazurier and Wernars (1992) obtained reproducibility in RAPD profiling of strains of *L. monocytogenes* by using suspensions of washed whole cells grown overnight in broth that were adjusted to an absorbance at 600 nm of 1.5 yielding 7.5×10^6 CFU/ml 5 μl of the cell suspension and that were then incorporated directly into a 50-μl PCR reaction volume without a prior cell lysis step. The same cell density among isolates

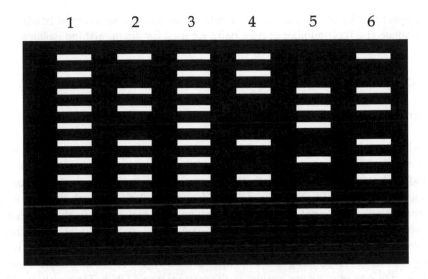

Figure 1.18 RAPD banding profiles of six bacterial isolates of the same species. Isolates corresponding to lanes 1 and 3 have identical banding patterns and are therefore considered of identical clonal origin. Isolates corresponding to lanes 2 through 6 are distinguishable from each of the other isolates.

being subjected to RAPD comparison is particularly important for reproducibility of faint bands.

VII. MULTILOCUS SEQUENCE TYPING (MLST)

A. Introduction

MLST makes use of the PCR to amplify sequences of several housekeeping genes (see Table 11.2 in Chapter 11). The resulting amplicons are then subjected to sequence analysis, and the sequences derived from the various isolates are compared with respect to each individual gene for differences.

VIII. RESTRICTION FRAGMENT LENGTH POLYMORPHISM (RFLP) AND PCR-RFLP

A. Introduction

RFLP refers to the application of one or more restriction nucleases directly to the genomic DNA of a bacterium and the resulting products are separated electrophoretically as bands in an agarose gel. PCR-RFLP refers to the initial amplification of a specific gene sequence by PCR followed by restriction of the resulting amplicons with one or more restriction nucleases and separation of the products electrophoretically in an agarose. With both techniques the numbers of bands from each isolate and their number of base pairs are used for distinguishing isolates.

IX. AMPLIFIED FRAGMENT LENGTH POLYMORPHISM (AFLP)

A. Introduction

AFLP involves digestion of purified genomic DNA using a restriction nuclease such as *Hind*III, followed by the ligation of the resulting fragments to a double-stranded oligonucleotide adapter that is complementary to the base sequence of the restriction site using T4 DNA ligase. The adapters are designed such that the original restriction sites are not restored after ligation, thus preventing further restriction digestion. Because the adapters are not phosphorylated, adapter-to-adapter ligation is prevented. Selective amplification by PCR of sets of these fragments is achieved using primers corresponding to the contiguous base sequences in the adapter restriction site plus one or more nucleotides in the original target DNA. The resulting DNA fragments amplified by PCR are then resolved by gel electrophoresis.

REFERENCES

Abolmaaty, A., Vu, C., Oliver, J., Levin, R. 2000. Development of a new lysis solution for releasing genomic DNA from bacterial cells for DNA amplification by polymerase chain reaction. *Microbios.* 101:181–189.

Albert, J., Fenyo, E. 1990. Simple, sensitive and specific detection of human immunodeficiency virus type 1 in clinical specimens by polymerase chain reactions with nested primers. *J. Clin. Microbiol.* 28:1560–1564.

Arbeit, R., Slutsky, A., Barber, T., Maslow, J., Niemczyk, S., Falkinham III, J., O'Connor, G., von Reyn, C. 1993. Genetic diversity: Among strains of *Mycobacterium avium* causing monoclonal and polyclonal bacteremia in patients with AIDS. *J. Infect. Dis.* 167:1384–1390.

Barrett, T., Lior, H., Green, J., Khakhria, R., Wells, J., Bell, B., Greene, K., Lewis, J., Griffin, P. 1995. Laboratory investigation of a multistate food-borne outbreak of *Escherichia coli* O157:H7 by using pulsed-field gel electrophoresis and phage typing. *J. Clin. Microbiol.* 32:3013–3017.

Birren, B., Lai, E., Clark, S., Hood, L., Simon, M. 1988. Optimized conditions for pulsed field gel electrophoretic separations of DNA. *Nucleic Acids Res.* 16:7563–7582.

Boerlin, P., Bannerman, E., Ischer, F., Rocurt, J., Bille, J. 1995. Typing of *Listeria monocytogenes:* A comparison of random amplification of polymorphic DNA with 5 other methods. *Res. Microbol.* 146:35–49.

Böttger, E. 1990. Frequent contamination of *Taq* DNA polymerase with DNA. *Clin. Chem.* 36:1258–1259.

Breslauer, K., Ronald, F., Blicker, H., Marky, L. 1986. Predicting DNA duplex stability from the base sequence. *Proc. Natl. Acad. Sci.* 83: 3746–3750.

Brett, M., Short, P., McLauchlin, J. 1998. A small outbreak of listeriosis associated with smoked mussels. *Int. J. Food Microbiol.* 43:223–229.

Cockerill, III, F., Uhl, J. 2002. Applications and challenges of real-time PCR for the clinical microbiology laboratory. In: *Rapid Cycle Real-Time PCR—Methods and Application*. Reischl, U., Wittwer, C., Cockerill, F., Eds., Springer Verlag: New York, pp. 3–27.

Corless, C., Guiver, M., Borrow, R., Edwards-Jones, V., Kaczmarski, E., Fox, A. 2000. Contamination and sensitivity issues with a real-time universal 16S rRNA PCR. *J. Clin. Microbiol.* 38:1747–1752.

Czajka, J., Bsat, N., Piani, M., Russ, W., Sultana, K., Wiedman, M. Whitaker, J., Batt. 1993. Differentiation of *Listeria monocytogens* and *Listeria innocua* by 16S rRNA genes and intraspecies discrimination of *Listeria monocytogenes* strains by random amplified polymorphic DNA polymorphisms. *Appl. Environ. Microbiol.* 59:304–308.

del Tufo, J., Tingey, S. 1994. RAPD assay. In: *Protocols for Nucleic Acid Analysis by Nonradioactive Probes*. P.G. Isaac, Ed., Humana Press: Totowa, NJ, pp. 237–241.

Diaco, R. 1995. Practical considerations for the design of quantitative PCR assays. In: *PCR Protocols: A Guide to Methods and Applications*. M.A. Innis, D.H. Gelfand, J.J. Sinsky, Eds., Academic Press: New York, pp. 84–108.

Freier, S., Kierzek, R., Jaeger, J., Sugimoto, N., Caruthers, M., Neilson, T., Turner, D. 1986. Improved free-energy parameters for predictions of RNA duplex stability. *Proc. Natl. Acad. Sci.* 83:9373–9377.

Gilgen, M., Hubner, P., Hofelein, C., Luthy, J., Candrian, U. 1998. PCR-based detection of verotoxin-producing *Excherichia coli* (VTEC) in ground beef. *Res. Microbiol.* 149:145–154.

Guan, J., Levin, R. 2002a. Sensitive and rapid detection of *Escherichia coli* O157:H7 in ground beef by nested PCR incorporating immunomagnetic separation. *J. Food. Biotechnol.* 16:145–154.

Guan, J., Levin, R. 2002b. Quantitative detection of *Escherichia coli* O157:H7 in ground beef by immunomagnetic separation and competitive polymerase chain reaction. *J. Food Biotechnol.* 16:155–156.

Gurrieri, S., Rizzarelli, E., Beach, D., Bustamante, C. 1990. Imaging of kinked configurations of DNA molecules undergoing orthogonal field alternating gel electrophoresis by fluorescene microscopy. *Biochemistry* 29:3396–3401.

Heid, C., Stevens, J., Livak, K. Williams, P. 1996. Real time quantitative PCR. *Genome Res.* 6:986–994.

Innis, M., Gelfand, D. 1990. Optimization of PCRs. In: *PCR Protocols: A Guide to Methods and Applications*. M.A. Innis, D.H. Gelfand, J.J. Sninsky, T.J. White, Eds., Academic Press: New York, pp. 3–20.

Jackson, D., Hayden, J., Quirke, P. 1991. Improving the sensitivity and specificity of PCR amplification. In: *PCR: A Practical Approach*. M.J. McPherson, P. Quirke, G.R. Taylor, Eds., Oxford Press: Oxford, UK, pp. 42–50.

Johansson, T., Rantala, L., Palmu, L., Honkanen-Buzalski, T. 1999. Occurrence and typing of *Listeria monocytogenes* strains in retail vacuum-packed fish products and in a production plant. *Int. J. Food Microbiol.* 47:111–119.

Kapperud, G., Vardund, T., Skjerve, E., Horres, E., Michaelsen, T. 1993. Detection of pathogenic *Yersinia enterocolitica* in foods and water by immunomagnetic separation, nested polymerase chain reactions, and colorimetric detection of amplified DNA. *Appl. Environ. Microbiol.* 59:2938–2944.

Lawrence, L., Gilmour, A. 1995. Characterization of *Listeria monocytogenes* isolated from poultry products and from poultry-processing environment by random amplification of polymorphic DNA and multilocus enzyme electrophoresis. *Appl. Environ. Microbiol.* 61:2139–2144.

Lawrence, L., Harvey, J., Gilmour, A. 1993. Development of a random amplification of poly-morphic DNA typing method for *Listeria monocytogenes*. *Appl. Environ. Microbiol.* 59:3117–311.

Lindqvist, R. 1999. Detection of *Shigella* spp. in food with a nested PCR method—sensitivity and performance compared with a conventional culture method. *J. Appl. Microbiol.* 86:971–978.

Mazurier, S., Wernars, K. 1992. Typing of *Listeria* strains by random amplification of poly-morphic DNA. *Res. Microbiol.* 143:499–505.

McConaughy, B., Laird, C., McCarthy, B. 1969. Nucleic acid reassociation in formamide. *Biochemistry* 8:3289–3295.

Meuner, J., Grimont, P. 1993. Factors affecting reproducibility of random amplified polymor-phic DNA fingerprinting. *Res. Microbiol.* 144: 373–379.

Mori, Y., Nagamine, K., Tomita, N., Notomi, T. 2001. Detection of loop-mediated isother-mal reaction by turbidity derived from magnesium pyrophosphate formation. *Biochem. Biophys. Res. Commun.* 289:150–154.

Nagamine, K., Hase, T., Notomi, T. 2002. Accelerated reaction by loop-mediated isothermal amplification using loop primers. *Mol. Cell. Probes* 16:223–229.

Nazarenko, I., Lowe, B., Darfler, M., Ikonomi, P., Scuster, D., Rashtchian, A. 2002. Multiplex quantitative PCR using self-quenched primers labeled with a single fluorophore. *Nucleic Acids Res.* 30:e37.

Notomi, T., Okayama, H., Masubuchi, H., Yonekawa, T., Watanabe, K., Amino, N., Hase, T. 2000. Loop-mediated isothermal amplification of DNA. *Nucleic Acids Res.* 28:e63.

Ozbas, Z., Lehner, A., Wagner, M. 2000. Development of a multiplex and semi-nested PCR assay for detection of *Yersinia enterocolitica* and *Aeromonas hydrophilia* in raw milk. *Food Microbiol.* 17:197–203.

Rand, K.H., Houck, H. 1990. *Taq* polymerase contains bacterial DNA of unknown origin. *Mol. Cell. Probes* 4:445–450.

Rupf, S., Merte, K., Eschrich, K. 1997. Quantification of bacteria in oral samples by competi-tive polymerase chain reaction. *J. Dent. Res.* 78:850–856.

Rychlik, W., Spencer, W., Rhoads, R. 1990. Optimization of annealing temperature for DNA amplification in vitro. *Nucleic Acids Res.* 18:6409–6412.

Sader, H., Pignatari, A., Leme, I., Burattini, M., Tancresi, R., Hollis, R., Jones, R. 1993. Epidemiologic typing of multiply drug-resistant *Pseudomonas aeruginosa* isolated from an outbreak in an intensive care unit. *Diag. Microbiol. Infect. Dis.* 17:13–18.

Struelens, M., De Ryck, R., Deplano, A. 2001. Analysis of microbial genomic macrorestric-tion patterns by pulsed-field gel electrophoresis (PFGE) typing. In: *New Approaches for the Generation and Analysis of Microbial Typing Data*. L. Diikshoorn, K.J. Towner, M. Struelens, Eds., Elsevier: New York, pp. 159–176.

Suggs, S., Hirose, T., Myake, E., Kawashima, M., Johnson, K., Wallace, R. 1981. Using puri-fied genes. ICN-UCLA Symp. *Mol. Cell. Biol.* 23:683–693.

Tenover, F., Arbeit, R., Goering, R., Mickelsen, P., Murray, B., Persing, D., Swaminathan, B. 1995. Interpreting chromosomal DNA restriction patterns produced by pulsed-field gel electrophoresis: Criteria for bacterial strain typing. *J. Clin. Microbiol.* 33:2233–2239.

Waage, A., Vardund, T., Lund, V., Kapperud, G. 1999. Detection of low numbers of *Salmonella* in environmental water, sewage and food samples by a nested polymerase chain reaction assay. *J. Appl. Microbiol.* 87:418–428.

Wiedman, M., Barany, F., Batt, C. 1995. Detection of *Listeria monocytogenes* by PCR-coupled ligase chain reaction. In: *PCR Strategies*. M.A. Innis, D.H. Gelfand, and J.J. Sninsky, Eds., Academic Press: New York, pp. 347–361.

Wilhelm, J., Hahn, M., Pingoud, A. 2000. Temperature non-homogeneity in rapid airflow-based cycler significantly affects real-time PCR. *Biotechniques* 33:508–512.

Yap, E., Mcgee, J. 1991. Short PCR product yields improved by lower denaturation temperatures. *Nucleic Acids Res.* 19:1713.

Zarchar, V., Thomas, A., Goustin, A. 1993. Absolute quantification of target DNA: A simple competitive PCR for efficient analysis of multiple samples. *Nucleic Acids Res.* 8:2917–2018.

CHAPTER 2

Escherichia coli O157:H7

I. CHARACTERISTICS OF THE ORGANISM

Enterohemorrhagic *E. coli* (EHEC) of serotype O157:H7 causes hemorrhagic colitis, which is characterized by microangiopathic haemolytic anaemia, thrombo-cytopenia, and central nervous system symptoms (Karmali, 1989) that may develop into life-threatening renal failure involving hemolytic-uremic-syndrome (HUS), particularly with children. *E. coli* isolates are serologically distinguished on the basis of three major surface antigens: somatic (O), flagella (H), and capsule (K) anti-gen. Although more than 100 *E. coli* serotypes produce Shiga-like toxins (SLTs), O157:H7 is the predominant serotype implicated in foodborne diseases. *E. coli* O157:H7 strains are sorbitol negative, β-D-glucuronidase negative, exhibit poor or no growth at 44.5°C, possess an attaching and effacing gene (*eae*), possess a 66-MDa plasmid, and express an uncommon 5000 to 8000 mol. wt. outer membrane protein (OMP; Padhye and Doyle, 1991). The infectious dose is considered to be about 1 CFU/g of raw food (Centers for Disease Control and Prevention, 1995; Griffin et al., 1994; Zhao, Doyle, and Wang, 1994).

II. VIRULENCE FACTORS

A. Hemolysins

Two different plasmid-encoded hemolysins have been described for Shiga toxin-producing *E. coli* (STEC). Alpha hemolysin encoded by the *hlyA* gene is formed by porcine edema disease-causing STEC strains that produce Stx variant 2e (Imberechts, De Greve, and Lintermans, 1992) and by *E. coli* causing urinary tract infections and septicemia (Korhonen et al., 1985; Orskov and Orskov, 1985). The alpha hemolysin produces a broad zone of hemolysis on enterohemolysin agar (Beutin et al., 1989). The second hemolysin encoded by the *elyA* gene is secreted exclusively by human STEC strains and produces a narrow, turbid, hemolytic halo on enterohemolysin agar (Beutin et al., 1989) and is 62% to 64% homologous to the *hlyA* gene (Kuhnert et al., 1997; Schmidt and Karch, 1996). The two genes *hlyA* and *elyA* harbored by STEC were found to be readily distinguished by PCR-restriction fragment length polymorphism (PCR-RFLP) analysis (Lehmacher et al., 1998). *Alu*I, *eco*RI, and *Mlu*I PCR-RFLP showed that all 93 human isolates of STEC examined harbored only the *elyA* gene and not the *hlyA* gene. Primers *hlyA*-start/*hlyA*-end homologous to both genes were used to amplify a 202-bp sequence of the *hlyA* gene and a 199-bp sequence of the *elyA* gene.

B. Intimin

Intimate adherence to intestinal cells is mediated by intimin, a 94-kda OMP encoded by the *eaeA* locus and results in effacement of underlying microvilli resulting in adhesion and effacement (AEA) lesions. The *eae* gene locus houses the *eaeA*, *eaeB*, and *sep* genes termed LEE for locus of enterocyte effacement and is absent in commensulate strains of *E. coli*.

C. Shiga-Like Toxins

Virulence of *E. coli* O157:H7 involves a number of factors including the production of one or more bacteriophage encoded SLTs, also called verotoxins (VTXs), that are lethal to Vero cells. VT1 and VT2 are synonymous with SLT-I and SLT-II and have been renamed Stx1 and Stx2. Stx1 from various strains are homogeneous whereas Stx2 exhibits heterogeneity among toxin-producing strains resulting in subgroups designated Stx2b, Stx2c, Stx2d, Stx2e, and Stx2f. Stxs consist of linked A and B subunits. The B subunits bind specifically to galactose moieties of globoglycolipids on the eukaryote cell surface. The A subunit is then cleaved from the B subunit, enters the cell, and proceeds to enzymatically rupture an N-glycoside bond of ribosomes so as to prevent the binding of tRNA resulting in the inhibition of protein synthesis.

D. Locus for Enterocyte Effacement

Among the various mechanisms involved in the virulence of *E. coli* O157:H7 is the ability to intimately adhere to enterocytes by an attaching and effacing mechanism. Attaching and effacing lesions are encoded on a pathogenicity island termed the locus for enterocyte effacement (LEE). LEE encodes a type III secretion system and *E. coli* secreted proteins that deliver effector molecules to the host cell and disrupt the host cytoskeleton (Donnenberg, Kaper, and Finlay, 1997; Elliot et al., 1998; Perna et al., 1998). LEE also carries the *eae* gene, which encodes an outer membrane protein (intimin) required for intimate attachment to epithelial cells (Yu and Kaper, 1992) and has been used as a convenient diagnostic marker for LEE-positive STEC strains (Gannon et al., 1993; Louie et al., 1994; Paton and Paton, 1998a). However, the presence of *eae* is not absolutely linked to human virulence in that HUS can be caused by LEE-negative strains (Paton and Paton, 1998b). SLT-II is associated with an increased risk of developing HUS (Boerlin et al., 1999; Kleanthous et al., 1990; Ostroff et al., 1989). Paton et al. (2001) described a gene designated *saa*, which is carried on a large plasmid of certain LEE-negative but not LEE-positive strains. This gene encodes a novel outer membrane protein that appears to function as an autoagglutinating adhesin.

E. Extracellular Serine Protease

Brunder, Schmidt, and Karch (1997) identified and characterized a novel extracellular protease designated EspP encoded by the large ~90-kb plasmid that cleaves human coagulation factor V required for prothrombin and clot formation. A specific immune response against EspP was detected in sera from patients suffering from EHEC infections. The degradation of factor V is thought to contribute to the prolonged mucosal hemorrhage observed in patients with hemorrhagic colitis.

F. Additional Virulence Factors

All *E. coli* O157:H7 strains harbor a large plasmid of about 66 MDa. However, several other SLT-producing *E. coli* serotypes also harbor this plasmid (Fratamico et al., 1995). An enterohemorrhagic toxin is encoded by this plasmid. In addition, a 90-kb plasmid encodes potential virulence factors, such as an enterohaemolysin (Schmidt, Kernbach, and Karch, 1996), a catalase-peroxidase (Brunder, Schmidt, and Karch, 1996), a serine protease (Brunder, Schmidt, and Karch, 1997), and a type II secretion pathway system (Schmidt, Henkel, and Karch, 1997). PCR detection relying solely on sequences specific for these plasmids will not guarantee detection of EHEC O157:H7 strains.

III. PHENOTYPIC VARIATION OF *E. COLI* O157:H7

Phenotypic variation has been found to occur in strains of *E. coli* O157:H7. Feng (1995) reported the recovery of sorbitol-fermenting isolates of *E. coli* O157:H7 from sorbitol-containing foods. A case of blood diarrhea in 1995 was reported to yield a β-D-glucuronidase-producing strain of *E. coli* O157:H7 (Hays et al., 1995). Ware, Abbott, and Janda (2000) reported on the percentage of phenotypically aberrant strains of *E. coli* O157:H7 among a total of 657 isolates. Typical reactions observed were as follows: no acid from rhamnose (1.2%), no acid from sucrose (0.8%), no acid from lactose (0.8%), no acid from glucose (0.8%), positive hydrolysis of urea (0.5%), negative production of lysine decarboxylase (0.3%), negative production of ornithine decarboxylase (0.3%), acid from D-sorbitol (0.3%), and negative production of indole (0.3%). None of the strains analyzed possessed multiple aberrant properties, nor were aberrant phenotypes associated with outbreaks with the exception of six strains that failed to ferment L-rhamnos linked to a daycare center outbreak. In addition, 21 strains were found to be nonmotile after incubation at 35°C for 96 hr in motility deeps. Many nonmotile isolates required multiple passages (4–17) before the presence of H7 antigen could be detected. In some instances, the H7 gene could only be demonstrated using the PCR. The authors concluded that such aberrant reactions could result in misidentification of O157:H7 isolates as some other species and discarded.

Hara-Kudo, Miyahara, and Kumagai (2000) found that strains of *E. coli* O157:H7 were culturable on agar media after being left in dH₂O at 18°C for 21 months. However, a number of cells had lost their O157 O antigenicity. The lost O157 O antigenicity was not recovered by growth in tryptic soy broth (TSB). Other phenotypic characteristics of O157:H7 strains were retained along with the SLT gene.

Wetzel and LeJeune (2007) reported on the isolation of a number of *E. coli* O157:H7 isolates that were PCR-negative for the presence of *slt-I* and various derivatives of *slt-II* but positive for the presence of the *eae* and *ehxA* genes. These isolates were experimentally lysogenized by Stx2-converting bacteriophage to a positive SLT-II state.

A. Conventional PCR

Genes used for PCR detection of *E. coli* O157:H7 and various phenotypic properties of isolates are presented in Table 2.1. Table 2.2 presents primer and probe sequences utilized with *E. coli* O157:H7 PCR assays.

Gannon et al. (1992) developed a PCR assay for detection of *slt-I* and *slt-II* genes in SLT-I- and SLT-II-producing strains of *E. coli* seeded into ground beef. When 25 g of ground beef were cultured in 225 ml of mTSB for 4 hr, 8 CFU/g were detected. With an enrichment time of 6 hr, 0.8 CFU/g was detected. DNA extraction and purification were essential to achieve these levels of detection.

Gilgen et al. (1998) developed a PCR assay for detection of verotoxin-producing *Escherichia coli* (VTEC) in ground beef. Ground beef (10 g) was stomached with 75 ml of homogenization buffer (10 mmolar Tris-HCl at pH 8.0, 150 mmolar NaCl, 2 mmolar ethylenediaminetetraacetic acid (EDTA), 0.1% sodium dodecyl sulfide (SDS), and 0.5 mg/ml pronase with A mesh insert. The bags were then incubated for 30 min at 37°C to release bacterial cells adsorbed to the meat surface. The sample (50 ml) was then centrifuged at 30,000 × g for 30 min. The pellet was then lysed in 2 ml of lysis buffer (10 mmolar Tris-HCl at pH 8.0, 50 mmolar KCl, 1.5 mmolar MgCl₂ containing 4 mg/ml of lysozyme, and 0.2 mg/ml proteinase K for 20 min at room temperature followed by 60 min at 60°C. The lysate (1 ml) was then centrifuged for 10 min at 14,500 × g and 450 µl used for nucleic acid isolation with the "Wizard" DNA resin. Ten µl of the 50-µl eluate were then incorporated into a 100-µl PCR reaction volume. Two nested sequential assays were used to detect the genes encoding VT1 and VT2 in various strains of *E. coli* regardless of serotype. The primers I-1/I-2 (Table 2.2) amplified a 614-bp sequence of the *vt1* gene. The internally nested primers I-3/I-4 (Table 2.2) amplified a 347-bp sequence of the *vt1* gene. The primers II-1/II-2 (Table 2.2) amplified a 779-bp sequence of the *vt2* gene. The internally nested primers II-3/II-4 (Table 2.2) amplified a 372-bp sequence of the *vt2* gene. The detection limit in ground beef was 110 CFU/10 g. The use of a 6 hr enrichment resulted in a detection limit of 1 CFU/10 g.

Karch and Meyer (1989) developed a single pair of primers MK1/MK2 (Table 2.2) capable of amplifying a 227-bp and a 224-bp sequence from *slt-I* and

Table 2.1   Genes Used for the PCR Identification of _E. coli_ O157:H7

Gene	Description
eaeA	Chromosomal gene encoding intimin, which is involved in attaching and effacing lesions of enterocytes with LEE-positive strains
saa	Large plasmid-bearing gene that encodes an outer membrane protein that functions as an autoagglutinating adhesin in LEE-negative strains but not LEE-positive strains
stx (_slt_)	Encodes the Stx toxin
stx$_1$ (_slt-I_)	Encodes the Stx$_1$ toxin
stx$_2$ (_slt-II_)	Encodes the Stx$_2$ toxin
stx$_{2b}$ (_slt-IIb_)	Encodes subtype of corresponding Stx$_2$ toxin
stx$_{2c}$ (_slt-IIc_)	Encodes subtype of corresponding Stx$_2$ toxin
stx$_{2d}$ (_slt-IId_)	Encodes subtype of corresponding Stx$_2$ toxin
stx$_{2e}$ (_slt-IIe_)	Encodes subtype of corresponding Stx$_2$ toxin
stx$_{2f}$ (_slt-IIf_)	Encodes subtype of corresponding Stx$_2$ toxin
flic$_{H7}$	Encodes the flagellin of H7 serotype
rfbE	Encodes an enzyme necessary for O-antigen synthesis
uidA	Encodes β-glucuronidase
Sil$_{O157}$	A small 2634 inserted locus (SIL) present in O157 serotypes that encodes an OMP designated IHP1
hly$_{933}$	Encodes the enterohemolysin of human O157:H7 strains
hly$_{21}$	Encodes the allelic enterohemolysin of bovine O157:H7 strains
ehxA	Encodes the enterohemolysin
orfU$_{leeO157}$	Encodes the origin of replication for the locus of enterocyte effacement with O157 strains
EspP	Encodes an extracellular serine protease produced by O157:H7 isolates that cleaves human coagulation factor V and thought to result in prolonged hemorrhagic colitis

slt-II, respectively. The primers amplified sequences from a conserved region of both _slt-I_ and _slt-II_ genes.

Pollard et al. (1990) developed primers for the detection of the _slt-I_ and _slt-II_ genes harbored by strains of _E. coli_. The primers VT1a/VT1b (Table 2.2) amplified a 130-bp sequence of the _slt-I_ gene and primers VT2a/VT2b (Table 2.2) amplified a 346-bp sequence of the _slt-II_ gene. Detection of these genes was restricted to strains of _E. coli_ with the exception that _slt-I_ was detected in five strains of _Shigella dysenteriae_.

Li and Drake (2001) developed a quantitative competitive PCR assay for detection and quantification of _E. coli_ O157:H7 in skim milk. The advantage of a competitive PCR assay is that it automatically corrects for partial inhibition of the PCR by components from a food. The primers TXAF/TXAR (Table 2.2) amplified a 401-bp sequence of the _slt-II_ gene. The primer TXAF1 in conjunction with TXAR amplified a competitive 275-bp sequence of the _slt-II_ gene. Target cop numbers were obtained from the standard curve where the log of the ratio of fluorescence intensities of the competitive band and the target band were plotted against the log of the concentration of competitive molecules and the use of equivalent fluorescent

Table 2.2   PCR Primers and DNA Probes[a]

Primer or Probe	Sequence (5' → 3')[a]	Size of Sequence (bp)	Amplified Gene or DNA Target Sequence	Reference
MFS1F MFS1R	ACG-ATG-TGG-TTT-ATT-CTG-GA CT-CAC-GTC-ACC-ATA-CAT-AT	166	60-MDa plasmid	Fratamico et al. (1995)
MK1 MK2	TTT-ACG-ATA-GAC-TTC-TCG-AC CAC-ATA-TAA-ATT-ATT-TCG-GTC	227 and 224	*slt-I* & *slt-II*	Fratamico et al. (1995) from Karch and Meyer (1989)
AE19 AE20	CAG-GTC-GTC-GTG-TCT-GCT-AAA TCA-GCG-TGG-TTG-GAT-CAA-CCT	1087	*eaeA*	Fratamico et al. (1995) from Gannon et al. (1993)
LEE LEE	CCA-TAA-TCA-TTT-TAT-TTA-GAG-GGA GAG-AAA-TAA-ATT-ATA-TTA-ATA-GAT-CGG-A	633	upstream of *eaeA*	Meng et al. (1997)
SLTI SLTI	TGT-AAC-TGG-AAA-GGT-GGA-GTA-TAC-A GCT-ATT-CTG-AGT-CAA-CGA-AAA-ATA-AC	210	*slt-I*	Meng et al. (1997)
SLTII SLTII	GTT-TTT-CTT-CGG-TAT-CCT-ATT-CC GAT-GCA-TCT-CTG-GTC-ATT-GTA-TTA-C	484	*slt-II*	Meng et al. (1997)
LP30 LP31	CAG-TTA-AAG-TGG-TGG-CGA-AGG CAC-CAG-ACA-ATG-TAA-CCG-CTG	348	*slt-I*	Cebula, Payne, and Feng (1995)
LP43	ATC-CTA-TTC-CCG-GGA-GTT-TAC-G	584	*slt-II*	Cebula, Payne, and Feng (1995)

LP44	GCG-TCA-TCG-TAT-ACA-CAG-GAG-C			
PT-2 PT-3	GCG-AAA-ACT-GTG-GAA-TTG-GG TGA-TGC-TCC-ATA-ACT-TCC-TG	252	*uidA* allele	Cebula, Payne, and Feng (1995)
O157BF O157Br O157rfbe	AAA-TAT-AAA-GGT-AAA-TAT-GTG-GGA-ACA-TTT-GG TGG-CCT-TTA-AAA-TGT-AAA-CAA-CGG-TCA-T FAM-CGC-TAT-GGT-GAA-GGT-GGA-ATG-GTT-GTC-ACG-AAT-AGC-DABCYL	149	*rfbe*	Fortin, Mulchandani, and Chen (2001)
TXAF TXAR	TTA-AAT-GGG-TAC-TGT-GCC-T CAG-AGT-GGT-ATA-ACT-GCT-GTC	401	*slt-II*	Li and Drake (2001)
TXAF1 TXAR	TTA-AAT-GGG-TAC-TGT-GCC-TTC-AGG-GGA-CCA-CAT-CGG-T CAG-AGT-GGT-ATA-ACT-GCT-GTC	275	*slt-II*	Li and Drake (2001)
MK1 MK2	CAG-CTC-TTC-AGA-TAG-CAT-TT CAG-CTC-TTC-AGA-TAG-CAT-TT	227 and 224	*slt-I* and *slt-II*	Karch and Meyer (1989)
SZ-I SZ-II SZI-97	CCA-TAA-TCA-TTT-TAT-TTA-GAG-GGA GAG-AAA-TAA-ATT-ATA-TTA-ATA-GAT-CGG-A FAM-TTG-CTG-CAG-GAT-GGG-CAA-CTC-TTG-TAMRA	632	*eaeA*	Oberst et al. (1998)

Continued

Table 2.2 PCR Primers and DNA Probes[a] (Continued)

Primer or Probe	Sequence (5′ → 3′)[a]	Size of Sequence (bp)	Amplified Gene or DNA Target Sequence	Reference
Forward	TTA-AAT-GGG-TAC-TGT-GCC-T	401	slt-II	McKillip and Drake (2000)
Reverse	CAG-AGT-GGT-ATA-ACT-GCT-GTC			
Probe	FAM-GCG-AGT-TGA-CCA-TCT-TCG-TCC-TCG-C-DABCYL			
VT1-A	CGC-TGA-ATG-TCA-TTC-GCT-CTG-C	302	slt-I	García-Sánchez et al. (2007) from Blanco et al. (2003)
VT1-B	CGT-GGT-ATA-GCT-ACT-GTC-ACC			
VT2-A	CTT-CGG-TAT-CCT-ATT-CCC-GG	516	slt-II	García-Sánchez et al. (2007) from Blanco et al. (2003)
VT2-B	CTG-CTG-TGA-CAG-TGA-CAA-AAC-GC			
HlyA1	GGT-GCA-GCA-GAA-AAA-GTT-GTA	1551	ehxA	García-Sánchez et al. (2007) from Schmidt, Beutin, and Karch (1995)
HlyA4	TCT-CGC-CTG-ATA-GTG-TTT-GGT-A			
EAE-1	GAG-AAT-GAA-ATA-GAA-GTC-GT	775	eaeA	García-Sánchez et al. (2007) from Blanco et al. (2003)
EAE-2	GCG-GTA-TCT-TTC-GCG-TAA-TCG-CC			
EAE-19	CAG-GTC-GTC-GTG-TCT-GCT-AAA	1087	eae_1	García-Sánchez et al. (2007) from Desmarchelier et al. (1998)
EAE-20	TCA-GCG-TGG-TTG-GAT-CAA-CCT			
O157-AF	AAG-ATT-GCG-CTG-AAG-CCT-TTG	497	rfbe$_{O157}$	

Primer	Sequence	Size	Target	Reference
O157-AR	CAT-TGG-CAT-CGT-GTG-GAC-AG			
H7-F H7-R	GCG-CTG-TCG-AGT-TCT-ATC-GAG-C CAA-CGG-TGA-CTT-TAT-CGC-CAT-TCC	625	$fliC_{h7}$	Garcia-Sánchez et al. (2007) from Gannon et al. (1997)
H7-F H7-R	TAC-CAC-CAA-ATC-TAC-TGC-TG TAC-CAC-CTT-TAT-CAT-CCA-CA	560	$fliC_{h7}$	Nagano et al. (1998)
O157-F O157-R	AAC-GGT-TGC-TCT-TCA-TTT-AG GAG-ACC-ATC-CAA-TAA-GTG-TG	678	O157	Nagano et al. (1998)
O157 A-F O157 A-R	AAG-ATT-GCG-CTG-AAG-GCT-TTG CAT-TGG-CAT-CGT-GTG-GAC-AG	497	$rfbe_{O157}$	Fach et al. (2003) from Desmarchelier et al. (1998)
O157 P-F8 O157 P-R8	CGT-GAT-GAT-GTT-GAG-TTG AGA-TTG-GTT-GGC-ATT-ACT-G	420	$rfbe_{O157}$	Fach et al. (2003) from Maurer et al. (1999)
P1EH P2EH	AGG-CGA-CTG-AGG-TCA-CT ACG-CTG-CTC-ACT-AGA-TGT	476	$eaeA_{O157}$	Fach et al. (2003) from Louie et al. (1994)
FLIC H7-F FLIC H7-R	GCG-CTG-TCG-AGT-TCT-ATC-GAG-C CAA-CGG-TGA-CTT-TAT-CGC-CAT-TCC	625	$fliC_{H7}$	Fach et al. (2003) from Gannon et al. (1997)

Continued

Table 2.2   PCR Primers and DNA Probes[a] (Continued)

Primer or Probe	Sequence (5' → 3')[a]	Size of Sequence (bp)	Amplified Gene or DNA Target Sequence	Reference
Sz I	CCA-TAA-TCA-TTT-TAT-TTA-GAG-GGA	632	$Orfb_{LEEO157}$	Fach et al. (2003) from Meng et al. (1996)
SZII	GAG-AAA-TAA-ATT-ATA-TTA-ATA-GAT-CGG-A			
RJD3	TTA-AAA-CCG-GTG-ACG-TGA-TGA-TGG-TG	125	SIL_{O157}	Fach et al. (2003) from Perelle et al. (2002)
SG7	AGC-AAC-AGG-CGC-AGA-TCG-TAG-CCA-C	578	SIL_{O157}	Perelle et al. (2002)
SF6	CGC-AGA-AAT-ACC-GGC-TTT-AAG-TAC-C			
VS8	GGC-GGA-TTA-GAC-TTC-GGC-TA	120	eaeA	Kawasaki et al. (2005) from Sharma, Dean-Nystrom, and Casey (1999)
VS9	CGT-TTT-GGC-ACT-ATT-TGC-CC			
SLTI-F3	GGC-ATT-AAT-ACT-GAA-TTG-TCA-TC	416	slt-I	Witham et al. (1996) from Gannon et al. (1992)
SLTI-R	CTG-AAT-CCC-CCT-CCA-TTA-TG			
SLT 102	FAM-CAG-AAT-GGC-ATC-QTGA-TGA-GTT-TCC-TAMRA			
VT1a	GAA-GAG-TCC-GTG-GGA-TTA-CG	130	slt-I	Pollard et al. (1990)
VT1b	AGC-GAT-GCA-GCT-ATT-AAT-AA			
VT2a	TTA-ACC-ACA-CCC-ACG-GCA-GT	346	slt-II	Pollard et al. (1990)
VT2b	GCT-CTG-GAT-GCA-TCT-CTG-GT			
RfbF	GTG-TCC-ATT-TAT-ACG-GAC-ATC-CAT-G	292	rfb	Zang and Meitzler (1999)
RfbR	CCT-ATA-ACG-TCA-TGC-CAA-TAT-TGC-C			

Primer	Sequence	Size	Gene	Reference
IntF	GAC-TGT-CGA-TGC-ATC-AGG-CAA-AG	368	*int*	Zang and Meitzler (1999)
IntR	TTG-GAG-TAT-TAA-CAT-TAA-CCC-CAG-G			
RAPD 1247	AAG-AGC-CCG-T	—	—	Grif et al. (1998) from Houvelink et al. (1995)
RAPD M13	GAG-GGT-GGC-GGT-TCT	—	—	Grif et al. (1998) from Grundmann et al. (1995)
SLT-F	GAA-CGA-AAT-AAT-TTA-TAT-GT	~900	*sltI* and *II*	Hopkins and Hilton (2000) from Lin et al. (1993)
SLT-R	AAA-TTA-CCA-ATG-TCA-GTA			
RAPD 1290	GTG-GAT-GCG-A	—	—	Hopkins and Hilton (2000) from Akopyanz et al. (1992)
RAPD 1247	AAG-AGC-CCG-T	—	—	
SAADF	CGT-GAT-GAA-CAG-GCT-ATT-GC	119	*saa*	Paton and Paton (2002)
SAADR	ATG-GAC-ATG-CCT-GTG-GCA-AC			
stxIF	ATA-AAT-CGC-CAT-TCG-TTG-ACT-AC			
stxIR	AGA-ACG-CCC-ACT-GAG-ATC-ATC			
stx2F	GGC-ACT-GTC-TGA-AAC-TGC-TCC			
stx2R	TCG-CCA-GTT-ATC-TGA-CAT-TCT-G			
eaeAF	GAC-CCG-GCA-CAA-GCA-TAA-GC			
eaeAr	CCA-CCT-GCA-GCA-ACA-AGA-GG			
hlyAf	CGA-TCA-TCA-AGC-GTA-CGT-TCC			
hlyAr	AAT-GAG-CCA-AGC-TGG-TTA-AGC-T			

Continued

Table 2.2   PCR Primers and DNA Probes[a] (Continued)

Primer or Probe	Sequence (5' → 3')[a]	Size of Sequence (bp)	Amplified Gene or DNA Target Sequence	Reference
595	CCG-AAG-AAA-AAC-CCA-GTA-ACA-G	~400	Q_{933}	LeJeune et al. (2004) from Unkmeir and Schmidt (2000)
Q_{933}	CGG-AGG-GGA-TTG-TTG-AAG-GC			
595	CCG-AAG-AAA-AAC-CCA-GTA-ACA-G	~400	Q_{933}	LeJeune et al. (2004) from Unkmeir and Schmidt (2000)
Q_{21}	GAA-ATC-CTC-AAT-GCC-TCG-TTG	~330	Q_{21}	
Primer Set A				
Stx1-a	TCT-CAG-TGG-GCG-TTC	338	stx_1	Wang, Clark, and Rodgers (2002)
Stx1-b	TAC-CCC-CTC-AAC-TGC-TAA-TA			
Stx2f-a	TGT-CTT-CAG-CAT-CTT-ATG-CAG	150	stx_{2f}	Wang, Clark, and Rodgers (2002)
Stx2f-b	CAT-GAT-TAA-TTA-CTG-AAA-CAG-AAA-C			
Stx2-a	GCG-GTT-TTA-TTT-GCA-TTA-GC	115	stx_2	Wang, Clark, and Rodgers (2002)
Stx2-b	TCC-CGT-CAA-CCT-TCA-CTG-TA			
Primer Set B				
Stx2c-a	GCG-GTT-TTA-TTT-GCA-TTA-GT	124	stx_{2c}	Wang, Clark, and Rodgers (2002)
Stx2c-b	AGT-ACT-CTT-TTC-CGG-CCA-CT			
Stx2e-a	ATG-AAG-TGT-ATA-TTG-TTA-AAG-TGG-A	303	stx_{2e}	Wang, Clark, and Rodgers (2002)
Stx2e-b	AGC-CAC-ATA-TAA-ATT-ATT-TCG-T			

Primer	Sequence	Product size	Gene target	Reference
EAE-a	ATG-CTT-AGT-GCT-GGT-TTA-GG	248	*eaeA*	Wang, Clark, and Rodgers (2002)
EAE-b	GCC-TTC-ATC-ATT-TCG-CTT-TC			
Primer Set C				
Stx2d-a	GGT-AAA-ATT-GAG-TTC-TCT-AAG-TAT	175	*stx*$_{2d}$	Wang, Clark, and Rodgers (2002)
Stx2d-b	CAG-CAA-ATC-CTG-AAC-CTG-ACG			
HlyA-a	AGC-TGC-AAG-TGC-GGG-TCT-G	569	EHEC *hlyA*	Wang, Clark, and Rodgers (2002)
HlyA-b	TAC-GGG-TTA-TGC-CTG-CAA-GTT-CAC			
RfbE-a	CTA-CAG-GTG-AAG-GTG-GAA-TGG	327	*rfbE*$_{O157}$	Wang, Clark, and Rodgers (2002)
RfbE-b	ATT-CCT-CTC-TTT-CCT-CTG-CGG			
FliC-a	TAC-CAT-CGC-AAA-AGC-AAC-TCC	247	*fliC*$_{H7}$	Wang, Clark, and Rodgers (2002)
FliC-b	GTC-GCC-AAC-GTT-AGT-GAT-ACC			
E16S-a	CCC-CCT-GGA-CGA-AGA-CTG-AC	401	*16S tRNA*	Wang, Clark, and Rodgers (2002)
E16S-b	ACC-GCT-GCC-AAC-AAA-GGA-TA			
stx1F934	GTG-GCA-TTA-ATA-CTG-AAT-TGT-CAT-CA	—	*stx1*	Jinneman, Yoshitoma, and Weagant (2003)
stxR104	GCG-TAA-TCC-CAC-GGA-CTC-TTC			
stx1P990	ROX-TGA-TGA-GTT-TCC-TTC-TAT-GTG-TCC-GGC-AGA-T-BHQ2			
stx2F1218	GAT-GTT-TAT-GGC-GGT-TTT-ATT-TGC	—	*stx2*	Jinneman, Yoshitoma, and Weagant (2003)
stx2R1300	TGG-AAA-ACT-CAA-TTT-TAC-CTT-TAG-CA			
stx2P1249	6FAM-TCT-GTT-AAT-GCA-ATG-GCG-GCG-GAT-T-BHQ1			

Continued

Table 2.2   PCR Primers and DNA Probes[a] (Continued)

Primer or Probe	Sequence (5′ → 3′)[a]	Size of Sequence (bp)	Amplified Gene or DNA Target Sequence	Reference
uidAF241	CAG-TCT-GGA-TCG-CGA-AAA-CTG	—	*uidA* allele	Jinneman et al. (2003)
uidAR383	ACC-AGA-CGT-TGC-CCA-CAT-AAT-T			
uidAP1249	TET-ATT-GAG-CAG-CGT-TGG-MGB/NFQ			
I-1	ACA-CTG-GAT-GAT-CTC-AGT-GG	614	*stx1*	Gilgen et al. (1998)
I-2	CTG-AAT-CCC-CCT-CCA-TTA-TG			
I-3	TTG-TCA-TCA-TCA-TGC-ATC-GC	247	*stx1*	
I-4	AGT-TAC-ACA-ATC-AGG-CGT-CG			
II-1	CCA-TGA-CRA-CGG-ACA-GCA-GTT	779	*stx2*	Gilgen et al. (1998)
II-2	CCT-GTC-ARC-TGA-GCA-CTT-TG			
II-3	GTT-CTG-CGT-TTT-GTC-ACT-GT	372	*stx2*	
II-4	AGC-TGT-CGT-TTT-GTC-ACT-GT			
SLTI-5	AGC-TGA-AGC-TTT-ACG-TTT-TCG-G	590	*sltI*	Tsen and Jian (1998)
SLTI-3	TTT-GCG-CAC-TGA-GAA-GAA-GAG-A			
SLTII-5	TTT-CCA-TGA-CAA-CGG-ACA-GCA-GTT	694	*sltII*	
SLTII-3	ATC-CTC-ATT-ATA-CTT-GGA-AAA-CTC-A			
VT1F probe	ATT-CAT-CCA-CTC-TGG-GGG-CA	63	*vt1*	Chen, Johnson, and Griffiths (1998)
VT1R probe	TCA-TTT-TAC-CCC-CTC-AAC-TG			
VT2F probe	TTG-CTG-TGG-ATA-TAC-GAG-GG	191	*vt2*	
VT2R probe	ACT-GCT-GTC-CGT-TGT-CAT-GG			

Primer	Sequence	Size	Target	Reference
SLTI-F	ACA-CTG-GAT-GAT-CTC-AGT-GG	614	*sltI*	Gannon et al. (1992)
SLTI-R	CTG-AAT-CCC-CCT-CCA-TTA-TG			
SLTII-F	CCA-TGA-CAA-CGG-ACA-GCA-GTT	779	*sltII*	
SLTII-R	CCT-GTC-AAC-TGA-GCA-CTT-TG			
VT1-FIP	GCT-CTT-GCC-ACA-GAC-TGC-ACA-TTC-GTT-GAC-TAC-TTC-TTA-TCT-GG	—	*vt1*	Hara-Kudo et al. (2007)
VT1-BIP	CTG-TGA-CAG-CTG-AAG-CTT-TAC-GCG-AAA-TCC-CCT-CTG-AAT-TTG-GC	—	*vt1*	
VT1-F3	GCT-ATA-CCA-CGT-TAC-AGC-GTG	—	*vt1*	
VT1-B3	ACT-ACT-CAA-CCT-TCC-CCA-GTT-C			
VT1-Lp F	AGG-TTC-CGC-TAT-GCG-ACA-TTA-AAT	—	*vt1*	
VT2-FIP	GCT-GTT-GAT-GCA-TCT-CTG-GTA-CAC-TCA-CTG-GTT-TCA-TCA-TAT-CTG-G	—	*vt2*	Hara-Kudo et al. (2007)
VT2-BIP	CTG-TCA-CAG-CAG-AAG-CCT-TAC-GGA-CGA-AAT-TCT-CCC-TGT-ATC-TGC-C	—	*vt2*	
VT2-F3	CAG-TTA-TAC-CAC-TCT-GCA-ACG-TG	—	*vt2*	
VT2-B3	CTG-ATT-ACC-ACT-GAA-CTC-CAT-TAA-CG	—	*vt2*	
VT2-Lp F1	TGT-ATT-ACC-ACT-GAA-CTC-CAT-TAA-CG	—	*vt2*	
VT2-Lp F2	GGC-ATT-TCC-ACT-AAA-CTC-CAT-TAA-CG	—	*vt2*	
AE22	ATT-ACC-ATC-CAC-ACA-GAC-GGT	397	*eae-A*	Fratamico, Bagi, and Pepe (2000) from Fratamico and Strobaugh (1998)
AE20-2	ACA-GCG-TGG-TTG-GAT-CAA-CCT			
HIFS1-F	ACG-ATG-TGG-TTT-ATT-CTG-GA	166	*hly*$_{933}$	Fratamico, Bagi, and Pepe (2000) from Fratamico et al. (1995)
Hlfs1-R	CTT-CAC-GTC-ACC-ATA-CAT-AT			

[a] Q, TAMRA; ROX, 6 carboxy-X-rhodamine; BHQ2, Black Hole Quencher; FAM, 6 carboxyfluorescene; BHQ1, Black Hole Quencher; MGB/NFQ, Black Hole Quencher; R = A+G.

intensities. The assay allowed the detection of 1×10^3 to 1×10^8 CFU per ml of skim milk.

Perelle et al. (2002) identified a chromosomal 2634-bp small inserted locus designated SIL_{O157} uniquely present in STEC O157:H7 strains. Two pairs of primers out of six were found to amplify SIL_{O157} sequences primarily present in *E. coli* O157:H7 serotypes. These primer pairs, RJD3/SF6 and SG7/SF6 (Table 2.2), detected all 34 STEC *E. coli* O257:H7 strains in addition to 22 O157 strains non-H7 and four O55:H7 strains known to be closely related genetically to O157:H7 strains, and one O127:H6 strain. Among 44 nontoxic O157 strains of several H serotypes, all were negative with both primer pairs. The authors concluded that no definitive conclusion was possible regarding the use of the SIL_{O157} locus for detecting STEC O157 in foods.

Although O157:H7 isolates do not exhibit β-glucuronidase activity, they carry the *uidA* gene. Sequencing has revealed that the *uidA* gene of O157:H7 strains has a G in place of the T present in commensurate *E. coli* in position 92 (Feng and Lampel, 1994). This conserved base change in the *uidA* allele is a powerful and coincidental marker of O157:H7 strains. Cebula, Payne, and Feng (1995) developed a multiplex PCR involving a set of primers in a mismatch amplification format to preferentially amplify the *uidA* allele of O157:H7 strains. A 20-bp, upstream, allele-specific primer designated PT-2 was designed that carried the conserved G (rather than T) at the 3′ end and also a G (rather than A) at position 19 (Figure 2.1). The double mismatch was designed to ensure that PT-2 would not anneal with the *uidA* gene of commensurate *E. coli*. In the multiplex PCR PT-2/PT-3 amplified a 252-bp sequence of the *uidA* allele of O157:H7 strains (Table 2.2). Primers LP30/LP31 amplified a 348-bp sequence of the *slt-I* gene (Table 2.2). Primers LP43/LP44 amplified a 584-bp sequence of the *slt-II* gene (Table 2.2).

All 42 O157:H7 strains examined yielded the *uidA* allelic amplicon plus the corresponding *slt-I* or *slt-II* amplicons known to be present in the specific strains. None of the 26 non-O157:H7 *E. coli* serotype strains yielded the *uidA* allelic amplicon although some did yield the *slt-I* or *slt-II* amplicons. The authors concluded that this multiplex PCR appeared to be highly specific for EHEC O157:H7 strains isolated from foods and clinical specimens.

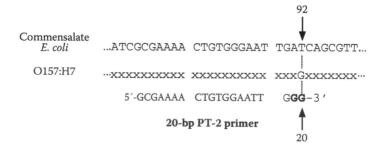

Figure 2.1   Partial sequence of the *uidA* gene of commensurate *E. coli* and *E. coli* O157:H7, illustrating the position of the PT-2 primer. The consensus bases are shown with x's, and the mismatched bases in the primer are boldface. (Adapted from Cebula, T., Payne, W., Feng, F. *J. Clin. Microbiol.* 33:248–250, 1995.)

Fach et al. (2003) compared the use of six pairs of primers that amplified sequences of six genes ($rfbE_{O157}$, $rfbB_{O157}$, $eaeA_{O157}$, $fliC_{H7}$, $orfU_{leeO157}$, and sil_{O157}; Table 2.2). Results indicated that among various O157:H serotypes that did not produce a Stx all were positive for the presence of the $rfbE_{O157}$ and $rfbB_{O157}$ genes and that amplification of the $fliC_{H}7$ gene of STEC O157 allowed the detection of toxigenic *E. coli* O157 of H type and strains of STEC or EPEC belonging to the O55:H7 serogroup, whereas the nontoxigenic *E. coli* O157 of H types other than H7 such as HND, H19, or H45 were not detected. Monoplex PCR assays were found relatively specific to STEC O157, but gave cross-reactions with O55:H7 and to a lesser extent with several other *E. coli* serotypes. The authors concluded that when both *stx* and *rfb* O157 genes are detected in a mixture of bacteria, it is uncertain that the signal is due to STEC O157 and that only isolation of clones and individual testing of isolates by PCR can demonstrate the presence or absence of STEC O157 in the sample. The sil_{O157} gene exhibited the highest level of specificity.

The phage encoded *stx* gene variant stx_2 is governed by the interaction of the transcription anti-terminator Q with the late promoter P_R' (5). LeJeune et al. (2004) identified the anti-terminator Q gene of phage 933W (Q_{933}) upstream of the stx_2 gene in 90% of human clinical isolates of *E. coli* O157:H7 and in 44.5% of bovine isolates. In addition, some bovine isolates were found to harbor only the Q_{21} allele, whereas others were found to harbor both genes, which appeared to be dependent on the geographic area of origin of the strains. Primers 595/Q_{933} (Table 2.2) amplified a ~400-bp sequence of the Q_{933} and a ~330-bp sequence of the bovine allelic Q_{21} gene. The presence of the Q_{933} gene was correlated with higher levels of *slt-II*. The authors concluded that the Q_{933} gene may be a useful molecule for epidemiological studies.

Auvray et al. (2007) subjected 164 minced beef samples to PCR enzyme-linked immunosorbent assay (ELISA) for detection of *stx* genes and five major serogroups after enrichment. Twenty-seven of the samples were *stx*-positive, none of which were also positive by Rti-PCR for at least one marker of the five main serogroups tested (O26, O103, O111, O145, and O157). The authors concluded that PCR techniques are applicable for rapid screening of samples containing both an *stx* gene and an O-group marker of the five major pathogenic STEC serogroups and that isolation of STC strains belonging to the main non-O157 serogroups remains difficult.

B. Multiplex PCR

Fratamico et al. (1995) developed a multiplex PCR for specifically identifying EHEC of serotype O157. Primers MFS1F/MFS1R (Table 2.2) amplified a 166-bp sequence of the 60-MDa plasmid. Primers MK1/MK2 (Table 2.2) amplified a 227-bp sequence of the *slt-I* gene and a 224-bp sequence of the *slt-II* gene. Primers Ae19/AE20 (Table 2.2) amplified a 1087-bp sequence of the *eaeA* gene. Although the 166-bp product in monoplex PCR was generated with as little as 1.2 CFU, in the multiplex PCRs, the detection limit of the *eaeA* gene amplicon was about 100 CFU and the detection limit of the *slt* gene was about 1000 CFU. The authors concluded that apparently the multiplex PCR conditions were more favorable for amplification of the plasmid and *eaeA* sequences than for the *slt* sequences. In addition, the plasmid may be present in multiple copies, whereas the *eaeA* and *slt* genes may be present as single copies. The multiplex PCR described gave a positive signal with all three primer pairs with only toxigenic O157 strains, whereas other *E. coli* strains were negative for at least one of the three amplicons.

Because there are other *E. coli* serotypes, such as *Citrobacter*, and additional AE-causing bacteria that possess the *eaeA* gene, there is a lack of absolute O157:H7 specificity with the PCR detection of just this gene alone. In addition, because the *slt* genes are also harbored by other *E. coli* serotypes (Fratamico et al., 1995), PCR detection of sequences specific for only *slt* genes does not guarantee specific detection of E. coli O157:H7, hence, the significance of employing primers for detection of the *slt-I, slt-II,* and *eaeA* genes in addition to primers for the detection of other genes in a multiplex format.

Meng et al. (1997) developed a multiplex PCR for identifying SLT-producing *E. coli* O157:H7. Three pairs of primers were used. Primers SZ-F/SZ-R (Table 2.2) amplified a 633-bp sequence immediately upstream of the *eaeA* gene of *E. coli* O157:H7, which is located in the locus of enterocyte effacement. Primers SLTI-F/SLTI-R (Table 2.2) amplified a 210-bp sequence of th *slt-I* gene. Primers SLTII-F /SLTII-R amplified a 484-bp sequence of the *slt-II* gene. This multiplex PCR readily distinguished *E. coli* O157:H7 from O55:H7 and other non-O157 serotype SLTEC strains.

Tsen and Jian (1998) developed a multiplex PCR for detection of *slt-I* and *slt-II* genes. The primers SLTI-5/SLTI-3 (Table 2.2) amplified a 590-bp sequence of *slt-I* and primers SLTII-5/SLTII-3 (Table 2.2) amplified a 694-bp sequence of *slt-II*. This multiplex system was additionally used in combination with pairs of primers for simultaneous detection of the heat-labile toxin gene *ltI* and the heat-stabile toxin gene *stII*.

Nagano et al. (1998) developed a multiplex PCR for detection of VTEC O157:H7 strains. The primers H7-F/H7-R (Table 2.2) amplified a 560-bp sequence of the *sliC* gene that encodes the flagellin structural gene of H7 type flagella. Primers O157-F/ O157-R (Table 2.2) amplified a 678-bp sequence of the *rfbE* gene that encodes an enzyme required for O157 antigen synthesis. Primers MK1/MK2 from Karch and Meyer (1989) amplified 227- and 224-bp sequences of the *slt-I and slt-II* genes, respectively (Table 2.2). The multiplex PCR generated three bands with the DNA from all 18 SLT-producing *E. coli* O57:H7 strains tested in addition to one O57:H⁻ strain. None of the 37 non–SLT-producing strains of *E. coli* examined yielded more than two bands. An O18:H7 strain did yield the H7 serotype amplicon. All five SLT negative strains of *E. coli* O.157:H7 yielded amplicons from the *rfbE* gene and the *fliC* gene but not from the *slt* gene. The multiplex PCR was therefore highly specific for detecting SLT-producing *E. coli* O157:H7 strains. The sensitivity of detection was the DNA from 3×10^3 CFU. This low level of detection sensitivity may have been due to the fact that DNA was extracted from cells by boiling cell suspensions in the absence of a detergent, which may not have released all the DNA from the cells.

Zang and Meitzler (1999) developed a multiplex PCR for detection of *E. coli* O157:H7 that involved five pairs of primers. Primers RfbF/RFbR (Table 2.2) amplified a 292-bp sequence of the *Rfb* gene required for O157 antigen synthesis. Primers FLIC H7-F/FLIC H7-R (Table 2.2) from Gannon et al. (1997) amplified a 625-bp sequence of the *flic* gene encoding the flagellin of the H7 serotype. Primers IntF/IntR

(Table 2.2) amplified a 368-bp sequence of the *int* gene encoding intimin. Primers SLTI-F/SLTI-R (Table 2.2) from Meng et al. (1997) amplified a 210-bp sequence of the *slt-I* gene. Primers SLTII-F/SLTII-R (Table 2.2) from Meng et al. (1997) amplified a 484-bp sequence of the *slt-II* gene. Only O157:H7 strains of *E. coli* were positive for the H7, O157, intimin, and one or both of the *slt* genes.

Fratamico, Bagi, and Pepe (2000) developed a multiplex PCR for detection of *E. coli* O157:H7 and to identify the H serogroup and the type of shiga toxin produced. The primers HlFSl-F/HlFI-R (Table 2.2) from Fratamico et al. (1995) amplified a 166-bp sequence of a plasmid-encoded hemolysin gene hly_{933}. The primers FLIC H7-F/FLIC H7-R (Table 2.2) from Gannon et al. (1997) amplified a 625-bp sequence of the chromosomal flagellar structural gene $FliC_{H7}$ of the H7 serogroup. The primers AE22/AE20-2 (Table 2.2) from Fratamico and Strobaugh (1998) amplified a 397-bp sequence of the *eaeA* gene. Primers SLTI-F/SLTI-R (Table 2.2) from Meng et al. (1997) amplified a 210-bp sequence of the *slt-I* gene. Primers SLTII-F/SLTII-R (Table 2.2) from Meng et al. (1997) amplified a 484-bp sequence of the *slt-II* gene. Following a 6-hr enrichment of seeded ground beef in modified *E. coli* broth containing 0.02 mg/ml of novobiocin, all five amplicons were detected with samples inoculated with 100 CFU/g but not with samples containing 10 CFu/g. An 8-hr enrichment allowed detection of 10 CFU/g and a 12-hr enrichment allowed detection of 1 CFU/g.

Hopkins and Hilton (2000) developed an RAPD-PCR multiplex assay for the combined epidemiological typing and *slt* detection of clinical SLT-producing O157 and non-O157 isolates of *E. coli*. The primers upstream and downstream (Table 2.2) from Lin et al. (1993) amplified a ~900-bp sequence of the *stx* gene. The two RAPD (random amplification of polymorphic DNA) primers 1290 and 1297 (Table 2.2) from Akopyanz et al. (1992) yielded similar profiles and allowed differentiation between epidemiologically unrelated STE-producing strains.

Paton and Paton (2002) developed a multiplex PCR assay for simultaneous detection of *stx-I*, *stx-II*, *eae*, *ehxA*, and *saa*. The primers SAADF/SAADR (Table 2.2) amplified a 119-bp sequence of the *saa* gene. The primers stx1F/stx1R (Table 2.2) amplified a 180-bp sequence of the *stx-I* gene. The primers stx2F/stx2R (Table 2.2) amplified a 255-bp sequence of the *stx-II* gene. The primers eaeAF/eaeAR (Table 2.2) amplified a 384-bp sequence of the *eae* gene. Primers hlyAF/hlyAR amplified a 534-bp sequence of the *ehxA* gene.

Kawasaki et al. (2005) developed a multiplex PCR for identifying *Salmonella*, *Listeria monocytogenes*, and *E. coli* O157:H7 directly from enrichment cultures of meat samples. Primers VS8/VS9 from Shama, Dean-Nystrom, and Casey (1999) amplified a 120-bp sequence of the *eaeA* gene. The limit of detection for *E. coli* O157:H7 was the DNA from one CFU per 25 g of spiked pork in agarose gels stained with SYBR green, which is 10 times more sensitive than ethidium bromide (Rengarajan et al., 2002).

Wang, Clark, and Rodgers (2002) developed multiplex PCR assays for simultaneous detection of stx_1, stx_2, stx_{2c}, stx_{2d}, stx_{2e}, stx_{2f}, *hlyA*, *rfbE*, *flic*, and *eaeA* among STEC strains of *E. coli* (Table 2.2). The assay involved running three primer sets simultaneously to detect the 10 *E. coli* O157:H7 genes. Primers (Table 2.2) for detection of the 16s tRNA of *E. coli* were also included as an internal control with each of the three multiplex assays. A total of 129 pathogenic strains was studied, which included 81 O157:H7 strains, 10 O157:non-H7, and 38 that were non-O157. Among the total of 129 strains, 101 (78.3%) were *stx* positive and 28 (21.7%) lacked *stx*, 92 (71.3%) were *hlyA* positive, and 96 (74.4%) were *eaeA* positive. A major advantage of the assay was the identification of the major virulence genes of *E. coli* without the need for restriction enzyme digestion and the differentiation of O157:H7 isolates from non-O157:H7 isolates.

IV. PCR ASSAYS INVOLVING MOLECULAR PROBES AND REAL-TIME PCR (RTI-PCR)

Witham et al. (1996) developed a PCR involving a dual-labeled probe for detection of the *slt-I* gene derived from SLT-producing strains of *E. coli* seeded into ground beef. DNA extraction with guanidium thiocyanate was superior to phenol extraction. Fluorescence analysis was performed with a luminescence spectrometer. The primers SLTI-F3/SLTI-R (Table 2.2) amplified a 416-bp sequence of the *slt-I* gene. The SLT 102 probe (Table 2.2) was labeled at the 5′ end with FAM and with TAMRA as the quencher. As few as 0.5 CFU/g of ground beef could be detected after a 12-hr enrichment in modified EC broth.

Oberst et al. (1998) developed a TaqMan assay for presumptive identification of *E. coli* O157:H7 in ground beef. Primers SZ-I/SZ-II (Table 2.2) amplified a 632-bp sequence of the *aeaA* gene. The SZI-97 probe (Table 2.2) was labeled at the 5′ end with FAM and at the 3′ end with the quencher dye TAMRA. Fluorescence intensity was monitored using a luminescence spectrometer. With the use of the QIA amp tissue DNA kit (Qiagen) the limit of detection was about 1×10^5 CFU/g of ground beef.

A molecular beacon consists of a short probe sequence that is complementary to the target DNA. Flanking the probe region are two GC-rich arm sequences complementary to each other. A fluorescent reporter molecule is conjugated to the end of one arm and a quencher is attached to the end of the other arm. In the absence of homologous DNA sequences, the molecular beacon assumes a hairpin configuration, with the two complementary arm regions hybridizing and the probe sequence forming a loop. In this state, the reporter dye does not fluoresce because of its close proximity to the quencher. McKillip and Drake (2000) applied a molecular beacon probe (Table 2.2) to detect *E. coli* O157:H7 in artificially contaminated skim milk following the PCR amplification of a 401-bp sequence of the *slt-IIA* subunit gene with a selected pair of primers (Table 2.2). 6FAM was the fluorescent reporter at the 5′ end, and DABCYL served as the quencher at the 3′ end (Figure 2.2). Monitoring of fluorescence was done every two cycles with a fluorescent microplate reader. The minimum level of detection was 1×10^3 CFU per ml of milk starting with 10 ml of skim milk.

Fortin, Mulchandani, and Chen (2001) developed an Rti-PCR assay for EHEC O157 serotypes utilizing a molecular beacon designated O157rfbe (Table 2.2) that

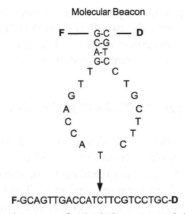

Figure 2.2   Linearization of the molecular beacon on annealing to the target sequence results in the fluorophore and quencher separating, resulting in fluorescence. **F**: 5′(6-FAM); **D**: (C6-NH)-(dABCYL)-3′. Adapted from McKillip and Drake (2000).

recognized a 26-bp sequence of the *rfbe* gene that encodes an enzyme necessary for O-antigen synthesis. The primers O157BF/O157BR (Table 2.2) amplified a 149-bp sequence of the *rfbE* gene. All eleven O157:H7 reference strains were positively recognized. Four O157 strains that were non-H7 were also detected. All 10 non-O157 STEC strains were negative. *E. coli* O55:H7, a serotype of EHEC that is known to be closely related genetically to O157:H7, was also negative. Following a 6-hr selective enrichment, 1 CFU/ml of raw milk was detected. However, with apple juice samples, a minimum of 11 hr of enrichment was required to detect 1 CFU/ml. PCR inhibition by apple juice was attributed to polyphenolic compounds in the juice. The limit of detection with a pure culture was 5×10^3 CFU/ml.

Jothikumar and Griffiths (2002) developed a SYBR green mediated Rti-PCR assay for simultaneous detection of SLT-I and SLT-II *E. coli* O157:H7. The primers JMs1F/JMS1R (Table 2.2) from Karch and Meyer (1989) amplified a 227-bp amplicon of the *slt-I* gene and a 224-bp amplicon from the *slt-II* gene. DNA was extracted by boiling. The limit of detection for cells from a pure culture suspended in water was 8.4×10^3 CFU/ml obtained from 1-ml samples. Because 11 ml was added to RTi-PCR mixtures, this equates to detecting the DNA from 8 CFU per reaction. The use of a larger volume of water (100 ml) and centrifugation to concentrate the cells and the incorporation of 10 ml of sample DNA into Rti-PCR mixtures would have increased the level of detection to 8 CFU per ml of water.

All genotypic O157:H7 strains are either O157:H7/H⁺ (produce H7 serotype flagella) or are phenotypically O157:H7/H⁻ (do not produce flagella but possess the *fli*$_{\mathrm{H7}}$ gene). Jinneman, Yoshitoma, and Weagant (2003) developed Rti-PCr multiplex PCR primers (Table 2.2) for the detection of the *stx1* and *stx2* genes and the allelic *uidA* (β-glucuronidase) gene responsible for the inability of O157:H7 and O157:H7/H⁻ strains to utilize β-glucuronidase due to a highly conserved point mutation at position 93 of the *uidA* gene. The cryptic presence of the nonexpressed *uidA* allelic gene was readily detected with an appropriate pair of PCR primers (Table 2.2). All 52 *E. coli* O157:H7 strains examined plus the two *E. coli* O157:H7/H⁻ were found to be *stx*-positive and *uidA*-positive with respect to the highly conserved single nucleotide point mutation detected with a specific dual-labeled probe uidAP266 (Table 2.2).

V. LOOP-MEDIATED ISOTHERMAL AMPLIFICATION (LAMP) OF DNA

Hara-Kudo et al. (2007) developed a LAMP assay for detection of the *vt1* and *vt2* genes in strains of *E. coli* O157:H7. Enrichment in mEc broth for 18 hr at 42°C allowed LAMP detection of 9 CFU/25 g of ground beef. The sensitivity of the LAMP assay was 100-fold higher that that of conventional PCR. The LAMP assay for *vt1* used five primers VT1-FIP, VT1-BIP, VT1-F3, VT1-B3, and VT1-loop F (Table 2.2). To ensure detection of several VT2 subtypes the LAMP assay for *vt2* used six primers VT2-FIP, VT2-BIP, VT2-F3, VT2-B3, VT2-loop F1, and VT2-loop F2 (Table 2.2). LAMP exhibited 100-fold greater detections sensitivity than conventional PCR.

VI. IMMUNOMAGNETIC SEPARATION AND PCR

Chen, Johnson, and Griffiths (1998) developed a magnetic capture-hybridization PCR assay for detection of the *vt1* and *vt2* genes of *E. coli* seeded into ground beef. After enrichment at 37°C of 25 g in 225 ml of brain heart infusion broth (BHI) a 1-ml sample was centrifuged at 12,000 g for 2 min and the pellet washed twice with 1 ml of dH₂O. The final preparation (100 µl) was boiled for 10 min to release DNA. Capture probes for *vt1* and *vt2* (100 µl each) were added and the preparation boiled for 10 min and cooled rapidly on ice. DNA capture probes for the *vt1* gene consisted of two 20-mer nucleotides that flanked a specific 63-bp sequence of the *vt2* gene, each labeled at the 5′-end with biotin (Table 2.2). DNA capture probes for the *vt2* gene consisted of two 20-mer nucleotides that flanked a specific 191-bp sequence of the *vt2* gene (Table 2.2). To the hybridization mix was added 3 µl of streptavidin-coated magnetic beads. After incubation at room temperature for 1 hr on a rotator, the beads were collected by a magnetic particle separator and washed twice with dH₂O and suspended in 50 µl of dH₂O and added directly to PCR reaction mixtures. The *vt1* and *vt2* primer pairs (Table 2.2) of Pollard et al. (1990) were used for PCR amplification. After a 15-hr enrichment the assay detected 1 CFU/g of ground beef.

Natural domestic hosts for STEC are ruminants such as sheep, goats, and cattle, in addition to swine and dogs. However, feral ruminants have also been found to harbor *E. coli* O157:H7. García-Sánchez et al. (2007) isolated the organism from various strains of wild deer. Retroanal mucosal swab samples from hunted deer were enriched for 18 hr at 37°C in 5 m of buffered peptone water supplemented with vancomycin (8 mg/l), cefixime (0.005 mg/l), and cefsulodine (10 mg/l). One ml of enrichment broth was added to 20 µl of magnetic beads coated with O157:H7 antibodies. The beads were then inoculated onto sorbitol MacConkey Agar containing sefixime (0.05 mg/l) and potassium tellurite (2.5 mg/l) and incubated at 37°C for 18 to 24 hr. A loop of cells was then suspended in 0.5 ml of sterilized dH$_2$O and boiled for 5 min to release the DNA. PCR assays for *stx-I*, *stx-II*, *ehxA*, *eae*, *eae-y1*, *O157rfbE*, and *fliCh7* were then performed with the primer pairs listed in Table 2.2. From each PCR-positive culture, 10 colonies were selected and biochemically confirmed as *E. coli*. β-D-glucuronidase activity was confirmed on Chromocult Agar.

   E. coli O157 cells tend to adsorb onto solid surfaces nonspecifically so as to interfere with immunomagntic separation. Tomoyashi (1998) developed a low-ionic strength solution by treating 100 ml of Milli-Q water with 5 g of Chelex 100 cation exchange resin overnight at room temperature. When the resulting supernatant was used for suspending cells with immunomagnetic beads, the proportion of *E. coli* O157 cells to non-O157 cells captured increased 31.4-fold compared to conventional immunomagnetic separation. These results suggest that the addition of the cation chelator EDTA to food homogenates may enhance the recovery of *E. coli* O157:H7 cells following differential centrifugation.

   Parham et al. (2003) undertook studies to optimize the immunomagnetic capture of O157:H7 cells from bovine feces. An increased magnetite content of magnetic beads from 12% to 80% had a positive effect on recovery. A bead size of 6.2 μ was also found superior to beads of 2.5 to 4.5 μ. PBS plus 0.01% Tween 20 for homogenization (10% w/v) of fecal samples was found superior to several other buffers for maximum recovery.

VII. RESTRICTION FRAGMENT LENGTH POLYMORPHISM

Samadpour et al. (1993) subjected genomic DNA prepared from 1568 isolates of *E. coli* O157:H7 to RFLP with the *Pvu*II restriction nuclease on Southern blots probed with ^{32}P-labeled bacteriophage λDNA. The isolates analyzed included strains from a recent large multistate outbreak of *E. coli* O157:H7 infection associated with undercooked beef in restaurants and a daycare center, and unrelated isolates. The isolates recovered from the incriminated meat and from 61 (96.8%) of 63 patients from Washington and Nevada possessed identical λRFLP patterns. The λRFLP patterns observed with 1 (91.7%) of 12 daycare center patients were identical with the multistate outbreak. Isolates from 42 patients unrelated to either infectious cluster possessed unique and different λRFLP patterns. λDNA was chosen as a [32]p-labeled probe because preliminary data demonstrated its superiority to toxin genes, rRNA, plasmid profiles, and the fact that SLTs are encoded by lambdoid bacteriophage.

VIII. SUBSPECIES TYPING OF *E. COLI* O157:H7 ISOLATES

   Although pulsed-field gel electrophorsis (PFGE) is considered the gold standard for subspecies typing of bacterial isolates, its inherent limitations must always be considered. Böhm and Karch (1992) found that PFGE with the use of the restriction nucleases *Xba*I, *Pac*I, *Sfi*I, and *Not*I was unable to distinguish between epidemiologically unrelated isolates of *E. coli* O157:H7.

Grif et al. (1998) made use of 17 human, 27 food, and 8 veterinary isolates of *E. coli* O157:H7 isolates to evaluate five different epidemiological typing methods. Ribotyping with the combined use of *ecoR*1 and *Pvu*II restriction nucleases was found unsuitable. All but one strain could be phage typed. Random amplified polymorphism performed with primers 1247 and M13 (Table 2.2) was found to be useful for epidemiological purposes. PFGE performed with the single restriction nuclease II was found to be of value as a subtyping system but was also found to have limitations when phylogenetically highly related strains were analyzed. The authors concluded that it is difficult to unequivocally prove by PFGE alone that a given clonal strain is responsible for the spread of an outbreak and that more than one subtyping method should be used for epidemiological investigations.

In an epidemiological study, Gupta et al. (2004) reported that among 38 clinical *E. coli* O157:H7 isolates confirmed as having a relatively uncommon but identical *Xba*I PFGE pattern. In contrast, *Bln*I PFGE yielded patterns with the same isolates that sorted into multiple distinct clusters. The authors concluded that the epidemiological identification of a single clonal cluster of *E. coli* O157:H7 using only one restriction nuclease with PFGE can be misleading and that two or more restriction nucleases are necessary.

REFERENCES

Akopyanz, N., Bukanov, N., Westblom, T., Kresovich, S., Berg, D. 1992. DNA diversity among clinical isolates of *Helicobacter pylori* detected by PCR-based RAPD-fingerprinting. *Nucleic Acids Res.* 20:5137–5142.

Auvray, F., Lecueuil, C., Taché, J., Leclerc, V., Deperrios, V., Lombard, B. 2007. Detection, isolation and characterization of Shiga toxin-producing *Escherichia coli* retail-minced beef using PCR-based techniques, immunoassays and colon hybridization. *Lett. Appl. Microbiol.* 45:646–651.

Beutin, L., Montenegro, M., Orskov, I., Orskov, F., Zimmerman, S., Stephan, R. 1989. Close association of verotoxin (Shiga-like toxin) production with enterohemolysin production in strains of *Escherichia coli*. *J. Clin. Microbiol.* 27:2559–2564.

Blanco, M., Blanco, J.E., Mora, A., Rey, J., Alonso, J., Hermoso, M., Hermoso, J., Alonso, M., Dahbi, G., Gonzalez, E., Bernardez, M., Blanco, J. 2003. Serotypes, virulence genes, and intimin types of Shiga toxin (verotoxin)-producing *Escherichia coli* isolates from healthy sheep in Spain. *J. Clin. Microbiol.* 41:1351–1356.

Boerlin, P., McEwen, S., Boerlin-Petzold, F., Wilson, J., Johnson, R., Gyles, C. 1999. Association between virulence factors of Shiga toxin-producing *Escherichia coli* and disease in humans. *J. Clin. Microbiol.* 37:497–503.

Böhm, H., Karch, H. 1992. DNA fingerprinting of *Escherichia coli* O157:H7 strains by pulsed-field gel electrophoresis. *J. Clin. Microbiol.* 30:2169–2172.

Brunder, W., Schmidt, H., Karch, H. 1996. KatP, a novel catalase-peroxidase encoded by the large plasmid of enterohaemorrhagic *Escherichia coli* O157:H7. *Microbiol.* 142:3305–3315.

Brunder, W., Schmidt, H., Karch, H. 1997. EsoP, a novel extracellular serine protease of enterohemorrhagic *Escherichia coli* O157:H7 cleaves human coagulation factor. *Mol. Microbol.* 24:767–778.

Cebula, T., Payne, W., Feng, F. 1995. Simultaneous identification of strains of *Esherichia coli* serotype O157:H7 and their Shiga-like toxin type by mismatch amplification mutation assay-multiplex PCRS. *J. Clin. Microbiol.* 33:248–250.

Centers for Disease Control and Prevention. 1995. *Escherichia coli* O157:H7 outbreak linked to commercially distributed dry-cured salami—Washington and California. 1994 *Morbid. Mortal. Weekly Rep.* 44:157–160.

Chen, J., Johnson, R., Griffiths, M. 1998. Detection of verotoxigenic *Escherichia coli* by magnetic capture-hybridization PCR. *Appl. Environ. Microbiol.* 64:147–152.

Desmarchelier, P., Bilge, S., Fegan, N., Mills, L., Vary Jr., J., Tarr, P. 1998. A PCR specific for *Escherichia coli* O157 based on the *rfb* locus encoding O157 lipopolysaccharide. *J. Clin. Microbiol.* 36:1801–1894.

Donnenberg, M., Kaper, J., Finlay, B. 1997. Interactions between enteropathogenic *Escherichia coli* and host epithelial cells. *Trends Microbiol.* 5:109–114.

Elliot, S., Wainwright, L., Mcdaniel, T., Jarvis, K., Deng, Y., Lai, L., McNamara, P., Donnenberg, M., Kaper. 1998. The complete sequence of the locus of enterocyte effacement (LEEE) from enteropathogenic *Escherichia coli* E2348/69. *Mol. Microbiol.* 28:1–4.

Fach, P., Perelle, S., Rout, J., Kilasser, F. 2003. Comparison of different PCR tests for detecting Shiga toxin-producing *Escherichia coli* O157:H7 and development of an ELISA-PCR assay for specific identification of the bacteria. *J. Microbiol. Methods* 55:383–0392.

Feng, P. 1995. *Escherichia coli* serotype O157:H7: A novel vehicle of infection and emergence of phenotypic variants. *Energ. Infect. Dis.* 1:47–52.

Feng, P., Lampel, K. 1994. Genetic analysis of *uidA* gene expression in enterohemorrhagic *Escherichia coli* serotype O157:H7. *Appl. Environ. Microbiol.* 57:320–323.

Fortin, N., Mulchandani, A., Chen, W. 2001. Use of real-time polymerase chain reaction and molecular beacons for the detection of *Escherichia coli* O157:H7. *Anal. Biochem.* 289:281–288.

Fratamico P., Bagi, I., Pepe, T. 2000. A multiplex PCR assay for rapid detection an identification of *Escherichia coli* O157:H7 in foods and bovine feces. *J. Food Protect.* 63:1032–1037.

Fratamico, P., Sackitey, S., Wiedman, M., Deng, M. 1995. Detection of *Esherichia coli* O157:H7 by multiplex PCR. *J. Clin. Microbiol.* 33:2188–2191.

Fratamico, P., Strobaugh, T. 1998. Simultaneous detection of *Salmonella* spp. and *Escherichia coli* O157:H7 by multiplex PCR. *J. Ind. Microbiol. Biotechnol.* 21:92–98.

Gannon, V., De'Souza, S., Graham, T., King, R., Rahn, K., Read, S. 1997. Use of the flagellar H7 gene as a target in multiplex PCR assays and improved specificity in identification of enterohemorrhagic *Escherichia coli* strains. *J. Clin. Microbiol.* 35:655–662.

Gannon, V., King, R., Kim, J., Thomas, E. 1992. Rapid and sensitive method for detection of Shiga-like toxin-producing *Escherichia coli* in ground beef using the polymerase chain reaction. *Appl. Environ. Microbol.* 58:3809–3815.

Gannon, V., Rashed, M., King, R., Thomas, E. 1993. Detection and characterization of the *eae* gene of Shiga-like toxin-producing *Escherichia coli* using polymerase chain reaction *J. Clin. Microbiol.* 31:1268–1274.

García-Sánchez, A., Sánchez, S., Rubio, R., Pereira, G., Alonso, J., de Mendoza, J., Rey, J. 2007. Presence of Shiga toxin-producing *E. coli* O157:H7 in a survey of wild artiodactyls. *Vet. Microbiol.* 121:373–377.

Gilgen, M., Hübner, P., Höfelein, C., Lüthy, J., Candrian, U. 1998. PCR-based detection of verotoxin producing *Escherichia coli* (VTEC) in ground beef. *Res. Microbiol.* 149:145–154.

Grif, K., Karch, H., Schneider, C., Daschner, F., Beutin, L., Cheasty, T., Smith, H., Rowe, B., Dierich, M., Allerberger, F. 1998. Comparative study of five different techniques for epidemiological typing of *Escherichia coli* O157. *Diag. Microbiol. Infect. Dis.* 32:165–176.

Griffin, P., Bell, B., Cieslak, P., Tuttle, J., Barrett, T., Doyle, M., McNamara, A., Shefer, A., Wells, J. 1994. Large outbreak of *Escherichia coli* O157:H7 in western United States: The big picture. In: M.A. Karmali, A. Goglio (Eds.), *Recent Advances in Verocytotoxin-Producing* Escherichia coli *Infections. Proceedings of the Second International Symposium and Workshop on Verotoxin (Shiga-Like Toxin)-Producing* Esherichia coli *Infections*. Elsevier: Amsterdam, pp. 7–12.

Grundmann, H., Schneider, S., Tichy, H., Simon, R., Klare, I., Hartung, D., Daschner, F. 1995. Automated laser fluorescence analysis of randomly amplified polymorphic DNA: A rapid method for investigation of nosocomial transmission of *Acinetobacter baumanii*. *J. Med. Microbiol.* 43:446–451.

Gupta, A., Hunter, S., Bidol, S., Dietrich, S., Kincaid, J., Salehi, E., Nicholson, L., Genese, C., Todd-Weinstein, S., Marengo, L., Kimura, A., Brooks, J. 2004. *Escherichia coli* O157 cluster evaluation. *Emerg. Infect. Dis.* 10:1856–1858.

Hara-Kudo, Y., Miyahara, M., Kumagai, S. 2000. Loss of O157 O antigenicity of verotoxin-producing *Escherichia coli* O157:H7 surviving under starvation conditions. *Appl. Environ. Microbiol.* 66:5540–5543.

Hara-Kudo, Y., Nmoto, J., Ohtsuka, K., Segawa, Y., Takatori, K., Kojima, T., Ikedo, M. 2007. Sensitive and rapid detection of vero toxin-producing *Escherichia coli* using loop-mediated isothermal amplification. *J. Med. Microbiol.* 56:398–406.

Hayes, P., Blom, K., Eng, P., Lewis, J., Srockbine, N., Swaminthan, B. 1995. Isolation and characterization of a β-D-gucuronidase-producing strain of *Escherichia coli* serotype O157:H7 in the United States. *J. Clin. Microbiol.* 33:3347–3348.

Hopkins, K., Hilton, A. 2000. Simultaneous molecular subtyping and Shiga toxin gene detection in *Escherichia coli* using multiplex polymerase chain reaction. *Lett. Appl. Microbiol.* 2000:122–125.

Houvelink, A.J., van de Kar, C., Meiss, J., Monnens, L., Melchers, W. 1995. Characterization of verotoxin-producing producing *Escherichia coli* O157 isolates from patients with haemolytic-uremic syndrome in Western Europe. *Epidemiol. Infect.* 115:1–14.

Imberechts, H., De Greve, H., Lintermans, P. 1992. The pathogenesis of edema disease in pigs. A review. *Vet. Microbiol.* 31:221–233.

Jinneman, K., Yoshitoma, K., Weagant, S. 2003. Multiplex real-time PCR method to identify Shiga toxins *stx1* and *stx2* and *Escherichia coli* O157:H7/H⁻ serotype. *Appl. Environ. Microbiol.* 69:6327–6333.

Jothikumar, N., Griffiths, M. 2002. Rapid detection of *Escherichia coli* O157:H7 with multiplex real-time PCR assays. *Appl. Environ. Microbiol.* 68:3169–3171.

Karch, H., Meyer, T. 1989. Single primer pair for amplifying segments of distinct Shiga-like-toxin-producing genes by polymerase chain reaction. *J. Clin. Microbiol.* 27:2751–2757.

Karmali, M. 1989. Infection by verocytotoxin-producing *Escherichia coli*. *Clin. Microbiol. Rev.* 2:15–38.

Kawasaki, S., Horikoshi, N., Okada, Y., Takeshita, K., Sameshima, S., Kawamoto, S. 2005. Multiplex PCR for simultaneous detection of *Salmonella* spp., *Listeria monocytogenes,* and *Escherichia coli* O157:H7 in meat samples. *J. Food Protect.* 68:551–556.

Kleanthous, H., Smith, H., Scotland, S., Gross, R., Rowe, B., Taylor, C., Milford, D. 1990. Haemolytic uraemic syndromes in British Isles, 1985–8: Association with Verocytotoxin producing *Escherichia coli*. Part 2: Microbiological aspects. *Arch. Dis. Child.* 65:722–727.

Korhonen, T., Valtonen, M., Parkkinen, J., Väisänen-Rhen, V., Finne, J., Orskov, F., Orskov, L., Svenson, S., Mäkelä, P. 1985. Serotypes, hemolysin production, and receptor recognition of *Escherichia coli* strains associated with neonatal sepsis and meningitis. *Infect. Immun.* 48:486–491.

Kuhnert, P., Heyberger-Meyer, B., Burnens, A., Nicolet, J., Frey, J. 1997. Detection of RTX toxin genes in Gram-negative bacteria with a set of specific probes. *Appl. Environ. Microbiol.* 63:2258–2265.

Lehmacher, A., Meier, H., Aleksic, S., Bockemühl, J. 1998. Detection of hemolysin variants of Shiga toxin-producing *Escherichia coli* by PCR and culture on vancomycin-cefixime-cefsulodin blood agar. *Appl. Environ. Microbiol.* 64:2449–2453.

LeJeune, J., Abedon, S., Takemura, K., Christie, N., Sreevatsan, S. 2004. Human *Escherichia coli* O157:H7 genetic marker in isolates of bovine origin. *Emerg. Infect. Dis.* 10:1482–1485.

Li, W., Drake, M. 2001. Development of a quantitative competitive PCR assay for detection and quantification of *Escherichia coli* O157:H7 cells. *Appl. Environ. Microbiol.* 67:3291–3294.

Lin, Z., Kurazono, H., Yamasaki, S., Takeda, Y. 1993. Detection of various variant verotoxin genes in *Escherichia coli* by polymerase chain reaction. *Microbiol. Immunol.* 37:543–548.

Louie, M., De Azavdo, J., Clark, R., Borezyk, A., Lior, H., Richter, M., Brunton, J. 1994. Sequence heterogeneity of the *eae* gene and detection of verotoxin-producing *Escherichia coli* using serotype-specific primers. *Epidemiol. Infect.* 112:449–461.

Maurer, J., Schmidt, D., Petrosko, P., Sanchez, S., Bolton, L., Lee, M. 1999. Development of primers to O-antigen biosynthesis genes for specific detection of *Escherichia coli* O157 by PCR. *Appl. Environ. Microbiol.* 65:2954–2960.

McKillip, J., Drake, M. 2000. Molecular beacon polymerase chain reaction detection of *Escherichia coli* O157:H7 in milk. *J. Food Protect.* 63:855–859.

Meng, J., Zhao, S., Doyle, M., Mitchell, S., Kresovich, S. 1996. Polymerase chain reaction for detecting *Escherichia coli* O157:H7. *Int. J. Food Microbiol.* 32:103–113.

Meng, J., Zhao, S., Doyle, M., Mitchell, S., Kresovich, S. 1997. A multiplex PCR for identifying Shiga-like toxin-producing *Escherichia coli* O157:H7. *Lett. Appl. Microbiol.* 24:172–176.

Nagano, I., Kunishima M., Itoh, Y., Wu, Z., Takahashi, Y. 1998. Detection of verotoxin-producing *Escherichia coli* O157:H7 by multiplex polymerase chain reaction. *Microbiol. Immunol.* 42:371–376.

Oberst, R., Hays, M., Bohra, L., Phebus, R., Yamashiro, C., Paszko-Kolva, C., Flood, S., Sargent, J., Gillespie, J. 1998. PCR-based DNA amplification and presumptive detection of *Escherichia coli* O157:H7 with an internal fluorogenic probe and the 5′ nuclease (Taqman) assay. *Appl. Environ. Microbiol.* 64:3389–3396.

Orskov, L., Orskov, F. 1985. *Escherichia coli* in extra-intestinal infections. *J. Hyg.* 95:551–575.

Ostroff, S., Tarr, M., Neill, M., Lewis, J., Hargrett-Bean, N., Kobayashi, J. 1989. Toxin genotypes and plasmid profiles as determinants of systemic sequelae in *Echerichia coli* O157:H7 infections. *J. Infect. Dis.* 160:994–999.

Padhye, N., Doyle, M. 1991. Production and characterization of monoclonal antibody specific for entrohaemorrhagic *Escherichia coli* serotypes O157:H7 and O26:H11. *J. Clin. Microbiol.* 29:99–103.

Parham, N., Spencer, J., Taylor, D., Ternent, H., Innocent, G., Mellor, D., Roberts, M., Williams, A. 2003. An adapted immunomagnetic cell separation method for use in quantification of *Escherichia coli* O157:H7 from bovine faeces. *J. Microbiol. Methods* 53:1–9.

Paton, A., Paton, J. 1998a. Detection and characterization of Shiga toxigenic Escherichia *coli* by using multiplex PCR assays for stx_1, stx_2, *eaeA,* enterohemorragic *E. coli hlyA,* rib_{O111}, and rfb_{O157}. *J. Clin. Microbiol.* 36:598–602.

Paton, A., Paton, J. 1998b. Pathogenesis and diagnosis of shiga toxin-producing *Escherichia coli* infections. *Clin. Microbiol. Rev.* 11:450–479.

Paton, A., Paton, J. 2002. Direct detection and characterization of Shiga toxigenic *Escherichia coli* by multiplex PCR for stx_1, stx_2, *eae, ehxa, and saa. J. Clin. Microbiol.* 40:271–274.

Paton, A., Srimanote, P., Woodrow, M., Paton, J. 2001. Characterization of Saa, a novel autoagglutinating adhesin produced by locus of enterocyte effacement-negative Shiga toxigenic *Escherichia coli* strains that are virulent for humans. *Infect. Immun.* 69:6999–7009.

Perelle, S., Fach, P., Dilasser, F., Grout, J. 2002. A PCR test for detecting *Escherichia coli* O57:H7 based on the identification of the small inserted locus (SIL_{O157}). *J. Appl. Microbiol.* 93:250–260.

Perna, N., Mayhew, G., Posfai, G., Elliott, S., Bonnenberg, M., Kaper, J., Blattner, F. 1998. Molecular evolution of a pathogenicity island from enterohemorrhagic *Escherichia coli* O157:H7. *Infect. Immun.* 66:3810–3817.

Pollard, D., Johnson, W., Lior, H., Tyler, S., Rozee, K. 1990. Rapid and specific detection of verotoxingenes in *Escherichia coli* by the polymerase chain reacion. *J. Clin. Microbiol.* 28:540–545.

Rengarajan, K., Cristol, S., Mehta, M., Nickrson, J. 2002. Quantifying DNA concentration using fluorometry: A comparison of fluorophors. *Mol. Vis.* 8:416–421.

Samadpour, M., Grimm, L., Desai, B., Alfi, D., Ongerth, J., Tarr, P. 1993. Molecular epidemiology of *Escherichia coli* O157:H7 strains by bacteriophage λ restriction fragment length polymorphism analysis: Application to a multistate foodborne outbreak and a day-care center cluster. *J. Clin. Microbiol.* 33:3179–3183.

Schmidt, H., Beutin, L., Karch, H. 1995. Molecular analysis of the plasmid-encoded hemolysin of *Escherichia coli* strain EDL 933. *Infect. Immun.* 63:1055–1061.

Schmidt, H., Henkel, B., Karch, H. 1997. A gene cluster closely related to the type II secretion pathway operons of gram-negative bacteria is located on the large plasmid of enterohemorrhagic *Escherichia coli* O157:H7 strains. *FEMS Microbiol. Lett.* 148:265–272.

Schmidt, H., Karch, H. 1996. Enterohemolytic phenotypes and genotypes of Shiga toxin-producing *Escherichia coli* O111 strains from patients with diarrhea and hemolytic-uremic syndrome. *J. Clin. Microbiol.* 34:2364–2367.

Schmidt, H., Kernbach, C., Karch, H. 1996. Analysis of the EHEAC *hyl* operon and its location in the physical map of the large plasmid of enterohemorrhagic *Escherichia coli* O157:H7. *Microbiol.* 142:907–914.

Shama, V., Dean-Nystrom, E., Casey, T. 1999. Semiautomated fluoreogenic PCR assays (Taqman) for rapid detection of *Escherichia coli* 157:H7 and other Shiga toxigenic *E. coli. Mol. Cell. Probes* 13:291–302.

Tomoyashu, T. 1998. Improvement of the immunomagnetic separation method selective for *Escherichia coli* O157 strains. *Appl. Environ. Microbiol.* 64:376–382.

Tsen, H., Jian, L. 1998. Development and use of a multiplex PCR system for the rapid screening of heat labile toxin I, heat stable toxin II and shiga-like toxin I and II genes of *Escherichia coli* in water. *J. Appl. Microbiol.* 84:585–592.

Unkmeir, A., Schmidt, H. 2000. Structural analysis of phage-borne *stx* genes and their flanking sequences in shiga toxin-producing *Escherichia coli* and *Shigella dysenteriae* type I strains. *Infect. Immun.* 68:4856–4864.

Wang, G., Clark, C., Rodgers, F. 2002. Detection in *Escherichia coli* of the genes encoding the major virulence factors, the genes defining the O157:H7 serotype, and components of the type 2 Shiga toxin family by multiplex PCR. *J. Clin. Microbiol.* 40:3613–3619.

Ware, J., Abbott, S., Janda, J. 2000. A new diagnostic problem: Isolation of *Escherichia coli* O157:H7 strains with aberrant biochemical properties. *Diag. Microbiol. Inf. Dis.* 38:185–187.

Wetzel, A., LeJeune, J. 2007. Isolation of *Escherichia coli* O157:H7 strains that do not produce Shiga toxin from bovine, avian and environmental sources. *Lett. Appl. Microbiol.* 45:504–507.

Witham, P., Yamashiro, C., Livak, K., Batt, C. 1996. A PCR-based assay for the detection of *Escherichia coli* Shiga-like toxin genes in ground beef. *Appl. Environ. Microbiol.* 672:1347–1353.

Yu, J., Kaper, J. 1992. Cloning and characterization of the *eae* gene of enterohemorrhagic *Escherichia coli* O157:H7. *Mol. Microbiol.* 6:411–417.

Zang, Y., Meitzler, J. 1999. Rapid and sensitive detection of *Escherichia coli* O157:H7 in bovine faeces by a multiplex PCR. *J. Appl. Microbiol.* 87:867–876.

Zhao, T., Doyle, M., Wang, G. 1994. Emerging pathogens and rapid detection methods: *Escherichia coli, Listeria, Salmonella, Vibrio cholerae*, pp. 386–403. In: *Proceedings of the Second Asian Conference on Food Safety.* Washington, DC: International Life Sciences Institute.

CHAPTER 3

Shigella

I. CHARACTERISTICS OF THE GENUS

Shigellosis, otherwise known as bacillary dysentery, is characterized by the presence of blood and mucus in the stools. There are four species of *Shigella*: *S. dysenteriae*, *S. flexneri*, *S. boydii*, and *S. sonnei*. Members of the genus *Shigella* are facultatively anaerobic, gram-negative, cytochrome oxidase negative, citrate negative, lysine decarboxylase negative, nonmotile rods that are either lactose negative or ferment lactose slowly. Although gastrointestinal infections by shigellae and EICC occur most frequently in developing nations, the responsible *E. coli* strains are indistinguishable from shigellae by PCR. In addition, with the exception of *S. boydii*, the other three species of *Shigella* are completely indistinguishable from one another by the PCR. The most fruitful approach to date in attempting to use the PCR for distinguishing the four species of *Shigella* from one another has been to first use immuno-capture based on somatic O antigenicity immediately prior to the use of the PCR to impart species selectivity to a coupled ELISA-PCR assay system. Most PCR assays involving the detection of shigellae have been designed to usually detect shigellae plus the *E. coli* pathotype EIEC as a collective and indistinguishable group of enteroinvasive organisms, or to distinguish *Shigella* from pathogenic members of other pathogenic genera.

Members of the genus *Shigella* are considered to have evolved from *E. coli* about 10,000 years ago (Stephens and Murray, 2001) by horizontal transfer of virulence plasmids and islands of pathogenicity (Lan and Reeves, 2002) by conjugation and transduction of certain virulence factors by tempered bacteriophage. There is some thought that *Shigella* spp. (except for *S. boydii* serotype 13) and *E. coli* constitute a single DNA relatedness group (Coimbra et al., 2001). DNA sequencing suggests that the four species of *Shigella* should be considered as clones of *E. coli*, which are much less biochemically active, host-restrictive, and carry plasmid encoding invasiveness data (Coimbra et al., 2001). This close genetic relationship between *Shigella* spp. and virulent *E. coli* strains, resulting in the common presence of virulence genes, makes it very difficult or impossible to distinguish *Shigella* and certain virulent types of *E. coli* by the PCR. This problem is presented in Table 3.1 and has resulted in the

Table 3.1   Genes Present in *Shigella* spp. and Virulent *E. coli* Used in
PCR Assays[a]

Organism	stx1	stx2	eaeA	StIb	ipaH	aggR	LTI	rrs[b]
E.coli K12								+
EHEC	+		+		+			+
EPEC			+					+
ETEC				+			+	+
EIEC					+			+
EAEC					+	+		+
S. dysenteriae	+				+			+
S flexneri					+			+
S. sonnei					+			+
S. boydii			+		+			+

[a] Compiled from Brandal et al. (2007).
[b] Positive control gene.

inability of most PCR assays to distinguish between shigellae and EIEC and to identify individual *Shigella* species.

II. VIRULENCE FACTORS

Virulence among the four species of *Shigella* involves both chromosomal and
plasmid-encoded genes. A large 120- to 230-kb plasmid is essential for invasion
(Theron et al., 2001). The invasion plasmid antigen gene (*ipaH*) is present in multiple copies on the invasion plasmid (Hartman et al., 1990). A 37-kb region of the
invasion plasmid of *S. flexneri* 2a contains all the genes essential for penetration into
mammalian cells and encodes about 33 genes. The genes comprising the ipaBCDA
(invasion plasmid antigens) cluster encode the immunodominant antigens. With the
exception of *ipaA*, these genes are required for invasion of mammalian cells. The
Ipa products are associated with the outer membrane of *Shigella*, with IpaB and IpaC
forming a complex on the cell surface and are responsible for transducing the signal
that results in entry into the host cell (Ménard et al., 1994). IpaB is also thought to be
the hemolysin (High et al., 1992) and the cause of apoptosis of infected macrophages
via lysis of the phagocytic vacuole (Zychlinsky et al., 1994).

From Table 3.1 it is obvious that all *Shigella* species examined in addition to
all virulent types of *E. coli* and the nonvirulent *E. coli* K12 possess the *rrs* gene.
It is shown that *Shigella* spp. contain the *ipaH* gene and therefore cannot be distinguished from EIEC strains. The *aggR* gene encodes a transcriptional activator on
the AA plasmid and its presence solely in enteroaggregative *E. coli* (EAEC) strains
allows such isolates to be distinguished from shigellae. The presence of LT1 or ST1b
in enterotoxigenic *E. coli* (ETEC) strains allows their distinction from all other virulent *E. coli* and shigellae. Although *stx1* occurs in both enterohemorrhagic *E. coli*
(EHEC) and *S. dysenteria*, *stx2* occurs only in EHEC strains.

The *eaeA* gene is located on the genomic DNA as part of the LEE (locus of enterocyte effacement) pathogenicity island and encodes intimin, which is involved in attachment of bacteria to enterocytes. Intimin is produced by strains of EHEC, enteropathogenic *E. coli* (EPEC), and *S. boydii* (Table 3.1). ETEC colonize the surface of the small bowel mucosa and cause diarrhea due to production of the LT (heat-labile) or ST (heat-stable) enterotoxin. The *virG* or *icsA* (intracellular spread) gene encodes a protein that catalyzes the polymerization of actin in the cytoplasm of an infected cell (Bernardoni et al., 1989; Makino et al., 1989). The Icsa protein is expressed at one pole (Goldberg et al., 1993). The polymerization of actin by IcsA forms a filamentous extension from the pole and propels the infecting bacterial cell through the cytoplasm. Synthesis of LPS, which is critical for correct unipolar localization of IcsA, requires *rfa* and *rfb,* which are chromosomal genes. Aerobactin is a hydroxamate siderophore that shigellae use to bind iron. The iuC locus contains the genes for aerobactin synthesis. The *stx1* and *stx2* genes in *S. dysenteriae* encode shiga toxins Stx1 and Stx2, which are readily transmitted to *Escherichia coli* via transducing bacteriophage, where they are known as shiga-like toxins 1 and 2 (Slt1 and Slt2) or vero toxins 1 and 2 (VT1 and VT2; Paton and Paton, 1998) causing lethal hemolytic uremic syndrome and bloody diarrhea. Sandvig (2001) has reviewed in detail the functionality of these *Shigella* toxins and their effects on mammalian cells.

In recent years, clusters of virulence genes have been identified in *Shigella* and other virulent organisms and are termed pathogenicity islands (PAIs). PAIs are usually large in size (20–200 kb) and contain numerous virulence factors in combination with transmissible genetic elements such as transposons, plasmids, or bacteriophages reflecting probable horizontal gene transfer. Four chromosomal PAIs have been identified on the *S. flexneri* chromosome (Walker and Vertma, 2002). The first designated SHI-1 is 46 kb in size and encodes an enterotoxin and two autotransporters thought to be involved in virulence (Al-Hasani et al., 2001; Rajakumar, Sasakawa, and Adler, 1997). The second designated SHI-2 encodes the siderophore aerobactin and colicin V immunity (Moss et al., 1999). The third PAI-designated SRL carries the resistance locus (RL) encoding resistance to streptomycin, ampicillin, chloramphenicaol, and tetracycline (Turner et al., 2001). The fourth PAI consists of a 42-kb sequence encoding insertion sequence elements, bacteriophage genes, and the *Shigella* virulence *criR* gene.

III. PCR DETECTION OF *SHIGELLAE*

A. Conventional PCR Assays

Lampel et al. (1990) used the PCR with primers KL1/KL8 (Table 3.2) to amplify a 760-bp sequence from the 220-kbp invasive plasmids of enteroinvasive *S. flexneri, S. dysenteriae, S. boydii, S. sonnei,* and *E. coli.* A nonvirulent mutant of *S. flexneri* lacking the virulence plasmid failed to yield the amplicon.

Lüscher and Altwegg (1994) made use of the PCR to detect shigellae, EIEC, and enterotoxic *E. coli* in stool samples of diarrheaic patients returning to Switzerland from tropical countries. Colonies were first isolated on MacConkey Agar. Primers

Table 3.2 PCR Primers and DNA Probes

Primer or Probe	Sequence (5' → 3')[a]	Size of Sequence (bp)	Amplified Gene or DNA Target Sequence	References
P1	ACC-AGG-ATT-TTG-GCT-TAG-AAG	~900	23S rDNA	Hong et al. (2004)
P2	CAC-TTA-CCC-CGA-CAA-GGA-AT			
Probe	AAG-CGA-CTT-GCT-CGT-GGA			
phoP-1	ATG-CAA-AGC-CCG-ACC-ATG-ACG	302	*phoP/phoQ*	Li and Mustapha (2004a)
phoP-2	GTA-TCG-ACC-ACC-ACG-ATG-GTT			
ipaH-1	GTT-CCT-TGA-CCG-GCT-TTC-CGA-TAC-CGT-C	620	*ipaH*	Li and Mustapha (2004b)
ipaH-2	GCC-GGT-CAG-CCA-CCC-TCT-GAG-AGT-AC			
KL1	TAA-TAC-TCC-TGA-ACG-GCG	760	*invap*	Lampel et al. (1990)
KL8	TTA-GGT-GTC-GGC-TTT-TCT-G			
HS9	ATC-AGG-TGT-CGT-AAT-TTT-AQ	952	*rfc*	Houng, Sethabutr, and Excheverria (1997)
HS57	GGG-CTA-AGT-TCC-CTC			
SF1	ATT-GGT-GGT-GGT-GGA-AGA-TTA-CTG-G	1096	*rfc*	
SF2	TTT-TGC-TCC-AGA-AGT-GAG-G			
SF5	AGC-TAA-TGC-GTT-TTG-GGG-AAT	~829	*rfc*	
SF6	TCC-CAA-TGA-CTG-ATA-CCA-TGG			
HS61	GTA-CAG-AAC-TGC-TGG-CAA-T	561	*rfc*	
HS62	CAA-CAT-GAT-TCA-TCC-ATG-G			

H8	GTT-CCT-TGA-CCG-CCT-TTC-CGA-TAC	620	*ipaH*	Theron et al. (2001) from Hartman et al. (1990)
H15	GCC-GGT-CAG-CCA-CCC-TC			
H8	GTT-CCT-TGA-CCG-CCT-TTC-CGA-TAC	401	*ipaH*	
H10	CAT-TTC--CTT-CAC-GGC-AGT-GGA			
STG-F	GAC-AGG-ATT-TGT-TAA-CAG-G	740	*stx*	Luo, Wang, and Peng (2002)
STG-R	TTC-CAG-TTA-CAC-AAT-CAG-GC			
IpaH-F	CCT-TGA-CCG-CCT-TTC-CGA-TA	606	*ipaH*	Kong et al. (2002)
IpaH-R	CAG-CCA-CCC-TCT-GAG-GTA-CT			
Probe	GGA-AAT-GCG-TTT-CTA-TGG-CGT-GTC		*ipaH*	Sethabutr et al. (2000)
Shig-1	TGG-AAA-AAC-TCA-GTG-CCT-CT	422	*ipaH*	Lüscher and Altwegg (1994)
Shig-2	CCA-GTC-CGT-AAA-TTC-ATT-CT			Lüscher and Altwegg (1994)
ial-Pr1	CTG-GAT-GGT-ATG-GTG-AGG	320	*ial*	Lüscher and Altwegg (1994) from Frankel et al. (1989)
ial-Pr2	GGA-GGC-CAA-CAA-TTA-TTT-CC			
St-Pr1	TCT-GTA-TTG-TCT-TTT-TCA-CC	186	*ST*	
St-Pr2	TTA-ATA-GCA-CCC-GGT-ACA-AGC			
Lt-Pr1	GGC-GAC-AGA-TTA-TAC-CGT-GC	750	*Lt*	
Lt-Pr2	CCG-AAT-TCT-GTT-ATA-TAT-GTC			

Continued

Table 3.2 PCR Primers and DNA Probes (Continued)

Primer or Probe	Sequence (5' → 3')[a]	Size of Sequence (bp)	Amplified Gene or DNA Target Sequence	References
Primer I	CTG-GTA-GGT-ATG-GTG-AGG	320	ial	Lindqvist et al. (1997) from Frankel et al. (1990)
Primer II	CCA-GGC-CAA-CAA-TTA-TTT-CC			
Stx1-F	AAA-TCG-CCA-TTC-GTT-GAC-TAC-TTC-T	370	Stx1	Brandal et al. (2007) from Brian et al. (1992)
Stx1-R	TGC-CAT-TCT-GGC-AAC-TCG-GGA-TGC-A			
Stx2-F	CAG-TCG-TCA-CTC-ACT-GGT-TTC-ATC-A	283	Stx2	Brandal et al. (2007) from Brian et al. (1992)
Stx2-R	TGC-CAT-TCT-GGC-AAC-TCG-GGA-TGC-A			
eaeA-F	TCA-ATG-CAG-TTC-CGT-TAT-CAG-TT	482	eaeA	Brandal et al. (2007) from Vidal et al. (2004)
eaeA-R	GGA-TAT-TCT-CCC-CAC-TCT-GAC-ACC			
STIb-F	GTA-AAG-TCC-GTT-ACC-CCA-ACC-TG	190	STIAb	Brandal et al. (2007) from Lopez-Saucedo et al. (2003)
STIb-R	ATT-TTT-CTT-TCT-GTA-TGT-CTT			
LTI-F	TCT-CTA-TGT-GCA-TAC-GGA-GC	322	LT1	Brandal et al. (2007) from Rappelli et al. (2001)

LTI-R	CAC-CCG-GTA-CAA-GCA-GGA-TT			
ipaH-F	GTT-CCT-TGA-CCG-CCT-TTC-CGA-TAC-CGT-C	619	*ipaH*	Brandal et al. (2007) from Toma et al. (2003)
ipaH-R	CCA-TAC-TGA-TTG-CCG-CAA-T			
ipaH-R	GCC-GGT-CAG-CCA-CCC-TCT-GAG-AGT-AC			
aggR-F	GTA-TAC-ACA-AAA-GAA-GGA-AGC	254	*aggR*	Brandal et al. (2007) from Toma et al. (2003)
aggR-R	ACA-GAA-TCG-TCA-GCA-TCA-GC			
rrs-F	CCC-CCT-GGA-CGA-AGA-CTG-AC	401	*rrs*	Brandal et al. (2007) from Wang, Clark, and Rodgers (2002)
rrs-R	ACC-GCT-GGC-AAC-AAA-GGA-TA			
virA-F	CTG-CAT-TCT-GGC-AAT-CTC-TTC-ACA-TC	215	*virA*	Vantarakis et al. (2000) from Villalobo and Torres (1998)
virA-R	TGA-TGA-GCT-AAC-TTC-GTA-AGC-CCT-CC			
Eco-malB-F	CAG-ACG-CTG-ACG-CTG-ACC-A	585	*malB*	Wang et al. (1997)
Eco-malB-R	GAC-CTC-GGT-TTA-GTT-GAC-AGA			

ial-Pr1/ial-Pr2 (Table 3.2) amplified a 320-bp sequence of the invasion plasmid antigen gene *ial*. Primers Shig-1/Shig-2 (Table 3.2) amplified a 422-bp sequence of the invasion plasmid antigen H gene *ipaH*. ETEC were identified using primers St-Pr1/ St-Pr2 that amplified a 186-bp sequence of the heat stable toxin gene *st* and primers Lt-Pr1/Lt-Pr2 (Table 3.2) that amplified a 750-bp sequence of the heat-labile toxin gene *lt* as described by Frankel et al. (1989). Among a total of 124 patients' stool samples, *Shigella*/EIEC and ETEC were detected in 19.8% and 18.5%, respectively.

Soft cheese samples represent a particularly difficult menstruum from which to extract low numbers of pathogenic bacteria for PCR detection. Lindqvist et al. (1997) described a buoyant density centrifugation (BDC) procedure to prepare samples for PCR analysis. The primers I/II (Table 3.2) amplified a 320-bp sequence of the *Shigella* spp. invasion plasmid (Frankel et al., 1990).

Blue cheese samples were seeded with *S. flexneri* and layered on top of Percoll media. Standard isotonic Percoll (SIP) was prepared by adding 100 mg of peptone and 850 mg of NaCl to 100 ml of Percoll and diluting with peptone water (8.5 g NaCl and 1.0 g peptone in 1 liter of dH_2O) to obtain the required discontinuous Percoll gradients. Eight milliliters of 50% SIP were added to 38 ml centrifuge tubes and 2 ml of 100% SIP were placed below the first layer. Blue cheese homogenates (1:0 diln.) were obtained by stomaching for 2 min and 12 ml layered onto the top of the Percoll and covered with paraffin oil. The tubes were centrifuged in an SW-27 Beckman rotor at 15,000 rpm for 10 min. Cheese homogenates subjected directly to PCR analysis required a 10^3-fold dilution to allow detection of 2×10^6 cells/g of cheese. In contrast, BDC allowed the PCR detection of 1×10^4 cells/g of cheese.

Sethabutr et al. (2000) compared conventional PCR-agarose, a PCR-ELISA, and conventional culture techniques for the ability to detect *Shigella* and enteroinvasive *E. coli* in stool samples from 89 children under the age of nine years with diarrheal stools in Kenya. DNA from stool samples was purified with SiO_2. The primers H8/H15 (Table 3.2) amplified a 620-bp sequence of the invasion plasmid antigen gene (*ipaH*) specific for *Shigella* species and EIEC. An ipaH probe H10 (Table 3.2) was biotin-labeled at the 5′-end and used as a capture probe. Digoxygenin-labeled amplicons were denatured and the ipaH–biotin probe added in microtiter plate wells coated with streptavidin hybridization. They were then allowed to incubate for 1.5 hr at 37°C. The wells were then thoroughly washed and the hybrid product–probe detected with anti-digoxigenin horseradish peroxidase (DIG-HRP) conjugate and color developed with 2,2′-azino-bis(3-ethylbenzthiazoline-6-sulfonic acid; ABTS). Conventional culture techniques yielded 25 (28%) *Shigella* positive samples, and 2 (2.2%) EIEC positive samples whereas PCR-ELISA yielded 42 (47%) positive and the conventional PCR-agarose-ethidium bromide (PCR-AGE) technique yielded 40 (45%) positive samples. However, 25 of the positive cultivation samples were PCR negative due presumably to PCR inhibitors.

Theron et al. (2001) developed a semi-nested PCR specific for virulent *Shigella* and EIEC. The first PCR was performed with primers H8/H15 (Table 3.2) that amplified a 620-bp sequence of the *ipaH* gene during the course of 10 thermal cycles. The second PCR was performed with 1 μl of the first PCR utilizing primers H8/H10 and yielded a 401-bp sequence. Using a pure culture of *S. flexneri*, cells were suspended in water at a level of 1×10^6 CFU/ml and lysed by heating at 100°C for 10 min and the preparation clarified by centrifugation. A volume of 10 μl of the cell lysate was used for the first PCR of 50 μl. The sensitivity of detection of the nested PCR was 1.6×10^5 CFU/ml of lysate (1.6×10^3 CFU/PCR). Enrichment for 6 hr at 37°C in gram-negative broth resulted in detection of 160 CFU/ml (1.6 CFU per PCR). The relatively low level of sensitivity without enrichment was attributed to the presence of PCR inhibitors.

Li and Mustapha (2004a) developed a PCR for simultaneously detecting *E. coli, Salmonella, Citrobacter*, and *Shigella* in ground beef. The primers phoP-1/phoP-2 (Table 3.2) amplified a 302-bp sequence of the *phoP/phoQ* gene locus, which is a two-component regulator system that regulates the expression of genes involved in virulence and macrophage survival. Because homologous sequences of this gene locus are present in all four bacterial groups, the primers used to amplify the *pho*P sequences allowed indiscriminate detection of all four bacterial groups. Ground beef (10 g) was homogenized with 90 ml of tryptic soy broth containing 0.5% yeast extract (TSBYE) and seeded with the individual organisms at various levels. DNA was purified with the use of a commercial kit. A 10-hr enrichment at 37°C allowed the detection of 1 CFU/g. The authors indicated that the application of species-specific primers to the amplicons obtained from the use of these homologous primers would conceivably allow species discrimination.

Brandal et al. (2007) attempted to clarify the virulence gene relationships among the five major pathotypes of *E. coli* that cause diarrhea and shigellae with the use of a multiplex PCR. PCR primers (Table 3.2) were utilized for seven virulence genes plus the positive control gene *rrs*. The authors concluded that with the use of such an octiplex PCR assay the five pathotypes of *E. coli* could be distinguished from one another, and that with the exception of EIEC strains, they could also be distinguished from *Shigella* spp. However, only strains of *S. boydii* could be distinguished from other species of *Shigella*.

B. Microarrays

Microarray technology is ideally suited for simultaneous detection of a number of pathogens present in a given sample. To select a common gene fragment for identification of the presence of multiple pathogens, such a gene must contain conserved regions common to these pathogens. In addition, sufficient sequence diversity must be present for genus and species identification. 16S rDNA and 23s rDNA have frequently been used for such a purpose. Hong et al. (2004) developed an oligonucleotide array assay for detection of 14 pathogenic bacterial species encompassing 13 bacterial genera including the genus *Shigella* using *S. dysenteriae*. The primers P1/P2 (Table 3.2) amplified a ~900-bp sequence of the 23S rDNA gene from all 13 species. Digoxygenin (DIG) was incorporated at the 5'-end of the P2 primer for color development. A different probe was used for each species. Probe no. 6 (Table 3.2) was used for detection of *Shigella*. All probes were immobilized onto nylon membranes that were then subjected to hybridization with thermally denatured PCR products labeled with DIG. The level of sensitivity was 10 CFU/ml. An advantage of DIG labeling of the probe is that it allows visual evaluation of results.

C. Multiplex PCR

Multiplex PCR refers to the application of the PCR for simultaneous detection of two or more different microorganisms in the same PCR reaction vial, making use of different pairs of primers that usually amplify the respective amplicons of significantly different size to facilitate agarose detection of each.

The *rfc* gene encodes the O-antigen polymerase that polymerizes the O-antigen subunit into lipopolysaccharide (LPS) chains. Houng, Sethabutr, and Escheverria (1997) reported on the development of a multiplex PCR using primers derived from the *rfc* genes of *S. sonnei*, *S. flexneri*, and *S. dysenteriae* I. The protocol developed was based on the assumption that homologous *rfc* sequences are present only in organisms that assemble identical, or even closely related, O-antigen saccharide into LPS chains of closely related serotypes. The assay system was able to detect and differentiate the three *Shigella* species in addition to EIEC. Primers HS9/HS51 yielded a 952-bp amplicon from *S. sonnei* (Table 3.2). Primers SF1/SF2 yielded a 1096-bp amplicon from *S. flexneri* (Table 3.2). Primers SF5/SF6 yielded a ~829-bp amplicon from *S. dysenteriae* (Table 3.2). Primers HS61/HS62 yielded a 561-bp amplicon from EIEC. The lowest *Shigella* level detected directly in feces was 7.4×10^4 CFU/g. The detection of *S. boydii* was not addressed.

Numerous studies have involved the detection of *Shigella* spp. simultaneously with other pathogens in multiplex PCR assays. Li and Mustapha (2004b) developed a multiplex PCR for detection of *Salmonella*, *E. coli* O157:H7, and *Shigella* spp. The primers ipaH-1/ipaH-2 (Table 3.2) were used to amplify a 620-bp sequence of the invasive plasmid encoding an antigen specific for the genus *Shigella*. The assay was applied to seeded homogenates of fruits and vegetables prepared with brain heart infusion broth. The detection level after 24 hr of enrichment at 37°C was 0.8 CFU per gram of produce or per ml of apple cider. Li, Zhuang, and Mustapha (2005) in a subsequent multiplex PCR study applied the same set of primers ipaH-1/ipaH-2 for detection of *Shigella* in conjunction with independent pairs of primers for simultaneously detecting *Salmonella* and *E. coli* O157:H7. A commercial kit was used for DNA purification. The limit of detection with ground beef was 1.6 CFU/g. However, the minimum level of detection with ground pork was 16 CFU/g.

Vantarakis et al. (2000) developed a multiplex PCR detection assay for *Salmonella* spp. and *Shigella* spp. in mussels. The primers virA-F/virA-R (Table 3.2) from Villalobo and Torres (1998) amplified a 215-bp sequence from the *virA* gene of *Shigella* spp. Following enrichment in buffered peptone water for 22 hr at 37°C, 1 to 10 CFU of *Shigella* spp. were detected per gram of tissue. This pair of primers, however, also amplifies a 215-bp sequence from the *virA* gene of EIEC (Villalobo and Torres, 1998).

Kong et al. (2002) developed a multiplex PCR for simultaneous detection of five pathogenic bacterial species representing six genera including *S. flexneri* as a representative of all four species of *Shigella*. The primers IpaH-F/IpaH-R amplified a 606-bp sequence of the gene encoding the outer antigen H on the invasive plasmid of *Shigella* spp. The primers were highly specific for all four *Shigella* species, although the authors did not address the issue of the primers possibly detecting

EIEC. Digestion of the 606-bp amplicon with the restriction nuclease *Sma*l yielded two bands of 289- and 317-bp, which presumably could be used for specific confirmation of *Shigella*.

D. Immunocapture PCR

In order to circumvent the lack of species specificity in PCR targeting of the shiga toxin gene, Luo, Wang, and Peng (2002) developed an immunocapture-PCR technique for specific detection of *S. dysenteriae* in sewage samples. The primers STG-F/STG-R (Table 3.2) amplified a 740-bp sequence of the shiga toxin gene. The wells of microtiter plates were coated with a monoclonal antibody specific for *S. dysenteriae* I. Following blocking, 20 µl of a bacterial suspension were added. After binding of the cells, the wells were thoroughly washed and the plates held at 100°C for 10 min for template preparation. When the amplicon was digested with *Hind*III characteristic fragments of 267- and 473-bp resulted. The limit of detection was the DNA from 5 CFU. This high level of sensitivity was attributable to the removal of PCR inhibitors by the affinity enrichment with the monoclonal antibody.

E. Molecular Typing

Shortly before 1985 an epidemic of bacillary dysentery occurred in West Bengal, India and adjacent Bangladesh involving multiple antibiotic-resistant isolates of *S. dysenteriae* I. All isolates were ampicillin-resistant. Palchaudhuri et al. (1985) undertook an extensive study of 300 such strains involving plasmid profile analysis. Results yielded six common plasmids in all strains isolated from West Bengal: 2.5, 4.5, 6.5, 10, 70, and 120 kb. The isolates from Bangladesh were identical except for the absence of the 12-kb plasmid. *Eco*R1 fragments generated from the total plasmid DNA content of each strain supported the view that the plasmids present in the West Bengal isolates were identical to the Bangladesh plasmids. Interestingly, growth at 42°C resulted in loss of the 70- and 120-kb plasmids resulting in conversion to sensitivity to chloramphenicol, tetracycline, and ampicillin. Ampicillin resistance was always found to be associated with the 120-kb plasmid. In addition, when the 70- and 120-kb plasmids of the West Bengal isolates were lost on growth at 41°C, they showed poor growth in tryptic soy broth reflecting the plasmid influence on metabolism and growth.

In an epidemic of shigellosis just prior to 1987 in southern Bangladesh, 19 out of 32 isolates of the causal organism *S. dysenteriae* I were resistant to nalidixic acid. All isolates harbored a large plasmid characteristic of invasive shigellae. In addition, all 19 nalidixic-acid-resistant isolates harbored a plasmid of about 20 MDa, which was not detected in any of the 13 nalidixic-acid-sensitive isolates (Munshi et al., 1987). Conjugation involving *E. coli* K12 (F⁻) as recipient resulted in transconjugants having acquired nalidixic acid resistance in conjunction with acquisition of the 20-MDa plasmid.

Just prior to 1996, an outbreak of dysentery due to multidrug-resistant *S. dysenteriae* I occurred along the coast of Kenya and shortly thereafter in the outskirts of adjacent Nairobi. All isolates were resistant to ampicillin, trimethoprim, sulphame-thoxazole, chloramphenicol, tetracycline, and streptomycin. Molecular typing of 22 isolates from this outbreak by Kariuki et al. (1996) utilized plasmid DNA profiles and pulsed-field gel electrophoresis (PFGE) of *Xba*I-restricted genomic DNA. Each of the 22 *S. dysenteriae* isolates had three plasmids of 2, 5, and 55 MDa. One isolate had an additional 24-MDa plasmid. Seven of the 22 isolates transferred all their antibiotic resistance characteristics, which were associated with the self-transmissi-ble 55-MDa plasmid, to a conjugation recipient of *E. coli* K12. Each of the isolates yielded a similar *Xba*I-PFGE banding pattern consisting of 18 to 19 bands between 10 and 1000 kbp indicating that all 22 strains were clonally related.

Wang, Cao, and Cerniglia (1997), in attempting to distinguish *Shigella* spp. from enteroinfectious and noninfectious *E. coli* strains, made use of several PCR primer pairs and 16S rDNA sequencing for studying the relationship among these organisms. The primers Eco-malB-F/Eco-malB-R (Table 3.2) amplified a 585-bp sequence of the *E. coli mal*B operon. The 585-bp amplicon was produced by all six strains of *E. coli* (three enteroinvasive and three nonenteroinvasive) in addition to *S. flexneri* and *S. sonnie* but not by *S. boydii* or *S. dysenteriae*, indicating that *S. flexneri* and *S. son-nei* are more closely related to *E. coli* than to *S. boydii* or *S. dystenteriae*. The prim-ers shig-IpaH-F/shig-IpaH-R (Table 3.2) amplified a 610-bp sequence of the *Shigella* Ipp-H invasion plasmid antigen H gene. None of the six enteroinfectious strains of *E. coli* yielded an amplicon, whereas all four species of *Shigella* yielded the 610-bp PCR product.

The IpaH sequences were present at multiple sites on both the large invasive plasmid and on the genomic DNA in *Shigella* and enteroinvasive *E. coli* (EIEC), and therefore EIEC (not included among the six *E. coli* strains examined) should be positive with this method. The primers Shig-set1A-F/Sig-set1A-R amplified a 258-bp sequence of the *S. flexneri* enterotoxin 1 (*set1A*) gene and yielded amplicons with two of the six *E. coli* strains and with *S. flexneri*, but not with the other three species of *Shigella,* indicating that *E. coli* is more closely related to *S. flexneri* than to the other three *Shigella* species. rDNA sequencing indicated that the species of *Shigella* form a cluster with *E. coli* and that *S. flexneri* and *S. sonnei* are closer to *E. coli* than to *S. boydii* and *S. dystenteriae*.

Because *Shigella* strains produce neither flagella nor capsular antigens, their antigenic characterization is completely dependent on their somatic (O) antigens. *S. dysteneriae, S. boydii,* and *S. flexneri* are comprised of 15, 18, and 8 serotypes, respectively. *S. flexneri* serotypes 1–6 are further subdivided into 12 subserotypes. *S. sonnei* exhibits only one serotype. A total of 59 serotypes constitutes the genus *Shigella* (Coimbra et al., 2001). In addition to cross-reactions confounding *Shigella* serotyping, occasional mutations in one of the multiple genes controlling O-antigen synthesis and polymerization can cause transition from smooth (S) forms to rough (R) forms that do not produce O-antigen and consequently are serologically untypeable. A total of 17 reference strains and clinical isolates of *Shigella* representing all sero-types were used by Coimbra et al. (2001) to establish a comparative database uti-lizing 16S and 23S rRNA gene restriction patterns generated with the restriction nuclease *Mlu*I. A total of 51 distinct ribotypes was obtained. The number of bands comprising each ribotype varied from 9 to 15, with fragment sizes ranging from 1.6 to 18.8 kbp. Fifteen ribotypes were shared by two or more serotypes or biotypes. With few exceptions, most isolates of the four *Shigella* species were easily identified as to species.

Navia, Gascon, and Vila (2005) subjected 43 strains of *S. sonnei* and 34 strains of *S. flexneri* to PFGE isolated in Barcelona, Spain from patients with persistent diarrhea acquired during widespread global travel abroad. Digestion of genomic DNA prior to PFGE was with *Xba* I. In contrast to the clonally related strains involved in an outbreak from a common geographic area as described above (Kariuki et al., 1996), the 77 isolates yielded a great variety of genetically diverse profiles with an average of 20–22 DNA bands reflecting the widespread intercontinental origins of the isolates and a high level of genomic rearrangements.

From 2001 to 2003 *S. sonnei* replaced *S. flexneri* to become the predominant species causing shigellosis in central Taiwan. A total of 425 *S. sonnei* isolates collected from 1996 to 2004 were subjected to *Not*I-PFGE, inter-ISI spacer typing (IST), and antimicrobial testing by Wei et al. (2007). Results indicated that at least 21 IST clones had emerged from the S. *sonnei* infections from 1996 to 2004. Most IST clones lasted for a short time; some circulated for two to three years. An IST 1 clone that was detected for the first time in 2000 was the most prevalent and responsible for the epidemic in 2001 to 2003. This IST 1 clone evolved into many strains with different PFGE genotypes and antibiograms. The ancestor PFGE genotype was found to have remained the predominant circulating strain during the period studies and constituted 139 of the 425 total strains examined.

REFERENCES

Al-Hasani, K.K., Rajakumar, K., Bulach, D., Robinson-Browne, R., Adler, B., Sakellaris, H. 2001. Genetic organization of the *she* pathogenicity island in *Shigella flexneri* 2a. *Microb. Pathogen.* 30:1–8.

Bernardoni, M., Mounier, J., d'Hauteville, H., Coquis-Rondon, M., Sansonetti, P. 1989. Identification of icsA, a plasmid locus in *Shigella flexneri* that governs bacterial intra- and intercellular spread through interaction with F-actin. *Proc. Natl. Acad. Sci. USA* 86:3867–3871.

Brandal, L., Lindstedt, B., Aas, L., Stavnes, T., Lassen, J., Kapperud, G. 2007. Octaplex PCR and fluorescence-based capillary electrophoresis for identification of human diarrheagenic *Escherichia coli* and *Shigella* spp. *J. Microbiol. Meth.* 68:331–341.

Brian, M., Rosolono, M., Muray, B., Mirandea, A., Lopez, E., Gomez, H., Cleary, T. 1992. Polymerase chain reaction for diagmosis of enterohemorrhagic *Escherichia coli* infection and hemolytic-uremic syndrome. *J. Clin. Microbiol.* 30:1801–1806.

Coimbra, R., Nicastro, G., Grimont, P., Grimont, F. 2001. Computer identification of *Shigella* species by rRNA restriction patterns. *Res. Microbiol.* 152:47–55.

Frankel, G., Giron, J., Valmassoi, J., Schoolnik, G. 1989. Multi-gene amplification: Simultaneous detection of three virulence genes in diarrhoeal stool. *Mol. Microbiol.* 3:1729–1734.

Frankel, G., Riley, L., Giron, J., Valmassoi, J., Friedman, A., Strockbine, N., Falkow, S., Schoolnik, G. 1990. Detection of *Shigella* in feces using DNA amplification. *J. Infect. Dis.* 161:1252–1256.

Goldberg, M., Baru, O., Parsot, C., Sansonetti, J. 1993. Unipolar localization and ATPase activity of IcsA a *Shigella flexneri* protein involved in intracellular movement. *J. Bacteriol.* 175:2189–2196.

Hartman, A., Venkatesan, M., Oaks, E., van Buysse, J. 1990. Sequence and molecular characterization of multicopy invasion plasmid antigen gene, *ipa*H, of *Shigella flexneri*. *J. Bacteriol.* 172:1905–1915.

High, N., Mounier, J., Prvost, C., Sansoneatti, P. 1992. IpaB of Shigella flexneri causes entry into epithelial dells an escape from tehphagoytaic vacuole. *EMBO J.* 11:1991–1999.

Hong, B., Jiang, L., Hu, Y., Fang, D., Guo, H. 2004. Application of oligonucleotide array technology for the rapid detection of pathogenic bacteria of food borne infections. *J. Microbiol. Meth.* 58:403–411.

Houng, H., Sethabutr, O., Escheverria, P. 1997. A simple polymerase chain reaction technique to detect and differentiate *Shigella* and enteroinvasive *Escherichia coli* in human feces. *Diagn. Microbiol. Infect. Dis.* 28:189–125.

Kariuki, S., Muthotho, N., Kimari, J., Waiyaki, P., Hart, C., Gilks, C. 1996. Molecular typing of multi-drug resistant *Shigella* dysenteriae type 1 by plasmid analysis and pulsed field gel electrophoresis. *Trans. Royal Soc. Trop. Med. Hyg.* 90:712–714.

Kong, R., Lee, S., Law, T., Law, R.S. 2002. Rapid detection of six types of bacterial pathogens in marine waters by multiplex PCR. *Water Res.* 36:2802–2812.

Lampel, K., Jagow, J., Truckness, M., Hill, W. 1990. Polymerase chain reaction for detection of invasive *Shigella flexne*ri in food. *Appl. Environ. Microbiol.* 56:1536–1540.

Lan, R., Reeves, P. 2002. *Escherichia coli* in disguise: Molecular origins of *Shigella. Microbes Infect.* 4:1125–1132.

Li, Y., Mustapha, A. 2004a. Development of a polymerase chain reaction assay to detect enteric bacteria in ground beef. *Food Microbiol.* 21:369–375.

Li, Y., Mustapha, A. 2004b. Simultaneous detection of *Escherichia coli* O157:H7, *Salmonella*, and *Shigella* in apple cider and produce by a multiplex PCR. *J. Food Protect.* 67:27–33.

Li, Y., Zhuang, S., Mustapha, A. 2005. Application of a multiplex PCR for the simultaneous detection of *Escherichia coli* O157:H7, *Salmonella* and *Shigella* in raw and ready-to-eat meat products. *Meat Sci.* 71:402–406.

Lindqvist, R., Norling, B., Lambertz, S. T. 1997. A rapid sample preparation methods for detection of food pathogens based on buoyant density centrifugation. *Appl. Microbiol.* 24:306–310.

Lopez-Saucedo, C., Cema, J., Vilegas-Sepulved, N., Thompson, R., Velaxquez, F., Torres, J., Tarr, P., Estrada-Garcia, T. 2003. Single multiplex polymerase chain reaction to detect diverse loci associated with diarrheagenic *Escherichia coli. Emerg. Infect. Dis.* 9:127–131.

Luo, W., Wang, S., Peng, X. 2002. Identification of shiga toxin-producing bacteria by a new immuno-capture toxin gene PCR. *FEMS Microbiol. Lett.* 216:39–42.

Lüscher, D., Altwegg, M. 1994. Detection of shigellae, enteroinvasive and enterotoxigenic *Escherichia coli* using the polymerase chain reaction (PCR) in patients returning from tropical countries. *Mol. Cell. Probes.* 8:285–290.

Makino, S. C., Kamata, K., Kurata, T., Yoshikawa, M. 1986. A genetic determinant required for continuous reinfection of adjacent cells on large plasmid in *Shigella flexneri* 2a. *Cell.* 46:551–555.

Ménard, R., Sansonetti, P., Parsot, C., Vasselon, R. 1994. The ipaB and IpaC invasins of *Shigella flexneri* associate in the extracellular medium and are partitioned in the cytoplasm by a specific chaperon. *Cell* 76:829–839.

Moss, J., Cardoz, T., Zychlinsky, A., Groisman, E. 1999. The *selC*-associated SHI-2 pathogenicity island of *Shigella flexneri. Mol. Microbiol.* 33:74–83.

Munshi, M., Haider, K., Rahaman, M., Sack, D., Ahmed, Z., Oroshed, M. 1987. Plasmid-mediated resistance to nalidixic acid in *Shigella dysenteriae* type 1. *Lancet* August 22:419–423.

Navia, M., Gascon, J., Vila, J. 2005. Genetic diversity *of Shigella* species from different intercontinental sources. *Infect. Gen. Evol.* 5:349–353.

Palchaudhuri, S., Kumar, R., Sen, D., Pal, R., Ghosh, S., Sarkar, B., Bhattaharya, S., Pal, S. 1985. Molecular epidemiology of plasmid patterns in *Shigella dysenteriae* type I obtained from an outbreak in West Bengal (India*). FEMS Microbiol. Lett.* 30:187–191.

Paton, J., Paton, A. 1998. Pathogenesis and diagnosis of Shiga toxin-producing *Escherichia coli* infections. *Clin. Microbiol. Rev.* 11:450–479.

Rajakumar, K., Sasakawa, C., Adler, B. 1997. Use of a novel approach, termed island probing, identifies the *Shigella flexneri she* pathogenicity island which encodes a homolog of the immunoglobulin A protease-like family of proteins. *Infect. Immun.* 65:4606–5614.

Rappelli, P., Maddau, G., Mannu, F., Colombo, M., Fiori, P., Cappuccinelli, P. 2001. Develeopment of a set of multiplex PCR assays for the simultaneous identification of enterotoxigenic, enteropathogenic, enterohemorrhagic and enteroinvasive *Escherichia coli. New Microbiol.* 24:77–83.

Sandvig, K. 2001. Shiga toxins. *Toxicon.* 39:1629–1635.

Sethabutr, O., Venkatesan, M., Yam, S., Pang, L., Smoak, B., Sang, W., Echeverria, P., Taylor, D., Isenbarger, D. 2000. Detection of PCR products of the *ipaH* gene from *Shigella* and enteroinvasive *Escherichia coli* by enzyme linked immunosorbent assay. *Diag. Microbiol. Infect. Dis.* 37:11–16.

Stephens, C., Murray, W. 2001. Pathogen evolution: How good bacteria go bad. *Curr. Microbiol.* 11:R53–R56.

Theron, J., Morar, D., Preez, M.K., Brözel, V., Venter, S. 2001. A sensitive seminested PCR method for the detection of *Shigella* in spiked environmental water samples. *Water Res.* 35:869–874.

Toma, C., Lu, Y., Higa, N., Nakasone, N., Chinen, I., Baschkier, A., Rivas, M., Iwanaga, M. 2003. Multiplex PCR assay for identification of human diarrheagenic *Escherichia coli. J. Clin. Microbiol.* 41:2660–2671.

Turner, S., Luck, S., Skellaris, H., Rajakumar, K., Adler, B. 2001. Nested deletions of the SRL pathogenicity island of *Shigella flexneri* 2a. *J. Bacteriol.* 183:5535–5543.

Vantarakis, A., Komninou, G., Venieri, D., Papapetropoulou, M. 2000. Development of a multiplex PCR detection of *Salmonella* spp. and *Shigella* spp. in mussels. *Lett. Appl. Microbiol.* 31:105–109.

Vidal, R., Vidal, M., Lagos, R., Levine, M., Prado, V. 2004. Multiplex PCR for diagnosis of enteric infections associated with diarrheagenic *Escherichia coli. J. Clin. Microbiol.* 42:1787–1789.

Villalobo, E., Torres, A. 1998. PCR for detection of *Shigella* spp. in mayonnaise. *Appl. Environ. Microbiol.* 64:1242–1245.

Walker, J., Vertma, N. 2002. Identification of a putative pathogenicity island in *Shigella flexneri* using subtractive hybridization of the *S. flexneri* and *Escherichia coli* genomes. *FEMS Microbiol. Lett.* 213:257–264.

Wang, G., Clark, C., Rodgers, F. 2002. Detection in *Escherichia coli* of the genes encoding the major virulence factors, the genes defining the O157:H7 serotype, and components of the type 2 Shiga toxin family by multiplex PCR. *J. Clin. Microbiol.* 40:3613–3619.

Wang, R., Cao, W., Cerniglia, C. 1997. Phylogenetic analysis and identification of *Shigella* spp. by molecular probes. *Mol. Cell. Probes.* 11:427–432.

Wei, H., Wang, Y., Li, C., Tung, S., Chiou, C. 2007. Epidemiology and evolution of genotype and antimicrobial resistance of an imported *Shigella sonnei* clone circulating in central Taiwan. *Diag. Microbiol. Infect. Dis.* 58:469–475.

Zychlinsky, A., Kenny, B., Menard, M., Provost, C., Holland, B., Sansonetti, P. 1994. IpaB mediates macrophage apoptosis induced by *Shigella flexneri. Mol. Microbiol.* 11:619–617.

CHAPTER 4

Salmonella

I. CHARACTERISTICS OF THE GENUS

All members of the genus *Salmonella* are gram-negative facultatively anaerobic peritrichously flagellated rods. The natural habitat of *Salmonella* spp. is the intestinal tract of mammals and avian species, most notably including poultry. Major metabolic characteristics of most isolates involve the fermentation of glucose with acid production and the ability to produce H2S. The genus *Salmonella* is presently composed of two species, *Salmonella enterica* and *Salmonella bongori*. *S. enterica* consists of six subgroups, groups I (*S. enterica* subsp. *enterica*) II, IIa, IIIb, IV, and VI. Only *S. enterica* subsp. *enterica* is considered of clinical relevance, and this subspecies includes the pathogen associated with typhoid fever *S. enterica* serovar Typhi. The other subspecies and *S. bongori* are usually associated with the environment or reptiles and are not regarded as clinically important.

Cellular invasion is an important factor influencing the virulence of *Salmonella* serotypes. The invasive phenotype is determined by a large cluster of genes in the *Salmonella* pathogenicity island 1 (SPI1), which is present in all invasive strains of *Salmonella*.

A large number of virulence genes and virulence-enhancing genes have been described for *Salmonella*. There are presently over 30 *Salmonella* specific genes that have been used for the polymerase chain reaction to detect and characterize *Salmonella*. These include *invA* gene sequences that are highly conserved among all *Salmonella* serotypes in addition to the amplification of *his* gene sequences also present throughout the genus as well as fimbriae protein-encoding genes and antibiotic-resistance genes (Table 4.1). The sensitivity of detection of *Salmonella* from complex matrices such as food and feces by PCR is invariably enhanced using nonselective or selective enrichment, particularly if followed by immunomagnetic separation in addition to coupling the PCR with ELISA formats. R-plasmids are considered to be the main factors responsible for the horizontal transfer of antibiotic resistance genes in *Salmonella*. A sizeable number of primer pairs are available for determining by

Table 4.1   Genes Used for the PCR Identification of *Salmonella* spp., Serotypes, and Antimicrobial Resistance

Gene	Description
Virulence Enhancing Genes	
invA	Triggers internalization required for invasion of deep tissue cells
InvE/A	Invasivity proteins
phoP/Q	Intramacrophage survival and enhanced bile resistance
stnB	Salmonella enterotoxin gene
irob	Regulation by iron
slyA	Salmolysin
hin/H2	Flagellar phase variation
afgA	Thin aggregative fimbriae
fimC	Pathogen relate fimbrae gene of *S. enterica*
sefA	Encodes major subunit fimbrial protein of serotype Enterica strains
pefA	Fimbrial virulence gene of *S.* Typhimurium
spvA	Virulence plasmid region
spvB	Virulence plasmid region
spvC	Virulence plasmid region that interacts with the host immune system and is responsible for an increased growth rate in host cells
rep-FIIA	Plasmid incompatibility group
sprC	Virulence gene
sipB-sipC	Junction of virulence genes *sipB–sipC*
himA	Encodes a binding protein
his	*Salmonella* genus specific histidine transport operon
prot6e	Located on 60-kb virulence plasmid and specific for *S.* Enteritidis
STM3357	Regulatory protein whose start codon sequence determines the dt⁻ phenotype exhibiting enhanced virulence
Integron Genes in Cassettes Encoding Antibiotic Resistance	
aadAla	Ampicillin resistance
pse1	Ampicillin resistance
bla$_{oxa-1\ like}$	Ampicillin resistance
bla$_{PSE-1}$	Ampicillin resistance
aadA2	Streptomycin resistance
aadA1a	Streptomycin resistance
dfrA1	Trimethoprim
dfrA12	Trimethoprim
qacEΔ1	Resistance to quarternary ammonium compounds/EB
sul1	Resistance to sulfonamides
floR	Resistance to chloramphenicol and florfenicol (fluorinated derivative of chloramphenicol)
gyrA	Nalidixic acid resistance
tetG	Tetracycline resistance

Table 4.1   Genes Used for the PCR Identification of *Salmonella* spp., Serotypes, and Antimicrobial Resistance (*Continued*)

Gene	Description
	Nonintegron Genes Encoding Antibiotic Resistance (Not Necessarily Specific for *Salmonella*)
bla$_{TEM}$	Ampicillin resistance
catA1	Chloramphenicol resistance
cm1A1	Choramphenicol resistance
aphaA1 -1AB	Kanamycin resistance
tet(A)	Tetracycline resistance
tet(B)	Tetracycline resistance
tet(G)	Tetracycline resistance
aac(3)-IV	Gentamycin resistance
sul1	Sulfonamide
sul2	Sulfonamide
aadA2	Strepomycin/spectinomycin resistance
dfrA14	Triethoprim resistance
	Other Genes
16S rRNA	16S rRNA gene
oriC	Origin of replication of *Salmonella*
spaQ	Required for protein secretion and for the ability of *Salmonella* to gain access to epithelial cells
fliC	Involved in the synthesis of phase 1 and 2 flagellin proteins

the PCR the presence of many antibiotic resistance genes in *Salmonella* isolates that are not necessarily specific for *Salmonella* (Table 4.1).

Amplification of 16S rDNA sequences has also been found useful for genus-specific detection of *Salmonella*. *d*-tartrate (dT+) fermenting strains have been found to result in less severe gastrointestinal infections than d-tartrate-nonfermenting (dT–) strains. Primers have therefore been developed for distinguishing between (dT+) and (dT–) strains. Among the molecular techniques available for strain discrimination of *Salmonella* isolates, pulsed field gel electrophoresis, random amplified polymporphic DNA analysis, ribotyping, multilocus sequence typing, subtracted fingerprinting, and enterobacterial repetitive intergeneric consensus typing have been found useful. Multiplex PCR has been found effective for simultaneously detecting *Salmonella* and other pathogens in foods, particularly with real-time PCR.

Currently, the method of serotyping *S. enterica* subsp. *enterica* is to discriminate isolates on the basis of O (surface polysaccharide) and H (flagellar) antigenic properties. Typing the O antigen denotes the serogroup, and typing the flagella denotes the serotype. The Vi or capsular antigens are specific to *S. enterica* serovar Typhi. Serotyping presently employs more than 150 O and H antigens for the characterization of over 2500 *Salmonella* serovars, of which 1478 are *S. enterica* subsp. *enterica*. Various molecular typing methods have been used to distinguish a number of prominent serotypes that are considered in this chapter. The manner in which the original authors designated species and serotypes has been retained throughout the text.

II. MOLECULAR TECHNIQUES

A. Conventional PCR

Rahn et al. (1992) established a PCR for detecting *Salmonella enterica* subsp. *typhimurium*. A collection of 630 strains of *Salmonella* comprising over 100 serovars including the 20 most prevalent serovars isolated from animals and humans was examined. Controls consisted of 142 non-*Salmonella* strains derived from 21 bacterial genera. The primers SalinvA139/SalinvA141 (Table 4.2) amplified a 284-bp sequence of the *invA* gene required for invasion of epithelial cells. With the exception of two S. Lichfield and two S. Senftenberg strains, all *Salmonella* strains were detected, and none of the non-*Salmonella* isolates yielded the 184-bp amplicon, although a few non-*Salmonella* strains yielded amplicons distinctly different in size from 184-bp. This pair of primers has been used by numerous investigators for detection and enumeration of *Salmonella* by the PCR in various food and environmental samples.

Cohen et al. (1993) developed a pair of primers Salm-U/Salm-L (Table 4.2) that amplified a 496-bp sequence of the *his* gene that encodes the histidine transport operon of salmonellae. The assay was found to be highly specific for members of the genus *Salmonella*.

Aabo et al. (1993) developed a PCR for specific detection of members of the genus *Salmonella*. The primers ST11/ST15 (Table 4.2) were derived from a randomly cloned fragment of the *S. typhimurium* genomic DNA and generated a 429-bp amplicon from 144 of 146 *Salmonella* strains tested (116 of 118 serovars). The two false-negative strains belonged to two different serovars of the rarely isolated subspecies IIIa (monophasic *S. arizonae*). No amplicons were produced with 86 non-*Salmonella* Enterobactriaceae strains tested, covering 41 species from 21 genera. These primers have also been used by numerous investigators for detection and enumeration of *Salmonella* under various PCR formats.

Aabo, Anderson, and Olsen (1995) developed a PCR assay for detection of *Salmonella* in enrichment broths derived from minced meat. Forty-eight unseeded minced pork and 48 unseeded minced beef samples were pre-enriched overnight in buffered peptone water (BPW) at 37°C. One ml of pre-enrichment culture was then transferred to 9 ml of tetrathionate broth (TB) and 0.1 ml was transferred to 9.9 ml of Rappaport–Vassiliadis (RV) medium and incubated at 41.5°C. After 7 hr of selective enrichment, 1 ml of TB culture and 0.05 ml of RV were transferred to 9.0 ml and 9.95 ml of Luria–Bertani (LB) broth, respectively, and incubated for 14 to 16 hr at 37°C. Cells in 5 ml of LB culture were then lysed and used for the PCR with the ST11/ST15 primers from Aabo et al. (1993). The sensitivity of the PCR was 99%, and the sensitivity of routine cultivation methodology was 50%.

Table 4.2 PCR Primers and DNA Probes

Primer or Probe	Sequence (5′ → 3′)[a]	Size of Sequence (bp)	Amplified Gene or DNA Target Sequence	References
probeST14	Alk.Phosph.- NH-TTT-GCG-ACT-ATC-AGG-TTA-CCG-TGG		*inva*	Li, Boudjeliab, and Zhao (2000). from Aabo et al. (1993)
Forward	ACC-ACG-CTC-TTT-CGT-CTG-G	941	*inva*	Lampel, Orlandi, and Kornegay (2000) from Galan et al. 1992)
Reverse	GAA-CTG-ACT-ACG-TAG-ACG-CTC			
HILA2-U	CTG-CCG-CAG-TGT-TAA-GGA-TA	497	*hilA*	Guo et al. (2000)
HILA2-D	CTG-TCG-CCT-TAA-TCG-CAT-GT			
Virulence Enhancing Genes				
iroB-F	TGC-GTA-TTC-TGT-TTG-TCG-GTC-C	606	*iroB* del	del Cerro, Soto, and Mendoza (2003) from Baümler, Heffron, and Reissbrodt (1997)
iroB-R	TAC-GTT-CCC-ACC-ATT-CTT-CCC			
agfA-F	TCC-GGC-CCG-GAC-TCA-ACG	261	*agfA*	Doran et al. (1993)
agfA-R	CAG-CGC-GGC-GTT-ATA-CCG			
hin/H2-F	CTA-GTG-AAA-TTG-TGA-CCG-CA	236	*hin/H2*	Way et al. (1993)
hin/H2-R	CCC-ATC-GCG-CTA-CTG-GTA-TC			
Cryptic DNA-F	AGC-CAA-CCA-TTG-CTA-AAT-TGG-CGC-A	429	Cryptic DNA	Aabo et al. (1993)

Continued

Table 4.2   PCR Primers and DNA Probes (Continued)

Primer or Probe	Sequence (5′ → 3′)[a]	Size of Sequence (bp)	Amplified Gene or DNA Target Sequence	References
Cryptic DNA-R	GGT-AGA-AAT-TCC-CAG-CGG-GTA-CTG			
phoP/Q-F	ATG-CAA-AGC-CCG-ACC-ATG-ACG	299	phoP/Q	Way et al. (1993)
phoP/Q-R	GTA-TCG-ACC-ACC-ACG-ATG-GTT			
stn-F	TTA-GGT-TGA-TGC-TTA-TGA-TGG-ACA-CCC		stn	Prager, Fruth, and Tschäpe (1995)
stn-R	CGT-GAT-GAA-TAA-AGA-TAC-TCA-TAG-G			
invE/invA-F	GCC-CGA--ACG-TGG-CGA-TAA-TT	457	invE/A	Stone et al. (1995)
invE/invA-R	TCA-CCG-GCA-TCG-GCT-TCA-AT			
slyA-F	GCC-AAA-ACT-GAA-GCT-ACA-GGT-G		slyA	Soto (2000)
slyA-R	CGG-CAG-GTC-AGC-GTG-TCG-TGC			
spvA-F	GTC-AGA-CCC-GTA-AAC-AGT	604	spvA	Guerra et al. (2001)
spvA-R	GCA-CGC-AGA-GTA-CCC-GCA			
spvB-F	ACG-CCT-CAG-CGA-TCC-GCA	1063	spvB	Guerra et al. (2001)
spvB-R	GTA-CAA-CAT-CTC-CGA-GTA			
spvC-F	ACT-CCT-TGA-ACA-ACC-AAA-TGC-CGA	298–1060	spvC	Chiu and Ou (1996)
spvC-R	TGT-CTT-CTG-CAT-TTC-GCC-ACC-ATC-A			

Primer	Sequence	Gene	Size	Reference
rep-FIIA-F rep-FIIA-R	CTG-TCG-TAA-GCT-GAT-GGC CTC-TGC-CAC-AAC-CTT-CAG-C	rep-FIIA-F		Guerra et al. (2002)
Class 1 integrons-F Class 1 integrons-R	GGC-ATC-CAA-GCA-AGC AAG-CAG-ACT-TGA-CCT-GAC		Class 1 integrons	Levèsque et al. (1995)
Salm3 salm4	GCT-GCG-CGC-GAA-CGG-GGA-AG TCC-CGG-CAG-AGT-TCC-CAT-T	*invA*	389	Ferretti et al. (2001)
Sal139-F Sal141-R	GTG-AAA-TTA-TCG-CCA-CGT-TCG-GGC-AA TCA-TCG-CAC-CGT-CAA-AGG-AAC-C	*invA*	285	Löfström et al. (2004) from Rahn et al. (1992)
Sal3 Sal4 Probe Salp	GCT-GCG-CGC-GAA-CGG-CGA-AG TCC-CGG-CAG-AGT-TCC-CAT-T TTT-GTG-AAC-TTT-ATT-GGC-GG	*invA*	389	Johnston et al. (2005) from Manzano et al. (1998)
SHIMA-L SHIMA-R Probe SHIMA-P	CGT-GCT-CTG-GAA-AAC-GGT-GAG CGT-GCT-GTA-ATA-GGA-ATA-TCT-TCA GGT-AAC-TTC-GGT-CTG-CGT-GAT-AAA-A	*himA*	122	Bej et al. (1994)
R-Factor Genes				
sul1-F sul1-R	CTT-CGA-TGA-GAG-CCG-GCG-GC GCA-AGG-CGG-AAA-CCC-GCG-GC	sul1	437	del Cerro et al. (2003) from Sandvang, Aerestrup, and Jensen (1998)

Continued

Table 4.2 PCR Primers and DNA Probes (Continued)

Primer or Probe	Sequence (5' → 3')[a]	Size of Sequence (bp)	Amplified Gene or DNA Target Sequence	References
qacEΔ1-F	ATC-GCA-ATA-GTT-GGC-GAA-GT	226	qacEΔ1	del Cerro, Soto, and Mendoza (2003) from Sandvang, Aerestrup, and Jensen (1997)
qacEΔ1-R	CAA-GCT-TTT-GCC-CAT-GAA-GC			
aadA1a-F	GTG-GAT-GGC-GGC-CTG-AAG-CC	527	aadA1a	del Cerro, Soto, and Mendoza (2003) from Sandvang et al. (1997)
aadA1a-R	ATT-GCC-CAG-TCG-GCA-GCG			
dfrA1-F	GTG-AAA-CTA-TCA-CTA-ATG-G	~450	dfrA1	Guerra et al. (2000a,b)
dfrA1-R	CCC-TTT--TGC--CAG-ATT-TGG			
dfra12-F	ACT-CGG-AAT-CAG-TAC-GCA	462	dfra12	Guerra et al. (2001)
dfrA12-R	GTG-TAC-GGA-ATT-ACA-GCT			
bla$_{TEM}$-F	TTG-GGT-GCA-CGA-GTG-GGT	504	bla$_{TE}$	Arlet and Phillippon (1991)
bla$_{TEM}$-R	TAA-TTG-TTG-CCG-GGA-AGC			
bla$_{OXA-1\ like}$-F	AGC-AGC-GCC-AGT-GCA-TCA	—	bla$_{OXA-1}$	Guerra et al. (2001)
bla$_{OXA-1\ like}$-R	ATT-CGA-CCC-CAA-GTT-TCC			
blA$_{PSE-1}$-F	CGC-TTC-CCG-TTA-ACA-AGT-AC	420	blA$_{PSE-1}$	Sandvang, Aerestrup, and Jensen (1998)
blA$_{PSE-1}$-R	CTG-GTT-CAT-TTC-AGA-TAG-CG			

Primer	Sequence	Product size	Gene	Reference
aac (3)-IV-F aac (3)-IV-R	GTT-ACA-CCG-GAC-CTT-GGA AAC-GGC-ATT-GAG-CGT-CAG	674	aac (3)-IV	Guerra et al. (2001)
aphA1-1AB-F aphA1-1AB-R	AAA-CGT-CTT-GCT-TTC-GAG-GC CAA-ACC-GTT-ATT-CAT-TCG-TGA	~500	aphA1-1AB	Frana, Carlson, and Griffith (2001)
catA1-F catA1-R	CCT-GCC-ACT-CAT-CGC-AGT CCA-CCG-TTG-ATA-TAT-CCC	623	catA1	Guerra et al. (2001)
cmlA1-F cmlA1-R	TGT-CAT-TTA-CGG-CAT-ACT-CG ATC-AGG-CAT-CCC-ATT-CCC-AT	435	cmlA1	Guerra et al. (2001)
floR-F floR-R	CAC-GTT-GAG-CCT-CTA-TAT ATG-CAG-AAG-TAG-AAC-GCG	868	floR gyrA-F	Ng et al. (1999)
tet (A)-F tet (A)-R	GCT-ACA-TCC-TGC-TTG-CCT CAT-AGA-TCG-CCG-TGA-AGA	210	tet (A)	Ng et al. (1999)
tet (B)-F tet (B)-R	TTG-GTT-AGG-GGC-AAG-TTT-TG GTA-ATG-GGC-CAA-TAA-CAC-CG	659	tet (B)	Ng et al. (1999)
tet (C)-F tet (C)-R	GCT-CGG-TGG-TAT-CTC-TGC AGC-AAC-AGA-ATC-GGG-AAC	468	tet (C)	Ng et al. (1999)
gyrA-F gyrA-R	AAA-TCT-GCC-CGT-GTC-GTT-GGT GCC-ATA-CCT-ACG-GCG-ATA-CC	343	gyrA	Vila et al. (1995)
Salm-U Salm-L	ACT-GGC-GTT-ATC-CCT-TTC-TCT-GGT-G ATG-TTG-TCC-TGC-CCC-TGG-TAA-GAG-A	496	hto	Cohen et al. (1993)

Continued

Table 4.2 PCR Primers and DNA Probes (Continued)

Primer or Probe	Sequence (5′ → 3′)[a]	Size of Sequence (bp)	Amplified Gene or DNA Target Sequence	References
FloF	ACC-CGC-CCT-CTG-GAT-CAA-GTC-AAG	584	*flo*	Khan et al. (2000)
FloR	CAA-ATC-ACG-GGC-CAC-GCT-GTA-TC			
VirF	GGG-GCG-GAA-ATA-CCA-TCT-ACA	392	*sprC*	Khan et al. (2000)
VirR	GCG-CCC-AGG-CTA-ACA-CG			
InvF	CGC-GGC-CCG-ATT-TTC-TCT-GGA	321	*invA*	Khan et al. (2000)
InVR	AAT-GCG-GGG-ATC-TGG-GCG-ACA-AG			
intF	GCC-CTC-CCG-CAC-GAT-GAT	265	*int*	Khan et al. (2000)
intR	ATT-GGC-GGC-CTT-GCT-GTT-CTT-CTA			
Pse-F	GGC-AAT-CAC-ACT-CGA-TGA-TGC-GT	156	*pse*	Chiu et al. (2006)
Pse-R	GGC-TCA-ATA-CGG-TCT-AGA-CGA-GT			
Flor-F	CTT-TGG-CTA-TAC-TGG-CGA-ATG	266	*FloR*	Chiu et al. (2006)
Flor-R	GAT-CAT-TAC-AAG-CGC-GAC-AG			
STR-F1	AGA-CGC-TCC-GCG-CTA-TAG-AAG-T	203	*str*	Chiu et al. (2006)
STR-R1	CGG-ACC-TAC-CAA-GGC-AAC-GCT			
Sul I-F	CGG-ATC-AGA-CGT-CGT-GGA-TGT	351	*sulI*	Chiu et al. (2006)
Sul I-R	TCG-AAG-AAC-CGC-ACA-ATC-TCG-T			

Primer	Sequence		Size	Gene	Reference
TetG-F	AGC-AGC-CTC-AAC-CAT-TGC-CGA-T		391	*tetG*	Chiu et al. (2006)
TetG-R	GGT-GTT-CCA-CTG-AAA-ACG-GTC-CT				
SpvC-1	ACT-CCT-TGC-ACA-ACC-AAA-TGC-GGA		447	*spvC*	Chiu et al. (2006)
spvC-2	TGT-CTC-TGC-ATT-TCG-CCA-CCA-TCA				
Other Genes					
Salm. spp. ST11	AGC-CAA-CCA-TTG-CTA-AAT-TGG-CGC-A		429	—	Aabo et al. (1993)
Aalm. spp. ST15	GGT-AGA-AAT-TCC-CAG-CGG-GTA-CTG				
16S rRNA-F	TGT-TGT-GGT-TAA-TAA-CCG-CA		571	*16S rRNA*	Aslam, Hogan, and Smith (2003) from Lin and Tsen (1995)
16S rRNA-R	CAC-AAT-CCA-TCT-CTG-GA				
SipB/C-F	ACA-GCA-AAA-TGC-GGA-TGC-TT		252	*sipB-sipC*	Sharma and Carlson (2000) from Carlson et al. (1999)
SipB/C-R	GCG-CGC-TCA-GTG-TAG-GAC-TC				
MINf	ACG-GTA-ACA-GGA-AGM		~402	*16S rRNA*	Trkov and Avguštin (2003)
MINr	TAT-TAA-CCA-CAA-CAC-CT				
TS-11	GTC-ACG-GAA-GAA-GAG-AAA-TCC-GTA-CG		375	—	Kawasaki et al. (2005) from Tsen, Liou, and Lin (1994)
TS-5	GGG-AGT-CCA-GGT-TGA-CGG-AAA-ATT-T				
STN-1	TTG-TCT-CGC-TAT-CAC-TGG-CAA-CC		617	*stn*	del Cerro et al. (2002) from Prager, Fruth, and Tschäpe (1995)
STN-2	ATT-CGT-AAC-CCG-CTC-TCG-TCC				

Continued

Table 4.2 PCR Primers and DNA Probes (Continued)

Primer or Probe	Sequence (5′ → 3′)[a]	Size of Sequence (bp)	Amplified Gene or DNA Target Sequence	References
RAPD-S	TCA-CGA-TGC-A			del Cerro et al. (2002) from Williams et al. (1990)
RAPD-C	AGG-GAA-CGA-G			
Ribotyping RIB-1	TTG-TAC-ACA-CCG-CCC-GTC-A	700–1000	—	del Cerro et al. (2002) from Kostman et al. (1992)
Ribotyping RIB-2	GGT-ACC-TTA-GAT-GTT-TCA-GTT-C			
RAPD ERIC	AAG-TAA-GTG-ACT-GGG-GTG-AGC-G			Garaizar et al. (2000)
RAPD M13	GTA-AAA-CGA-CGG-CCA-GT			
RAPD OPS-19	GAG-TCA-GCA-G			
IRS-PCR adapter AX				
AX1	CTA-GTA-CTG-GCA-GAC-TCT			
AX2	GCC-AGT-A			
IRS-PCR adapter AT				
AT1	CCT-GAT-GAG-TCC-TGA-C			
AT2	CGG-TCA-G			

Name	Sequence	Size (bp)	Gene	Reference
IRS-PCR primer PX	AGA-GTC-TGC-CAG-TAC-TAG-A			
IRS-PCR primer AH1	AGA-ACT-GAC-CTC-GAC-TCG-CAC-G			
IRS-PCR primer AT1	CCT-GAT-GAG-TCC-TGA-C			
InvE primer	TGC-CTA-CAA-GCA-TGA-AAT-GG	457	invE-invA	Notzon, Helmuch, and Bauer (2006) from Stone et al. (1994)
INVA primer	AAA-CTG-GAC-CAC-GGT-GAC-AA			
Probe 259-285	CGT-CTT-ATC-TTG-ATT-GAA-GCC-GAT-GCC-Fluo			
Probe 288-311 LCRed-640-	TGA-AAT-TAT-CGC-CAC-GTT-CGG-GCA-Pho			
Primer SEFA-1	GCA-GCG-GTT-ACT-ATT-GCA-GC	310	sefA	De Medici et al. (2003)
Primer SEFA-2	CTG-TGA-CAG-GGA-CAT-TTA-GCG			
Forward primer	GGC-TTC-GGT-ATC-TGG-TGG-TGT-A	98	sefA	Seo et al. (2004)
Reverse primer	GGT-CAT-TAA-TAT-TGG-CCC-TGA-ATA			Seo et al. (2004)
Probe SEF14	6FAM-CCA-CTG-TCC-CGT-TCG-TTG-ATG-GAC-A-TAMRA			
Forward Prot6e-5	ATA-TCG-TCG-TTG-CTG-CTT-CC	206	prot6e	Malorny, Bunge, and Helmuth (2007)
Reverse Prot6E-6	CAT-TGT-TCC-ACC-GTC-ACT-TTG			
Prot6e Probe	FAM-AGG-CGC-TCA-TCG-GTC-CTG-CTG-T-DQ			

Continued

Table 4.2 PCR Primers and DNA Probes (Continued)

Primer or Probe	Sequence (5′ → 3′)[a]	Size of Sequence (bps)	Amplified Gene or DNA Target Sequence	References
Forward primer 139	GTG-AAA-TTA-TCG-CCA-CGT-TCG-GGC-AA	110	invA	Malorny, Bunge, and Helmuth (2007)
Reverse primer 141	TCA-TCG-CAC-CGT-CAA-AGG-AAC-C			
Probe invA	Yakima Yelow-CTC-TGG-ATG-GTA-TGC-CCG-GTA-AAC-A-DQ			
Forward 167	CAC-ATT-ATT-CGC-TCA-ATG-GAG	290	STM3357	Malorny, Bunge, and Helmuth (2003)
Reverse 166	GTA-AGG-GTA-ATG-GGT-TCC			
Forward SalinvA139	GTG-AAA-TTA-TCG-GCA-CGT-TCG-GGC-AA	284	invA	Rahn et al. (1992)
Reverse SalinvA141	TCA-TCG-CAC-CGT-CAA-AGG-AAC-C			
Forward invE	TGC-CTA-CAA-GCA-TGA-AAT-GG	457	invE-invA	Notzon, Helmuch, and Bauer (2006)
Reverse invA	AAA-CTG-GAC-GAC-CAC-GGT-GAC-AA			
Fret probe 259-285	CGT-CTT-ATC-TTG-ATT-GAA-GCC-GAT-GCC-Fluo			
Fret probe 288-311	Red-640-TGA-AAT-TAT-CGC-CAC-GTT-CGG-GCA-Pho			
INV1F	GGT-CAT-TCC-ATT-ACC-TAC-CT	485	invA	Rychlik et al. (1999)
INV1R	CAA-TAG-CGT-CAC-CTT-TGA-TA			
INV2F	TGG-TGT-TTA-TGG-GGT-CGT-TCT-A	284	invA	
INV2R	CTT-TCA-AAT-CGG-CAT-CAA-TAC-TC			

S. Typhimurium Fli15	CGG-TGT-TGC-CCA-GGT-TGG-TAA-T	559	FliC	Rychlik et al. (1999)
S. Typhimurium Tym	ACT-CTT-GCT-GGC-GGT-GCG-ACT-T			
S. Enteritidis Sef167	AGG-TTC-AGG-CAG-CGG-TTA-CT	312	sefA	
S. Enteritidis Sef478	GGG-ACA-TTT-AGC-GTT-TCT-TG			
S. Typhimurium Primer 1	TTA-TTA-GGA-TCG-CGC-CAG-GC	164	oriC	Fluit et al. (1993)
S. Typhimurium Primer 2	AAA-GAA-TAA-CCG-TTG-TTC-AC			
S. typhi Forward primer	TGC-CGG-AAA-CGA-ATC-T	300	5S–23s spacer	Zhu, Lim, and Chan (1996)
S. typhi Reverse primer	AGT-GCA-TTG-GCA-TGA-CAA-CC			
Salm. spp. invA-1	TTG-TTA-CGG-CTA-TTT-TGA-CCA	521	invA	Cortez et al. (2006)
Salm. spp. invA-2	CTG-ACT-GCT-ACC-TTG-CTG-ATG			
S. Enteritidis sefA-1	GCA-GCG-GTT-ACT-ATT-GCA-GC	330	sefA	
S. Enteritidis sefA-2	TGT-GAC-AGG-GAC-ATT-TAG-CG			
S. Typhimurium pefA-1	TTC-CAT-TAT-TGC-ACT-GCG-TG	497	pefA	

Continued

Table 4.2 PCR Primers and DNA Probes (Continued)

Primer or Probe	Sequence (5′ → 3′)[a]	Size of Sequence (bp)	Amplified Gene or DNA Target Sequence	References
S. Typhimurium pefA-2	GGC-ATC-TTT-CGC-TGT-GGC-TT			
S. enterica Srt2F	ATA-AAT-CCG-GCG-GCC-TGA	102	fimC	Piknova et al. (2005)
S. enterica Srt2R	TGG-TAT-CGA-CGC-CTT-TAT-CT			
S. enterica Srt2P	6-FAM-TTA-CAC-CGG-AGT-GGA-TTA-AAC-GGC-TGG-G-TAMRA			
Salm. spp. invA139	GTG-AAA-TTA-TCG-CCA-CGT-TCG-GGC-AA[a]	284	invA	Jofré et al. (2005) from Rahn et al. (1992)
Salm. spp. invA141	TCA-TCG-CAC-CGT-CAA-AGG-AAC-C[b]			
Salmonella iac139	A-TCG-CCA-CGT-TCG-GGC-AA-TCT-ATT-AAC 120-ACT-GCT-GC[c]			
Salmonella iac141	TCG-CAC-CGT-CAA-AGG-AAC-C-CTC-TCA-GCT-TAT-CCT-G[c]			
Salm. spp. Sal1	TTA-TTA-GGA-TCG-CGC-CAG-GC	163	oriC	Espinoza-Medina et al. (2006)
Salm. spp. Sal2	AAA-GAA-TAA-CCG-TTG-TTC-AC			
Salm. spp. 16S-Sal	GTG-TTG-TGG-TTA-ATA-ACC-GCA-GCA	324	16S rDNA	Lin et al. (2004)
Salm. spp. 16S-CCR	TGT-TBG-MTC-CCC-ACG-CTT-TCG[d]			

Name	Sequence	Size	Gene	Reference
RAPD DG100	CCA-GCA-GCT-T	—	—	Lim et al. (2005) from Maré, Dicks, and van der Walt (2001)
RAPD DG102	GGT-GCG-GGA-A	—	—	Lim et al. (2005) from Chansiripornchai et al. (2000)
RAPD DG93	AGC-AGC-GCC-TCA	—	—	Lim et al. (2005) from Miyata et al. (1995)
ERIC DG111	ATG-TAA-GCT-CCT-GGG-GAT-TCA-C	—	—	Lim et al. (2005) from Millemann et al. (1996)
ERIC DG112	AAG-TAA-GTG-ACT-GGG-GTG-AGC-G	—	—	Lim et al. (2005) from Millemann et al. (1996)
Ribotyping-PCR DG109	TTG-TAC-ACA-CCG-CCC-GTC-A	—	—	Lim et al. (2005) from Kostman et al. (1992)
Ribotyping-PCR DG110	GGT-ACC-TTA-GAT-GTT-TCA-GTT-C	—	—	Lim et al. (2005) from Kostman et al. (1992)
SSCP DG146	GTG-AAA-TTA-TCG-CCA-CGT-TCG-GGC-AA	284	invA	Lim et al. (2005) from Rahn et al. (1992)
SSCP DG147	TCA-TCG-CAC-CGT-CAA-AGG-AAC-C			Lim et al. (2005) from Rahn et al. (1992)
RAPD 1254	CCG-CAG-CCA-A	—	—	Eriksson et al. (2005) from Akopyanz et al. (1992)
RAPD S	TCA-CGA-TGC-A	—	—	Soto et al. (1999) from Williams et al. (1990)
RAPD OPB-6	TGC-TCT-GCC-C	—	—	Soto et al. (1999) from Lin et al. (1996)
RAPD OPB-7	AGG-GAA-CGA-G	—		

Continued

Table 4.2   PCR Primers and DNA Probes (Continued)

Primer or Probe	Sequence (5′ → 3′)[a]	Size of Sequence (bp)	Amplified Gene or DNA Target Sequence	References
RAPD OPL-2	TGG-GCG-TCA-A	—	—	Maré, Dicks, and van der Walt (2001)
RAPD OPL-3	CCA-GCA-GCT-T	—	—	—
RAPD OPL-12	GGG-CGG-TAC-T	—	—	—
MLVA SE-1F	AGA-CGT-GGC-AAG-GAA-CAG-TAG	—	—	Boxrud et al. 2007
MLVA SE-1R	CCA-GCC-ATC-CAT-ACC-AAG-AC	—	—	—
MLVA SE-2F	CTT-CGG-ATT-ATA-CCT-GGA-TTG	—	—	
MLVA SE-2R	TGG-ACG-GAG-GCG-ATA-G	—	—	
MLVA SE-3F	CAA-CAA-AAC-AAC-AGC-AGC-AT –	–	—	
MLVA SE-3R	GGG-AAA-CGG-TAA-TCA-GAA-AGT	—	—	
MLVA SE-4F	ACT-TTA-GAA-AAT-GCG-TTG-AC	—	—	
MLVA SE-4R	AAG-TCA-ACT-GCT-CTA-CCA-AC	—	—	
MLVA SE-5F	CGG-GAA-ACC-ACC-ATC-AC	—	—	
MLVA SE-5R	CAG-GCC-GAA-TAG-CAG-GAT	—	—	
MLVA SE-6F	CCC-CTA-AGC-CCG-ATA-ATG	—	—	
MLVA SE-6R	GCC-GTT-GCT-GAA-GGT	—	—	
MLVA SE-7F	GAT-AAT-GCT-GCC-GTT-GGT-AA	—	—	
MLVA SE-7R	ACT-GCG-TTT-GGT-TTC-TTT-TCT	—	—	
MLVA SE-8F	TTG-CCG-CAT-AGC-AGC-AGA-AGT	—	—	
MLVA SE-8R	GCC-TGA-ACA-CGC-TTT-TTA-ATA-GGC-T	—	—	
MLVA SE-9F	CGT-AGC-CAA-TCA-GAT-TCA-TCC-CGC-G	—	—	
MLVA SE-9R	TTT-GAA-ACG-GGG-TGT-GGC-GCT-G	—	—	
MLVA SE-10F	GCT-GAG-ATC-GCC-AAG-CAG-ATC-GTC-G	—	—	
MLVA SE-10R	ACT-GGC-GCA-ACA-GCA-GCA-ACA-G	—	—	

Primer	Sequence	Size	Gene	Reference
RAPD	CGA-CGC-CCT			Shabarinath et al. (2006)
invE F	CAG-GAT-ACC-TAT-AGT-GCT-GC	166	*invE*	Shabarinath et al. (2006) from Chen and Griffiths (2001)
invE R	CAC-CAA-TAT-CGC-CAG-TAC-GA			
hns LHNS-531	TAC-CAA-AGC-TAA-ACG-CGC-AGC-T	152	*hns*	Jones, Law, and Bej (1993)
hns RHNS-682	GCA-ACT-GGA-AGA-TTT-CCT-GAT-CA			
RAPD 23L	AGG-GAA-CGA-G			Heyndrickx et al. (2007)
RAPD OPB	CCG-AAG-CTG-C			
RAPD P1254	AAT-CGG-GCT-G			
RAPD OPA4	CCG-CAG-CCA-A			
MLST Forward	CGC-TGT-ACG-GTA-TTT-CAT-T	394	*spaM*	Fakhr, Nolan, and Logue (2005) from Hudson et al. (2001)
MLST Reverse	CTG-ACT-CGG-CCT-CTT-CCT-G			
MLST Forward	CCG-GCA-CCG-AAG-AGA	893	*manB*	Fakhr, Nolan, and Logue (2005) from Kotetishvili et al. (2002)
MLST Reverse	CGC-CGC-CAT-CCG-GTC			
MLST Forward	CTC-AAA-GTC-GCY-GGY-GC[e]	518	*pduF*	Fakhr, Nolan, and Logue (2005) from Kotetishvili et al. (2002)
MLST Reverse	GGG-TTC-ATT-GCA-AAA-CC			
MLST Forward	CCG-CGA-CCT-TTA-TGC-CAA-AAC-CG	474	*glnA*	Fakhr, Nolan, and Logue (2005) from Kotetishvili et al. (2002)
MLST Reverse	CCT-GTG-GGA-TCT-CTT-TCG-CT			

Continued

Table 4.2 PCR Primers and DNA Probes (Continued)

Primer or Probe	Sequence (5′ → 3′)[a]	Size of Sequence (bp)	Amplified Gene or DNA Target Sequence	References
Styinva-JHO-2-left	TCG-TCA-TTC-CAT-TAC-CTA-CC	119	*invA*	Nam et al. (2005) from Hoorfar, Ahrens, and Rådström (2000)
Styinva-JHO-2-right	AAA-CGT-TGA-AAA-ACT-GAG-GA			
Salm. spp. Sal-F	GCG-TTC-TGA-ACC-TTT-GGT-AAT-AA	102	*invA*	Daum et al. (2002)
Salm. spp. Sal-R	CGT-TCG-GGC-AAT-TCG-TTA			Daum et al. (2002)
Salm. spp. probe	FAM-TGG-CGG-TGG-GTT-TTG-TTG-TCT-TCT-TAMPA			Daum et al. (2002)
Salm spp INVAF	ACA-GTG-CTC-GTT-TAC-GAC-CTG-AAT	243	invA	Catarame et al. (2006) from Rahn et al. (1992)
Salm. Spp. INVAR	AGA-CGA-CTG-GTA-CTG-ATC-GAT-AAT			
Salm. spp. 16SF	ACG-GTA-ACA-GGA-AG(AC)A-G	402	16S *rRNA*	Trkov and Avguštin (2003)
Salm. spp. 16SR	TAT-TAA-CCA-CAA-CAC-CT			
Salm. spp. Salm16S-F	CGG-GGA-GGA-AGG-TGT-TGT-G	178	16S *rRNA*	Fukushima et al. (2007) from Fey et al. (2004)
Salm. spp. Salm16S-R1	GAG-CCC-GGG-GAT-TTC-ACA-TC			

Name	Sequence	Product size	Target gene	Reference
S. senftenberg SAO1	TAT-CGT-ACT-GGC-GAT-ATT-GGT-GTT-TA	540	*invA*	Vázquez-Novelle et al. (2005)
S. senftenberg SAO2	GGA-CAA-ATC-CAT-ACC-ATG-GCG-AGT-CA			
S. senftenberg SAO3	GAA-ATT-ATC-GCC-ACG-TTC-GGG	281	*invA*	
S. senftenberg SAO4	TCA-TCG-CAC-CGT-CAA-AGG-AAC			
Salm. spp. invAF	CGC-TCT-TTC-GTC-TGG-CAT-TAT-C	408	*invA*	Hong et al. (2003)
Salm. spp. invAR	CCG-CCA-ATA-AAG-TTC-ACA-AAG			
Salm. spp. invar probe	TTT-CTC-TGG-ATG-GTA-TGC-CC-biotin	*invA*		
Salm. spp. L17	CAA-GGC-ATC-CAC-CGT-GT	—	16S-23S rRNA	Baudart et al. (2000)
Salm. spp. G17	GTG-AAG-TCG-TAA-CAA-GG			
Salm. spp. Fim1A	CCT-TTC-TCC-ATC-GTC-CTG-AA	85	*fimA*	Cohen, Mechanda, and Lin (1996)
Salm. spp. Fim1A	TGG-TGT-TAT-CTG-CCT-GAC-CA			
Salm. spp. S18	ACC-GCT-AAC-GCT-CGC-CTG-TAT	159	*ompC*	Kwang, Littledike, and Keen (1996)
Salm. spp. S19	AGA-GGT-GGA-CGG-GTT-GCT-GCC-GTT			

Continued

Table 4.2 PCR Primers and DNA Probes (Continued)

Primer or Probe	Sequence (5′ → 3′)[a]	Size of Sequence (bp)	Amplified Gene or DNA Target Sequence	References
Salm. spp. M1	TTA-TTA-GGA-TCG-CGC-GAG-GC	163	Oric	Mahon et al. (1994) from Widjojoatmodjo et al. (1991)
Salm. spp. M2	AAA-GAA-TAA-CCG-TTG-TTC-AC			
S. Enteritidis S1	GCC-GTA-CAC-GAG-CTT-ATA-GA	620	vir plasmid	
S. Enteritidis S2	AA-GTG-ATG-CCT-TCT-GCA-TC			
Salm. spp. SAL-1F	GTA-GAA-ATT-CCC-AGC-GGG-TAC-TG	438	—	Waage et al. (1999)
Salm. spp. SAL-2R	GTA-TCC-ATC-TAG-CCA-ACC-ATT-GC			
Salm. spp. SAL-3F	TTT-GCG-ACT-ATC-AGG-TTA-CCG-TGG	312	—	
Salm. spp. SAL-4R	AGC-CAA-CCA-TTG-CTA-AAT-TGC-CGC			

[a] Underlined sequence homologous to underlined sequence of iac139 primer.
[b] Underlined sequence homologous to underlined sequence of iac141 primer.
[c] Bold sequence homologous to *B. napus Accg8* gene.
[d] B = G, C, or T; M = A or C.
[e] Y = C or T.

To overcome problems associated with application of the PCR to clinical samples, Stone et al. (1994) combined a nonselective short enrichment with a *Salmonella*-specific PCR-hybridization assay to specifically identify *Salmonella* serovars from clinical samples of various animal species. Fecal samples were seeded with known numbers of *Salmonella* followed by nonselective and selective enrichment for 2 hr. The PCR was performed with the use of the primers invA/invE (Table 4.2) that amplified a 457-bp sequence traversing portions of both the *invA* and *invE* genes. Confirmation of the identity of resulting amplicons was achieved with a ^{32}P-labeled internal oligonucleotide probe via Southern hybridization. Forty-seven *Salmonella* isolates representing 32 serovars were evaluated, and all yielded the 457-bp amplicon with confirmation by the probe. No hybridization occurred with 53 non-*Salmonella* organisms. The assay detected as few as 9 CFU in pure culture and 80 and 100 CFU per gram of seeded feces after 2-hr incubation in brain–heart infusion and selenite-cystine broth (SC), respectively. Selective enrichment in tetrathionate broth and Rappaport–Vassiliadis broth was inhibitory to the PCR.

Food products can contain a variety of PCR inhibitors including DNAses, proteases, and hemin compounds. Soumet et al. (1994) evaluated six different DNA extraction procedures for the detection of *Salmonella* from chicken products by the PCR. The primers ST11/ST15 (Table 4.2) from Aabo et al. (1993) were utilized. No PCR product was directly detected from poultry tissue homogenates seeded with 170 CFU/25 g of tissue (28 CFU/g) and incubated for 12 to 24 hr. However, an initial inoculum of 1.0 CFU/25 g of tissue (0.5 CFU/g) was readily detected after 10 hr of enrichment and required the presence of 1×10^3 to 1×10^5 CFU/ml of enrichment broth for reliable results. Nonselective enrichment consisted of chicken fillets (25 g) homogenized with 225 ml of BPW and seeded with 1, 10, and 170 *Salmonella enterica* subsp. *enterica* serotype Typhimurium LT2 (*S. typhimurium*) and incubated at 37°C for 10 hr. Three of six DNA extraction procedures yielded reliable, rapid, and sensitive results. Successful protocols used centrifugation to concentrate the cells from 1.5 ml of enrichment broth and Proteinase K.

Typhoid fever is still a serious public health problem in many geographic areas of the world. An ideal primer pair for specific detection of the causative organism *S. typhi* should not cross-react with other *Salmonella*. The ribosomal RNA gene is genetically stable and consists of conserved and variable regions. The variable regions may differ considerably among different bacterial species, and therefore such target sequences can be used for differential detection. Sequence data from the 5S part of the 23S rRNA genes and the 5S-23S spacer region of *S. typhi* indicated that the 5S-23S spacer was a variable region specific for *S. typhi*. A pair of primers Forward/Reverse (Table 4.2) amplifying a 300-bp sequence of the 5S-23S spacer region yielded the expected amplicon from all 54 strains of *S. typhi* tested. None of the non–*S. typhi Salmonella* strains, which included *S. paratyphi* A, B, C plus *enteritidis*, and *cholerasuis*, yielded amplicons, nor did strains representing six non-*Salmonella* species. However, *S. typhimurium* did yield a 450-bp amplicon that was readily distinguished from the 300-bp amplicon of *S. typhi*.

Salmonellosis is known to result occasionally from the consumption of contaminated shellfish, particularly raw oysters. Bej et al. (1994) developed a PCR assay for the detection of *Salmonella* spp. in raw oysters. The primers SHIMA-L/SHIMA-R (Table 4.2) amplified a 122-bp sequence derived from the *himA* gene that encodes a binding protein in S. typhimurium and *E. coli*. The primers were designed from sequence data specific for *Salmonella*. A 3'-labeled 25-nucleotide probe SHIMA-P (Table 4.2) was also developed for confirmation. DNA was extracted with guanidine isothiocyanate, purified with chloroform, and precipitated with ethanol. Nonselective enrichment was in BPW or Luria–Bertani broth for 3 hr at 37°C and allowed the PCR to detect 1 to 10 CFU/g of oyster tissue seeded with S. typhimurium.

Recognizing that poultry, poultry products, cattle, and dairy products are the major sources of *Salmonella*-contaminated food products that cause human salmonellosis, Rychlik et al. (1999) developed a universal protocol for the nested PCR detection of *Salmonella* spp. in seeded minced meat and chicken feces. The protocol involved enrichment in BPW for 18 hr followed by centrifugal layering of 20 μl of the enrichment broth onto 300 μl of a 60% sucrose solution with centrifugation at 4500 g for 10 min. The interface containing *Salmonella* was transferred to a fresh tube and centrifuged at 14,000 g. The pellet was washed twice with 500 μl of sterile water. DNA was released by suspending the pellet in 50 μl of lysing solution (0.05% proteinase K plus 0.1% Triton X-100), followed by incubation at 50°C for 30 min and then heating at 95°C for 10 min to inactivate the proteinase K.

The initial PCR utilized primers INV1F/INV1R (Table 4.2) that yielded a 485-bp amplicon derived from the *invA* gene. The second PCR involved the use of primers INV2F/INV2R (Table 4.2) that amplified an internally nested sequence of 284 bp. Without enrichment of fecal samples, the level of detection was 1×10^5 CFU of S. typhimurium per gram. Following enrichment for 18 hr in BPW, 1×10^2 CFU/g were detected. With seeded minced meats, following enrichment, less then 10 CFU/g were detected. When applied to 159 enriched field samples, 11 samples were both PCR and culture positive for *Salmonella*, 13 samples were positive by PCR but negative by culture methodology, and four samples (liver) were negative in PCR but positive by culture methodology, presumably due to the potent inhibition of the PCR by heme. The remaining 131 samples were negative with both methods. The authors concluded that the PCR was more sensitive in detection of *Salmonella* than standard culture methods.

Fluit et al. (1993) found that when chicken meat was inoculated with S. typhimurium the PCR without prior enrichment yielded a detection limit of 1×10^7 CFU/g of tissue. Separation of target cells using immunomagnetic beads (IMB) reduced the level of detection to 1×10^5 CFU/g. Seeded tissue samples (25 g) enriched for 24 hr in either SC or BPW resulted in 0.1 CFU being detected in BPW, whereas enrichment in SC failed to result in amplification with 1.0 CFU/g. The use of IMB following enrichment for 24 hr in SC resulted in the detection of 0.01 CFU/g, whereas with BPW, 0.1 CFU/g of tissue was detected. Primer 1/Primer 2 (Table 4.2) amplified a 164-bp sequence derived from the origin of DNA replication (*oriC*) of *Salmonella*.

Li, Boudjeliab, and Zhao (2000) utilized the primers ST11/ST15 of Aabo et al. (1993) to develop a combined magnetic bead separation PCR and slot blot (MBS-PCR-SB) assay for detection of *Salmonella* and *L. monocytogenes* in milk. The ST14 probe (Table 4.2) of Aabo et al. (1993) was labeled at the 5'-end with alkaline phosphatase. Membrane hybridization was followed by the addition of a chemilumi-nescent alkaline phosphatase substrate and exposure of imaging film to the membrane for 5 to 60 min. With milk samples, the combined MBS-PCR-SB technique was about 10 times more sensitive than conventional PCR-electrophoresis.

Salmonella produce aggregative type 1 fimbrae. A single gene, *fimA*, encodes the major fimbrial unit. Cohen, Mechanda, and Lin (1996) made use of the primers Fim1A/Fim2A (Table 4.2) that amplified an 85-bp sequence of the *fimA* gene for identifying members of the genus *Salmonella*. A 10% polyacrylamide gel was used for resolution of the amplicons. The method readily distinguished *Salmonella* from non-*Salmonella* isolates and was found useful for identifying *Salmonella* at the genus level but not at the species level. It is interesting that nonfimbrial *Salmonella* such as S. Gallinarum and S. Pullorum which form type 2 fimbrae and which lack adhesiveness, still yielded the *fimA* amplicon. This indicates that the gene is still present, although not expressed. The authors concluded that all *Salmonella* serovars process a closely related *fimA* gene, even though some strains produce antigenically unrelated fimbrae.

The outer membrane proteins of *Salmonella* and other gram-negative bacteria are considered to be major structural proteins. Kwang, Littledike, and Keen (1996) developed a qualitative PCR assay for detection of members of the genus *Salmonella* using the primers S18/S19 (Table 4.2) that amplified a 159-bp sequence of the outer membrane protein C gene *ompC*. The primer pair successfully amplified DNA from 40 *Salmonella* serovars (60 isolates) but not that from 24 non-*Salmonella* species. Detection of amplicons in ground beef required a 4–6 hr enrichment with an initial count of 20 CFU/g.

Fresh tomatoes have been associated with several multistate infectious outbreaks of salmonellosis during the past 15 years. Guo et al. (2000) developed a PCR assay involving a pair of primers HILA2-U/HILA2-D (Table 4.2) that amplified a 497-bp sequence from the *hilA* virulence gene of *Salmonella* serovars, which is a positive transcriptional regulator of several invasive genes. This primer pair was found to be highly specific for 83 *Salmonella* strains representing 38 serotypes and did not yield amplicons from 12 non-*Salmonella* bacterial species. Seeded tomatoes (~75 g) were rinsed with 20 ml of 0.1% peptone water (PW), which was centrifuged at 12,000 × g for 10 min and the pellets suspended in 5 ml of PW, which was then added to 5 ml of brain–heart infusion (BHI) broth for 6-hr enrichment at 37°C. Crude DNA, released by boiling from cells without enrichment derived from the surface of tomatoes, yielded a detection limit of 1×10^5 CFU/g, whereas with a 6-hr enrichment in BHI the detection limit was 1×10^2 and 1×10^3 CFU/g on and in tomatoes, respectively. Tomatoes are well recognized to contain one or more potent inhibitors, as yet unidentified, of the PCR.

Waage et al. (1999) developed a PCR assay with two nested primers selected from conserved sequences within a 2.3-kb randomly cloned DNA fragment from *S. typhimurium* genomic DNA (Aabo, Anderson, and Olsen, 1995). A total of 129 *Salmonella* strains representing 27 different serovars and 31 strains of other genera were included in the study. Minced beef seeded with 280 CFU/g was detected as positive without enrichment, whereas seeding with 28 CFU/g yielded negative results. After overnight enrichment of food samples an initial level of <10 CFU/g was detected. Extracted DNA was not purified.

In an attempt to circumvent the problems associated with the inhibition of the PCR by complex food matrices, Lampel, Orlandi, and Kornegay (2000) developed a protocol using FTA filters to prepare DNA templates from foods without enrichment. The FTA filter is a fibrous matrix impregnated with chelators and denaturants that effectively traps and lyses micro-organisms on contact. Released DNA is sequestered and preserved intact within the membrane. Various fruits, vegetables, and ground beef (10 g) were first rinsed with 10 ml of Butterfield's phosphate buffer. The rinse buffer was then passed through glass wool, the filtrate centrifuged at 8000 g for 5 min, and the pellet resuspended in 100 µl of phosphate buffered saline (PBS). The sample (100 µl) was then applied to an FTA filter (Fitz Co., Inc.). The filter was air-dried at 56°C and washed with FTA purification buffer (Tris-EDTA, pH 8.0). Spotted areas were then removed with a 6-mm diam. hole punch and directly used as sources of DNA template in 100- to 200-µl PCR reaction volumes. Without the FTA filter, ground beef yielded no *Salmonella invA* amplicons when seeded with 1×10^7 CFU/g. In contrast, the FTA filter allowed detection of 40 CFU/g. It is interesting that the limit of detection with sliced tomato tissue (10 g) for *Salmonella* was 5.0×10^2/g without and with the FTA filter.

Some multiple antibiotic resistance clones have been found to display similar antibiotic resistance profiles (R-phenotypes) but present different R-gene profiles that are frequently inserted in integron/transposon structures that facilitate their intracellular movement (Caratoli, 2001; Guerra et al., 2000a, 2001). Both integron-borne and nonintegronborne R-genes can be chromosome or plasmid located, and the R-plasmids are the main factors responsible for the horizontal transfer of the R-genes in *Salmonella*. Del Cerro, Soto, and Mendoza (2003) undertook an extensive study involving the application and evaluation of PCR protocols for the characterization of *Salmonella* isolates by determination of the presence of virulence genes and their relation to class 1 integrons and plasmids. The presence of R-genes contained in the integrons was determined using nested-PCR protocols. In these assays, the PCR products obtained after the integron sequence amplification were used as template DNA in a second PCR reaction that used primers for amplification of specific R-gene sequences. A total of 30 genes (13 for enhancement of virulence and 17 for antibiotic resistance) were targeted (Table 4.2). All 56 *Salmonella* isolates of animal origin involving meat, poultry, feces, and eggs and all 20 human clinical isolates were found to possess the *invE/A* genes (encoding invasivity genes), *phoP/Q* (intramacrophage survival and enhanced bile resistance), and *stn* (enterotoxin) genes. Other genes enhancing virulence were present at frequencies from 66% to 98%. The presence of antibiotic resistance genes determined by PCR and disk assays were in agreement and varied between strains. In addition, a correlation between *V-gene* profile and serotype was observed for both the food and human clinical isolates.

Ferretti et al. (2001) developed a 12-hr PCR-based assay for detection of *Salmonella* serovars in Italian salami. Enrichment was in BPW (10 g plus 90 ml) for 6 hr. Cell lysis involved Proteinase K. DNA was not purified. The primers Salm3/ Salm4 (Table 4.2) amplified a 389-bp sequence within the conserved *invA* gene of *Salmonella* serovars. The protocol detected 1 CFU/100 ml of homogenate (1 CFU/10 g).

Baudart et al. (2000) determined the diversity of *Salmonella* strains isolated from different natural aquatic systems within a Mediterranean coastal watershed (river, wastewater, and marine coastal areas). A total of 574 strains were identified by both serotyping and ribosomal spacer-heteroduplex polymorphism (RS-HP). More than 40 different serotypes were found. The RS-HP method was based on the PCR amplification of the intergeneric spacer region between the 16S and 23S rRNA genes and size polymorphism of the resulting products generated with the primers L17/G17 (Table 4.2). These sequences are generally found in multiple copies in most bacterial genomes and consist of homologous 3′ and 5′ sequences flanking heterologous intervening sequences. RS-HP produced amplicon banding profiles that allowed the discrimination of species at both serotype and intraserotype levels.

Nucera et al. (2006) compared the API 20E metabolic system and an *invA*-based PCR analysis for identification of *Salmonella enterica* isolates from swine farms. The primers spvCF/spvCR (Table 4.2) from Chiu and Ou (1996) amplified a 244-bp sequence of the *invA* gene of *S. enterica*. API 20E had the highest agreement with other tests at the 99.9% likelihood level. Both tests had 100% sensitivity and 96% specificity compared to 16 s rRNA sequencing. Compared to serotyping, both tests had 96% sensitivity; specificity was 86% for API 20E and 79% for *invA* PCR.

Hong et al. (2003) developed a direct nonenrichment PCR-ELISA assay for rapid detection of *S. enterica* from poultry carcass rinse samples. A commercial DNA purification kit was used involving a DNA affinity minicolumn. The primers used, invAF/invAR (Table 4.2), amplified a 408-bp sequence of the *invA* gene. PCR mixes contained digoxygenin (DIG)-labeled deoxynucleotides. A capture probe (Table 4.2) was designed to anneal to the central region of the PCR amplicon with a biotin molecule added to its 3′-end. The wells of ELISA plates were first coated with streptavidin (SA) followed by the addition of amplicon–probe duplexes to facilitate binding of the amplicons to the bottom of the plates via the bound biotin-labeled probe and the addition of blocking buffer. Anti-DIG antibody–peroxidase conjugate was then added and finally the peroxidase substrate (ABTS plus H_2O). The limit of detection for *Salmonella* without enrichment was 2×10^2 CFU/ml of rinse samples. The ELISA assay increased the sensitivity of the conventional PCR by 100-fold.

Strains of *S. enterica* subsp. *enterica* differing in intensity of pathogenicity, and resulting symptoms can be differentiated by fermentation of dextrorotary [L(+)]-tartrate (*d*-tartrate). *d*-Tartrate-nonfermenting (dT−) strains exhibit enhanced human pathogenicity causing typhoidlike disease, whereas d-tartrate fermenting strains (dT+) result in less severe gastrointestinal infections. Anaerobic fermentation of tartrate usually proceeds via stereospecific dehydratases to oxaloacetic acid. The two genes *ttdA* and *ttdB* encode the subunits of the L-tartrate dehydratase. Upstream of the *ttdA* and *ttdB* genes are two open reading frames encoding the regulatory proteins STM3357 and STM3358. Sequencing of the intragenic region of the STM3357 and STM3356 regulatory genes revealed that only a single nucleotide located in the ATG start codon of gene STM3356 was different between dT+ and dT− strains. AdT+ strains possessed a regular ATG start codon, whereas a dT− strain possessed an A nucleotide instead of the G within the start codon (Malorny, Bunge, and Helmuth, 2003). Malorny, Bunge, and Helmuth (2003) developed a multiplex PCR assay for distinguishing dT⁻ from dT⁺ strains based on the presence of these two different start codons. The primers 167/166 were designed to yield a 290-bp amplicon when the ATG start codon is present in the STM3356 gene and no amplicon when the start codon is absent. The 167/166 primers were combined with a second pair of primers ST11/ST15 (Table 4.2) from Aabo et al. (1993) simultaneously generating a 429-bp amplicon when *Salmonella* DNA was present.

Foods and feeds are known to contain a number of PCR inhibitors. Liao and Schollenberger (2003) found that alfalfa seed homogenates markedly reduced the sensitivity of *Salmonella* detection using a commercial PCR kit. The minimal number of CFU of *Salmonella* was determined to be 1 to 10 and 100 to 1000 in the absence and presence of seed homogenates, respectively. Alfalfa homogenates in BPW incubated for 24 or 48 hr exhibited higher levels of PCR inhibition compared to those not subjected to enrichment. The use of IMS eliminated the PCR inhibition. The level of detection was reduced to 2 to 3 CFU of *Salmonella* in 25 g of seeds following 8 hr of enrichment and IMS.

Löfström et al. (2004) examined the inhibition of the PCR for detection of *Salmonella* by animal feeds. Feeds examined were soy, rapeseed, wheat, oats, palm kernal expeller, maize (gluten and pellets), mixed feeds, whey, draff (distiller's waste), meat meal, and fish meal. Samples (25 g) were stomached with 225 ml of BPW. The primers SalinvA139/SalinvA141 (Table 4.2) from Rahn et al. (1992) amplified a 285-bp sequence of the *invA* gene. DNA polymerases examined for determination of the extent of inhibition were DyNAzyme II, Fast Start Tach, Platinum *Taq*, *Pw*, *rTth*, *Taq*, and *Tfl*. The homogenized feed samples were all found to be inhibitory to the PCR, but the level of inhibition varied among the different DNA polymerases. Rapeseed and rapeseed meal were found to be highly inhibitory for the DNA polymerases except for *rTth*. *rTth* allowed the incorporation of 0.4% homogenates directly to PCR reaction mixes. The optimum protocol developed with nonselective enrichment in BPW for 18 hr at 37°C resulted in a 0.81 probability of detection of 1 CFU/25 g of feed. Among 55 unseeded feed samples, 8% were positive by PCR and 3% by conventional bacteriological techniques.

Raw seed sprouts are presently recognized as an important cause of foodborne disease. The conditions that promote sprout growth (high temperature, high humidity, and abundant nutrients) also promote the growth of bacteria including pathogens such as *Salmonella* and therefore represent a unique challenge in terms of producing pathogen-free raw product. Johnston et al. (2005) developed a PCR for the direct detection of *Salmonella* and *E. coli* O157:H7 in seeded raw alfalfa sprouts. Alfalfa sprouts (25 g) were seeded with *S. typhimurium* and stomached with 225 ml of BPW containing 2.5% tween 80 plus 4% polyethylene glycol (PEG) 8000 (to prevent bacterial cell clumping) with a filter containing stomacher bag. The filtrate was centrifuged at 9100 × g for 10 min. The resulting pellet was extracted with a guanidine-detergent lysing solution and then extracted with chloroform and the DNA precipitated with ethanol. The DNA was additionally purified with a commercial kit. The primers Sal3/Sal4 (Table 4.2) from Manzano et al. (1998) amplified a 389-bp sequence from the *invA* gene. An internal probe (Table 4.2) labeled with digoxygenin (DIG) was used for confirmation in a Southern blot hybridization assay. The detection limit for *Salmonella* without enrichment was 10 CFU/g of alfalfa sprouts. With seeded irrigation water, the detection limit was 0.1 CFU/ml.

The 16S *rRNA* genes are ubiquitous sequences in bacteria and are now widely used for species identification. Relatively conserved in addition to relatively variable regions are known to be associated with 16S rRNA sequences. The variable regions have been found to be amenable to use for the specific detection of various bacterial species. Lin et al. (2004) developed a novel pair of PCR primers 16S-Sal/16S-CCR (Table 4.2) from sequence alignment of the V3 and V6 regions of 16S rRNA genes of various *Salmonella* serovars for generation of a 324-bp amplicon. 16S-CCR is a degenerate primer because it matches rDNAs from most members of the *Enterobacteriaceae*. The specificity of the primer pair is therefore based on the specificity of the primer 16S-Sal. With the use of an 8-hr enrichment in selenite cystine broth (SCB) at 37°C, 1 to 9 CFU of *Salmonella* per gram could be detected in chicken tissue. Without enrichment, detection sensitivity was 100 to 900 CFU/g.

It is possible to detect low numbers of *Salmonella* in homogenized meat samples with a short enrichment of 5 hr. Croci et al. (2004) compared an ELISA-flow injection assay (ELISA-FIA) and a conventional PCR assay using the ST11/ST15 primers (Table 4.2) of Aabo et al. (1993). Seeded meat samples (25 g) were stomached in 225 ml of BPW and enriched for 5 hr at 37°C. DNA for the PCR was extracted by boiling and was not purified. A sandwich ELISA was directed against intracellular *Salmonella* soluble proteins. After the addition of polyclonal antibody conjugated to horseradish peroxidase (HRP), a fluorescent substrate was added and then injected into a fluorescence-detecting flow injection unit. The sensitivity of detection of both methods was 1 to 10 CFU/25 g after 5 hr of enrichment, which allowed the detection of ~5 × 10^3 CFU/g with the ELISA-FIA and 1 × 10^3 CFU/g with the PCR. Among 30 samples of various meat products, 19 were positive after 5 hr of enrichment by both methods.

There is relatively little information regarding marine environments as reservoirs of salmonellosis. Martinez-Urtaza, Echeita, and Liebana (2006) undertook phenotypic and genotypic characterization of *S. enterica* serotpe Paratyphi B isolates from environmental and human sources in Galacia, Spain. Isolates from enteric fever and gastroenteritis cases are normally discriminated by their capacity to ferment *d*-tartrate. A total of 16 *S.* Paratyphi B isolates were investigated. Four of these represented all of the *S.* Paratyphi B isolates derived from over 6000 samples of mollusks and seawater, eight isolates were of human clinical origin from the area, and four were reference strains. These isolates were investigated for *d*-tartrate fermentation, the presence of genes encoding the effector proteins sopE1 and avrA, pulsed field gel electrophoresis (PFGE), and antimicrobial susceptibility. The *d*-tartrate fermenting strains (dT⁺) were designated as *S. enterica* serotype Java and those failing to use *d*-tartrate (dT⁻) were designated *S. enterica* serotype Paratyphi B sensu stricto. Systemic variant strains (dT⁻) were dominant among the marine environment isolates. All dT⁻ isolates were *sopE1* positive and *avrA* negative, presented an indistinguishable *Xba*I PFGE profile, and included three isolates from the marine environment and two human isolates. The authors concluded that the systemic variant isolates of *S.* Paratyphi B in the marine environment are of notable public health significance as a result of the potential of acquiring enteric fever linked to the consumption of raw shellfish.

Vázquez-Novelle et al. (2005) developed an 8-hr nested PCR assay for the detection of *Salmonella* Senftenberg in raw oysters. The outer primers SAO1/SAO2 (Table 4.2) amplified a 540-bp sequence of the *invA* gene. The inner primers SAO3/SAO4 (Table 4.2) amplified a 281-bp sequence. Four different DNA extraction and purification methods were compared. Samples were enriched for 3.5 hr at 37°C in BPW followed by Chelex 100 sample treatment, and the released DNA purified by chloroform-isoamyl alcohol extraction and ethanol precipitation. This protocol allowed the detection of <1 CFU/g of oyster tissue.

Pork is considered a significant contributor to salmonellosis. Nowak et al. (2006) studied *Salmonella* contamination in swine at slaughter in a single slaughterhouse and on farms in Germany using an antibody ELISA assay and the PCR. Meat juice samples from the diaphragm, pillor muscle, jejunal lymph nodes, and tonsils were obtained and assayed for antibody titers and PCR detection of *Salmonella* spp. A total of 383 animals from 32 different swine farms were involved. The PCR primers ST11/ST15 (Table 4.2) amplified a 429-bp sequence from a randomly cloned genomic DNA fragment and are specific for *Salmonella* spp. (Aabo et al., 1993). A total of 27 (7%) of the 383 slaughtered animals from 6 of the 32 fattening farms was ELISA positive. *Salmonella* were found in 16.4% of the jejunum lymph nodes and in 15% of the tonsils. The percentage of ELISA-positive animals on the farms ranged from 7% to 50%. The results of the PCR analysis of the jejunum with its lymph nodes indicated that 11 farms delivered a total of 63 (16.5%) *Salmonella*-positive animals to slaughter, notably higher than the 7% detected with the ELISA assay for serum antibodies. The authors concluded that the risk of cross-contamination increases when animals are slaughtered after animals that are positive.

Given the fact that the genus *Salmonella* is an extremely polymorphic and diverse group, comprising approximately 2500 serotypes, Bansal, Gray, and Mcdonell (2006) undertook an extensive validation of the PCR for the routine detection of *Salmonella* in a variety of foods that included raw meats, poultry, cooked meat products, seafood, vegetables, and dairy products. The primers ST11/ST15 (Table 4.2) from Aabo et al. (1993) were used. A total of 100 different *Salmonella* serovars (125 isolates) and 34 non-*Salmonella* (39 isolates) were used for validation. All 125 *Salmonella* isolates yielded the appropriate amplicon, whereas none of the 34 non-*Salmonella* isolates did. Among 503 naturally contaminated food samples, 21% contained *Salmonella*. The incidence with raw poultry was 100% following nonselective pre-enrichment of 25 g in BPW for 16 to 20 hr at 37°C followed by selective enrichment in mannitol selenite cystine broth (MSC) and Rappaport–Vassiliadis Broth. There was 100% agreement between bacterial culture methods and the PCR.

The contamination of fruits such as cantaloupes that develop on the ground and are eaten raw constitute a potential source of *Salmonella* contamination from the soil and irrigation water. Espinoza-Medina et al. (2006) examined the incidence of

Salmonella associated with the farming and harvesting of cantaloupes in Sonora, Mexico. Samples (25 g) were pre-enriched in 225 ml of lactose broth for 24 hr at 37°C. One ml was then transferred to 10 ml of tetrathionate broth and also to selenite-cystine broth followed by incubation at 37°C for 24 hr. Standard bacteriological methods and the PCR were then applied to the selective enrichment broths. Released DNA was purified with phenol-chloroform-isoamyl alcohol. The primers Sal1/Sal2 (Table 4.2) amplified a 163-bp fragment of the *oriC* gene encoding the origin of replication of *Salmonella* spp. A total of 190 samples was collected consisting of irrigation water (17), groundwater (11), chlorinated water (21), crop soil (24), in-field cantaloupes (35), in-field workers' hands (24), packed cantaloupes (34), and packing workers' hands (24). Among the 190 samples, only 6 (3.2%) were positive using standard bacteriological methods (4/17 from irrigation water and 2/24 from crop soil). In contrast, 26 samples (13.7%) were positive by the PCR. The authors concluded that this frequency of contamination was indicative of a high incidence of fecal contamination.

There have been few studies comparing various enrichment procedures regarding PCR sensitivity of detection for *Salmonella* in foods. Myint et al. (2006) studied the effect of several enrichment protocols on the sensitivity and specificity of the PCR for detection of poultry naturally contaminated with *Salmonella* spp. Poultry tissue was enriched in BPW and the primers ST11/ST15 (Table 4.2) from Aabo et al. (1993) were utilized. Among 90 tissue samples, none were positive for *Salmonella* without enrichment. Enrichment with BPW yielded 20 positive samples, and pre-enrichment in BPW followed by enrichment in Rappaport–Vassiliadis (RV) broth yielded 22 positive samples, and pre-enrichment in BPW followed by selective enrichment in tetrathionate-Hajna broth (TT-H) yielded 28 positive samples. Culture procedures also yielded 28 positive samples resulting in a sensitivity of 100% for the BPW-TT-H enrichment followed by the PCR.

In Europe, as a result of the E.U. *Salmonella* control program involving the vaccination of poultry stocks against *S.* Enteritidis and *S.* Typhimurium, the prevalence of these serovars has decreased significantly and *S.* Infantis has become the predominant serovar in poultry (van Duijkeren et al., 2002). This is reflected in the observation that the proportion of human salmonellosis due to *S.* Infantis has slowly increased in Europe and is now the third most frequent serovar infecting humans (Galanis et al., 2006). Kardos et al. (2007) developed a novel PCR assay for specific identification of *S. enterica* serovar Infantis. Chelex 100 was used for purification of DNA from broth cultures. The primer pairs 558f/1275r and 878f/1275r (Table 4.2) amplified 727-bp and 413-bp sequences of the *fliB* gene encoding the flagella protein flagellin. The detection limit of the assay with poultry fecal samples was low and determined to be 10^5 CFU/ml from pure broth cultures and 10^6/g from spiked fecal samples, which may have been due to the method of lysing the cells that was not indicated. A major advantage of the assay is that it is capable of identifying *S.* Infantis even when a given strain is not producing flagella.

Oikonomou, Halatsi, and Kyriacou (2008) developed a unique PCR assay that permits independent amplification of an internal amplification control (IAC) and a target sequence using the same set of primers to improve the sensitivity for detection of *Salmonella* spp. The major purpose of the IAC is to prevent false-negative results due to PCR inhibition. The assay targeted the quorum-sensing gene *sdiA* and is based on a large size difference between the IAC (3196 bp) and the target sequence (274 bp) and therefore relies on their different extension times. The assay involves noncompetitive amplification of the IAC and target sequence and is the first PCR assay to utilize extension time as a critical parameter for enhancing sensitivity in a PCR assay with an internal amplification control. An internal portion of the *sdiA* gene (274 bp) of *S. enterica* was amplified with primers sdiA1/sdiA2 (Table 4.2) and the resulting amplicon inserted into a plasmid that was then transformed into an *Escherichia coli* recipient, which served as a reservoir for the resulting 3196 plasmid. A linearized form of the plasmid was then used as the IAC. Thermal programming consisted in part of an initial stage of cycling involving 10 cycles of extension at 72°C for 30 s and a second stage of 25 cycles involving extension at 72°C for 3.5 min. The first stage of amplification involving a 30 s extension time was used for the selective amplification of the target sequence. The second stage of amplification involving a 3.5 min extension time allowed amplification of the IAC. Consequently, if a target sequence were present in an environmental or clinical sample, it would be amplified in both stages and the IAC would not be amplified. If there were no target sequence in the sample only the IAC would be amplified during the second stage.

B. Real-Time PCR (Rti-PCR)

In June of 2001, an outbreak of gastroenteritis among individuals at a picnic in Texas, USA, was reported. Nine food items were screened for *Salmonella* using a Rti-PCR protocol (Daum et al. 2002). Barbecued chicken was the only food item found positive for *Salmonella*. The primers Sal-F/Sal-R (Table 4.2) amplified a 102-bp sequence of the *invA* gene of almost all *Salmonella* serotypes. A dual-labeled probe (Table 4.2) with Fam at the 5′-end and TAMPA at the 3′-end was utilized.

Eyigdor, Carlik, and Unal (2002) applied Rti-PCR to tetrathionate (TTB) selective enrichments of 492 intestinal homogenates and 27 drag swabs from 47 poultry flocks in Turkey. InvA139/invA141 (Table 4.2) from Rahn et al. (1992) amplified a 281-bp sequence of the *invA* gene with SYBR Green. The number of positive individual samples by Rti-PCR and standard bacteria cultivation was 65% and 35%, respectively. The level of sensitivity following 18 hr of TTB enrichment at 37°C was 6 CFU/ml.

The use of SYBR green for multiplex Rti-PCR has certain limitations, in that the fluorescent signal for the amplicons derived from two or more different organisms has the same wavelength and therefore cannot be distinguished. However, if the resulting T_m values (thermal denaturation temperature) of the different amplicons are sufficiently different, they can then be used to distinguish the two amplicons. Jothikumar et al. (2003) utilized the primers Fim1A/Fim2A (Table 4.2) from Cohen, Mechanda, and Lin (1996) that amplified an 85-bp sequence of the *fimA* gene of *Salmonella* encoding a fimbrinlike protein. The resulting amplicon had a T_m of 86°C. The primers used for *L. monocytogenes* in the multiplex Rti-PCR assay yielded an amplicon with a T_m value of 80°C. Samples of seeded milk were enriched in nutrient broth for 16 hr at 37°C. The detection limit for *Salmonella* serovars was 2.5 CFU/10 ml of milk.

Meat foods are invariably complex matrices that contain numerous PCR inhibitors. De Medici et al. (2003) evaluated four DNA extraction and purification methods in conjunction with SYBR Green I mediated Rti-PCR for detection of *Salmonella enterica* serotype Enteritidis in poultry from BPW enrichment cultures of homogenized tissue. Boiling, alkaline lysis, and Nucleospin purification yielded equal levels of detection sensitivity. Immunomagnetic beads yielded notably lower detection sensitivity. Boiling was therefore selected as the method of choice because of its simplicity and rapidity. The primers used, SEFA-1/SEFA-2 (Table 4.2), amplified a 310-bp sequence of the *sefA* gene that encodes a major subunit protein of a novel fimbrial structure on the surface of serotype Enterica that is found in few other *Salmonella* serotypes. A standard curve was generated from 1×10^3 to 1×10^8 CFU/ml. Sensitivity of detection was less than 1×10^3 CFU/ml.

Wang, Jothikumar, and Griffiths (2004) developed a novel method of DNA extraction and purification in conjunction with a multiplex Rti-PCR assay for simultaneous detection of *Salmonella* and *L. monocytogenes* in raw sausage meat. The PCR primers for *Salmonella* SF/SR (Table 4.2) from Cohen, Mechanda, and Lin (1996) were used in conjunction with SYBR green. Meat samples were enriched in tryptic soy broth (TSB) for 6 to 8 hr. Lysis buffer (400 µl) was added to 200 µl of the enrichment broth, mixed, and centrifuged at $12,000 \times g$ for 5 min. The pellet was resuspended in 200 µl of lysis buffer containing glycogen (0.03 mg/ml) and then heated in a boiling water bath for 10 min and iced. Sodium iodide solution (300 µl) consisting of 6 M NaI in 50 mM Tris-HCl and 25 mM EDTA at pH 8.0) was added followed by 500 µl of isopropanol and the DNA was then purified with a commercial kit. The detection limit for *Salmonella* was 4 cells/g of sausage.

Kessel, Kars, and Perdue (2003) reported on the efficiency of a portable Rti-PCR system for detecting *Salmonella* serovars in raw milk utilizing SYBR green as a direct fluorophore. Enrichment was in tetrathionate broth for 24 hr at 37°C. The commercial primers used targeted a sequence of the *spaQ* gene in *Salmonella*, which is part of the chromosomal *inv-spa* complex responsible for virulence. The limit of detection was 10 to 20 CFU/100 ml of milk.

Salmonella contamination of bulk milk from dairies is thought to occur through fecal contamination, and therefore the incidence of *Salmonella* in raw milk is of public health importance. Samples from 854 farms in 21 states in the United States were collected in 2002 from bulk tank milk and subjected to enrichment in tetrathionate broth by Karns et al. (2005). With the use of real-time PCR 101 samples (11.8%) were found to contain *Salmonella enterica*, whereas conventional culture techniques detected only 22 (2.6%) positive samples. A commercial real-time PCR kit targeting the *spaQ* gene of genomic *Salmonella* DNA was used. Because T_m values of the resulting amplicons were determined following amplification, SYBR green was presumably the fluorescent reporter dye. The authors concluded that the real-time PCR assay indicated that the presence of *S. enterica* in U.S. bulk tank milk is substantially higher than previously reported.

Mercanoglu and Griffiths (2005) developed an IMB-Rti-PCR assay for rapid detection of *Salmonella* in milk, ground beef, and alfalfa sprouts. SYBR green was used as the fluorogenic reporter molecule in conjunction with the primers SalinvA139/SalinvA141 (Table 4.2) from Rahn et al. (1992) that amplified a 284-bp sequence of the *invA* gene. Following a 10-hr enrichment in BPW, cells were removed by IMB and DNA was extracted with a commercial kit. A detection limit of 1 CFU/ml of milk, 25 CFU/25 g of beef, and 1.5 CFU/25 g of alfalfa sprouts was achieved.

Notzon, Helmuch, and Bauer (2006) evaluated an immunomagnetic bead Rti-PCR (IMB-Rti-PCR) assay for the rapid detection of *Salmonella* in meat. Homogenized samples were nonselectively enriched in BPW for 12 hr at 37°C. The primers inVE/inVA (Table 4.2) from Stone et al. (1994) amplified a 457-bp sequence spanning the *invE* and *invA* genes and were used in conjunction with dual-labeled fluorescent resonance energy transfer (FRET) probes (Table 4.2). One probe was labeled with fluorescein at the 3'-end and the other with Light Cycler-red 640 at the 5'-end. The FRET probes hybridized at a distance of 3 nucleotides from one another. Proteinase K was used for cell lysis in lysis buffer, and the DNA released after boiling was purified with a commercial kit. Among a total of 491 naturally contaminated meat samples assayed, 43 were positive with the standard bacteriological cultivation method and IMB-Rti-PCR. Seven samples were false-negatives and three were false-positives by the IMB-Rti-PCR assay in comparison to the standard culture method. These results yielded a sensitivity of 83.7% (false-negative rate of 16.3%) and a specificity of 99.3% (false-positive rate of 0.7%) for the IMS-Rti-PCR assay. The authors concluded that the IMB-Rti-PCR assay is apparently more sensitive than the standard culture method used.

Seo et al. (2004) developed an Rti-PCR assay for specific detection of *Salmonella* Enteritidis. A dual-labeled fluorogenic probe SEF14 (Table 4.2) was used, labeled with FAM at the 5'-end and TAMRA at the 3'-end. The primers forward/reverse (Table 4.2) amplified a 98-bp sequence of the *sefA* gene that encodes a major subunit fimbrial protein specific for *Salmonella* group D strains such as *S.* Enteritidis. Without enrichment, the level of detection was 1×10^3 CFU/ml of pooled eggs. Enrichment of pooled egg samples was in TSB at 37°C for 24 hr and yielded a detection level of <1 CFU/600 g of pooled eggs.

Seo, Valentin-Bon, and Brackett (2006) using the same pair of primers and probe previously utilized (Seo et al., 2004) found that the Rti-PCR assay was as sensitive as the conventional plate count method in the frequency of detection of *S.* Enteritidis in homemade ice cream involved in an outbreak. However, populations of *S.* Enteritidis derived from the Rti-quantitative PCR were approximately 1 log higher than obtained by MPN and CFU values obtained by conventional culture methods. The higher counts obtained by Rti-PCR compared to viable counts were attributable to die-off of *S.* Enteritidis during frozen storage of the ice cream and therefore to the PCR detection of the targeted DNA sequence from both viable and dead cells.

S. enterica subspecies *enterica* serovar Enteritidis is the world's leading cause of salmonellosis. The reservoir for *S.* Enteritidis is mainly poultry. Malorny, Bunge, and Helmuth (2007) developed a duplex Rti-PCR assay for detection of *S.* Enteritidis in whole chicken carcass rinses and eggs. Enrichment was in BPW for 20 hr at

37°C, which was followed by Chelex 100 purification and then cell lysis by heating at 100°C for 8 min. The primers Prot6e-5/Prot6e-6 (Table 4.2) amplified a 206-bp sequence of the Prot6e gene located on the *S*. Enteritidis specific 60-kb virulence plasmid in conjunction with a Prot6e-probe labeled at the 5′-end with 6-carboxyfluorescein (FAM) and at the 3′-end with the Eclipse-Dark-Quencher (DQ; Table 4.2). A second primer pair 139/141 (Table 4.2) was used in conjunction with an invA-probe labeled at the 5′-end with Yakima-Yellow and at the 3′-end with DQ (Table 4.2), as an internal amplification control, and amplified a 110-bp sequence of the *invA* gene present in all *Salmonella*. The detection limit was less than 3 CFU per 50 ml of carcass BPW rinses or 10 ml of egg.

Rti-PCR is ideally suited for quantitative PCR. Piknova et al. (2005) developed an Rti-PCR assay system for the quantification of *Salmonella enterica* from 38 serovars. The primers Srt2F/Srt2R (Table 4.2) amplified a 102-bp sequence from the pathogen-related *fimC* gene. A dual-labeled probe Srt2P (Table 4.2) was labeled with 6-FAM at the 5′-end and with TAMRA at the 3′-end. A standard curve was linear when the C_T values were plotted against the log of CFU from 1×10^3 to 1×10^7 CFU/ml.

Salmonella are frequently found in association with animal manure, which is often applied as fertilizer for vegetable production. Environmental substrates such as manure and soil are therefore presently a major concern regarding food safety. The reliability of the PCR is dependent on the quality of the DNA. DNA extracted from soil, manure, or compost can possess coextracted PCR inhibitors such as heme and fulvic acids in addition to other inhibitors from soil. Klerks et al. (2006) evaluated five commercial DNA extraction methods with respect to the sensitivity of detection of *S. enterica* serovar Enteritidis by real-time PCR. The primers and probe were from Hoorfar, Ahrens, and Rådström (2000). Notable differences were observed with respect to the efficiency of DNA extraction between the various kits. Soil and manure samples were more inhibitory to the PCR than compost samples.

Nam et al. (2005) developed an Rti-PCR assay for detection of *Salmonella* in dairy farm environmental samples. The primers Styinva-JHO-2-left-5′/Styinva-JHO-2-right-5′ from Hoorfar, Ahrens, and Rådström (2000) amplified a 119-bp sequence of the *invA* gene in conjunction with the use of SYBR green. Samples derived from lagoon water, feed/silage, bedding soil, and bulk tank milk samples were seeded with *S*. Enteritidis. Sensitivity of detection was 10^3 to 10^4 CFU/ml of inoculated samples. Enrichment in TSB for 18 hr at 37°C resulted in the PCR detection of *Salmonella* in all samples seeded with 1 CFU/ml.

Hein et al. (2006) developed an Rti-PCR assay for detection and quantification of *Salmonella* spp. The assay is based on an internationally validated conventional PCR system, which was suggested as a standard method for detection of *Salmonella* spp. The primers Sal139/Sal141 (Table 4.2) from Rahn et al. (1992) amplified a 284-bp sequence of the *invA* gene. Each of 92 *Salmonella* isolates representing 30 serovars was subjected to Rti-PCR analysis with SYBR green as the reporter molecule, which allowed the T_m value (thermal denaturation temperature) for the amplicon generated by each isolate to be determined for verification of identity. In addition, two isolates were seeded into various food products and enrichment in BPW for 16 hr at

37°C yielded detection of 2.5 CFU/25 g of salmon and minced meat, 5 CFU/25 g of chicken meat, and 5 CFU/25 ml of raw milk.

Aqueous samples from the food processing industry can be expected to contain fewer PCR inhibitors than homogenized meat samples and to be more amenable to detection of low numbers of *Salmonella* by the PCR without the necessity of enrichment. Wolffs et al. (2006) applied the primers SalinvA139/SalinvA141 of Rahn et al. (1992) that amplified a 284-bp sequence of the *invA* gene in conjunction with SYBR green to the Rti-PCR detection of *Salmonella* in poultry rinse water and spent mungbean irrigation water (used for sprouting). A two-step filtration protocol was used in place of enrichment. *Tth* DNA polymerase was used to generate amplicons in PCR reaction mixes, because this enzyme is recognized to be less sensitive to PCR inhibitors than conventional *Taq* polymerase. In the two-step filtration protocol, cheesecloth was found to retain significantly fewer cells than filter paper. The second filtration step consisted of passing the filtrate from the cheese cloth through a 0.22 μ porosity Duropore filter. The cells captured onto the surface of the membrane filter were released by vortexing with 1 ml of saline and 4 μl added to Rti-PCR reaction mixes. A detection level of 2×10^2 CFU/100 ml of rinse water equivalent to the DNA from 8 CFU per PCR reaction was achieved.

The primer pair and sequence to be amplified can influence the level of detection sensitivity for *Salmonella* in meats. Catarame et al. (2006) developed an Rti-PCR assay for the detection of *Salmonella* in 100 meat samples naturally contaminated, such as beef, chicken, pork, and turkey. The primers invaf/invar of Rahn et al. (1992) amplified a 243-bp sequence of the *invA* gene, and the primers 16SF/16SR of Trkov and Avguštin (2003) amplified a 402-bp sequence of the 16S rRNA gene. SYBR green was the fluorescent reporter dye. Meat samples were enriched in RV broth for 24 hr at 37°C prior to PCR. A commercial kit was used for DNA extraction and purification. Results indicated that with the 16S rRNA primers, 5/100 samples were positive for *Salmonella*. The invA primers yielded only two positive samples. Standard bacteriological methods using enrichment in BPW for 96 hr at 37°C confirmed the five positive samples. The authors concluded that the selection of primers must be done with considerable care to achieve maximum detection.

Fukushima et al. (2007) made use of buoyant density gradient centrifugation (DC) to separate bacterial cells of various genera including *Salmonella* from complex food matrices and to remove PCR inhibitors. This protocol allowed target organisms in food samples to be concentrated 250-fold and to be detected at levels down to 10 CFU/g without enrichment. Samples were analyzed by Rti-PCR utilizing the primers Salm16S-F/Salm16S-R1 (Table 4.2) from Fey et al. (2004). Meat (25 g) was stomached with 225 ml of 0.02% Tween 20 in BPW with a filter mesh. The Tween 20 was used to homogenize the fat. The filtrate was centrifuged at 1880 × g for 5 min and the upper portion removed and centrifuged at 16,000 × g for 5 min. The pellet was suspended in 0.5 ml of 1.5 M NaCl and mixed with 1 ml of Percoll solution (1.050 g/ml) and centrifuged at 4500 × g for 15 min to achieve flotation of cellular debris. The upper portion was removed. The bottom portion (about 0.5 ml) containing target cells and food particles was homogenized and placed on top of two layers (0.6 ml of 1.050 g/ml Percoll solution layered on top of a 1.123 g/ml Percoll solution in a microcentrifuge tube with two colored density markers (orange for 1.033 g/ml and green for 1.098 g/ml) or red for 1.121 g/ml). The preparation was centrifuged at 14,500 × g for 5 min, and then about 1 ml was taken from the interface between the two density markers and mixed with 1 ml of 0.15 M NaCl and centrifuged at 14,500 × g for 5 min. The pellet was suspended in 100 μl of a commercial lysing solution. The level of detection sensitivity with Rti-PCR was 10 CFU/g of tissue, which was equivalent to the DNA from 3 CFU per Rti-PCR reaction.

Bohaychuk et al. (2007) developed an Rti-PCR assay for the detection of *Salmonella* in a variety of food and food-animal matrices. The primers used LC-Sal1/LC-Sal2 (Table 4.2) amplified an unspecified bp sequence from the *invA* gene. A pair of fluorescent resonance energy transfer (FRET) probes LC-Sal probe 1/LC-sal probe 2 (Table 4.2) was used for detection of the specific amplicon. An internal control (IC) targeting a partial sequence of the pcDNA 3.1 plasmid was synthesized and utilized in conjunction with a second set of FRET probes LC-Sal IC probe 1/ LC-Sal IC probe 2 (Table 4.2). Both IC and target DNA were amplified by the same set of primers (Table 4.2) to monitor each PCR reaction for inhibition and accuracy of reagent preparation. A titration of IC template against target genomic DNA was used to determine the amount of IC to be added to each Rti-PCR reaction that would minimally interfere competitively with amplification of the target DNA. Samples were first subjected to pre-enrichment cultivation in buffered peptone water (BPW) and 0.1 ml of BPW pre-enrichments and then transferred to 10 ml of tetrathionate broth (TB) for selective enrichment at 35°C for 24 hr and to Rappaport–Vasschiades (RV) broth with incubation at 42°C for 24 hr. Each selective enrichment culture was streaked onto xylose-lysine-tergitol 4 agar and Ramback agar plates and incubated at 35°C. Aliquots (150 μl) of each of the TB- and RV-selective broth enrichments were combined and subjected to immunomagnetic separation followed by the use of a commercial DNA extraction kit. When the Rti-PCR assay was applied to field samples and to conventional selective enrichment and isolation, sensitivity ranged from 97.1% to 100% for the various matrices. The estimated cost of the Rti-PCR assay was 62% of conventional cultivaton when labor and materials were included.

C. Multiplex PCR (mPCR)

The ability of the PCR to detect all serotypes of *Salmonella* in addition to the distinguishable presence of a specific species of *Salmonella* has certain advantages over the cultural detection of all *Salmonella*. Mahon et al. (1994) developed an mPCR for simultaneous detection of all *Salmonella* and *S.* Enteritidis on chicken skin. The primers M1/M2 (Table 4.2) amplified a 163-bp sequence specific to the origin of replication (OriC) of *Salmonella* genomic DNA, which is specific for all *Salmonella*. Primers S1/S2 (Table 4.2) amplified a 620-bp sequence derived from a *S.* Enteritidis virulence plasmid. Nonselective enrichment was in BPW followed by the PCR. Among 68 samples tested, the total number positive for *Salmonella* by the PCR was 10 (15%) in contrast to only 5 (7.4%) positive samples obtained with routine bacteriological cultivation.

The most common serotypes of *Salmonella* isolated from humans are *S.* Typhimurium and *S.* Enteritidis, derived frequently from the consumption of eggs or egg products. Soumet et al. (1999) undertook an extensive study of 35 broiler poultry houses in France involving 22 swabs each from a variety of physical locations before chicks were introduced and nine swabs from each after 42 days of breeding. A total of 1078 swab samples were subjected to mPCR detection of *S.* Typhimurium and *S.* Enteritidis. Gauze swabs were incubated in 150 ml of phosphate-buffered peptone water (BPW) for 18–20 hr at 37°C. For multiplex PCR, three primer pairs were utilized. Primers ST11/ST15 (Table 4.2) amplified a 429-bp sequence from a randomly cloned genomic DNA fragment and are specific for *Salmonella* spp. (Aabo et al. 1993). Primers Sef167/Sef478 (Table 4.2) from Rychlik et al. (1999) amplified a 312-bp sequence derived from the *sefA* gene that encodes the fimbrial antigen Sef14 and are specific for *S.* Enteritidis. The occurrence of this gene sequence among other *Salmonella* serotypes is restricted to the serotypes Enteridis, Blegdam, Dublin, Galinarum, Pullorum, Rostock, Seremban, and Typhi (Turcotte and Woodward, 1993). Primers Fli15/Tym (Table 4.2) from Rychlik at al. 1999) amplified a 559-bp sequence of the *fliC* gene encoding flagellin H1, which is specific for *S.* Typhimurium and for no other *Salmonella* serotype. Among a total of 1078 swabs from the 35 poultry houses, 330 were positive via m-PCR and routine bacteriological techniques. Only 19 swabs were bacteriologically positive and PCR negative (false PCR negative) resulting in a total of 349 positive samples. Among 729 bacteriologically negative samples 28 were PCR-positive indicating a smaller percentage of false-negatives from m-PCR than from bacteriological techniques. Among a total of 406 PCR positive samples, 38 were *S.* Enteritidis and 31 were *S.* Typhimurium. The remaining 337 were other *Salmonella* species and serotypes. A comparison of results derived from both m-PCR and bacteriological techniques indicated that the m-PCR had a sensitivity of 95% and a specificity of 96.2.

Dual-labeled probes can be used with the PCR without the necessity of a costly real-time Rti-PCR unit. Sharma and Carlson (2000) developed a multiplex fluorogenic PCR assay for simultaneous detection of *Salmonella* and *E. coli* O157:H7. The primers for *Salmonella* SipB/C-F/SipB/C-R (Table 4.2) from Carlson et al. (1999) amplified a 250-bp sequence of a junctional segment of the virulence genes *sipB* and *sipC* in *Salmonella* serovars. The primers for *E. coli* O157:H7 amplified a 150-bp sequence of an intragenic fragment of the *eae* gene. DNA probes labeled at the 5′-end with FAM (for *Salmonella*) or HEX (for *E. coli* O157:H7) as the reporter dye and TAMRA at the 3′-end as a quencher were used. Detection of amplified products was achieved by reading a 90-well plate in a computer-controlled dual-scanning microplate spectrofluorometer. The assay system could detect <10 CFU/g of meat following 18 hr of enrichment of 1 g of meat in 9 ml of gram-negative tryptic soy broth (GNTSB).

Vantarakas et al. (2000) developed an mPCR detection assay for *Salmonella* spp. and *Shigella* spp. in mussels. The primers Sal3/Sal4 (Table 4.2) from Rahn et al. (1992) amplified a 275-bp sequence from the *invA* gene of *Salmonella* serovars and the *Shigella* primers amplified a 215-bp sequence from the *virA* gene of *Shigella* spp. Following enrichment in BPW for 22 hr, 1 to 100 CFU per ml of tissue homogenate were detected.

Aslam, Hogan, and Smith (2003) reported on the development of an m-PCR for simultaneous detection of *Salmonella*, *E. coli* O157:H7, and *L. monocytogenes* in milk. The *Salmonella* primers 16S rRNA-F/16S rRNA-R (Table 4.2) from Lin and Tsen (1995) amplified a 571-bp sequence of the 16S rRNA *Salmonella* gene. Boiling for 15 min was used to extract DNA. When *Salmonella* were added to milk without enrichment, the limit of detection was 5 CFU/ml of milk. However, when *Salmonella* were grown in milk overnight and DNA extracted by boiling, the detection limit was 1×10^2 CFU/ml of milk. The coating of cells by milk fat may have prevented efficient cell lysis after growth in milk. It was previously shown that certain inhibitors present in milk may interfere with the PCR when bacteria are grown in milk (Khan et al., 1998; Kim et al., 2001).

An mPCR was developed by Kim et al. (2006) to differentiate between the most common serotypes of *S. enterica* subsp. *enterica* in the United States. Six genetic loci (Table 4.2) from *S. enterica* serovar Typhimurium and four from *S. enterica* serovar Typhi were used to create a PCR assay system consisting of two five-plex PCRs. The assays gave reproducible results with 30 different serotypes representing 75% of the most common clinical serotypes of *S. enterica* subsp. *enterica* encountered.

Multidrug-resistant *Salmonella enterica* serovar Typhimurium phage type DT104 first emerged in the 1990s and is now globally distributed. Most strains of this phage type are characterized by chromosomal resistance to ampicillin (A), chloramphenicol (C), streptomycin (S), sulfonamides (Su), and tetracycline (T) and are referred to as ACSSuT-type. The ACSSuT-type serovar Typhimurium is considered to be derived from two separate evolutionary events. One is the integration of a 43-kb *Salmonella* genomic island 1 (SGI1), which carries the above ACSSuT antimicrobial resistance genes. The other is the integration of a P22-like phage into the genomic DNA to form prophage PDT17 or ST104. In serovar Typhimurium DT104, the SGI1 island (harboring the five antibiotic resistance genes) is thought to be disseminated among *Salmonella* serovars through a mechanism of mobilization of the IncC plasmid R55. In the presence of a helper plasmid, the SGI1 can spread between different serovar Typhimurium phage types or between serovars. Khan et al. (2000) developed an m-PCR assay for detection of *S.* Typhimurium DT104 (ACSSuT-type) strains. Four pairs of primers were used: FloF/FloR (Table 4.2) amplified a 584-bp sequence from the *flost* gene, VirF/VirR (Table 4.2) amplified a 392-bp sequence of the *sprC* virulence gene, InvF/InvR (Table 4.2) amplified a 321-bp sequence of the *invA* invasive gene, and IntF/IntR (Table 4.2) amplified a 265-bp sequence of the *int* integron gene. Twenty-two ACSSuT-type DT104 strains and two U302 phage ACSSuT-type strains tested yielded amplification of all four genes. All strains examined were PCR positive for the *invA* gene. The PCR indicated that the *spvC* gene was present in 31 of 32 (97%) of the *S.* Typhimurium strains examined and that 30 of the 32 strains (94%) were positive for the *int* gene. All multidrug ACSSuT-resistant DT104 strains were PCR-positive for *spvC*.

Variable-number tandem repeats (VNTR) are short sequence repeats that consist of unique DNA elements that are repeated in tandem. Individual strains within a bacterial species often maintain the same sequence element but with different copy numbers. Because sequence homology exists between strains in the flanking region of the VNTR locus, PCR amplification with flanking-sequence-specific primers can be used to determine the variations in copy numbers of repeat units that reflect diversity among strains of the same species or serotype. This allows individual strains to be distinguished on the basis of amplicon sizes. Liu et al. (2003) developed an mPCR for typing 59 clinical strains of *S. typhi* from five Asian countries using three different VNTR loci designated TR1, TR2, and TR3 as molecular markers. The primer pairs involved for amplifying sequences from each locus were: TR1F1/TR1R1, TR2F1/TR2F2, and TR3F1/TR3R1 (Table 4.2). Each strain yielded one amplicon of a characteristic size with each of the three primer pairs. The molecular size of each of the three resulting amplicons (mPCR banding pattern) for each strain was then used to distinguish a total of 49 VNTR types out of the 59 strains. Substantial genetic heterogeneity at the VNTR loci was found to exist among the *S.* Typhi strains from within the same country and among different countries. Most of the isolates from each country yielded distinctive banding patterns that were different from the isolates from the other countries. When the VNTR mPCR was applied to strains of *S.* Typhimurium and *S.* Paratyphi two amplicons of ~200- and 30-bp were observed. This was in sharp contrast to the highly diversified banding patterns obtained with strains of *S.* Typhi. Individual PCR assays indicated that the 200-bp amplicon was produced by the TR1 primers whereas the 300-bp amplicon was produced by the TR2 primers. Interestingly, no amplicon was generated from these serotypes by the TR3 primers, suggesting that sequences of the TR3 primer sites are highly conserved in S. *typhi*.

The *fliC* gene is involved in the synthesis of flagellin and encodes the phase 1 flagellar protein (H1 antigen) and is expressed alternately with the *fliB* gene encoding the phase 2 flagellar protein (H2 antigen) according to a phase variation process with both genes located on the genomic DNA. Touron et al. (2005) developed a nested mPCR (nmPCR) involving four external primers and three internal primers. This nmPCR was based on amplification with primers targeting sequences of the conserved regions of the *fliC* gene present in all *Salmonella* serovars, even the nonmotile ones located on both sides of a variable central region of the *fliC* gene (Figure 4.1). DNA was extracted and purified from aquatic and sediment samples using commercial kits. The external primer pairs Sal345/Sal1312 and Sal345d/Sal1312d (Figure 4.1) initially target a 948-bp sequence that can vary by about 10 nucleotides depending on the different serovars. The internal primer pairs Salnes3d/Salnes1d and Salnes3d/Salnes1d amplify an 888–892-bp sequence that can also vary by a few nucleotides depending on the different serovars (see the legend to Figure 4.1 for precise target amplification). Following initial PCR amplification with all four external primers for 35 cycles, 1 µl was then used for the nested PCR with the three internal primers for 35 cycles. Restriction analysis was then performed using the *Hha*I endonuclease and the RFLP products visualized after electrophoresis in 2% agarose with ethidium bromide.

Outbreaks of gastroenteritis due to consumption of contaminated produce have more than doubled in the United States since 1987. Most of the reported outbreaks were caused by *E. coli* O157:H7, *Salmonella*, and *Shigella*. Li and Mustapha (2004) developed an mPCR for detection of *Salmonella*, *E. coli* O157:H7, and *Shigella*. The

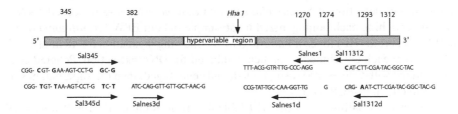

Figure 4.1 Nested mPCR involving four external primers and three internal primers for detection of *Salmonella* spp. Assay is based on amplification with primers targeting sequences of the conserved regions of the *fliC* gene present in all *Salmonella* serovars, even the nonmotile ones located on both sides of a variable central region of the *fliC* gene. Bold font indicates differing aligned nucleotides of primers. R = A or G. Sal1312/Sal345 used in conjunction with Salnes1/Salnes3d amplify fliC gene sequences specific to diphasic serovars, whereas Sal1312d/ Sal345d and Salnes1d/Salnes3d amplify equivalent *fliC* gene sequences specific to the monophasic serovars: *S. enterica* Montevideo, Enteritidis, Dublin, Derby, Agona, and Senftenberg. (Adapted from Touron, A. et al., *Res. In Microbiol.* 156:541–553, 2005.)

assay yielded a 429-bp amplicon from *S. typhimurium* utilizing the primers ST11/ ST15 (Table 4.2) from Aabo et al. (1993), a 252-bp amplicon from *E. coli* O157:H7, and a 620-bp amplicon from *Shigella flexneri*. The multiplex assay was applied to apple cider, fruits, and vegetables. DNA was extracted and purified using the method of Pitcher, Saunders, and Owen (1989). The detection level for all three pathogens was 0.8 CFU/g or CFU/ml after 24 hr of nonselective enrichment in brain–heart infusion (BHI) broth.

Grain is considered to be a product with a low risk of contamination with pathogenic bacteria due to its low water activity. However, several studies have indicated the presence of low numbers of bacterial pathogens in wheat and flour (Eyles, Moss, and Hocking, 1989; Richter et al., 1993; Berghofer et al., 2003). Kim et al. (2006) established an mPCR for simultaneous detection of *E. coli, L. monocytgenes,* and *S. typhimurium* in artificially inoculated wheat grain. Dried wheat grain (25 g) was enriched in 225 ml of BPW and incubated for 24 hr at 32°C. The primers for *S. typhimurium* were SalinvA139/SalinvA141 (Table 4.2) from Rahn et al. (1992) and amplified a 284-bp sequence of the *invA* gene. *S. typhimurium* was detected at a level of <54 CFU/ml in the multiplex PCR. Singleplex PCR resulted in detection of 3 CFU/ml.

The simultaneous detection of *Salmonella* spp. and *L. monocytogenes* in foods by PCR following enrichment poses certain critical problems stemming from fundamental differences between the two genera. These differences are reflected in part by the fact that *Salmonella* spp. are gram-negative, whereas *L. monocytogenes* is gram-positive. In addition, *L. monocytogenes* is nutritionally fastidious, whereas *Salmonella* spp. are capable of growth in a mineral salts–glucose medium. Jofré et al. (2005) found that simultaneous enrichment in rich media such as tryptic soy broth yeast extract (TSBYE) or BPW resulted in at least a one log predominance of *Salmonella* over *L. monocytogenes*, even when the initial ratio of the former to the latter was 1:100. However, Jofré et al. (2005) successfully developed an mPCR for simultaneous detection of *Salmonella* spp. and *L. monocytogenes* in seeded cooked ham by enriching each homogenized meat sample in both BPW for *Salmonella* and half Frazer broth (HF) for *L. monocytogenes*. Following incubation for 48 hr at 37°C, 1.0 ml from each enrichment was then combined for DNA extraction. Pelleted cells were mixed with Chelex 100 resin, boiled, and centrifuged; then the supernatant was used directly in m-PCR assays.

The primers Salinv139/Salinv141 (Table 4.2) from Rahn et al. (1992) were used to amplify a 284-bp sequence of the *invA* gene located on pathogenicity island 1 of *Salmonella* spp. encoding proteins of the type III secretion system, essential for the invasion of epithelial cells by *Salmonella* serovars. A pair of primers Lip2/Lip3 amplified a 215-bp sequence of the *prfA* gene of *L. monocytogenes*. A separate internal amplification control (iac) was used for each genus. The primers iac139/iac141 (Table 4.2) amplified a 120-bp fragment of the *Brassica napus ACCy8* gene flanked by a sequence of the *Salmonella* primers. A similar *L. monocytogens* internal control was synthesized consisting of a 132-bp fragment of the *Brassica napus ACCy8* gene flanked by a sequence of the *L. monoctogenes* primers. The resulting chimeric amplicons (iac-Salm and iac-Lmono) were added to m-PCR reactions as dual internal amplification controls. Identical primer-binding sites allowed coamplification of the *prfA* or *invA* gene fragment and the corresponding IAC. The difference in the amplicon size between the *Salmonella invA* or *L. monocytogenes prfA* specific amplimers (284- and 215-bp, respectively) and the corresponding IAC (120- and 132-bp, respectively) readily allowed discrimination in a 2.0% agarose gel. The use of an IAC internal control greatly facilitates the interpretation of negative PCR results, because a control signal will always be produced when there is no target sequence present and can thus reveal the failure or inhibition of a PCR reaction.

The sanitation of chicken abattoirs is of considerable concern in terms of the incidence and levels of *Salmonella*-contaminated poultry carcasses. Cortez et al. (2006) developed a multiplex PCR that was applied to samples from chicken abattoirs in São Paulo State, Brazil. A total of 288 samples derived from feces, feathers, water (scald, evisceration, and chiller), and rinse water of noneviscerated, eviscerated, and chilled carcasses were collected from the processing lines of three chicken abattoirs. Three sets of primers were utilized. Primers invA-1/invA-2 (Table 4.2) amplified a 521-bp sequence of the invasion gene *invA* of *Salmonella* serovars. Primers sefA-1/sefA-2 amplified a 330-bp sequence of the fimbrial *sefA* gene of *S.* Enteritidis. Primers pefA-1/pefA-2 yielded a 497-bp amplicon with both *S.* Enteritidis and *S.* Typhimurium derived from the *pefA* fimbrial virulence gene. However, differentiation was achieved by the presence of a restriction site for the enzyme *Kpn*I in *S.* Typhimurium that does not exist in *S.* Enteritidis. Results yielded *Salmonella* from 52 out of the 288 collected samples. From these, 29 (10%) were positive for *Salmonella* serovars, 16 (5.6%) for *S.* Enteritidis, and 7 (2.4%) for *S.* Typhimurium. The highest incidence of *Salmonella* serovars was found to occur at a level of 6 out of 36 feces samples (16.7%) and from 11 out of 36 rinse water samples of noneviscerated carcasses, most probably reflecting fecal contamination of the carcasses. The highest incidence of *S.* Enteritidis was from 4 out of 36 (11.1%) chilled water samples and 4 of 36 rinse water samples of chilled carcasses. The highest incidence of *S.*

Typhimurium was from 3 out of 36 (8.3%) rinse water samples of noneviscerated carcasses. The authors concluded that there was a significant difference in the occurrence of *Salmonella* among the three abattoirs and that there is a need to improve quality control measures for *Salmonella* in chicken abattoirs.

Chiu et al. (2006) collected a total of 104 clinical isolates of *Salmonella* serogroup B from three major hospitals in Taiwan during 1997 to 2003, which were examined by an mPCR targeting the resistance genes *pse*, *floR*, *str*, *tetG*, and *sulI*, and the *spv* gene of the virulence plasmid (Table 4.2). The presence of the *spv* plasmid was used to identify serovar Typhimurium because 90% of this serovar harbor the virulence plasmid (Borrego et al., 1992; Chiu, Lin, and Ou, 1999). Six primer pairs were used in an mPCR assay. A total of 51 isolates (49%) were resistant to all five antibiotics (ampicillin, chloramphenicol, streptomycin, sulfonamide, and tetracycline) tested and all contained a 1.25-kb PCR fragment of integron that is part of the 43-kb *Salmonella* genomic island 1. A second group of 29 isolates (28%) was resistant to only streptomycin and sulfonamide; a third group of 13 isolates (13%) was susceptible to all five drugs. A majority (94.1%) of the *Salmonella* serogroup B isolates resistant to all five antibiotics harbored a virulence plasmid. Phage typing identified three major phage types, DT104, DT120, and U302 with two genotypes among the DT104 strains. Analysis by PFGE revealed six genotypes indicating genetic variations among isolates within the same phage type suggesting diversification of serovar Typhimurium in Taiwan.

Kawasaki et al. (2005) developed an mPCR for simultaneous detection of *Salmonella* serovars, *L. monocytogenes*, and *E. coli* O157:H7 in enriched meat samples. Enrichment in one of 20 broths tested was determined to yield maximum numbers of all three organisms after 20 hr at 35°C. This enrichment broth consisted of tryptone, 1.0%; beef extract, 0.5%; yeast extract, 0.5%; KH_2PO_4, 1.5%; Na_2HPO_4, 0.7%; NaCl, 0.5%; and dextrose, 0.05%. The primers TS-11/TS-5 (Table 4.2) amplified a 375-bp sequence from the *Salmonella*-specific sequence reported by Tsen, Liou, and Lin (1994). Boiling of cells of *Salmonella* and *E. coli* O157:H7 was effective for lysis and release of DNA, but failed with *L. monocytogenes*. Among five DNA extraction methods examined, treatment of cells with achromopeptidase and lysozyme followed by guanidine thiocyanate was optimum for all three organisms. The PCR sensitivity of detection after optimization of a number of variables was 1 CFU/15 g of meat.

The sensitivity of PCR detection of a target pathogen in foods is notably dependent on the composition of the food product and the presence of PCR inhibitors. Li, Zuang, and Mustapha (2005) developed an mPCR for simultaneous detection of *Salmonella*, *Shigella*, and *E. coli* O157:H7 in raw and ready-to-eat meat products. Samples (25 g) were homogenized with 225 ml of BHI Broth and incubated at 37°C for 24 hr. DNA was extracted with a commercial kit. Primers ST11/ST15 (Table 4.2) from Aabo et al. (1993) were used to amplify a 429-bp amplicon from members of the genus *Salmonella*. With samples of ground beef, following enrichment, an initial inoculum of 1.5 CFU/g was detectable with each of the three pathogens. However, amplification was not successful from enriched homogenates of ground pork seeded with each organism at 1.5 CFU/g. Higher inoculation levels of 15 to 15,000 CFU/g yielded the expected amplicons.

An mPCR method was developed by Kim et al. (2006b) to differentiate between the most common serotypes of *Salmonella enterica* subsp. *enterica* in the United States. The technique is based on the PCR detection of genes present in specific serotypes but not others. The gene sequences selected for amplification were based on previous work including whole-genome sequencing of various serotypes. Six genetic loci from *S. enterica* serovar Typhimurium and four from *S. enterica* serovar Typhi were used to create a PCR assay system consisting of two fiveplex PCRs. The five primer pairs consisting of STM1, STM2, STM3, STM4, and STM5 (Table 4.2) were combined in one fiveplex. The primer pairs STY1, STY2, STY3, STM6, and STY4 (Table 4.2) were combined into a second fiveplex PCR. Resulting banding profiles were found to correlate with specific serotypes. To further discriminate between serotypes with the same amplicon patterns, two additional primer sets PT4 and STM7 (Table 4.2) were developed. The assays gave reproducible results with 30 different serotypes representing 75% of the most common clinical serotypes of *S. enterica* subsp. *enterica* encountered. PCR serotyping was nearly as discriminating as conventional serotyping.

D. Pulsed Field Gel Electrophoresis (PFGE)

Garaizar et al. (2000) undertook an extensive interlaboratory comparison of PFGE, random amplified polymorphic DNA (RAPD), and infrequent restriction-site PCR (IRS-PCR), involving 212 isolates of *S. enterica*. RAPD analysis utilized the three random primers ERIC-2, M13, and OPS-19 (Table 4.2). With IRS-PCR, DNA was first restricted with combinations of two restriction nucleases: *Xba*I-*Hha*I and *Xba*I-*Taq*I. Appropriate adapters (Table 4.2) were then ligated to the restriction products and PCR undertaken. With the enzyme combination *Xba*I-*Hha*I the number of bands ranged from 7 to 16, and with *Xba*I-*Taq*I the number of bands ranged from 3 to 8. With IRS-PCR, serovars were efficiently discriminated; however, discrimination between strains was poor. With RAPD, some banding profiles were found to vary, depending on the thermocycler used. The authors concluded that RAPD could be useful as an epidemiological marker when isolates are processed in a single batch with identical reagents and the same thermocycler, but interlaboratory variations prevented the establishment of a library-definitive typing system. PFGE utilizing the restriction nucleases *Xba*I, *Bln*I, and *Spe*I yielded high interlaboratory reproducibility values and high levels of discrimination and therefore was selected as the method of choice for establishing a computerized library for *S.* serovar Enteritidis isolates utilizing the combined banding profiles derived from all three restriction nucleases.

Contamination of pork products by *Salmonella* is often the result of feces being spread onto carcasses during slaughter and processing of swine. Wonderling et al. (2003) recovered *S. enterica* isolates from a processing plant in Ireland over a 2-month period in the spring of 2000. PFGE analysis was performed on 581 confirmed *Salmonella* isolates from 84 *Salmonella*-positive samples. A total of 32 *Xba*I visually distinguishable PFGE pattern types were obtained with 12 PFGE cluster groups. Three of the groups predominated throughout the sampling period. Both carcass and fecal isolates of *Salmonella* were recovered from 13 swine resulting in matched samples. PFGE typing of the 252 isolates recovered from the matched samples revealed that 7 (54%) of the 13 carcasses were contaminated with *Salmonella* PFGE types that were not isolated from the feces of the same animal. Conversely, from 6 of the 13 (46%) of the matched animals, *Salmonella* clonal types were isolated from the feces that were not isolated from the carcasses of the same animals. The authors concluded that swine feces from one animal can contaminate many carcasses during slaughter and processing.

Vieira-Pinto, Tenreiro, and Martins (2006) obtained 69 isolates of *Salmonella* spp. from the ileum, tonsils, carcass, and mandibular and ileocolic lymph nodes of individual swine slaughtered in a single Portuguese abattoir. The isolates were subjected to PFGE and serotping. *Xba*I macrorestriction revealed 18 genotypes among the 8 serotypes. Eighty percent of the animals with ileum and ileocolic lymph node positive samples also presented the same PFGE pattern type in the corresponding tonsils and among swine with positive tonsils. PFGE analysis suggested three different sources of animal infection: the farm of origin, the transportation, and the lairage or holding area.

S. enterica subsp. *enterica* serovar Newport has re-emerged as a public and animal health problem. In 1987, a chloramphenicol-resistant strain of this serovar was identified in a Southern California outbreak. Contaminated hamburger derived from dairy cattle from dairies using chloramphenicol was implicated. During the late 1980s serovar Newport was the most common *Salmonella* isolate from dairy cattle in California. From 1991 to 1998, this serovar accounted for only 1.2% of salmonella isolates in California from dairy cattle. Beginning in 1999, serovar Newport again emerged clinically in dairy cattle. These isolates were resistant to multiple antibiotics and exhibited resistance toward the new cephalosporin ceftiofur used in the treatment of salmonellosis. The multidrug-resistant (MDR) serovar Newport affected young and adult cattle and humans in California. Many of these cases were linked to the consumption of soft cheese products. In 2000, serovar Newport was reported by the Center for Disease Control (CDC) to be the third most common serovar.

Berge, Adaska, and Sischo (2004) undertook a study comparing PFGE patterns and antibiotic resistance profiles of MDR serovar Newport isolates from the 1980s with contemporary 1999–2002 isolates. *Xba*I was used to generate PFGE patterns for 198 isolates from a wide variety of dairy environments and sources in addition to human clinical isolates and resistance to 18 antibiotics was determined. Although the antibiotic resistance patterns of historic (1980 to 1995) and contemporary (1999 to 2001) isolates were similar, the contemporary isolates differed from the historic isolates by being resistant to cephalosporins and florfenicol and in their general sensitivity to kanamycin and neomycin. With few exceptions, the contemporary isolates clustered together and were clearly separated from the historic isolates. One PFGE-antibiogram cluster combination was predominant for the recent isolates, which were of human origin from all areas of the United States including California, indicating a rapid dissemination of this phenotypic strain. The authors concluded that the data are consistent with the hypothesis that the re-emergence of MDR serovar Newport is not simply an acquisition of further antibiotic resistance genes by the historic isolates but reflects a different genetic lineage.

Chung et al. (2004) analyzed a total of 81 isolates of *S*. Enteritidis from Seoul, Korea, for antibiotic susceptibility, phage typing, and PFGE. Thirty-two isolates were from broiler carcasses and pig feces and 49 were from humans. Antibiotic resistance was most prevalent among human isolates, 89.9% of which were resistant to more than two antibiotics, whereas 44.7% of poultry and 13.3% of pig isolates were resistant to more than two antibiotics. The isolates yielded six PFGE patterns with *Xba*I and *Spe*I digestion and five PFGE patterns with *Not*I digestion. A single PFGE pattern X1, S1, or N1 was dominant, and the remainder of the PFGE patterns differed by only one or two bands. Results indicated the spread of a genetically related clone of *S*. Enteritidis in foods and humans in Korea and that phage typing as well as PFGE may offer an improved level of discrimination for the epidemiological investigation of *S*. Enteritidis.

Although PFGE is considered superior to other subspecies molecular typing methods, typeability may not be ideal for some bacterial species due to DNA degradation. Chen et al. (2005) reported on the first case of neonatal salmonellosis resulting in meningitis caused by the transmission of *S. enterica* serotype Panama from contaminated breast milk in Taiwan. This serotype is one of several that are known to cause disease in children and is more invasive than other serotypes. Asymptomatic colonization of the mammary gland was indicated. Ribotyping of isolates from the infant's blood, cerebral spinal fluid, and from the mother's breast milk collected aseptically revealed identical banding patterns. PFGE was not used because the serotype Panama isolates possessed intracellular DNase activity that degraded the genomic DNA. Silbert et al. (2003) found that the addition of 50 µg/l of thiourea to the gel buffer prevented DNA degradation and greatly improved band resolution and typeability of *Salmonella* serovars.

Five different serotypes of *S. enterica* were implicated in a large outbreak of salmonellosis linked to fresh Roma tomatoes served at gas station deli counters in Pennsylvania and adjoining states during July 2004 (Sandt et al., 2006). One of these serotypes (Anatum) was isolated from both tomatoes and patients. PFGE patterns utilizing the restriction nucleases *Xba*I, *Bol*nI, and *Spe*I indicated that the Anatum serotype isolates from tomatoes were clonally identical to Anatum isolates from patients. In addition, isolates of serotype Javiana were also involved and were found to be the main cause of the outbreak. Out of 146 Javiana isolates from patients, 132 had an identical PFGE pattern. The authors concluded that PFGE was essential for discriminating between outbreak-related isolates and unrelated sporadic isolates for each of the five serotypes epidemically linked to this multistate outbreak.

Although PFGE is considered the "gold standard" for subtyping *Salmonella* isolates, PFGE also exhibits limited discriminatory power. Two PFGE patterns comprise nearly 48% of the *S*. serotype Enteritidis isolates in the Pulse Net National Database. Multiple-locus variable-number tandem repeat analysis (MLVA) is a subtyping technique that involves amplification and size analysis of polymorphic regions of DNA containing variable numbers of tandemly repeated sequences. Boxrud et al. (2007) developed an MLVA method for subtyping *Salmonella* serotype Enteritidis. The discrimination ability and epidemiological concordance of MLVA were compared with PFGE and phage typing. A total of 153 *S*. serotype Enteritidis isolates from Minnesota residents, comprised of 40 isolates from four separate foodborne disease outbreaks and 113 isolates from sporadic cases, were typed. Tandem repeat sequences were identified from genomic sequence data. PCR primers targeting the regions flanking tandem repeat loci were designed (Table 4.2). *Xba*I and *Bln*I were used in conjunction with PFGE. MLVA provided greater discrimination among non-epidemiologically linked isolates than did PFGE or phage typing. MLVA was better able to differentiate isolates between the individual outbreaks than either PFGE or phage typing. MLVA typing was found to have enhanced resolution, good reproducibility, and good epidemiological concordance.

Gaul et al. (2007) subjected 674 isolates of *Salmonella* isolated from swine from various locations throughout the United States and Canada to PFGE with *Xba*I. A total of 66 subtypes were differentiated, which when subjected to cluster analysis could be separated into the 11 serotypes based on their PFGE banding profiles. The authors concluded that if an unserotyped isolate had a PFGE pattern which exactly matched one found in the database, the serotype of the unknown isolate would be identical to the serotype of the matching pattern. If the isolate had a new PFGE pattern, the closest related pattern would be of the same serotype.

The precise origin of *Salmonella* contamination of poultry is an ongoing concern globally. Heyndrickx et al. (2007) undertook an extensive multiple typing study of 18 Belgian broiler flocks from the hatchery to the slaughterhouse. The study involved serotyping, genotyping, and phage typing. Genotyping encompassed PFGE, RAPD, and plasmid profiling. With PFGE, the restriction nucleases *Xba*I, *Spe*I, *Not*I, and *Bln*I were used. RAPD was performed with the primers 23L, OPB-17, P1254, and OPA4 (Table 4.2). For 12 of the 18 flocks there was no correlation between the serotypes found preharvest and those isolated from the transport crates and on the carcasses in the slaughterhouse. Serotypes found in the crates were usually also found on the carcasses. In 5 of the 10 flocks with *Salmonella*-positive broilers, complex contamination patterns with the involvement of different serotypes, genotypes, or both were revealed. In two of these flocks, the *S.* Enteritidis contamination could be traced to the hatchery. In one flock, evidence was found for the acquisition, during rearing, of a megaplasmid in the *S.* Enteritidis strains. In three other positive flocks the environment and movable material (e.g., footwear) influenced the shedding pattern of the broilers. With two flocks reared consecutively in the same broiler house, a persistent *Salmonella* Hadar geno/phage type predominated in the preharvest period, whereas another *Salmonella* Hadar geno/phage type was found in the broiler house or the environment but never in the broilers. The authors concluded that most of the time, *Salmonella* strains that contaminate Belgium broiler carcasses do not predominate in the preharvest environment.

Harbottle et al. (2006) undertook a comparison of multilocus sequence typing (MLST), PFGE, and antimicrobial susceptibility typing for characterization of 81 *S. enterica* serotype Newport isolates from humans, food animals, and retail foods presumably from Maryland. Fifty-nine percent of the isolates were resistant to nine or more antibiotics. The restriction nuclease *Xba*I generated 43 PFGE patterns indicating a genetically diverse population. MLST resulted in 12 sequence types with one type encompassing 62% of the strains. Strain discrimination was enhanced by combining PFGE, antimicrobial susceptibility, and MLST results.

Chandel and Malhotra (2006) subjected 16 strains of *S. enterica* serotype Typhi to PFGE and insertional sequence probe IS200 analysis. The IS200 is a 707- to 710-bp sequence that is widely distributed among conserved loci of most *Salmonella* serotypes. The apparent copy numbers are 10 to 25 in serotype Typhi isolates. The actual copy number and positions of IS200 indicate the evolutionary origin of such strains. PFGE analysis using *Xba*I revealed two major banding profiles. Both the IS200 and PFGE analysis discerned two unique clones that were nearly identical. The IS200 probe detected strains with either 11 to 14 IS200 copies or only 2 copies.

Fakhr, Nolan, and Logue (2005) applied PFGE and multilocus sequence typing (MLST) to 85 *S. enterica* serovar Typhimurium isolates from cattle. PFGE utilizing *Xba*I generated 50 profiles. MLST was based on determination of the DNA sequence of three selected housekeeping genes *manB, pduF, glnA* and one virulence gene *spaM* (Table 4.1) utilizing the appropriate PCR primers for these genes (Table 4.2). DNA sequencing of all four genes showed no genetic diversity among the 85 isolates. The authors concluded that MLST, using these genes, lacks the discriminatory power of PFGE for typing *S. enterica* serovar Typhimurium.

Typhoid fever is a significant cause of morbidity and mortality worldwide, causing an estimated 16 million cases and 600,000 deaths annually. Drug resistance among *S. enterica* serotype Typhi strains has emerged globally. Kubota et al. (2005) examined the distribution of PFGE patterns of *S. enterica* serotype Typhi isolates from patients with a history of international travel. PFGE subtyping utilized the restriction nuclease *Xba*I. Isolates indistinguishable with *Xba*I were further characterized using the restriction nuclease *Bln*I. A total of 139 isolates were typed, representing travel to 31 countries. *Xba*I yielded 79 PFGE patterns. Among the 139 isolates, 46 (33%) were resistant to one or more antimicrobial agents (ampicillin, chloramphenicol, trimethoprim, and sulfomethoxazole) traditionally used in the treatment of typhoid fever. Twenty-seven (59%) of the 46 multidrug resistant (MDR) isolates had indistinguishable PFGE patterns with both *Xba*I and *Bln*I. The authors concluded that MDR serotype Typhi has emerged as a predominant clone in Southeast Asia and the Indian subcontinent.

S. Senftenberg has been described as an uncommon human pathogen with most reported cases being of nosocomial origin. Recently, *S.* Senftenberg has been revealed as one of the predominant serovars isolated from marine environments and seafood, especially in temperate and tropical zones. Contamination of the marine environment by this serovar has been associated with its ability to survive in brines with salt concentrations of 30% (Martinez-Urtaza and Liebana, 2005). Mussels to be frozen are processed in three steps: (1) live mussels are steamed for a few minutes to open their shells, (2) they then descend into a pool of brine (30% NaCl) to separate the shell from the tissue, and (3) finally the mussel tissue is transported to a freeze tunnel. The use of brine was linked to the first contamination events in 1998 and 1999 in Spain. Martinez-Urtaza and Liebana (2005) reported on the persistent detection of *S.* Senftenberg in mussel-processing facilities between 1998 and 2002 in Spain. A total of 110 isolates from eight processing plants were subjected to PFGE utilizing *Xba*I with 21 different PFGE patterns resulting. Processing plants that used brine in their processing lines had greater diversity among their *S*. Senftenberg population, which supports the hypothesis that imported salt used for brine preparation could have been the origin of contamination. Isolates from mussel-processing plants exhibited PFGE typing patterns clearly distinct from those of clinical origin except for one human isolate that was identical to an isolate from a frozen mussel sample.

Nde et al. (2006) examined whole carcasses at eight stages on a turkey-processing line, and *Salmonella* prevalence was determined via enrichments. Recovered *Salmonella* were characterized using serotyping and PFGE analysis with *Aba*I. Contamination rates varied along the line and were greatest after defeathering and after chilling. Analysis of contamination in relation to serotype and PFGE profiles found that on some visits the same serotype was present all along the processing line whereas on other days, additional serotypes were recovered that were not detected earlier on the line, suggesting that birds harbored more than one serotype of *Salmonella* or there was cross-contamination occurring during processing. Following washing, *Salmonella* prevalence was significantly reduced.

E. Subtracted Restriction Fingerprinting (SFP)

S. enterica subsp. *enterica* serovar Derby is frequently isolated from pork, is among the 20 most frequently isolated serotypes from nonhuman sources, and is among the 10 most frequent serotypes from humans in Germany and Brazil. Sixty-two isolates of this serovar from slaughter pigs and minced pork products (from 395 samples) from Southern Brazil were analyzed for their genomic relationships and for the presence of antimicrobial resistance genes by Michael et al. (2006). Twenty-four isolates were indistinguishable on the basis of their SFP patterns generated with *EcoRI/Pau*I. Macrorestriction with *Xba*I yielded 24 strains with the same pattern, and *Bin*I yielded 25 strains with identical patterns. In contrast to the *Bin*I-macrorestriction patterns, the *Xba*I-macrorestriction patterns were in good agreement with the results of SFP analysis and phage typing. The resistance genes found (Table 4.1) were *sul1* or *sul2* (sulfonamide resistance), and *aadA2* (streptomycin/spectinomycin resistance), *tet(A)* (tetracycline resistance), *tet(B)* (tetracycline/minocycline resistance), bla_{TEM} (ampicillin resistance), and *dfrA14* (trimethoprim resistance). The authors concluded that the large number of isolates indicates the potential resistant risk that *S*. Derby can pose to human health when such organisms enter the food chain.

F. Random Amplified Polymorphic DNA (RAPD) Analysis

del Cerro et al. (2002) evaluated a number of PCR protocols for the confirmation of 46 salmonella isolates from 117 samples of animal origin (17 raw minced meat, 27 raw chicken meat, 8 raw sausages, and 25 egg samples in addition to 18 poultry fecal samples, and cecal swab samples). Positive control strains consisted of 40 serotypes representing 20 serogroups. Ten pairs of primers amplifying sequences of the virulence enhancing genes were assessed for specificity. All *Salmonella* strains and no non-*Salmonella* strains generated amplicons with the *stn* and *phoP/Q*-primers (Table 4.2). One or more non-*Salmonella* strains among 21 yielded amplicons with primers for *iroB*, *st*, *himA*, *invE/A*, and *agfA*. These results indicate a significant level of conservation of these gene sequences amplified among Eubacteria. The primer pair STN-1/STN-2 (Table 4.2) from Prager, Fruth, and Tschäpe (1995) were utilized to screen for *Salmonella* from the foods. PCR-ribotyping with primers RIB-1/RIB-2 (Table 4.2) from Kostman et al. (1992) yielded two to three bands for each food isolate, which ranged from 700 to 1000 bp. In addition, PCR-ribotyping resulted in a very low level of discrimination, differentiating only three profiles among the 46 isolates. In contrast, RAPD analysis with random primers S and C (Table 4.2) yielded 8 and 11 profiles, respectively, which when combined yielded a total of 15 profile types. IMS was found to be more efficient than centrifugation for detection of *Salmonella* in enrichment cultures. This is most probably due to enhanced removal of PCR inhibitors resulting from washing the beads after capture of target cells. A sensitivity of DNA from 5×10^2 to 1×10^3 CFU per PCR reaction was obtained with pure cultures.

Lim et al. (2005) compared four molecular typing methods for the differentiation of 57 *Salmonella* isolates using RAPD with three random primers DG100, DG102, and DG93 (Table 4.2) by enterobacterial repetitive intergeneric consensus (ERIC) using primers DG111 and DG112 (Table 4.2), by ribotyping-PCR with primers DG109 and DG110 (Table 4.2), and by single strand conformational polymorphism (SSCP) using primers DG146 and DG147 (Table 4.2). RAPD with primers DG100, DG102, and DG93 produced 42, 51, and 54 typing patterns, respectively. ERIC typing produced 50 patterns, ribotyping-PCR produced four patterns, and SCCP produced 11 patterns. Discrimination using a combination of RAPD (primer DG102 or primer DG93) and ERIC was superior to combinations and differentiation of all 57 *Salmonella* strains; it was also superior to the combination of results with two RAPD primers.

Hilton, Banks, and Penn (1996) applied primer 1254 (Table 4.2) for distinguishing 37 isolates of *Salmonella* representing 20 different serotypes. This primer was found suitable for production of profiles yielding four to ten bands for *Salmonella* isolates. Primer 1254 was found capable of discriminating between some but not all isolates of *S*. Typhimurium and *S*. Enteritidis. Further discrimination was possible using alternative primers to distinguish serotypes that fell into the same RAPD profile group with one particular primer. Phage typing proved to be more discriminating for both of these serogroups compared to the use of just one RAPD primer.

Soto et al. (1999) applied three independent RAPD primers S, OPB-6, and OPB-7 (Table 4.2) to 326 isolates of *Salmonella* from clinical sources in Spain encompassing 12 serotypes. With primer S, 21 amplified DNA band profiles were differentiated; with primer OPB-6, 14 profiles were differentiated; and with primer OPB-7, 40 profiles were differentiated. Combining the results obtained with all three primers yielded 57 RAPD types with a discriminating index (DI) of 0.94. Combining RAPD typing and phage typing allowed further discrimination. The concordance of RAPD typing and serotyping was 100%.

Maré, Dick, and van der Walt (2001) undertook the characterization of 33 strains of *S*. Enteritidis by RAPD, phage typing, plasmid profiles, and LD_{50} values in mice. Three RAPD primers were used: OPL-2, OPL-3, and OPL-12 (Table 4.2). Their results indicated that strains of the same phage type are not always genetically related and that strains of a high genetic relatedness were of different phage types. No specific plasmid profile could be linked to any of the 11 phage types. Based on results obtained from LD_{50} virulence tests, strains containing the 38 MDa plasmid were more virulent than strains that did not carry the plasmid.

During 2000 and 2001 an outbreak of human salmonellosis occurred in Sweden and Norway caused by *Salmonella* Livingstone. Eriksson et al. (2005) undertook a comparative study of three strains from food sources during the outbreak, two human strains, plus 27 more or less unrelated strains. PFGE distinguished 10 types; RAPD, 4 types; and automated ribotyping 4 types. The combination of all three typing methods yielded 12 types. Simpson's discriminating index calculated for each method yielded 0.766 for PFGE, 0.556 for ribotyping, and 0.236 for RAPD. It is well recognized that at least two separate RAPD primers are required for achieving the same level of discrimination as the use of one restriction nuclease with PFGE. In this study PFGE was performed after restriction with *Xba*I and RAPD was performed with primer 1254 (Table 4.2).

Shabarinath et al. (2006) made use of the PCR for detection of *Salmonella* from seafood in addition to employing RAPD and ERIC for discriminating *Salmonella* isolates from tropical seafood in India. The PCR employed three pairs of primers for detection of *Salmonella*, the *invA* primers Salinv139/Salinv141 of Rahn et al. (1992), the *invE* primers F/R of Chen and Griffiths (2001), and the *hns* primers LHNS-531/RHNS-682 of Jones, Law, and Bej (1993; Table 4.2). The primers of Millemann et al. (1996) were used for ERIC (Table 4.2). Among RAPD primers initially screened, one PM5 (Table 4.2) was selected for differentiating the strains. Among 100 samples of seafood, 20 were positive for *Salmonella* by enrichment cultivation and routine bacteriological methods and 52 by PCR. Among the three different primer pairs used, the *hns* primers yielded the highest number of positive samples compared to primer pairs *invE* and *invA*. Selenite cystine broth yielded the highest number of positive samples compared to tetrathionate broth and Rappaport–Vassiliadis broth. RAPD and ERIC individually distinguished four typing patterns, the combination of both methods discriminated six different types.

REFERENCES

Aabo, S., Anderson, J., Olsen, J. 1995. Research note: Detection of *Salmonella* in minced meat by the polymerase chain reaction method. *Lett. Appl. Microbiol.* 21:180–182.

Aabo, S., Rasmussen, O., Rosen, L., Sorensen, P., Olsen, J., 1993. *Salmonella* identification by the polymerase chain reaction. *Mol. Cell Prob.* 7:171–178.

Akopyanz, N., Bukanaov, N., Westblom, T., Kresovich, S., Berg, D. 1992. DNA diversity among clinical isolates of *Helicobacter pylori* detected by PCR-based RAPD fingerprinting. *Nucleic Acids Res.* 20:5137–5142.

Arlet, G., Phillippon, A. 1991. Construction by polymerase chain reaction and intragenic DNA probes for three main types of transferable β-lactamases (TEM, SHV, CARB). *FEMS Microbiol. Lett.* 82:19–26.

Aslam, M., Hogan, J., Smith, L. 2003. Development of a PCR-based assay to detect Shiga toxin-producing *Escherichia coli*, *Listeria monocytogenes*, and *Salmonella* in milk. *Food Microbiol.* 20:345–350.

Bansal, N., Gray, V., Mcdonell, F. 2006. Validated PCR assay for the routine detection of *Salmonella* in food. *J. Food Prot.* 69:282–287.

Baudart, J., Lemarchand, K., Brisabois, A., Lebaron, P. 2000. Diversity of *Salmonella* strains isolated from the aquatic environment as determined by serotyping and amplification of the ribosomal DNA spacer regions. *Appl. Environ. Microbiol.* 66:1544–1552.

Baümler, A., Heffron, F., Reissbrodt, R. 1997. Rapid detection of *Salmonella enterica* with primers specific for *iroB. J. Clin. Microbiol.* 35:1224–1230.

Bej, A., Mahbubani, M., Boyce, M., Atlas, R. 1994. Detection of *Salmonella* spp. in oysters by PCR. *Appl. Environ. Microbiol.* 60:368–373.

Berge, A., Adaska, J., Sischo, W. 2004. Use of antibiotic susceptibility patterns and pulsed-field gel electrophoresis to compare historic and contemporary isolates of multi-drug resistant *Salmonella enterica* subsp. *enterica* serovar Newport. *Appl. Environ. Microbiol.* 70:318–323.

Berghofer, L., Hocking, A., Miskelly, D., Jansson, E. 2003. Microbiology of wheat and flour milling in Australia. *Int. J. Food Microbiol.* 85:137–149.

Bohaychuk, V., Gensler, G., McFall, M., King, R., Renter, D. 2007. A real-time PCR assay for the detection of *Salmonella* in a wide variety of food and food-animal matrices. *J. Food Prot.* 70:1080–1087.

Borrego, J., Castro, D., Jimenez-Notario, M., Luque, E., Martinez-Manzanares, E., Rodriquez-Avial, C., Picazo, J. 1992. Comparison of epidemiological markers of *Salmonella* strains isolated from different sources in Spain. *J. Clin. Microbiol.* 30:3058–3064.

Boxrud, D., Pederson-Gulrud, K., Wotton, J., Medus, C., Lyszkowicz, E., Besser, J., Bartkus, J. 2007. Comparison of multiple-locus variable-number tandem repeat analysis, pulsed-field gel electrophoresis, and phage typing for subtype analysis of *Salmonella enterica* serotype Enteritidis. *J. Clin. Microbiol.* 45:536–543.

Caratoli, A. 2001. Importance of interons in the diffusion of resistance. *Vet. Rev.* 32:197–396.

Carlson, S., Bolton, L., Briggs, C., Hurd, H., Sharma, V., Fedorka-Cray, P., Jones, B., 1999. Detection of multiresistant *Salmonella typhimurium* DT104 using multiplex and fluoro-genic PCR. *Mol. Cell Probes.* 13:213–222.

Catarame, T., O'Hanlon, K., Mcdowel, D., Blair, I., Duffy, G. 2006. Comparison of a real-time polymerase chain reaction assay with a culture method for the detection of *Salmonella* in retail meat samples. *J. Food Saf.* 26:1–9.

Chandel, D., Malhotra, P. 2006. Molecular typing reveals a unique clone of *Salmonella enterica* serotype Typhi among Indian strains. *J. Clin. Microbiol.* 44:2673–2675.

Chansiripornchai, N., Ramasoota, P., Bangtrakulnonth, A., Sasipreeyajan, J., Svenson, S. 2000. Application of randomly amplified polymorphic DNA (RAPD) analysis for typing avian *Salmonella enterica* subsp. *enterica. REMS Immunol. Med. Microbiol.* 29:221–225.

Chen, J., Griffiths, M., 2001. Detection of *Salmonella* and simultaneous detection of *Salmonella* and Shiga-like toxin producing *Escherichia coli* using the magnetic capture hybridization polymerase chain reaction. *Lett. Appl. Microbiol.* 32:7–11.

Chen, T., Thien, P., Liaw, S., Fung, C., Siu, L. 2005. First report of *Salmonella enterica* serotype Panama meningitis associated with consumption of contaminated breast milk by a neonate. *J. Clin. Microbiol.* 43:5400–5402.

Chiu, C., Lin, T., Ou, J. 1999. Prevalence of the virulence plasmids of nontyphoid *Salmonella* in the serovars isolated from humans and their association with bacteremia. *Microbiol. Immunol.* 43:899–903.

Chiu, C., Ou, J. 1996. Rapid identification of *Salmonella* serovars in feces by specific detection of virulence genes, *invA* and *spvC*, by an enrichment broth culture-multiplex PCR combination assay. *J. Clin. Microbiol.* 34:2619–2622.

Chiu, C., Su, L., Chu, C., Wang, M., Yeh, C., Weill, F., Chu, C. 2006. Detection of multidrug-resistance *Salmonella enterica* serovar Typhimurium phage types DT02, DT04, and U302 by multiplex PCR. *J. Clin. Microbiol.* 44:2354–2358.

Chung, Y., Kwon, Y., Kim, S., Kim, S., Lee, B., Change, Y. 2004. Antimicrobial susceptibilities and epidemiological analysis of *Salmonella* Enteritidis isolates in Korea by phage typing and pulsed-field gel analysis. *J. Food. Prot.* 67:264–270.

Cohen, H., Mechanda, S., Lin, W. 1996. PCR amplification of the *fimA* gene sequence of *Samonella* Typhimurium, a specific method for detection of *Salmonella* spp. *Appl. Environ. Microbiol.* 62:4303–4308.

Cohen, N., Neibergs, H., McGruder, E., Witford, H., Behle, R., Ray, P., Hargis, B. 1993. Genus-specific detection of salmonellae using the polymerase chain reaction (PCR). *J. Vet. Diagn. Invest.* 5: 368–371.

Cortez, A., Carvalho, A., Ikuno, A., Bürger, K., Vidal-Martins, A. 2006. Identification of *Salmonella* spp. isolates from chicken abattoirs by multiplex-PCR. *Res. Vet. Sci.* 81:340–344.

Croci, L., Delibato, E., Volpe, G., De Medici, D., Palleschi, G. 2004. Comparison of PCR, electrochemical enzyme-linked immunosorbent assays, and the standard culture method for detecting *Salmonella* in meat products. *Appl. Environ. Microbiol.* 70:1393–1396.

Daum, L., Barnes, W., McAvin, J., Neidert, M., Cooper, L., Huf, W., Gaul, L., Riggins, W., Morris, S., Salmen, A., Lohman, K. 2002. Real-time PCR detection of *Salmonella* in suspect foods from a gastroenteritis outbreak in Kerr county, Texas. *J. Clin. Microbiol.* 40:3050–3052.

del Cerro, A., Soto, S., Landeras, E., Gonzalez-Hevia, M., Guijarro, J., Mendoza, M. 2002. PCR-based procedures in detection and DNA-fingerprinting of *Salmonella* from samples of animal origin. *Food Microbiol.* 19:567–575.

del Cerro, A., Soto, S., Mendoza, M. 2003. Virulence and antimicrobial-resistance gene profiles determined by PCR-based procedures for *Salmonella* isolated from samples of animal origin. *Food Microbiol.* 20:431–438.

De Medici, D., Croci, L., Delibato, E., Pasquale, S., Filetici, E., Toti, L. 2003. Evaluation of DNA extraction methods for use in combination with SYBR Green I real-time PCR to detect *Salmonella enterica* serotype Enteritidis in poultry. *Appl. Environ. Microbiol.* 69:3456–3461.

Doran, J., Collinson, S., Burian, J., Sarlos, G., Tood, E., Munro, C., Kay, C., Bauser, P., Peterkin, P., Kay W. 1993. DNA-based diagnostic test for *Salmonella* species targeting agf-a, the structural gene for thin, aggregative fimbriae. *J. Clin. Microbiol.* 31:2263–2273.

Eriksson, J., Löfström, C., Aspán, A., Gunnarsson, A., Karisson, I., Borch, E., de Jong, B., Rådström, P. 2005. Comparison of genotyping methods by application to *Salmonella* Livingstone strains associated with an outbreak of human salmonellosis. *Int. J. Food Microbiol.* 104:93–103.

Espinoza-Medina, I., Rodríguez-Leyva, F., Vargas-Arispuro, I., Islas-Osuna, M., Acedo-Félix, E., Martínez-Téllez, M. 2006. PCR identification of *Salmonella*: Potential contamination sources from production and postharvest handling of cantaloupes. *J. Food Prot.* 69:1422–1425.

Eyigdor, A., Carlik, K., Unal, C. 2002. Implementation of real-time PCR to tetrathionate broth enrichment step of *Salmonella* detection in poultry. *Lett. Appl. Microbiol.* 34:37–41.

Eyles, M., Moss, R., Hocking, A. 1989. Microbiological status of Australian flour and the effects of milling procedures on the microflora of wheat and flour. *Food Aust.* 1989:704–708.

Fakhr, M., Nolan, L., Logue, C. 2005. Multilocus sequence typing lacks the discriminatory ability of pulsed-field gel electrophoresis for typing *Salmonella enterica* serovar Typhimurium. *J. Clin. Microbiol.* 43:2215–2219.

Ferretti, R., Mannazzu, I., Cocolin, L., Comi, G., Clementi, F. 2001. Twelve-hour PCR-based method for detection of *Salmonella* spp. in food. *Appl. Environ. Microbiol.* 67:977–978.

Fey, A., Eichleer, S., Flavier, S., Christen, R., Hofle, M., Guzman, C. 2004. Establishment of a real-time PCR-based approach for accurate quantification of bacterial RNA targets in water, using *Salmonella* as a model organism. *Appl. Environ. Microbiol.* 70:3618–3623.

Fluit, A., Widjojoatmodjo, M., Box, A., Torensma, R., Verhoef, J. 1993. Rapid detection of *Salmonella* in poultry with the magnetic immuno-polymerase chain reaction assay. *Appl. Environ. Microbiol.* 59:1342–1346.

Frana, T., Carlson, S., Griffith, R. 2001. Relative distribution and conservation of genes encoding aminoglycoside-modifying enzymes in *Salmonella enterica* serotype *typhimurium* phage type DT104. *Appl. Environ. Microbiol.* 67:445–448.

Fukushima, H., Katsube, K., Hata, Y., Ryoko, K., Fujiwara, S. 2007. Rapid separation and concentration of food-borne pathogens in food samples prior to quantification by viable-cell counting and real-time PCR. *Appl. Environ. Microbiol.* 73:92–100.

Galanis, E., Danilo, M., Wong, L., Patrick, D., Binsztein, M., Cieslik, N., Chalermchaikit, T., Aidara-Kane, A., Ellis, A., Angulo, F., Wegener, H., 2006. Web-based surveillance and global *Salmonella* distribution, 2000–2002. *Emerg. Infect. Dis.* 12:381–388.

Garaizar, J., López-Molina, N., Laconcha, I., Baggesen, D., Rementeria, A., Vivanco, A., Audicana, A., Perales, I. 2000. Suitability of PCRS fingerprinting, infrequent-restriction-site PCRS, and pulsed-field gel electrophoresis, combined with computerized gel analysis, in library typing of *Salmonella enterica* serovar Enteritidis. *Appl. Environ. Microbiol.* 66:5273–5281.

Gaul, S., Wedel, S., Erdman, M., Harris, D., Harris, I., Ferris, K., Hofman, L. 2007. Use of pulsed-field gel electrophoresis of conserved XbaI fragments for identification of swine *Salmonella* stereotypes. *J. Clin. Microbiol.* 45:472–476.

Guerra, B., Laconcha, I., Soto, S., González-Hevia, M., Mendoza, M. 2000b. Molecular characterization of emergent multiresistant *Salmonella enterica* serotype [4,5,12:i:-] organisms causing salmonellosis. *FEMS Microbiol. Lett.* 190:341–347.

Guerra, B., Soto, S., Argueles, J., Mendoza, M. 2001. Multidrug resistance is mediated by large plasmids carrying a class integron 1 in the emergent *Salmonella enterica* serotype [4,5,12:i:-] *Antimicrob. Agents Chemother.* 45:1305–1308.

Guerra, B., Soto, S., Cal, S., Mendoza, M. 2000a. Antimicrobial resistance and spread of class 1 integrons among *Salmonella* serotypes. *Antimicrob. Agents Chemother.* 44:2166–2169.

Guerra, B., Soto, S., Helmuth, R., Mendoza, M. 2002. Characterization of a self-transferable plasmid from *Salmonella enterica* serotype Typhimurium clinical isolates carrying two integron-borne gene cassettes together with virulence and drug resistance genes. *Antimicrob. Agents Chemother.* 46:2977–2981

Guo, X., Chen, J., Beuchat, L., Brackett, R. 2000. PCR detection of *Salmonella enterica* serotype Montevideo in and on raw tomatoes using primers derived from *hilA*. *Appl. Environ. Microbiol.* 66:5248–5252.

Harbottle, H., White, D., McDermott, P., Walker, R., Zhao, S. 2006. Comparison of multilocus sequence typing, pulsed-field gel electrophoresis, and antimicrobial susceptibility typing for characterization of *Salmonella enterica* serotype Newport isolates. *J. Clin. Microbiol.* 44:2449–2457.

Hein, I., Flekna, G., Krassnig, M., Wagner, M. 2006. Real-time PCR for the detection of *Salmonella* spp. in food: An alternative approach to a conventional PCR system suggested by the FOOD-PCR project. *J. Microbiol. Meth.* 66:538–547.

Heyndrickx, M., Herman, L., Vlaes, L., Butzler, J., Wildemauwe, C., Godard, C., De Zutter, L. 2007. Multiple typing for the epidemiological study of the contamination of broilers with *Salmonella* from the hatchery to the slaughterhouse. *J. Food Prot.* 70:323–334.

Hilton, A., Banks, J., Penn, C. 1996. Random amplification of polymorphic DNA (RAPD) of *Salmonella*: Strain differentiation and characterization of amplified sequences. *J. Appl. Bacteriol.* 81:575–584.

Hong, Y., Berang, M., Liu, T., Hofacre, C., Sanchez, S., Wang, L., Maurer, J. 2003. Rapid detection of *Campylobacter coli*, *C. jejuni*, and *Salmonella enterica* on poultry carcasses by using PCR-enzyme-linked immunosorbent assay. *Appl. Environ. Microbiol.* 69:3492–3499.

Hoorfar, J., Ahrens, P., Rådström, P. 2000. Automated 5' nuclease assay for identification of *Salmonella enterica*. *J. Clin. Microbiol.* 38:3429–3435.

Hudson, C., Garcia, M., Gast, R., Maurer, J. 2001. Determination of close genetic relatedness of the major *Salmonella enteritidis* phage types by pulse-field gel electrophoresis and DNA sequence analysis of several *Salmonella* virulence genes. *Avian Dis.* 45:875–886.

Jofré, A., Martin, B., Garriga, M., Hugas, M., Pla, M., Rodríquez-Lázaro, D., Aymerich, T. 2005. Simultaneous detection of *Listeria monocytogenes* and *Salmonella* by multiplex PCR in cooked ham. *Food Microbiol.* 22:109–115.

Johnston, L., Elhanafi, D., Drake, M., Jaykus, L. 2005. A simple method for the direct detection of *Salmonella* and *Escherichia coli* O157:H7 from raw alfalfa sprouts and spent irrigation water using PCR. *J. Food Prot.* 68:2256–2263.

Jones, J., Law, R., Bej, A. 1993. Detection of *Salmonella* spp. in oysters using polymerase chain reaction (PCR) and gene probes. *J. Food Sci.* 58:1191–1197.

Kahn, M., Kim, C., Kakoma, I., Morin, D., Hansen, R., Hurley, D., Jothikumar, N., Wang, X., Griffiths, M. 2003. Real-time multiplex SYBR Green I-based PCR assay for simultaneous detection of *Salmonella* serovars and *Listeria monocytogens*. *J. Food Prot.* 66:2141–2145.

Kardos, G., Farkas, T., Antal, M., Nógrády, N., Kiss, I. 2007. Novel PCR assay for identification of *Salmonella enterica* serovar infantis. *Lett. Appl. Microbiol.* 45:421–425.

Karns, J., Kessel, J., Mccluskey, B., Perdue, M. 2005. Prevalence of *Salmonella enterica* in bulk tank milk from US dairies as determined by polymerase chain reaction. *J. Dairy Sci.* 88:3475–3479.

Kawasaki, S., Horikoshi, N., Okada, Y., Takeshita, K., Sameshima, T., and Kawamoto, S. 2005. Multiplex PCR for simultaneous detection of *Salmonella* spp., *Listeria monocytogenes*, and *Escherichia coli* O157:H7 in meat samples. *J. Food Prot.* 68:551–556.

Kessel, J., Kars, J., Perdue, M. 2003. Using a portable real-time PCR assay to detect *Salmonella* in raw milk. *J. Food Prot.* 66:1762–1767.

Khan, A., Nawaz, M., Khan, S., Cerniglia, C. 2000. Detection of multidrug-resistant *Salmonella typhimurium* DT104 by multiplex polymerase chain reaction. *FEMS Microbiol. Lett.* 182:355–360.

Khan, M., Kim, C., Kakoma, I., Morin, D., Hansen, R., Hurley, W., Tripathy, D., Baek, B. 1998. Detection of *Staphylococcus aureus* in milk by use of polymerase chain reaction analysis. *Am. J. Vet. Res.* 59:807–813.

Kim, C., Khan, M., Morin, D., Hurley, W., Tripathy, D., Kehrli, Jr., M., Oluoch, A., Kakoma, I. 2001. Optimization of the PCR for detection of *Staphylococcus aureus nuc* gene in bovine milk. *J. Dairy Sci.* 84:74–83.

Kim, J., Demeke, T., Clear, R., Patrick, S. 2006a. Simultaneous detection of PCR of *Escherichia coli*, *Listeria monocytogenes* and *Salmonella typhimurium* in artificially inoculated wheat grain. *Int. J. Food Microbiol.* 111:21–25.

Kim, S., Frye, J., Hu, J., Fedorka-Cray, P., Gautom, R., Boyle, D. 2006b. Multiplex PCR-based method for identification of common clinical serotypes of *Salmonella enterica* subsp. *enterica*. *J. Clin. Microbiol.* 44:3608–3615.

Klerks, M., van Bruggen, A., Zijlstra, C., Donnikov, M. 2006. Comparison of methods of extracting *Salmonella enterica* serovar Enteritidis DNA from environmental subsrates and quantification of organisms by using a general internal procedural control. *Appl. Environ. Microbiol.* 72:3879–3886.

Kostman, J., Edlind, T., LiPuma, J., Stull, T. 1992. Molecular epidemiology of *Pseudomonas cepacia* determined by polymerase chain reaction ribotyping. *J. Clin. Microbiol.* 30:2084–2087.

Kotetishvili, M., Stine, O., Kreger, A., Morris, J., Jr., Sulakvelidze, A.' 2002. Multilocus sequence typing for characterization of clinical and environmental *Salmonella* strains. *J. Clin. Microbiol.* 40:1626–1635.

Kubota, K., Barrett, T., Ackers, M., Brachman, P., Mintz, E. 2005. Analysis of *Salmonella enterica* serotype Typhi pulsed-field gel electrophoresis patterns associated with international travel. *J. Clin. Microbiol.* 43:1205–1209.

Kwang, J., Littledike, E., Keen, J. 1996. Use of the polymerase chain reaction for *Salmonella* detection. *Lett. Appl. Microbiol.* 22:46–51.

Lampel, K., Orlandi, P., Kornegay, L. 2000. Improved template preparation for PCR-based assays for detection of food-borne bacterial pathogens. *Appl. Environ. Microbiol.* 66:4539–4542.

Levèsque, C., Pyche, L., Larose, C., Roy, P. 1995. PCR mapping integrons reveals several novel combinations of resistance genes. *Antimicrob. Agents Chemother.* 39:185–191.

Li, X., Boudjeliab, N., Zhao, X. 2000. Combined PCR and slot blot assay for detection of *Salmonella* and *Listeria monocytogenes*. *Int. J. Food Microbiol.* 56:167–177.

Li, Y., Mustapha, A. 2004. Simultaneous detection of *Escherichia coli* O157:H7, *Salmonella*, and *Shigella* in apple cider and produce by a multiplex PCR. *J. Food Prot.* 67:27–33.

Li, Y., Zuang, S., Mustapha. 2005. Application of a multiplex PCR for the simultaneous detection of *Escherichia coli* O157:H7, *Salmonella* and *Shigella* in raw and ready-to-eat meat products. *Meat Sci.* 71:402–406.

Liao, C., Shollenberger, L. 2003. Detection of *Salmonella* by indicator agar media and PCR as affected by alfalfa seed homogenates and native bacteria. *Lett. Appl. Microbiol.* 36:152–156.

Lim, H., Lee, K., Hong, C., Bahk, G., Choi, W. 2005. Comparison of four molecular typing methods for the differentiation of *Salmonella* spp. *Int. J. Food Microbiol.* 105:411–418.

Lin, A., Usera, M., Barret, T., Goldsby, R. 1996. Application of random amplified polymorphic DNA analysis to differentiate strains of *Salmonella enteritidis*. *J. Clin. Microbiol.* 34:870–876.

Lin, C., Hung, C., Hsu, S., Tsai, C., Tsen, H. 2004. An improved PCR primer pair based on 16S rDNA for the specific detection of *Salmonella* serovars in food samples. *J. Food Prot.* 67:1335–1343.

Lin, C., Tsen, H. 1995. Development and evaluation of two novel oligonucleotide probes based on 16S rRNA sequence for the identification of *Salmonella* in food. *J. Appl. Bacteriol.* 78:507–520.

Liu, Y., Lee, M., Ooi, E., Mavis, Y., Tan, A., Quek, H. 2003. Molecular typing of *Salmonella enterica* serovar Typhi isolates from various countries in Asia by a multiplex PCR assay on variable-number tandem repeats. *J. Clin. Microbiol.* 41:4388–4394.

Löfström, C., Knutsson, R., Axelsson, C., Rådström, P. 2004. Rapid and specific detection of *Salmonella* spp. in animal feed samples by PCR after culture enrichment. *Appl. Environ. Microbiol.* 70:69–75.

Mahon, J., Murphy, C., Jones, P., Barrow, P. 1994. Comparison of multiplex PCR and standard bacteriological methods of detecting *Salmonella* on chicken skin. *Lett. Appl. Microbiol.* 19:169–172.

Malorny, B., Bunge, C., Helmuth, R. 2003. Discrimination of d-tartrate-fermenting and non-fermenting *Salmonella enterica* subsp. *enterica* isolates by genotypic and phenotypic methods. *J. Clin. Microbiol.* 41:4292–4297.

Malorny, B., Bunge, C., Helmuth, R. 2007. A real-time PCR for detection of *Salmonella* Enteritidis in poultry meat and consumption eggs. *J. Microbiol. Meth.* 70:245–251.

Manzano, M., Cocolin, I., Astori, G., Pipan, C., Botta, G., Cantoni, C., Comi, G. 1998. Development of a PCR microplate-capture hybridization method for simple, fast and sensitive detection of *Salmonella* serovars in food. *Mol. Cell Probes* 12:227–234.

Maré, L., Dick, L., van der Walt, M. 2001. Characterization of South African isolates of *Salmonella enteritidis* by phage typing, numerical analysis of RAPD-PCR banding patterns and plasmid profiles. *Int. J. Food Microbiol.* 64:237–245.

Martinez-Urtaza, J., Echeita, A., Liebana, E. 2006. Phenotypic and genotypic characterization of *Salmonella enterica* serotype Paratyphi B isolates from environmental and human sources in Galicia, Spain. *J. Food Prot.* 69:1280–1285.

Martinez-Urtaza, J., Liebana, E. 2005. Use of pulsed-field gel electrophoresis to characterize the genetic diversity and clonal persistence of *Salmonella senftenberg* in mussel processing facilities. *Int. J. Food Microbiol.* 105:153–163.

Mercanoglu, B., Griffiths, M. 2005. Combination of immumomagnetic separation with real-time PCR for rapid detection of *Salmonella* in milk, ground beef, and alfalfa sprouts. *J. Food Prot.* 68:557–561.

Michael, G., Cardosa, M., Rabsch, W., Schwarz, S. 2006. Phenotpic and genotypic differentiation of porcine *Salmonella enterica* subsp. *enterica* serovar Derby isolates. *Vet. Microbiol.* 118:312–318.

Millemann, Y., Lesage-Descauses, M., Lafont, J., Chaslus-Dancia, E. 1996. Comparison of random amplified polymorphic DNA analysis and enterobacterial repetitive intergenic consensus-PCR for epidemiological studies of *Salmonella*. *FEMS Immunol. Med. Microbiol.* 14:129–134.

Miyata, M., Acki, T., Inglis, V., Yoshida, T., Endo, M. 1995. RAPD analysis of *Aeromonas salmonicida* and *Aeromonas hydrophila*. *J. Appl. Bacteriol.* 79:181–185.

Myint, M., Johnson, Y., Tablante, N., Heckert, R. 2006. The effect of pre-enrichment protocol on the sensitivity and specificity of PCR for detection of naturally contaminated *Salmonella* in raw poultry compared to conventional culture. *Food Microbiol.* 23:599–604.

Nam, H., Srinivasan, V., Gillespie, B., Murinda, S., Oliver, S. 2005. Application of SYBR green real-time PCR assay for specific detection of *Salmonella* spp. in dairy farm environmental samples. *Int. J. Food Microbiol.* 102:161–171.

Nde, C., Sherwwod, J., Doetkott, C., Logue, C. 2006. Prevalence and molecular pro-files of *Salmonella* collected at a commercial turkey processing plant. *J. Food Prot.* 69:1794–1801.

Ng, L., Mulvey, M., Martin, I., Petters, G., Johnson, W. 1999. Genetic characterization of anti-microbial resistance in Canadian isolates of *Salmonella* serovar Typhimurium DT104. *Antimicrob. Agents Chemother.* 43:3018–3021.

Notzon, A., Helmuch, R., Bauer, J. 2006. Evaluation of an immunomagnetic separation-real-time PCR assay for the rapid detection of *Salmonella* in meat. *J. Food Prot.* 69:2896–2901.

Nowak, B., Müffling, T., Chaunchom, S., Hartung, J. 2006. *Salmonella* contamination in pigs at slaughter and on the farm: A field study using an antibody ELISA test and PCR tech-nique. *Int. J. Food Microbiol.* 20:259–267.

Nucera, D., Maddox, C., Hoien-Dalen, P., Weigel, R. 2006. Comparison of API 20E and *invA* PCRf or identification of *Salmonella enterica* isolates from swine production units. *J. Clin. Microbiol.* 44:3388–3390.

Oikonomou, I., Halatsi, K., Kyriacou, A. 2008. Selective PCR: A novel internal amplification control strategy for enhanced sensitivity in *Salmonella* diagnosis. *Lett. Appl. Microbiol.* 46:456–461.

Piknova, L., Kaclikova, E., Pangallo, D., Polek, B., Kuchta, T. 2005. Quantification of *Salmonella* by 5′ nuclease real-time polymerase chain reaction targeted to *fimC* gene. *Curr. Microbiol.* 50:38–42.

Pitcher, D., Saunders, N., Owen, R. 1989. Rapid extraction of bacterial genomic DNA with guanidium thiocyanate. *Lett. Appl. Microbiol.* 8:151–156.

Prager, R., Fruth, A., Tschäpe, H. 1995. *Salmonella* enterotoxin (stn) gene is prevalent among strains of *Salmonella enterica*, but not among *Salmonella* bangdori and other Enterobacteriaceae. *FEMS Immunol. Med. Microbiol.* 12:47–50.

Rahn, K., De Grandis, S., Clarke, R., McEwen, S., Galan, J., Ginocchio, C., Curtis, III, R., Gyles, C. 1992. Amplification of an *invA* gene sequence of *Salmonella enterica* by poly-merase chain reaction as a specific method of detection of *Salmonella. Mol. Cell. Probes* 6: 271–279.

Richter, K., Dorneanu, E., Eskridge, K., Rao, C. 1993. Microbiological quality of flours. *Cereal Foods World* 1993:367–369.

Rychlik, I., van Kesteren, L., Cardová, L., Sveskova, A., Martinkova, R., Sisák, F. 1999. Rapid detection of *Salmonella* in field samples by nested polymerase chain reaction. Lett. Appl. Microbiol. 29:269–272.

Sandt, C., Krouse, D., Cook, C., Hackman, A., Chmielecki, W., Warren, N. 2006. The key role of pulsed-field gel electrophoresis in investigation of a large multiserotype and multi-state food-borne outbreak of *Salmonella* infections centered in Pennsylvania. *J. Clin. Microbiol.* 44:3208–3212.

Sandvang, D., Aerestrup, F., Jensen, L. 1997. Characterization of integrons and antibiotic resis-tance genes in Danish multiresistant *Salmonella typhimurium* DT104. *FEMS Microbiol. Lett.* 160:37–41.

Seo, K., Valentin-Bon, I., Brackett, R. 2006. Detection and enumeration of *Salmonella* Enteritidis in home made ice cream associated with an outbreak: Comparison of con-ventional and real-time PCR methods. *J. Food Prot.* 69:639–643.

Seo, K., Valentin-Bon, I., Brackett, R., Holt, P. 2004. Rapid, specific detection of *Salmonella* Enteritidis in pooled eggs by real-time PCR. *J. Food Prot.* 67:864–869.

Shabarinath, S., Kumar, H., Khushiramani, R., Karunasagar, I. 2006. Detection and characteriza-tion of *Salmonella* associated with tropical seafood. *Int. J. Food Microbiol.* 114:227–233.

Sharma, V., Carlson, S. 2000. Simultaneous detection of *Salmonella* strains and *Escherichia coli* O157:H7 with fluorogenic PCR and single enrichment-broth culture. *Appl. Environ. Microbiol.* 66:5472–5476.

Silbert, S., Oyken, L., Hollis, R., Pfaller, M. 2003. Improving typeability of multiple bacterial species using pulsed-field gel electrophoresis and thiourea. *Diag. Microbiol. Infect. Dis.* 47:619–621.

Soto, S., Guerra, B., González-Hevia, M., Mendoza, M. 1999. Potential of three-way randomly amplified polymorphic DNA analysis as a typing method for twelve *Salmonella* serotypes. *Appl. Environ. Microbiol.* 65:4830–4836.

Soto, S., Martinez, N., Guerra, B., González-Hevia, M., Mendoza, M. 2000. Usefulness of genetic typing methods to trace epidemiologically *Salmonella* serotype Ohio. *Epidemiol. Infect.* 125:481–489.

Soumet, C., Ermel, G., Fach, P., Colin, P. 1994. Evaluation of different DNA extraction procedures for the detection of *Salmonella* from chicken products by polymerase chain reaction. *Lett. Appl. Microbiol.* 19:294–298.

Soumet, C., Ermel, G., Rose, V., Rose, N., Drouin, P., Salvat, G., Colin, P. 1999. Identification by a multiplex PCR-based assay of *Salmonella* Typhimurium and *Salmonella* Enteritidis strains from environmental swabs of poultry houses. *Lett. Appl. Microbiol.* 29:1–6.

Stone, G., Oberst, R., Hays, M., McVey, S., Chengappa, M. 1994. Detection of *Salmonella* serovars from clinical samples by enrichment broth cultivation PCR procedure. *J. Clin. Microbiol.* 32:1742–1749.

Stone, G., Oberst, R., Hays, M., McVey, S., Chengappa, M. 1995. Combined PCR-oligonucleotide ligation assay for rapid detection of *Salmonella* serovars. *J. Clin. Microbiol.* 33:2888–2893.

Touron, A., Berthe, T., Pawlak, B., Petit, F. 2005. Detection of *Salmonella* in environmental water and sediment by a nested polymerase chain reaction assay. *Res. Microbiol.* 156:541–553.

Trkov, M., Avguštin, G. 2003. An improved 16S rRNA based PCR method for the specific detection of *Salmonella enterica*. *Int. J. Food Microbiol.* 80:67–75.

Tsen, H., Liou, J., Lin, C. 1994. Possible use of a polymerase chain reaction method for specific detection of *Salmonella* in beef. *J. Ferment. Bioeng.* 77:137–143.

Turcotte, C., Woodward, J. 1993. Cloning, DNA nucleotide sequence and distribution of the gene coding the SEF14 fimbrial antigen of *Salmonella* Enteritidis. *J. Gen. Microbiol.* 139:1477–1485.

Van Duijkeren, E., Wannet, W., Houwers, D., van Pelt, W. 2002. Serotype and phage type distribution of *Salmonella* strains isolated from humans, cattle, pigs, and chickens in the Netherlands from 1984 to 2001. *J. Clin. Microbiol.* 40:3980–3985.

Vantarakas, A., Komninou, G., Venieri, D.K., Papapetropoulou, M. 2000. Development of a multiplex PCR detection of *Salmonella* spp. and *Shigella* spp. in mussels. *Lett. App. Microbiol.* 31:105–109.

Vázquez-Novelle, M., Pazos, A., Abad, M., Sánchez, J., Pérez-Parallé, M. 2005. Eight-hour PCR-based procedure for the detection of *Salmonella* in raw oysters. *FEMS Microbiol. Lett.* 243:279–283.

Vieira-Pinto, M., Tenreiro, R., Martins, C. 2006. Unveiling contamination sources and dissemination routes of Salmonella spp. in pigs at a Portuguese slaughterhouse through macrorestriction profiling by pulsed-field gel electrophoresis. *Intl. J. Food Microbiol.* 110:77–84.

Vila, J., Ruiz, J., Goñi, P., Marcos, A., Jiménez de Anta, T. 1995. Mutations in the *gyrA* gene of quinolone-resistant clinical isolates of *Acinetobacter baumanii*. *Antimicrob. Agents Chemother.* 39:1201–1203.

Waage, A., Vardund, T., Lund, V., Kapperud, G. 1999. Detection of low numbers of *Salmonella* in environmental water, sewage, and food samples by a nested polymerase chain reaction assay. *J. Appl. Microbiol.* 87:418–428.

Wang, X., Jothikumar, N., Griffiths, M. 2004. Enrichment and DNA extraction protocols for the simultaneous detection of *Salmonella* and *Listeria monocytogenes* in raw sausage meat with multiplex real-time PCRS. *J. Food Prot.* 67:189–192.

Way, J., Josephson, K., Pillai, S., Abbaszadagan, M., Gesba, C., Pepper, I. 1993. Specific detection of *Salmonella* spp. by multiplex polymerase chain reaction. *Appl. Environ. Microbiol.* 59:1473–1479.

Widjojoatmodjo, M., Fluit, A., Torensma, R., Keller, B., Verhoef, J. 1991. Evaluation of the magnetic immuno PCR assay for rapid detection of *Salmonella*. *Eur. J. Clin. Microbiol. Infect. Dis.* 10:935–938.

Williams, J., Kubelin, A., Livak, K., Rafalski, J., Tingey, S. 1990. DNA polymorphisms amplified by arbitrary primers useful as genetic markers. *Nucl. Acid Res.* 18:6531–6535.

Wolffs, P., Glencross, K., Thibaudeau, R., Griffiths, M. 2006. Direct quantification and detection of salmonellae in biological samples without enrichment, using two-step filtration and real-time PCR. *Appl. Environ. Microbiol.* 72:3896–3900.

Wonderling, L., Pearce, R., Wallace, R., Call, J., Feder, I., Tamplin, M., Luchansky, J. 2003. Use of pulsed-field gel electrophoresis to characterize the heterogeneity and clonality of *Salmonella* isolates obtained from the carcasses and feces of swine at slaughter. *Appl. Environ. Microbiol.* 69:4177–4182.

Zhu, Q., Lim, C., Chan, Y. 1996. Detection of *Salmonella typhi* by polymerase chain reaction. *J. Appl. Bacteriol.* 80:244–251.

CHAPTER 5

Vibrio vulnificus

I. CHARACTERISTICS OF THE ORGANISM

Vibrio vulnificus is a gram-negative, nonspore-forming, polarly flagellated, facultatively anaerobic short rod, capable of fermenting glucose to acid but not gas. The organism is considered the most serious and invasive of all human pathogenic vibrios in the United States, accounting for 95% of all seafood-related deaths in this country (Oliver, 1989). Three biotypes are presently recognized and distinguished on the basis of biochemical characteristics, serology, and molecular typing. *V. vulnificus* requires at least 0.5% NaCl for growth and has been found to be a natural inhabitant of marine coastal waters and to be globally ubiquitous (Kaysner et al., 1987; Kelly, 1982; Oliver, 1989, O'Neill, Jones, and Grimes, 1992; Tamplin et al., 1982). It has been isolated from coastal waters and shellfish of Massachusetts and Florida (Oliver, Warner, and Cleland, 1983) in addition to U.S. Pacific coastal areas. An average of 6 × 10⁴ CFU/g has been reported for oysters (Oliver and Kaper, 1997).

Elevated counts have been found to be correlated with elevated coastal water temperature during the summer months (Kelly, 1982; Oliver, 1989; Tilton and Ryan, 1987; Pfeffer, Hite, and Oliver, 2003). The organism is of particular concern along the warm coastal waters of the United States Gulf coast. Primary septicemic infections due to this organism resulting from the consumption of raw oysters can result in fatality rates as high as 60% (Oliver, 1989). The most frequent symptoms include fever, chills, nausea, hypotension, and endotoxic shock, which are usually associated with endotoxicity derived from gram-negative lipopolysaccharides (LPS; McPherson et al., 1991). Symptoms of gastroenteritis (abdominal pain, vomiting, and diarrhea) occur in less than half the cases (Oliver and Kaper, 1997). The incubation period has been found to vary from several hours to several days with a median of 26 hours (Oliver, 1989). The organism was originally referred to in 1976 as the "lactose fermenting vibrio" (Hollis et al., 1976). However, Bisharat et al. (1999) reported that lactose fermentation can vary. Liver damage or cirrhosis, such as arising from chronic alcoholism, chronic renal disease, diabetes, and immunocompromising diseases, are considered major factors in susceptibility (Desenclos et al., 1991; Oliver,

1989; Johnston, Becker, and Mcfarland, 1986) and are thought to be responsible for the phenomenon wherein outbreaks involving the consumption of oysters from a specific lot usually involve only a single susceptible individual developing symptoms (Oliver and Kaper, 1997). Secondary necrotic lesions of the extremities frequently occur (69%) often necessitating surgical debridement or limb amputation (Oliver and Kaper, 1997).

The organism produces an unusually large number of extracellular virulence factors. A number of selective agar media have been developed for isolation of the organism incorporating various levels of the antibiotics colistin and polymyxin-B, in addition to bile salts and K tellurite. The molecular techniques applied to detecting, identifying, and characterizing the organism include the polymerase chain reaction (PCR), real-time PCR, random amplified polymorphic DNA analysis (RAPD), multiplex PCR, ribotyping, multilocus enzyme electrophoresis (MLEE), PCR-restriction fragment length polymorphism (PCR-RFLP), and gene sequence typing. The heat-stable hemolysin-cytolysin gene *cth* has been used most frequently for the specific PCR detection and identification of *V. vulnificus*. In addition, semi-nested reverse transcription-PCR (RT-PCR) has been used to detect viable but nonculturable (VBNC) cells of *V. vulnificus* utilizing primers that amplified a sequence of the *vvhA* gene, which encodes the cytolysin structural gene product.

II. VIRULENCE FACTORS

A. Capsule Production

All virulent strains of *V. vulnificus* were found by Simpson et al. (1987) to produce opaque colonies derived from encapsulated cells, whereas nonencapsulated spontaneous mutants of the same strains were found to produce translucent (noncapsular) colonies that were avirulent. Nonencapsulated translucent colony types were found not to revert to the capsular-virulent type. Biosca et al. (1993b) studied spontaneous translucent colony mutants of *V. vulnificus* derived from opaque colonies from infected eels. The rate of spontaneous mutation from opaque to translucent colonies was about 100-fold higher than that observed for translucent to opaque colonies. The LD_{50} values resulting from intraperitoneal (I.P.) injection of elver (young) eels was 1 to 3 log units higher with translucent colony types than with opaque mutants. The authors concluded that the capsule of the opaque form increases pathogenicity for eels by performing a protective function.

Yoshida, Ogawa, and Mizaguchi (1985) observed that the frequency of spontaneous mutation of opaque to translucent colony mutation was higher than for mutation from the translucent to the opaque colony form. Encapsulated opaque colony-forming strains were found to be notably more resistant to the bactericidal action of antiserum. Opaque colony types exhibited LD_{50} values for intravenous (I.V.) injection of mice that were 1 to 3 log cycles lower than those obtained with translucent colony producing strains.

B. Extracellular Virulence Factors

V. vulnificus produces a large number of extracellular factors that are considered to contribute to its virulence including hemolysins (Moreno and Landgraf, 1998), proteases (Desmond et al. 1984; Kreger and Lockwood, 1981; Oliver, Thomas, and Wear, 1986), elastase (Kothary and Kreger, 1985; Oliver, Thomas, and Wear, 1986; Moreno and Landgraf, 1998), collagenase (Poole, Bowdre, and Klapper, 1982; Smith and Merkel, 1982; Oliver, Thomas, and Wear, 1986), DNase (Desmond et al., 1984; Kreger and Lockwood, 1981; Oliver, Thomas, and Wear, 1986; Moreno and Landgraf, 1998), lipase (Desmond et al., 1984; Oliver, Thomas, and Wear, 1986; Tison et al., 1982; Moreno and Landgraf, 1998), phospholipase (Desmond et al., 1984; Testa, Daniel, and Kreger, 1984; Tison et al., 1982), mucinase (Oliver, Thomas, and Wear, 1986; Moreno and Landgraf, 1998), fibrinolysin (Oliver, Thomas, and Wear, 1986), chondroitin sulfatase (Oliver, Thomas, and Wear, 1986), hyaluronidase (Oliver, Thomas, and Wear, 1986), elastase (Moreno and Landgraf, 1998), and an alkaline sulfatase (Kitaura et al., 1983; Oliver, Thomas, and Wear, 1986). A heat-stable hemolysin is produced that exhibits cytolytic activity against a variety of mammalian erythrocytes, cytotoxic activity to CHO cells, vascular permeability to guinea pig skin, and lethality for mice (Kreger and Lockwood, 1981) and has been utilized for molecular identification of the organism. This cytolysin–hemolysin toxin is encoded by the *cth* gene, has a molecular weight of about 56,000 Da (Gray and Kreger, 1985a) but by itself cannot account entirely for the organism's pathogenicity. Strains have been studied that produce the hemolysin–cytolysin toxin that are not virulent (Johnson and Calia, 1981; Johnson et al., 1984). Glucose represses the production of the hemolysin-cytotoxin (Oliver, 1989); however, its activity is notably enhanced by sphingomyelinase on sheep blood agar plates (Levin, unpublished observations). The elastase is a protease that degrades elastin, casein, albumin, immunoglobulin G, and complement factors C3 and C4 (Poole, Bowdre, and Klapper, 1982; Kothary and Kreger, 1985).

Wright et al. (1995) cloned the *cth* gene of *V. vulnificus* into *E. coli* HB101 and found filtered sonic extracts to be toxic to CHO cells and hemolytic for rabbit erythrocytes. Both activities were neutralized by antisera to the crude *V. vulnificus* toxin and to purified cytolysin. The cloned toxin appeared identical to the 56,000-Da cytolysin described by Gray and Kreger (1985a).

The cytolysin gene of *V. vulnificus* was sequenced in 1990 by Yamamoto et al. (1990) and was found to contain two open reading frames designated vvhA and vvhB. VvhA was found to encode the cytolysin structural gene, which included a region that encodes the sequence for the 10 N-terminal amino acids of the ca. 56-kDa cytolysin reported by Gray and Kreger (1985a), whereas no known gene product was associated with vvhB. Certain sequence regions of the structural gene of the *V. vulnificus* cytolysin were found to greatly resemble sequence regions of the *V. cholerae* El Tor hemolysin structural gene, suggesting that the two genes possess a common evolutionary origin.

C. Serum-Iron Availability

Liver disease such as cirrhosis arising from chronic alcoholism is considered a high risk factor for infection by this organism (Tacket, Brenner, and Blake, 1984), presumably due to increased levels of serum iron released by damaged hepatocytes. The major iron-binding protein in serum is transferrin, normally resulting in about 10^{-18} M free iron in serum, which is far below the 10^{-6} M required for growth of *V. vulnificus* (Oliver, 1989). Chronic liver damage may result in an iron overload of the iron-binding capacity of serum transferrin, resulting in sufficiently elevated levels of free serum iron to allow septicemic growth of *V. vulnificus*. Liver damage resulting in elevated levels of iron in blood serum has been found to reduce the LD_{50} dose in mice from about 10^6 to 1 CFU (Wright, Simpson, and Oliver 1981). This observation was later confirmed by Stelma et al. (1992), who simultaneously injected noncompromised mice with iron compounds and strains of *V. vulnificus*.

III. ENRICHMENT AND ISOLATION MEDIA

A. Alkaline Peptone Salt Broth (APS)

Alkaline peptone salt broth is widely used for enrichment cultivation of *V. vulnificus* and is useful for MPN enumeration followed by streaking onto one of the selective agar media described below. The medium consists of 1% peptone plus 3.0% NaCl. The pH is adjusted to 8.5 and it therefore imparts some degree of selectivity for alkaline pH tolerant marine vibrios such as *V. vulnificus*.

B. Colistin-Polymyxin-β-Cellobiose (CPC) Agar

Massad and Oliver (1987) found CPC Agar (Table 5.3) to be highly selective for *V. vulnificus*. Among 17 species of *Vibrio* tested, only *V. vulnificus* and *V. cholerae* were able to grow on this medium. Cellobiose utilization by *V. vulnifiucus* resulting in yellow colonies distinguishes the organism from *V. cholerae*. The mCPC agar has the same composition as CPC agar except that colistin methanesulphonate is reduced from 1.36×10^6 U to 4×10^5 U/L. Cellobiose-colistin (CC) agar (Høi, Dalsgaard, and Dalsgaard, 1998) has the same composition as CPC agar except that it contains no polymyxin B and the concentration of colistin is reduced to 4×10^5 U/L.

C. Thiosulfate-Citrate-Bile-Salts Sucrose (TCBS) Agar

Most isolation studies from marine sources have involved the use of TCBS agar (Table 5.3) to select for *V. vulnificus,* although it is widely recognized that significant numbers of nonvibrios are also capable of growth on this medium (Brayton et al., 1983; Oliver, Warner, and Cleland, 1983; West et al., 1982). Because TCBS agar does not distinguish *V. vulnificus* from other sucrose-negative vibrios, additional selective media have been developed for its isolation. The identification of lactose fermenting vibrios from seafood with TCBS is much more difficult than from clinical sources. This is due to the large number of sucrose-negative, lactose-positive vibrios in marine environments that yield colonies on selective media as TCBS agar that are similar or identical in appearance to those produced by *V. vulnificus* (Oliver, Warner, and Cleland, 1983).

D. *Vibrio vulnificus* (VV) Agar

Brayton et al. (1983) proposed the medium known as *Vibrio vulnificus* or VV agar (Table 5.3) employing tellurite as a selective agent and indicated its superiority over TCBS agar for selective isolation of *V. vulnificus*. However, other pathogenic vibrios produce colonies similar to those of *V. vulnificus* on VV agar and members of several other genera other than *Vibrio* grow well on this medium (Oliver, 1989).

E. Sodium Dodecyl Sulfate-Polymyxin-Sucrose (SPS) Agar

Kitaura et al. (1983) developed SPS agar (Table 5.3) based on the ability of *V. vulnificus* to produce an alkyl sulfatase capable of hydrolyzing sodium dodecyl sulfate and resistance to polymyxin B. Sulfatase-positive isolates produce a halo around the colony. However, many other species of *Vibrio* also produce this enzyme and resulting halo.

F. Cellobiose-Colistin (CC) Agar

Høi, Dalsgaard, and Dalsgaard (1998) developed an improved selective medium for *V. vulnificus* designated CC agar having the same composition as CPC agar except that it contains no polymyxin B and colistin is reduced to 4×10^5 U/L (Table 5.3). These authors based the composition of CC agar on the observation that concentrations of colistin and polymyxin B in CPC agar inhibit the growth of a proportion of *V. vulnificus* strains. CC agar gave a higher plating efficiency of *V. vulnificus* than did CPC agar, mCPC agar, or TCBS agar.

G. *Vibrio vulnificus* Medium (VVM)

Cerdà-Cuéllar, Jofre, and Blanch (2000) proposed a more selective medium for isolation of *V. vulnificus* known as *Vibrio vulnificus* medium (VVM). This medium (Table 5.3) contains the same level of polymyxin B and colistin as CPC agar but in addition has a pH of 8.5 compared to CPC that has a pH of 7.6.

IV. IDENTIFICATION OF *V. VULNIFICUS* ISOLATES

A. Biochemical Characteristics

Identification of isolates of *V. vulnificus* from human infections, environmental sources, and infected eels is usually dependent on biochemical characteristics (Table 5.1). Fermentation of lactose distinguishes this organism from *V. parahaemolyticus*. API-20E strips may incorrectly identify *V. vulnificus* unless a 2.0% marine salts solution is used as the diluent instead of 0.85% saline (Ghosh and Bowen, 1980). Errors in identification of *V. vulnificus* have also been reported with other

Table 5.1   Major Diagnostic Characteristics of *V. vulnificus*

Characteristic	Biotype 1	Biotype 2	Biotype 3
Growth in 0% NaCl	–	–	–
Growth in 0.5% NaCl	+	+	+
Growth in 6.0% NaCl	+	+	+
Growth in 8.0% NaCl	+	–	–
Acid but no gas from glucose	+	+	+
β-Hemolysis of sheep red blood cells	+	+	+
Inhibition by O/125[a]	+	+	+
Lysine decarboxylase	+	+	+
Ornithine decarboxylase	+	–	+
Arginine dihydrolase	–	–	–
Production of indole	+	–	+
Acid from			
Cellobiose	+	+	–
D-galactose			
Lactose[b]	+	+	–
Sucrose[c]	–	–	–
Maltose	+	+	+
D-mannitol	+	–	–
D-sorbitol	–	+	–
Salicin	+	+	–
Citrate	+	+	–

[a] 50 mg/disk (2,4-diamino-6,7-diisopropyl pteridine phosphate).
[b] Fermentation of lactose may be delayed several days or have a delayed ONPG reaction.
[c] 15% of isolates may ferment sucrose (Farmer, Hickman-Brenner, and Kelly, 1985).

commercial systems such as the API Rapid E, API Rapid NFT, and Minitek systems (Chester and Cleary, 1980; Overman, Kessler, and Seabolt, 1985).

B. Serotyping and Immunological Techniques

V. vulnificus isolates possess unique H-antigens (Tassin et al., 1983) allowing flagellar antiserum in slide agglutination assays to distinguish the organism from all other vibrios. *V. vulnificus* possesses a species-specific H antigen located on the flagellum core that is shrouded by the flagellum sheath. Polyclonal antibodies raised to the extracted and purified flagellum core were used by Simonson and Siebeling (1993) for the development of a species-specific agglutination assay for *V. vulnificus* by coating formalized cells of *Staphylococcus aureus* with the anti-H antibody. The detection rate was 99.3%. Distinctive cell surface antigens have also been used to identify *V. vulnificus* with a 100% detection rate and a false-positive incidence of only 0.9% (Gray and Kreger, 1985b).

Tamplin et al. (1991) described the use of a monoclonal antibody (mAb) to a species-specific intracellular protein derived from the injection of mice with whole cells of *V. vulnificus*. Specificity was assessed with 72 bacterial strains representing 15 genera and 34 species. The minimum level of detection was 2000 *V. vulnificus* cells per well.

Simonson and Siebeling (1993) found that purified capsular polysaccharide (CPS) was poorly immunogenic for rabbits and mice. However, conjugation of CPS from three strains of *V. vulnificus* to keyhole limpet hemocyanin (KLH) and coating of *S. aureus* with the resulting CPS antibodies resulted in only three cross-reacting strains among 32 via coagglutination, indicating that numerous capsular serotypes exist.

Marco-Noales et al. (2000) reported on the development of an indirect membrane-based microscopic immunofluorescent LPS-antibody assay for detection and enumeration of *V. vulnificus* serovar E that is able to also detect "viable but non-culturable (VBNC) cells." A minimum of 10^4 cells per black membrane filter could be detected. The assay was considered to have two major advantages over previously developed immunoassays for *V. vulnificus* in that it allows direct cell enumeration and rapid detection of the organism in tissue.

V. TYPING OF *V. VULNIFICUS* ISOLATES BELOW THE SPECIES LEVEL

The major concepts regarding the subspecies classification of *V. vulnificus* isolates have changed considerably during the past two decades and are still in a process of evolution as additional studies are published on new isolates. The major emphasis on subspecies designations has been based on species pathogenicity (humans versus eels), and biochemical and serological observations that have led to present conclusions which contradict certain original subspecies concepts regarding the organism.

A. Biogroups

Muroga, Jo, and Nishibushi (1976) were the first to describe *Vibrio* isolates pathogenic for eels. However, Tison et al. (1982) were the first to allocate *Vibrio* strains pathogenic for eels to the species *V. vulnificus*, which at that time were not associated with pathogenicity for humans. They performed a comparative study of human clinical, environmental, and eel pathogenic isolates of *V. vulnificus* using phenotypic comparison, eel and mouse pathogenicity, and DNA–DNA hybridization studies and concluded that human clinical isolates should be designated as belonging to biogroup 1 and that eel pathogen isolates be designated as belonging to biogroup 2 of the species *V. vulnificus*. Biogroup 2 was phenotypically defined as differing biochemically from biogroup 1 in being negative for indole production, ornithine decarboxylase activity, acid production from mannitol and sorbitol, and growth at 42°C (Table 5.1). Interestingly, they also reported that neither human clinical nor environmental isolates exhibited pathogenicity for eels, whereas all human clinical, environmental, and eel isolates tested resulted in mortality in suckling mice, with eel isolates yielding the highest level of mortality.

Tison and Seidler (1981) found several environmental and human clinical isolates of *V. vulnificus* to be phenotypically indistinguishable and to possess 85% or greater DNA–DNA homology and 45.7% to 47.8% mols% G+C values.

Shimada and Sakazaki (1984) proposed a serovar O serological agglutination typing scheme that recognized seven O serovars. Martin and Siebeling (1991) developed monoclonal antibodies (mAb) specific for epitopes in the LPS fraction of the cell wall using an ELISA assay. The resulting five mAbs reacted with 10 of 17 clinical strains of *V. vulnificus,* indicting that at least six LPS serovars exist. In contrast, polyclonal antisera, produced in rabbits immunized with whole cells, resulted in indiscriminate agglutination of all 17 strains examined. All biotype 2 strains examined were found by Amaro et al. (1992) to possess the same LPS profiles that were immunologically distinct from those of biotype 1 strains. Biosca et al. (1993a) found that the electrophoretic outer membrane profiles (OMP) of Japanese and European biotype 2 strains were almost identical except for slight differences in the relative amounts of some minor bands. Protein bands stained with anti-biotype 2 OMP antisera resulted in identical immunoblots with both biotype 1 and 2 strains, clearly indicating the presence of antigenically related OMP in both biotypes.

Amaro and Biosca (1996) found that a human clinical strain of *V. vulnificus,* originally isolated from a human leg wound, was a biotype 2 isolate and belonged to a recently proposed serogroup E, characteristic of biotype E strains (Biosca, Oliver, and Amaro, 1996). The strain was pathogenic for both eels and mice and was negative for indole production and mannitol utilization. The authors concluded that strains of biotype 2 (eel pathogens) are also opportunistic pathogens for humans.

The LPS of *V. vulnificus* biotype 2 isolates has been found to be immunologically distinct from biotype 1 isolates and from other *Vibrio* species (Biosca, Oliver, and Amaro, 1996). Biosca, Oliver, and Amaro (1996) characterized 17 biotype 1 and 83 biotype 2 strains biochemically and serologically (based on LPS and agglutination) and found the biotype 2 strains to be biochemically homogeneous and distinguishable from biotype 1 strains based on their inability to produce indole. These authors emphasized that the indole reaction is the only reliable biochemical reaction distinguishing between biotypes 1 and 2. The biotype 2 stains were also found to possess a common LPS profile that is immunologically distinct from the LPS of biotype 1 strains. The authors concluded that biotype 2 constitutes a LPS-based O serogroup that is phenotypically homogeneous and pathogenic for eels, which they designated serogroup E (for eels). In contrast, biotype 1 strains were previously found to exhibit a high degree of seroheterogeneity (Martin and Siebeling, 1991; Shimada and Sakazaki, 1984).

Nucleic acid probes have been found unable to differentiate between biotypes 1 and 2 (Wright et al., 1993; Aznar et al., 1994). However, Høi et al. (1997) reported that ribotype profiles can be used to distinguish biotype 1 and biotype 2 isolates. Biosca et al. (1997a) reported on a serogroup E strain virulent for eels but indole-positive and an additional isolate not belonging to serogroup E but pathogenic for eels and also indole-positive. They also reported that ribotyping with *Hind*III indicated that pathogenicity for eels is not associated with a specific ribotype in contrast to the conclusion of Høi et al. (1997). Because at this point in time it was evident that no biochemical test or specific serogroup could with certainty be associated with eel virulence, they proposed that *V. vulnificus* strains be classified into serovars instead of biotypes and that biotype E strains be designated as belonging to serotype E. Unfortunately, their recommendation did not allocate biotype 1 isolates into any specific serotype.

Biosca et al. (1997b) developed an indirect ELISA assay specific for biotype 2 (eel pathogens) of *V. vulnificus* based on polyclonal antibodies raised against LPS in rabbits. Weak cross-reactions were observed with antisera to crude LPS by other *Vibrio* species and biotype 1 isolates but not with antisera to purified LPS. The detection limit with antisera to purified LPS was between 10^4 and 10^5 cells/well. The assay was successful in detecting the organism in infected eel tissue and in detecting cells in the nonculturable state.

Amaro et al. (1999) reported on the isolation of serovar E isolates from seawater and oysters in Taiwan that were avirulent for eels. These isolates differed from eel pathogenic strains in (1) positive ability to ferment mannitol, (2) *Hind*III ribotyping profiles, (3) notable susceptibility to eel serum, and (4) immunoreactive LPS components. The authors concluded that these Taiwanese strains had clonal origins different from those of eel serovar E strains and that ribotyping is more related to the source of the strain than to serovar. Similar results were obtained by Arias et al. (1997) with other serovar E strains, exhibiting different ribotype patterns depending on their source.

Dalsgaard et al. (1999) reported on the isolation of indole-positive *V. vulnificus* isolates of serovar E from infected eels at a Danish eel farm and concluded that indole production may not be a reliable marker to identify strains virulent for eels, and that the inability to produce acid from mannitol might be a more useful phenotypic characteristic of such isolates. The authors also agreed with Arias et al. (1997) that the division of *V. vulnificus* into two biotypes based on phenotypic criteria originally established by Tison et al. (1982) is no longer tenable and leads to taxonomic confusion.

During 1996–1997, 62 cases of wound infections and bacteremia due to *V. vulnificus* were found to result from contact with purchased inland pond-raised tilapia in Israel (Bisharat et al., 1999). The outbreak was due to a new marketing policy of selling live fish instead of marketing them packed in ice postmortem. The isolates exhibited five atypical biochemical test results (Table 5.1), were nontypable by pulsed field gel electrophoresis (PFGE), and all had the same polymerase chain reaction–restriction fragment length polymorphism (PCR-RFLP) pattern derived from a 388-bp DNA fragment of the *cth* gene. Following PCR amplification of this fragment, digestion was with the three restriction nucleases *Dde*I, *Hha*I, and *Hpa*II. These isolates were distinguishable from biogroups 1 and 2 on the basis of negative reactions with respect to the utilization of citrate, salicin, cellobiose, and lactose (Table 5.1). The authors designated these isolates as belonging to a newly established subspecies group biotype 3.

Colodner et al. (2002) reported a biotype 3 isolate resulting in an unusual case of an individual developing bacteremia and extensive necrosis surrounding a puncture wound to a finger by a wire that occurred 24 hours after handling a tilipia followed by hand washing, suggesting colonization of the skin by the organism. Interestingly, the individual suffered from uncontrolled non-insulin-dependent diabetes, which is considered a predisposing factor to infection by the organism.

Gutacker et al. (2003) made use of multilocus enzyme electrophoresis (MLEE) involving the electrophoretic resolution of 15 intracellular enzymes, RAPD (Aznar, Ludwig, and Schleifer, 1993) and sequence typing of two genes, *recA* (encodes RecA protein involved in homologous recombination, recombinant DNA repair, and the SOS response) and *glnA* (encodes glutamine synthetase) in attempting to establish the phylogenetic relationship between strains of the three *V. vulnificus* biotypes. A pair of primers designated recA-1 and recA-2 (Table 5.2) were designed to amplify a 543-bp fragment of the *recA* gene for sequence analysis. A pair of primers designated glnA-1 and glnA-2 were also designed to amplify a 402-bp fragment of the *gln-A* gene for sequence analysis. Eleven of the 15 enzymes from all 62 *V. vulnificus* isolates were polymorphic, and four were monomorphic. A total of 43 electrophoretic types (ETs) were identified from the 62 isolates. Cluster analysis of the 43 ETs showed no significant clone associated with clinical or environmental origin and separated the isolates into two major divisions, I and II.

Alignment of the 543-bp amplified sequence of the *recA* gene yielded 35 nucleotide substitutions, 28 of which were shared by more than one sequence. Only two nucleotide substitutions resulted in amino acid substitutions. Alignment of the 402-bp amplified sequence of the *glnA* gene revealed 32 nucleotide substitutions, 25 of which were shared by more than one sequence. Only one amino acid substitution resulted. Phylogenetic analysis of the *recA* and *glnA* gene sequences separated the 62 strains into two major heterogeneous divisions (I and II) that were not correlated with any particular phenotypic trait.

Several clusters were identified; one consisted of all four Israeli human clinical isolates of tilapia origin (biotype 3), and another comprised indole-negative eel-pathogenic isolates (biotype 2), which formed a single cluster in the *recA* gene tree but were divided into two groups within division II of the *glnA* gene tree. It is interesting that biotype 1 strains were found to be distributed throughout the *recA* and *glnA* phylogenies, reflecting the heterogeneous nature of these isolates. Indole-positive eel-pathogenic strains did not form a monophylogenetic group in either the *recA* or *glnA* gene trees and in general did not cluster together with indole-negative eel-pathogenic isolates.

Primer M13 (Table 5.2) applied to all 62 strains yielded a total of 28 different RAPD profiles with isolates falling into two divisions, I and II. One cluster within division II included all 11 strains from diseased eels derived from several different geographic areas plus isolates not associated with eel pathogenicity and exhibiting a positive indole reaction. Another cluster within division II comprised all four human clinical isolates from Israel (biotype 3) with identical RAPD profiles. MLEE, RAPD, and sequence typing separated the isolates into two distinct divisions with 80% of the isolates included in the same genetic groups. MLEE, RAPD, and sequence typing indicated that indole-negative eel-pathogenic strains from different geographic origins tended to cluster as a separate genotype, in contrast to a variable phylogeny with the indole-positive eel-pathogenic isolates. The authors concluded that the designation biotype 2 should not be limited to the indole-negative isolates from diseased

Table 5.2 PCR Primers and DNA Probes

Primer or Probe	Sequence (5′ → 3′)	Size of Amplified Sequence (bp)	Gene	Reference
VVp1	CCG-GCG-GTA-CAG-GTT-GGC-C	519	*cth*	Hill et al. (1991)
VVp2	CGC-CAC-CCA-CTT-TCG-GGC-C			
RAPD - Gen 1-50-03	AGG-AYA-CGT-G	—	—	Radu et al. (1998)
RAPD Gen 1-50-09	AGA-AGC-GAT-G	—	—	
RAPD primer	GGA-TCT-GAA-C	—	—	Høi et al. (1997)
Choi-1	GAC-TAT-CGC-ATC-AAC-AAC-CG	704	*vvhA*	Lee, Eun, and Choi (1997)
Choi-2	AGG-TAG-CGA-GTA-TTA-CTG-CC			
P1	GAC-TAT-CGC-ATC-AAC-AAC-CG	704	*vvh*	Lee et al. (1998)
P2	AGG-TAG-CGA-GTA-TTA-CTG-CC			
P3	GCT-ATT-TCA-CCG-CCG-CTC-AC	222	*vvh*	
P4	CCG-CAG-AGC-CGT-AAA-CCG-AA			
Vv oligo 1	CGC-CGC-TCA-CTG-GGG-CAG-TGG-CTG	386	*cth*	Brauns, Hudson, and Oliver (1991)
Vv oligo 3	CCA-GCC-GTT-AAC-CGA-ACC-ACC-CGC			
VV1	GAC-TAT-CGC-ATC-AAC-AAC-CG	704	*vvh*	Fischer-Le Saux et al. (2002) from Lee, Eun, Choi (1997)
VV2R	AGG-TAG-CGA-GTA-TTA-CTG-CC			

Continued

Table 5.2 PCR Primers and DNA Probes (Continued)

Primer or Probe	Sequence (5′ → 3′)	Size of Amplified Sequence (bp)	Gene	Reference
VV3	GCT-ATT-TCA-CCG-CCG	604		Fischer-Le Saux et al. (2002) from Lee et al. (1998)
L-CTH	TTC-CAA-CTT-CAA-ACC-GAA-CTA-TGA-C	205	vvh	Brasher et al. (1998)
R-CTH	GCT-ACT-TTC-TAG-CAT-TTT-CTC-GC			
P-CTH probe	GAA-GCG-CCC-GTG-TCT-GAA-ACT-GGC-GTA-ACG			
L-vvh	TTC-CAA-CTT-CAA-ACC-GAA-CTA-TGA-C	205	vvh	Lee, Panicker, and Bej (2003)
R-vvh	GCT-ACT-TTC-TAG-CAT-TTT-CTC-GC			
PP-vvh probe	GAA-GCG-CCC-GTG-TCT-GAA-ACT-GGC-GTA-ACG-GAT-TT	—		
BP-vvh probe	GTT-CTT-CCT-TCA-GCG-CTG-TTT-TCG-GTT-TAC	—		
RAPD primer	TAT-CAG-GCT-GAA-AAT-CTT	—	—	Vickery et al. (1998)
R-PSE420				
Probe 610	A(K)A-(R)TT-GGC-GCC-GAC-GA	—	16S rDNA	Aznar, Ludwig, and Schleifer (1993)
Probe 1038	GCT-GTT-CCT-TTA-AGC-GAT-G	—	23S rDNA	
M13	GAA-ACA-GCT-ATG-ACC-ATG	—	—	
T3	ATT-AAC-CCT-CAC-TAA-AGG	—	—	
T7	AAT-ACG-ACT-CAC-TAT-AGG	—	—	

Name	Sequence	Size (bp)	Target	Reference
M13	GAA-ACA-GCT-ATG-ACC-ATG	—	—	Arias et al. (1998)
T3	ATT-AAC-CCT-CAC-TAA-AGG	—	—	
1038 probe	UAG-CGA-AAU-UCC-UUG-UCG	—	—	
Vvu1 probe	CAT-AGA-ACA-TTG-CCG-CAG	—	—	Aznar et al. (1994)
Vvu2 probe	ACT-CAA-TGA-TAC-TGG-CTT-A	—	—	
Vvu3 probe	ACC-GTT-CGT-CTA-ACA-CAT	—	—	
Vvu4 probe	TCA-AAG-AAC-ATT-GCC-GCA	—	—	
L-vvh	TTC-CAA-CTT-CAA-ACC-GAA-CTA-TGA	205	*vvh*	Panicker, Myers, and Bej (2004)
R-vvh	ATT-CCA-GTC-GAT-GCG-AAT-ACG-TTG			
VVAP probe	GAG-CTG-TCA-CGG-CAG-TTG-GAA-CCA	—	*vvha*	Wright et al. (1993)
recA-1	GAC-GAG-AAT-AAA-CAG-AAG-GC	543	*recA*	Gutacker et al. (2003)
recA-2	TCG-CCG-TTA-TAG-CTG-TAC-C			
glnA-1	TGA-CCC-ACG-CTC-TAT-CGC	402	*glnA*	
glnA-2	GCG-TGT-GCA-ACG-TTG-TG			
UtoxF	GAS-TTT-GTT-TGG-GGY-GAR-CAA-GGT-T[a]	—	*toxR*	Bauer and Rørvik (2007)
vptoxR	GGT-TCA-ACG-ATT-GCG-TCA-GAA-G	297	*toxR*	
vctoxR	GGT-TAG-CAA-CGA-TGC-GTA-AG	640	*toxR*	
vvttoxR	AAC-GGA-ACT-TAG-ACT-CCG-AC	436	*toxR*	
RAPD	GGA-TCT-GAA-C	—	—	Warner and Oliver (1998)

[a] S = , Y = C + T, R = A + G.

eels and that the designation serovar E, which presently includes biotype 2 strains plus other eel-pathogenic isolates, is not supported by the phylogenetic results.

VI. VIABLE BUT NONCULTURABLE (VBNC) *V. VULNIFICUS*

A rapid decrease in cell viability was observed when the organism was held at 4°C in an oyster homogenate (Oliver, 1981). Reduction of temperature has been found to trigger a response resulting in the VBNC state. VBNC cells cannot be cultured on routine media, and their viability can be ascertained only by the use of various direct assays for detection of metabolic activity within the cell. Resuscitation of VBNC cells of *V. vulnificus* occurs when the temperature stress is removed for a number of hours by an upward shift in temperature. Warner and Oliver (1998) used the fluorescent redox probe cyanoditolyltetrazolium chloride (CTC) and fluorescent microscopy to detect low-temperature-induced VBNC cells in artificial seawater. VBNC cells were resuscitated by raising the temperature from 5°C to 22°C for 12 hr. RAPD profiles were generated with the 10-mer primer of Høi et al. (1997; Table 5.2) specific for vibrios. *V. vulnificus* cells made with VBNC lost their RAPD signal after an average of 7 days. The authors hypothesized that either newly synthesized DNA binding proteins or chromosomal supercoiling, or a combination of both, prevents detection of VBNC cells by RAPD. Their preliminary studies using the antibiotic ciprofloxacin, a supercoiling inhibitor, resulted in maintenance of the RAPD signal under these conditions of low-temperature stress.

Bryan et al. (1999) observed that when a culture of *V. vulnificus* was shifted from 35°C to 6°C it became converted to a nonculturable state. Cells adapted to 15°C prior to shifting to 6°C, however, remained viable and culturable. In addition, cultures adapted to 15°C exhibited greater survival when frozen at –78°C compared to being frozen directly from 35°C. De novo protein synthesis was shown to be required for low-temperature survival. It is interesting that removal of iron from the growth medium by adding 2,2′-Dipyridyl prior to cold adaptation at 15°C decreased the ability of the culture to adapt to 6°C.

Detection of mRNA is thought to be a reliable marker for viability due to its short half-life. With this in mind, Fischer Le-Saux et al. (2002) detected VBNC cells of *V. vulnificus* by applying seminested reverse transcription-PCR (RT-PCR) targeting the *vvhA* gene of VBNC populations induced by holding cells at 4°C in artificial seawater. The *VvhA* nested system of Lee et al. (1998) was used. Following RT, two external primers designated VV1 and VV2R (Table 5.2) delineating a 704-bp sequence within the open reading frame of the *vvhA* gene were used in conjunction with an internal primer designated VV3 (Table 5.2) delineating a 604-bp fragment in conjunction with the VV2R in the seminested PCR. Transcripts were shown to persist in nonculturable populations for over 4.5 months, with a progressive decline of the signal over time. The methodology not only detected VBNC cells but also ensured that only viable cells were detected.

In a PCR technique developed for *V. vulnificus*, higher levels of DNA were required to detect VBNC than growing cells (Brauns, Hudson, and Oliver, 1991; Coleman and

Oliver, 1996). In the same way, using an RAPD method to detect grown cells, the loss of a signal of RAPD amplification products was observed with starved and VBNC cells of this pathogen, until it became undetectable (Warner and Oliver, 1998).

VII. MOLECULAR METHODS FOR DETECTION AND TYPING

A. Conventional PCR

Hill et al. (1991) were the first to develop a PCR procedure for detection of *V. vulnificus*. They seeded the organism into oyster homogenates and found that among several DNA extraction procedures examined, DNA recovered from cells in homogenates by lysing with guanidine isothyocyanate (GITC) followed by extraction with chloroform and precipitation with ethanol was most suitable for use as a PCR template. In contrast, extraction of the homogenates with GITC alone notably inhibited the PCR. The two primers used VVp1 and VVp2 (Table 5.2) targeted a 519-pb sequence of the *cth* gene. With the GITC-chloroform DNA extraction method, 20 µl of oyster extract were found to be noninhibitory when added directly to PCR reactions. The limit of detection was 10^2 CFU/g of oyster tissue following a 24-hr enrichment at 33–35°C in a 1:10 homogenate of oyster tissue in APW.

Aono et al. (1997) used the two primers VVp1 and VVp2 (Table 5.2) developed by Hill et al. (1991) for evaluating the effectiveness of the PCR in identifying isolates of *V. vulnificus* from marine environments. A total of 13,325 bacteria isolates from seawater, sediments, oysters, and goby specimens collected along the coastal regions of Tokyo Bay were metabolically screened. Among these, 713 grew at 40°C, required NaCl for growth, formed greenish colonies on TCBS agar, and were presumptively identified as *V. vulnificus*. The PCR amplified the targeted 519-bp sequence of the *cth* gene with 61 of these isolates. DNA–DNA hybridization with the type strain of *V. vulnificus* and the API 20E system confirmed the PCR results. The authors concluded that the PCR method is useful for rapid and accurate identification of *V. vulnificus* from marine sediments.

Hervio-Heath et al. (2002) examined French coastal water and mussels for the presence of several pathogenic vibrios including *V. vulnificus*. The primers of Lee, Eun, and Choi (1997) VV-1 and VV-2R targeting the 704-bp sequence of the *vvhA* cytolysin gene were used to confirm the identity of the presumptive *V. vulnificus* isolates as well as the primers Vv oligo 1 and Vv oligo 3 of Brauns, Hudson, and Oliver (1991) delineating a 386-bp fragment of the 704 *cth* sequence. A total of 242 strains was isolated on TCBS or mCPC agar medium, of which 190 (78.5%) were identified as members of the genus *Vibrio*. Among these 190 *Vibrio* isolates, 20 were identified as *V. vulnificus*, with 16 derived from estuarine water samples and 4 from mussels. One of these isolates of *V. vulnificus* was found to be indole-positive and ornithine decarboxlase-negative, distinguishing it from *V. vulnificus* biogroups 1, 2, and 3 (Table 5.1). The other 19 strains belonged to biotype 1. The authors also found that mCPC agar yielded notably more *V. vulnificus* isolates than TCBS agar and recommended that TCBS agar not be used as an isolation medium for *V. vulnificus*.

These observations confirmed those of Høi et al. (1998) that TCBS agar has a low plating efficiency for clinical and environmental strains of *V. vulnificus*.

Lee et al. (1997) made use of a primer pair designated Choi-1 and Choi-2 for PCR amplification of a 704-bp sequence of the *vvhA* gene following enrichment of seeded homogenates of octopus tissue. Enrichments consisted of incubation at 30°C of samples for 4 hr in modified brain–heart infusion broth containing 2.0% NaCl at pH 8.0, which was found superior to APW. Sensitivity of detection was 10 CFU/ml of homogenates.

Lee et al. (1998) established a nested PCR for direct identification of *V. vulnificus* in blood serum and bulbous lesions (bullae) from patients. The external primers were identical to those used by Lee et al. (1997) but were designated P1 and P2 (Table 5.2) and again amplified a 704-bp sequence of the *vvh* gene. The nested primer pair P3 and P4 (Table 5.2) amplified an internal 222-bp sequence. The fluid samples were pelleted by centrifugation and the pellets lysed by boiling in 1 mM EDTA-0.5% Triton X-100. Sensitivity was 1 CFU per PCR reaction tube. Among 18 culture-positive specimens, 17 (94.4%) were positive with the nested PCR. The one negative result was presumably due to antibiotic treatment of the patient. Among 19 culture-negative specimens, 8 (42.1%) were positive with the nested PCR. Amplicons were detected in agarose gels stained with ethidim bromide. The assay time was 6 hr.

Wang and Levin (2005a) developed a quantitative competitive PCR assay for *V. vulnificus* in shellfish based on the hemolysin (*vvh*) gene. The specificity of the primers for *V. vulnificus* was proven by gel electrophoresis. The detection sensitivity for *V. vulnificus* was 220 CFU per PCR reaction with cells from a pure culture and 270 CFU/g of tissue without enrichment. After 10 hr of enrichment at 37°C the minimum detection level was reduced to 7 CFU/g of tissue.

An efficient procedure was developed for the quantitative PCR detection of *Vibrio vulnificus* in pure culture by Wang and Levin (2005b). The procedure involved boiling cell suspensions in TZ solution (2% Triton X-100 and 2.5 mg/ml NaN3 in 0.1 M Tris-HCl buffer at pH 8.0). Serial dilutions of lysed cell suspensions were then used for PCR. The method of visualizing amplified bands in agarose gels was evaluated by comparing the nucleic acid dyes GelStar and ethidium bromide (EB). GelStar stain was found to yield discernible amplified bands with lower levels of target DNA than could be achieved with EB. The minimum detection level with the GelStar stain was 16 CFU per PCR reaction compared to 40 CFU with EB. The relative fluorescence intensity of the DNA bands was analyzed with the NIH Image 1.61 software program. Calibration curves relating fluorescence intensity of amplified bands to lysed cells were obtained and the method was found suitable for the quantification of genomic DNA derived from $\sim10^1$–10^3 CFU per PCR reaction.

B. Multiplex PCR

Brasher et al. (1998) developed a multiplex PCR method for simultaneous amplification of targeted gene segments of five gram-negative pathogens including *V. vulnificus* in shellfish tissue homogenized in APW. The primers L-CTH and R-CTH

(Table 5.2) targeted a 205-bp sequence of the *cth* gene of *V. vulnificus*. A 6-hr enrichment at 35°C was used prior to DNA purification and the PCR. The sensitivity of detection was 10^1–10^2 CFU following a double multiplex PCR. Amplicons were detected by agarose gel electrophoresis and ethidium bromide staining of DNA bands. Alternatively, the amplicon was detected using a biotinylated oligonucleotide probe (Table 5.2), using a dot blot assay and a streptavidin–alkaline phosphatase conjugate and a chromogenic substrate.

Lee, Panicker, and Bej (2003) developed a similar multiplex PCR method for simultaneous amplification of targeted gene segments of the same five pathogens including *V. vulnificus*, followed by DNA–DNA sandwich hybridization. The pair of primers used with *V. vulnificus* L-vvh and R-vvh amplified a 205-bp sequence of the *vvh* gene (Table 5.2). The method used a primary phosphorylated oligonucleotide probe (Table 5.2) bound to a novel microwell surface, followed by addition of PCR amplified DNA, and then the addition of a secondary biotinylated oligonucleotide probe (Table 5.2) to facilitate detection. The addition of an alkaline phosphatase–avidin conjugate was followed by the phosphatase substrate para-nitrophenyl phosphate in diethanolamine and incubation for 2 hr with the absorbance at 405 nm determined with a microplate reader. With a preliminary 3-hr enrichment in APW the level of detection was 10^2 cells/g of seeded oyster tissue. A major advantage of this technique is that once the primary oligonucleotide probes are immobilized onto the surface of the wells the plates can be stored at 5°C for up to 2 months. Disadvantages of the method are (1) difficulties with respect to quantitation, and (2) the extended overnight period of time for DNA hybridization.

Bauer and Rørvik (2007) developed a novel multiplex PCR for the simultaneous identification of *V. parahaemolyticus*, *V. cholerae*, and *V. vulnificus* involving the *toxR* gene. The forward primer UtoxF (Table 5.2) was universal for all three species, and the reverse primers were species-specific. The primers UtoxF/vptoxR (Table 5.2) amplified a 297-bp sequence of the *toxR* gene from *V. parahaemolyticus*. The primers UtoxF/vctoxR (Table 5.2) amplified a 640-bp sequence of the *toxR* gene from *V. cholerae*. The primers UtoxF/vvtoxR (Table 5.2) amplified a 435-bp sequence of the *toxR* gene from *V. vulnificus*. The primer pairs for *V. cholerae* and *V. vulnificus* yielded a sensitivity of 100% with no false-positive or negative results yielding a specificity of 100%. The primer pair for *V. parahaemolyticus* yielded a specificity of 90% resulting from 18 of 114 *V. alginolyticus* isolates being positive.

C. Real-Time PCR (Rti-PCR)

Panicker, Myers, and Bej (2004) described a SYBR green I-based real-time PCR assay for detection of *V. vulnificus* in oyster tissue homogenate. A pair of primers designated L-vvh and R-vvh (Table 5.2) was used to amplify a 205-bp sequence of the *vvh* gene. The minimum level of detection was 100 CFU per PCR tube. A 5-hr enrichment allowed detection of 1 CFU per ml of tissue homogenate, which is equivalent to 10 CFU/g of tissue. The assay required 8 hr for completion.

Wang and Levin (2006) developed an Rti-PCR assay for quantification of *Vibrio vulnificus* seeded into clam tissue homogenates. Without enrichment, the limit of detection was 1×10^2 CFU/g of tissue with a linear detection range of 1×10^2 to 1×10^8 CFU/g. With a 5-hr nonselective enrichment, the limit of detection was 1 CFU/g of tissue with a linear detection range of 1 to 1×10^6 CFU/g of tissue. A tenfold higher detection limit with seeded clam tissue homogenates occurred compared to a pure culture. After 5 hr of nonselective enrichment the detection limits with a pure broth culture and seeded tissue homogenates were identical at 1 CFU/ml; however, the C_T value with tissue homogenates was about three cycles higher than with a pure culture, reflecting some level of PCR inhibition from the tissue.

Five thermal factors, including initial denaturation temperature, cycling denaturation temperature, annealing temperature, extension temperature, and the temperature at which the intensity of the fluorescent signal is read, were evaluated by Wang and Levin (2007) for their effects on the detection of *Vibrio vulnificus* by Rti-PCR. Fluorescent signal detection after extension was set between the T_m value of the primer–dimers (79°C) and that of the PCR target amplicons (84°C). This effectively eliminated the overestimation of the yield of PCR amplicons due to the presence of primer–dimers that otherwise led to erroneously lower C_T values (1.9 cycles lower). The annealing and extension steps were combined to convert a three-step PCR to a two-step PCR. This consisted of initial denaturation at 95°C for 3 min, cycling denaturation at 94°C for 15 s, and a combined annealing and extension step at 60°C for 5 s in each PCR cycle. One genomic target per real-time PCR reaction was detected with the simplified two-step PCR.

Ethidium bromide monoazide (EMA) is a DNA intercollating agent that with visible light activation cross-links the two strands of DNA so as to prevent PCR amplification. EMA penetrates only membrane-damaged cells. This has allowed the selective Rti-PCR amplification of DNA from viable bacterial pathogens (Nogva et al., 2003). EMA has been utilized to selectively allow the Rti-PCR amplification of a targeted DNA sequence in viable cells of *Vibrio vulnificus* (Wang and Levin, 2005c). The optimized light exposure time to achieve cross-linking of DNA by the EMA in dead cells and to photolyse the free EMA in solution was at least 15 min. The use of 3.0 µg/ml or less of EMA did not inhibit the PCR amplification of DNA derived from viable cells of *V. vulnificus*. The minimum amount of EMA to completely inhibit the Rti-PCR amplification of DNA derived from heat-killed cells was 2.5 µg/ml. Amplification of DNA from dead cells in a mixture with viable cells was successfully inhibited by 2.5 µg/ml of EMA, whereas the DNA from viable cells present was successfully amplified by Rti-PCR.

EMA Rti-PCR was used by Lee and Levin (2007) to discriminate between DNA from γ-irradiated and nonirradiated cells. γ-irradiation at 1.08 KGy and above of cell suspensions containing 1×10^6 CFU/ml resulted in 0 CFU/ml. RTi-PCR was able to detect 86.6% destruction by 1.08 KGy, and EMA real-time PCR was able to detect 93.2% destruction at this dose. With 3.0 and 5.0 KGy, EMA real-time PCR was able to detect 99.3% and 100% destruction, respectively. These results reflected the fact that higher doses of γ-irradiation resulted in greater membrane damage allowing more EMA to enter the cells so as to prevent DNA amplification from cells destroyed by the γ-irradiation.

Lee and Levin (2008) developed a novel method for discriminating *Vibrio vulnificus* by Rti-PCR before and after γ-irradiation based on the observation that γ-irradiation results in extensive reduction in the molecular size of DNA. Irradiation of viable cells (1×10^6 CFU/ml) at 1.08 KGy resulted in 100% destruction determined by plate counts, with most of the DNA from the irradiated cells having a bp length of less than 1000. The use of a pair of primers to amplify a 1000-bp sequence of DNA from cells exposed to 1.08 KGy failed to yield amplification. In contrast, primers designed to amplify sequences of 700-, 300-, and 70-bp yielded amplification with C_T values resulting in 13.4%, 27.6%, and 45.4% detection of genomic targets. When viable cells of *V. vulnificus* were exposed to 1.08, 3.0, and 5.0 kGY, the average molecular size of genomic DNA visualized in an agarose gel decreased with increasing dose, corresponding to an increased probability of amplification with primers targeting sequences of decreasing size.

D. RAPD

Radu et al. (1998) subjected 26 biotype 1 and 10 biotype 2 isolates to RAPD analysis using two random primers designated Gen 1-50-03 and Gen 1-50-09 (Table 5.2). A total of six RAPD types was distinguished with primer Gen 1-50-03, with all six RAPD types represented by one or more strains of biotype 1. With biotype 2 strains, only three of these RAPD types were distinguished. With primer Gen 1-50-09, a total of five RAPD types was distinguished, with all five RAPD types represented by one or more strains of biotype 1. With biotype 2 strains, only four of these RAPD types were generated. Results also indicated that certain biotype 1 and biotype 2 strains yielded identical RAPD profiles with both RAPD primers, indicating a high degree of DNA sequence similarity between such strains of the two biotypes.

Vickery et al. (1998) made use of a random primer designated R-PSE420 (Table 5.2) for generating RAPD profiles of *V. vulnificus* strains. The primer yielded 15 different DNA banding profiles with 16 strains. A great deal of genomic heterogeneity was observed with strains derived from different oyster samples and even from strains derived from the same patient with wound infections.

In a subsequent RAPD study, Vickery, Harold, and Bej (2000) used the same primer for RAPD profiling of 10 *V. vulnificus* isolates from patients who succumbed as a result of infections derived from consuming raw oysters. Analysis of the DNA band profiles revealed significant genetic heterogeneity among these strains also.

Høi et al. (1997) screened 10 RAPD primers for analysis of *V. vulnificus* isolates and found that one primer (Table 5.2) was superior and yielded 10–15 bands. Use of this primer for RAPD analysis of isolates failed to distinguish between Danish and U.S. strains and to separate biotype 1 and 2 strains due to excessive heterogeneity of the RAPD profiles. In contrast, ribotyping differentiated Danish and U.S. strains and distinguished between biotypes 1 and 2.

Warner and Oliver (1999) developed an RAPD protocol for detecting *V. vulnificus* and for distinguishing this organism from other members of the genus *Vibrio*. A 10-mer primer (Table 5.2) previously described by Warner and Oliver (1998) was used. Each of 70 *V. vulnificus* strains examined produced a unique banding pattern, indicating that members of this species are highly heterogeneous. All of the clinical isolates yielded a unique band (200 to 178 bp) that was only occasionally found with environmental strains. The authors concluded that this band may be correlated with human pathogenicity. Subsequent observations by DePaola et al. (2003) with this primer indicated that only 70% of clinical isolates possessed this amplicon and that a band of about 460-bp was present in 86% of these same strains.

DePaola et al. (2003) examined strains of *V. vulnificus* from market oysters and oyster-associated primary septicemia cases (25 strains from each group) for potential virulence markers that could possibly distinguish strains from these two sources. The isolates were analyzed for plasmid content, the presence of the 460-bp amplicon by RAPD using the primer of Warner and Oliver (1998), and for virulence in inoculated mice with serum-iron overload. Both groups of isolates yielded strains with similar results. About half of both oyster and clinical isolates produced the 460-bp band. The authors concluded that nearly all *V. vulnificus* strains in oysters are virulent and that these assay methods cannot distinguish between fully virulent and less virulent strains or between clinical and environmental isolates.

E. Ribotyping

Aznar, Ludwig, and Schleifer (1993) subjected strains of *V. vulnificus* to ribotyping by digesting DNA with the restriction nucleases *Hind*III, *Sal*I, *Kpn*I, and *Sph*I followed by hybridization with the oligonucleotide probe 610 (Table 5.2) reactive with 16S rDNA and with probe 1038, reactive with 23S rDNA (Table 5.2). The highest degree of discrimination by ribotyping was when DNA was digested with *hpn*I and hybridized to probe 1038, which resulted in discrimination of biotypes 1 and 2 in addition to individual strains. These authors also found that strains of *V. vulnificus* biotypes 1 and 2 could be differentiated by RAPD with the universal primer M13, T3, or T7 (Table 5.2). Compared with riboytping, RAPD appeared to be a faster method for diagnosing the identity of *V. vulnificus* biotypes.

Arias et al. (1997) determined the intraspecies genomic relatedness of 44 biotype 1 and 36 biotype 2 isolates from different geographic origins by ribotyping and with the use of amplified fragment length polymorphism (AFLP). Ribopatterns of DNAs digested with *Kpn*I and hybridized with labeled olidgonucleotide probe 1038 (Table 5.2) revealed up to 19 ribotypes that were different for the two biotypes. Sixteen different ribotypes were found within biotype 1 strains from clinical and environmental sources, and only three, mainly from diseased eels, were found among biotype 2 strains. Within biotype, 96% of the strains yielded the same ribopattern. AFLP fingerprinting obtained by selective PCR amplification of *Hind*II-*Taq*I double-restricted DNA fragments following ligation to specific oligonucleotide adapters exhibited a strain-specific pattern that allowed the finest differentiation of subgroups within the eel-pathogenic isolates sharing the same ribotypes. Ribotyping clearly separated the eel-pathogenic strains from the clinical and environmental isolates, whereas AFLP distinguished individual strains and therefore constitutes one of the most discriminative methods for epidemiological and ecological studies.

Arias et al. (1998) determined the genetic relationships among 132 strains of *V. vulnificus* derived from human infections, diseased eels, seawater, and shellfish with the use of ribotyping and RAPD. For ribotyping, genomic DNA was digested with *Kpn*I and hybridized with the 18-mer universal digoxigenin-labeled 1038 oligonucleotide probe (Table 5.2) complementary to a highly conserved sequence in the 23S rRNA gene. RAPD was performed with the universal primers M13 and T3 (Table 5.2). Both ribotyping and RAPD revealed a high level of homogeneity of diseased-eel isolates in contrast to the genetic heterogeneity of seawater-shellfish isolates of the Mediterranean. Although differentiation within diseased-eel isolates was only possible by ribotyping, the authors proposed that RAPD is a better technique than ribotyping for less laborious and rapid typing of new *V. vulnificus* isolates.

F. 16S rRNA Sequencing

Aznar et al. (1994) compared several 16S rRNA sequences of pathogenic *Vibrio* species and found that *V. vulnificus* represents a group that is not closely related to the core organisms of the genus *Vibrio*. Among four oligonucleotide probes (Table 5.2) specific for *V. vulnificus*, two (Vvv2 and Vvu3) hybridized with all 21 *V. vulnificus* strains tested, and the other two probes (Vvu1 and Vvu4) distinguished *V. vulnificus* biotype 1 strains from all other organisms examined.

G. Oligonucleotide Probe

Wright et al. (1993) developed an oligonucleotide probe designated VVAP constructed from a portion of the *V. vulnificus* cytolysin gene *vvhA* and labeled with alkaline phosphatase. Naturally occurring *V. vulnificus* were detected without enrichment or selective media by plating dilutions of oyster homogenates and seawater directly onto Luria agar followed by incubation overnight at 35°C or at room temperature for 72 hr. Plates with colonies were overlaid with membrane filters or filter-paper disks, which were then microwaved and treated with proteinase K to remove background alkaline phosphatase activity prior to hybridization. The VVAP probe was then added and hybridization allowed to occur for 30 min at 50°C or 1 hr at 56°C. After washing, alkaline phosphatase activity was then assayed with nitroblue tetrazolium plus 5-bromo-4-chloro-3-indolyl-phosphate. Color development of *V. vulnificus* colony blots was complete within 60 min. The method has the advantage of not requiring purification of colonies or metabolic characterization of isolates.

Table 5.3   Selective Agar Media for Isolation of *V. vulnificus*[a]

Cellobiose-Polymyxin B–Colistin (CPC) Agar (Massad and Oliver, 1987)

Solution 1:   Peptone ------------------- 10 g
                Beef extract ------------- 5 g
                NaCl----------------------- 20 g
                Bromothymol blue ------ 40 mg
                Cresol red---------------- 40 mg
                Agar ----------------------- 15 g
                d.H$_2$O --------------------- 900 ml
                pH 7.6

Solution 2:   Cellobiose --------------- 15 g
                Colistin -------------------- 1.36 x 10^6 U
                Polymyxin B-------------- 10^5 U
                d.H$_2$O --------------------- 100 ml

Solution 1 is adjusted to a pH of 7.6, autoclaved, cooled to 55°C, and solution 2 added after filter sterilization. When solidified, CPC agar is olive green to light brown. *V. vulnificus* produces yellow colonies surrounded by a yellow zone due to fermentation of cellobiose.

Modified CPC (mCPC) Agar (Tamplin et al., 1991)

Same composition as CPC agar except that colistin methanesulphonate is reduced to 4 x 10^5 U/L.

Cellobiose-Colistin (CC) Agar (Høi, Dalsgaard, and Dalsgaard, 1998)

Same composition as CPC agar except that it contains no polymyxin B and the concentration of colistin is reduced to 4 x 10^5 U/L.

Vibrio vulnificus (VV) agar (Brayton et al., 1983)

Peptone--- 2 g/L
Oxgall --- 8 g
Casamino acids ----------------------------- 0.5 g
NaCl--- 10 g
MgCl.6H$_2$0 ----------------------------------- 2.0 g
KCl--- 1.0 g
Crystal violet (0.15% aq. soln.) ----------- 1.0 ml
Tween 80 (10% aq. soln.) ----------------- 5.0 ml
Salicin (20% aq. soln.)---------------------- 100 ml
K tellurite (0.5% aq. soln.) ---------------- 1 ml
Agar -- 15 g

The medium is boiled, adjusted to a pH of 8.6, autoclaved for 10 min, cooled to 55°C and then filter sterilized salicin (20% soln.) is added to a final concentration of 2%. Growth of *V. vulnificus* appears as large gray colonies with black centers. Recovery and growth of *V. vulnificus* on VV agar was reported by Brayton et al. (1983) to be superior to that on TCBS agar.

Sodium Dodecyl Sulfate-Polymyxin-Sucrose (SPS) Agar (Kitaura et al., 1983)

Proteose peptone -------------------------- 10 g/L
Beef extract----------------------------------- 5 g
Sucrose--- 15 g
Sodium chloride ----------------------------- 20 g
SDA -- 1 g
Polymyxin B ----------------------------------- 100,000 U
Bromthymol blue ---------------------------- 0.04 g
Cresol red-------------------------------------- 0.04 g
Agar --- 15 g
pH 7.6

Table 5.3   Selective Agar Media for Isolation of V. vulnificus[a] (Continued)

Thiosulfate-Citrate-Bile-Salts Sucrose (TCBS) Agar (Difco)

Yeast extract---------------------------------- 5 g/L
Proteose peptone No. 3 ------------------- 10 g
Sodium citrate ------------------------------ 10 g
Sodium thiosulphate ----------------------- 10 g
Oxgall --- 8 g
Sucrose-- 20 g
Sodium chloride ----------------------------- 10 g
Ferric citrate -------------------------------- 1 g
Brom Thymol.blue-------------------------- 0.04 g
Thymol blue --------------------------------- 0.04 g
Agar -- 15 g

Final pH is 8.6. Originally developed as a selective medium for V. cholarae El Tor, which produces yellow colonies, whereas nonsucrose-fermenting V. parahaemolyticus and V. vulnificus produce blue-green colonies.

Vibrio vulnificus Medium (VVM; Cerdà-Cuéllar, Jofre, and Blanch, 2000)

Cellobiose----------------------------------- 15 g/L
NaCl--- 15 g
Yeast extract-------------------------------- 4 g
MgCl$_2$.6H$_2$0 ---------------------------------- 4 g
KCl-- 4 g
Cresol red------------------------------------ 40 mg
Bromothymol blue-------------------------- 40 mg
Polymyxin B --------------------------------- 10^5 U
Colistin methanesulfonate --------------- 10^5
Agar --- 15 g

All of the components are added together, heated to boiling, cooled to 50°C, and then the pH is adjusted to 8.5 with 5 M NaOH. VVM does not require boiling.

[a] Contents per liter.

REFERENCES

Amaro, C., Biosca, E. 1996. *Vibrio vulnificus* biotype 2 pathogenic for eels, is also an opportunistic pathogen for humans. *Appl. Environ. Microbiol.* 62:1454–1457.

Amaro, C., Biosca, E., Fouz, B., Garay, E. 1992. Electrophoretic analysis of heterogeneous lipopolysaccharides from various strains of *Vibrio vulnificus* biotypes 1 and 2 using silver staining and immunoblotting. *Curr. Microbiol.* 25:99–104.

Amaro, C., Hor, L., Marco-Noales, E., Bosque, T., Fouz, B., Alcaide, E. 1999. Isolation of *Vibrio vulnificus* serovar E from aquatic habitats in Taiwan. *Appl. Environ. Microbiol.* 65:1352–1355.

Aono, E., Sugita, H., Kawasaki, J., Sakakibara, H., Takahashi, T., Endo, K., Deguchi, Y. 1997. Evaluation of the polymerase chain reaction method for identification of *Vibrio vulnificus* isolated from marine environments. *J. Food Prot.* 60:81–83.

Arias, C., Pujalte, M., Garay, E., Aznar, R. 1998. Genetic relatedness among environmental, clinical, and diseased-eel *Vibrio vulnificus* isolates from different geographic regions by ribotyping and randomly amplified polymorphic DNA PCR. *Appl. Environ. Microbiol.* 64:3403–3410.

Arias, C., Verdonck, L., Swings, J., Garay, E., Aznar, R. 1997. Intraspecific differentiation of *Vibrio vulnificus* biotypes by amplified fragment length polymorphism and ribotyping. *Appl. Environ. Microbiol.* 63:2600–2606.

Aznar, R., Ludwig, W., Amann, R., Schleifer, K. 1994. Sequence determination of rRNA genes of pathogenic *Vibrio* species and whole-cell identification of *Vibrio vulnificus* with rRNA-targeted oligonucleotide probes. *Int. J. Syst. Bacteriol.* 44:330–337.

Aznar, R., Ludwig, W., Schleifer, K. 1993. Ribotyping and randomly amplified polymorphic DNA analysis of *Vibrio vulnificus* biotypes. *Syst. Appl. Microbiol.* 16:303–309.

Bauer, A., Rørvik, L. 2007. A novel multiplex PCR for the identification of *Vibrio parahaemolyticus, Vibrio cholerae,* and *Vibrio vulnificus. Lett. Appl. Microbiol.* 45:371–375.

Biosca, E., Amaro, C., Larsen, J., Pederson, K. 1997a. Phenotypic and genotypic characterization of *Vibrio vulnificus:* Proposal for the substitution of the subspecific taxon biotype for serovar. *Appl. Environ. Microbiol.* 63:1460–1466.

Biosca, E., Garay, E., Toranzo, A., Amaro, C. 1993a. Comparison of outer membrane protein profiles of *Vibrio vulnificus* biotypes 1 and 2. *FEMS Microbiol. Lett.* 107:217–222.

Biosca, E., Llorens, H., Garay, E., Amaro, C. 1993b. Presence of a capsule in *Vibrio vulnificus* biotype 2 and its relationship to virulence for eels. *Infect. Immun.* 61:1611–1618.

Biosca, E., Oliver, J., Amaro, C. 1996. Phenotypic characterization of *Vibrio vulnificus* biotype 2, a lipopolysaccharide-based homogeneous O serogroup within *Vibrio vulnificus. Appl. Environ. Microbiol.* 62:918–927.

Biosca, E.G., Marco-Noales, E., Amaro, C., Alcaide, E. 1997b. An enzyme-linked immunosorbent assay for detection of *Vibrio vulnificus* biotype 2: Development and field studies. *Appl. Environ. Microbiol.* 63:537–542.

Bisharat, N., Agmon, V., Finkelstein, R., Raz, R., Ben-Dror, G., Memer, L., Soboh, S., Colodner, R., Cameron, D., Wykstra, D., Swerdlow, D., Farmer J., III. 1999. Clinical, epidemiological, and microbiological features of *Vibrio vulnificus* biogroup 3 causing outbreaks of wound infection and bacteremia in Israel. *Lancet* 354:1421–1424.

Brasher, C., DePaola, A., Jones, D., Bej, A. 1998. Detection of microbial pathogens in shellfish with multiplex PCR. *Current Microbiol.* 37:101–107.

Brauns, L.A., Hudson, M., Oliver, M. 1991. Use of the polymerase chain reaction in detection of culturable and nonculturable *Vibrio vulnificus* cells. *Appl. Environ. Microbiol.* 57:2651–2655.

Brayton, P., West, P., Russek, E., Colwell, R. 1983. New selective plating medium for isolation of *V. vulnificus* biogroup 1. *J. Clin. Microbiol.* 17:1039–1044.

Bryan, P., Sefan, R.J., DePaola, A., Foster, J., Bej, A. 1999. Adaptive response to cold temperatures in *Vibrio vulnificus. Current Microbiol.* 38:168–175.

Cerdà-Cuéllar, M., Jofre, J., Blanch, A. 2000. A selective medium and a specific probe for detection of *Vibrio vulnificus. Appl. Environ. Microbiol.* 66:855–859.

Chester, B., Cleary, T. 1980. Evaluation of the Minitek system for identification of nonfermentative and nonenteric fermentative Gram-negative bacteria. *J. Clin. Microbiol.* 12:509–516.

Coleman S., Oliver, J. 1996. Optimization of conditions for the polymerase chain reaction amplification of DNA from culturable and nonculturable cells of *Vibrio vulnificus. FEMS Microbiol. Ecol.* 19:127–132.

Colodner, R., Chazan, B., Kipelowitz, J., Keness, Y., Raz, R. 2002. Unusual portal of entry of *Vibrio vulnificus:* evidence of its prolonged survival on the skin. *Clin. Infect. Dis.* 34:714–715.

Dalsgaard, I., Hoi, L., Siebeling, R.J., Dalsgaard, A. 1999. Indole-positive *Vibrio vulnificus* isolated from disease outbreaks on a Danish eel farm. *Dis. Aquat. Org.* 35:187–194.

DePaola, A., Nordstrom, J., Dalsgaard, A., Forslund, A., Oliver, J., Bates, T., Bourdage, K., Gulig., P. 2003. Analysis of *Vibrio vulnificus* from market oysters and septicemia cases for virulence markers. *Appl. Environ. Microbiol.* 69:4006–4011.

Desenclos, J., Klontxz, K., Wolfe, L., Hoecherl, S. 1991. The risk of *Vibrio* illness in the Forida raw oyster eating population, 1981–1988. *Am. J. Epidemiol.* 134:290–297.

Desmond, E., Janda, J., Adams, F., Bottone, E. 1984. Comparative studies and laboratory diagnosis of *Vibrio vulnificus* and invasive *Vibrio* sp. *J. Clin. Microbiol.* 19:122–125.

Farmer, J., III, Hickman-Brenner, F., Kelly, M. 1985. *Manual of Clinical Microbology*, 4th ed. ASM Press: Washington, DC, pp. 282–301.

Fischer-Le Saux, M., Hervio-Heath, D., Loaec, S., Colwell, R., Pommeypuy, M. 2002. Detection of cytolysin-hemolysin mRNA in conculturable populations of environmental and clinical *Vibrio vulnificus* strains in artificial seawater. *Appl. Environ. Microbiol.* 68:5641–5646.

Ghosh, H., Bowen, T. 1980. Halophilic vibrios from human tissue infections on the Pacific coast of Australia. *Pathology* 12:397–402.

Gray, L., Kreger A. 1985a. Purification and characterization of an extracellular cytolysin produced by *Vibrio vulnificus*. *Infect. Immun.* 48:62–72.

Gray, L. Kreger, A. 1985b. Identification of *Vibrio vulnificus* by indirect immunofluorescence. *Diagn. Microbiol. Infect. Dis.* 3:461–468.

Gutacker, M., Conza, N., Benagli, C., Pedroli, A., Bernasconi, M.V., Permin, L., Aznar, R., Piffaretti, J. 2003. Population genetics of *Vibrio vulnificus*: Identification of two divisions and a distinct eel-pathogen clone. *Appl. Environ. Microbiol.* 69:3203–3212.

Hervio-Heath, D., Colwell, R., Derrien, A., Robert-Pillot, A., Fournier, J.M., Pommepuy, M. 2002. Occurrence of pathogenic vibrios in coastal areas of France. *J. Appl. Microbiol.* 92:1123–1135.

Hill, W., Keasler, S., Trucksess, M., Feng, P., Kaysner, C., Lampel, K. 1991. Polymerase chain reaction identification of *Vibrio vulnificus* in artificially contaminated oysters. *Appl. Environ. Microbiol.* 57:707–711.

Høi, L., Dalgaard, A., Larsen, J., Warner, J., Oliver, J. 1997. Comparison of ribotyping and randomly amplified polymorphic DNA PCR for characterization of *Vibrio vulnificus*. *Appl. Environ. Microbiol.* 63:1674–1678.

Høi, L., Dalsgaard, I., Dalsgaard, A. 1998. Improved isolation of *Vibrio vulnificus* from seawater and sediment with cellobiose-colistin agar. *Appl. Environ. Microbiol.* 64:1721–1724.

Hollis, D., Weaver, R., Baker, C., Thornsberry, C. 1976. Halophilic vibrio species isolated from blood cultures. *J. Clin. Microbiol.* 3:425–431.

Johnson, D., Calia, F. 1981 Hemolytic reaction of clinical and environmental strains of *Vibrio vulnificus*. *J. Clin. Microbol.* 14:457–459.

Johnson, D., Calia, F., Musher, D., Coree, A. 1984. Resistance of *Vibrio vulnificus* to serum and opsonizing factors: Relation to virulence in suckling mice and humans. *J. Infect. Dis.* 150:413–418.

Johnston, J., Becker, S., Mcfarland, L. 1986. Gastroenteritis in patients with stool isolations of *Vibrio vulnificus*. *Am. J. Med.* 80:336–338.

Kaysner, C., Abeyta, C. Jr., Wekell, M., DePaola, A., Stott, R., Jr., Leitch, J. 1987. Virulent strains of *Vibrio vulnificus* from estuaries of the United States West Coast. *Appl. Environ. Microbiol.* 53:1349–1351.

Kelly, M.T. 1982. Effect of temperature and salinity on *Vibrio (Beneckea) vulnificus* occurrence in a Gulf Coast environment. *Appl. Environ. Microbiol.* 44:820–824.

Kitaura, T., Doke, S., Azuma, I., Imaida, M., Miyano, K., Harada, K., Yabuuchi, E. 1983. Halo production by sulfatase activity of *Vibrio vulnificus* and *Vibrio cholerae* O1 on a new selective sodium dodecyl sufate-containing agar medium: A screening marker in environmental surveillance. *FEMS Microbol. Lett.* 17:205–209.

Kothary, M., Kreger, A. 1985. Production and partial characterization of an elastolytic protease of *Vibrio vulnificus*. *Infect. Immun.* 50:534–540.

Kreger, A., Lockwood, D. 1981. Detection of extracellular toxin (s) produced by *Vibrio vulnificus*. *Infect. Immun.* 33:583–590.

Lee, C., Panicker, R., Bej, A.K. 2003. Detection of pathogenic bacteria in shellfish using multiplex PCR followed by CovaLink™ NH microwell plate sandwich hybridization. *J. Microbiol. Meth.* 53:199–209.

Lee, J., Levin, R. 2007. Discrimination of γ-irradiated and nonirradiated *Vibrio vulnificus* by using real-time polymerase chain reaction. *J. Appl. Microbiol.* 104:728–734.

Lee, J., Levin, R. 2008. New approach for discrimination *of Vibrio vulnificus by real-time PCR before and after γ-irradiation*. *J. Microbiol. Meth.* 73:1–6.

Lee, J.Y., Eun, J.B., Choi, S.H. 1997. Improving detection of *Vibrio vulnificus* in *Octopus variabilis* by PCR. *J. Food Sci.* 62:179–182.

Lee, S.E., Kim, S.O., Kim, S.J., Shin, J.H., Choi, S.H., Chung, S.S., Rhee, J.H. 1998. Direct identification of *Vibrio vulnificus* in clinical specimens by nested PCR. *J. Clin. Microbiol.* 36:2887–2892.

Marco-Noales, E., Biosca, E.G., Milán, M., Amaro, C. 2000. An indirect immunoflorescent antibody technique for detection and enumeration of *Vibrio vulnificus* serovar E (biotype 2): Development and applications. *J. Appl. Microbiol.* 89:599–606.

Martin, S.J., Siebeling, R.J. 1991. Identification of *Vibrio vulnificus* O serovars with anti-lipopolysaccharide monoclonal antibody. *J. Clin. Microbiol.* 29:1684–1688.

Massad, G., Oliver, J.D. 1987. New selective and differential medium for *Vibrio cholerae* and *Vibrio vulnificus*. *Appl. Environ. Microbiol.* 53:2262–2264.

McPherson, V.L., Watts, J.A., Simpson, L.M., Oliver, J.D. 1991. Physiological effects of the lipopolysaccharide of *Vibrio vulnificus* on mice and rats. *Microbios* 67:141–149.

Moreno, M.L.G., Landgraf, M. 1998. Virulence factors and pathogenicity of *Vibrio vulnificus* strains isolated from seafood. *J. Appl. Microbiol.* 84:747–751.

Muroga, K., Jo, Y., Nishibushi, M. 1976. Pathogenic *Vibrio* isolated from cultured eels. I. Characteristics and taxonomic status. *Fish. Pathol.* 11:141–145.

Nogva, H., Drømtorp, S., Nissen, H., Rudi, K. 2003. Ethidium monoazide for DNA-based differentiation of viable and dead bacteria by 5′-nuclease PCR. *BioTechniques.* 34:804–813.

Oliver, J. 1981. Lethal cold stress of *Vibrio vulnificus* in oysters. *Appl. Environ. Microbiol.* 41:710–717.

Oliver, J., Thomas, M., Wear, J. 1986. Production of extracellular enzymes and cytotoxicity by *Vibrio vulnificus*. *Microbiol. Infect. Dis.* 5:99–111.

Oliver, J., Warner, R., Cleland, D. 1983. Distribution of *Vibrio vulnificus* and other lactose-fermenting vibrios in the marine environment. *Appl. Environ. Microbiol.* 45:985–998.

Oliver, J.D. 1989. *Vibrio vulnificus*. In: *Foodborne Bacterial Pathogens*. M.P. Doyle, Ed., Marcel Dekker, New York, pp. 569–600.

Oliver, J.D., Kaper, J.B. 1997. *Vibrio* species. In: *Food Microbiology Fundamentals and Frontiers*. M.P. Doyle, L.R. Beuchat, T.J. Montville, Eds., ASM Press: Washington, DC, pp. 228–264.

O'Neill, K., Jones, S., Grimes, D. 1992. Seasonal incidence of *Vibrio vulnificus* in the Great Bay estuary of New Hampshire and Maine. *Appl. Environ. Microbiol.* 58:3257–3262.

Overman, T., Kessler, J., Seabolt, J. 1985. Comparison of API 20E, API Rapid E, and API Rapid NFT for identification of members of the family *Vibrionaceae*. *J. Clin. Microbiol.* 22:778–781.

Panicker, G., Myers, M., Bej, A. 2004. Rapid detection of *Vibrio vulnificus* in shellfish and Gulf of Mexico water by real-time PCR. *Appl. Environ. Microbiol.* 70:498–507.

Pfeffer, C., Hite, M., Oliver, J. 2003. Ecology of *Vibrio vulnificus* in estuarine waters of Western North Carolina. *Appl. Environ. Microbiol.* 69:3526–3531.

Poole, M., Bowdre, J., Klapper, D. 1982. Elastase produced by *Vibrio vulnificus*: *In vitro* and *in vivo* effects. *Abstracts, Annual Meeting of the American Society for Microbiology*, B155, p. 43.

Radu, S., Elhadi, N., Hassan, Z., Rusul, G., Lihan, S., Fifadara, N., Yuherman, R., Purwati, E. 1998. Characterization of *Vibrio vulnificus* isolated from cockles (*Anadara granosa*): Antimicrobial resistance, plasmid profiles and random amplification of polymorphic DNA analysis. *FEMS Microbiol. Lett.* 165:139–143.

Shimada, T., Sakazaki, R. 1984. On the serology of *Vibrio vulnificus*. *Jpn. J. Med. Sci. Biol.* 37:241–246.

Simonson, J., Siebeling, R. 1986. Rapid serological identification of *Vibrio vulnificus* by anti-H coagglutination. *Appl. Environ. Microbiol.* 52:1299–1304.

Simonson, J., Siebeling, R. 1993. Immunogencity of *Vibrio vulnificus* capsular polysaccharides and polysaccharide-protein conjugates. *Infect. Immun.* 61:2053–2058.

Simpson, L., White, V., Zande, S., Oliver, J. 1987. Correlation between virulence and colony morphology in *Vibrio vulnificus*. *Infect. Immun.* 55:269–272.

Smith, G., Merkel, J. 1982. Collagenolytic activity of *Vibrio vulnificus*: Potential contribution to its invasiveness. *Infect. Immun.* 35:1155–1156.

Stelma, G., Jr., Reyes, A., Peeler, J., Johnson, C., Spaulding, P. 1992. Virulence characteristics of clinical and environmental isolates of *Vibrio vulnificus*. *Appl. Environ. Microbiol.* 58:2776–2782.

Tacket, C., Brenner, F., Blake, P. 1984. Clinical features and an epidemiological study of *Vibrio vulnificus* infections. *J. Infect. Dis.* 149:558–561.

Tamplin, M., Martin, A., Ruple, A., Cook, D., Kaspar, C. 1991. Enzyme immunoassay for identification of *Vibrio vulnificus* in seawater sediment and oysters. *Appl. Environ. Microbol.* 57:1235–1240.

Tamplin, M., Rodrick, G., Blake, N., Cuba, T. 1982. Isolation and characterization of *Vibrio vulnificus* from two Florida estuaries. *Appl. Environ. Microbiol.* 44:1466–1470.

Tassin, M., Siebeling, R., Roberts, N., Larson, A. 1983. Presumptive identification of *Vibrio* species with H antiserum. *J. Clin. Microbiol.* 18:400–407.

Testa, J., Daniel, L., Kreger, A. 1984. Extracellular phospholipase A_2 and lysophospolipase produced by *Vibrio vulnificus*. *Infect. Immun.* 45:458–463.

Tilton, R., Ryan, R. 1987. Clinical and ecological characteristics of *Vibrio vulnificus* in the Northeastern United States. *Diagn. Microbiol. Infect. Dis.* 6:109–117.

Tison, D., Nishibuchi, J., Greenwood, J., Seidler, R. 1982. *Vibrio vulnificus* biogroup 2: New biogroup pathogenic for eels. *Appl. Environ. Microbiol.* 44:640–646.

Tison, D., Seidler, R. 1981. Genetic relatedness of clinical and environmental isolates of the lactose-positive *Vibrio vulnificus*. *Curr. Microbiol.* 6:181–184.

Vickery, M., Harold, N., Bej, A. 2000. Cluster analysis of AP-PCR generated DNA fingerprints of *Vibrio vulnificus* isolates from patients fatally infected after consumption of raw oysters. *Lett. Appl. Microbiol.* 30:258–262.

Vickery, M., Smith, A., DePaola, A., Jones, D., Steffan, R., Bej, A. 1998. Optimization of the arbitrary-primed polymerase chain reaction (AP-PCR) for intra-species differentiation of *Vibrio vulnificus*. *J. Microbiol. Meth.* 33:181–189.

Wang, S., Levin, R. 2005a. Quantitative determination of *Vibrio parahaemolyticus* by polymerase chain reaction. *Food Biotechnol.* 18:279–287.

Wang, S., Levin, R. 2005b. Quantification of *Vibrio vulnificus* using the polymerase chain reaction. *Food Biotechnol.* 19:27–35.

Wang, S., Levin, R. 2005c. Discrimination of viable *Vibrio vulnificus* cells from dead cells in real-time PCR. *J. Microbiol. Meth.* 64:1–8.

Wang, S., Levin, R. 2006. Rapid quantification of *Vibrio vulnificus* in clams (*protochaca staminea*) using real-time PCR. *Food Microbiol.* 23:757–761.

Wang, S., Levin, R. 2007. Thermal factors influencing detection of *Vibrio vulnificus* using real-time PCR. *J. Microbiol. Meth.* 69:358–363.

Warner, J., Oliver, J. 1998. Randomly amplified polymorphic DNA analysis of starved and viable but nonculturable *Vibrio vulnificus* cells. *Appl. Environ. Microbiol.* 64:3025–3028.

Warner, J., Oliver, J. 1999. Randomly amplified polymorphic DNA analysis of clinical and environmental isolates of *Vibrio vulnificus* and other *Vibrio* species. *Appl. Environ. Microbiol.* 65:1141–1144.

West, P., Russek, E., Brayton, P., Colwell, R. 1982. Statistical evaluation of a quality control method for isolation of pathogenic *Vibrio* species on selected thiosulfate-citrate-bile salts-sucrose agars. *J. Clin. Microbiol.* 16:1110–1116.

Wright, A., Micelli, G., Landry, W., Christy, J., Watkins, W., Morris, J., Jr. 1993. Rapid identification of *Vibrio vulnificus* on nonselective media with an alkaline phosphatase-labeled oligonucleotide probe. *Appl. Environ. Microbiol.* 59:541–546.

Wright, A., Morris, J., Jr., Maneval, D., Jr., Richardson, K., Kaper, J. 1985. Cloning of the cytotoxin-hemolysin gene of *Vibrio vulnificus*. *Infect. Immun.* 50:922–924.

Wright, A., Simpson, L., Oliver, J. 1981. Role of iron in pathogenesis of *Vibrio vulnificus* infections. *Infect. Immun.* 34:503–507.

Yamamoto, K., Wright, A.C., Kaper, J., Morris, J., Jr. 1990. The cytolysin gene of *Vibrio vulnificus*: Sequence and relationship to the *Vibrio cholerae* El Tor hemolysin gene. *Infect. Immun.* 58:2706–2709.

Yoshida, S., Ogawa, M., Mizaguchi, Y. 1985. Relation of capsular materials and colony opacity to virulence of *Vibrio vulnificus*. *Infect. Immun.* 47:446–451.

CHAPTER 6

Vibrio parahaemolyticus

I. CHARACTERISTICS OF THE ORGANISM

Vibrio parahaemolyticus is a gram-negative, polarly flagellated, facultatively anaerobic enteropathogenic rod indigenous to coastal marine environments and shellfish that is capable of causing mild gastroenteritis to severe debilitating dysentery. Infections of the gastrointestinal tract are usually due to consumption of raw shellfish. In addition, extraintestinal infections have also been reported to be due to the organism, such as eye and ear infections and wound infections of the extremities. The first recorded outbreak of seafood infection due to *V. parahaemolyticus* occurred in Japan in 1950 (Miwatani and Takeda, 1976). Among 272 patients with acute gastroenteritis, 20 succumbed. The incubation period with most cases was 2 to 6 hr. The symptoms included acute abdominal pain, vomiting, and diarrhea, with watery and in some cases bloody diarrhea. *V. parahaemolyticus* was discovered to be the causative agent in this initial outbreak by Fujino (1951). The organism is considered to be halophilic with an optimum NaCl concentration of about 3.0% (Takikawa, 1958).

The characteristics of the organism were first described in detail by Fujino et al. (1953). A major distinction between *V. parahaemolyticus* and members of the genera *Aeromonas* and *Pseudomonas* is the formation of spheroplasts (Fujino et al., 1965). In 1963, Sakazaki, Iwanami, and Fukumi (1963) defined two biotypes of *V. parahaemolyticus*. Biotypes 1 and 2 can be differentiated by the ability of biotype 2 to grow in alkaline peptone water (APW) containing 10% NaCl, a positive Voges–Proskauer reaction, and fermentation of sucrose. Biotype 1 is negative for these attributes. Sakazaki, Iwanami, and Fukumi (1963) named the biotype 2 group *Vibrio alginolyticus*. Anderson and Ordal (1972) found that a reference strain of *V. parahaemolyticus* showed more than 91% DNA–DNA homology to other *V. parahaemolyticus* strains and only 63% to 70% homology to *V. alginolyticus* strains.

All strains of *V. parahaemolyticus* have been found to possess a thermolabile hemolysin (Fujino et al., 1969; Miwatani et al., 1972) encoded by the *lht* gene that is not directly related to virulence. PCR primer pairs have therefore been developed utilizing the resulting amplicons from the *lht* gene for identification of all isolates of *V. parahaemolyticus*. Virulence has been found to be associated with two principal genes that code for (1) a thermally stable direct acting hemolysin (*tdh*) and (2) a thermally stable direct acting-related hemolysin (*trh*). However, not all clinical strains have been found to possess the *trh* gene. Primer pairs targeting sequences of the *tdh* gene are therefore used to distinguish virulent from nonvirulent strains. Virulent strains are usually characterized as Kanagawa phenomenon (KP) positive, which refers to β-hemolysis on a special blood agar known as Wagatsuma Blood agar (Table 6.3). Epidemiological studies have indicated that specific clones of certain serotypes, notably O3:K6 having enhanced virulence, have become endemically established in certain global locales.

II. CLINICAL SYMPTOMS DUE TO INFECTIONS BY *V. PARAHAEMOLYTICUS*

Gastrointestinal infections due to *V. parahaemolyticus* resulting from the ingestion of raw seafood are usually mild with duration of 2–3 days (Twedt, 1989). Clinical symptoms accompanying diarrhea may include abdominal cramps, nausea, vomiting, headache, low-grade fever, and chills. The incubation period can range from 4 to 96 hr (Twedt, 1989). A more severe and debilitating dysenteric form of gastrointestinal infection with bloody stools and marked leucocytosis requiring hospitalization has been observed in Southeast Asia, India, and Bangladesh, with several outbreaks in the United States, due particularly to strains of the serotype O3:K6 (Bolen, Zamiska, and Greenough, 1974; Hughes et al., 1978; Daniels et al., 2000). Extraintestinal infections due to *V. parahaemolyticus* are also recognized involving wounds of the extremities (Roland, 1970; Bonner et al., 1983; Hollis et al., 1976), ear (Ghosh and Bowen, 1980; Olsen, 1978) and eye infections (Steinkuller et al., 1980; Tacket et al., 1982), and bacteremia (Bonner et al., 1983; Hollis et al., 1976). Symptoms of gastroenteritis were found by Sanyal and Sen (1974) to occur rapidly after volunteers ingested 2×10^5 to 3×10^7 CFU of KP$^+$ *V. parahaemolyticus*. They concluded that the infectious dose for *V. parahaemolyticus* is ~10^5 CFU. In contrast, volunteers receiving 4×10^9 to 1.6×10^{10} CFU of KP$^-$ *V. parahaemolyticus* did not encounter symptoms of diarrhea. However, more recent studies have suggested that unusually virulent strains may be infectious at lower cell numbers (DePaola et al., 2000).

III. ECOLOGY OF *V. PARAHAEMOLYTICUS*

V. parahaemolyticus is widespread along marine coastal waters globally. The organism is considered a common inhabitant of estuaries, being found throughout estuarine environments including water, sediment, suspended particles, plankton, fish, and shellfish and is infrequently found in freshwater or full-strength seawater (Joseph, Colwell, and Kaper, 1982). Higher numbers are usually encountered during warm summer months. The organism is cold-sensitive and, in addition, is unable to grow at hydrostatic pressures of 200 to 1000 atm. encountered in the deep sea environment (Schwartz and Colwell, 1974). For these reasons, the organism appears to be limited to inshore coastal and estuarine areas.

Sutton (1974) reported that in oysters from the coast of Sydney, Australia, the viable count increased with increasing water temperature, but that even with a water temperature of 25°C, the count of *V. parahaemolyticus* never exceeded 12 per 100 g of oyster tissue.

Fishbein et al. (1974) reported the isolation of *V. parahaemolyticus* from 30 different marine species, including eel, crab, clams, oysters, lobsters, scallops, sardines, shrimp, and squid. From 1969 to 1972, 546/635 (85%) of seafood samples in the United States were found by the U.S. Food and Drug Administration (FDA). to be positive for the organism (Fishbein et al., 1974). Counts of *V. parahaemolyticus* have been found to be as high as 1300 per gram in oysters (Felsenfeld and Cabirac, 1977), which is of considerable public health importance considering the widespread practice of consuming shellfish raw or undercooked (Abbott et al., 1989; CDC, 1998; Daniels et al., 2000; Myers, Panicker, and Bej, 2003).

Asakawa, Akahane, and Noguchi (1994) reported that recently landed horse mackerel yielded no more than 10^2 CFU of *V. parahaemolyticus* on the entire surface of individual fish. However, holding at 21–25°C for 10 hr resulted in 10^5 to 10^6 *V. parahaemolyticus*.

Sircar et al. (1976) reported that 33.3% of patients stricken in Calcutta with gastroenteritis infections due to *V. parahaemolyticus* had not eaten seafood for at least 7 days prior to the onset of symptoms. About 15% of family contacts of these patients were found to be healthy carriers who shed the organism.

IV. PHENOTYPIC CHARACTERISTICS OF *V. PARAHAEMOLYTICUS*

A. General

Hugh and Sakazaki (1972) were the first to describe the minimal characteristics for identification of *V. parahaemolyticus* (Table 6.1). A more extensive list of characteristics is presented in the eighth edition of *Bergy's Manual of Determinative Bacteriology* (Shewan and Véron, 1974) and by Joseph, Colwell, and Kaper (1982). The DNA base composition of isolates ranges from 44 to 46 mols% G+C (Joseph, Colwell, and Kaper, 1982). The organism is considered to have a minimum growth temperature of 9 to 10°C and a maximum growth temperature of about 44°C (Horie et al., 1966; Jackson, 1974). Beuchat (1973), however, reported moderate growth at 5°C. Most strains are indole-positive; however, indole-negative strains have been found associated with several food infection outbreaks. Sakai, Kudoh, and Zen-Yoji (1974) reported on *V. parahaemolyticus* gastroenteritis outbreaks from 1970 to 1972 in Tokyo due to indole-negative strains and mixed infections with indole-positive and negative strains. A total of 64 indole-negative strains were serologically typed as O1:K1 and the remainder as O1:TNK11. All were KP+.

Table 6.1   Minimal Characters for Identification of
V. parahaemolyticus[a]

Character	Reaction	% +
Gram-negative, asporogenous rod	+	100
Indophenol oxidase	+	100
Glucose, acid under a seal of petroleum	+	100
Glucose, gas	–	0
D-Mannitol, acid	+	99.6
Sucrose, acid	–	5.3
Acetylmethylcarbinol	–	0
Hydrogen sulfide	+	97[b]
L-Lysine decarboxylase	+	100
L-Arginine dihydrolase	–	0
L-Ornithine decarboxylasae	+	97.3
Indole production	+	98.9
Growth in 1% tryptone broth	–	0
Growth in 1% tryptone broth with 8% NaCl	+	—
Growth in 1% tryptone broth with 10% NaCl	–	100

[a] Modified from Hugh and Sakazaki (1972) and Twedt (1989).
[b] From Colwell (1970).

B. Flagellation

V. parahaemolyticus produces unsheathed peritrichous or lateral flagella when grown on solid media and a single sheathed polar flagellum when grown in a liquid medium (Baumann, Baumann, and Mandel, 1971; Allen and Baumann, 1971; Yabuuchi et al., 1974a,b; Belas, Simon, and Silverman, 1986). The peritrichous flagella are curled and atypical, whereas the polar flagellum is considered the normal type (Yabuuchi et al., 1974b). Kimura, Tateiri, and Iida (1979) reported that the formation of peritrichous, but not polar flagella, was inhibited in media of pH 8.5 and higher. The single polar flagellum propels the cell in liquid for swimming motility and is powered by the sodium-motive force, and the lateral flagella are driven by the proton-motive force and enable swarming over surfaces (Atsumi, McCarter, and Imae, 1992). The polar flagellum is produced continuously, whereas production of the lateral flagella is induced when the organism is grown on surfaces (Belas, Simon, and Silverman, 1986; McCarter, Hillmen, and Silverman, 1988).

Genes responsible for formation of the lateral flagella (*laf*) resulting in swarmer cells are induced when *V. parahaemolyticus* is grown not only on the surface of solid media, but also when suspended in viscous media, or agglutinated with antibody against the polar flagellum (Belas, Simon, and Silverman, 1986; McCarter, Hilmen, and Silverman, 1988). These conditions have in common the constraint of the polar flagellum, which induces the swarmer *laf* genes and lateral flagella formation (McCarter, Hillmen, and Silverman, 1988). Mutations in genes encoding components of the polar flagellum (*fla*) have been found to result in the constitutive expression of the *laf* genes and constitutive formation of lateral flagella (McCarter, Hillmen, and Silverman, 1988). Because the polar flagellum appears capable of sensing external forces influencing its motion, the term dynamometer was applied to it by McCarter, Hillmen, and Silverman (1988). Belas, Simon, and Silverman (1986) found that insertional mutations in the *laf* genes resulted in the inability to form lateral flagella resulting from defects in the swarming phenotype.

Stewart and McCarter (2003) studied the regulatory gene cascade controlling the formation of the lateral flagella (*Laf* gene system). The lateral flagella system was found to involve 38 genes. Swarming and swimming motility were entirely dependent on the sigma factor (σ^{54}), whereas the *FliAl* gene (lateral σ^{28}) and *LafK* gene appeared to be swarming-specific regulators in that only swarming but not swimming was abolished by induced mutations of these two genes.

C. Antigenic Properties

Three principal categories of outer antigens are produced by strains of *V. parahaemolyticus*: thermostable somatic O antigens, thermolabile K antigens, and flagellar antigens. Antigenicity of the protein flagellin, from the polar flagellum, is common to all isolates of *V. parahaemolyticus* in addition to many other *Vibrio* species (Shinoda et al., 1976). The polar flagellin of *V. parahaemolyticus* differs antigenically from the flagellin of the peritrichous (lateral) flagella (Shinoda et al., 1974). The use of antiserum against lateral flagella in an agglutination assay has been found useful in identifying *V. parahaemolyticus* isolates from marine samples (Shinoda et al., 1983).

The K antigens are acidic polysaccharides that are released from the cell surface by heating at 100°C for 1–2 hr (Omori et al., 1966) resulting in exposure of the somatic O antigens. The O antigens are lipopolysaccharides (Torii et al., 1969; Torrii, 1974). With few exceptions, each K antigen is associated with a single somatic O antigen (Sakazaki et al., 1968). The complete antigenic scheme for *V. parahaemolyticus* encompassing both O and K antigens is given in Table 6.2.

Table 6.2   Antigenic Scheme of *V. parahaemolyticus*[a]

O Group	K Antigen
1	1, 25, 26, 32, 38, 41, 56, 58, 64, 69
2	3, 28
3	4, 5, 6, 7, 29, 30, 31, 33, 37, 43, 45, 48, 54, 57, 58, 59, 65
4	4, 8, 9, 10, 11, 12, 13, 34, 42, 49, 53, 55, 63, 67
5	15, 17, 30, 47, 60, 61, 68
6	18, 46
7	19
8	20, 21, 22, 39, 70
9	23, 44
10	19, 24, 52, 66, 71
11	36, 40, 50, 51, 61

[a] From Twedt (1989), personal communication from R. Sakazaki in 1986.

D. Hemolysins

1. General

At least five hemolytic components have been found to be produced by *V. parahaemolyticus:* (1) a heat-stable direct hemolysin (Obara, 1971; Sakurai, Matsuzaki, and Miwatani, 1973; Douet et al., 1992; Taniguchi et al., 1986; Nishibuchi and Kaper, 1985, 1990); (2) a heat-stable direct-related hemolysin that is heat labile (Bej et al., 1999; Kishishita et al., 1992; Nishibuchi et al., 1989); (3) a heat-labile direct hemolysin (Taniguchi et al., 1986; Bej et al., 1999); (4) a phospholipase A (Yanagase et al., 1968); and (5) a lysophospholipase (Yanagase et al., 1970; Misaki and Matsumoto, 1978). *V. parahaemolyticus* also produces a lecithinase (Yanagase et al., 1968) and a glycerophosphorylcholine diesterase (Yanagase et al., 1970). The latter four function indirectly by enhancing hemolysis.

2. Direct Acting Hemolysins

Hemolysis on Wagatsuma's Agar (Wagatsuma, 1968) is referred to as the Kanagawa phenomenon (KP) and has been found to correlate well with human pathogenicity. The KP phenomenon is characterized by the appearance on Wagatsuma agar of a sizeable clear halo of hemolysis after 18 to 24 hr of incubation at 37°C. The presence of 7% NaCl in Wagatsuma agar (Table 6.3) is thought to stress the cells resulting in enhanced production of hemolysin. The use of serological detection of the hemolysin has resulted in some KP⁻ strains being designated weak KP⁺ (Ohashi et al., 1977). Lu (2003) found that the incorporation of sphingomyelinase (10 units/L) to blood agar base containing 5.0% defibrinated sheep blood resulted in most shellfish isolates of *V. parahaemolyticus* exhibiting readily discernible but weak β-hemolysis.

Sakazaki et al. (1968) reported that 2655/2720 (96.6%) of human clinical isolates were KP⁺ and that only 7/650 (1%) of environmental isolates were KP⁺. Thompson, Vanderzant, and Ray (1976) found only 4/2218 environmental isolates to be KP⁺. Other investigators have reported similar results (Joseph, Colwell, and Kaper, 1982). The relationship between KP⁺ strains and human pathogenicity has been suggested by Sakazaki et al. (1974) as resulting from selective multiplication of KP⁺ strains in the

Table 6.3   Composition of Wagatsuma's Medium[a,b]

Yeast extract	5.0 g
Peptone	10.0
Mannitol	5.0
K_2HPO_4	5.0
NaCl	70.0
Agar	15.0
dH_2O	1.0 L

[a] Wagatsuma (1968).
[b] Final pH is 7.5; 5% defibrinated and washed rabbit or human red blood cells are added after sterilization of medium.

human intestine. The absence of KP+ strains in the environment has raised the question of whether KP+ strains actually originate from the natural environment. The predominance of KP− strains in the marine environment is thought to result from the greater survival of KP− strains than KP+ strains in seawater (Joseph, Colwell, and Kaper, 1982).

Miyamoto et al. (1969) were the first to observe that strains of *V. parahaemolyticus* from human sources (infected individuals and healthy carriers) exhibited hemolysis, whereas those from natural environments or from seafood were nonhaemolytic. Hemolysis was found to be enhanced on Wagatsuma blood agar. In addition, if the NaCl concentration was lower than 5%, nonhemolytic strains were found to also exhibit hemolysis and the distinction was lost.

Chun, Chung, and Tak (1974) found that more than one-third of marine isolates of *V. parahaemolyticus* were KP+, which is contrary to the observations of Sakazaki et al. (1968). KP+ stains showed noticeable hemolysis after 16–20 hr at 37°C on Wagatsuma blood agar, whereas KP− strains required 32–40 hr incubation for detectable hemolysis to appear. When 0.01 M CaCl₂ was added to Wagatsuma blood agar, KP− strains became strongly hemolytic after 24 hr incubation and the hemolysis of KP+ strains was greatly intensified.

The purified thermostabile hemolysin (TDH) is a dimeric protein of 44,000 Da (Miyamoto et al., 1980), has been found to be responsible for the Kanagawa phenomenon, causes fluid accumulation in the ligated mouse ileum, is cytolytic against cultured mammalian cells, and lyses erythrocytes of various animal spieces (Douet et al., 1992). It is stable to heating at 100°C for 10 min (Sakurai, Matsazaki, and Miwatani, 1973; Douet et al., 1992) and is activated by Ca++ (Chun, Chung, and Tak, 1974). The heat-labile hemolysin is usually found in KP− strains, but not consistently in KP+ strains, which characteristically possess the thermostabile hemolysin (Miyamoto et al., 1980). At least 10 direct-acting hemolysins have been found in various strains of *V. parahaemolyticus* (Table 6.4).

Table 6.4 Thermally Resistant Direct Acting (TDH), TDH-Related (TRH), and Thermally Labile (TL) Hemolysins of *V. parahaemolyticus*

Hemolysin	Gene	Reference
Vp-TDH1	Vp-*tdh1*	Nishibushi and Kaper (1985), Taniguchi et al. (1986)
Vp-TDH2	Vp-*tdh2*	Nishibuchi and Kaper (1990)
Vp-TDH3	Vp-*tdh3*	Nishibuchi and Kaper (1990)
Vp-TDH4	Vp-*tdh4*	Nishibuchi and Kaper (1990)
Vp-TDH5	Vp-*tdh5*	Baba et al. (1991)
TRH1	*trh1*	Kishishita et al. (1992)
TRH2	*trh2*	Kishishita et al. (1992)
TDHA	*tdhA*	Iida and Yamamoto (1990)
TDHS	*tdhS*	Iida and Yamamoto (1990)
TDHx	*tdhX*	Xu et al. (1994)
TRHx	*trhX*	Xu et al. (1994)
TL	*tl*	Taniguchi et al. (1986)

Cherwongrodzky and Clark (1981) found that hemolysin production in peptone broth cultures of *V. parahaemolyticus* occurred only when the pH was between 6.5 and 5.5. Mannitol, present at a 1.0% level in Wagatsuma agar, was found to reduce the pH to within this pH range. Glucose and mannose, although readily metabolized, lowered the pH below this critical range, inhibiting hemolysin production. The authors concluded that production of the hemolysin was pH controlled.

Barrow and Miller (1974) observed that both KP+ and KP- strains survive well in seawater below 25°C, but at 37°C KP+ strains lose their ability to cause hemolysis of blood cells and would thus be regarded as KP-. The authors also noted that several strains that were KP+ on primary isolation from seafoods subsequently became KP- almost immediately after subculture at 37°C. Repeated attempts to obtain KP+ revertants failed.

Sakurai et al. (1973) purified the TDH from a strain of *V. parahaemolyticus* to electrophoretic homogeneity with the use of exclusion chromatography with a SephadexG-200 column from which they derived a mol. wt. of 118,000 Da for the hemolysin that was stable to heating at 100°C for 10 min.

Miwatani et al. (1974) found that heating the crude TDH from a KP+ strain at 60°C for 10 min resulted in partial inactivation of hemolytic activity. In contrast, heating at 80 to 90°C for 10 min did not result in significant inactivation. The "Arrhenius effect" was found to disappear when the hemolysin was purified and was attributed to a protein that inhibited the TDH at 60°C. An alternate explanation is that the hemolysin heated at 60°C may have activated a proteinase that was rapidly inactivated at 80 to 90°C and which was removed on purification of the TDH. They reported that the purified TDH had a mol. wt. of 118,000 and 114,000 by gel filtration and sedimentation velocity, respectively. These values are notably higher than those obtained in subsequent investigations, which may have been due to aggregation of the purified TDH (Takeda, Thaga, and Miwatani, 1978).

Zen-Yoji et al. (1974) distinguished two hemolysins from strains of *V. parahaemolyticus* that were immunologically distinguishable and were designated "a" and "a¹". The former was considered identical to the TDH.

Takeda, Thaga, and Miwatani (1978) purified the TDH from *V. parahaemolyticus* and found it to have a mol. wt. of ~42,000 and to consist of two subunits of ~21,000 Da. Miyamoto et al. (1980) purified the hemolysin of *V. parahaemolyticus* associated with the Kanagawa phenomenon. The purified enzyme was stable to heating at 100°C for 60 min at pH 6.0, was found to have an estimated mol. wt. of 44,000 Da, to consist of two subunits of 22,000 Da each, and have an isoelectric point of 4.9.

Douet et al. (1992) purified and characterized the Kanagawa haemolysin from a strain of *V. parahaemolyticus* and found it to be stable to heating at 100°C for 10 min and to be a monomeric protein with a mol. wt. of about 29,000 Da. Unlike the previous studies (Zen-Yoji et al., 1974; Miyamoto et al., 1980), Douet et al. (1992) found the haemolysin to lack dimeric subunits with PAGE. They suggested that a possible explanation for the apparent higher molecular weight reported in previous studies may have been due to the retarding interaction of the haemolysin on the dextran-based gel filtration columns used to estimate the molecular weight.

Honda, Ni, and Miwatani (1988) purified the hemolysin from a KP⁻ strain of
V. parahaemolyticus of human clinical origin. The clinical symptoms of gastroenteritis due to this strain were indistinguishable from those due to KP⁺ strains. The
hemolysin was designated "thermostable direct hemolysin (Vp-TDH)-related hemolysin" or Vp-TRH. The hemolysin had an estimated molecular weight of 48,000 Da
and was found to consist of two subunits of 23,000 Da each. The isoelectric point
was determined to be 4.6. The hemolytic activity of VP-TRH was labile on heating
at 60°C for 10 min at pH 7.0, unlike that of Vp-TDH. The hemolytic activities of
Vp-TRH on sheep, calf, and chicken erythrocytes were notably less than those of
Vp-TDH. The *tdh* and *trh* genes have been found to have 68.6% nucleotide sequence
homology (Nishibuchi et al., 1989).

Burstyn, McNicol, and Voll (1980) used nalidixic acid to enrich for spontaneous
auxotrophic and KP⁻ mutants of *V. parahaemolyticus*. Some Cyst⁻ and Arg⁻ mutants
of a KP⁺ strain were found to be KP⁻. Reversion to prototrophy by these strains was
not accompanied by a return to the parental KP⁺ phenotype. With few exceptions,
only KP⁻ strains are usually recovered from food involved in gastroenteritis by *V.
parahaemolyticus*. Barker and Gangarosa (1974) earlier suggested that KP⁻ strains
associated with seafood may undergo conversion to KP⁺ pathogenic strains during
transit through the intestine, and that such conversion may be due to the result of
DNA transfer from resident intestinal microorganisms. An alternative mechanism
may involve environmental stimuli that influence insertion/excision events of an
IS-like element governing the expression of the hemolysin gene (Burstyn, McNicol,
and Voll, 1980).

Nishibuchi et al. (1985) developed a 406-bp gene probe specific for the Vp-TDH
gene. All 66 KP⁺ strains examined were detected with the probe, in addition to 12/14
weakly KP⁺ strains and 10/61 KPO⁻ strains. Among 121 other *Vibrio* isolates exclu-
sive of *V. parahaemolyticus*, only *V. hollisae* strains reacted with the probe under
stringent conditions.

Molecular epidemiological studies have revealed that not only strains carrying
the *tdh* gene but strains carrying a *trh* gene or both genes are strongly associated with
gastroenteritis (Okuda et al., 1997a; Shirai et al., 1990).

Shirai et al. (1990) used *tdh* and *trh* gene probes to detect the TDH and TRH
producing genes in strains of *V. parahaemolyticus*. Among a total of 214 clinical
strains, 112 (52.3%) had the *tdh* gene only, 52 strains (24.3%) had the *trh* gene, and 24
strains (11.2%) carried both the *tdh* and *trh* genes. Among 71 environmental strains,
5 (7.0%) exhibited weak hybridization with the *trh* gene probe and none hybridized
with the *tdh* gene probe. These results suggest that the TRH as well as the TDH is
an important virulence factor for *V. parahaemolyticus*. A particularly interesting
observation in this study was the absence of both the *tdh* and *trh* genes in 26 of the
clinical isolates, which may reflect significant *tdh* or *trh* nucleotide sequence variation among certain clinical strains. It is interesting that among 48 *trh* gene-positive
strains of *V. parahaemolyticus*, only 18 (37.5%) were found to produce TRH in culture media when examined by a sensitive ELISA assay. Some discrepancies have
been noted between the KP, as defined by β-hemolysis on Wagatsuma agar, and the
actual production of Vp-TDH, as determined by immunological methods, in that

immunological methods have been found useful for detecting low levels of TDH in certain strains that are KP⁻ (Honda et al., 1980).

Taniguchi et al. (1985) cloned the *tdh* and *trh* genes. Colony hybridization indicated that the *tdh* gene was present in all 15 KP⁺ strains, but not in 14 KP⁻ strains. It is interesting that the *trh* gene was present in all 29 strains.

KP⁺ strains have been found to carry two copies of nonidentical *tdh* genes (*tdh1* and *tdh2*) but not the *trh1* or the *trh2* genes (Okuda et al., 1997a). Strains carrying both *tdh* and *trh* genes are KP⁻ and produce TDH in much smaller amounts than KP⁺ strains (Shirai et al., 1990). In the KP⁺ strain the expression level of the *tdh2* gene is higher than that of the *tdh1* gene (Nakaguchi et al., 2003). Both genes possess 97.2% sequence homology to each other (Nishibuchi and Kaper, 1990).

Kelly and Stroh (1989) studied the phenotypic characteristics of 13 clinical and 221 environmental isolates of *V. parahaemolyticus* from the Pacific Northwest coast of Canada. Of the environmental isolates 1.45% were KP⁺, and of the clinical isolates 23% were KP⁺. Clinical isolates from locally acquired extraintestinal infections were KP⁻ and urease-negative (UH⁻). Isolates from locally acquired gastrointestinal cases were KP⁻ and UH⁺. Isolates from travelers' diarrhea were KP⁺ and UH⁻. Eight percent of the environmental isolates were KP⁻ and Uh⁻. The expression of the Kanagawa hemolysin was concluded not to be essential for pathogenesis of *V. parahaemolyticus* and that gastroenteritis in the Pacific Northwest may be due to strains that are KP⁻ and UH⁺.

Burstyn, McNicol, and Voll (1980) isolated auxotrophic and KP⁻ mutants. Cys⁻ and Arg⁻ mutants of KP⁺ strains were found to be KP⁻. Reversion to prototrophy was not accompanied by a concomitant return to the KP⁺ phenotype. The authors suggested that environmental stimuli might influence insertion/excision events of an intrastrand-like element governing the expression of the hemolysin gene, which might explain the organism's ability to respond to changes in the environment.

Kaper et al. (1984) were the first to clone the *tdh* gene. The cloned *tdh* gene was then labeled and used as a colony probe for confirmation of KP⁺ and KP⁻ strains. The *tdh* gene was first sequenced by Nishibuchi and Kaper (1985) and found to encode a protein of 165 amino acids preceded by a 24-amino acid peptide for an overall calculated molecular weight (mol. wt.) of 21,140 Da, which is consistent with previous mol. wt. determinations performed on the purified TDH (Takeda, Thaga, and Miwatani, 1978). In contrast to *Escherichia coli,* where the hemolysin after introduction was detected only in cell lysates, introduction of the cloned gene into a KP⁻ strain of *V. parahaemolyticus* resulted in the production of extracellular hemolysin.

Taniguchi et al. (1986) determined the nucleotide sequences of the *tdh* and *tl* genes and deduced the mature TDH and TL to have mol. wts. of 18,500 and 45,300, respectively. Maxicell analysis indicated the mature TDH and TL proteins to have mol. wts. of 19,500 and 45,500 Da, respectively.

Xu et al. (1994) studied the unusual simultaneous production of the TDH and TRH by a KP⁺ strain of *V. parahaemolyticus* designated TH3766. The two hemolysins were differentially purified. The mol. wt. was estimated to be 23,000 for both. The purified TDHx was indistinguishable from the previously reported TDH/I from strain TH102 (Honda et al., 1991) but was different from the authentic TDH of a KP⁺ strain (T4750) physicochemically. The mobility of TRHx in nondenaturing

PAGE differed from all known TDHs and TRHs. The genes for both hemolysins were cloned and nucleotide homologies between *tdhX* and *tdhA* (a gene for authentic TDH) and between *trhX* and *trhA* (a gene for authentic *trh*) were 98.1% and 99.1%, respectively. Homology between *tdhX* and *trhX* was 68.1%. The amino acid sequence of TDHx was identical to that of TDH/I. Two amino acid differences were found between TRHX and TRH (Nishibuchi et al., 1989).

Iida and Yamamoto (1990) cloned and sequenced the gene encoding the TDH from a KP+ strain of *V. parahaemolyticus* designated T4750. The strain contained two DNA sequences designated *tdhA* and *tdhS* homologous to the *tdh* gene previously reported by Nishibuchi and Kaper (1985) and Taniguchi et al. (1986). DNA sequence homology between *tdhA* and *tdhS* was 97.2%. The deduced amino acid sequence of the TDHA was identical to that of the TDH protein purified from *V. parahaemolyticus* by Tsunasawa et al. (1987) except for glutamic acid at position 118 in place of glycine. The amino acid sequence deduced from the second gene, *tdhS*, differed in eight amino acids from the TDH protein. However, it agreed with the sequence of TDH deduced from the previously cloned *tdh* gene. The *tdha* gene was concluded to be the structural gene for the TDH found in the culture supernatant of *V. parahaemolyticus* T4750. Little if any expression of the *tdhS* gene occurred with strain T4750. TDHS and TDHA are therefore considered synonymous with Vp-TDH2 and Vp-TDH1 (Table 6.4), respectively.

Yoh et al. (1991) characterized four representative *tdh* genes of *V. parahaemolyticus* derived from two representative strains. *tdh1* and *tdh2* were derived from a KP+ strain (WP-1) and *tdh3* (chromosomal) and *tdh4* (plasmid located) were from a KP− strain (AQ3776). The four *tdh* gene products exhibited different electrophoretic mobilities under nondenaturing conditions, stimulated vascular permeability in the rabbit skin, and were lethal to mice. The *tdh2* gene product was found to be the major extracellular TDH of the KP+ strain. Antigenicities of the four gene products (TDHs) were indistinguishable. Maximum nucleotide sequence divergence among the four *tdh* genes was 3.3%. The authors concluded that the four *tdh* genes have evolved to maintain a fundamental molecular structure and that biological activities of the gene products and minor structural or charge differences of the molecules are perhaps responsible for the slight divergence of their biological activities.

Kishishita et al. (1992) assessed the public health significance of rare environmental strains of *V. parahaemolyticus* that exhibited very weak hybridization signals with a *trh* gene-specific DNA probe. The *trh*-like gene was cloned from an environmental strain and the nucleotide sequence determined. They designated the original *trh* gene *trh1*. The very weak hybridization signal was found due to a variant of the *trh1* gene that they designated *trh2*, which was found to possess 84% homology to the *trh1* gene and 54.8% to 68.8% homology to the various *tdh* genes (Table 6.4). The nucleotide sequences of the *trd1* and *trd2* genes in various strains of *V. parahaemolyticus* were found to exhibit significant strain-to-strain variation. The authors concluded that both the *trh1*- and *trh2*-carrying strains should be considered potentially virulent.

Nishibuchi et al. (1992) converted a TDH+ strain of *V. parahaemolyticus* to an isogenic TDH− stain by specifically inactivating the two copies of the *tdh* gene encoding TDH1 and TDH2. Cells of the parent strain AQ3815 resulted in fluid accumulation in the rabbit ileal loop assay, whereas the TDH⁻ mutant did not. The culture filtrate of the parent strain produced a striking alteration in the ion-flux rabbit ileum in vitro assay, whereas the TDH⁻ strain produced little or no alteration in the ion-flux. Conversion of the TDH⁻ strain to the TDH2 state restored the ability of the culture filtrate to alter the ion-flux in the ileum in vitro assay. These results indicated that TDH⁺ and not TDH⁻ cultures were highly associated with the diarrhea resulting from *V. parahaemolyticus* infection of the gastrointestinal tract.

Okuda and Nishibuchi (1998) determined the importance of variation in the promoter sequence for the expression level of the *tdh* genes. *Tdh1* and *tdh2*-lacZ fusions were used to determine the effect of promoter sequences on expression of both genes. Two bases (positions 24 and 34) within the *tdh2* promoter sequence were shown to be primarily responsible for the difference in the promoter strength and expression of the *tdh2* and *tdh1* genes. The sequence of the promoters of KP⁻ strains was found to be similar to the *tdh2* promoter of KP⁺ strains. They differ, however, at position 34 but have the same base at position 24 as the *tdh2* promoter. Base substitution of the *tdh* promoter of KP⁻ strains only at position 34 was sufficient to increase expression of these genes to the KP⁺ level. Introduction of point mutations at positions 24 and 34 of a *tdh1* promoter converted a KP⁻ strain to a KP⁺ strain with a 65-fold increase in intracellular TDH.

In addition to the hemolysins as virulence factors, Lee et al. (2002) purified and characterized an extracellular serine protease designated protease A from a clinical strain of *V. parahaemolyticus* carrying neither the *tdh* nor the *trh* genes. The purified protease was cytotoxic, heat labile (>50°C), lysed erythrocytes, caused tissue hemorrhage and death in mice when injected intraperitoneally and intravenously, and represents an additional virulence factor.

McCarthy et al. (1999) developed *tlh* gene probes labeled with alkaline phosphatase and digoxigenin (DIG) to facilitate rapid identification of *V. parahaemolyticus* isolates. Both probes functioned with 98% agreement and were effectively equivalent. In a similar manner *tdh* gene probes labeled with alkaline phosphatase and DIG were subsequently developed and found to be highly specific for the *tdh* gene of *V. parahaemolyticus* (McCarthy et al., 2000).

DePaola et al. (2000) determined the density of total *V. parahaemolyticus* and the occurrence of pathogenic (*tdh⁺* or *trh⁺*) strains in oysters following outbreaks in Washington, Texas, and New York in 1997 and 1988 with the use of *tdh* and *trh* digoxigenin colony probes. *V. parahaemoltycus* was recovered from 77%, 100%, and 93% of the samples from the Pacific Northwest, Texas, and New York, respectively. Only two samples (46,000 and 23,000 CFU/g) exceeded the level of concern of 10,000 CFU/g. Most positive counts were between 10 and 1000 CFU/g. The authors concluded from their data that findings of more than 10,000 CFU/g of total *V. parahaemolyticus* or >10/g of *tdh⁺* or *trh⁺* *V. parahaemolyticus* in environmental oysters should be considered extraordinary.

DePaola et al. (2003a) examined oysters for the presence and density of *V. parahaemolyticus* harvested from March 1999 through September 2000 in Mobile Bay, Alabama, a coastal area known to endemically harbor the organism and to be involved in outbreaks at that time. DNA probes targeting the *tlh* and *tdh* genes were used for confirmation of total and pathogenic (*tdh⁺* strains). *V. parahaemoltyicus* was detected in all 156 samples with densities ranging from <10 to 12,000/g. Higher densities of the organism were associated with higher water temperatures. Pathogenic strains (*tdh*+) were detected in 34 (21.8%) of the 156 samples; 97% of the *tdh*+ strains produced urease and were also *trh*+. The O3:K6 serotype was not detected. The efficient screening of numerous isolates by colony lift and DNA probe procedures was thought to account for the higher prevalence of *tdh*+ samples than previously reported.

E. H₂S Production

Early observations regarding H_2S production were negative (Fujino, Sakazaki, and Tamura, 1974; Joseph, Colwell, and Kaper, 1982). Sakazaki, Iwanami, and Fukumi (1963) reported that none of 1702 strains examined produced H_2S using triple sugar iron (TSI) or sulfide indole motility (SIM) media. Twedt, Spaulding, and Hall (1969), however, found that nearly all cultures of *V. parahaemolyticus* examined produced H_2S using SIM and lead acetate agar. Colwell (1970), using more sensitive methods for detection, also found that nearly all of the strains examined produced H_2S. Fujino, Sakazaki, and Tamura (1974) indicated that *V. parahaemolyticus* was H_2S negative with Kligler iron agar. Jegatheson and Paramasivam (1976) found that the use of Russel's triple sugar agar containing 0.18% lead acetate but not TSI was able to detect H_2S production by all strains of *V. parahaemolyticus* examined. The present author found that peptone iron agar is highly reliable for the detection of H_2S production by strains of *V. parahaemolyticus* but frequently requires up to 5 days incubation at 37°C.

F. Urease (Uh) Production

The majority of clinical and environmental isolates are usually found to be urease-negative (Uh⁻; Sakazaki, Iwanami, and Fukumi, 1963; Osawa et al., 1996). Sakazaki, Iwanami, and Fukumi (1963) reported only 4% of strains tested to be urease-positive. Colwell (1970), however, found 97% of isolates examined to be Uh⁺ and Chitu, Cufedu, and Nacescu (1977) found 6/8 strains to be Uh⁺, whereas Joseph, Colwell, and Kaper (1982) reported 13/19 strains to be positive. Abbott et al. (1989) were the first to report the isolation of Uh⁺ clinical strains from the West Coast of California and Mexico and indicated that by 1983, Uh⁺ *V. parahaemolyticus* had become the predominant biotype in California outbreaks. Between 1979 and 1987, out of 45 isolated *V. parahaemolyticus* strains, 32 were found to be Uh⁺. Among these 32 Uh⁺ strains, 19 were found to belong to serovar O4:K12. It is interesting that this serovar was not found among any of the 13 Uh⁻ isolates. Uh⁺ strains cured of their plasmids still retained the Uh⁺ phenotype.

A correlation has been found between the presence of the *trh* gene and urease production among clinical isolates, which is considered to be an unusual characteristic of *V. parahaemolyticus* (Magalhães et al., 1992; Okuda et al., 1997a; Osawa et al., 1996; Suthienkul et al., 1995). Cal and Ni (1996) found that the purified urease from clinical isolates that were KP⁻ and urease-positive resulted in a weak accumulation of intestinal fluid in the rabbit ileal loops test and a weak positive response with the suckling mouse test. High concentrations of ammonium ions are thought to exert toxic effects on the permeability of the intestinal mucosa. The authors concluded that urease is an important etiological factor in the development of gastrointestinal inflammatory lesions.

Osawa et al. (1996) found that among 132 strains of *V. parahaemolyticus* isolated from patients and suspected food items of former foodborne cases occurring in Kanagawa Prefecture, Japan, 10 strains were Uh+ and 4 of these were *tdh*-. In addition, 106/132 strains (12%) were *tdh*+ but less than 6% were Uh+, whereas all 5 *trh*+ strains were Uh+. The authors concluded that urea hydrolysis may not be a reliable marker for identifying *tdh*+ strains in Japan but may be a marker for *trh*+ stains.

Okuda et al. (1997a) analyzed Uh+ and Uh- strains of *V. parahaemolyticus* from patients on the West Coast of the United States isolated between 1979 and 1995 for the presence of the *tdh*, *trh1*, and *trh2* genes. Among 60 Uh+ strains, 59 (98%) possessed either the *trh1* or *trh2* gene and 54 strains (90%) carried the *tdh* gene. The stronger correlation of Uh+ with the *trh* gene than with the *tdh* gene was mostly attributed to strains possessing only the *trh2* gene. Among 25 Uh- strains, 20 (80%) possessed the *tdh* gene and none had the *trh* gene.

Nakaguchi et al. (2003) found that TDH and TRH are produced at low levels by Uh+ strains of *V. parahaemolyticus* and that the urease gene cluster has no influence on the regulation of *tdh* and *trh* expression.

V. SENSITIVITY OF *V. PARAHAEMOLYTICUS* TO LOW TEMPERATURES

The sensitivity of *V. parahaemolyticus* to low temperatures is well documented. Rapid rates of destruction of the organism at refrigerator temperatures has been reported by a number of workers. Temyo (1966) observed a 3-log reduction of *V. parahaemolyticus* at 4°C in peptone broth containing 3% NaCl. Similar observations have been reported when the organism was held at low temperatures in shrimp (Bradshaw, Francis, and Twedt, 1974; Vanderzant and Nickelson, 1972), oysters (Johnson and Liston, 1973; Thomson and Thacker, 1973; Goatcher et al., 1974), homogenized fish fillets (Matches, Liston, and Daneault, 1971; Covert and Woodburn, 1972; Johnson and Liston, 1973), and crabmeat (Beuchat, 1977; Johnson and Liston, 1973). The recovery of *V. parahaemolyticus* from refrigerated shrimp homogenates was found to be more effective when enrichments in tryptic soy broth (TSB) containing 7% NaCl were streaked onto a selective agar compared to enrichment in a selective broth prior to isolation (Vanderzant, Nickelson, and Hazelwood, 1974).

Ray, Hawkins, and Hackney (1978) studied the effect of refrigerated (5°C) and frozen storage (−20°C) on *V. parahaemolyticus* in various seafood homogenates. Cells were sensitive to both storage temperatures, with many cells dying, and many sublethally injured. Refrigerated storage appeared more injurious than frozen storage. Initial recovery of sublethally injured cells was highest in a nutritionally rich, nonselective liquid medium (TSB). A repair-detection method was developed for maximum recovery of injured cells in commercial seafoods that consisted of allowing cells to repair in a nonselective broth, such as TSB, for about 2 hr followed by adding NaCl to a final concentration of 3%, and then incubating further for maximum growth. This procedure was then followed by selective enrichment in glucose salt Teepol broth (GSTB), followed by streaking onto thiosulfate citrate bile salts sucrose (TCBS) agar plates.

Boutin et al. (1985) studied the survival of several strains of *V. parahaemolyticus* in shrimp homogenates at 4°, –20°, and at –80°C. The process of freezing and thawing was found to destroy 1 log cycle of the cells. Thereafter, the counts remained unchanged at –80°C, but continued to decline an additional 5 logs during 25 days at –20°C, and an additional 6 logs at 4°C. Cells suspended in 9% dimethylsulfoxide exhibited no loss in cell viability at –80°C.

Jiang and Chai (1996) found that *V. parahaemolyticus* could reach the nonculturable stage in 50 to ~80 days during starvation at 3.5°C. The critical factor was the low temperature. A KP⁻ strain lost culturability more slowly than a KP⁺ strain at low temperature. The initial difference in CFU between TCBS agar and a nonselective agar was about 1 log. With increased time at 3.5°C the CFU on TCBS agar declined to below 10 CFU/ml, whereas the CFU on the nonselective agar were still 10^6/ml.

VI. ISOLATION AND CULTIVATION OF *V. PARAHAEMOLYTICUS*

The U.S. FDA manual (Elliot et al., 1998) recommends blending food samples in 3% NaCl or PBS, enriching into alkaline peptone water (1% NaCl) or APS (3% NaCl), and then streaking onto TCBS agar (Table 6.5). Typical colonies are bluish green. Isolates such as *V. alginolyticus* produce yellow colonies due to the fermentation of sucrose. TCBS agar is known to allow the growth of other gram-negative

Table 6.5   Thiosulfate Citrate Bile Salts Sucrose (TCBS) Agar (Difco)[a,b]

Yeast extract	5 g/L
Proteose peptone No. 3	10 g
Sodium citrate	10 g
Sodium thiosulphate	10 g
Oxgall	8 g
Sucrose	20 g
Sodium chloride	10 g
Ferric citrate	1 g
Brom Thymol blue	0.04 g
Thymol blue	0.04 g
Agar	15 g

[a] Final pH is adjusted to 8.6. Originally developed as a selective medium for *Vibrio cholarae* El Tor, which produces yellow colonies, whereas nonsucrose fermenting *V. parahaemolyticus* and *V. vulnificus* produce blue-green colonies.

[b] Difco formulation above replaces original 0.3% sodium cholate (Kobayashi et al., 1963) with an additional 0.3% oxgal totaling 0.8%.

organisms (Joseph, Colwell, and Kaper, 1982). The inability of *V. parahaemolyticus* isolates to ferment sucrose is a primary differential characteristic. Colwell (1970), however, found 6% of *V. parahaemolyticus* strains examined to be sucrose positive. Fujino, Sakazaki, and Tamura (1974) found that 2% of marine isolates were sucrose positive. Twedt (1989) listed 13 liquid media and 11 agar media for selective cultivation of *V. parahaemolyticus*.

VII. PRESERVATION OF *V. PARAHAEMOLYTICUS* ISOLATES

The author's experience has shown that cultures of *V. parahaemolyticus* can be maintained for up to a year by washing off Peptone iron agar slants with heavy growth into tubes of 1.0% cooked meat particles overlaid with mineral oil and stored at 20°C. For long term storage, –80°C in TSB containing 3% NaCl plus 20% glycerol has been found effective (Wong et al., 1999). Storage at –60°C in 2% NaCl plus 10% dimethyl-sulfoxide (DMSO) has been found effective for long-term maintenance of cultures in the author's laboratory.

VIII. BACTERIOPHAGE FOR *V. PARAHAEMOLYTICUS*

Nakanishi et al. (1966) isolated three phages from a fecal sample of a patient suffering from infection by *V. parahaemolyticus*. These phages did not attack other *Vibrio* species, but attacked only strains of *V. parahaemolyticus*. The phages exhibited varying plaque morphology on different strains of *V. parahaemolyticus*, possibly reflecting virulent and lysogenic infections. Baross, Liston, and Morita (1978) reported that phages were frequently isolated from shellfish, with a total of 117/355 enrichments yielding phage infectious for *V. parahaemolyticus*. They found that the titers of phage increased dramatically with increase in water temperature. The possibility exists that human clinical isolates may differ from environmental isolates with respect to the acquisition of the *trh* gene via transduction and lysogenization.

IX. USE OF PCR FOR DETECTION OF *V. PARAHAEMOLYTICUS*

Tada et al. (1992) established PCR protocols for the specific detection of the *tdh* and *trh* genes of *V. parahaemolyticus*. The selection of primers took into consideration that the *tdh* and *trh* genes are known to have sequence divergence of up to 3.3% and 16%, respectively. An annealing temperature of 55°C was required with the three primer pairs D1–D2, D5–D2, and D5–D3 (Table 6.6) for detection of the *tdh* gene. An annealing temperature of 60°C was required with the primer pair R3–R5 (Table 6.6) for high-specificity detection of the *trh1* gene. The R2–R6 primer pair (Table 6.6) detected both *trh1* and *trh2* genes with an annealing temperature of 55°C. The limit of sensitivity was 400 fg of cellular DNA in each PCR reaction tube derived from 100 cells.

Table 6.6   PCR Primers and DNA Probes

Primer or Probe	Sequence (5' → 3')	Size of Amplified Sequence (bp)	Gene or DNA Target Sequence	References
D1	CCA-TCT-GTC-CCT-TTT-CCT-GC	373	*tdh*	Tada et al. (1992)
D2	CCA-AAT-ACA-TTT-TAC-TTG-G			
D5	GGT-ACT-AAA-TGG-CTG-ACA-TC	199	*tdh*	Tada et al. (1992)
D2	CCA-AAT-ACA-TTT-TAC-TTG-G			
D5	GGT-ACT-AAA-TGG-CTG-ACA-TC	251	*tdh*	Tada et al. (1992)
D3	CCA-CTA-CCA-CTC-TCA-TAT-GC			
R3	GCC-TCA-AAA-TGG-TTA-AGC-GC	210	*trh1*	Tada et al. (1992)
R5	TGG-CGT-TTC-ATC-CAA-ATA-CG			
R2	GGC-TCA-AAA-TGG-TTA-AGC-G	250	*trh1, trh2*	Tada et al. (1992)
R6	CAT-TTC-CGC-TCT-CAT-ATG-C			
VP-1	CGG-CGT-GGG-TGT-TTC-GGT-AGT	285	*gyrB*	Venkateswaran, Dohmoto, and Harayama (1998)
VP-2r	TCC-GCT-TCG-CGC-TCA-TCA-ATA			
toxR-F1	GTC-TTC-TGA-CGC-AAT-CGT-TG	350	*toxR*	Kim et al. (1999)
toxR-R1	ATA-CGA-GTG-GTT-GCT-GTC-ATG			
toxR-F2	AGC-CCG-CTT-TCT-TCA-GAC-TC	390	*toxR*	Kim et al. (1999)
toxR-R2	AAC-GAG-TCT-TCT-GCA-TGG-TG			

Continued

Table 6.6 PCR Primers and DNA Probes (Continued)

Primer or Probe	Sequence (5' → 3')	Size of Amplified Sequence (bp)	Gene or DNA Target Sequence	References
toxR-F3	CGC-TTT-CTT-CAG-ACT-CAA-GC	394	toxR	Kim et al. (1999)
toxR-R2	AAC-GAG-TCT-TCT-GCA-TGG-TG			
2 RAPD	GTT-TCG-CTC-C	—	—	Hara-Kudo et al. (2003)
4 RAPD	AAG-AGC-CCG-T	—	—	
284-RAPD	CAG-GCG-CAC-A	—	—	Wong et al. (1999)
1 RAPD	GGT-GCG-GGA-A	—	—	Okuda et al. (1997a)
2 RAPD	GTT-TCG-CTC-C			
VPF2	CGC-TTA-GAT-TTG-GGG-GTG-TG	327	—	Khan et al. (2002)
VPR2	GTT-GGT-TGA-GGC-ATA-GGT-AGC			
L-tl	AAA-GCG-GAT-TAT-GCA-GAA-GCA-CTG	450	tl	Brasher et al. (1998)
R-tl	GCT-ACT-TTC-TAG-CAT-TTT-CTC-TGC			
P-tl	ACG-GAC-GCA-GGT-GCG-AAG-AAC-TTC-ATG-TTG			

Primer	Sequence	Size	Gene	Reference
L-tdh	GTA-AAG-GTC-TCT-GAC-TTT-TGG-AC	269	*tdh*	Bej et al. (1999)
R-tdh	TGG-AAT-AGA-ACC-TTC-ATC-TTC-ACC			
L-trh	TTG-GCT-TCG-ATA-TTT-TCA-GTA-TCT	500	*trh*	Bej et al. (1999)
R-trh	CAT-AAC-AAA-CAT-ATG-CCC-ATT-TCC-G			
ERIC 1R	ATG-TAA-GCT-CCT-GGG-GAT-TCA-C	—	—	Marshall et al. (1999)
GS-VP.1	TAA-TGA-GGT-AGA-AAC-A	651	*toxRS*	Matsumoto et al. (2000)
GS-VP.2	ACG-TAA-CGG-GCC-TAC-A			
F-03MM824	AGG-ACG-CAG-TTA-CGC-TTG-ATG	369	*ORF8*	Meyers, Panicker, and Bej (2003)
R-03MM1192	CTA-ACG-CAT-TGT-CCC-TTT-GTA-G			
FP	AAA-CAT-CTG-CTT-TTG-AGC-TTC-CA	75	*tdh*	Blackstone et al. (2003)
RP	CTC-GAA-CAA-CAA-ACA-ATA-TCT-CAT-CAG			
P	FAM-TGT-CCC-TTT-TCC-TGC-CCC-CGG-TAMRA			

Brasher et al. (1998) developed a double multiplex PCR assay enabling the simultaneous detection in shellfish of *Escherichia coli, Salmonella typhimurium, V. vulnificus, V. cholerae,* and *V. parahaemolyticus.* The sensitivity of detection for each species was 10^1–10^2 cells following a double multiplex PCR protocol. For detection of *V. parahaemolyticus* a 450-bp sequence of the *tlh* gene was amplified (Table 6.6). A colorimetric biotinylated probe in addition to conventional agarose gel electrophoresis followed by staining with ethidium bromide was used for detection of amplicons. PCR reaction volumes were 100 μl. The primary PCR consisted of 30 cycles followed by a similar secondary PCR of 30 cycles involving the use of 5 μl of reaction volume from the primary PCR transferred to a final PCR reaction volume of 100 μl.

Bej et al. (1999) developed a multiplex PCR assay for total and virulent strains of *V. parahaemolyticus* based on the amplification of a 450-bp sequence (Brasher et al., 1998) of the thermolabile hemolysin gene (*tlh*), a 269-bp sequence of the thermostable direct hemolysin gene (*tdh*), and a 500-bp sequence of the thermostable direct hemolysin-related (*trh*) gene (Table 6.6). All 111 *V. parahaemolyticus* isolates studied yielded the *tlh* amplicons. However, only 60 isolates yielded the *tdh* amplicon, and 43 yielded the *trh* amplicon. Sensitivity of detection for all three amplicons was between 10^1 and 10^2 CFU/gram of oyster tissue following homogenization in alkaline peptone water and incubation at 35°C for 6 hr.

The first European incident of gastroenteritis due to *V. parahaemolyticus* was derived from the consumption of live oysters in Spain and was reported by Lozano-León et al. (2003). The two isolates studied in detail were KP+ and produced TDH, possessed the *tlh* and *tdh* but not the *trh* gene as evidenced by PCR (Brasher et al., 1998; Bej et al., 1999; Table 6.6).

RNA sequences are frequently used for identification and to establish phylogenetic relationships. However, the rRNA sequence homologies between *V. parahaemolyticus* and related species are too high to allow their use for such purposes with *V. parahaemolyticus* (Kim et al., 1999). For example, the 16S rRNA sequences of *V. parahaemolyticus* and *V. alginolyticus* are >99% identical (Kita-Tsukamoto et al., 1993; Ruimy et al., 1994). The *gyrB* gene encodes the B subunit of DNA gyrase, which is essential for DNA replication. The homology of the *gyrB* sequences between *V. parahaemolyticus* and *V. alginolyticus* is 86.8% (Venkateswaran, Dohmoto, and Harayama, 1998). For this reason, Venkateswaran, Dohmoto, and Harayama (1998) developed a PCR procedure using primers VP-1/VP-2r (Table 6.6) targeting a 285-bp sequence of the *gyrB* gene for specific detection of *V. parahaemolyticus.*

The *toxR* gene was first discovered as the regulatory gene of the cholera toxin operon and was later found to regulate many other genes in *V. cholerae* (DiRita, 1992; Miller, Taylor, and Mekalanos, 1987). The *toxR* gene is well conserved among species of *Vibrio.* The degree of homology of the *toxR* gene between *V. parahaemolyticus* and *V. cholerae* is 52%, which is much lower than the value of 91–92% for the rRNA gene (Kita-Tsukamoto et al., 1993; Lin et al., 1993). Based on these earlier observations, Kim et al. (1999) developed a DNA colony hybridization test with the use of a 678-bp polynucleotide probe (Lin et al., 1993) for the *toxR* gene of *V. parahaemolyticus,* to confirm the identity of isolates. Kim et al.

(1999) also developed a specific PCR assay for the identification of *V. parahaemolyticus* based on amplifying amplicons of the *toxR* gene. Three effective primer pairs were identified (Table 6.6). These primer sequences were selected from the regions of the *toxR* gene not conserved between *V. parahaemolyticus* and *V. cholerae* (Lin et al., 1993). A total of 373 strains of *V. parahaemolyticus* were all found to carry the *toxR* gene.

Dileep et al. (2003) studied the incidence of *V. parahaemolyticus* in seafoods, water, and sediment samples using PCR and conventional microbiological methods. Among 86 samples analyzed, 28 (32.6%) were positive for *V. parahaemolyticus* by conventional microbiological methods, and 53 (61.6%) were positive by the *toxR*-targeted PCR, performed directly on enrichment broth lysates.

Blackstone et al. (2003) developed a real-time PCR assay for detection of *V. parahaemolyticus* in oysters with the use of a pair of primers amplifying a 75-bp sequence of the *tdh* gene (Table 6.6) in conjunction with a dual-labeled fluorogenic probe. Their procedure involved homogenizing oyster tissue at a 1:10 dilution in alkaline peptone water (pH 8.5) followed by overnight enrichment incubation at 35°C. The assay detected target DNA from 1 CFU per PCR reaction.

Kaufman et al. (2002) examined eight clinical and nine oyster isolates of *V. parahaemolyticus* isolated during the Pacific Northwest outbreak in 1997 and an additional three clinical isolates from the 1994 outbreak. A multiplex PCR assay for simultaneous detection of the *tdh*, *trh*, and *tlh* genes was used. All isolates of *V. parahaemolyticus* studied were tl^+, which is considered a specific marker for all isolates of *V. parahaemolyicus* (Bej et al., 1999). All 11 clinical isolates harbored both the *tdh* and *trh* genes; five out of eight clinical isolates from the 1997 outbreak were of serogroup 04 and two others were of serogroup 01. All three isolates from the 1994 outbreak belonged to serogroup 04. In contrast, among the nine oyster isolates, only four were of serogroup 034, and three others belonged to serogroup 01. The authors concluded that *V. parahaemolyticus* strains potentially virulent to humans belong primarily to serogroups 01 and 04. All 11 clinical isolates and 9 out of 11 shellfish isolates in the study produced urease. Two oyster isolates lacking the *tdh* and *trh* genes were urease-negative. The authors suggested that the *tdh*, *trh*, and urease test can be used to identify and track potentially virulent strains in oysters.

A PCR method for quantitative detection of *Vibrio parahaemolyticus* was developed by Wang and Levin (2004). The primers L-tdh/R-tdh (Table 6.6) from Bej et al. (1999) were used. Several lysis methods were compared, and a lysis solution designated TZ developed by Abolmaaty et al. (2000) proved effective. The PCR amplification products were visualized after agarose gel electrophoresis with the nucleic acid stains GelStar and ethidium bromide. The relative fluorescent intensity of the DNA bands was analyzed using the NIH Image 1.61 software program. The GelStar stain was found to be more sensitive than ethidium bromide. The limit of detection by staining DNA bands with GelStar was 16 CFU per PCR and 48 with ethidium bromide. A log-linear relationship between the number of CFU per PCR reaction and the fluorescent intensity of the DNA bands was obtained, and calibration curves were generated.

X. MOLECULAR TYPING OF *V. PARAHAEMOLYTICUS* ISOLATES BELOW THE SPECIES LEVEL

Wong et al. (1996) screened 16 restriction nucleases for use in conjunction with pulsed-field gel electrophoresis (PFGE) analysis of *V. parahaemolyticus* strains. The restriction nuclease *Sfi* I was found to yield 17 clear and discernible bands and was applied to 130 clinical strains from Thailand. These 130 isolates were grouped into 14 PFGE types and each type was subdivided into one to six patterns, resulting in a total of 39 discernible PFGE patterns. PFGE patterns were found to be unrelated to serotype or to KP+ or KP- isolates. Strains from Thailand clustered into major groups, and these major groups showed low degrees of similarity to strains from the United States and the Netherlands.

Wong et al. (1999) subjected 308 clinical isolates of *V. parahaemolyticus* derived from food outbreaks in Taiwan between 1993 and 1995 to RAPD analysis. The 10-mer primer designated 284 (Table 6.6) was used and generated 41 RAPD patterns. The patterns were grouped into 16 RAPD types, the first four of which were the major patterns and accounted for 91.25% of the domestic clinical isolates. The major RAPD types were phylogenetically more closely related to each other than to any of the minor RAPD types. The RAPD typing patterns were correlated with previously reported PFGE typing patterns (Wong et al., 1996) of these isolates.

Marshall et al. (1999) compared various molecular typing methods for differentiating clinical strains of *V. parahaemolyticus* below the species level. During the 1997 outbreak along the West Coast of Canada, 38 clinical and 16 environmental samples were collected and subjected to analysis by (1) enterobacterial repetitive intergeneric sequence (ERIC) PCR, (2) detection of restriction fragment length polymorphism (RFLP) in rRNA genes (ribotyping), (3) PFGE, and (4) RFLP analysis of the genetic locus encoding the polar flagellum (Fla locus RFLP analysis). ERIC PCR and ribotyping were the most informative and discriminating methods, especially when used together. Fla locus RFLP analysis was the least discriminating.

XI. THE O3:K6 PANDEMIC CLONE

Honda et al. (1987) were the first to report on the isolation of KP-clinical isolates of *V. parahaemolyticus* belonging to the serovar O3:K6. Among 51 individuals with travelers' diarrhea at Osaka Airport in Japan who had visited the Maldives, 12 (24%) were found to be infected with *V. parahaemolyticus*. Among these 12 isolates, 11 (92%) were KP- and also *tdh-*. All 11 strains belonged to serovar O3:K6 and caused fluid accumulation in the rabbit ileal loop assay.

The occurrence of foodborne disease outbreaks in Taiwan increased dramatically in 1996 (Chiou et al., 2000). This increase was correlated with a high rate of isolation of *V. parahaemolyticus*, which caused 69% to 71% of the total outbreaks from 1996 to 1999. Serotyping of 3743 *V. parahaemolyticus* isolates yielded 40 serovars, the most frequent of which was O3:K6 (Chiou et al., 2000).

Strains of the O3:K6 serovar were found to have appeared for the first time in Calcutta, India, in February of 1996 when the incidence of clinical cases of *V. para-haemolyticus* infection was suddenly found to increase dramatically at that time. Okuda et al. (1997b) examined 134 strains isolated from 1994 to 1996 in Calcutta with respect to serovar, the presence of the *tdh* gene, and the *tdh*-related hemolysin genes *trh1* and *trh2*. All of the serovar O3:K6 strains carried the *tdh* gene but not the *trh* genes and did not produce urease. RAPD analysis indicated that the O3:K6 serovar strains belonged to a unique clone. Clinical O3:K6 strains isolated between 1982 and 1993 from travelers arriving in Japan from Southeast Asia were found to be RAPD distinct from the Calcuttta O3:K6 clone, and strains isolated in 1995 and 1996 were indistinguishable from the Calcutta O3:K6 strains. These results suggested that the unique O3:K6 clone may have become prevalent not only in Calcutta but also in Southeast Asia. The O3:K6 serovar is considered more highly infectious than other serovars with up to 75% of exposed individuals becoming infected compared to 56% with other serovars (Daniels et al., 2000).

Bag et al. (1999) further examined the genomes of 30 clinical isolates of the O3:K6 strains isolated in Calcutta from February 1996 to June 1998 with the use of ribotyping, *tdh* genotyping, and PFGE. No restriction fragment length polymorphism (RFLP) in the *tdh* gene was observed. Five ribotypes were, however, distinguished. O3:K6 strains isolated between June and August 1996 exhibited different PFGE types compared to the PFGE-type strains isolated before and after this period, indicating genetic reassortment among these strains occurred during this brief period. Strains isolated between August 1999 and March 1998 showed identical, or nearly identical, PFGE patterns. The authors concluded that there is a certain degree of genomic reassortment among the O3:K6 clones but that these strains are predominantly one clone.

Chowdbury et al. (2000) subjected 35 pandemic isolates to ribotyping and PFGE analysis. These isolates consisted of 21 strains of O3:K6, 10 of O4:K68, and 4 of O1:KUT isolated from 1996 to 1999. In addition, a total of 13 nonpandemic strains of O3:K6, O1:KUT, and other serotypes isolated before 1996 were also examined. The nonpandemic strains exhibited heterogeneous ribotype and PFGE profiles. In contrast, the ribotypes of the 35 pandemic strains were indistinguishable with two exceptions and their PFGE profiles were all nearly identical and distinct from the nonpandemic strains. The authors concluded that the O4:K68 and O1:KUT strains most likely originated from the pandemic O3:K6 clone.

Nasu et al. (2000) found that a phage designated f237 was specifically and exclusively associated with O3:K6 serovar strains isolated since 1996. This phage has 10 open reading frames (ORFs), including a unique ORF designated ORF8, that exhibits CTX, which carries the cholera enterotoxin genes *ctxA* and *ctxB* (Iida et al., 2001). It has also been speculated that the ORF8 may encode an adhesion protein that might also contribute to the enhanced virulence of the O3:K6 serotype (Chiou et al., 2000). In *V. parahaemolyticus* O3:K6 strains, the f237 phage has ORF8 in place of *ctxAB* in *Vibrio cholerae* (Waldor and Mekalanos, 1996). It is interesting that Iida et al. (2001) found that 53/96 *V. parahaemolyticus* strains isolated in 1999 in Japan from recently arrived Asian travelers with dysentery carried the ORF8 sequence; 34 of these ORF8 carrying strains were of serotype O3:K6, 11 were of serotype O4:K68, and 8 were of serotype O1:KUT (UT = untypeable). PFGE analysis indicated identical genotypes. DNA probe analysis indicated that phage f237 had integrated into the bacterial chromosome of strains of all three serotypes carrying f237.

Vuddhakul et al. (2000) used immunomagnetic beads coated with K6 antibody to isolate a strain of *V. parahaemolyticus* serovar O3:K6 from fresh shellfish in Thailand. This environmental isolate was indistinguishable from several clinical isolates with the use of six RAPD primers. These results indicated that the clinical O3:K6 and environmental isolates belonged to the same O3:K6 pandemic clone.

Meyers, Panicker, and Bej (2003) described the development and use of a set of primers specific for a 369-bp sequence of ORF8 designated F-03MM824 and R-03MM1192 (Table 6.6). This set of primers was found to be highly effective in rapidly screening and detecting newly acquired isolates of *V. parahaemolyticus* from marine waters from the Gulf of Mexico for the 03:K6 serovar. These authors concluded that inasmuch as all newly emerged O3:K6 isolates are derived from a single clone, it is likely that this strain has been transported from one geographic locale to another via ship ballast water.

Nasu et al. (2000) studied 99 clinical *V. parahaemolyticus* strains isolated from January to August 1996 in Osaka, Japan, from travelers with diarrhea returning from Southeast Asian countries. Among these strains, 24 were of the O3:K6 serovar. All 24 of these O3:K6 strains possessed a common plasmid designated pO3K6 whose genome consisted of 8782-bp, with 10 ORFs. The gene organization of pO3K6 was similar to the filamentous phage VF33 reported earlier (Taniguchi et al., 1984; Chang et al., 1998). A single-stranded DNA phage was obtained from the culture supernatant of an O3:K6 strain. The double-stranded DNA obtained by treatment of the genome with DNA polymerase was identical to that of pO3K6 as evidenced by *Hind*III restriction fragment analysis. The authors concluded that pO3K6 is a replicative form of f237.

The first reported outbreak of gastroenteritis due to *V. parahaemolyticus* serovar O3:K6 in North America occurred between May 31 and July 10, 1998. The outbreak involved 416 individuals in 13 states who had eaten raw oysters harvested from Galveston Bay, Texas, and Long Island Sound (Daniels et al., 2000; Gendel et al., 2001). Gendel et al. (2001) used automated ribotyping to determine the genetic relationships between *V. parahaemolyticus* O3:K6 isolates derived from the 1998 outbreaks in Texas and the northeastern United States. The patterns resulting from the use of the restriction enzymes *Eco*RI and *Pst*I suggested that the outbreak in the northeastern United States was caused by a single strain closely related to the Asian clone. In contrast, multiple strains were involved in the Texas outbreak, and the predominant type was genetically distinct from the northeastern U.S. and Asian clones.

Khan et al. (2002) reported that O3:K6 strains possessed a specific 850-bp sequence that was absent in other *Vibrio* species and related organisms. A set of primers VPF2/VPR2 (Table 6.6) was then developed that amplified a 327-bp segment of this unique sequence. A total of 25 strains of *V. parahaemolyticus* isolated during 1997 to 1998 were then subjected to PCR analysis. All seven Pacific Northwest strains of *V. parahaemolyticus* lacked this 327-bp sequence and were not serovar O3:K6. In contrast, all eight New York and all ten Texas isolates possessed this specific sequence and were serovar O3:K6. This study clearly indicated that the pandemic serovar O3:K6 had spread to the Eastern and Southern coastal areas of North America and suggested that the Pacific Northwest coastal areas may not at that time have been in contact with this serovar.

Matsumoto et al. (2000) showed with RAPD analysis O3:K6 strains from six countries, including the United States isolated from 1997, and later, belonging to the same clone. The *toxR* and *toxS* genes in the *toxRS* operon encode transmembrane proteins involved in the regulation of the virulence-associated genes and are well conserved in the genus *Vibrio*. The nucleotide sequences of a 1364-bp region covering 95% of the *toxRS* coding region of five O3:K6 pandemic strains were found to be 100% identical. In contrast, sequence variation was found among all prepandemic strains. The sequences of the two groups differed invariably at seven base positions. The authors then utilized the consistency of the bases at two of these positions in the *toxRS* sequence of the pandemic strains to develop a group-specific PCR for convenient detection and identification of strains belonging to this new O3:K6 pandemic clone, designated GS-PCR, for group-specific PCR. The pair of primers used, GS-VP.1 and GS-VP.2 (Table 6.6), amplify a 651-bp sequence of the toxRS operon. Application of the GS-PCR technique showed that the *toxRS* sequences of pandemic O4:K68 and O1:KUT serotypes were identical to the 1995 and later pandemic O3:K6 strains. The authors concluded that the GS-PCR-positive O4:K68 and O1:KUT strains may have diverged from the pandemic O3:K6 clone by alteration of the genes associated with the O:K antigens.

Bhuiyan et al. (2002) studied 66 strains of *V. parahaemolyticus* belonging to 14 serotypes isolated from hospitalized patients in Bangladesh in 1998, 1999, and 2000. Among these 66 strains, 60 were positive for the *tdh* gene but negative for the *trh* gene. The *trh* gene was carried only by four strains, and two strains were negative for both the *tdh* and *trh* genes, even though these two strains were isolated from hospitalized patients with acute diarrhea. Among the 66 strains, 48 belonged to four serotypes having the O3:K6 genotype (GS-PCR+) and possessed the *tdh* gene. Among these 48 GS-PCR+ strains, 39 were positive for the presence of the ORF8 amplicon. Eight O3:K6 strains and one O4:K68 strain were GS-PCR+ but ORF8−. RAPD analysis of the ORF8+ and ORF8− strains from both serotypes yielded identical profiles. Ribotyping yielded similar results. The authors concluded that GS-PCR+ strains lacking ORF8 may be due to a loss of all or part of the phage f237 genome and that ORF8 is not always a stable marker for the pandemic clone. The authors also observed that as of 1999, a fourth serotype O1:K25 has emerged, possessing the GS-PCR genotype, in addition to the presence of the ORF8 marker for phage f237. The fact that the strains of *V. parahaemolyticus* belonging to the O3:K6 serotype isolated from 1996 onward are more virulent than strains isolated before 1996 was taken to indicate that the O3:K6 serotype is enriching its gene pool to a higher level of virulence.

The *tdh* gene was detected by PCR in 33 of 329 shellfish samples (10%) from four geographic regions of Japan in 2001 by Hara-Kudo et al. (2003). The authors used the D5 and D3 pair of 20-mer primers from Tada et al. (1992) that amplify a 251-bp sequence of the *tdh* gene (Table 6.6). RAPD was performed with the use of primers

2 and 4 of Okuda et al. (1997a; Table 6.6). Analysis of 19 strains from shellfish and marine sediments indicated that pandemic O3:K6 strains were distributed throughout the Japanese coastal environment; 14/19 strains were *tdh⁻* and GS-PCR-positive and yielded the same RAPD profiles as the reference strains of the O3:K6 pandemic clones. High-resolution PFGE analysis showed that three of the O3:K6 strains isolated from the Japanese coastal environment were indistinguishable from the O3:K6 pandemic strains of other countries.

DePaola et al. (2003b) determined the serotypes, potential pathogenicity (based on possession of *tdh* gene), and ribotypes for 178 pathogenic *V. parahaemolyticus* isolates from clinical, environmental, and food sources on the Pacific, Atlantic, and Gulf coasts of the United States and from clinical sources in Asia. Multiplex PCR was used to confirm the identity of the isolates as *V. parahaemolyticus* and the presence of the *tdh* and *trh* genes. Most of the environmental, food, and clinical isolates from the United States were positive for *tdh*, *trh*, and urease production. Clinical isolates from Texas, New York, and Asia were predominantly of serotype O3:K6 and possessed only *tdh*. In contrast, none of the 26 clinical isolates from Washington State were of the O3:K6 serotype. The O3:K6 serotype was not detected among any of the 94 environmental and 16 food isolates from the United States. The combination of serotyping and ribotyping indicated that isolates from the Pacific coast are genetically distinct from the Atlantic and Gulf coast isolates.

REFERENCES

Abbott, S., Powers, C., Kaysner, C., Takeda, Y., Ishibashi, M., Joseph, S., Ianda, J. 1989. Emergence of a restricted bioserovar of *Vibrio parahaemolyticus* as the predominant cause of *Vibrio*- associated gastroenteritis on the West Coast of the United States and Mexico. *J. Clin. Microbiol.* 27:2891–2893.

Abolmaaty, A., Vu, C., Oliver, J., and Levin, R.E. 2000. Development of a new lysis solution for releasing genomic DNA from bacterial cells for DNA amplification by polymerase chain reaction. *Microbios.* 101:181–189.

Allen, R., Baumann, P. 1971. Structure and arrangement of flagella in species of the genus *Beneckea* and *Photobacterium fisheri*. *J. Bacteriol.* 107:295–302.

Anderson, R., Ordal, E. 1972. Deoxyribonucleic acid relationships among marine vibrios. *J. Bacteriol.* 109:696–706.

Asakawa, Y., Akahane, S., Noguchi, M. 1974. Quantitative studies on pollution with *Vibrio parahaemolyticus* during distribution of fish. In: *International Symposium on* Vibrio parahaemolyticus, T. Fujino, G. Sakaguchi, R. Sakazaki, Y. Takeda, Eds., Saikon: Tokyo, pp. 97–103.

Atsumi, T., McCarter, L., Imae, Y. 1992. Polar and lateral flagellar motors of marine *Vibrio* are driven by different ion-motive forces. *Nature* 355:182–184.

Baba, K., Shirai, H., Terai, A., Takeda, Y., Nishibuchi, M. 1991. Analysis of the *tdh* gene cloned from a *tdh* gene- and *trh* gene-positive strains of *Vibrio parahaemolyticus*. *Microbiol. Immunol.* 35: 253–258.

Bag, P., Nandi, S., Bhadra, R., Ramamurthy, T., Bhattacharya, S., Nishibushi, M., Hamabata, T., Yamasaki, S., Takeda, Y., Nair, G. 1999. Clonal diversity among recently emerged strains of *Vibrio parahaemolyticus* O3:K6 associated with pandemic strains. *J. Clin. Microbiol.* 37:2354–2357.

Barker, W., Gangarosa, E. 1974. Food poisoning due to *Vibrio parahaemolyticus. Annu. Rev. Med.* 25:75–81.

Baross, J., Liston, J., Morita, R. 1978. Ecological relationship between *Vibrio parahaemolyticus* and agar-digesting vibrios as evidenced by bacteriophage susceptibility patterns. *Appl. Environ. Microbiol.* 36:500–505.

Barrow, G., Miller, D. 1974. Growth studies on *Vibrio parahaemolyticus* in relation to pathogenicity. In: *International Symposium on Vibrio parahaemolyticus*, T. Fujino, G. Sakaguchi, R. Sakazaki, Y. Takeda, Eds., Saikon:Tokyo, pp. 205–209.

Baumann, P., Baumann, L., Mandel, M. 1971. Taxonomy of marine bacteria: The genus *Beneckea. J. Bacteriol.* 107:268–294.

Bej, A., Patterson, D., Brasher, C., Vickery, M., Jones, D., Kaysner, C. 1999. Detection of total and hemolysin-producing *Vibrio parahaemolyticus* in shellfish using multiplex PCR amplification of *tlh*, *tdh*, and *trh*. *J. Microbiol. Meth.* 36:215–225.

Belas, R., Simon, M., Silverman, M. 1986. Regulation of lateral flagella gene transcription in *Vibrio parahaemolyticus. J. Bacteriol.* 167:210–218.

Beuchat, L.R. 1973. Interacting effects of pH, temperature, and salt concentration on growth and survival of *Vibrio parahaemolyticus. Appl. Microbiol.* 25:844–846.

Beuchat, L. 1977. Evaluation of enrichment broths for enumerating *Vibrio parahaemolyticus* in chilled and frozen crab meat. *J. Food Prot.* 40:592–595.

Bhuiyan, N., Ansurazzaman, M., Kmruzzaman, M., Alam, K., Chowdhury, N., Nishibushi, M., Faruque, S., Sack, D., Takeda, Y., Nair, G. 2002. Prevalence of the pandemic genotype of *Vibrio parahaemolyticus* in Dhaka, Bangladesh, and significance of its distribution across different serotypes. *J. Clin. Microbiol.* 40:284–286.

Blackstone, G., Nordstrom, J., Vickery, M., Bowen, T., Meyer, R., DePaola, A. 2003. Detection of pathogenic *Vibrio parahaemolyticus* in oyster enrichments by real time PCR. *J. Microbiol. Meth.* 53:149–155.

Bolen, J., Zamiska, A., Greenough, W. 1974. Clinical features in enteritis due to *Vibrio parahaemolyticus. Am. J. Med.* 57:638–641.

Bonner, J., Coker, A., Berryman, C., Polock, H. 1983. Spectrum of *Vibrio* infections in a Gulf Coast community. *Ann. Intern. Med.* 99:464–469.

Boutin, B., Reyes, A., Peeler, J., Twedt, R. 1985. Effect of temperature and suspending vehicle on survival of *Vibrio parahaemolytcus* and *Vibrio vulnificus. J. Food Prot.* 48: 875–878.

Bradshaw, J., Francis, D., Twedt, R. 1974. Survival of *Vibrio parahaemolyticus* in cooked seafood at refrigeration temperatures. *Appl. Microbiol.* 27:657–661.

Brasher, C., DePaola, A., Jones, D., Bej, A. 1998. Detection of microbial pathogens in shellfish with multiplex PCR. *Current Microbiol.* 37:101–107.

Burstyn, D., McNicol, L., Voll, M. 1980. Isolation and characterization of spontaneously arising auxotrophic and Kanagawa phenomenon-negative mutants of *Vibrio parahaemolyticus. Infect. Immun.* 27:889–896.

Cal, Y., Ni, Y. 1996. Purification, characterization, and pathogenicity of urease produced by *Vibrio parahaemolyticus. J. Clin. Lab. Anal.* 10:70–73.

CDC. 1998. Outbreak *of Vibrio parahaemolyticus* infections associated with eating raw oysters and clams harvested from Long Island Sound-Connecticut, New Jersey, and New York, 1998. *Morbid. Mortal. Wkly. Rep.* 48:48–51.

Chang, B., Taniguchi, H., Miyamoto, H., Yoshida, S. 1998. Filamentous bacteriophages of *Vibrio parahaemolyticus* as a possible clue to genetic transmission. *J. Bacteriol.* 180:5094–5101.

Cherwongrodzky, J., Clark, A. 1981. Effect of pH on the production of the Kanagawa hemolysin by *Vibrio parahaemolyticus*. *Infect. Immun.* 34:115–119.

Chiou, C., Hsu, S., Chiu, S., Wang, S., Chao, C. 2000. *Vibrio parahaemolyticus* serovar O3:K6 as cause of unusually high incidence of food-borne disease outbreaks in Taiwan from 1996 to 1999. *J. Clin. Microbiol.* 38:4621–4625.

Chitu, M., Cufedu, C., Nacescu, N. 1977. The isolation and characterization of some *Vibrio parahaemolyticus* strains isolated from salted herring and roe. *Zbl. Bakt. Hyg. 1. Abt. Orig. A.* 238:59–65.

Chowdhury, N., Chakraborty, S., Ramamurthy, T., Nishibuchi, M., Yamasaki, S., Takeda, Y., Nair, G. 2000. Molecular evidence of clonal *Vibrio parahaemolyticus* pandemic strains. *Emerg. Inf. Dis.* 6:631–636.

Chun, D., Chung, J., Tak, R. 1974. Some observations on Kanagawa type hemolysis of *Vibrio parahaemolyticus*. In: *International Symposium on Vibrio parahaemolyticus*, T. Fujino, G. Sakaguchi, R. Sakazaki, Y. Takeda, Eds., Saikon: Tokyo, pp. 199–204.

Colwell, R.R. 1970. Polyphasic taxonomy of the genus *Vibrio*: Numerical taxonomy of *Vibrio cholerae, Vibrio parahaemolyticus*, and related *Vibrio* species. *J. Bacteriol.* 104:410–433.

Covert, D., Woodburn, M. 1972. Relationships of temperature and sodium concentration to the survival of *Vibrio parahaemolyticus* in broth and fish homogenate. *Appl. Microbiol.* 23:321–325.

Daniels, N., Ray, B., Easton, A., Maaranao, N., Kahn, E., McShan, A., Del Rosario, L., Baldwin, T., Kingsleyk, M., Puhr, N., Wells, J., Angulo, F. 2000. Emergence of a new *Vibrio parahaemolyticus* serotype in raw oysters. *J. Am. Med. Assoc.* 284:1541–1545.

DePaola, A., Kaysner, C., Bowers, J., Cook, D. 2000. Environmental investigations of *Vibrio parahaemolyticus* in oysters after outbreaks in Washington, Texas, and New York (1997 and 1998). *Appl. Environ. Microbiol.* 66:4649–4654.

DePaola, A., Nordstrom, J., Bowers, J., Wells, J., Cook, D. 2003a. Seasonal abundance of total and pathogenic *Vibrio parahaemolyticus* in Alabama oysters. *Appl. Environ. Microbiol.* 69:1521–1526.

DePaola, A., Ulaszek, J., Kaysner, C., Tenge, B., Nordsrom, J., Wells, J., Puhr, N., Gendel, S. 2003b. Molecular, serological, and virulence characteristics of *Vibrio parahaemoltyicus* isolated from environmental, food, and clinical sources in North America and Asia. *Appl. Environ. Microbiol.* 69:3999–4005.

Dileep, V., Kumar, H., Kumar, Y., Nishubichi, M., Karunasagar, I., Karunasagar, I. 2003. Application of polymerase chain reaction for detection of *Vibrio parahaemolyicus* associated with tropical seafoods and coastal environment. *Lett. Appl. Microbiol.* 36:423–427.

DiRita, V., 1992. Co-ordinate expression of virulence genes by ToxR in *Vibrio cholerae. Mol. Microbiol.* 6:451–458.

Douet, J., Castroviejo, M., Dodin, A., Bebear, C. 1992. Purification and characterization of Kanagawa haemolysin from *Vibrio parahaemolyticus*. *Res. Microbiol.* 143:569–577.

Elliot, E., Kaysner, C., Jackson, L., Tamplin, M. 1998. *Vibrio cholerae, V. parahaemolyticus, V. vulnificus*, and other *Vibrio* spp. *U.S. FDA Bacteriological Analytical Manual.* AOAC International: Gaithersburg, MD, Chapter 9, pp. 9.01–9.27.

Felsenfeld, O., Cabirac, H.B. 1977. A study of the ecology of *Vibrio parahaemolyticus* and *Vibrio alginolyticus* in Southeast Louisiana USA with special consideration of seafood consumption. *J. Appl. Nutr.* 29:17–28.

Fishbein, M., Wenz, B., Landry, W., MacFachern, B. 1974. *Vibrio parahaemolyticus* isolates in the U.S.: 1969–1972. In: *International Symposium on* Vibrio parahaemolyticus, T. Fujino, G. Sakaguchi, R. Sakazaki, Y. Takeda, Eds., Saikon: Tokyo, p. 53.

Fujino, T. 1951. Bacterial food poisoning. *Saishin Igaku.* 6: 263–271. In Japanese.

Fujino, T., Miwatani, T., Takeda, Y. Tomaru, A. 1969. A thermolabile direct hemolysin of *Vibrio parahaemolyticus. Biken J.* 12:145–148.

Fujino, T., Miwatani, T., Yasuda, J., Kondo, M., Takeda, Y., Akita, Y., Kotera, K., Okada, M., Nishimune, H., Shimizu, Y., Tamura, T., Tamura, Y. 1965. Taxonomic studies on the bacterial strains isolated from cases of "shirasu" food poisoning (*Pasteurella parahaemolytica*) and related organisms. *Biken J.* 8:63–71.

Fujino, T., Okuno, Y., Nakada, D., Aoyama, A., Fukai, K., Mukai, T., Ueho, T. 1953. On the bacteriological examination of shirasu food poisoning. *Med. J. Osaka Univ.* 4:299–304.

Fujino, T., Sakazaki, R., Tamura, K. 1974. Designation of the type strain of *Vibrio parahaemolyticus* and description of 200 strains of the species. *Int. J. Syst. Bacteriol.* 24:447–449.

Gendel, S., Ulaszek, J., Nishibuchi, M., DePaola, A. 2001. Automated ribotyping differentiates *Vibrio parahaemolyticus* O3:K6 strains associated with a Texas outbreak from other clinical strains. *J. Food Prot.* 64:1617–1620.

Ghosh, H., Bowen, T. 1980. Halophilic vibrios from human tissue infections on the Pacific coast of Australia. *Pathology* 12:397–402.

Goatcher, L., Engler, S., Wagner, D., Westhoff, D. 1974. Effect of storage at 5 C on survival of *Vibrio parahaemolyticus* in processed Maryland oysters (*Crassostrea virginica*). *J. Milk Food Technol.* 37:74–77.

Hara-Kudo, Y., Sugiyama, K., Nishibushi, M., Showdhury, A., Yatsuyanagi, J., Ohtomo, Y., Saito, A., Hidetoshi, N., Nishina, T., Nakagawa, H., Konuma, H., Miyara, M., Kumagail, S. 2003. Prevalence of pandemic thermostable direct hemolysin producing *Vibrio parahaemolyticus* O3:K6 in seafood and the coastal environment of Japan. *Appl. Environ. Microbiol.* 69:3883–3891.

Hollis, D., Weaver, R., Baker, C., Thornsberry, C. 1976. Halophilic *Vibrio* species isolated from blood cultures. *J. Clin. Microbiol.* 3:425–431.

Honda, S., Goto, I., Minematsu, I., Ikeda, N., Ishibashi, M., Kinoshita, Y., Nishibushi, M., Honda, T., Miwatani, T. 1987. Gastroenteritis due to Kanagawa negative *Vibio parahaemolyticus. Lancet* 1(8528):331–332.

Honda, T., Abad-Lapuebla, M., Ni, Y., Yamamoto, K., Miwatani, T. 1991. Characterization of a new thermostable direct haemolysin produced by a Kanagawa-phenomenon-negative clinical isolate of *Vibrio parahaemolyticus. J. Gen. Microbiol.* 137:253–259.

Honda, T., Chearskul, S., Takeda, Y., and Miwatani, T. 1980. Immunological methods for detection of Kanagawa phenomenon of *Vibrio parahaemolyticus. J. Clin. Microbiol.* 11:600–603.

Honda, T., Ni, Y., Miwatani, T. 1988. Purification and characterization of a hemolysin produced by a clinical isolate of Kanagawa phenomenon-negative *Vibrio parahaemolyticus* and related to the thermostable direct hemolysin. *Infect. Immun.* 56:961–965.

Horie, S., Okuzuma, M., Kato, N., Saito, K. 1966. Comparative observation on the range of growth temperature among three biotypes of *Vibrio parahaemolyticus. Bull. Jap. Soc. Fish.* 32:424–426.

Hugh, R., Sakazaki, R. 1972. Minimal number of characters for the identification of *Vibrio* species, *Vibrio cholerae*, and *Vibrio parahaemolyticus. Conf. Public Health Lab. Directors.* 30:133–137.

Hughes, J., Boyce, J., Alem, A., Wells, A., Rhaman, A., and Curlin, G. 1978. *Vibrio parahaemoltcus* enterocolitis in Bangdladesh: Report of an outbreak. *Am. J. Trop. Med. Hyg.* 27:106–112.

Iida, T., Hattori, A., Tagomori, K., Nasu, H., Naim, R., Honda, T. 2001. Filamentous phage associated with recent pandemic strains of *Vibrio parahaemolyticus. Emerg. Infec. Dis.* 7:477–478.

Iida, T., Yamamoto, K. 1990. Cloning and expression of two genes encoding highly homologous hemolysins from a Kanagawa phenomenon-positive *Vibrio parahaemolyticus* T4750 strain. *Gene*. 93:9–15

Jackson, H. 1974. Temperature relationships of *Vibrio parahaemolyticus*. In: *International Symposium on* Vibrio parahaemolyticus, T. Fujino, G. Sakaguchi, R. Sakazaki, Y. Takeda, Eds., Saikon: Tokyo, pp. 139–145.

Jegatheson, M., Paramasivam, T. 1976. Hydrogen sulphide production as an aid to the identification of *Vibrio parahaemolyticus*. *S.W. Asian J. Trop. Med. P. H.* 7:377–379.

Jiang, X., Chai, T. 1996. Survival of *Vibrio parahaemolyticus* at low temperatures under starvation conditions and subsequent resuscitation of viable, nonculturable cells. *Appl. Environ. Microbiol.* 62:1300–1305.

Johnson, H.C., Liston, J. 1973. Sensitivity of *Vibrio parahaemolyticus* to cold in oysters, fish fillets, and crab meat. *J. Food Sci.* 38:437–441.

Joseph, S.W., Colwell, R.R., Kaper, J.B. 1982. *Vibrio parahaemolyticus* and related halophilic vibrios. *Crit. Rev. Microbiol.* 10:77–124.

Kaper, J. , Campen, R., Seidler, R., Baldini, M., Falkow, S. 1984. Cloning of the thermostable direct hemolysin or Kanagawa phenomenon-associated hemolysin of *Vibrio parahaemolyticus*. *Inf. Immun.* 45:290–292.

Kaufman, G., Myers, M., Pass, C., Bej, A., Kaysner, C. 2002. Molecular analysis of *Vibrio parahaemolyticus* isolated from human patients and shellfish during US Pacific northwest outbreaks. *Lett. Appl. Microbiol.* 34:155–161.

Kelly, M., Stroh, E. 1989. Urease-positive, Kanagawa-negative *Vibrio parahaemolyticus* from patients and the environment in the Pacific Northwest. *J. Clin. Microbol.* 27:2820–2822.

Khan, A., McCarthy, S., Wang, R., Cerniglia, C. 2002. Characterization of United States outbreak isolates of *Vibrio parahaemolyticus* using enterobacterial repetitive intergenic consensus (ERIC) PCR and development of a rapid PCR method for detection of O3:K6 isolates. *FEMS Microbiol. Lett.* 206:209–214.

Kim, Y., Okuda, J., Matsumoto, C., Takahashi, N., Hashimoto, S., Nishibushi, M. 1999. Identification of *Vibrio parahaemolyticus* strains at the species level by PCR targeted to the *toxrR* gene. *J. Clin. Microbiol.* 37:1173–1177.

Kimura, K., Tateiri, S., Iida, H. 1979. Effects of pH of the medium on flagellation of *Vibrio parahaemolyticus*. *Appl. Environ. Microbiol.* 37:1248–1249.

Kishishita, M., Matsuoka, N., Kumagai, K., Yamasaki, S., Takeda, Y., Nishibuchi, M. 1992. Sequence variation in the thermostable direct hemolysin-related hemolysin (*trh*) gene of *Vibrio parahaemolyticus*. *Appl. Environ. Microbiol.* 58:2449–2457.

Kita-Tsukamoto, K., Oyaizu, H., Namba, K., Shimidu, U. 1993. Phylogenic relationships of marine bacteria, mainly members of the family *Vibrionaceae*, determined on the basis of 16S rRNA sequences. *Int. J. Syst. Bacteriol.* 43:8–19.

Kobayashi, T., Enomoto, S., Sakazaki, R., Kuwabaara, S. 1963. A new selective isolation medium for the *Vibrio* group (modified Nakanishi's medium—TCBS agar). *Jpn. J. Bacteriol.* 18:387–392.

Lee, C., Cheng, M., Yu, M., Pan, M. 2002. Isolation and characterization of a putative virulence factor, serine protease, from *Vibrio parahaemolyticus*. *FEMS Microbiol. Lett.* 209:31–37.

Lin, Z., Kumagai, K.K, Baba, K., Mekalanos, J., Nishibushi, M. 1993. *Vibrio parahaemolyticus* has a homolog of the *Vibrio cholerae toxRS* operon that mediates environmentally induced regulation of the thermostable direct hemolysin gene. *J. Bacteriol.* 175:3844–3855.

Lozano-León, A., Tores, J., Osorio, C., Martinex-Urtaza, J. 2003. Identification of *tdh*-positive *Vibrio parahaemolyticus* from an outbreak associated with raw oyster consumption in Spain. *FEMS Microbiol. Lett.* 226:281–284.

Lu, B. 2003. The isolation, enumeration, biochemical and molecular identification of *Vibrio vulnificus* and *Vibrio parahaemolyticus* from shellfish. M.S. thesis, University of Massachusetts, Amherst, MA.

Magalháes, M., Takeda, Y., Magalháes, V., Tateno, S. 1992. Brazilian urease-positive strains of *Vibrio parahaemolyticus* carry genetic potential to produce TDH-related hemolysin. *Mem. Inst. Oswaldo Cruz* 87:167–168.

Marshall, S., Clark, S.G., Wang, G., Mulvey, M., Kelly, M.T., Johnson, W.M. 1999. Comparison of molecular methods for typing *Vibrio parahaemolyticus*. *J. Clin. Microbiol.* 37:2473–2478.

Matches, J., Liston, J., Daneault, L. 1971. Survival of *Vibrio parahaemolyticus* in fish homogenate during storage at low temperature. *Appl. Environ. Microbiol.* 21:951–952.

Matsumoto, C., Okuda, J., Ishibashi, M., Iwanaga, M., Garg, P., Rammamurthy, T., Wong, H., DePaola, A., Kim. Y., Albert, M., Nishibuchi, M. 2000. Pandemic spread of an O3:K6 clone of *Vibrio parahaemolyticus* and emergence of related strains evidenced by arbitrarily primed PCR and *toxRS* sequence analyses. *J. Clin. Microbiol.* 38:578–585.

McCarter, I., Hilmen, M., Silverman, M. 1988. Flagellar dynamometer controls swarmer cell differentiation of *Vibrio parahaemolyticus*. *Cell.* 54:345–351.

McCarthy, A., DePaola, A., Cook, D., Kaysner, C., Hill, W. 1999. Evaluation of alkaline phosphatase- and digoxigenin-labeled probes for detection of the thermolabile hemolysin (*tlh*) gene of *Vibrio parahaemolyticus*. *Lett. Appl. Microbiol.* 28:66–70.

McCarthy, S., DePaola, A., Kaysner, C., Hill, W., Cook. D. 2000. Evaluation of nonitotopic DNA hybridization methods for detection of the *tdh* gene of *Vibrio parahaemolyticus*. *J. Food Prot.* 63:1660–1664.

Meyers, M., Panicker, G., Bej, A. 2003. PCR detection of a newly emerged pandemic *Vibrio parahaemolyticus* O3:K6 pathogen in pure cultures and seeded waters from the Gulf of Mexico. *Appl. Environ. Microbiol.* 69:2194–2200.

Miller, Y., Taylor, R., Mekalanos, J. 1987. Cholera toxin transcriptional activator ToxR is a transmembrane DNA binding protein. *Cell* 48:271–279.

Misaki, H., Matsumoto, M. 1978. Purification of lysophospholipase of *Vibrio parahaemolyticus* and its properties. *J. Biochem.* 83:1395–1405.

Miwatani, T., Sakurai, J., Takeda, Y., Shinoda, S. 1974. Studies on direct hemolysins of *Vibrio parahaemolyticus*. In: *International Symposium on Vibrio parahaemolyticus*, T. Fujino, G. Sakaguchi, R. Sakazaki, T. Takeda, Eds., Saikon: Tokyo, pp. 245–251.

Miwatani, T., Sakurai, J., Yoshihara, A., Takeda, Y. 1972. Isolation and partial purification of thermolabile direct hemolysin of *Vibrio parahaemolyticus*. *Biken J.* 15:61–66.

Miwatani, T., Takeda, Y. 1976. *Vibrio parahaemolyticus—A Causative Bacterium of Food Poisoning*. Saikon: Tokyo.

Miyamoto, Y., Kato, T., Obara, Y., Akiyama, S., Takizawa, K., Yamai, S. 1969. In vitro hemolytic characteristic of *Vibrio parahaemolyticus*: Its close correlation with human pathogenicity. *J. Bacteriol.* 100:1147–1149.

Miyamoto, Y., Obara, Y., Nikkawa, T., Yamai, S., Kato, T., Yamada, Y., Ohashi, M. 1980. Simplified purification and biophysicochemical characteristics of Kanagawa phenomenon-associated hemolysin of *Vibrio parahaemolyticus*. *Inf. Immun.* 28:567–576.

Nakaguchi, Y., Okuda, J., Iida, T., Nishibuchi, M. 2003. The urease gene cluster of *Vibrio parahaemolyhticus* does not influence the expression of the thermostable direct hemolysin gene or the TDH-related hemolysin gene. *Microbiol. Immunol.* 47:233–239.

Nakanishi, H., Iida, Y., Maeshima, K., Teramoto, T., Hosaka, Y., Ozaki, M. 1966. Isolation and properties of bacteriophages of *Vibrio parahaemolyticus*. *Biken J.* 9:149–157.

Nasu, H., Iida, T., Sugahara, T., Yamaichi, Y., Park, K., Yokoyama, K., Makino, H., Shinagawa, H., Honda, T. 2000. A filamentous phage associated with recent pandemic *Vibrio parahaemolyticus* O3:K6 strains. *J. Clin. Microbiol.* 38:2156–2161.

Nishibuchi, M., Fasano, A., Russell, R., Kaper, J. 1992. Enterotoxigenicity of *Vibrio parahaemolyticus* with and without genes encoding thermostable direct hemolysin. *Infect. Immun.* 60:3539–3545.

Nishibuchi, M., Ishibashi, M., Takeda, Y., Kaper, J. 1985. Detection of the thermostable direct hemolysin gene and related DNA sequences in *Vibrio parahaemolyticus* and other *Vibrio* species by the DNA colony hybridization test. *Infect. Immun.* 49:481–486.

Nishibuchi, M., Kaper, J.B. 1985. Nucleotide sequence of the thermostable direct hemolysin gene of *Vibrio parahaemolyticus*. *J. Bacteriol.* 162:558–564.

Nishibuchi, M., Kaper, J. 1990. Duplication and variation of the thermostable direct hemolysin (*tdh*) gene in *Vibrio parahaemoltyicus*. *Mol. Microbiol.* 4:87–99.

Nishibuchi, M., Taniguchi, T., Misawa, T., Khaeomanee-Iam, V., Honda, T., Miwatani, T. 1989. Cloning and nucleotide sequence of the gene (*trh*) encoding the hemolysin related to the thermostable direct hemolysin of *Vibrio parahaemoticus*. *Inf. Immun.* 57:2691–2697.

Obara, Y. 1971. Studies on hemolytic factors of *Vibrio parahaemolyticus*. II. The extraction, and its properties. *J. Jpn. Assoc. Infect. Dis.* 45:392–398.

Ohashi, M., Ohta, K., Tsuno, M., Zen-Yoji, H. 1977. Development of a sensitive serological assay, reversed passive hemagglutination test, for detection of enteropathogenic toxin (Kanagawa hemolysin) of *Vibrio parahaemolyticus*, and re-evaluation of the toxin producibility of the isolates from various sources. *Proceedings of the 13th Joint Conference on Cholera*, DEHW Publ No. (NIH) 78-1590, pp. 403–413.

Okuda, J., Ishibashi, M., Abbott, I., Janda, J., Nishibuchi, M. 1997a. Analysis of the thermostable direct hemolysin (*tdh*) gene and the *tdh*-related hemolysin (*trh*) genes in the urease-positive strains of *Vibrio parahaemolyticus* isolated on the West Coast of the United States. *J. Clin. Microbiol.* 35:1965–1971.

Okuda, J., Ishibashi, Hayakawa, E., Nishino, T., Takeda, Y., Mukhopadhyay, A., Garg, S., Bhattacharya, S., Nair, G., Nishibuchi, M. 1997b. Emergence of a unique O3:K6 clone of *Vibrio parahaemolyticus* in Calcutta, India and isolation of strains from the same clonal group from Southeast Asian travelers arriving in Japan. *J. Clin. Microbiol.* 35:3150–3155.

Okuda, J., Nishibuchi, M. 1998. Manifestation of the Kanagawa phenomenon, the virulence-associated phenotype, of *Vibrio parahaemolyicus* depends on a particular single base change in the promoter of the thermostable direct haemolysin. *Mol. Microbiol.* 39:499–511.

Olsen, H. 1978. *Vibrio parahaemolyticus* isolated from discharge from the ear in two patients exposed to sea water. *Acta Pathol. Microbiol. Scand. Sect. B.* 86:247–248.

Omori, G., Iwao, M., Iida, S., Kuroda, K. 1966. Studies on K antigen of *Vibrio parahaemolyticus*. I. Isolation and purification of K antigen from *Vibrio parahaemolyticus* A55 and some of its biological properties. *Biken J.* 9:33–43.

Osawa, R., Okitsu, T., Morozumi, H., Yamai, S. 1996. Occurrence of urease-positive *Vibrio parahaemolyticus* in Kanagawa, Japan with specific reference to presence of thermostable direct hemolysin (TDH) and the TDH-related hemolysin genes. *Appl. Environ. Microbiol.* 62:725–727.

Ray, B., Hawkins, S., Hackney, C. 1978., Method for the detection of injured *Vibrio parahaemolyticus* in seafoods. *Appl. Environ. Microbol.* 35:1121–1127.

Roland, F. 1970. Leg gangrene and endotoxin shock due to *Vibrio parahaemolyticus*: An infection acquired in New England coastal waters. *N. Engl. J. Med.* 282:1306.

Ruimy, R., Breittmayer, V., Elbaze, P., Lafay, B., Boussemart, O., Gauthier, M., Christine, R. 1994. Phylogenetic analysis and assessment of the genera *Vibrio, Photobacterium, Aeromonas,* and *Pleisiomonas* deduced from small-subunit rRNA sequences. *Int. J. Syst. Bacteriol.* 44:416–426.

Sakai, S., Kudoh, Y., Zen-Yoji, H. 1974. Food poisoning caused by indole-negative strains of *Vibrio parahaemolyticus* in Tokyo. In: *International Symposium on Vibrio parahaemolyticus,* T. Fugino, G. Sakaguchi, R. Sakazaki, Y. Takeda, Eds., Saikon: Tokyo, pp. 59–62.

Sakazaki, R., Iwanami, S., Fukumi, H. 1963. Studies on the enteropathogenic, facultatively halophilic bacteria, *Vibrio parahaemolyticus.* I. Morphological, cultural, and biochemical properties and its taxonomic position. *Jpn. J. Med. Sci. Biol.* 16:161–188.

Sakazaki, R., Iwanami, S., Fukumi, H. 1968. Studies on the enteropathogenic, facultatively halophilic bacteria, *Vibrio parahaemolyticus.* II. Serological characteristics. *Jpn. J. Med. Sci. Biol.* 21:313–324.

Sakazaki, R., Tamura, K., Kato, T., Obara, Y., Yamai, S., Hobo, K. 1968. Studies on the enteropathogenic, facultatively halophilic bacterium *Vibrio parahaemolyticus.* III. Enteropathogenicity. *Jpn. J. Med. Sci. Biol.* 21:325–331.

Sakazaki, R., Tamura, K., Nakamura, K., Kurata, T., Gohda, A., Kazuno, Y. 1974. Studies on enterophathogenic activity of *Vibrio parahaemolyticus* using ligated gut loop model in rabbits. *Jpn. J. Med. Sci. Biol.* 27:35–43.

Sakurai, J., Matsuzaki, A., Miwatani, T. 1973. Purification and characterization of thermostable direct hemolysin of *Vibrio parahaemolyticus. Infect. Immun.* 8:775–780.

Sanyal, S.C., Sen, P.C. 1974. Human volunteer study on the pathogenicity of *Vibrio parahaemolyticus.* In: *International Symposium on* Vibrio parahaemolyticus, T.G. Sakaguchi, R. Sakazaki, Y. Takeda, Eds., Saikon: Tokyo, pp. 227–230.

Schwartz, J.R., Colwell, R.R. 1974. Effect of hydrostatic pressure on growth and viability of *Vibrio parahaemolyticus. Appl. Microbiol.* 28:977–981.

Shewan, J., Véron, M. 1974. *Genus I.* Vibrio *Pacini 1854, 411.* In: *Bergy's Manual of Determinative Bacteriology,* 8th ed. R.E. Buchanan, N.E. Gibbons, Eds., Williams and Wilkins: Baltimore, MD, p. 342.

Shinoda, S., Honda, T., Takeda, Y., Miwatani, T. 1974. Antigenic difference between polar monotrichous and peritrichous flagella of *Vibrio parahaemolyticus. J. Bacteriol.* 120:923–928.

Shinoda, S., Kariyama, R., Ogawa, M., Takeda, Y., Miwatani, T. 1976. Flagellar antigens of various species of the genus *Vibrio* and related genera. *Int. J. Syst. Bacteriol.* 26:97–101

Shinoda, S., Nakahara, N., Nonomiya, Y., Itoh, K., Haruaki, K. 1983. Serological method for identification of *Vibrio parahaemolyticus* from marine sediments. *Appl. Environ. Microbiol.* 45:148–152.

Shirai, H., Ito, H., Hirayama, T., Nakabayashi, Y., Kumagai, K., Takeda, Y., Nishibuchi, M. 1990. Molecular epidemiological evidence for association of thermostable direct hemolysin (TDH) and TDH-related hemolysin of *Vibrio parahaemolyticus* with gastroenteritis. *Infect. Immun.* 58:3568–3573.

Sircar, B., Deb, B., De, S., Ghosh, A., Pal, S. 1976. Clinical and epidemiological studies on *V. parahaemolyticus* infection in Calcutta. *Ind. J. Med. Res.* 64:1576–1580.

Steinkuller, P., Kelly, M., Sands, S., Barber, J. 1980. *Vibrio parahaemolyticus* endophthalmitis. *J. Pediatr. Ophthalmol. Strabismus.* 17:150–153.

Stewart, B., McCarter, L. 2003. Lateral flagellar gene system of *Vibrio parahaemolyticus*. *J. Bacteriol.* 185:4508–4518.

Suthienkul, O., Ishibashi, M., Iida, T., Nettip, N., Supavej, S., Eampokalap, B., Makino, M., Honda, T. 1995. Urease production correlates with possession of the *trh* gene in *Vibrio parahaemolyticus* strains isolated in Thailand. *J. Infect. Dis.* 172:1405–1408.

Sutton, R.G. 1974. Some quantitative aspects of *Vibrio parahaemolyticus* in oysters in the Sydney area. In: *International Symposium on* Vibrio parahaemolyticus, T. Fugino, G. Sakaguchi, R. Sakazaki, Y. Takeda, Eds., Saikon: Tokyo, pp. 71–76.

Tacket, C., Barret, T., Sanders, G., Blake, P. 1982. Panophthalmitis caused by *Vibrio parahaemolyticus*. *J. Clin. Microbiol.* 16:195–196.

Tada, J., Ohashi, T., Nishimura, N., Shirasaki, Y., Ozaki, H., Fukushima, S., Takano, J., Nishibushi, M., Takeda, Y. 1992. Detection of the thermostable direct hemolysin gene (*tdh*) and the thermostable direct hemolysin-related gene (*trh*) of *Vibrio parahaemolyticus* by polymerase chain reaction. *Mol. Cell Probes.* 6:477–487.

Takeda, Y., Thaga, S., Miwatani, T. 1978. Evidence that thermostable direct hemolysin of *Vibrio parahaemolyticus* is composed of two subunits. *FEMS Microbiol. Lett.* 4:271–274.

Takikawa, I. 1958. Studies on pathogenic halophilic bacteria. *Yokohama M. Bull.* 2:313–322.

Taniguchi, H., Hirano, H., Kubomura, S., Higashi, K., Misuguchi, Y. 1986. Comparison of the nucleotide sequences of the genes for the thermostable direct hemolysin and the thermolabile hemolysin from *Virbio parahaemolyticus*. *Microb. Pathol.* 1:425–432.

Taniguchi, H., Ohta, H., Ogawa, M., Mizuguchi, Y. 1985. Cloning and expression in *Escherichia coli* of *Vibrio parahaemolyticus* thermostable hemolysin and thermolabile hemolysin genes. *J. Bacteriol.* 162:510–515.

Taniguchi, H., Sato, K., Ogawa, M., Udou, T., Mizuguchi, Y. 1984. Isolation and characterization of a filamentous phage, Vf33, specific for *Vibrio parahaemolyticus*. *Microbiol. Immumol.* 28:327–337.

Temyo, R. 1966. Studies on the prevention of outbreaks of food poisoning caused by *Vibrio parahaemolyticus*. *Bull. Tokyo Med. Dent. Univ.* 13:89–510.

Thompson, C., Jr., Vanderzant, C., Ray, S. 1976. Serological and hemolytic characteristics of *Vibrio parahaemolyticus* from marine sources. *J. Food Sci.* 41:204–205.

Thomson, W., Thacker, C. 1973. Effect of temperature on *Vibrio parahaemolyticus* in oysters at refrigerator and deep freeze temperatures. *Can. Inst. Food Sci. Technol. J.* 6:156–158.

Torii, M. 1974. Extraction and antigenic specificity of o-antigens of *Vibrio parahaemolyticus*. In: *International Symposium on Vibrio parahaemolyticus,* T. Fugino, G. Sakaguchi, R. Sakazaki, Y. Takeda, Eds., Saikon: Tokyo, pp. 187–192.

Torii, M., An, T., Igarashi, K., Sakai, K., Kuroda, K. 1969. Immunochemical studies on O-antigens of *Vibrio parahaemolyticus*. 1. Preparation, specificity and chemical nature of the antigens. *Biken J.* 12:77–84.

Tsunasawa, S., Sugihara, A., Masaki, T., Sakiyama, T., Takeda, F., Miwatani, T., Narita, K. 1987. Amino acid sequence of thermostable direct hemolysin produced by *Vibrio parahaemolyticus*. *J. Biochem.* 101:111–121.

Twedt, R.M. 1989. *Vibrio parahaemolyticus*. In: *Foodborne Bacterial Pathogens*, M.P. Doyle, Ed., Marcel Dekker: New York, pp. 543–600.

Twedt, R.M., Spaulding, P.L., Hall, H.E. 1969. Morphological, cultural, biochemical, and serological comparison of Japanese strains of *Vibrio parahaemolyticus* with related cultures isolated in the United States. *J. Bacteriol.* 98:511–518.

Vanderzant, C., Nickelson, R. 1972. Survival of *Vibrio parahaemoyticus* in shrimp tissue under various environmental conditions. *Appl. Microbiol.* 23:34–37.

Vanderzant, C., Nickelson, R., Hazelwood, R.W. 1974. Effect of isolation-enumeration procedures on the recovery of normal and stressed cells of *Vibrio parahaemolyticus*. In: *International Symposium on Vibrio parahaemolyticus*, T. Fugino, G. Sakaguchi, R. Sakazaki, Y. Takeda, Eds., Saikon: Tokyo, pp. 111–116.

Venkateswaran, K., Dohmoto, N., Harayama, S. 1998. Cloning and nucleotide sequence of the *gyrB* gene of *Vibrio parahaemolyticus* and its application in detection of this pathogen in shrimp. *Appl. Environ. Microbiol.* 64:681–687.

Vuddhakul, V., Chowdbury A., Laohaprertthisan, V., Pungrasamee, P., Patararungrong, N., Thianmonyri, P., Ishibasi, M., Matsumoto, C., Nishibuchi, M. 2000. Isolation of a pandemic O3:K6 clone of a *Vibrio parahaemolyticus* strain from environment and clinical sources in Thailand. *Appl. Environ. Microbiol.* 66:2685–2689.

Wagatsuma, S. 1968. On a medium for hemolytic reaction. *Media Circle.* 13:159–162 (in Japanese).

Waldor, M., Mekalanos, J. 1996. Lysogenic conversion by a filamentous phage encoding cholera toxin. *Science* 272:1910–1914.

Wang, S., Levin, R.E. 2004. Quantitative determination of *Vibrio parahaemolyticus* by polymerase chain reaction. *Food Biotechnol.* 18:279–287.

Wong, H., Liu, C., Pan, T., Wang T., Lee, D., Shih, Y. 1999. Molecular typing of *Vibrio parahaemolyticus* isolates obtained from patients involved in food poisoning outbreaks in Taiwan by random amplified polymorphic DNA analysis. *J. Clin. Microbiol.* 37: 1809–1812.

Wong, H., Liu, K., Pan, T., Lee, D., Shih, Y. 1996. Subspecies typing of *Vibrio parahaemolyticus* by pulsed-field gel electrophoresis. *J. Clin. Microbiol.* 34:1536–1539.

Xu, M., Iida, T., Yamamoto, K., Takarada, Y., Miwatani, T., Honda, T. 1994. Demonstration and characterization of simultaneous production of a thermostable direct hemolysin (TDHI) and a TDH-related hemolysin (TRHx) by a clinically isolated *Vibrio parahaemolyticus* strain, TH3766. *Infect. Immun.* 62:166–171.

Yabuuchi, E., Miwatani, T., Takeda, Y. Arita, M. 1974a. Flagellar morphology of *Vibrio parahaemolyticus*. In: *International Symposium on* Vibrio parahaemolyticus, T. Fujino, G. Sakaguchi, R. Sakazaki, Y. Takeda, Eds., Saikon: Tokyo, pp. 163–168.

Yabuuchi, E., Miwatani, T., Takeda, Y. Arita, M. 1974b. Flagellar morphology of *Vibrio parahaemolyticus* (Fujino et al.) Sakazaki, Iwanami, and Fukumi 1963. *Jpn. J. Microbol.* 18:295–305.

Yanagase, Y., Inoue, K., Ozaki, M., Ochi, T., Amano, T., Chazano, M. 1970. Hemolysins and related enzymes of *Vibrio parahaemolyticus*. I. Identification and partial purification of enzymes. *Biken J.* 13:77–92.

Yanagase, Y., Ozaki, M., Ochi, T., Amano, T. 1968. Lecithinase of *Vibrio parahaemolyticus*. *Jpn. J. Bacteriol.* 23: 35–36.

Yoh, M., Honda, T., Miwatani, T., Nishibushi, M. 1991. Characterization of thermostable direct hemolysins encoded by four representative *tdh* genes of *Vibrio parahaemolyticus*. *Microb. Patholog.* 10:165–172.

Zen-Yoji, H., Kudoh, Y., Igarashi, H., Ohta, K., Fukai, K. 1974. Purification and identification of enteropathogenic toxins "a" and "a" produced by *Vibrio parahaemolyticus,* and their biological and pathological activities. In: *International Symposium on* Vibrio parahaemolyticus, T. Fugino, G. Sakaguchi, R. Sakazaki, Y. Takeda, Eds., Saikon: Tokyo. pp. 237–243.

CHAPTER 7

Vibrio cholerae

I. CHARACTERISTICS OF THE ORGANISM

Vibrio cholerae is a gram-negative, facultatively anaerobic, curved rod that is polarly flagellated, cytochrome oxidase-positive, and characteristically ferments glucose to acid without gas production. The generally accepted method for its isolation involves enrichment in alkaline peptone water (APW) followed by culture on selective thiosulfate-citrate-bile salts-sucrose (TCBS) agar (Table 6.5, Chapter 6). Sucrose-positive, smooth yellow colonies on TCBS agar are submitted to the oxidase test. Positive colonies are then purified on a nonselective agar and subjected to biochemical and serological tests. Detection of cholera toxin (CTX) is usually by an ELISA assay.

V. cholera is considered a heterogeneous species with 206 serotypes presently recognized. However, only two serotypes are associated with epidemic infections, O1 and O139. The O139 strains have been shown to be genetically similar to O1 strains and are hypothesized as having evolved from strains of the early seventh pandemic by a mechanism involving insertion of exogenous DNA encoding the O139 LPS (Bik et al. 1995; Bik, Gouw, and Mooi, 1996; Karaolis, Lan, and Reeves, 1995; Dumontier and Berche, 1998). There are strains of these two serogroups, however, that do not produce the cholera enterotoxin and are not infectious. In addition, CTX negative *V. cholera* O1 strains have been implicated in occasional cases of diarrhea and extraintestinal infections. Some non-O1/non-O139 strains produce a heat stabile enterotoxin designated NAG-ST (nonagglutinable *Vibrio* ST) encoded by the *sto* gene. *V. cholerae* O1 is divided into two biotypes, classical and El Tor. The classical biotype has been gradually replaced by the El Tor biotype since 1961, and presently the classical is considered extinct after not reappearing after 1983. The El Tor biotype is therefore presently considered the most significant. In the United States, crabs, shrimp, and oysters have been the most frequently implicated vehicles.

Seven pandemics of cholera have been recorded since 1817, the ongoing seventh pandemic having started in 1961 in Asia. There have been two major upsurges of the seventh pandemic: one in the 1970s spread to Africa and the other in 1991 spread cholera to South America. Both continents had been free of pandemic cholera for over a century. The majority of published molecular studies on *V. cholerae* isolates during the past two decades have involved strains from the seventh pandemic.

II. FACTORS ASSOCIATED WITH THE VIRULENCE OF *V. CHOLERAE*

Cholera toxin (CTX) consists of two polypeptides; the single polypeptide A is usually proteolytically nicked to form two disulphide-linked polypeptides named A1 and A2. The A subunits are responsible for toxic activity. The A subunit encoded by *ctxA* is responsible for adenylate cyclase activation in enterocytes of the small intestine, inducing extensive secretion resulting in the tremendous loss of water and electrolytes associated with cholera. The B subunit encoded by *ctxB* consists of five identical peptides responsible for binding to the epithelial cell surface receptor GM_1. In addition to CTX, the pathogenicity of *V. cholera* O1 and O139 strains also depends on the ability to adhere and colonize the small intestine via production of pili encoded by *tcpA*.

There are a large number of genes associated with the virulence of *V. cholera* that have been used to detect and characterize isolates. The major virulence-associated factors are present in clusters within at least three regions in the *V. cholera* genome. The first consists of the *ctxA/ctxB* genes that reside on a lysogenized phage (Waldor and Mekalanos, 1996). The second is a large pathogenicity island that encodes a toxin coregulated pilus gene cluster involving a type IV pilus that is the primary adhesion and colonizing factor (Taylor et al., 1987) and acts as a phage receptor. The third gene cluster, the RTX gene cluster, encodes a cytotoxin active against cultured Hep-2 cells in *V. cholera* El Tor strains (Lin et al., 1999). Table 7.1 lists the genes that have been used to identify and characterize isolates of *V. cholerae*.

III. PCR DETECTION, IDENTIFICATION, AND CHARACTERIZATION OF *V. CHOLERAE*

In January 1991, an outbreak of cholera started in Peru and rapidly spread throughout most of Latin America. Within 15 months over 450,000 cases occurred, with about 4000 deaths. The causative organism was toxigenic *V. cholerae* O1 of the El Tor biotype, which is distinct from the U.S. Gulf Coast strains. Fields et al. (1992) reported on the use of primers CTX2/CTX3 (Table 7.2) that amplified a 564-pb sequence of the *ctxA* gene for its detection in 150 *V. cholerae* isolates derived from patients, food, and water from the 1991–1992 outbreak. There were 140 isolates found to be toxigenic both by PCR and immunoassay.

Koch et al. (1993) reported on the development of a PCR assay for detection of *V. cholerae* with seeded oysters, crabmeat, shrimp, and lettuce. The primers P1/P3 (Table 7.2) amplified a 778-bp sequence of the *ctxB* gene from a *V. cholerae* O1 strain. Seeded foods were homogenized or rinsed with APW followed by a 6–8 hr enrichment incubation at 37°C. One milliliter enrichments were boiled and 2 to 5

Table 7.1   Genes Used for the PCR Identification of *V. cholerae*

Gene	Description
ace	Encodes the accessory cholerae enterotoxin
ctxA	Encodes cholera toxin CTXA that is phage encoded and responsible for cholera symptoms, functions intracellularly to activate adenyl cyclase resulting in diarrhea
ctxB	Encodes ctxB peptide that binds to enterocyte receptors (GM$_1$)
dnaJ	A housekeeping gene that encodes heat shock protein 40 used for the identification of all *Vibrio* species
epsM	Encodes an extracellular secretory protein
hlyA	Encodes a hemolysin
hsp60	Encodes a 60-kDa heat shock protein
lolB	Previously called the *hemM* gene that encodes an outer membrane lipoprotein
mshA	Encodes mannose sensitive pili
nanH	Encodes a neuraminidase
nint	Encodes a phagelike integrase
O1-*rfb*	Encodes the O antigen of O1 strains
O139-*rfb*	Encodes the O antigen of O139 strains
ompU	Encodes an outer membrane protein that confers bile resistance and enhanced survival in the environment
ompW	Encodes an outer membrane protein found exclusively in all strains of *V. cholerae* and used for species confirmation
pgm	Encodes phosphoglucomutase; used as a housekeeping gene
rfbN	Encodes the O antigen of O1 strains
rtx	A gene cluster encoding a presumptive cytotoxin rtxA, an acyl transferase, rtxC, and proteins RtxB and RStxD involved in toxin transport
rs1	Encodes repetitive sequence 1
sto	Encodes thermotolerant enterotoxin
toxR	Regulates expression of *ctx*, *ctp*, and outer membrane protein genes
tcpA (El Tor)	Encodes toxin-coregulated pilus involved in intestinal colonization
tcpA (Classical)	Encodes toxin-coregulated pilus involved in intestinal colonization
wbeT	Encodes the somatic antigen synthesis region
zot	Encodes the zonula occludans toxin (ZOT) that results in increased permeability of the intestinal mucosa enhancing water secretion

µl added to 100 µl PCR reaction volumes. A detection limit of 1 CFU/10 g of food was obtained.

Varela et al. (1993) subjected 43 *V. cholerae* strains isolated from human infections, water, and fish in Argentina during an outbreak just prior to 1993 to ELISA and PCR assays for ctxA1 expression and detection of the *ctxA1* gene. The primers ctxA1-F/ctxA1-R (Table 7.2) amplified a 431-bp sequence of the *ctxA1* gene. The primers ctxA2-B-F/ctxA2-B-R (Table 7.2) amplified a 566-bp portion of the *ctxA2-B* gene sequence. Twenty-four out of twenty-six Ogawa and eight out of ten Inaba isolates were positive for ctxA1 by ELISA and PCR. Two Ogawa and two Inaba strains were negative for ctxA1 by both methods. Seven non-O1 isolates were ctxA1 negative by ELISA; however, one of these seven non-O1 isolates was positive for *ctxA1* by the PCR. This isolate was designated S563. To determine whether strain S563 had a mutation in a region other than *ctxA1*, primers to detect the *ctxA2-B* region were developed. The S563 isolate failed to yield the expected 566-bp amplicon, whereas positive controls that were both ELISA- and PCR-positive for ctxA1 yielded the 566-bp amplicon. The PCR detection of *ctxA* therefore does not guarantee the production of the CTX toxin. The authors concluded that the mutation present in the S563 strain was different from those present in other nontoxigenic *V. cholerae* isolates.

Guglielmetti et al. (1994) developed a PCR assay for detection of the *sto* gene encoding the heat-stabile enterotoxin of non-O1 strains of *V. cholerae* (ST). The primers STO-F/STO-R (Table 7.2) amplified a 238-bp sequence of the *sto* gene. Among 22 *V. cholerae* non-O1 isolates from sporadic cases of cholera from Cuba, four strains (18.2%) were found to carry the *sto* gene.

Shangkuan, Sow, and Wang (1995) developed a multiplex PCR to detect *V. cholerae* and to biotype O1 isolates. The study involved a total of 96 strains. Two pairs of primers (Table 7.2) were used to amplify a *ctxA2-B* gene fragment of *V. cholerae*. The first pair of primers C2F/C2R amplified a 385-bp fragment of *ctxA2-B*. The second pair of primers C4F/C4R is nested within the first PCR product and amplifies a 103-bp fragment. Three primers were used to amplify the *hlyA* gene for detection of classical and El Tor isolates. The outer primer pair H4F/H3R (Table 7.2) was designed to amplify a well-conserved region of the *hlyA* gene among both classical and El Tor strains and yields amplicons of 486- and 497-bp, respectively. The internal primer H3F (Table 7.2) was chosen to be specific for El Tor strains and to yield a 243-bp amplicon. The internal primer pair H3F/H3R yielded a 243-bp amplicon derived from the *hlyA* gene. The primers C2F, C2R, H4F, H3R, and H3F were used in a single multiplex PCR reaction to simultaneously amplify the *ctxA2-B* and *hlyA* gene fragments. The nested PCR for the *hlyA* gene used 5 µl of the multiplex PCR product for template DNA with internal primers C4F and C4R and 20 thermal cycles. The second nested PCR with *ctxA2-B* internal primers allowed detection of less than 3 CFU/g with both 1:10 and 1:100 enriched homogenates. Classical strains yielded the 486-bp amplicon from the *hlyA* gene, whereas El Tor strains yielded both the 497- and 243-bp amplicons from the hlyA gene. All 33 CT toxin-producing strains produced the 385-bp fragment derived from the *ctxA2-B* gene. In contrast, no 385-bp amplicon was produced from any of the 12 nontoxigenic strains. Seeded oyster tissue homogenates prepared with 10 g or 1 g of tissue in 90 or 99 ml of APW, respectively, were incubated statically at 37°C for 6 hr. The 6-hr enrichment in APW allowed detection of 30 CFU/g from the *ctxA2-B* amplicon with a 1:100 APW dilution enrichment, whereas with 1:10 enrichments, the 385-bp *ctxA2-B* amplicon was just barely visible at the same level of sensitivity because of the presence of PCR inhibitors derived from the tissue.

DePaola and Hwang (1995) determined the optimum conditions of enrichment for detection of *V. cholera* by the PCR. Recovery and PCR detection were significantly greater from oyster homogenates diluted 1:100 in alkaline peptone water and incubated at 42°C for 18–21 hr. The primers used (Table 7.2) were from Fields et al. (1992) and amplified a 564-bp sequence of the *ctxA* gene.

Damian et al. (1998) subjected 89 *V. cholerae* clinical strains isolated in Romania from 1977 to 1994 and six strains from Moldavia to ribotyping, toxin gene restriction pattern analysis (toxinogenotype), and analysis of the distribution of the genes *ctx*, *ace*, and *zot* (Table 7.2). After *Bgl*I restriction of genomic DNA, a total of 18 ribotypes and 21 toxinogenotypes were distinguished. PCR amplicons were transferred to a nylon membrane and hybridized with corresponding digoxygenin (DIG) probes followed by DIG detection. Although 94% of the 95 strains harbored the *ctx* gene, 20% of the strains had an incomplete "virulence cassette."

Hoshino et al. (1998) developed a multiplex PCR assay for rapid detection of toxigenic *V. cholerae* O1 and O139. The assay targeted the *rfb* sequence specific for the O1 and O139 serogroups and the *ctxA* gene. Primers O139-F2/O139-R2 (Table 7.2) amplified a 449-bp sequence of the *rfb* gene from O139 strains. The primers O1F2-1/O1F2-2 (Table 7.2) amplified a 192-bp sequence of the *rfb* gene from O1 strains. The primers VCT1/VCT2 amplified a 308-bp sequence of the *ctxA* gene. Among 121 stool samples from hospitalized patients in Calcutta exhibiting acute secretory diarrhea, 38 were both culture and PCR positive. An additional four samples that were culture-negative were PCR positive. Direct PCR detection without enrichment involved boiling a 250 µl portion of each stool sample for 10 min followed by 1:10 dilution of the samples using Tris-HCl buffer (pH 8.0). Bovine serum albumin (BSA) was added at a concentration of 400 ng/ml to the stool samples to circumvent the presence of PCR inhibitors. The samples were then centrifuged at 2000 rpm for 2 min and 3 µl of the supernatants added to PCR mixes to final reaction volumes of 30 µl.

Ripabelli et al. (1999) reported on the occurrence of various pathogenic *Vibrio* species from mussels harvested from approved shellfish waters in the Adriatic Sea. *V. cholera* O1 and O139 serotypes were not detected. However, *V. cholera* non-O1/non-O139 was found in 1.6% of the samples compared to 32.2% for *V. alginolyticus*. PCR with primers STO-F/STO-R (Table 7.2) were used to amplify a 238-bp sequence of the *sto* gene that encodes the thermotolerant enterotoxin of *V. cholerae* and revealed the absence of the *sto* gene in all of these environmental isolates of *V. cholerae*.

Nandi et al. (2000) assessed the distribution of genes for an outer membrane protein (OmpW) and a regulatory protein (ToxR) among 254 *V. cholerae* isolates. The primers ompW-F/ompW-R (Table 7.2) amplified a 588-bp sequence of the *ompW* gene. The primers toxR-F/toxR-R (Table 7.2) amplified an 883-bp sequence of the *toxR* gene. The primers ctxA-F/ctxA-R (Table 7.2) amplified a 301-bp sequence of the *ctxA* gene. All 254 isolates were found to harbor the *ompW* gene and 229 (98%) were found to harbor the *toxR* gene. None of the other 40 strains belonging to other *Vibrio* species produced amplicons with either *ompW*- or *toxR*-specific primers, nor did 80 strains from other bacterial genera. Restriction fragment length polymorphism (RFLP) analysis and nucleotide sequence data revealed that the *ompW* gene sequence is highly conserved among *V. cholerae* strains belonging to different biotypes or serogroups. The authors concluded that their observations suggested that the *ompW* gene can be targeted for the species-specific identification of *V. cholerae* strains and is more suitable than the *toxR* gene. They then developed a multiplex PCR involving both the *ompW* and *ctxA* genes for screening both toxigenic and nontoxigenic strains of clinical and environmental isolates of *V. cholerae*.

Table 7.2 PCR Primers and DNA Probes for *V. cholerae*

Primer or Probe	Sequence (5' → 3')	Sequence (bp)	Gene	Reference
CTX2	CGG-GCA-GAT-TCT-AGA-CCT-CCT-G	564	*ctxA*	Fields et al. (1992)
CTX3	CGA-TGA-TCT-TGG-AGC-ATT-CCC-AC			
P1	TGA-AAT-AAA-GCA-GTC-AGG-TG	778	*ctxB*	Koch et al. (1993)
P3	GGT-ATT-CTG-CAC-ACA-AAT-CAG			
STO-F	CAT-GAG-AAA-CCT-ATT-CAT-TGC	238	*sto*	Ripabelli et al. (1999)
STO-R	TTA-ATT-TAA-ACA-TCC-AAA-GCA-AG			
rtxA-F	CTG-AAT-ATG-AGT-GGG-TGA-CTT-ACG	417	*rtxA*	Chow et al. (2001)
rtxA-R	GTG-TAT-TGT-(C/T)CG-ATA-TCC-GCT-ACG			
rtxC-F	CGA-CGA-AGA-TCA-TTG-ACG-AC	263	*rtxC*	Chow et al. (2001)
rtxC-R	CAT-CGT-CGT-TAT-GTG-GTT-GC			
ctxB$_2$	GAT-ACA-CAT-AAT-AGA-ATT-AAG-GAT	460	*ctxB*	Chow et al. (2001)
ctxB$_3$	GGT-TGC-TTC-TCA-TCA-AAC-CAC			
ompW-F	CAC-CAA-GAA-GGT-GAC-TTT-ATT-GTG	588	*ompW*	Nandi et al. (2000)
ompW-R	GAA-CTT-ATA-ACC-ACC-CGC-G			
toxR-F	ATG-TTC-GGA-TTA-GGA-CAC	883	*toxR*	Nandi et al. (2000) from Miller, Taylor, and Mekalanos (1987)
toxR-R	TAC-TCA-CAC-ACT-TTG-ATG-GC			

ctxA-F	CTC-AGA-CGG-GAT-TTG-TTA-GGC-ACG	301	ctxA	Nandi et al. (2000) from Keasler and Hall (1993)
ctxA-R	TCT-ATC-TCT-GTA-GCC-CCT-ATT-ACG			
VHMF	TGG-GAG-CAG-CGT-CCA-TTG-TG	519	lolB	Lalitha et al. (2008)
VHA-AS5	CAA-TCA-CAC-CAA-GTC-ACT-C			
ctxA-F	TTT-GTT-AGG-CAC-GAT-GAT-GGA-T	—	ctx	Blackstone et al. (2007)
ctxA-R	ACC-AGA-CAA-TAT-AGT-TTG-ACC-CAC-G			
ctxAP	FAM/TET-TGT-TTC-CAC-CTC-AAT-AGT-TTG-AGA-AGT-GCC-C-BHQ			
Outer ctxA-F	TCT-ATC-TCT-GTA-GCC-CCT-ATT-ACG	—	ctxA	Mendes, Abath, and Leal (2008)
Outer ctxA-R	ATA-CCA-TCC-ATA-TAT-TTG-GGA-G			Mendes, Abath, and Leal (2008) from Li et al. (2002)
Inner ctxA-F	CTC-AGA-CGG-GAT-TTG-TTA-GGC-ACG	302	ctxA	Mendes, Abath, and Leal (2008) from Keasler and Hall (1993)
Inner ctxA-R	TCT-ATC-TCT-GTA-GCC-CCT-ATT-ACG			
rfbN-F	GTT-TCA-CTG-AAC-AGA-TGG-G	198	rfbN	Mendes, Abath, and Leal (2008) from Islam et al. (2004)
rfbN-R	GGT-CAT-CTG-TAA-GTA-CAA-C			
rtxC-F	CGA-CGA-AGA-TCA-TTG-ACG-AC	255	rtx	Fukushima, Tsunomori, and Seki (2003) from Chow et al. (2001)
rtxC-r	CAT-CGT-CGT-TAT-GTG-GTT-GC			

Continued

Table 7.2   PCR Primers and DNA Probes for *V. cholerae* (Continued)

Primer or Probe	Sequence (5′ → 3′)	Sequence (bp)	Gene	Reference
CT-F	ACA-GAG-TGA-GTA-CTT-TGA-CC	308	*ctx*	Fukushima, Tsunomori, and Seki (2003) from Nair et al. (1994)
CT-R	ATA-CCA-TCC-ATA-TAT-TTG-GGA-G			
L-ctx	CTC-AGA-CGG-GAT-TTG-TTA-GGC-ACG	302	*ctxA*	Lee, Panicker, and Bej (2003)
R-ctx	TCT-ATC-TCT-GTA-GCC-CCT-ATT-ACG			
PP-ctx	ATT-AGT-TTG-AGA-AGT-GCC-CAC-TTA-GTG-GGT-CAA-AC	—	*ctxA*	
BP-ctx	GTT-TCT-GCT-TTA-GGT-GGG-ATT-CCA-TAC-TCC	—	*ctxA*	
V1F	AGC-AAG-AGC-ATT-GTT-GTT-CCT-ACC	120	*rtxA*	Gubula (2006)
V1R	ACT-TCC-CTG-TAC-CGC-ACT-TAG-AC			
V3F	TGG-TTG-ATC-GCT-TGG-CGC-ATC	145	*epsM*	
V3R	ATG-GCA-GCC-TTT-GAG-TGA-G			
V6F	ACA-CCT-GGA-ACA-GTT-ATT-GAT-GGC	113	*mshA*	
V6R	TCA-CTC-GAA-GTA-TCT-AGC-GTT-TGC			
V7F	TGC-AAT-GAC-ACA-AAC-TTA-TCC-TAG	147	*tcpA*	
V7R	CCC-ATA-GCT-GTA-CCA-GTG-AA-G			
94F	CGG-GCA-GAT-TCT-AGA-CCT-CCT-G	564	*ctxA*	Rivera et al. (2001) from Fields et al. (1992)
614R	CGA-TGA-TCT-TGG-AGC-ATT-CCC-AC			
80F	ACG-CTG-ACG-GAA-TCA-ACC-AAA-G	869	*ompU*	
906R	GCG-GAA-GTT-TGG-CTT-GAA-GTA-G			

Primer	Sequence	Size	Gene	Reference
25F	TCG-CTT-AAC-GAT-GGC-GCG-TTT-T	947	*zot*	Rivera et al. (2001) from Fields et al. (1992)
1129R	AAC-CCC-GTT-TCA-CTT-CTA-CCC-A			
101F	CCT-TCG-ATC-CCC-TAA-GCA-ATA-C	779	*toxR*	
837R	AGG-GTT-AGC-AAC-GAT-GCG-TAA-G			
72F	CAC-GAT-AAG-AAA-ACC-GGT-CAA-GAG	451	*tcpA* El Tor	Rivera et al. (2001) from Fields et al. (1992)
477R	CGA-AAG-CAC-CTT-CTT-TCA-CGT-TG			
647R	TTA-CCA-AAT-GCA-ACG-CCG-AAT-G	620	*tcpA* Classical	
132F	TAG-CCT-TAG-TTC-TCA-GCA-GGC-A	862	*tcpI*	Rivera et al. (2001) from Fields et al. (1992)
951R	GGC-AAT-AGT-GTC-GAG-CTC-GTT-A			
489F	GGC-AAA-CAG-CGA-AAC-AAA-TAC-C	481	*hlyA* El Tor	Rivera et al. (2001) from Fields et al. (1992)
744F	GAG-CCG-GCA-TTC-ATC-TGA-AT	738/727	*hlyA* Classical	
1184R	CTC-AGC-GGG-CTA-ATA-CGG-TTT-A			
67F	TCG-CAT-TTA-GCC-AAA-CAG-TAG-AAA	172	*stn/sto*	Rivera et al. (2001) from Fields et al. (1992)
194R	GCT-GGA-TTG-CAA-CAT-ATT-TCG-C			
C2F	AGG-TGT-AAA-ATT-CCT-TGA-CGA	335	*ctx*	Radu et al. (2002) from Shangkuan, Sow, and Wang (1995)
C2R	TCC-TCA-GGG-TAT-CCT-TCA-TC			

Continued

Table 7.2   PCR Primers and DNA Probes for *V. cholerae* (Continued)

Primer or Probe	Sequence (5′ → 3′)	Sequence (bp)	Gene	Reference
zot1-F	TGG-CTT-CGT-CTG-CTG-CCG-GCG-ATT	—	zot	Radu et al. (2002) from Shangkuan, Sow, and Wang (1995)
zot2-R	CAC-TTC-TAC-CCA-CAG-CGC-TTG-CGC			
ace1-F	TAA-GGA-TGT-GCT-TAT-GAT-GGA-CAC-CC	—	ace	Radu et al. (2002) from Shangkuan, Sow, and Wang (1995)
ace2-R	CGT-GAT-GAA-TAA-AGA-TAC-TCA-TAG-G			
RAPD GEN15005	CGG-ATA-ACT-G			Radu et al. (2002) from Shangkuan, Sow, and Wang (1995)
ctxA1-F ctxA1-R	AGA-CGG-GAT-TTG-TTA-GGC-ACG-AT AGA-ACC-TCG-TAA-GGG-TGT-TGG-GC	431	ctxA1	Valera et al. (1993)
ctxA2-B-F ctxA2-B-R	TAG-AGC-TTG-GAG-GGA-AGA-GCC-GT ATT-GCG-GCA-ATC-GCA-TGA-GGC-GT	566	ctxA2-B	Valera et al. (1993)
ctx1	CGG-GCA-GAT-TCT-AGA-CCT-CCT-G	564	ctx	Damian et al. (1998) from Fields et al. (1992)
ctx2 ctx probe	CGA-TGA-TCT-TGG-AGC-ATT-CCC-AC CGT-TAA-TGA-TGT-ATT-AGG-GGC-ATA			Damian et al. (1998) from Tamayo et al. (1997)
ace1	TAA-GGA-TGT-GCT-TAT-GAT-GGA-CAC-CC	314	ace	Damian et al. (1998) from Trucksis et al. (1993)
ace2	CGT-GAT-GAA-TAA-AGA-TAC-TCA-TAG			

Primer	Sequence	Size	Target	Reference
ace probe	CCG-CTT-ATC-CAA-CAG-GCT-ATC			
zot1	TGG-CTT-CGT-CTG-CTG-CCG-GCG-ATT	1C83	zot	Damian et al. (1998) from Tamayo et al. (1997)
zot2	CAC-TTC-TAC-CCA-CAG-GGC-TTG-GGC			
zot probe	GCC-ACT-TTA-ACC-GCG-CCA-C			
RAPD 1	GGT-GCG-GGA-A	—	—	Coelho et al. (1995)
RAPD 2	AAC-GGT-GAC-C	—	—	
RAPD 3	CCA-GAT-GCA-C	—	—	
RAPD 4	AAG-ACC-CCT-C	—	—	
RAPD 5	CTT-CAG-AGT-AGA-ACG-CAA-TG	—	—	
RAPD 6	GAC-ATA-AGA-ACA-AGT-TAC-AG	—	—	
RAPD 7	CGC-TAG-CAA-TTA-ATG-TGC-ATC	—	—	
RAPD 8	GCT-CTA-GAT-AAG-AAC-AAG-TTA-CAG-ACG	—	—	
RAPD 9	CGC-GGA-TCC-ATA-AGA-ACA-AGT-TAC-AGA-CG	—	—	
RAPD 10	AGA-GGG-CAC-A	—	—	
NA1	GGA-TCA-GAA-TGC-CAC-GGT-G	—	16S/23S rRNA	Coelho et al. (1995)
NB2	TCG-CTC-GCC-GCT-ACT-GG			
Vib-L2	TAT-CCC-GGG-TCC-CTC-TTG-AGG-CGT-TTG-TTA	300–1000	—	Castañeda et al. (2005)
Vib-R2	CGT-TGC-TAG-CCC-CTT-AGG-CGG-GCG-TTA			
REP1-I	III-ICG-ICG-ICA-TCI-GGC	750–6000	—	Castañeda et al. (2005)
REP2-I	ICG-ICT-TAT-CIG-GCC-TAC			

Continued

Table 7.2 PCR Primers and DNA Probes for *V. cholerae* (Continued)

Primer or Probe	Sequence (5′ → 3′)	Sequence (bp)	Gene	Reference
ctxB-F	GGT-TGC-TTC-TCA-TCA-TCG-AAC-CAC	460	*ctxB*	Tapchaisri et al. (2007) from Mitra et al. (2001)
ctxB-R	GAT-ACA-CAT-AAT-AGA-ATT-AAG-GAT			
Ace-F	TAA-GGA-TGT-GCT-TAT-GAT-GGA-CAC-CC	316	*ace*	Tapchaisri et al. (2007) from Shi et al. (1998)
Ace-R	CGT-GAT-GAA-TAA-AGA-TAC-TCA-TAG-G			
nanH-F	GAC-AGT-CCA-GCC-AAA-CAG	1900	*nanh*	Tapchaisri et al. (2007) from Jermyn and Boyd (2002)
nanH-R	CGT-TAG-CGT-TGT-TAG-CCT-C			
ninT-F	ATC-TGA-TGG-CGG-CAA-TC	1000	*ninh*	Tapchaisri et al. (2007) from Jermyn and Boyd (2002)
ninT-R	GCG-GCT-TCA-ATG-ACA-TC			
O139-F2	AGC-CTC-TTT-ATT-ACG-GGT-GG	449	*rfb* O139	Hoshino et al. (1998)
O139-R2	GTC-AAA-CCC-GAT-CGT-AAA-CG			
O1F2-1	GTT-TCA-CTG-AAC-AGA-TGG-G	192	*rfb* O1	Hoshino et al. (1998)
O1F2-2	GGT-CAT-CTG-TAA-GTA-CAA-C			
VCT1	ACA-GAG-TGA-GTA-CTT-TGA-CC	308	*ctxA*	Hoshino et al. (1998)
VCT2	ATA-CCA-TCC-ATA-TAT-TTG-GGA-G			
RAPD ERIC1	ATG-TAA-GCT-CCT-GGG-GAT-TCA-C	—	—	Pazzani et al. (2006)

Primer	Sequence	Size (bp)	Gene	Reference
RAPD ERIC2	AAG-TAA-GTG-ACT-GGG-GTG-AGC-G	—		
RAPD VCR1	CAG-CTC-CTT-AGG-CGG-GCG-TTA-G	—		
RAPD VCR2	ACA-GTC-CCT-CTT-GAG-GCG-TTT-G	—		
RAPD ATX1	AAG-CGA-TTG-AAA-GGA-TGA	—		
RAPD ATX2	CCG-CGA-GTG-CTT-GGT-TAG	—		
STO-F	CAT-GA(G/A)-AAA-CCT-ATT-CAT-GC	238	*sto*	Guglielmetti et al. (1994)
STO-R	GAA-CGA-AAC-CTA-CAA-ATT-TAA-TT			
VM-F	CAG-GTT-TGY-TGC-ACG-GCG-AAG-A	—	*dnaJ*	Nhung et al. (2007)
VC-Rmm	AGC-AGC-TTA-TGA-CCA-ATA-CGC-C	375	*dnaJ*	
VP-MmR	TGC-GAA-GAA-AGG-CTC-ATC-AGA-G	96	*dnaJ*	
VV-Rmm	GTA-CGA-AAT-TCT-GAC-CGA-TCA-A	412	*dnaJ*	
VM-Rmm	YCT-TGA-AGA-AGC-GGT-TCG-TGC-A	177	*dnaJ*	
V.al2-mmR	GAT-CGA-AGT-(A/G)CC-(A/G)AC-ACT-(A/C)GG-A	144	*dnaJ*	
RAPD 1281	AAC-GCG-CA-C	—	—	Bhowmick et al. (2007)
NST-F	CCT-ATT-CAT-TGC-ATT-AAT-G	215	*stn*	Sharma and Chaturvedi (2006) from Ogawa et al. (1990)
NST-R	CCA-AAG-CAA-GCT-GGA-TTG-C			
TcpA-F (El Tor)	GAA-GAA-GTT-TGT-AAA-AGA-AGA-ACA-C	417	*tcpA*	Sharma and Chaturvedi (2006) from Ogawa et al. (1990)
TcpA-R (El Tor)	GAA-AGG-ACC-TTC-TTT-CAC-GTT-G			
OmpW-F	CAC-CAA-GAA-GGT-GAC-TTT-ATT-GTG	588	*ompW*	Sharma and Chaturvedi (2006) from Nandi et al. (2000)
OmpW-F	GAA-CTT-ATA-ACC-ACC-CGC-G			

Continued

Table 7.2 PCR Primers and DNA Probes for *V. cholerae* (Continued)

Primer or Probe	Sequence (5' → 3')	Sequence (bp)	Gene	Reference
ctxA-F	CTC-AGA-CGG-GAT-TTG-TTA-GGC-ACG			
ctxA-R	TCT-ATC-TCT-GTA-GCC-CCT-ATT-ACG			
tcpA-F (El Tor)	GAA-GAA-GTT-TGT-AAA-AGA-ACA-C	472	*tcpA*	Khuntia, Bhusan, and Chhotray (2008) from Keasler and Hall (1993)
tcpA-R (El Tor)	GAA-AGG-ACC-TTC-TTT-CAC-GTT-G			
ctxA-F	GGC-TGT-GGG-TAG-AAG-TGA-AAC-GG	1140	*ctxA*	Kotetishvili et al. (2003)
ctxA-R	CTA-AGG-ATG-TGG-AAT-AAA-AAC-ATC			
ctxB-F	GGC-TGT-GGG-TAG-AAG-TGA-AAC-GG	1152	*ctxB*	Kotetishvili et al. (2003) from Li et al. (2002)
ctxB-R	CTA-AGG-ATG-TGG-AAT-AAA-AAC-ATC			
tcpA-F	AAA-ACC-GGT-CAA-GAG-GG	600	*tcpA*	Kotetishvili et al. (2003) from Karaolis et al. (1998)
tcpA-R	CAA-AAG-CTA-CTG-TGA-ATG-G			
tcpA-R	CAA-ATG-CAA-CGC-CGA-ATG-G			
gyrB-F	GAA-GG(G/T/C)-GGT-ATT-CAA-GC	560	*gyrB*	Kotetishvili et al. (2003)
gyrB-R	GAG-TCA-CCC-TCC-AC(A/T)-ATG-TA			
pgmA-F	AAA-GAT-ACT-CA(C+T)-GCS-CTG-TC	730	*pgm*	Kotetishvili et al. (2003)
pgmA-R	AAC-CAG-CGT-TTT-ACC-GAC-GGC-AAC-A			

recA-F	GAA-ACC-ATT-TCG-ACC-GGT-TC	700	*recA*	Kotetishvili et al. (2003)
recA-R	CCG-TTA-TAG-CTG-TAC-CAA-GCG-CCC			
Vc.sodB-F	AAG-ACC-TCA-ACT-GGC-GGT-A	248	*sodB*	Tarr et al. (2007)
Vc.sodB-R	GAA-GTG-TTA-GTG-ATC-GCC-AGA-GT			
Vm.sodB-F	CAT-TCG-GTT-GTT-TCG-CTG-AT	121	*sodB*	Tarr et al. (2007)
Vm.sodB-R	GAA-GTG-TTA-GTG-ATT-GCT-AGA-GAT			
Vp.flaE-79F	GCA-GCT-GAT-CAA-AAC-GTT-GAG-T	897	*flaE*	Tarr et al. (2007)
Vp.flaE-934R	ATT-ATC-GAT-CGT-GCC-ACT-CAC			
Vv.hsp-326-F	GTC-TTA-AAG-CGG-TTG-CTG-C	410	*hsp*	Tarr et al. (2007)
Vv.hsp-697-R	CGC-TTC-AAG-TGC-TGG-TAG-AAG			
V.16S-700F	CGG-TGA-AAT-GCG-TAG-AGA-T	663	*16S rDNA*	Tarr et al. (2007)
V.16S-1325R	TTA-CTA-GCG-ATT-CCG-AGT-TC			
C2F	AGG-TGT-AAA-ATT-CCT-TGA-CGA	335	*ctxA2-B*	Shangkuan, Sow, and Wang (1995)
C2R	TCC-TCA-GGG-TAT-CCT-TCA-TC			
C4F	ACT-GAT-TTG-TGT-GCA-GAA-TAC-CAC	103	*ctxA2-B*	
C4R	TGA-TAG-CCA-TCT-CTC-TTT-TTC-CAG			
H4F	GCA-AAC-AGC-GAA-ACA-AAT-ACC	497/486	*hlyA*	Shangkuan, Sow, and Wang (1995)
H3R	TCC-ACC-CCA-CCA-GTC-ACC			
H3F	GGG-AGC-CGG-CAT-TCA-TCT	243		

Continued

Table 7.2   PCR Primers and DNA Probes for *V. cholerae* (Continued)

Primer or Probe	Sequence (5′ → 3′)	Sequence (bp)	Gene	Reference
Pvc55-1[a]	aat-tct-aat-acg-act-cac-tat-agg-gAA-TCT-CTT-CGG-TCC-GAT-CAA	135	hlyA	Fykse et al. (2007)
Pvc56-2[a]	TGA-TGC-TGA-AGG-TCA-AGC-AG			
MBvc[a]	ccg-atc-TCA-GAA-AGG-CTT-ATG-GGG-TGg-atc-gg			
Pvc62-1[a]	aat-tct-aat-acg-act-cac-tat-agg-gCG-CTG-AGA-CCA-CAC-CCA-TA	102	tcpA	Fykse et al. (2007)
Pvc60-2	GAA-GAA-GTT-TGT-AAA-AGA-AGA-ACA-CG			
MBvc11[a]	ccg-atc-AGA-AAA-CCG-GTC-AAG-AGG-GTg-atc-gg			
Pvc64-1[a]	aat-tct-agt-acg-act-cac-tat-agg-gAG-AAG-GTG-GGT-GCA-GTG-GCT-ATA-ACA	151	ctxa	Fykse et al. (2007)
Pvc61-2	TGA-TCA-TGC-AAG-AGG-AAC-TCA			
MBvc-12[a]	ccg-atc-TTG-TTA-GGC-ACG-ATG-ATG-GAg-atc-gg			
Pvc65-1[a]	aat-tct-aat-acg-act-cac-tat-agg-gAT-GAT-GTT-GCC-CAC-GCT-AGA	116	groEL	Fykse et al. (2007)
Pvc66-2	GGT-TAT-CGC-TGC-GGT-AGA-AG			
MBvc13[a]	ccg-atc-CTG-TCT-GTA-CCT-TGT-GCC-GAg-atc-gg			
Pvc69-1[a]	aat-tct-aat-acg-act-cac-tat-agg-gcG-GAA-CCG-TTT-TGA-CGT-ATT	139	toxR	Fykse et al. (2007)
Pvc72-2	CTC-GCA-ATG-ATT-TGC-ATG-AC			
MBvc14[a]	ccg-atc-TTA-ACC-CAA-GCC-ATT-TCG-ACg-atc-gg			

[a] Nucleotide sequences of primers are in capital letters; lowercase letters represent surrounding sequences. Lowercase letters of molecular beacons are stem sequences.

V. cholerae O1 and O139 serotypes are considered to cause noninvasive epidemic cholera in developing countries, but non-O1/non-O139 serotypes may be invasive and cause systemic bacteremia and septicemia. Namdari, Klaips, and Hughes (2000) reported on the consumption of raw clams by a healthy individual in Maryland, followed 18 hr later by the development of a severe profuse watery diarrhea, nausea, and vomiting with complete recovery after 72 hr. Blood cultures were positive for a non-O1/non-O139 strain of V. cholerae that was cytotoxic to Hep-2 cell cultures. PCR confirmed that the isolate did not harbor the ctxA gene.

Chow et al. (2001) developed a PCR assay for detection of the rtxA, rtxC (encoding the RTX repeat in toxin), and the ctxB toxin genes among 166 clinical and environmental isolates of V. cholerae. All 166 isolates were O1 El Tor, O139, or non-O1 serotypes, and all harbored the rtxA and rtxC genes that are considered specific for all V. cholera isolates. Only the non-O1 serogroups failed to harbor the ctxB gene.

The ctx genes of O1 and O139 strains of V. cholerae are carried by a filamentous phage (CTX phage), which is known to use the toxin-coregulated pili (TCP) as its receptor. In an effort to understand the mechanism of sporadic emergence of toxigenic O141 V. cholerae causing severe gastroenteritis, Dalsgaard et al. (2001) probed a collection of O041 clinical and environmental isolates from Denmark, Japan, and the United States for genes involving TCP production, toxigenicity, virulence regulation, and other phylogenetic markers. All eight clinical isolates hybridized to probes specific for the ctx, zot, rs1, tcpA, rtxA, and toxR genes. In contrast, all but one of the seven nonclinical O141 isolates were negative for the first five genes, although these nonclinical strains were positive for rtxA and toxR. The authors concluded that their data supported a model for emergence of toxigenic O141 that involves acquisition of the CTX phage sometime after these strains had acquired the pathogenicity island encoding TCP. The authors also concluded that the clonal nature of toxigenic O141 strains isolated from diverse global locations suggested that the emergence is a rare event but that once it occurs toxigenic strains are capable of regional and perhaps even global dissemination.

A total of 69 isolates of V. cholerae (19 toxigenic O1, clinical and environmental; 39 non-O1 environmental; 4 nontoxigenic O1; and seven O139) were examined by Rivera et al. (2001) using multiplex PCR with two primer pairs (Table 7.2) for the presence of eight genes associated with pathogenicity. Based on tcpA and hlyA gene comparison, the strains were grouped into classical and El Tor biotypes. The toxR, hlyA, and ompU genes were present in 100%, 98.6%, and 87% of the V. cholerae isolates, respectively. The ctx and toxin-coregulated pilus El Tor tcpA genes were present in all toxigenic V. cholerae O1 and O139 strains. ctxA and zot were present in all of the toxigenic O1 and O139 strains tested and were absent in nontoxignic O1 strains. It is interesting that among non-O1/non-O139 strains, four were found to be ctx positive and zot negative.

Singh, Isac, and Colwell (2002) developed a hexaplex PCR (Table 7.2) for rapid detection of the virulence and regulatory genes for cholera toxin subunit S (ctxA), zona occludence toxin (zot), accessory cholera enterotoxin (ace), toxin-coregulated pilus (tcpA), outer membrane protein (ompU), and central regulatory protein ToxR (toxR) in V. cholerae and Vibrio mimicus. The primer pairs of Fields et al. (1992) and of Shi et al. (1998) were used for *ctxA* and *ace,* respectively (Table 7.2). The remaining primer pairs (Table 7.2) were from Rivera et al. (2001). Thirty-one of 35 clinical and environmental isolates of *V. cholerae* O1 and 23 of 24 clinical and environmental isolates of O139 gave positive results for *ctxA, zot, ace, ctp* (El Tor specific), *ompU,* and *toxR* genes. One O139 isolate was positive for both the *ompU* and *toxR* genes plus the El Tor–specific *tcpA* gene, whereas four O1 isolates harbored only the two genes *ompU* and *toxR.* All of the 21 non-O1/non-O139 isolates were positive for the *toxR* gene, and none were positive for the other five genes. The authors concluded that environmental isolates of O1 and O139 like clinical isolates possess these six virulence and regulatory genes comprising the virulence cassette and genes encoding surface organelles required for intestinal adherence and colonization.

Radu et al. (2002) subjected 33 *V. cholerae* Malaysian clinical outbreak (1997–1998) strains to PCR detection of the *tcp, ctx, zot*, and *ace* genes (Table 7.2), RAPD, and antibiotic susceptibility. The primers for *tcp* (Table 7.2) were from Keasler and Hall (1993). The primers for *ctx* (Table 7.2) were from Shangkuan et al. (1995). For RAPD, a single random primer GEN15005 (Table 7.2) was used. All isolates were susceptible to chloramphenicol, and only a single isolate was resistant to cefoperazone. The isolates exhibited a high incidence of resistance to cephalothin (90.9%), bacitracin (97%), cefuroxim (97%), streptomycin (87.9%), rifampin (77.8%), tetracycline (77.6%), and carbenicilin (69.7%). Some strains exhibited resistance to 13 antibiotics. The *ctx, zot*, and *ace* genes are clustered with *cep* (core encoded pilus) and *orfU* (an open reading frame of unknown function). PCR results indicated the presence of the core toxin region among the isolates and confirmed the co-occurrence of the *tcp, ctx, zot*, and *ace* genes in 32 of the 33 *V. cholerae* O1 clinical isolates. One isolate was negative for all four genes, suggesting that this isolate may not have been the causative organism and indicating that more than one isolate should be obtained from each patient. RAPD analysis indicated that the isolates exhibited a high level of genetic heterogeneity.

Lee, Panicker, and Bej (2003) developed a multiplex PCR assay linked to a microwell sandwich assay for detection of *Salmonella* and three *Vibrio* species including *V. cholerae* in seeded oyster homogenates. The primers L-ctx/R-ctx (Table 7.2) amplified a 302-bp sequence of the *ctxA* gene of *V. cholerae.* Individual capture probes were then added and covalently bound to the wells. The phosphorylated capture probe PP-ctx (Table 7.2) was used for *V. cholerae.* Multiplex amplicons were denatured in the wells and incubated for hybridization to the immobilized probes. The biotinylated probe BP-ctx (Table 7.2) was then added followed by an alkaline phosphatase–avidin conjugate and then enzyme substrate added for color development. Enrichment of seeded oyster homogenates in alkaline peptone water (APW) allowed the detection of 10^2 CFU/g of tissue.

Fukushima, Tsunomori, and Seki (2003) developed a series of duplex Rti-PCR assays for detection of 17 species of food and waterborne pathogens, including *V. cholerae,* in stools utilizing SYBR green as the fluorescent reporter molecule. The primers rtxC-F/rtxC-R (Table 7.2) amplified a 265-bp sequence of the *rtx* toxin gene and were used to detect *V. cholera* strains O1 and O139 as well as non-O1 strains, except for the classical *V. cholerae* O1 strains. The primers CT-F/CT-R amplified a

308-bp sequence of the *ctx* gene and were used to detect *V. cholerae* O139 Bengal. Without enrichment of seeded stool samples, the detection level with DNA purification was about 10^5 CFU/g of stool. The protocol for detection of less than 10^4 CFU/g required overnight enrichment.

Gopal et al. (2005) reported the occurrence of various species of *Vibrio* that were predominant in water, sediment, and shrimp samples from shrimp farms in Japan. *V. cholerae* was not among the dominant pathogenic vibrios but was detected in some samples. All such isolates, however, were negative for the cholera-producing toxin gene *ctx*. The primers (Table 2) of Koch et al. (1993) were used to detect the *ctx* gene of the *V. cholerae* isolates.

A specific repetitive sequence for the genome of *V. cholerae* is known to consist of a repeated sequence of 123–126-bp (Barker, Clark, and Manning, 1994). This element is spread in the *V. cholerae* "super integron," which harbors nine copies in the locus encoding the mannose- and fucose-resistant hemagglutinin. These sequences, called VCR (for *V. cholerae* repetitive DNA sequence), are nearly uniform, and are interspersed between structural genes.

Castañeda et al. (2005) developed a VCR–PCR novel typing method for genetic analysis of *V. cholerae* O1 and non-O1 isolates utilizing the primers Vib-L2/Vib-R2 (Table 7.2). A total of 69 strains of *V. cholerae* were examined, 50 O1 and 19 non-O1 epidemic strains isolated from 1992 to 2000 in Argentina. The 69 isolates yielded a total of 29 patterns with five to eight bands in each ranging between 300 and 1000-bp. The most prevalent pattern included 28 (68%) of human and environmental origin and all were toxigenic O1 isolates. This pattern was found to have spread in distinct provinces of Argentina from 1992 to 1998 and suggested that a single clone with epidemic behavior was responsible for the epidemic in Latin America during this period. Repetitive extragenic palindromic sequence-PCR (REP-PCR) analysis of genomic DNA of the 69 isolates utilizing the primers REP1-I/REP2-I (Table 7.2) resulted in 14 profiles with each comprised of five to eight bands of 750–6000 bp. In near agreement with VCR-PCR, 36 isolates fell into a single REP-PCR pattern, most of which fell into the major VCR-PCR group.

Gubula (2006) developed a multiplex Rti-PCR assay for detection of four highly conserved genes characteristic of potentially toxigenic O1 and O139 strains of *V. cholera: rtxA, epsM, mshA*, and *tcpA*. The four pairs of primers (Table 7.2) were specifically designed to yield amplicons having notably different T_m values (85.3, 87.4, 80.8, and 84.1°C, respectively) for individual detection of the respective amplicons because SYBR green was used as a common fluorophore.

The virulence of *V. cholerae* is regulated primarily by the *ctxAB* and *tcpA* genes, which have been thought to be associated exclusively with clinical O1 and O139 serogroup isolates. Sharma and Chaturvedi (2006) examined the presence of the virulence genes *stn, ompW, ctxA,* and *tcpA* of classical and El Tor variants among 115 environmental strains of non-O1 *V. cholerae* from river water in India. The primers NST-F/NST-R (Table 7.2) amplified a 215-bp sequence of the *stn* gene. The primers (Table 7.2) for *ctxA* and *tcpA* (classical) were from Ogawa et al. (1990). The primers TcpA-F/TcpA-R (Table 7.2; El Tor) amplified a 417-bp sequence of the *tcpA* gene. The primers OmpW-F/OmpW-R (Table 7.2) amplified a 588-bp sequence of the *ompW* gene and were used to confirm the identity of all 115 *V. cholerae* isolates examined in the study. The *ctxA* and *tcpA* genes were present in 14 isolates (13%). The gene *stn* encoding the heat-stabile toxin was absent in all 115 isolates.

Blackstone et al. (2007) developed an Rti-PCR assay for detection of *V. cholerae* harboring the *ctxA* gene. The primers ctxA-F/ctxA-R (Table 7.2) amplified a sequence of the *ctxA* gene. The probe ctxAP (Table 7.2) was labeled at the 5′-end with FAM or TET and at the 3′-end with a black hole quencher (BHQ). Shellfish tissue from Mobile Bay (three oyster samples and three clam samples) were homogenized in APW and subjected to overnight enrichment incubation at 42°C. A 1-ml aliquot of enrichment was boiled for 10 min and 2 to 2.5 µl incorporated into Rti-PCR assays. All six shellfish samples were positive for *V. cholerae* and harbored the *ctxA* gene. The detection limit was 0.8 CFU per Rti-PCR reaction with clams and was less then 10 CFU per Rti-PCR with oysters.

V. cholerae was isolated by Saravanan et al. (2007) from 2 of 5 shrimp, 2 of 5 clams, and from 5 of 20 river water estuary samples in Southwest India. The presence of 11 genes associated with virulence was assessed using the PCR with respective primer pairs. All of the *V. cholerae* isolates were positive for the *toxR*, *ompU* (outer membrane protein), and *hlyA* (hemolysin A) genes. *V. cholerae* was most prevalent in samples from fresh fish markets and from a landing dock. All except one isolate belonged to the *V. cholerae* non-O1 serogroup and were therefore non-toxigenic (lacking the *ctxA* gene) but possessed the *hlyA* gene and therefore could potentially cause mild gastroenteritis. The remaining genes were only sparsely present in the isolates. It has been reported that most environmental strains harbor the *ompU* gene, which confers bile resistance and enhanced survival in the environment (Karunasagar et al., 2003).

Nhung et al. (2007) developed a multiplex PCR assay for detecting five human pathogenic *Vibrio* species including *V. cholerae*. Based on the differing sequence of the *dnaJ* gene for each of the five *Vibrio* species, pairs of primers including one universal forward primer (VM-F) and five reverse primers (VC-Rmm, VP-MmR, VV-Rmm, VM-Rmm, and V.al2-mmR) specific to the five species (*V. cholerae, V. parahaemolyticus, V. vulnificus, V. mimicus,* and *V. alginolyticus,* respectively) were designed (Table 7.2).

Tarr et al. (2007) developed a multiplex PCR for simultaneous detection of four human pathogenic *Vibrio* species (*V. cholerae, V. parahaemolyticus, V. vulnificus,* and *V. mimicus*). Included was a primer pair V.16S-700F/V.16S-1325R for amplification of a 663-bp sequence of a conserved portion of the 16S rDNA specific for confirming the identity of all 309 isolates examined as members of the genus *Vibrio*. Primers are listed in Table 7.2.

Fykse et al. (2007) developed a multitarget molecular beacon-based real-time nucleic acid sequence-based amplification (NASBA) assay specific for detection of *Vibrio cholerae*. The genes encoding the cholera toxin (*ctx*), the toxin-coregulated pilus (*tcpA*), the *ctaA* toxin regulator (*toxr*), hemolysin (*hlyA*), and the 60-kDa chaperonin product (*groEL*) were selected as the target sequences. Molecular beacon probes for the five different genetic targets were labeled with FAM at the 5′-ends and with dabsyl at the 3′-ends. The NASBA assay (Compton, 1991) is a sensitive, transcription-based amplification system specifically designed for detecting messenger RNA. The technology relies on the simultaneous activity of three different enzymes: RNase H, avian myoblastosis virus reverse transcriptase, and T7 RNA polymerase. The presence of a T7 promoter sequence at the 5′-end of the forward primer is essential and is used by the T7 RNA polymerase during synthesis of new RNA amplicons. In contrast to PCR and reverse transcriptase-PCR, the NASBA method is isothermal (41°C). In NASBA, single-stranded RNA amplicons are produced that can be readily detected by hybridization with a sequence-specific probe such as a molecular beacon. The entire assay including RNA extraction and NASBA was completed within 3 hr. A correlation between cell viability and NASBA was demonstrated for the *ctxA*, *toxR*, and *hlyA* targets.

Lalitha et al. (2008) developed a PCR assay specific for all strains of *V. cholerae* including O1, O139, and non-O1/non-O139 serogroups and biotypes. The primers VHMF/VHA-AS5 (Table 7.2) amplified a 519-bp sequence of the *lolB* gene that encodes an outer membrane lipoprotein. All 44 *V. cholerae* strains (34 O1 El Tor, one classical, four O139, and five non-1/non-O139) were found to harbor the *lolB* gene, whereas 40 other *Vibrio* species and 56 enteric gram-negative reference strains of other genera did not harbor the gene. The diagnostic sensitivity and specificity with 633 clinical rectal swab samples were 98.5% and 100%, respectively.

Mendes, Abath, and Leal (2008) developed a multiplex single-tube PCR assay for detection of the *V. cholerae* serotype. The *ctxA* gene was targeted with a pair of external primers and a pair of internally nested primers that yielded a final amplicon of 302 bp (Table 7.2). In addition, a pair of primers (Table 7.2) was added that amplified a 198-bp sequence of the *rfbN* gene that encodes the O1 serotype.

Khuntia, Bhusan, and Chhotray (2008) developed a quadruplex PCR for simultaneous detection of serotype, biotype, toxigenic potential, and control regulating factors of *V. cholerae*. The assay specifically targeted the *rfb* genes for O1 and O139 serotypes (primers from Hoshino et al., 1998; Table 7.2), *ctxA* (primers from Keasler and Hall, 1993; Table 7.2), *tcpA* (El Tor primers from Keasler and Hall, 1993; Table 7.2), *tcpA* (classical primers from Rivera et al. 2001; Table 7.2), and *toxR* (primers from Miller, Taylor, and Mekalanos, 1987; Table 7.2) genes.

IV. MOLECULAR TYPING OF *V. CHOLERAE* ISOLATES

Coelho et al. (1995) described five (1–5) random primers (Table 7.2) used for distinguishing 50 El Tor, four classical, and two Gulf Coast strains of *V. cholerae*. An additional five (6–10) random primers (Table 7.2) were described for distinguishing more closely related El Tor and five Bengal strains. In addition, primers NA1/NB2 (Table 7.2) were developed to amplify a 16S/23s rRNA spacer region of the strains. Under conditions of low stringency, a series of bands was obtained. The Bengal strains yielded a more prominent band of 0.35 kbp and weaker bands at the top of the pattern compared to the other biotypes. A 0.5-kbp band was produced from the El Tor, Bengal, and Gulf Coast strains and was absent from classical strains. A 0.8-kbp strong band was produced by all four groups.

In 1994 a cholera epidemic occurred in Italy and Albania after more than a decade of absence. Pazzani et al. (2006) examined 110 *V. cholerae* El Tor isolates from this epidemic using RAPD with six random primers (Table 7.2), *BgI*I ribotyping, and PFGE with *Sfi*I and *Not*I. All strains were of biotype 6, and the respective RAPD and PFGE patterns were identical as well. These findings indicated that the 1994 isolates belonged to the same clone and that the clone was part of the larger global spread of epidemic ribotype 6, which started in southern Asia in 1990.

Scrascia et al. (2006) examined 80 *V. cholerae* O1 strains, selected to represent the 1998–1999 history of the largest cholera epidemic in Kenya, with respect to ribotyping antimicrobial susceptibility, and RAPD using the six random primers from Pazzani et al. (2006). Sixty-one of the 80 isolates fell into a single ribotype designated B27 and were resistant to chloramphenicol, spectinomycin, streptomycin, sulfamethoxazole, and trimethoprim. The 61 ribotype B27 strains fell into a single RAPD cluster. Six of the 61 ribotype B27 isolates were also resistant to doxycycline and tetracycline resulting from the presence of a conjugative plasmid. These observations were interpreted to indicate that the predominant B27 ribotype strains had a common clonal origin that rapidly spread from West Africa to eastern Africa.

In 1997 an outbreak of cholera in north-central Bangladesh occurred. Farugue et al. (1999) examined 19 clinical isolates of *V. cholerae* from this region plus 39 strains from other regions of Bangladesh, in addition to 7 strains from India and other geographic regions of the globe. Ribotyping indicated that the recently isolated strains belonged to a new ribotype that was distinct from previously described ribotypes of toxigenic O139 isolates from the same area. All strains carried the genes for toxin coregulated pili (*tcpa* and *tcpI*) accessory colonization factor (*acfB*), the regulator gene *toxR*, and multiple copies of the lysogenic genome encoding cholerae toxin (CTXΦ) and belonged to a previously described *ctxA* genotype. Analysis of the *rfb* gene cluster by PCR revealed the absence of a large region of the O1-specific *rfb* operon and the presence of an O139-specific genomic region in all O139 strains. Southern hybridization analysis of the O139-specific genomic region produced identical restriction patterns in strains belonging to the new ribotype and those of previously described ribotypes. These results suggested that the new ribotype of Bengal *V. cholerae* possibly originated from an existing O139 strain by genetic changes in the rRNA operon. In contrast to previously isolated O139 strains, most of which were resistant to trimethoprim, sulfamethoxazol, and streptomycin, all of the 19 new ribotype strains were susceptible to these antibiotics. Molecular analysis revealed the possible deletion of a 3.6-kb region of the SXT element in strains that were susceptible to these antibiotics. The authors concluded that *V. cholerae* O139 strains were undergoing considerable reassortment in genetic elements encoding antimicrobial resistance with epidemic potential similar to previously described O139 *V. cholerae* isolates being retained.

In order to investigate the origin of *V. cholerae* O1 biotype El Tor isolates in Japan in 1997, Arakawa et al. (2000) subjected 67 strains to pulsed field gel electrophoresis (PFGE) after digestion with *Not*I and *Sfi*I. Thirty-six strains were from patients who had gone abroad, and 31 strains were isolates from patients just returning from India and various geographic regions in Asia. Sixty-six of the 67 O1 El

Tor isolates belonged to serotype Ogawa, and only one belonged to serotype Inaba. Among the 365 domestic isolates, 25 belonged to a single PFGE subtype. In contrast, PFGE analysis separated the 31 imported strains into 13 subtypes, with only one of these strains exhibiting the same PFGE pattern common to the major domestic PFGE subtype.

A collection of 45 isolates of *V. cholera* from the seventh pandemic sampled over a 3-year period was analyzed by Lan and Reeves (2002) using amplified fragment length polymorphism (AFLP) fingerprinting employing 16 EcoRI-*Mse*I primer pair combinations. Among the 45 isolates, all but 11 were distinguishable from one another by AFLP. AFLP revealed far more variations than ribotyping. AFLP grouped most of the 45 isolates into two clusters. Cluster I consisted mainly of strains from the 1960s and 1970s, and cluster II contained mainly strains from the 1980s and 1990s revealing a temporal pattern of change in the clone.

Based on the B subunit of the cholera toxin (CTX), two immunologically related but distinguishable epitypes have been described: CTX1 is the prototype produced by classical biotype strains, and CTX2 is produced by the El Tor biotype and O139 strains (Finkelstein et al. 1987). An alternate classification identifies three types of *ctxB* genes based on three nonrandom base changes resulting in deduced amino acid sequence changes. Genotype 1 is found in strains of the classical biotype globally and in U.S. Gulf Coast strains, genotype 2 is found in El Tor biotype strains from Australia, and genotype 3 is found in El Tor biotype strains from the seventh pandemic. Thus the *V. cholerae* O1 classical biotype CTX belongs to the CTX1 epitype and genotype 1, whereas the El Tor biotype of the seventh pandemic produces CTX of the CT2 epitype and genotype 3. Nair et al. (2006) examined a total of 185 strains of *V. cholerae* O1, consisting of 31 classical biotypes isolated between 1960 and 1990, 113 strains of El Tor biotype, and 41 hybrid strains (non-El Tor and non-classical) isolated between 1960 and 2005. Detection of the CTX subtypes was by ganglioside GM-specific ELISA assays using mouse monoclonal antibodies (MAbs) specific for the CTX subtypes. All 31 classical and all of the El Tor hybrid strains isolated between 2001 and 2000 produced CTX of the classical subtype. This is in contrast to the El Tor strains isolated from 1971 to 2000, which produced mostly CTX of the El Tor subtype. Nucleotide sequence analysis of the *ctxB* genes of eight representative El Tor strains isolated from 2001 to 2005 indicated that the DNA sequences were identical to that of the classical *ctxB* gene. These results indicate that the CTX produced by the El Tor strains isolated from Bangladesh between 2001 and 2005 have shifted from epitype CTX2 and genotype 3 to epitype CTX1 and genotype 1. Thus, present El Tor biotype strains produce CTX of the classical biotype. In addition, this study indicated that El Tor strains producing classical CTX have completely replaced the El Tor CTX in Bangladesh.

Kotetishvili et al. (2003) characterized 22 *V. cholerae* isolates including epidemic (O1 and O139) and nonepidemic serogroups by PFGE multilocus sequence typing (MLST) by using three housekeeping genes, *gyrB, pgm,* and *recA*. Sequence data were also obtained for the virulence associated genes *tcpA, ctxA,* and *ctxB*. PFGE was performed following restriction with *Ceu*I. Although the restriction nuclease *Not*I is most frequently used for PFGE analysis of *V. cholerae* isolates, the authors chose *Ceu*I because *Not*I restriction digestion of *V. cholerae* DNA generates numerous bands of similar sizes that are difficult to resolve. In contrast, *Ceu*I generates from six to ten bands at most from *V. cholerae* DNA that are readily resolved. Thirteen PFGE types were identified among the 22 isolates. Six gene loci (three housekeeping genes and three genes associated with virulence) were sequenced using the respective amplicons generated with pairs of primers listed in Table 7.2. For *tcpA* one forward and two reverse primers (Table 7.2) were simultaneously incorporated into the PCR reaction. Analysis of the six sequences revealed 21 unique sequence types among the 22 isolates examined. The authors concluded that this observation suggests that the discriminatory power of MLST is superior to that of PFGE. However, the reader is reminded that the discriminatory power of any method depends on the number of independent observations that the method allows. The use of two or three restriction nucleases separately or in combination with PFGE compared to the use of only two gene loci for sequence typing might be expected to yield more or a near equal number of PFGE types compared to the number of MLST types.

The presumably extinct classical biotype and the El Tor biotype are closely related in their O-antigen biosynthesis genes but differ in other genome regions. There are several distinct phenotypic and genotypic traits that distinguish these two O1 biotypes.

V. cholerae O1 isolates from Mozambique during an early 2004 epidemic were found to harbor the classical CTX prophage but were phenotypically identified to the El Tor biotype indicating that strains were genomic hybrids. Ansurazzaman et al. (2007) subjected such presumably hybrid isolates from the 2004–2005 Mozambique epidemic to PFGE analysis following digestion with *Not*I. A total of 18–21 bands yielded five closely related PFGE patterns that were distributed similarly in both years (2004–2005) among the 80 strains tested. The PFGE patterns indicated an overall El Tor lineage. The PFGE patterns grouped the hybrid strains from Mozambique into a separate cluster from Bangladesh clinical and environmental hybrid strains. A band of 398 bp that contained an *rstR* allele of the classical type was detected from all the hybrid O1 strains, which was absent in all conventional classical and El Tor strains. This band was considered useful for identifying the hybrid strains.

Twelve strains of *V. cholerae* El Tor that clustered into phage type 27, two untypeable El Tor strains plus twelve O139 strains of the Ogawa serotype that clustered into phage type 1 from different geographic areas of India were examined by Bhowmick et al. (2007) with PFGE and RAPD to determine the extent of their genetic relatedness. RAPD was performed with primer 1281 (Table 7.2). PFGE analysis was performed after digestion with *Not*I. The RAPD profiles of 11 isolates were identical, and the remaining 15 distinguishable. With PFGE, six strains yielded identical profiles. Only two strains yielded identical RAPD patterns and identical PFGE patterns and were of different phage types. The authors concluded that RAPD and PFGE typing indicated that the majority of isolates of the same phage type were genetically heterogeneous except for three sets of clonally identical strains involving six isolates that had both identical RAPD and PFGE profiles and were of the same phage type and serotype.

Tapchaisri et al. (2007) analyzed a total of 240 *V. cholerae* El Tor O1 strains from patients in Thailand during two different epidemic periods, 1999–2000 (193 Ogawa and seven Inaba strains) and 2001–2002 (40 Inaba strains), for the presence of nine virulence genes: *ctxA, ctxB, zot, ace, toxR, tcpA, hlyA, nanh,* and *ninh* by PCR (Table 7.2). The primers of Keasler and Hall (1993) were used for *ctxA,* those of Rivera et al. (2001) were used for *zot, toxR, tcpA* (classical and El Tor), and *hlyA* (classical and El Tor; Table 7.2). Amplicons from *ctxA, ctxB, zot, ace, toxR, tcpA, hlyA, nanh,* and *ninh* were obtained from all 10 *V. cholerae* reference strains and from 239 of 240 isolates studied. One Inaba isolate of 2001–2002 yielded only amplicons of *toxR* and *hlyA.* The isolates of 1999–2000 revealed ribotypes D, G, H, and I. In contrast, the majority of isolates from 2001–2002 were of ribotype G plus two new ribotypes J and K. The authors concluded that the clinical isolates of the two epidemics showed a sustained appearance of one epidemic *V. cholerae* ribotype clone (G) and a constant but gradual minor change in the genetic composition of the other *V. cholerae* strains as indicated by the change in ribotype of the two studies.

V. ACQUISITION OF ANTIBIOTIC RESISTANCE

In the 1996–1997 cholera epidemic in Guinea-Bissau, the emergence of a multidrug-resistant strain of *V. cholera* O1 was observed. Before the emergence of the multidrug-resistant strain, the case fatality rate was 1.0%, and after its emergence it was 5.3%. Dalsgaard et al. (2000) showed that the strain contained a 150-kb conjugative multiple antibiotic-resistance plasmid. Seven strains isolated from patients at different locations and at different times from the epidemic were found to have identical antibiotic-resistance profiles showing resistance to ampicillin, colistin, furazolidone, gentamyin, kanamycin, O129, streptomycin, sufamethoxazole, trimethoprim, and tetraycycline. A representative donor strain from this group carrying the plasmid was used to obtain transconjugants of *Escherichia coli* K12 that exhibited the same antibiotic-resistance pattern as the donor strain except that it retained sensitivity to colistin and furazolidine. RFLP analysis revealed that the 150-kb plasmid had been transferred and that the multiple antibiotic resistance, except for colistin and furazolidine, was plasmid encoded. The plasmid and multiple antibiotic resistance was subsequently transferred to a *V. cholera* recipient initially resistant only to colistin. The authors suggested that the finding of transferable resistance to all of the antimicrobials most commonly used to treat cholera may have contributed to the increased fatality rate.

VI. BIOTYPE CONVERSION BY PHAGE

V. cholerae serogroups O1 and O139 are presently the major causative agents of cholera. Within serogroups O1 several biotypes exist, of which the classical and El Tor biotypes have been responsible for all recent epidemics. *V. cholerae* O1 of the classical biotype was responsible for the first six cholera pandemics. The El Tor biotype, although isolated as early as 1906 (Gotschlich, 1906), was not considered of epidemic potential at that time. Beginning in 1961 the El Tor biotype dominated the classical biotype in global pandemics, and after 1966 all cholera epidemics were due to El Tor biotype strains. In 1992 the O139 serotype emerged. Molecular studies have shown that O139 strains originated from serogroup O1 El Tor strains by the acquisition of a novel DNA sequence inserted within the *rfb* gene encoding the O antigen (Bik et al., 1995). Mukhopadhyay and Ghosh (2000) presented evidence that phage PS166 when infecting a recipient El Tor strain can convert it to the classical biotype by integrating inside the *hlyA* gene locus so as to disrupt the gene. Evidence was also presented that phage integration can also result in complete loss of the *hlyA* gene in addition to multiple disruption or integration. In additon, 10–15% of lysogens became Cys⁻ or Cys⁻ Met⁻ being converted to the classical biotype with changes in all El Tor-specific determinants, namely, loss of (1) soluble hemolysin, (2) cell-associated chicken erythrocyte hemagglutination, (3) polymyxin B resistance, (4) resistance to phage Φ149, and (5) sensitivity to El Tor-specific phage e4. It is interesting that repeated culture of such lysogens was found to result in the loss of the prophage but the other phenotypic characteristics of the classic biotype remained unchanged.

At least five distinct *V. cholerae* prophage CTXΦ types have been described that differ in their *rstR* phage repressor genes and are designated CTX^{ET}Φ (El Tor), CTX^{Class}Φ (classical), CTX^{Var}Φ (pre-O139 El Tor), CTX^{Calc}Φ (O139), and CTX^{Env}Φ (non-O1/non-O139). Strains of the recently (1992) emerged O139 serogroup often carry two distinct CTXΦ prophage types arrayed in tandem. One corresponds to the previously recognized El Tor prophage CTX^{ET}Φ and the other to the CTXΦ type designated Calcutta (CTX^{Calc}Φ), found to date only in strains of the O139 serogroup. Restriction fragment length polymorphism analysis by Ledón et al. (2008) of the array of CTXΦ prophages in two strains of *V. cholerae* O139 revealed the presence of copies of complete CTXΦ and pre-CTXΦ prophages coexisting at a single chromosome locus in each strain. Restriction patterns and comparative nucleotide sequence analysis revealed pre-CTXΦ precursors of both El Tor and Calcutta lineages. The authors hypothesized that two precursor variants independently acquired cholera toxin genes and gave rise to the current El Tor and Calcutta CTXΦ prophages.

Reid and Mekalanos (1995) showed that the temperate phage K139 is involved in the emergence of O139 from an El Tor strain by the transfer of the *rfb* operon. In addition, the phage CTXΦ uses pili encoded by *tcp* as its receptor, which is encoded by another lysogenic filamentous phage, VPIΦ. In this manner, one phage encodes the receptor for another phage and thus phage–phage interaction produces the pathogenicity of the host bacteria.

The discovery of a lysogenic filamentous phage (CTXΦ) that encodes the three toxin genes *ctx, ace,* and *zot* and that is able to integrate into the *V. cholerae* chromosome provides evidence of horizontal transmission of toxigenicity among *V. cholerae* strains (Waldor and Mekalanos, 1996). Transmission of the CTXΦ requires that the recipient strain express a toxin-coregulated pilus (TCP) encoded by *tcp*, which is essential for colonization and also encodes a transcriptional regulator encoded by *toxR* that activates both *tcp* and *ctx* genes during infection.

REFERENCES

Ansurazzaman, M., Bhian, N., Ashrafus, S., Sultana, M., Mcuamule, A., Mondlane, C., Wang, X., Den, J., Seidlein, L., Clemens, J., Lucas, M., Sack, D., Nair, G. 2007. Genetic diversity of El Tor strains of *Vibrio cholerae* O1 with hybrid traits isolated from Bangladesh and Mozambique. *Int. J. Med. Microbol.* 297:443–449.

Arakawa, E., Toshiyuki, M., Matsushita, S., Shimada, T., Yamai, S., Ito, T., Watanabe, H. 2000. Pulsed-field gel electrophoresis-based molecular comparison of *Vibrio cholerae* O1 isolates from domestic and imported cases of cholera in Japan. *J. Clin. Microbiol.* 38:424–426.

Barker, A., Clark, C., Manning, P. 1994. Identification of VCR, a repeated sequence associated with a locus encoding a hemagglutinin in *Vibrio cholerae* O1. *J. Bacteriol.* 176:5450–5458.

Bhowmick, T., Das, M., Roy, N., Sarkar, B. 2007. Phenotypic and molecular typing of *Vibrio cholerae* O1 and O139 isolates from India. *J. Infect.* 54:475–482.

Bik, E., Bunschoten, A., Gouw, R., Mooi, F. 1995. Genesis of the novel epidemic *Vibrio cholerae* O139 strain: Evidence for horizontal transfer of genes involved in polysaccharide synthesis. *EMBO J.* 14:209–216.

Bik, E., Gouw, R., Mooi, F. 1996. DNA fingerprinting of *V. cholerae* strains with a novel insertion sequence element: A tool to identify epidemic strains. *J. Clin. Microbiol.* 34:1453–1461.

Blackstone, G., Nordstrom, J., Bowen, M., Myer, R., Imbro, P., DePaola, A. 2007. Use of real time PCR assay for detection of the *ctxA* gene of *Vibrio cholerae* in an environmental survey of Mobile Bay. *J. Microbiol. Meth.* 68:254–259.

Casteñada, N., Pichel, M., Orman, B., Binsztein, N., Roy, P., Centron, D. 2005. Genetic characterization of *Vibrio cholerae* isolates from Argentina by *V. cholerae* repeated sequences-polymerase chain reaction. *Diag. Microbiol. Infect. Dis.* 53:175–183.

Chow, K., Ng, T., Yuen, K., Yam, W. 2001. Detection of RTX toxin gene in *Vibrio cholerae* by PCR. *J. Clin. Microbiol.* 39:2594–2597.

Coelho, A., Vicente, A., Baptista, M., Momen, H., Santos, F., Salles, C. 1995. The distinction of pathogenic *Vibrio cholerae* groups using arbitrarily primed PCR fingerprints. *Res. Microbiol.* 146:671–683.

Compton, J. 1991. Nucleic acid sequence-based amplification. *Nature* 350:91–92.

Dalsgaard, A., Serichantalergs, O., Forslund, A., Lin, W., Mekalanos, J., Mintz, E., Shimada, T., Wells, J. 2001. Clinical and environmental isolates of *V. cholerae* serogroup O141 carry the CTX phage and the genes encoding the toxin-coregulated pili. *J. Clin. Microbiol.* 39:4086–4092.

Damian M., Koblaavi, S., Carle, I., Nacescu, N., Grimont, F., Ciufecu, C., Grimont, P. 1998. Molecular characterization of *Vibrio cholerae* O1 strains isolated from Romania. *Res. Microbiol.* 149:745–755.

DePaola, A., Hwang, G. 1995. Effect of dilution, incubation time, and temperature of enrichment on cultural and PCR detection of *Vibrio cholerae* obtained from the oyster *Crassostrea virginica. Mol. Cell. Probes.* 9:75–81.

Dumontier, S., Berche, P. 1998. *Vibrio cholerae* O22 might be a putative source of exogenous DNA resulting in the emergence of the new strain of *Vibrio cholerae* O139. *FEMS Microbiol. Lett.* 164:91–98.

Farugue, S., Siddique, A., Saha, M., Asadulghani, A., Rahman, M., Zahman, K., Albert, M., Sack, D., Sack, R. 1999. Molecular characterization of a new ribotype of *Vibrio cholerae* O139 Bengal associated with an outbreak of cholera in Bangladesh. *J. Clin. Microbiol.* 37:1313–1318.

Fields, P., Popovic, T., Wachsmuth, K., Olsvik, O. 1992. Use of polymerase chain reaction for detection of toxigenic *Vibrio cholera* O1 strains from the Latin American cholera epidemic. *J. Clin. Microbiol.* 30:2118–2121.

Finkelstein, R., Burks, F., Zupan, A., Dallas, W., Jacob, C., Ludwig, D. 1987. Epitopes of the cholera family of enterotoxins. *Rev. Infect. Dis.* 9:544–561.

Fukushima, H., Tsunomori, Y., Seki, R. 2003. Duplex real-time SYBR green PCR assays for detection of 17 species of food- or waterborne pathogens in stools. *J. Clin. Microbiol.* 41:5134–5146.

Fykse, E., Skogan, G., Davies, W., Olsen, J., Blatny, J. 2007. Detection of *Vibrio cholerae* by real-time nucleic acid sequence-based amplification. *Appl. Environ. Microbiol.* 73:1457–1466.

Gopal, S., Otta, S., Kumar, S., Karunasagar, O., Nishibuchi, M., Karanasagar, I. 2005. The occurrence of *Vibrio* species in tropical shrimp culture environments, implications for food safety. *Int. J. Food Microbiol.* 102:151–159.

Gotschlich, F. 1906. Uber Cholera- und choleraähnliche Vibrionen unter den aus Mekka zurückkehrenden Pilgern. *Z. Hyg. Infekt-Kr.* 53:281–304.

Gubula, A. 2006. Multiplex real-time PCR detection of *Vibrio cholerae. J. Microbiol. Meth.* 65:278–293.

Guglielmetti, P., Bravo, L., Zanchi, A., Monte, R., Lombardi, G., Rossolini, G. 1994. Detection of the *Vibrio cholerae* heat-stable enterotoxin gene by polymerase chain reaction. *Mol. Cell. Probes.* 8:39–44.

Hoshino, K., Yamasaki, S., Mukhopadhyay, A., Chakrabarty, S., Basu, A., Bhatachaarya, S., Nair, G., Shimada T., Takeda, Y. 1998. Development and evaluation of a multiplex PCR assay for rapid detection of toxigenic *Vibrio cholerae* O1 and O139. *FEMS Immunol. Med. Microbiol.* 20:201–207.

Islam, M., Ahsan, S., Khan, S., Ahmed, Q., Rashid, M., Islam, K., Sack, R. 2004. Virulence properties of rough and smooth strains of *Vibrio cholerae* O1. *Microbiol. Immunol.* 48:229–235.

Jermyn, W., Boyd, E. 2002. Characterizatin of a novel *Vibrio* pathogenicity island (VPI-2) encoding neuraminidase (nanH) among toxigenic *Vibrio cholerae* isolates. *Microbiol.* 148:3681–3693.

Karaolis, D., Johnson, J., Bailey, C., Boedeker, E., Kaper, J., Reeves, P. 1998. A *Vibrio cholerae* pathogenicity island associated with epidemic and pandemic isolates. *Proc. Natl. Acad. Sci. USA* 95:3134–3139.

Karaolis, D., Lan, R., Reeves, P. 1995. The sixth and seventh cholera pandemics are due to independent clones separately derived from environmental nontoxigenic, non-O1 *Vibrio cholerae. J. Bacteriol.* 17:3191–3198.

Karunasagar, I., Rivera, I., Joseph, B., Kennedy, B., Setty, V., Huq, A., Karnasagar, I., Colwell, R. 2003. OmpU genes in non toxigenic *Vibrio cholerae* associated with aquaculture. *J. Appl. Mirobiol.* 95:338–343.

Keasler, S., Hall, R. 1993. Detecting and biotyping *Vibrio cholerae* O1 with multiplex polymerase chain reaction. *Lancet* 341:1661.

Khuntia, H., Bhusan, B., Chhotray, P. 2008. Quadruplex PCR for simultaneous detection of serotype, biotype, toxigenic potential, and central regulating factor of *Vibrio cholerae. J. Clin. Microbiol.* 46:2399–2401.

Koch, W., Payne, W., Wentz, B., Cebula, T. 1993. Rapid polymerase chain reaction method in detection of *Vibrio cholerae* in foods. *Appl. Environ. Microbiol.* 59:556–560.

Kotetishvili, M., Stine, O., Chen, Y., Kreger, M., Sulakvelidze, A., Sozhamannan, S., Morris, J., Jr. 2003. Multilocus sequence typing has better discriminatory ability for typing *V. cholerae* than does pulsed-field gel electrophoresis and provides a measure of phylogenetic relatedness. *J. Clin. Microbiol.* 41:2191–2196.

Lalitha, P., Suraiya, M., Lim, K., Le, S., Halindawaty, A., Chan, Y., Ismail, A., Zainuddin, Z., Ravichandran, M. 2008. Analysis of *lolB* gene sequence and its use in the development of a PCR assay for the detection of *Vibrio cholerae*. *J. Microbiol. Meth.* 75:142–144.

Lan, R., Reeves, P. 2002. Pandemic spread of cholera: Genetic diversity and relationships within the seventh pandemic clone of *Vibrio cholerae* determined by amplification fragment length polymorphism. *J. Clin. Microbiol.* 40:172–181.

Ledón, T., Campos, J., Suzarte, E., Rodriguez, B., Mareno, K., Fando, R. 2008. El Tor and Calcutta CTXΦ precursors coexisting with intact CTXΦ copies in *Vibrio cholerae* O139 isolates. *Res. Microbiol.* 159:81–87.

Lee, C., Panicker, G., Bej, A. 2003. Detection of pathogenic bacteria in shellfish using multiplx PCR followed by Covalink™ NH microwell plate sandwich hybridization. *J. Microbiol. Meth.* 53:199–209.

Li, M., Shimada, T., Morris, J., Jr., Sulakvelidze, A., Sozhamannan, S. 2002. Evidence for the emergence of non-O1 and non-O139 *Vibrio cholerae* strains with pathogenic potential by exchange of O-antigen biosynthesis regions. *Infect. Immun.* 70:2441–2453.

Liu, W., Fullner, J., Clayton, R., Sexton, J., Rogers, M., Calia, K., Calderwood, S., Fraser, C., Mekalanos, J. 1999. Identification of a *Vibrio cholerae* RTX toxin gene cluster that is tightly linked to the cholera toxin prophage. *Proc. Natl. Acad. Sci. USA* 96:1071–1076.

Mendes, C., Abath, F., Leal, N. 2008. Development of a multiplex single-tube nested PCR(MSTNPCR) assay for *Vibrio cholerae* O1 detection. *J. Microbiol. Meth.* 72:191–196.

Miller, V., Taylor, R., Mekalanos, J. 1987. Cholera toxin transcriptional activator toxR is a transmembrane DNA binding protein. *Cell* 48:271–279.

Mitra, R., Nandy, R., Ramamurthy, T., Bhatacharya S., Yamasaki, S., Shimada, T., Takida, Y., Naiar, G., 2001. Molecular characterization of rough variants of *Vibrio cholerae* isolated from hospitalized patients with diarrhea. *J. Med. Microbiol.* 50:268–276.

Mukhopadhyay, R., Ghosh, R. 2000. Mechanism of phage PS166-mediated biotype conversion in *Vibrio cholerae*: Role of the *hlyA* locus. *Virology.* 273:44–51.

Nair, G., Qadri, F., Holmgren, J., Svennerholm, A., Safa, A., Bhuiyan, N., Ahmad, Q., Farugue, S., Farugue, A., Takeda, Y., Sack, D. 2006. Cholera due to altered El Tor strains of *Vibrio cholerae* O1 in Bangladesh. *J. Clin. Microbiol.* 44:4211–4213.

Nair, G., Shimada, T., Kurazono, H., Okuda, J., Pal, A., Karasawa, T., Mihara, T., Uesaka, Y., Shirai, H., Garg, S., Saha, P., Mukhopadhyay, A., Ohashi, T., Tada, J., Nakayama, T., Fukushima, S., Takeda, T., Yoshifumi, Y. 1994. Characterization of phenotypic, serological, and toxigenic traits of *Vibrio cholerae* O139 Bengal. *J. Clin. Microbiol.* 32:2775–2779.

Namdari, H., Klaips, C., Hughes, J. 2000. A cytotoxin-producing strain of *Vibrio cholerae* non-O1, non-O139 as a cause of cholera and bacteremia after consumption of raw clams. *J. Clin. Microbiol.* 38:3518–3519.

Nandi, B., Nandy, R., Mukhopadhyay, S., Nair, G., Shimada, T., Ghose, A. 2000. Rapid method for species-specific identification of *Vibrio cholerae* using primers targeted to the gene of outer membrane protein OmpW. *J. Clin. Microbiol.* 38:4145–4151.

Nhung, P., Ohkusu, J., Sun, X., Ezaki, T. 2007. Rapid and specific identification of 5 human pathogenic *Vibrio* species by multiplex polymerase chain reaction to *dnaJ* gene. *Diag. Microbiol. Infect. Dis.* 59:271–275.

Ogawa, A., Kato, J., Watanabe, H., Nair, G., Takeda, T. 1990. Cloning and nucleotide sequence of a heat stable enterotoxin gene from *Vibrio cholerae* non-O1 isolated from a patient with travelers diarrhea. *Infect. Immun.* 58:3325–3329.

Pazzani, C., Scrascia, M., Dionisi, A., Maimone, F., Luzzi, I. 2006. Molecular epidemiology and origin of cholera reemergence in Italy and Albania in the 1990s. *Res. Microbiol.* 157:508–512.

Radu, S., Vincent, M., Apun, K., Rahim, R., Benjamin, P., Rusul, G. 2002. Molecular characterization of *Vibrio cholerae* O1 outbreak strains in Miri, Sarawak (Malaysia). *Acta Tropica.* 83:169–176.

Reid, J., Mekalanos, J. 1995. Characterization of *Vibrio cholerae* bacteriophage K139 and use of a novel mini-transposon to identify a phage encoded virulence factor. *Mol. Microbiol.* 18:685–701.

Ripabelli, G., Sammarco, M., Grasso, G., Fanelli, I., Capriola, A., Luzzi, I. 1999. Occurrence of *Vibrio* and other pathogenic bacteria in *Mytilus galloprovinialis* (mussels) harvested from Adriatic Sea, Italy. *Int. J. Food Microbiol.* 49:43–48.

Rivera, I., Cun, J., Huq, A., Sack, R., Colwell, R. 2001. Genotypes associated with virulence in environmental isolates of *Vibrio cholerae. Appl. Environ. Microbiol.* 67:2421–2429.

Saravanan, V., Kumar, S., Karunasagar, I., Karunasagar, I. 2007. Putative virulence genes of *Vibrio cholerae* from seafoods and the coastal environment of Southwest India. *Int. J. Food Microbiol.* 119:329–333.

Scrascia, M., Maimone, F., Mohamu, K., Materu, S., Grimont, F., Grimont, P., Pazzani, C. 2006. Clonal relationship among *Vibrio cholerae* O1 El Tor strains causing the largest cholera epidemic in Kenya in the late 1990s. *J. Clin. Microbiol.* 44:3401–3404.

Shangkuan, Y., Sow, Y., Wang, T. 1995. Multiplex-polymerase chain reaction to detect toxigenic *Vibrio cholerae* and to biotype *Vibrio cholerae* O1. *J. Appl. Bacteriol.* 79:267–273.

Sharma, A., Chaturvedi, A. 2006. Prevalence of virulence genes (*ctxA, stn, OmpW* and *tcpA*) among non-O1 *Vibrio cholerae* isolated from fresh water environment. *Int. J. Hyg. Environ. Health.* 209:521–526.

Shi, L., Mioshi, S., Hiura, M., Tomochika, K., Shimada, T., Sinoda, S. 1998. Detection of genes encoding cholera toxin (CT), zonula occludans toxin (ZOT), accessory cholera enterotoxin (ACE) and heat-stable enterotoxin (ST) in *Vibrio mimicus* clinical strains. *Microbiol. Immunol.* 42:823–828.

Singh, D., Isac, S., Colwell, R. 2002. Development of a hexaplex PCR assay for rapid detection of virulence and regulatory genes in *Vibrio cholerae* and *Vibrio mimicus. Clin. Microbiol.* 40:4321–4324.

Tamayo, M., Koblavi, S., Grimont, F., Castañeda, E., Grimont, P. 1997. Molecular epidemiology of *Vibrio cholerae* O1 isolates from Colombia. *J. Med. Microbiol.* 46:611–616.

Tapchaisri, P., Na-Uol, M., Jaipaw, J., Srimanote, P., Chongsa-nguan, M., Yamasaki, S., Hayashi, H., Nair, G., Kurazono, H., Chaicumpa, W. 2007. Virulence genes of clinical *Vibrio cholerae* O1 isolates in Thailand and their ribotypes. *J. Infect. Dis.* 55:557–565.

Tarr, C., Patel, J., Puhr, N., Sowers, E., Bopp, C., Strockbine, N. 2007. Identification of *Vibrio* isolates by a multiplex PCR assay and *rpoB* sequence determination. *J. Clin. Microbiol.* 45:134–140.

Taylor, R., Miller, V., Furlong, D., Mekalanos, J. 1987. Use of *phoA* gene fusions to identify a pilus colonization factor coordinately regulated with cholera toxin. *Proc. Natl. Acad. Sci. USA* 84:2833–2837.

Trucksis, M., Galen, J., Michalski, J., Fasano, A., Kaper, J. 1993. Accessory cholera enterotoxin (Ace), the third toxin of *Vibrio cholerae* virulence cassette. *Proc. Natl. Acad. Sci. USA* 90:5267–5271.

Varela, P., Rivas, M., Binszrein, N., Cremona, M., Hermann, P., Burrone, O., Ugalde, R., Frasch, C. 1993. Identification of toxigenic *Vibrio cholerae* from the Argentine outbreak by PCR for *ctx* A1 and *ctx* A2-B. *FEBS* 315:74–76.

Waldor, M., Mekalanos, J. 1996. Lysogenic conversion by a filamentous phage encoding cholera toxin. *Science* 272:1910–1914 .

CHAPTER 8

Aeromonas hydrophila

I. CHARACTERISTICS OF THE ORGANISM

The genus *Aeromonas* presently consists of 14 species, five of which, *A. hydrophila*, *A. veronii* biovar *sobria*, *A. jandaei*, *A schubertii*, and *A. caviae*, have been implicated in human infections (Janda and Abbot, 1998). *A. hydrophila* is the species most frequently implicated in gastroenteritis (Daskalov, 2006). The organism is a gram-negative, facultatively anaerobic, nonspore-forming rod, mobile by a single polar flagellum, DNase positive, protease positive, catalase positive, cytochrome oxidase positive, and ferments glucose with acid and gas production. The optimum growth temperature is considered to be ~28°C, and most isolates are capable of psychotrophic growth at refrigerator temperatures, with some isolates exhibiting growth at –1°C (Daskalov, 2006). *A. hydrophila* is widespread in lakes and streams, seafood, shellfish, poultry, ground meat, raw milk, raw vegetables, and municipal wastewater. Human clinical isolates of *A. hydrophila* presently fall into three DNA hybridization groups: HG1, HG2, and HG3 (Kirov, 2003). Pathogenicity is associated with the ability to produce exotoxins (agglutinins and hemolysins), cytotoxins, endotoxins, siderophores, invasins, adhesins (pili), S-layer (surface array protein layer), and flagella (Daskalov, 2006).

Two types of gastroenteritis have been attributed to *A. hydrophila*; the first and most common is a cholera-like illness characterized by watery stools and a mild fever. The second is a dysentery-like illness characterized by the presence of blood and mucus in the stools. The organism has also been implicated in extraintestinal infections involving septicemia, meningitis, peritonitis, urinary tract infections, and severe muscle degeneration as well as respiratory and wound infections (Chopra and Houston, 1999). In more recent years, the involvement of *A. hydrophila* has been increasingly detected in infections of severe burn patients (Barillo et al., 1996; Chim and Song, 2007). In addition, the organism has been implicated in hemolytic-uremic syndrome (HUS) involving a cytotoxin with homology to Shiga toxin 1 (Haque et al., 1996).

Several genes have been utilized for the specific PCR detection of potentially virulent aeromonads in foods and environmental samples. These include the gene encoding the β-hemolysin, the *aero* gene encoding aerolysin that exhibits hemolytic and cytolytic properties, and the effector gene *aexU* involved in a type III secretion system. An additional approach for imparting PCR specificity for detecting the members of a specific species of *Aeromonas* has been to initially employ immunocapture of the organism utilizing antibodies specific for binding the targeted species followed by the PCR.

II. PCR DETECTION

PCR methodology in the detection of bacteria can be divided into three types depending on the specificity of the primer pairs. One involves specific primer PCR (SPPCR) utilizing a pair of primers highly specific for a DNA sequence unique to the bacterial species targeted; the second involves a pair of primers specific for a conserved sequence present only in a specific genus; and the third involves universal primer PCR (UPPCR) utilizing a pair of primers that amplify a specific sequence present in all bacteria. Pollard et al. (1990) developed a PCR for the rapid and specific detection of the aerolysin gene in hemolytic strains of *A. hydrophila* associated with human infections. The sequence of the *aero* gene from *A. sobria* was found to have only 77% homology to the *aero* gene from *A. hydrophila,* which allowed the design of a species-specific pair of primers. The primers Aerola/Aerolb (Table 8.1) amplified a 209-bp sequence of the *aero* gene derived only from strains of *A. hydrophila.*

Kingombe et al. (2004) assessed a total of 78 raw and 123 processed and ready-to-eat retail food samples for the presence of mobile *Aeromonas* spp. harboring virulence genes (cytotoxic enterotoxin and hemolysin genes) using primers AHCF1/AHCR1 (Table 8.1). Conventional cultivation indicated 65/201 (32.3%) of the samples were presumptively positive for *Aeromonas* spp. The PCR method following enrichment indicated that 51/201 (25.4%) of the samples harbored *Aeromonas* spp. with the virulence genes.

The pathogenicity of *A. hydrophila* depends in part on the production of aerolysin encoded by the *aero* gene. Aerolysin is a hydrophilic protein that exhibits both hemolytic and cytolytic properties. The toxin binds to eukaryotic cells and forms holes in the membrane resulting in the destruction of membrane permeability and osmotic lysis. Tombelli et al. (2000) developed a unique DNA PCR piezoelectric biosensor for identification of *A. hydrophila* based on the *aero* gene. The primers AERO1/AERO2 (Table 8.1) amplified a 205-bp sequence of the *aer* gene. A 233-mer biotinylated probe (Table 8.1) was immobilized onto a streptavidin-coated gold disc on the surface of a quartz crystal to achieve peizoelectric detection.

Özbas, Lehner, and Wagner (2000) developed a seminested PCR assay for detection of *A. hydrophila* in raw milk. The primers 1a/1b (Table 8.1) amplified a 209-bp sequence of the *aer* gene. A forward seminested primer Aero 2a (Table 8.1) in conjunction with primer 1b amplified a 150-bp internal sequence. After the first amplification with primers 1a/1b, a second amplification using the seminested primers

Aero 2a/1b was employed with a 1:100 dilution of the first amplification product. The primers were found to be highly specific for the species *A. hydrophila*. The sensitivity of detection was 1.5×10^2 CFU/ml of milk. Among 56 samples of raw milk, 13/56 (23%) were found to be positive by PCR, whereas culture methodology yielded 8/56 (14%) positive for *A. hydrophila*. The higher level of detection was ascribed to the PCR detection of dead as well as live cells.

Peng et al. (2002) developed an immunocapture PCR specific for *A. hydrophila* where purified mouse IgG to *A. hydrophila* was used to coat the wells of a microtiter plate. Twenty microliters of a pure bacterial suspension were added to each plate. After 1 hr at 37°C the plates were washed, the cells lysed, and the released DNA used for 16S rDNA amplification utilizing the universal primers UPF/UPR (Table 8.1) that generated a 500-bp amplicon. Confirmation of species identity was accomplished with the restriction nuclease *Hae*III applied to the 500-bp amplicon that yielded two bands, a 180-bp band and two-coincident 160-bp bands. The assay was capable of detecting five CFU per well with a 20-µl DNA sample.

Kong et al. (2002) amplified a 720-bp sequence of the *Aero* gene that encodes the cytolytic autolysin utilizing the primers *Aero*-F/*Aero*-R (Table 8.1) for detecting *Aeromonas* in marine waters. Restriction digestion with *Taq*1 yielded fragments of 44-, 310-, and 366-bp, which were found to be specific for *A. hydrophila*.

Fukushima, Tsunomori, and Seki (2003) developed a series of duplex Rti-PCR assays utilizing SYBR green as the fluorophore for detection of 17 species of food or waterborne pathogens in stools. DNA was extracted with a commercial kit. The primers AHCF1/AHCR2 (Table 8.1) amplified a 232-bp sequence of the cytolytic enterotoxin of *A. hydrophila*. The analytical sensitivity of the assay without enrichment was approximately $>10^3$ CFU/g of stool samples. The identity of amplicons derived from *A. hydrophila* was confirmed from their T_m of 88.9°C. The duplex nature of the Rti-PCR assays was based on the simultaneous amplification of target sequences from two pathogens yielding amplicons of notably different T_m values.

Xia et al. (2004) developed a species-specific PCR for *A. hydrophila*. A pair of primers AP1/AP2 (Table 8.1) amplified a 208-bp sequence of the β-hemolysin gene.

III. DISTRIBUTION OF TOXIN GENES

Kingombe et al. (1999) developed a PCR for detecting enterotoxin and aerolysin genes in *Aeromonas* spp. The *A. hydrophila* cytolytic enterotoxin gene (AHCYTOEN) served as the reference gene in that it has been described as a multivirulence gene resulting in lethality in mice, hemolysis, cytotoxicity, and enterotoxigenicity (Chakraborty et al., 1986). Some of these activities are part of the virulence factors of other *Aeromonas* species. The primers AHCF1/AHCR1 (Table 8.1) amplified a 232-bp sequence of the AHCYTOEN gene and were developed because of 100% homology between the AHCYTOEN gene and an extracellular hemolysin gene, which represented the two main groups of virulence factors in the genus *Aeromonas* (enterotoxin and hemolysin). Among 220 *Aeromonas* isolates from raw food samples (beef, fish, vegetables) 157 (71%) were PCR positive; among 59

Table 8.1   PCR Primers and DNA Probes[a]

Primer or Probe	Sequence (5′ → 3′)[a]	Size of Sequence (bp)	Amplified Gene or DNA Target Sequence	References
AP1	CAA-GGA-GGT-CTG-TGG-CGA-CA	208	β-hemolysin	Xia et al. (2004)
AP2	TTT-CAC-CGG-CGG-TAG-CAG-GAT-TG			
EUB f933	GCA-CAA-GCG-GTG-GAG-CAT-GTG-G	500	16S rDNA	Ji et al. (2004)
EUB r1387	GCC-CGG-GAA-CGT-ATT-CAC-CG			
16srDNA-F	AGG-TTG-ATG-CCT-AAT-ACG-TA	—	16S rDNA	Bi, Liu, and Liu (2007)
16SrDNA-R	CGT-GCT-GGC-AAC-AAA-GGA-CAG			
Aero-F	TGT-CGG-SGA-TGA-CAT-GGA-YGT-G	720	aero	Kong et al. (2002)
Aero-R	CCA-GTT-CCA-GTC-CCA-CCA-CTT-CA			
aexUCF	TTG-CCA-GCT-GTC-ACC-AGT-GC	—	aexU	Sha et al. (2007)
aexUCR	TTA-CAG-ATA-GTC-AGC-CCC-GAC			
AERO1	CCA-AGG-GGT-CTG-TGG-CGA-CA	—	aero	Tombelli et al. (2000)
AERO2	TTC-CAC-CGG-TAA-CAG-GAT-TG			
AERO probe	CAC-CAG-GTA-TTG-GAC-GCT-GTC-CC	—	aero	
1a	CCA-AGG-GGT-CTG-TGG-CGA-CA	209	aero	Özbas, Lehner, and Wagner (2000)
1b	TTT-CAC-CGG-TAA-CAG-GAT-TC			
Aero 2A	AAG-CAA-TAT-TGT-CGG-CAT-GA	150	aero	
Aero 1b	TTT-CAC-CGG-TAA-CAG-GAT-TC			

Primer	Sequence	Size (bp)	Target	Reference
Aerola Aerolb	CCA-AGG-GGT-CTG-TGG-CGA-CA TTT-CAC-CGG-TAA-CAG-GAT-TG	209	*aero*	Pollard et al. (1990)
UPF UPR	AAA-CTC-AAA-GGA-ATT-GAC GAC-GGG-CGG-TGT-GTA-CAA	500	16S rDNA	Peng et al. (2002)
AHCF1 AHCR1	GAG-AAG-GTG-ACC-ACC-AAG-AAC-A AAC-TGA-CAT-CGG-CCT-TGA-ACT-C	232	AHCYTOEN	Kingombe et al. (2004)
AHH1F AHH1R	GCC-GAG-CGC-CCA-GAA-GGT-GAG-TT GAG-CGG-CTG-GAT-GCG-GTT-GT	130	*Ahh1*	Wang et al. (2003)
AH-aerAF AH-aerAR	CAA-GAA-CAA-GTT-CAA-GTG-GCC-A ACG-AAG-GTG-TGG-TTC-CAG-T	309	*aerA*	Wang et al. (2003)
ASA1F ASA1R	TAA-AGG-GAA-ATA-ATG-ACG-GCG GGC-TGT-AGG-TAT-CGG-TTT-TCG	249	*asa1*	Wang et al. (2003)
A16SF A26SR	GGG-AGT-GCC-TTC-GGG-AAT-CAG-A TCA-CCG-CAA-CAT-TCT-GAT-TTG	356	16S rDNA	Wang et al. (2003)
AHCF1 AHCR1	GAG-AAG-GTG-ACC-ACC-AAG-AAC-A AAC-TGA-CAT-CGG-CCT-TGA-ACT-C	232	*ahc*	Fukushima, Tsunomori, and Seki (2003)
Aer-F Aer-R	AGA-GTT-TGA-TCA-TGG-CTC-AG GGT-TAC-CTT-GTT-ACG-ACT-T	1502	16S rDNA	Borrel et al. (1997)

[a] S = G or C; Y = C or T.

Aeromonas isolates from environmental water samples, 34 (74%) were PCR positive; and among 71 human clinical *Aeromonas* isolates, 36 (51%) were PCR positive. The characterization of the PCR products by restriction fragment length polymorphism (PCR-RFLP) using the endonuclease *Hpa*II and PCR-amplicon sequence analysis (PCR-ASA) revealed three types of amplicons indicating that the virulence genes classified into three main groups: (1) aerolysins-hemolysins, (2) cytolytic enterotoxins, and (3) cytotonic enterotoxins.

The pathogenicity of *A. hydrophila* depends in part on the production of aerolysin encoded by the *aero* gene. Aerolysin is a hydrophilic protein that exhibits both hemolytic and cytolytic properties. The toxin binds to eukaryotic cells and forms holes in the membrane resulting in the destruction of membrane permeability and osmotic lysis. Tombelli et al. (2000) developed a unique DNA PCR piezoelectric biosensor for identification of *A. hydrophila* based on the *aero* gene. The primers AERO1/ AERO2 (Table 8.1) amplified a 205-bp sequence of the *aer* gene. A 233-mer biotinylated probe (Table 8.1) was immobilized onto a streptavidin-coated gold disc on the surface of a quartz crystal to achieve peizoelectric detection.

Albert et al. (2000) studied the distribution of three toxin genes among 115 *Aeromonas* isolates from 1735 children with diarrhea. Alt is a heat-labile cytotonic enterotoxin, Ast is a heat-stable cytotonic enterotoxin, and Act is a cyctotoxic enterotoxin. In addition, 27 aeromonads isolated from 830 control children, and 120 randomly selected aeromonads from different components of surface water in Bangladesh were also examined for the distribution of these three toxin genes. The number of isolates positive only for the presence of the *ast* gene was significantly higher for the environmental samples than for samples from diarrheal children. Isolates positive only for the presence of the *act* gene were not found in any of the three sources. Note that the number of isolates positive for both the *alt* and *ast* genes was significantly higher for diarrheal children than control children and the environment. Among 11 *A. hydrophila* isolates from diarrhea children, none harbored the *alt* or *ast* genes individually, but six (54.6%) harbored both genes. In contrast, none of five isolates of *A. hydrophila* from diarrheal children harbored both the *alt* and *ast* genes, whereas among *A. hydrophila* isolates isolated from the environment, only two (11.1%) harbored both the *alt* and *ast* genes. The products of both the *alt* and *ast* genes may therefore synergistically act to induce severe diarrhea.

Ji et al. (2004) utilized a universal primer PCR (UPPCR) involving a pair of universal primers EUB f933/EUB r1387 (Table 8.1) to amplify a 500-bp sequence of the conserved 16S rDNA in conjunction with density gradient gel electrophoresis (DGGE) and single-stranded conformation polymorphism (SSCP) for rapid PCR detection of *A. hydrophila* among other pathogens. When the 500-bp amplicon was subjected to DGGE, the 500-bp amplicon from *A. hydrophila* exhibited a distinctly different migration location compared to other organisms. SSCP yielded three major bands with *A. hydrohila* that were distinct from the banding profiles of other organisms. The combination of UPPCR-DGGE was found to yield a highly sensitive method for identification of *A. hydrophila* that was somewhat superior to UPPCR-SCCP.

Recently the presence of a type III secretion system (T3SS) was reported for *A. hydrophila* (Burr et al., 2002; Yu et al., 2004; Sha et al., 2005, 2007) and its

contribution to virulence established. Sha et al. (2007) described a cytotoxin designated AexT associated with a T3SS having ADP-ribosomyltransferase activity. The
toxin was derived from a diarrheal isolate of A. hydrophila. A new T3SS effector
(aexU) was described and characterized, the COOH-terminus of which exhibited no
sequence similarity to any known functional proteins in the database. Forward and
reverse PCR primers aexUCF/aexUCR (Table 8.1) derived from the COOH terminus
were described. The authors reported the presence of the aexU gene in various isolates among 250 aeromonads from clinical and water sources.

Bi, Liu, and Liu (2007) used primers 16SrDNA-F/16SrDNA-R (Table 8.1) to amplify
a sequence of 16S rDNA for confirmation of the identity of A. hydrophila mutants with
reduced virulence. Evidence for species specificity, however, was not presented.

IV. GENOTYPING USING HEMOLYSIN AND AEROLYSIN GENES

Wang et al. (2003) developd a multiplex PCR (mPCR) assay for amplification
of the hemolysin and aerolysin genes of A. hydrophila and A. veronii bv. Sobria.
The assay was evaluated using 121 clinical isolates and seven reference strains of
Aeromonas. A hemolytic isolate of A. hydrophila was used as a positive control. The
primers used are listed in Table 8.1. One pair of primers A16SF/A16SR (Table 8.1)
amplified a 356-bp sequence of the 16S rDNA as an amplification control. Among
82 hemolytic isolates of A. hydrophila, 35 harbored only the ahh1 gene and were
allocated to genotype 1; none harbored the asa1 gene alone (genotype 2); one was
allocated to genotype 3, harbored both the ahh1 and asa genes; 46 harbored both the
ahh1 and aerA genes and were allocated to genotype 4; and none of the 82 isolates
was completely lacking in one of the four toxin genes (genotype 5). In contrast, none
of the other six species of Aeromonas species represented by 41 isolates possessed
both the ahh1 and the aerA genes, as did 46/82 (56%) of the hemolytic A. hydrophila
isolates. The most common single hemolysin gene carried among the total of 128
Aeromonas isolates (seven species) was ahh1, with 99/128 (77%) of isolates positive
for this gene either alone or in combination with one of the other hemolysin genes.
A statistically significant correlation was found between cytotoxin levels (from Vero
cell culture cytotoxicity assays) and the hemolysin genotype. Isolates belonging to
genotype 4 (carrying both ahh1 and aerA genes) expressed higher cytotoxin titers
than isolates of other genotypes, suggesting that genotype 4 isolates may have greater
clinical significance.

V. RESTRICTION FRAGMENT LENGTH POLYMORPHISM (RFLP)

Borrel et al. (1997) applied restriction fragment length polymorphism (RFLP) to
reference strains of all species of Aeromonas and 76 clinical isolates of diverse origin. Genomic DNA was extracted and purified. The primers AerF/AerR (Table 8.1)
were used to amplify a 1502-bp sequence of the 16S rDNA of all Aeromonas species.
The resulting amplicons were restricted with AluI and MboI, and the DNA fragments

produced (33- to 346-bp) resolved as banding patterns by agarose gel electrophoresis. The resulting banding patterns were used to identify isolates of *Aeromonas* at the species level. Most RFLP results were in agreement with biochemical identification. Three presumed *A. vaeronii* and a strain of *A. caviae* were identified as *A. hydrophila* on the basis of both RFLP and biochemical analyses.

REFERENCES

Albert, M., Ansaruzzaman, M., Talukder, K., Chopra, A., Kuhn, I., Rahman, M., Faruque, A., Islam, M., Sack, R., Mollby, R. 2000. Prevalence of enterotoxin genes in *Aeromonas* spp. isolated from children with diarrhea, healthy controls, and the environment. *J. Clin. Microbiol.* 38:3785–3790.

Barillo, D., McManus, A., McManus, W., Kim, S., Pruitt, Jr., B. 1996. *Aeromonas* bacteraemia in burn patients. *Burns* 22:48–52.

Bi, Z., Liu, Y., Liu, C. 2007. Contribution of AhyR to virulence of *Aeromonas hydrophila* J-1. *Res. Vet. Sci.* 83:150–156.

Borrell, N., Acinas, S., Figueras, M., Marinez-Murcia, A. 1997. Identification of *Aeromonas* clinical isolates by restriction fragment length polymorphism of PCR-amplified 16S rRNA genes. *J. Clin. Microbiol.* 35:1671–1674.

Burr, S., Stuber, K., Wahli, T., Frey, J. 2002. Evidence for a type III secretion system in *Aeromonas salmonicida* subsp. Salmonicida. *J. Bacteriol.* 184:5966–5970.

Chakraborty, T., Huhle, B., Bergbauer, H., Goebel, W. 1986. Identification of *Aeromonas hydrophila* hybridization group I by PCR assays. *Appl. Environ. Microbiol.* 62:1167–1170.

Chim, H., Song, C. 2007. *Aeromonas* infection in critically ill burn patients. *Burns* 33:756–759.

Chopra, A., Houston, C. 1999. Enterotoxins in *Aeromonas*-associated gastroenteritis. *Microbes Infect.* 1:1129–1137.

Daskalov, H. 2006. The importance of *Aeromonas hydrophila* in food safety. *Food Control* 17:474–483.

Fukushima, H., Tsunomori, T., Seki, R. 2003. Duplex real-time SYBR Green PCR assays for detection of 17 species of food-or waterborne pathogens in stools. *J. Clin. Microbiol.* 41:5134–5146.

Haque, Q., Sigoama, M., Iwade, Y., Miorikawa, Y., Yamauchi, T. 1996. Diarrheal and environmental isolates of *Aeromonas* spp. produce a toxin similar to Shiga-like toxin 1. *Curr. Microbiol.* 32:239–245.

Janda, J., Abbot, S. 1998. Evolving concepts regarding the genus *Aeromonas*: an expanding panorama of species, disease presentations, and unanswered questions. *Clin. Infect. Dis.* 27:332–344.

Ji, N., Peng, B., Wang, G., Wang, S., Peng, X. 2004. Universal primer PCR with DGGE for rapid detection of bacterial pathogens. *J. Microbiol. Meth.* 57:409–413.

Kingombe, C., Huys, G., Howald, D., Luthi, E., Swings, J., Jemmi, T. 2004. The usefulness of molecular techniques to assess the presence of *Aeromonas* spp. harboring virulence markers in foods. *Int. J. Food Microbiol.* 94:113–121.

Kingombe, C., Huys, G., Tonolla, M., Albert, M., Swings, J., Peduzzi, R., Jemmi, T. 1999. PCR detection, characterization, and distribution of virulence genes in *Aeromonas* spp. *Appl. Environ. Microbiol.* 65:5293–5302.

Kirov, S. 2003. *Aeromonas* species. In: *Foodborne Microorganisms of Public Health Significance,* 6th ed., A. D. Hocking, Ed., AIFST, NSW Branch Blackwell Publ., Milton, Queensland, Australia. pp. 553–575.

Kong, R., Lee, S., Law, S., Wu, R., 2002. Rapid detection of six types of bacterial pathogens in marine waters by multiplex PCR. *Water Res.* 36:2802–2812.

Özbas, Z., Lehner, A., Wagner, M. 2000. Development of a multiplex and semi-nested PCR assay for detection of *Yersinia enterocolitica* and *Aeromonas hydrophila* in raw milk. *Food Microbiol.* 17:197–203.

Peng, X., Zhang, J., Wang, S., Lin, Z., Zhang, W. 2002. Immuno-capture PCR for detection of *Aeromonas hydrophila. J. Microbiol. Meth.* 49:335–338.

Pollard, D., Johnson, W., Lior, H., Tyler, S., Rozee, K. 1990. Detection of the aerolysin gene in *Aeromonas hydrophila* by the polymerase chain reaction. *J. Clin. Microbiol.* 28:2477–2481.

Sha, J., Pillai, L., Fadl, A., Galindo, C., Erova, T., Chopra, A. 2005. The type II secretion system and cytotoxic enterotoxin alter the virulence of *Aeomonas hydrophila. Infect. Immun.* 73:6446–6457.

Sha, J., Wang, S., Suarez, G., Sierra, G., Sierra, J., Fadl, A., Erova, T., Foltz, S., Khajanchi, B., Silver, A., Graf, J., Schein, C., Chopra, A. 2007. Further characterization of a type III secretion system (T3SS) and of a new effector protein from a clinical isolate of *Aeromonas hydrophila*-Part I. *Microbial Path.* 43:127–146.

Tombelli, S., Mascini, M., Sacco, C., Turner, A., 2000. A DNA piezoelectric biosensor assay coupled with a polymerase chain reaction for bacterial toxicity determination in environmental samples. *Anal. Chim. Acta* 418:1–9.

Wang, G., Clark, C., Liu, C., Pucknell, C., Munro, C., Kruk, T., Caldeira, R., Woodward, D., Rodgers, F. 2003. Detection and characterization of the hemolysin genes in *Aeromonas hydrophila* and *Aeromonas sobria* by multiplex PCR. *J. Clin. Microbiol.* 41:1048–1054.

Xia, C., Ma, Z., Raman, H., Wu, R. 2004. PCR cloning and identification of the β-hemolysin gene of *Aeromonas hydrophila* from freshwater fishes in China. *Aquaculture* 229:45–53.

Yu, H., Rao, P., Lee, H., Vilches, S., Merino, S., Tomas, J., Leung, K. 2004. A type III secretion system is required for *Aeromonas hydrophila* AH-1 pathogenesis. *Infect. Immun.* 72:1248–1256.

CHAPTER 9

Plesiomonas shigelloides

I. CHARACTERISTICS OF THE ORGANISM

Plesiomonas shigelloides is a unique gram-negative, polarly flagellated pathogenic bacterium native to aquatic animals and environments. The genus *Plesiomonas* consists of a single homogeneous species. Its metabolism is similar to that of the genus *Vibrio* in that sugars are fermented with acid production but no gas. 5S rDNA sequencing has indicated the organism to be closely related to the genus *Proteus*. Diarrhea is the major symptom, although extraintestinal infections including septicemia are known to occur.

The organism also causes various extraintestinal infections of high mortality with septicemia and meningitis (Miller and Koburger, 1985; Ampofo et al., 2001), particularly with predisposed individuals. Oysters are the major food incriminated in outbreaks in the United States. A temperature of 42 to 44°C is recommended for isolation to eliminate aeromonads. The utilization of inositol with acid production is a unique characteristic of the organism that is exploited with several agar media developed for its selective and differential isolation. The organism is β-hemolytic and produces a cholera-like (CL) enterotoxin in addition to a thermostabile (TS) and a thermolabile labile (LT) enterotoxin. A large plasmid (>120 mDa) has also been found to facilitate invasion. PCR detection and identification of *P. shigelloides* is based on primers that yield a specific amplicon from the 23S rRNA gene or on primers that yield a specific amplicon from the DNA gyrase B gene (*gyrB*). A variety of factors has been found to influence results obtained from application of the polymerase chain reaction (PCR) to the DNA of the organism.

II. NOMENCLATURE AND TAXONOMY

The genus *Plesiomonas* consists of a single homogeneous species (*P. shigelloides*). The organism was first described in 1947 (Ferguson and Henderson, 1947) and was indicated to have certain properties in common with *Shigella*. The initial strain possessed the major somatic antigen of *S. sonnei* but differed biochemically. This strain, designated C27, and similar isolates were subsequently placed chronologically into the genera *Pseudomonas, Aeromonas,* and *Vibrio* (Koburger, 1989). Habs and Schubert (1962) initially established the genus *Plesiomonas* in the family *Vibrionaceae* based on a number of its unique features and its similarity to the genus *Aeromonas* (*plesio,* neighbor; *monas, Aeromonas*). In common with other members of the family *Vibrionaceae,* it is polarly flagellated, facultatively anaerobic, and cytochrome oxidase-positive. In addition, isolates have in common with members of the genus *Vibrio* sensitivity to O/129 (Kirov, 1997). 5S rDNA sequencing has indicated the organism to be closely related to the genus *Proteus* (MacDonnel and Colwell, 1985; Martinez-Mucia, Nenlock, and Collins, 1992).

III. PHYSIOLOGICAL AND BIOCHEMICAL CHARACTERISTICS

The minimum temperature range for growth is 8–10°C, and the maximum is 42–45°C depending on the specific strain (Miller and Koburger, 1986a). Most strains do not grow below 8°C; however, one strain has been reported to grow at 0°C (Rouf and Rigney, 1971). Most strains have an optimum growth temperature between 35 and 38°C. A temperature of 42 to 44°C is recommended for isolation to eliminate aeromonads (Huq et al., 1991). Although most isolates are able to grow from a pH of 4.0 to 9.0, strains have been reported to be killed rapidly at a pH of 4.0 and below (Kirov, 1997). Most isolates exhibit growth from 2.0% to 3.0% NaCl. Some strains have been found to grow in 5.0% NaCl (Miller and Koburger, 1986a). NaCl, however, is not an absolute requirement for growth (Janda and Abbot, 1999).

Identification of *P. shigelloides* is biochemically based on the organism being cytochrome oxidase positive, lysine and ornithone decarboxylase positive, and arginine dihydrolase positive, with production of acid from inositol (Table 9.1) and the fermentation of sugars without gas production (Kelly and Kain, 1991; Farmer, 2003).

IV. ECOLOGICAL DISTRIBUTION

The primary habitats of *P. shigelloides* are freshwater ecosystems (rivers, lakes, and surface waters) and marine estuaries in tropical and temperate climates (Monteil and Harf-Monteil, 1997). In aquatic systems, *P. shigelloides* occurs as free-living cells and in fish, crabs, shrimp, mussels, and oysters (Schubert, 1984; Huber et al., 2004; Oxley et al., 2002). In addition, the organism has been isolated from soil and a number of terrestrial animals (amphibians, monkeys, birds, polecats, and reptiles) in addition to domestic animals (sheep, swine, cattle, cats, dogs, and goats; Schubert, 1984; Khardori and Fainstein, 1988). The organism has also been reported to occur at a low incidence (0.0078%) in healthy humans (Arai et al., 1980). Oysters have been the major food incriminated in outbreaks in the United States (Farmer, Arduino, and Hickman-Brenner, 1992).

Table 9.1   Metabolic Characteristics of *P. shigelloides*[a]

Metabolic Property	% of Strains Positive
Phenylalanine deaminase	0–4
Arginine dihydrolase	100
Lysine decarboxylase	100
Ornithine decarboxylase	100
Cytochrome oxidase	100
NO_3 reduction to NO_2	100
Indole production from tryptophan	100
Methyl red test	90–100
Motility (36°C)	92–95
DNase	0–100
β-galactosidase	90–99
Gelatinase	0
β-glucosidase (aesculine hydrolysis)	0
Hydrogen sulfide	0
Urease	0
Voges–Proskauer test	0
Citrate utilization	0
Fermentation of:	
Adonitol	0
L-arabinose	0
Cellobiose	0
Dulcitol	0
Erythritol	0
Glucose	100
Glycerol	35–66
Lactose	76–81
Maltose	100
Mannitol	0
D-mannose	70–95
Melibiose	70–95
α-methyl-D-glucoside	0
Mucate	0
Inositol	95–100
Raffinose	0
Rhamnose	0
Sorbitol	0
Sucrose	0–5
Trehalose	95–100
D-xylose	0

[a] Compiled from Stock (2004), Kelly and Kain (1991), and Farmer (2003).

V. TOXINS AND INVASIVE FACTORS

P. shigelloides produces a cholera-like (CL) enterotoxin (Gardner, Fowlston, and George, 1987; Abbott, Kokka, and Janda, 1991), a thermostable (TS; Mathews, Douglas, and Guiney, 1988; Abbott, Kokka, and Janda, 1991) and a thermolabile (TL; Fálcon et al., 2003) enterotoxin. A large plasmid (>120 mDa) has also been found to facilitate invasion (Herrington et al., 1987).

Binns et al. (1984) reported HeLa cell invasion by 5 of 16 clinical isolates of *P. shigelloides*. However, Herrington et al. (1987) found that among five clinical strains of *P. shigelloides*, none invaded HeLa cells, none produced a positive keratoconjunctivitis reaction in guinea pigs, nor did any produce diarrhea in rabbits. Genetic probes for heat-stable enterotoxins related to enterotoxigenic *Escherichia coli* and for gene sequences common to the invasive plasmid of *Shigella* spp. and enteroinvasive *E. coli* were negative. Heat-labile enterotoxin was not detected using a GM1 ELISA asay. Rabbit ileal loop assays were negative. One strain cured of its plasmid was unable to invade the intestinal mucosa of genetobiotic piglets given 1×10^{10} CFU orally, whereas the wild-type exhibited invasion. Three piglets that received 1×10^{10} wild-type CFU developed severe illness involving shock with edema of the eyelids, conjunctivae, head, and neck. A fourth piglet that received a plasmid-cured strain remained healthy. Electron microscopy of tissue sections indicated that intestinal epithelial cells of infected piglets harbored *P. shigelloides*, primarily inside vacuoles. An oral dose of 1×10^9 CFU resulted in diarrhea in one of three piglets.

Gardner, Fowlston, and George (1987) grew 29 strains of *P. shigelloides*, mostly of clinical origin, in an iron-depleted medium (syncase broth) and found that sterile filtrates of 24 of the 29 isolates produced elongation of Chinese hamster ovary (CHO) cells. Heating of the sterile filtrates (100°C for 30 min) and addition of cholera antitoxin to the filtrates prevented this effect on CHO cells. The addition of iron to syncase broth eliminated the elongation effect on CHO cells by sterile culture filtrates (Gardner, Fowlston, and George, 1990).

Eight strains of *P. shigelloides* were assayed for enterotoxin production using the rabbit ileal loop assay by Mathews, Douglas, and Guiney (1988). It is interesting that seven of the strains required serial in vivo passages through the rabbit's intestine before enterotoxin activity was detected in cell filtrates. Enterotoxin production was readily lost with subculture of these toxigenic cells. Heat treatment (65°C for 10 min) of the culture filtrates from three strains that had never been passed in vivo led to detectable enterotoxin activity. No DNA homology was found to the cloned enterotoxin genes of *E. coli* and *Vibrio cholerae*.

Abbott, Kokka, and Janda (1991) found that sterile filtrates of all 16 strains of *P. shigelloides* studied contained a low level of cytolysin active on HEP-2 (human epithelial) and Y1 (mouse adrenalin tumor) cell monolayers but not on CHO cell monolayers. Susceptible tissue culture cells became rounded with internal granulation within 24 hr prior to eventual lysis. This cytolysin was not inactivated by heating at 100°C for 10 min. The median 50% lethal dose for Swiss Webster mice was determined to be 3.5×10^8 CFU with a range of 3.2×10^7 to >1×10^9 CFU.

Culture filtrates of four isolates of *P. shigelloides* from water were assayed by Fálcon et al. (2003) for cytotoxic activity in CHO, Vero (African green monkey), HeLa (human cervix), HT29 (human epithelial intestinal), and SK6 (swine epithelial kidney) cells. Intensive cytoplasmic vacuolation including cell rounding and swelling with massive cell death and destruction of the cell monolayers was observed. CHO, HeLa, and Vero cells were the most sensitive to the vacuolating activity, which was microscopically observed within 30 min of exposure. The toxic activity was inactivated by heating at 56°C and was neutralized with antiserum to the cytotoxin of *A. hydrophila*. Among the four isolates studied, three were found to be β-hemolytic. In the suckling mouse assay, two of the four isolates produced intense fluid accumulation. This observation may reflect the fact that none of the four isolates was of clinical origin.

VI. β-HEMOLYSIS

Although several reports (Kelly and Kain, 1991; Stock, 2004) have indicated the absence of β-hemolysin production, an agar overlay system has been found to yield β-hemolysis (Daskaleros, Stoebner, and Payne, 1991; Janda and Abbott, 1993). Janda et al. (1993) reported that over 90% of 36 isolates produced a cell-associated β-hemolysin detected with agar overlay and contact-hemolysis assays. It is interesting that the addition of the iron-chelating agent EDDA [ethylene-di (*o*-hydroxyphenyl acetic acid)] to L broth was reported to notably increase the levels of hemolysis from culture supernatants (Daskaleros, Stoebner, and Payne, 1991). The observation in the same study that surface colonies fail to yield hemolysis is of particular significance and suggests that either oxygen tension or viscosity may influence β-hemolysin production or release from cells.

Baratéla et al. (2001) found that all seven strains of *P. shigelloides* studied produced β-hemolysis on the surface of blood agar plates pepared with Luria agar (LA), but with trytic soy agar (TSA) surface growth yielded five positive strains and on brain–heart infusion agar (BHIA) only two positive strains. In contrast, using the agar overlay method, all seven isolates exhibited β-hemolysis with all three media. The greatest hemolysis was detected with LA and the weakest with BHIA. The addition of iron to the culture medium did not affect bacterial growth, but reduced hemolytic activity. In the presence of an iron chelator, growth was inhibited, but hemolytic activity was enhanced. Calcium ions stimulated and EDTA reduced hemolytic activity when added to assay mixtures. These observations indicate that the production of β-hemolysis clearly depends on the medium and method used for cultivation of the organism, which may explain the negative results obtained by several authors.

VII. ISOLATION

Enteric agars were first used for the isolation and identification of *P. shigelloides* but are not ideal (Koburger, 1989). Growth of most isolates of *P. shigelloides* is obtained on MacConkey, *Salmonella-Shigella*, desoxycholate, Hektoen enteric, and xylose lysine desoxycholate agars however, some strains may be inhibited (Koburger, 1989). Jeppesen (1995) has reviewed the various agar media for selective isolation of *P. shigelloides*. Schubert (1977) was the first to develop a selective medium for the specific isolation of *P. shigelloides* designated inositol brilliant green bile salts (IBB) agar (Table 9.2). Its differential property is based on the fermentation of inositol and its selective ability on bile salts. IBB agar was found by von Graevenitz and Bucher (1983) to be highly effective for selective isolation of *P. shigelloides* from human feces following enrichment in alkaline peptone water (APW). Miller and Koburger (1985) developed *Plesiomonas* agar (PL), which contains the nonfermented carbohydrates mannitol and arabinose at levels of 0.75% and 0.5%, respectively (Table 9.2). The fermentable carbohydrate inositol is present at a critically low level of 0.1%, along with 0.2% lysine and a notably low level of 0.1% bile salts No. 3. The initial pH is adjusted to 7.4 before autoclaving. Contaminating colonies fermenting mannitol and arabinose will be red. Fermentation of the low level of inositol and decarboxylation of lysine will result in a near-neutral pH so that typical colonies of *P. shigelloides* are pink. Miller and Koburger (1986b) found that with plating of aquatic samples, both IBB and PL agars should be used simultaneously at 35°C for maximum recovery. IBB was found to have a greater recovery rate, and PL agar is less inhibitory to injured cells.

Huq et al. (1991) compared the plating efficiency of *P. shigelloides* on *Plesiomonas* differential agar (PDA; Table 9.2), IBB agar, and modified *Salmonella-Shigella* (MSS) agar where lactose was replaced with inositol. Cell suspensions contained equal numbers of *P. shigelloides* and aeromonads derived from pure cultures. The largest number of *P. shigelloides* CFU appeared on PDA at 42°C. At this temperature, colonies of aeromonads also appeared in high numbers. At 44°C on PDA plates, there was a slight decrease in recovery of *P. shigelloides*, but complete inhibition of *A. hydrophila* and significant inhibtion of *A. caviae*. *A. sobria* appeared in high numbers on PDA at 44°C; however, the average colony size was drastically reduced to approximately 1 mm, making them clearly distinct from *P. shigelloides* colonies, which were about 4 mm in diameter. Although the recovered CFU of *P. shigelloides* on PDA at 44°C was about 10% less than at 44°C, the authors concluded that PDA at 44°C and with a 24-hr incubation period should be a suitable medium for isolating *P. shigelloides* from water samples and possibly also clinical samples where the presence of aeromonads poses a problem.

VIII. SEROLOGY

Serotyping of *P. shigeloides* isolates is based on the detection of O (somatic) and H (flagellar) antigens. Shimada and Sakasaki (1978) proposed the first antigenic typing scheme for *P. shigelloides*, which was based on 30 O antigenic groups and 11 H antigens. Subsequent studies indicated more than 100 serovars (Aldova and Schubert, 1996; Aldova and Shimada, 2000). One of the most frequently encountered serovars of *P. shigelloides* is O17 (Aldova, 1987), which is involved with protection against

Table 9.2 Agar Media for Selective Isolation of *P. shigelloides*

Inositol Brilliant Green Bile Salt (IBB) Agar (Schubert, 1977)	g/L	*Plesiomonas* (PL) Agar (Miller and Koburger, 1985)	g/L	*Plesiomonas* Differential Agar (PDA) (Huq et al., 1991)	g/L
Peptone	10.0	Peptone	10	Peptone	7.5
Beef extract	5.0	NaCl	5.0	Beef extract	7.5
NaCl	5.0	Yeast extract	2.0	NaCl	5.0
Bile salt mixture	8.5	Mannitol	7.5	Meso-inositol	10.0
Brilliant green	0.00033	Arabinose	5.0	Bile salt mixture	8.5
Neutral red	0.025	Inositol	1.0	Brilliant green	0.00033
Meso-inositol	10.0	Lysine	2.0	Neutral red	0.025
Agar	13.5	Bike salts No. 3	1.0	Agar	13.5
pH 7.2, 42°C and 44°C for 48 hr		Phenol red	0.08	pH 7.4, 44°C for 24 hr	
		Agar	15.0		
		pH 7.4, 42°C for 24 hr			

shigellosis due to *S. sonnei* via a common lipopolysaccharide (LPS) in the cell walls of both species (Sayeed, Sack, and Qadri, 1992).

IX. EPIDEMIOLOGY AND OUTBREAKS

P. shigelloides has been ranked third as a cause of travelers' diarrhea in Japan (Schubert and Holz-Bremer, 1999) and third as a cause of diarrhea among certain military units in China (Bai et al., 2004) and among civilians in Hong Kong (Chan et al., 2003).

The first outbreak of gastroenteritis due to *P. shigelloides* occurred in Japan in 1963 (Ueda, Yamazaki, and Hori, 1963) and was due to contaminated cuttlefish salad involving 275 cases of diarrheal infection out of 870 individuals who consumed the salad. Salted mackerel resulted in an outbreak in 1966 in Japan involving 53 cases (Hori and Hayashi, 1966). Subsequent outbreaks involved waterborne diarrhea affecting 978 out of 2141 persons in Japan (Tsukamoto et al., 1978) and various other waterborne outbreaks (Centers for Disease Control and Prevention, 1996; Medema and Schels, 1993), in addition to oyster consumption (Ratula et al., 1982). Uncooked shellfish have been found to be the most important sources of foodborne illnesses from *P. shigelloides* (Holmberg et al., 1986). More recent outbreaks have been discussed by Jagger (2000).

P. shigelloides is also responsible for a variety of extraintestinal infections, particularly among children and immunocompromised individuals with underlying maladies such as malignancy, splenectopy, alcoholic liver disease, cirrhosis, sickle-cell anemia, and primary haemochromatosis (excess deposition of iron throughout the body (Ampofo et al., 2001). Most of these predispositional factors suggest that the availability of iron may be a major factor preventing or limiting infections in healthy individuals due to *P. shigelloides*.

X. APPLICATION OF PCR

A. Conventional PCR

González-Rey et al. (2000) were the first to develop a PCR assay specific for *P. shigelloides*. The assay was used to confirm the identity of 25 isolates from aquatic environments, 10 isolates from human clinical cases of diarrhea, and 5 isolates from animals. The forward primers PS23FW3/PS23RV3 (Table 9.3) amplify a 284-bp sequence of the 23S rRNA gene (Table 9.3).

Gu et al. (2006a) subjected 26 isolates of *P. shigelloides* from Sweden (10 fresh water, 6 fish, 10 human clinical) to random amplified polymorphic DNA (RAPD) analysis with the use of two random primers (LMPB1 and LMPB4; Table 9.3). Prior to RAPD analysis, the identity of all isolates was confirmed via PCR utilizing the primer pair of González-Rey (Table 9.3). There was notable genetic variability among most of the isolates, and none of the isolates had the same composite RAPD

Table 9.3 PCR Primers and DNA Probes

Primer or Probe	Sequence (5' → 3')	Size of Amplified Sequence (bp)	Gene or Target Sequence	References
PS23FW3	CTC-CGA-ATA-CCG-TAG-AGT-GCT-ATC-C	284	23S rDNA	González-Rey et al. (2000)
PS23RV3	CTC-CCC-TAG-CCC-AAT-AAC-ACC-TAA-A			
PSG237-F	TTC-CAG-TAC-GAG-ATC-CTG-GCT-AA	68	*gyrB*	Fuchushima and Tsunomori (2005)
PAG110R	ACC-GTC-ACG-GCG-GAT-TAC-T			
Forward	AGC-GCC-TCG-GAC-GAA-CAC-CTA	112	23S rDNA	Loh and Yap (2002)
Reverse	GTG-TCT-CCC-GGA-TAG-CAG			
Probe	LCRed640-GGT-AGA-GCA-CTG-TTA-AGG-CTA-GGG-GGT-CAT-C-P			
PS-F	GCA-GGT-TGA-AGG-TTG-GGT-AA	628	23S rDNA	Gu and Levin (2006a)
PS-R	TTG-AAC-AGG-AAC-CCT-TGG-TC			
RAPD: LMPB1	GGA-ACT-GCT-A			Gu et al. (2006a) from Boerlin et al. 1995)
RapD: LMPB4	AAG-GAT-CAG-C			

profile. The results indicated that most of the isolates from the same source grouped together and that the isolates from fish had a closer linkage to the human clinical isolates than did the freshwater isolates, suggesting that fish may be the more serious source for potential risk of infection. It was also observed that certain human clinical isolates had almost the same RAPD profiles as certain of the isolates from freshwater and fish, indicating that *P. shigelloides* from both seafood and freshwater should be considered potential pathogens.

Various factors have been found to affect the quantitative PCR assays of *Plesiomonas shigelloides* (Gu and Levin, 2006a). Different *Taq* polymerase preparations, varying sets of primers, different DNA strains, and different cell lysing agents were found to significantly influence the linear relationship between the fluorescent intensities of DNA bands and the log of CFU per PCR. The primer–dimers formed in the PCR can be eliminated by using different *Taq* polymerase preparations and different sets of primers to run the PCR. The two sets of primers compared were those of González-Rey et al. (2000) and a pair developed in the study designated PS-F and PS-R that amplify a 628-bp sequence of the 23S rRNA gene of *P. shigelloides* (Table 9.3).

A rapid and efficient procedure for quantitative detection of *Plesiomonas shigelloides* in pure culture was developed by Gu, Cao, and Levin (2006a), which is the first report of quantitative detection of *P. shigelloides* by the polymerase chain reaction (PCR). A minimum of 4 CFU per PCR reaction could be detected. There was a linear relationship between the relative fluorescent intensities of the DNA bands and the log of the CFU from 4 to 1.2×10^3 CFU per PCR reaction. The effects on detection sensitivity of two different nucleic acid dyes, GelStar™ and ethidium bromide (EB), and the effects of two different staining methods with each dye were evaluated. Adding the dye to the agarose solution before gel formation proved to yield superior results compared to staining the DNA bands after electrophoresis. The GelStar™ stain was found superior to EB with minimum detection levels of 4 CFU and 12 CFU per PCR reaction, respectively.

A quantitative assay for *P. shigelloides* in clams and oysters based on the conventional polymerase chain reaction was developed by Gu and Levin (2006b). The primers used (Table 9.3) were those of González-Rey et al. (2000). The assay involved the treatment of homogenized tissue samples with 4.0% formaldehyde that presumably denatured DNases and proteases present in the tissue that would otherwise inactivate the PCR reaction. The level of detection of *P. shigelloides* in clam tissue without enrichment was 200 CFU/g. The addition of 0.1% bovine serum albumin (BSA) to PCR reactions or the DNA purification system reduced the level of detection to 60 CFU/g. Formaldehyde had no effect on the level of detection with clam tissue. The level of detection of *P. shigelloides* in oyster tissue without enrichment was 6×10^5 CFU/g. The addition of 4.0% formaldehyde to oyster tissue homogenates reduced the level of detection to 6×10^2 CFU/g in contrast to the addition of 0.1% BSA to PCR reactions or the DNA purification system, which reduced the level of detection to only 2×10^5 CFU/g. The combination of formaldehyde plus BSA, formaldehyde plus DNA purification, or formaldehyde plus BSA plus DNA purification all gave a detection level of 2×10^2 CFU/g of oyster tissue. With clam tissue, the linear range for detection of *P. shigelloides* was 60 to 2×10^4 CFU/g. With oyster tissue, the linear range for detection of *P. shigelloides* was 2×10^2 to 6×10^4 CFU/g.

Ethidium bromide monoazide (EMA) is a DNA intercollating agent that with visible light activation cross-links the two strands of DNA so as to prevent PCR amplification. EMA penetrates only membrane-damaged cells. This has allowed the selective Rti-PCR amplification of DNA from viable bacterial pathogens (Nogva et al., 2003). EMA has been utilized to allow selectively the rapid and efficient PCR quantitative detection of viable *P. shigelloides* (Gu and Levin, 2007a). The addition of EMA (1 µg/ml) to mixtures of viable and heat-killed cells of *P. shigelloides* inhibited the PCR amplification of DNA derived from the dead cells, but did not inhibit the PCR amplification of DNA derived from the viable cells. EMA at 5 µg/ml or less had little or no inhibition on the PCR amplification of DNA derived from viable cells of *P. shigelloides*. After EMA treatment, the DNA from viable *P. shigelloides* cells in varying ratios of viable to dead cells could be selectively quantified by PCR. The minimum level of detection was DNA from 24 CFU per PCR reaction. A linear relationship was found between the relative fluorescent intensity of the DNA bands and the log of genomic targets derived from the viable cells in mixtures of viable and dead cells in the range of 2.4×10^1 to 2.4×10^4 DNA targets from viable cells per PCR.

A quantitative assay for *Plesiomonas shigelloides* in pure culture and clams based on the competitive polymerase chain reaction was developed by Gu and Levin (2007b), which is the first report for quantitative detection of *P. shigelloides* by competitive PCR. A forward primer (PS-F) and reverse primer (PS23RV3) along with a hybrid primer were designed (Table 9.3), and the specificity of the forward and reverse primers for *P. shigelloides* was proven. An internal standard DNA sequence was synthesized by PCR with the hybrid primer as the forward primer, and PS23RV3 as the reverse primer. A single concentration (0.588 pg/PCR) of internal standard (IS) was used for competitive PCR. The lowest level of detection of *P. shigelloides* was 80 CFU per PCR in pure culture, 240 CFU/g of clam tissue without enrichment, and 40 CFU/g of clam tissue after 7 hr. Nonselective enrichment occurred at 37°C. There was a linear relationship between the log of the ratio of the relative fluorescent intensities of amplified target DNA bands to the internal standard DNA bands (IS) and the log of the CFU within a certain range in pure cultures and in clam tissue either with or without enrichment. The linear range with cells from a pure culture was the DNA derived from 8.0×10^1 to 8.0×10^4 CFU per PCR, whereas with clam tissue, the linear range was the DNA derived from 2.4×10^2 to 2.4×10^5 CFU/g of tissue (1.2×10^1 to 1.2×10^4 CFU per PCR) without enrichment, and 4.0×10^1 to 1.2×10^4 CFU/g of clam tissue with enrichment, respectively.

B. Real-Time PCR (Rti-PCR)

Loh, Peng, and Yap (2002) were the first to develop an Rti-PCR assay for *P. shigelloides*. The primer pair used (Table 9.3) amplifies a 112-bp sequence of the 23s rRNA gene. Their assay involved the use of SYBR green I in conjunction with a probe labeled at the 5′-end with LCRed640 (Table 9.3) for the establishment of a fluorescent resonance energy transfer (FRET) system. SYBR Green I binds to the minor groove of double-stranded DNA and in so doing undergoes an eightfold increase in fluorescence and is a general indicator of amplification. SYBR green I has an excitation maximum near that of fluorescence, which enables it to be a donor in FRET so as to excite LCRed640. Channel F1 of the light cycler was used to detect the intensity of SYBR green fluorescence as a function of amplification. Channel F2 was used to detect the increase in fluorescence observed from LCRed640 resulting from annealing of the probe to its homologous site, which is excited by SYBR green I. The assay time was 3 hr.

Fuchushima and Tsunomori (2005) developed an Rti-PCR assay using SYBR green for detecting *P. shigelloides* in stool samples. The forward primer PSG237-F and the reverse primer PAG110R (Table 9.3) amplified a 68-bp sequence of the *gyrB* (DNA gyrase B gene). A commercial stool extraction kit was used for DNA extraction and purification. The assay was completed within 2 hr.

A quantitative assay for *Plesiomonas shigelloides* in pure culture and oysters based on Rti-PCR and utilizing SYBR green was developed by Gu and Levin (2008). The primers PS23FW3/PS23RV3 (Table 9.3) from González-Rey et al. (2000) were used. The methodology involved the treatment of oyster tissue homogenates with formaldehyde, differential centrifugation, and treatment of samples with activated charcoal coated with *Pseudomonas fluorescens*. With seeded oyster tissue homogenates, without formaldehyde or coated charcoal treatments, the lowest level of detection for *P. shigelloides* was 1×10^7 CFU per gram of tissue, equivalent to the DNA from 2.5×10^5 CFU per Rti-PCR. The addition of 4% formaldehyde to tissue homogenates reduced the minimum level of detection of *P. shigelloides* to 1×10^5 genomic targets per gram of tissue, equivalent to the DNA from 2.5×10^3 CFU per Rti-PCR. The treatment of tissue homogenates with only activated charcoal coated with *P. fluorescens* reduced the minimum level of detection of *P. shigelloides* to 1×10^4 CFU per gram, equivalent to the DNA from 2.5×10^2 CFU per real-time PCR, without DNA purification or enrichment. The combination of adding 4.0% formaldehyde to oyster tissue homogenates and treatment with coated charcoal reduced the level of detection of *P. shigelloides* to 1×10^3 CFU per gram, equivalent to the DNA from 25 CFU per Rti-PCR. The linear range of detection was from 1×10^3 to 1×10^6 genomic targets per gram without enrichment.

REFERENCES

Abbott, S., Kokka, R., Janda, J. 1991. Laboratory investigations on the low pathogenic potential of *Plesiomonas shigelloides*. *J. Clin. Microbiol.* 29:148–153.

Aldova, E. 1987. Serotyping of *Plesiomonas shigelloides* strains with our own antigenic scheme: an attempted epidemiological study. *Zentralbl. Bacteriol. Hyg. A* 265:253–262.

Aldova, E., Schubert, R. 1996. Serotyping of *Plesiomonas shigelloides*: A tool for understanding of ecological relationship. *Med. Microbiol. Lett.* 5:33–39.

Aldova, E., Shimada, T. 2000. New O and H antigens of the international antigenic scheme for *Plesiomonas shigelloides*. *Folia Microbiol. (Praha)* 45:301–304.

Ampofo, K., Graham, P., Ratner, A., Rajagopalan, L., Della-Latta, P., Saiman, L. 2001. *Plesiomonas shigelloides* sepsis and splenic abscess in an adolescent with sickle-cell disease. *Pediatr. Infect. Dis.* 20:1178–1179.

Arai, T., Ikejima, N., Itoh, T., Sakai, S., Shimada, T., Sakazaki, R. 1980. A survey of *Plesiomonas shigelloides* from aquatic environment, domestic animals, pets, and humans. *J. Hyg. (Cambridge)* 84:203–211.

Bai, Y., Dai, Y., Li, J., Nie, J., Chen, Q., Wang, H., Rui, Y., Zhang, Y., Yu, S. 2004. Acute diarrhea during army field exercise in southern China. *World J. Gastroenterol.* 10:127–131.

Baratéla, K., Saridakis, H., Gaziri, L., Pelayo, J. 2001. Effects of medium composition, calcium, iron and oxygen on haemolysin production by *Plesiomonas shigelloides* isolated from water. *J. Appl. Microbiol.* 90:482–487.

Binns, M., Vaughan, M., Sanyal, S., Timmis, K. 1984. Invasive ability of *Plesiomonas shigelloides*. *Zentralbl. Bakteriol. Mikrobiol. Hyg. A* 257:343–347.

Boerlin, P., Bannerman, E., Ischer, F., Rocourt, J., Bille J. 1995. Typing *Listeria monocytogenes*: A comparison of random amplification of polymorphic DNA with 5 other methods. *Res. Microbiol.* 146:35–39.

Centers for Disease Control and Prevention. 1998. *Plesiomonas shigelloides* and *Salmonella* serotype Hartford infections associated with a contaminated water supply—Livingston County, New York. *Morb. Mortal. Wkly. Rep.* 47:394–396.

Chan, S., Ng, K., Lyon, D., Cheung, W., Cheng, A., Rainer, T. 2003. Acute bacterial gastroenteritis: A study of adult patients with positive stool cultures treated in the emergency department. *Emerg. Med.* 20:335–338.

Daskaleros, P., Stoebner, J., Payne, S. 1991. Iron uptake in *Plesiomonas shigelloides:* cloning of the genes for the heme-iron uptake system. *Inf. Immun.* 59:2706–2711.

Fálcon, R., Carbonell, G., Figueredo, P., Butião, F., Saridakis, H., Pelayo, J., Yano, T. 2003. Intracellular vacuolation induced by culture filtrates of *Plesiomonas shigelloides* isolated from environmental sources. *J. Appl. Microbiol.* 95:273–278.

Farmer, J., III. 2003. *Enterobacteriaceae*: Introduction and identification. In: *Manual of Clinical Microbiology*, 8th ed. P. Murray, E. Baron, J. Jorgensen, M. Pfaller, R. Yolken, Eds., ASM Press: Washington, DC, pp. 637–640.

Farmer, J., III, Arduino, M., Hickman-Brenner, F. 1992. The genera *Aeromonas* and *Plesiomonas*. In: *The Prokaryotes,* 2d ed. A. Balows, H. Trüper, M. Dworkin, W. Harder, K.-H. Schlcifer, Eds., Springer-Verlag: New York, pp. 3012–3028.

Fergusen, W., Henderson, N. 1947. Description of strain C27: A motile organism with the major antigen of *Shigella sonnei* phase I. *J. Bacteriol.* 54:179–181.

Fuchushima, H., Tsunomori, Y. 2005. Study of real-time PCR assays for rapid detection of food-borne pathogens. *J. Jpn. Assn. Infect. Dis.* 79:644–655.

Gardner, S., Fowlston, S., George, W. 1987. In vitro production of cholera toxin-like activity by *Plesiomonas shigelloides*. *J. Infect. Dis.* 156:720–722.

Gardner, S., Fowlston, S., George, W. 1990. Effect of iron on production of a possible virulence factor by *Plesiomonas shigelloides*. *J. Clin. Microbiol.* 28:811–813.

González-Rey, C., Svensonk, S., Bravo, L., Rosinsky, J., Ciznar, I., Krovacek. K. 2000. Specific detection of *Plesiomonas shigelloides* isolated from aquatic environments, animals and human diarrhoeal cases by PCR based on 23S rRNA gene. *FEMS Immunol. Med. Microbol.* 29:107–113.

Gu, W., Cao, J., Levin, R.E. 2006a. Quantification of *Plesiomonas shigelloides* using PCR based on 23S rRNA gene. *Food Biotechnol.* 20:211–218.

Gu, W., Gonzalez-Rey, C., Krovacek, K., Levin, R. 2006b. Genetic variability among isolates of *Plesiomonas shigelloides* from fish, human clinical sources and fresh water, determined by RAPD typing. *Food Biotechnol.* 20:1–12.

Gu, W., Levin, R. 2006a. Factors affecting quantitative PCR assay of *Plesiomonas shigelloides*. *Food Biotechnol.* 20:219–230.

Gu, W., Levin, R. 2006b. Quantitative detection of *Plesiomonas shigelloides* in clam and oyster tissue by PCR. *Int. J. Food Microbiol.* 111:81–86.

Gu, W., Levin, R. 2007a. Quantification of viable *Plesiomonas shigelloides* in a mixture of viable and dead cells using ethidium bromide monoazide and conventional PCR. *Food Biotechnol.* 21:145–159.

Gu, W., Levin, R. 2007b. Quantification *of Plesiomonas shigelloides* in pure culture and clams by competitive PCR. *Food Biotechnol.* 21:17–31.

Gu, W., Levin, R.E. 2008. Innovative methods for removal of PCR inhibitors for quantitative detection of *Plesiomonas shigelloides* in oysters by real-time PCR. *Food Biotechnol.* 22:98–113.

Habs, H., Schubert, R. 1962. Über die biochemischen merkmald and die taxonomische stellung von *Pseudomonas shigelloides* (Bader). *Zentralbl. Bacteriol.* 1 [original]. 186:316–327.

Herrington, D., Tzipori, S., Robins-Browne, R., Tall, B., Levine, M. 1987. In vitro and in vivo pathogenicity of *Plesiomonas shigelloides. Infect. Immun.* 55:979–985.

Holmberg, S., Wachsmuth, L., Hickman-Brenner, P., Blake, P., Farmer, J., III. 1986. *Plesiomonas* enteric infections in the United States. *Ann. Intern. Med.* 105:690–694.

Hori, M., Hayashi, K. 1966. Food poisoning caused by *Aeromonas shigelloides* with an antigen common to *Shigella dysenteriae. J. Jpn. Assoc. Infect. Dis.* 39:433–441.

Huber, I., Spanggard, B., Appel, K., Rossen, L., Nielson, T., Gram, L. 2004. Phylogenetic analysis and in situ identification of the intestinal microbial community of rainbow trout (*Oncorhynchus mykiss,* Walbaum). *J. Appl. Microbiol.* 96:117–132.

Huq, A., Akhtar, A., Chowdbury, M., Sack, D. 1991. Optimum growth temperature for the isolation of *Plesiomonas shigelloides* using various selective and differential agars. *Can. J. Microbiol.* 37:800–802.

Jagger, T. 2000. *Plesiomonas shigelloides*—A veterinary persective. *Infect. Dis. Rev.* 2:199–210.

Janda, J., Abbott, S. 1993. Expression of hemolytic activity of *Plesiomonas shigelloides. J. Clin. Microbiol.* 31:1206–1208.

Janda, J., Abbott, S. 1999. Unusual food-borne pathogens. *Listeria monocytogenes, Aeromonas, Plesiomonas,* and *Edwardsiella* species. *Clin. Lab. Med.* 19:553–582.

Jeppesen, C. 1995. Media for *Aeromonas* spp., *Plesiomonas shigelloides* and *Pseudomonas* spp. from food and environment. *Int. J. Food Microbiol.* 26:25–41.

Kelly, M., Kain, K. 1991. Biochemical characteristics and plasmids of clinical and environmental *Plesiomonas shigelloides. Experentia* 47:439–441.

Khardori, N., Fainstein, V. 1988. *Aeromonas* and *Plesiomonas* as etiological agents. *Ann. Rev. Microbiol.* 42:395–419.

Kirov, S. 1997. *Aeromonas* and *Plesiomonas* species. In: *Food Microbiology Fundamentals and Frontiers.* M.P. Doyle, L.R. Beuchat, T.J. Montville, Eds., ASM Press: Washington, DC, pp. 265–286.

Koburger, J. 1989. *Plesiomonas shigelloides* In: *Foodborne Bacterial Pathogens.* M.P. Doyle, Ed., Marcel Dekker: New York, pp. 311–325.

Loh, J., Peng, E., Yap, E. 2002. Rapid and specific detection of *Plesiomonas shigelloides* directly from stool by lightCycler PCR. In: *Rapid Cycle Real-Time PCR Methods and Applications: Microbiology and Food Analysis.* U. Reischl, C. Wittwer, F. Cockeriw, Eds., Springer Verlag: New York, pp. 161–169.

MacDonnel, M., Colwell, R. 1985. Phylogeny of the *Vibrionaceae* and recommendations for two new genera, *Listonella* and *Shewanella*. *Syst. Appl. Microbiol.* 6:171–182.

Martinez-Mucia, A., Nenlock, S., Collins, M. 1992. Phylogenetic interrelationships of members of the genera *Aeromonas* and *Plesiomonas* as determined by 16S ribosomal DNA sequencing: Lack of congruence with results of DNA-DNA hybridization. *Int. J. Syst. Bacteriol.* 42:412–421.

Mathews, B., Douglas, H., Guiney, D. 1988. Production of a heat stable enerotoxin by *Plesiomonas shigelloides*. *Microb. Pathogen.* 5:207–213.

Medema, G., Schels, C. 1993. Occurrence of *Plesiomonas shigelloides* in surface water: Relationship with faecal pollution and trophic state. *Zentralbl. Hyg. Umweldtmed.* 194:398–404.

Miller, M., Koburger, J. 1985. *Plesiomonas shigelloides:* An opportunistic food and water borne pathogen. *J. Food Prot.* 48:449–457.

Miller, M., Koburger, J. 1986a. Tolerance of *Plesiomonas shigelloides* to pH, sodium chloride, and temperature. *J. Food Prot.* 49:877–879.

Miller, M., Koburger, J. 1986b. Evaluation of inositol brilliant green bile salts and *Plesiomonas* agars for recovery of *Plesiomonas shigelloides* from aquatic samples in a seasonal survey of the Suwane river estuary. *J. Food. Prot.* 49:274–278.

Monteil, H., Harf-Monteil, H. 1997. *Plesiomonas shigelloides:* Une bacterie exotique. *La Lettre de l'infectiologue de la Microbiologie à la Clinique.* XII:255–262.

Nogva, H., Drømtorp, S., Nissen, H., Rudi, K. 2003. Ethidium monoazide for DNA based differentiation of viable and dead bacteria by 5'-nuclease PCR. *BioTechniques* 34:804–813.

Oxley, A., Shipton, W., Owens, L., McKay, D. 2002. Bacterial flora from the gut of wild and cultured banana prawn, *Prenaeus merginsis*. *J. Appl. Microbiol.* 93:214–223.

Ratula, W., Sarubi, F., Jr., Finch, C., McCormick, J., Steinkraus, G. 1982. Oyster-associated outbreak of diarrhoeal disease possibly caused by *Plesiomonas shigelloides*. *Lancet.* 1:739.

Rouf, M., Rigney, M. 1971. Growth temperatures and growth characteristics of *Aeromonas*. *Appl. Microbiol.* 22:503–506.

Sayeed, S., Sack, D., Qadri, F. 1992. Protection from *Shigella sonnei* infection by immunization of rabbits with *Plesiomonas shigelloides* (VC01). *J. Med. Microbiol.* 37:382–384.

Schubert, R. 1977. Ueber den Nachweis von *Plesiomonas shigelloides* Habs und Schubert, 1962, und ein Lectivemedium, den Inositol-Brillantgrün-Gallesalz-Agar. *E. Rodenwaldt-Arch.* 4:97–103.

Schubert, R. 1984. Genus IV. *Plesiomonas*. In: *Bergey's Manual of Systematic Bacteriology*. Vol I. N. Kreig, I. Holt, Eds., Williams and Wilkins: Baltimore, pp. 548–550.

Schubert, R., Holz-Bremer, A. 1999. Cell adhesion of *Plesiomonas shigelloides*. *Z. Hyg. Umweltmed.* 202:383–388. *Med. Sci. Biol.* 31:135–142.

Shimada, T., Sakasaki, R. 1978. On the serology of *Plesiomonas shigelloides*. *Jpn. J. Med. Sci. Biol.* 31:135–142.

Stock, I. 2004. *Plesiomonas shigelloides:* An emerging pathogen with unusual properties. *Rev. Med. Microbiol.* 15:129–139.

Tsukamoto, T., Kinoshita, Y., Shimada, T., Sakazaki, R. 1978. Two epidemics of diarrhoeal disease possibly caused by *Plesiomonas shigelloides*. *J. Hg. Camb.* 80:275–280.

Ueda, S., Yamazaki, S., Hori, M. 1963. Isolation of a paracolon C27 and halophilic organisms from an outbreak of food poisoning. *Jpn. J. Publ. Hlth.* 10:67–70.

Von Graevenitz, A., Bucher, C. 1983. Evaluation of differential and selective media for isolation of *Aeromonas* and *Plesiomonas* spp. from human feces. *J. Clin. Microbiol.* 17:16–21.

CHAPTER 10

Campylobacter jejuni

I. CHARACTERISTICS OF THE ORGANISM

Campylobacter jejuni is considered to be the leading cause of enteric illness in the United States and other industrialized nations (Stern and Kazmi, 1989; Wesley et al., 2000; Thomas et al., 1999) causing mild to severe symptoms including bloody diarrhea (Blaser et al., 1979). Occasionally invasive infections can result in reactive arthritis (Melby, Kvien, and Glennås, 1996), meningitis, pneumonia, miscarriage, and a severe form of Guillain–Barré syndrome involving neuromuscular paralysis of the extremities (Blaser et al., 1986). *C. jejuni* is carried in the intestine of many wild and domestic animals, particularly avian species including poultry, where the intestine is colonized and results in healthy animals becoming carriers. This ubiquitous distribution of *Campylobacter* spp. in animals and raw meat products and the contact of animal wastes with surface waters makes such vehicles primary risk factors in contracting campylobacteriosis. The organism can survive at refrigerator temperatures for several weeks under moist microaerobic conditions but only a few days at room temperature (Blaser et al. 1980; Castillo and Escartin, 1994; Clark and Bueschkens, 1986). The organism is a cytochrome oxidase-positive, microaerophilic, curved gram-negative rod (1.5–5 µ) exhibiting corkscrew motility. *C. jejuni* is considered closely related to *C. coli*, the former being distinguished by the utilization of hippuric acid, and the latter distinguished on the basis of propionic acid utilization. The infective dose for humans has been found to be as low as 2–3 cells/ml (Robinson, 1981).

There are presently 14 recognized species of *Campylobacter* (Snelling et al., 2005a). *C. jejuni* is the species most frequently associated with human campylobacteriosis. This review concerns itself primarily with *C. jejuni*. The reader, however, is reminded that many studies on *C. jejuni* also include parallel studies and comparative data obtained with *C. coli*, *C. lari*, and *C. upsaliensis*, which are the four human pathogenic species of *Campylobacter* and are often referred to as the thermophilic *Campylobacter* species, being able to grow at 43°C. The term "thermotolerant" is with reference to those *Campylobacter* species able to grow at 41 to 43°C. In this regard, the term "thermotolerant" is perhaps more appropriate in that these organisms do not exhibit true thermophily (growth at 55°C or above). The D value at 50°C for *C. jejuni* strains in skimmed milk has been found to be from 1.3 to 4.5 min (Christopher, Smith, and Vanderzant, 1982a). In the same study, inoculation of a heat-tolerant strain of *C. jejuni* into roast beef at a level of 5.9×10^6/g resulted in no survivors by the time the internal temperature had risen to 55°C.

A. Phenotypic Characteristics of Campylobacters

Identification of *Campylobacter* species is based on biochemical tests, antibiotic resistance patterns, and growth temperatures. Unique characteristics of *Campylobacter jejuni* include anaerobic/microaerophilic growth conditions, optimum growth at 42–43°C, and the inability to grow below 30°C, which is associated with the absence of cold shock protein genes that play a role in low-temperature adaptation with many bacteria (Hazeleger et al., 1998). Additional major phenotypic characteristics that assist in species identification of *C. jejuni* include nitrate reduction to nitrite, hippurate hydrolysis, absence of urease, nalidixic acid susceptibility, cephalothin resistance, and the inability to utilize carbohydrates (Table 10.1). Most isolates of *C. jejuni* produce catalase; however, atypical strains are known that are catalase-negative, so that it is not considered a common feature of all isolates of the species (Hernandez et al., 1991). Although most isolates are H_2S^- and resistant to cephalothin (30 µg/disk), atypical isolates *of C. jejuni* from patients with diarrhea have been described that are H_2S^+, sensitive to cephalothin, and grow optimally at 37°C (Tee et al., 1987). Hébert et al. (1982) proposed a biotyping scheme based on the observation that 47% of *C. jejuni* isolates do not grow on charcoal–yeast extract (CYE) agar, 46% hydrolyze DNA, and 81% hydrolyze sodium hippurate (Table 10.2).

II. ECOLOGICAL DISTRIBUTION OF CAMPYLOBACTERS

A. Environmental Factors

Human infectious campylobacters have been found to be environmentally ubiquitous, being readily isolated from numerous wild and domesticated mammals in addition to wild avian species, poultry, and freshwater and marine aquatic environments. Hudson et al. (1999) studied the seasonal variation (August versus February) of *Campylobacter* types in New Zealand from human cases, veterinary cases, raw poultry, milk, and untreated water. Differences were noted between August 1996 and February 1997 serotypes, with the most frequent serotype isolated in February being completely absent in August. In contrast to the serotyping data, one pulsed field gel electrophoresis (PFGE) restriction profile type was dominant in samples from both February and August. Another group that was absent in August dominated the February isolates.

In an attempt to explain the annual increase in the incidence of campylobacteriosis in England and Wales that begins in May and reaches a maximum in June,

Table 10.1   Phenotypic Characteristics of C. jejuni

Gram-negative	Nitrate reduced to nitrite (+)
Growth at 42°C	Nitrite reduced (−)
Microaerophilic	DNAse production (+)
Catalase production (+)	Cephalothin resistant
Urease production (−)	Nalidixic acid sensitive
Hippurate utilization (+)	Cytochrome oxidase positive
Sensitive to nalidixic acid	No growth below 30°C
Carbohydrates not utilized	No growth with 3.5% NaCl
Alk. phosphatase production (+)	Reduction of triphenyltetrazolium chloride
Citrate utilization (+)	H_2S production (−)
Succinate utilization (+)	Indoxyl acetate utilization (+)

Table 10.2   Biotyping Scheme for C. jejuni[a]

	Phenotypic Characteristic for Respective Biotype							
Test	1	2	3	4	5	6	7	8
Growth on CYE agar	+	−	+	−	+	−	+	−
DNA hydrolysis	+	+	−	−	+	+	−	−
Hippurate hydrolysis	+	+	+	+	−	−	−	−

[a] Adapted from Hébert et al. (1982).

Nichols (2005) presented the hypothesis that the seasonal increase is associated with flies. The seasonal increase in fly population during the warm summer months is thought to result from rainy weather and elevated temperatures that lead to rapid development of flies from eggs in a few days. The resulting contact of flies with human and animal feces is presumably responsible for observed seasonal outbreaks prevalent during the warm summer months. Most organically (76%) and conventionally (74%) raised poultry have been found to be contaminated with campylobacters (Cui et al., 2005). Patrick et al. (2004) found that as ambient temperature increases in Denmark, there is a corresponding increase in the incidence of human campylobacteriosis and the percentage of infected broiler flocks at slaughter. The largest increase in incidence of campylobacteriosis occurred between 13 and 20°C.

1. Aquatic Sources of Campylobacters

Thermophilic campylobacters have been historically recognized as urease-negative. Bolton, Holt, and Hutchinson (1985) were the first to report the isolation of urease-positive thermophilic campylobacters (UPTC) from Europe. Mégraud et al. (1988) subsequently established that such isolates are variants of C. lari by DNA hybridization dot-blot assay and biochemical methods. Matsuda et al. (1996) reported on the first isolation of two urease-positive thermophilic campylobacter isolates from the water of different rivers in Japan. With the applicaton of PFGE analysis using the restriction nucleases ApaI, salI, and SmaI, both isolates were found to be identical clones genotypically distinct from the previously reported UPTC isolates from Europe. Positive urease production in addition to positive arylsulphatase production metabolically distinguished these isolates from strains of *C. lari*, *C. jejuni*, *C. coli*, and *C. fetus*. Three strains of urease-positive thermophilic campylobacters were isolated from seagulls by Kaneko et al. (1999). PFGE analysis involving genomic DNA digestion with the restriction nucleases *Apa*I and *Sma*I yielded identical profiles. The isolates were phenotypically identical to a reference strain of *C. lari* except for urease production.

Hernández et al. (1996) subjected 32 strains of thermophilic campylobacters isolated from marine recreational waters to biotyping, ribotyping, and RAPD profiling. A 10-mer random primer (Table 10.6) was used for RAPD analysis. The majority of seawater isolates (29, 90%) were *C. coli*, and three strains (9%) were *C. jejuni*. Strains within each species were found to be genomically diverse.

Schönberg-Norio et al. (2004) examined the risk factors for *Campylobacter* infections in Finland. A total of 100 patients infected with *C. jejuni* or *C. coli* were involved in the survey. All cases were sporadic and not associated with known outbreaks. Three major risk factors were identified: (1) tasting or eating raw or undercooked meat, (2) drinking water from a dug well, and (3) swimming in water from natural sources. Private untreated water supplies were found to present a significant risk factor.

Michaud, Menard, and Arbeitt (2004) reported on an extensive epidemiological study in Quebec, Canada in 2000–2001 involving 158 cases of campylobacteriosis. Exposure to poultry accounted for less than half of the cases. Drinking untreated tap water at home from a deep well was the only risk factor identified for 53% of the cases. These results are consistent with the hypothesis that the waterborne route of infection may be the common factor linking infections in humans, poultry, other domestic animals, and wild birds.

C. jejuni has been found to remain viable when internalized by the waterborne protozoa *Tetrahymena pyreformis* and *Acanthamoeba castelani* for up to 36 hr longer than when they were in a purely planktonic state (Snelling et al., 2005b). Internalized campylobacters could be detected for up to 8 days at 25°C and were significantly more resistant to chemical disinfection than planktonic cells. The authors concluded that the internalization of campylobacters by protozoa can delay the decline of viability in broiler drinking water and increases the potential for *Campylobacter* colonization of poultry.

Kahn and Edge (2007) developed a novel triplex PCR assay for the detection and differentiation of thermophilic species of *Campylobacter* in water samples using a 16S-23s rDNA intrnal transcribed spacer region. The primers ICJ-UP/ICJ-DN (Table 10.6) amplified a 349-bp sequence from *C. jejuni*. Primers ICL-UP/ICL-DN (Table 10.6) amplified a 279-bp sequence from *C. lari*. Primers ICC-UP/ICC-DN (Table 10.6) amplified a 72-bp sequence from *C. coli*.

2. Wildlife as a Potential Reservoir for Infection by C. jejuni

Wildlife have long been considered an infectious reservoir for campylobacters because of their close association with surface waters. Pacha, Clark, and Williams

(1985) reported that among 189 muskrat fecal samples 90 (47.5%) were positive for *C. jejuni*. These animals characteristically inhabit areas adjacent to streams and ponds and deposit fecal material near or in such aqueous environments, greatly facilitating the dissemination of the organism.

Among five commonly encountered intestinal organisms, *C. jejuni* was the most frequently isolated from pigeons over a 12-month period in Barcelona, Spain. Among 105 cloacal samples, 28 (26%) were positive for *C. jejuni* (Casanovas et al., 1995).

Broman et al. (2000) obtained fecal samples from 140 albatross, 100 macaroni penguins, and 206 fur seals from Bird Island, a subantarctic region of south Georgia, which were plated onto *Campylobacter*-selective agar supplemented with 5% citrated horse blood, 10 mg/L of vancomycin, 500 IU/L of polymyxin B, and 5 mg/L of trimethoprim. Three isolates of *C. jejuni* were obtained solely from three penguins and exhibited low genetic diversity. PCR amplicons obtained with primers FLA4F and FLA1728R (Table 10.6) that amplify a 1702-bp sequence of the *flaA* gene and primers C16SF and B37 (Table 10.6) that amplify a 1464-bp sequence of the 16S rDNA gene were used for cloning and subsequent sequencing studies. Although the *flaA* gene is recognized to exhibit high intraspecies variability in *Campylobacter* (Meinersmann et al., 1997) all three isolates yielded identical nucleotide sequences for this gene with 99.9% homology to a 1993–1994 retail chicken isolate from Seattle, Washington. PFGE of whole-cell DNA was performed with the restriction nucleases *Sma*I and *Kpn*I and yielded identical restriction patterns for all three isolates, strongly indicating that all three strains were identical and were from a single clone.

Peterson et al. (2001) investigated the potential importance of wildlife as a source of infection in commercial poultry flocks and in humans by comparing the serotype distributions, PCR-restriction fragment length polymorphism (PCR-RFLP) profiles of the *fla* gene (using the restriction endonucleases *Dde*I and *Ala*I), and PFGE profiles (using the restriction endonucleases *Sma*I, *Klpn*I, and *Bam*HI) of *C. jejuni* isolates from different sources. Results indicated a relatively low number of wildlife strains with an inferred clonal relationship to human and poultry strains, suggesting that the importance of wildlife as a reservoir of infections is limited.

Waldenström et al. (2002) examined a total of 1794 migrating birds trapped at a coastal site in southern Sweden for detection of *Campylobacter* species. Isolates were identified to species by PCR directed at the 23S rRNA gene used to identify thermophilic *Campylobacter* isolates. PCR primers used were THERM1 and THERM4 (Table 10.6) of Fermér and Engvall (1999). Subsequent endonuclease digestion of the PCR products with *Alu*I was used to generate banding patterns for species identification of *C. jejuni*, *C. coli*, and *C. lari*. A second PCR method used for identification of *Campylobacter* species involved application of a multiplex PCR specific for *C. jejuni* and *C. coli* developed by Vandamme et al. (1997). *C. jejuni* was found in 5.0% of the birds, *C. lari* in 5.6%, and *C. coli* in 0.9%. An additional 10.7% of the tested birds were infected with hippurinate hydrolysis-negative *Campylobacter* species that were not identified to the species level. The prevalence of campylobacters differed significantly between ecological guilds of birds. Shore-line-foraging birds feeding on invertebrates and opportunistic feeders were most commonly infected (76.8% and 50%, respectively). High prevalence was also found in ground-foraging invertebrate feeders (11.0%), ground-foraging insectovores (20.3%), and plant-eating species (18.8%). Almost no campylobacters were found among ground-foraging granivores, arboreal insectovores, and aerial insectovores, or reed- and herbaceous plant-foraging insectovores.

C. fetus is known to be pathogenic for humans and animals (Tu et al. 2005). Two subspecies have been identified, *C. fetus* subsp. *fetus* and *C. fetus* subsp. *venerealis*, and there are two serotypes, A and B. With the use of RAPD involving 10 random primers, PFGE with *Sma*I and *Sal*I, DNA–DNA hybridization, and 16S rRNA sequence analysis, Tu et al. (2005) examined the genetic divergence of strains of *C. fetus* of mammalian and reptilian origin. Each of the RAPD primers was able to distinguish the mammalian from the reptilian strains. DNA–DNA hybridization revealed substantial genomic-homology differences between strains of mammalian and reptilian origin. The authors concluded that there is more extensive genetic divergence between reptile and mammal strains of *C. fetus* subsp. *fetus* strains than between the two subspecies and between the type A and type B serotype strains.

3. Campylobacters Associated with Farm and Domesticated Animals

C. jejuni has been found to be commonly present in the feces of 2- to 3-week-old unweaned calves and from sheep (Gill and Harris, 1982). Small numbers of the organism were recovered from equipment during the processing of unweaned calves but not after routine cleaning. *C. jejuni* was found in 82.9% of 94 chicken wing packages analyzed on the day of arrival at supermarkets and in 15.5% of 45 packages obtained from the same supermarket shelves a few days later (Kinde, Genigeorgis, and Pappaioanou, 1983). At day zero, the mean *C. jejuni* count per wing was 1×10^3, and after day six of refrigerated storage it was 1×10^2 CFU/wing. Piglets and mothers have been found to often harbor *Campylobacter* isolates with identical genetic subtyping profiles, suggesting that piglets become infected via their mothers (Weijtens et al., 1997). Weijtens et al. (1997) monitored the *Campylobacter* infection of 10 sows and their piglets. The sows were found to be infected with *Campylobacter* before litter. Half the piglets became infected with *Campylobacter* within 7 days after birth and 85% after 4 weeks. Alter et al. (2005) failed to detect campylobacters in the feces of piglets on the day of birth. However, within 1 week following birth, the incidence of *C. coli* was 32.8%. After transfer to the nursery on the fourth week, the incidence rose to 56.6%, which eventually reached 79.1%. No *Campylobacter* species other than *C. coli* were isolated in the study, which confirmed earlier reports that swine are mainly infected with *C. coli*.

It has been found possible to rear swine throughout many successive generations under highly controlled conditions that are free of campylobacters. Weijtens, Urlings, and Van der Plas (2000) described an experimental government farm in the Netherlands for the rearing of pathogen-free swine. The farm buildings are built air-tight with positive air pressure inside and filtration of all incoming air. Feed is sterilized by gamma irradiation. Staff shower and don new sterile clothing at the entrance to each building. The farm was started in 1980 with piglets obtained through caesarian section. Subsequent generations were reared and bred naturally, within the controlled environment of the farm. Such an experimental facility clearly indicates that it is possible to rear *Campylobacter*-free animals in a nonsterile state. However, the overall cost would probably be prohibitive for most swine-rearing farms. The study did show, however, that restocking a normal pig farm with *Campylobacter*-free swine and maintaining a vigorous sanitation regimen can reduce the number of infected animals from 98% to 22%.

Wesley et al. (2000) detected *C. jejuni* and *C. coli* in the feces of healthy dairy cows using the multiplex PCR assay of Harmon, Ransom, and Wesley (1997). Among 2085 cows examined from 31 dairy farms, 786 (37.7%) were positive for *C. jejuni*, whereas only 37 (1.8%) were positive for *C. coli*. The accessibility of feed to birds was positively correlated with herd prevalence of *C. jejuni*. Varma, Pushpa, and Subramanyam (2005) found that among 100 raw milk samples from lactating cows, 24 contained viable *C. jejuni*.

Isolates of *C. jejuni* colonizing cattle farms originating from distances less than 1.0 km from each other have been found to be genetically more similar than isolates originating from greater distances (French et al., 2005).

Cabrita ct al. (1992) determined the incidence of *C. jejuni* and *C. coli* in wild and farm animals. The organisms were isolated from 59 of 98 chickens (60.2%), 65 of 110 swine (59.1%), 31 of 54 black rats (57.4%), 61 of 134 sparrows (45.5%), 21 of 52 ducks (40.5%), 32 of 164 cows (19.5%), and 27 of 176 sheep (15.3%). Resistance to ampicillin was found in 5.5% of the strains, 5.5% were resistant to tetracycline, 12.6% to erythromycin, and 23.5% to streptomycin. Resistance to erythromycin (26.2%) and to streptomycin (58.4%) was notably high in isolates from swine. Tetracycline resistance was encoded by a 33 or a 41 mDa plasmid and was transferred by conjugation.

C. jejuni has been found to colonize primarily the lower gastrointestinal (GI) tract of chicks (Beery, Hugdahl, and Doyle, 1988). When 8-day-old chicks were orally administered *C. jejuni*, the principal sites of localization after 7 days were the ceca, large intestine, and cloacae where densely packed cells were observed in mucus within crypts. *C. jejuni* cells were found to freely pervade the lumina of crypts without attachment to crypt microvilli. Chemoattraction of *C. jejuni* to the fucose content of mucus presumably attracts the organism to mucus, in which it is highly motile. Within the crypts, mucin is thought to be used as a substrate that presumably promotes the establishment of the organism in the intestine.

Refrigerated storage at 4°C of seeded poultry skin for 3 days was found to result in a reduction in CFU of 0.31 log/g (Bhaduri and Cotrell, 2004). Frozen storage alone for 14 days at –20°C resulted in a reduction in CFU of 1.38 log/g. When skin was refrigerated for 3 days following seeding and then frozen for 14 days the reduction in CFU was 2.55 log/g. Counts were consistently lower when plating was performed using modified cefoperazone charcoal deoxycholate agar compared to tryptic soy agar containing 5% sheep blood. These results indicate that the extent of observed loss in CFU as a result of frozen storage is dependent on the extent of refrigerated storage damage and the plating medium.

Adark et al. (2005) reported on an extensive epidemiological study of 1,724,315 patients in England and Wales afflicted with foodborne infections from 1996 to

2000. The most important cause of foodborne disease was contaminated poultry involving species of *Campylobacter*.

Poultry on a broiler farm in southern England were found by Pearson et al. (1993) to be colonized by a single serotype of *C. jejuni*, previously implicated in an outbreak in 1984. The serotype persisted on the farm for at least 18 months after the outbreak. The predominant source of *C. jejuni* on the farm was found to be the water supply, which therefore excluded vertical transmission. Clarification of the water supply, shed cleaning and disinfection, and withdrawal of furazolidine from feed reduced the colonization of poultry by *C. jejuni* from 81% to 7% and resulted in a 1000 to 10,000-fold reduction in campylobacters recoverable from carcasses. Two months after the end of the intervention program 84% colonization occurred.

A combination of *Citrobacter diversus*, *Klebsiella pneumonia*, and *Escherichia coli* in conjunction with 2.5% mannose in drinking water was found by Schoeni and Wong (1994) to be highly effective for competitive exclusion of *C. jejuni* fed to young poultry. *C. jejuni* was not detected in the ceca of chicks receiving such competitive exclusion treatment.

The numbers of campylobacters in the ceca of poultry was found by Wallace et al. (1997) to vary considerably with season, with the highest numbers occurring in the months of June and July, reaching over 10^{12}/g of feces in the ceca. The lowest numbers (1×10^7)/g in the ceca occurred in November. Numbers were frequently several logs higher in the ceca than in the intestine.

Certain genotypes have been found capable of dominating poultry and poultry environments. Peterson and Wedderkopp (2001) studied the incidence of *C. jejuni* among 12 broiler houses located on ten farms in Denmark in 1998. PFGE typing of isolates revealed that the majority of houses carried identical strains in two or more broiler flocks. Seven houses carried persistent clones that spanned an interval of at least four broiler flock rotations involving at least one half year. One dominant RFLP type was represented by 44% of the 99 isolates examined, indicating that this genotype was well adapted for colonization of broiler flocks and for broiler house survival.

Transcriptional profiling of *C. jejuni* colonization of the chick cecum identified 59 genes that were differentially expressed in vivo compared with the genes in vitro (Woodall et al., 2005). Their data suggested that *C. jejuni* regulates electron transport and controls metabolic pathways to alter its physiological state during establishment in the chick cecum. Microarray data indicted that within chick ceca limited oxygen conditions result in an increase in the transcription of the C4-dicarboxylate transporter genes *dcuA*, *dcuB*, and *dctA*. Limited oxygen was found to have a greater effect on the increased expression levels of these genes than the presence of the C4-dicarboxylates fumarate and succinate.

In addition to gastroenteritis in humans caused by *C. jejuni* and to a lesser extent by *C. coli*, *C. upsaliensis* is also considered a human pathogen with household pets such as dogs and cats, suspected of transmitting the organism to humans. Steinhauserova, Fojtikova, and Klimes (2000) studied the incidence of *C. upsaliensis* among 225 dogs and cats suffering from intense diarrhea. A total of 16 of these animals (7%) were found to harbor the organism. A control group of 126 dogs and cats without clinical symptoms yielded 16 (12.6%) *C. jejuni/C. coli* strains and 4

(3.1%) *C. upsaliensis* strains. All of the *C. upsaliensis* isolates from both diarrheagenic and control animals were from dogs.

Devane et al. (2000) applied a double enrichment multiplex-PCR method for the detection of *C. jejuni* and *C. coli* in fecal, meat, and riverwater samples involving a total of 1450 samples from a defined geographic area in New Zealand. Serotyping and *Sma*I-PFGE revealed a high level of diversity among the isolates. *C. jejuni* and *C. coli* subtypes indistinguishable from those obtained from human cases of campylobacteriosis were detected from most of the environmental matrices.

The highest incidence of campylobacters on the interior packaging material of raw meat was found to be from game fowl (3.6%), followed by raw chicken (3.0%), lamb (1.6%), and beef (0.1%; Burgess et al., 2005). *C. coli* isolates from the external packaging were more multiresistant to antimicrobial drugs than were *C. jejuni* isolates. The authors concluded that the external packaging material for raw meats is a vehicle for potential cross-contamination of campylobacters in retail premises and consumers' homes.

B. Vertical versus Horizontal Transmission among Poultry

Pearson et al. (1996) reported on a 5-year broiler chicken study that yielded 3304 of 12,233 (27%) chickens positive for *C. jejuni*. The study encompassed 32 consecutively reared flocks. A low level of transmission was found to occur between flocks raised in the same sheds. They found that 85% of the *C. jejuni* isolates were of 10 serotype complexes but 58% were of 3 serotype complexes, indicating a high degree of strain similarity throughout the entire study. Data suggested that an intermittent common external *Campylobacter* source was involved. The incidence of *C. jejuni* isolation in broilers from two hatcheries was 17.6% and 42.9%, indicating that the presence of *C. jejuni* in the broiler flocks was attributable to the hatcheries and not to horizontal transmission on the rearing farm. The authors hypothesized that introduction of *C. jejuni* involved vertical transmission from broiler breed stock to chicks with amplification in the broiler sheds.

III. VIRULENCE FACTORS

A. Toxins

Some confusion presently exists regarding the number and specificity of toxic factors produced by campylobacters that are considered to be derived from a wide variety of strains, bioassay systems, and cultural conditions employed (Johnson and Lior, 1988). In addition, some toxin studies have been performed with crude culture filtrates and others with purified toxins. Friis et al. (2005) have reviewed in detail the use of in vitro cell culture methods for investigating *Campylobacter* invasion mechanisms, particularly adhesion to mammalian cells and cell penetration.

C. jejuni enterocolitis in humans is characterized by inflammatory infiltration of neutrophils and mononuclear cells, villus degeneration and atrophy, loss of mucus, crypt abscess, and ulceration of the mucosal epithelium (Everest et al., 1993). The infection is therefore histologically similar to ulcerative colitis associated with chronic inflammatory cells (Skirrow, 1986). Inflammation in ulcerative colitis is mediated in part by the release from leukocytes of leukotriene B_4 (LTB_4) and prostaglandin E_2 (PGE_2). In addition, PGE_2 decreases active sodium and chloride absorption and increases fluid secretion in the small intestine and colon by activation of adenylate cyclase (Everest et al., 1993). Bloody diarrhea associated with *Campylobacter* enteritis (Walker et al., 1986) is considered to reflect invasiveness (Klipstein and Engert, 1985).

The first report of the production of an extracellular mammalian cytotoxin produced by *C. jejuni* was by Yeen, Puthucheary, and Pang (1983). Culture filtrates were found to produce cytolethal effects on the human cell lines HeLa, MRC-5, and HEP-2, which included cell rounding, loss of adhesiveness, and cell death within 24–48 hr (Table 10.3). These morphological changes were observed with 8/11 strains of *C. jejuni* isolated from patients with either acute gastroenteritis or septicemia. The toxic factor did not retain its activity after heating at 100°C for 30 min. It is interesting that the culture filtrate did not exhibit toxic activity on monkey Vero, MK2, PMK, or mouse L929 cells.

Guerrant et al. (1987) reported on a unique heat-labile cytotoxin from culture supernatants of *C. jejuni* strains different from those previously described (Table 10.3). It was lethal to Chinese hamster ovary (CHO) and HeLa cells, had no effect on Vero cells, failed to elicit fluid accumulation in rat ileal loops and in the suckling mouse assay, and was not neutralized by anti-Shiga-like and cholera toxins.

Isolates of *C. jejuni* have been found to produce a heat-labile enterotoxin (CJT) that was observed to raise intracellular cyclic AMP levels, cause cytotonic changes in CHO cells, and induce fluid accumulation in ligated rat ileal loops (Ruiz-Palacios et al., 1983). Cell culture supernatants of *C. jejuni* have been found to induce a net sodium secretory flux and an impaired glucose transport in perfused jejunal segments of adult rats in vivo (Fernández et al., 1983). The secretory activity of the toxin is neutralized by antiserum to cholera toxin (CT; Ruiz-Palacios et al., 1983).

McCardell, Madden, and Lee (1984) screened cell-free supernatants from strains of *C. jejuni* and *C. coli* for ELISA binding to cholera toxin (CT) antibody and obtained positive binding results indicating the presence of an extracellular factor immunologically similar to CT. The extracellular factor had a mol. wt. of 70 kDa, was destroyed by trypsin, and caused (a) rounding of Y-1 mouse adrenalin cells, (b) rabbit skin permeability, (c) fluid accumulation in rabbit ileal loops, and (d) elongation of CHO cells. Antibody to *Campylobacter* cytotonic toxin (CCT) eliminated the cytotoxic effects of cholera toxin (CT). Rounding of Y-1 mouse adrenalin cells was not eliminated by heating cell-free concentrates at 100°C for 10 min.

Two distinct toxin types from *C. jejuni* and *C. coli* were reported by Johnson and Lior (1984): a heat-labile cytotoxin (CJT) active on Vero cells and a heat-stabile cholera-like cytotoxin that elongated CHO cells but exhibited no activity in the suckling mouse assay or with Y-1 cells.

CJT was found by Klipstein and Engert (1985) to contain a subunit that was purified and found to be immunologically related to the B subunits of cholera toxin (CT)

Table 10.3   Factors Involved in the Development of Gastroenteritis by *C. jejuni*

Toxin	Characteristics	References
Enterotoxin (CJT)	Results in fluid accumulation in animal intestinal loops, rounding of Y1 cells and elongation of CHO cells. Neutralized by antitoxin to cholera toxin (CT) and *E. coli* heat-labile enterotoxin (LT). Stimulates cyclic AMP accumulation.	Ruiz-Palacios et al. (1983) Johnson and Lior (1986) Klipstein and Engert (1984a)
Cytotoxins		
1. Cytotoxin (CT)	68-kDa protein. Causes rounding and lysis of CHO and INT407 cells. Lethal to Vero, Hela, MRC-5, and Hep-2 cells. Heat labile, noninvasive with chicken embryos, lethal to fertile chicken eggs, and toxic to chicken embryo fibroblasts. Does not bind to GM_1. Binds to amino terminal of N-acetylneuraminic acid. Immunologically unrelated to cholera and *E.coli* enterotoxins.	Yeen, Puthucheary, and Pang (1983) Tenover et al. (1985) Mahajan and Rodgers (1990)
2. Vero-negative cytotoxin	Cytotoxic and lethal to Hela and CHO cells but not to Vero cells.	Guerrant et al. (1987)
3. Cytolethal distending toxin (CLDT)	Induces a slowly developing swelling of CHO cells but not of Y1 cells. Toxin is heat labile. Mediates release of interleukin −8 from INT407 cells.	Johnson and Lior (1988) Hickey et al. (2000) Picket et al. (1996) Lara-Tejero and Galán (2000, 2001)
4. Cytolethal distending toxin (CLDT)	Causes swelling and elongation of Hela, Hep-2, and secondary embryonic calf gut cells. Vero cells markedly less sensitive. Toxin is heat labile.	Schulze, Hänel, and Borrmann (1996)
5. Cytolethal distending toxin (TF)	Causes elongation of Vero and Hep-2 cells rounding of CHO cells with a high percentage of death in the latter two cell types. Toxin is resistant to 80°C.	Fragoso et al. (1996)
Enterocyte Binding Factors		
1. *Campylobacter* adhesion factor	Required for binding to fibronectin receptor.	Konkel et al. (1997, 1999) Monteville and Konkel (2002)
2. Flagella (flagellin)	Required for GI colonization.	Nuijten et al. (1990) Nachamkin, Yang, and Stern (1993b)
3. PEB1	An outer membrane protein that is a major cell-binding factor for adherence to HeLa cells.	Fauchèr et al. (1989) Pei and Blaser (1993)
4. Lipopolysaccharide (LPS)	Required for maximum adherence and invasion of human intestinal embryonic INT407 cells.	Fry et al. (2000)

and *Escherichia coli* heat-labile toxin (LT). CJT was found to share with CT and LT their property of attachment to the GM_1 mammalian ganglioside receptor that is known to be a function of the B subunits (Donta and Viner, 1975).

The amount of CCT produced in broth cultures has been found to vary from none to large amounts, depending on the isolate (Klipstein and Engert, 1984a). The production of moderate or large amounts of CCT resulted in preparations that caused fluid secretion in rat ileal loops. GM_1 ganglioside eliminated the cytotoxic response in CHO cell assays. Antiserum to *E. coli* heat-labile enterotoxin (LT) neutralized the CCT secretory effect in rat ileal loop assays. These findings indicate that CCT is immunologically related to both CT and LT and shares with CT, LT, and *Salmonella* enterotoxin the property of binding to GM_1 ganglioside tissue receptor (Klipstein and Engert, 1984b).

Nonvirulent strains of *C. jejuni* have been documented. All eight isolates from asymptomatic carriers were found by Klipstein and Engert (1985) to produce neither enterotoxin nor cytotoxin nor caused fluid accumulation in rat ligated ileal loops. All six strains from patients with secretory-type diarrhea produced enterotoxin. The broth filtrate of one strain also had a cytotoxic effect on both Vero and Hela cells. Broth filtrates of all six strains evoked fluid secretion in rat ileal loops. All six isolates from individuals with invasive-bloody diarrhea failed to produce enterotoxin but all six produced a cytotoxin. Antiserum to *E. coli* enterotoxin did not neutralize the cytotoxin produced. None of the broth filtrates from these strains invoked a fluid response in rat ileal loops.

Johnson and Lior (1986) assessed 185 strains of *C. jejuni* for cytotonic and cytolethal toxin production. Cytotonic toxicity was determined on the basis of elongation of CHO cells, the rounding of Y-1 cells, or morphological changes in Vero cells. Out of 185 strains, 27 (15%) were toxin negative, 100 (54%) were both cytotoxic and cytolethal, 30 (16%) were cytotoxic but not cytolethal, and 28 (15%) were cytolethal and not cytotonic. Cyclic AMP did not increase in CHO cells by more than twofold, compared to 30-fold increases by *E. coli* and *V. cholerae* enterotoxins. Partial neutralization of *C. jejuni* enterotoxins occurred with antitoxin to cholera and *E. coli* heat-labile enterotoxins.

Johnson and Lior (1988) described a heat-labile toxin produced by *C. jejuni* and several other *Campylobacter* species detectable in culture filtrates. The toxin was cytolethal to CHO, Vero, Hela, and Hep-2 cells and negative with Y-1 cells (Table 10.3). The only in vivo bioassay responsive to the toxin was the ligated rat ileal loop test. Strains positive with the ileal rat loop assay were negative in suckling mice, adult rabbit ileal loops, and rabbit skin. This toxin was described as a cytolethal distending toxin (CLDT) reflecting the progressive cell distension and eventual cytotoxicity observed with all sensitive tissue cells. Tissue culture assays required 96 hr of observation to detect elongation and cytolethal events with CHO cells. The CLDT response in rat ileal loops was uniformly associated with hemorrhagic reactions, mucosal inflammation, and limited fluid accumulation. The toxin is distinct from *Campylobacter* enterotoxins, cytotoxins described earlier, and cholera-like enterotoxin produced by some *Campylobacter* species. Nucleotide sequencing of genes in *C. jejuni* encoding the CLDT have revealed three genes, *cdtA, cdtB,* and *cdtC*, encoding proteins related to the CDT protein of *E. coli* (Picket et al. 1996). All three genes were required for toxic activity with Hela cells. Strains of *C. coli* were found to harbor the *cdtB* gene, but most produced CLDT at less than 1/10 the level of the majority of *C. jejuni* strains in Hela cell assays.

Partially purified *C. jejuni* enterotoxin was found to yield three bands with sodium dodecyl sulfide–polyacrylamide gel electrophoresis (SDS-PAGE) of 68, 59, and 43 kDa (Daikoku et al., 1990). This fraction enhanced the adenylate cyclase activity of Hela cell membranes. Affinity chromatography with anti-cholera toxin immunoglobulin (IgG) and ganglioside revealed that the eluent from an anti-cholera toxin IgG column, after binding and washing, exhibited a single band (68 kDa) with SDS-PAGE, whereas the eluent from a ganglioside column exibited two bands (68 and 54 kDa) with SDS-PAGE. These results suggest that the 68 kDa polypeptide should have an immunological relationship with cholera toxin, and the 68- and 54-kDa polypeptides might be responsible for the recognition of ganglioside receptors in enterocyte membranes.

Florin and Antillon (1992) studied enterotoxin and cytotoxin production by *C. jejuni* human clinical and poultry isolates in Costa Rica. The production of enterotoxin was tested primarily in Y-1 cells to avoid confusion with the effect of the CLDT that is active on CHO cells but not on Y-1 cells. An enterotoxic effect was defined as rounding of Y-1 cells and elongation of CHO cells without lethality. The effect of the cytotoxin was similar with Vero, MRC-5, and Hela cells. After 24 hr the cells were rounded, wrinkled, and shrunken, and after several more days underwent lethal degeneration. With CHO and Y-1 cells, the cytotoxin resulted in permanent rounding without eventual growth, presumably reflecting cell death. Both toxins were heat labile. Enterotoxin production occurred with a frequency of 47% among 44 strains from children with diarrhea and 34% from 35 poultry isolates. Cytotoxic effects were not observed among 26 strains from children with watery diarrhea. Among 18 isolates from children with bloody or inflammatory diarrhea, 2 isolates produced cytotoxic effects on tissue cells, as did 6 of the 36 poultry strains. The authors suggested that watery diarrhea results when only enterotoxin is produced, whereas inflammatory or bloody diarrhea is due to a combined effect of enterotoxin and cytotoxin. They also considered that the cytotoxin observed in their study was most probably identical with the one active on Vero cells as observed by Johnson and Lior (1984) and Klipstein and Engert (1985).

Fragoso et al. (1996) reported on a partially heat-labile extracellular toxin of *C. jejuni* active on Vero, Hep-2, and CHO cells and negative on Y-1 cells (Table 10.3). This toxin provoked elongation of Vero and Hep-2 cells and a rounding of CHO cells and a high percentage of death in the latter two cell types. Among 63 strains of *C. jejuni*, 14 (22.2%) were positive for this toxin, whereas among 41 strains of *C. coli*, 11 (26.8%) were positive.

CLDT from a strain of *C. jejuni* was found to cause a rapid and specific cell cycle arrest in Hela and Caco-2 cells (Whitehouse et al., 1998). CLDT caused Hela cells to arrest at the G2 or early M phase of cell division and to cause accumulation of the inactive tyrosine-phosphorylated form of CDC2. These results indicated that CLDT treatment of cells results in failure to activate CDC2, which leads to cell cycle arrest at G_2/M. The authors suggested that CLDT produced in proximity to intestinal epithelial cells may cause these rapidly proliferating cells to become blocked in G_2 so as to prevent crypt cells from surviving and maturing into functional villus epithelium, resulting in temporary villus erosion and loss of absorptive function.

The CLDT is encoded by three conserved genes: *CdtA, CdtB,* and *CdtC. CdtA* bears homology to the ricin B chain, *CdtB* is homologous to enzymes having phosphoesterase activity including nucleases, and *CdtC* is most similar in homology to *CdtA* (Hassane et al., 2001). *CdtB* is the active component whereas *CdtA* and *CdtC* are involved in delivering the *CdtB* molecule into mammalian cells (Ohara, Oswald, and Sugai, 2004). Recent studies have suggested that the CLDT functions as an intracellular DNase (Dlakié, 2000; Elwell and Dreyfus, 2000; Lara-Tejero and Galan, 2000).

Purified *CdtB* has been found to exhibit in vitro nuclease activity and is considered to exert its effect by damaging cellular DNA, thereby resulting in G_2/M cell division arrest (Lara-Tejero and Galán, 2000). In addition, when *CdtB* was microinjected into host cells, it was capable by itself of causing all of the toxic effects of the CDT holotoxin including G_2/M arrest and cytoplasmic distention (Lara-Tejero and Galán, 2000). Purified *CdtA, CdtB,* or *CdtC* do not result in toxic activity when applied externally to susceptible cells individually (Lara-Tejero and Galán, 2001). In contrast, when *CdtA, CdtB,* or *CdtC* are combined, they interact to form an active tripartite holotoxin that exhibits full cellular toxicity. *CdtA* has a domain with shared similarity with the B chain of ricin-related toxins. Lara-Tejero and Galán (2001) therefore proposed that CDT is a tripartite toxin composed of *CdtB* as the enzymatically active subunit of *CdtA* and *CdtC,* which are required for delivery of *CdtB*.

Hassane et al. (2001) found that expression of the *CdtB* subunit cloned into *Saccharomyces cerevisiae* under the control of the GAL1 promoter resulted in G_2/M cellular arrest and cell death, similar to that observed with mammalian cells. By 10 hr postinduction, *CdtB*-associated DNA degradation was observed by pulsed field gel elecrophoresis (PFGE), reflecting presumed DNase activity by the *CdtB* gene product. Cloning of the genes *CdtA* and *CdtC* into *S. cerevisiae* had no effect on the cells.

The presence of the *cadF, cdtA, cdtB,* and *cdtC* genes, in addition to the *iam* sequence (having no known gene product) was determined in 115 *C. jejuni* and 57 *C. coli* isolates from children with diarrhea and from chicken carcasses in Poland (Rozynek et al., 2005). The *cadF* gene was found to be present in nearly 100% of *Campylobacter* isolates, regardless of origin or species. In contrast, the *iam* region was found in 83.3% and 100% of *C. coli* from children and chickens, respectively, but in only 1.0% and 54.7%, respectively, of *C. jejuni* isolates. All three *cdt* genes were found in nearly all *C. jejuni* isolates from both children and chickens, but only in 5.6% of human *C. coli* isolates as compared to 87.2% in chicken *C. coli* isolates.

One of the more definitive studies regarding the characterization of a purified cytotoxin from *C. jejuni* was performed by Mahajan and Rodgers (1990). A 68-kDa protein was isolated from a fully virulent strain of *C. jejuni*. The protein was heat-labile, sensitive to trypsin, and was lethal to fertile chicken eggs. It also exhibited toxic effects on chicken embryo fibroblasts and caused rounding and lysis of CHO and ENT407 cells. The protein did not bind to GM_1 ganglioside in contrast to cholera toxin, and was immunologically unrelated to *V. cholerae* and *E. coli* enterotoxins. The amino terminal of N-acetylneuraminic acid was found to function as the membrane receptor for the toxin. The authors considered the 68-kDa protein to be a cytotoxin rather than an enterotoxin.

Interleukin-8 (IL-8) is a proinflammatory cytokine, a potent chemotactic factor for many immune effector cells, and a mediator of localized inflammatory responses. Live cells of *C. jejuni* have been found to induce the release of IL-8 from intestinal epithelial INT407 cells by two independent mechanisms, one of which requires adherence or invasion and the second of which requires CLDT (Hickey et al., 1999, 2000).

Autoagglutination (AAG) is known to be a marker of virulence in certan gram-negative bacterial pathogens including *V. cholerae*. In *C. jejuni*, AAG has been found to be strongly associated with flagella expression. Strong AAG in *C. jejuni* has been found associated with enhanced adherence to INT407 cells compared to strains exhibiting weak AAG (Misawa and Blaser, 2000).

Infection of rabbit ileal loops with inflammatory *C. jejuni* strains was found by Everest et al. (1993) to cause elevation of cyclic AMP, PGE_2, and LTB_4 levels in tissue fluids. Incubation of cultured Caco-2 cells with loop fluids caused elevated cellular cyclic AMP levels, which were inhibited by antiserum to PGE_2. These observations suggest that the mechanism of *C. jejuni* caused enterocolitis is similar to that of inflammatory bowel disease in which active secretion is stimulated by inflammation of the tissue. The infection of human colonic epithelial cells (Caco-2 cell monolayers) by *C. jejuni* in vitro has been found to increase the activity of mitogen-activated protein kinases and their phosphorylation, suggesting that these pathways are important in inflammatory responses induced by *C. jejuni* (MacCallum, Haddock, and Everest, 2005).

B. Cell Adhesion Factors

The flagella of *C. jejuni* have been found to be involved in colonization of the intestine. Nachamkin, Bohachick, and Patton (1993a) orally administered a wild-type flagellated, a nonflagellated mutant, and a truncated flagella mutant to 3-day-old chicks at levels of 6.6×10^8 CFU per chick. Only the fully motile, wild-type strain colonized the chick ceca, indicating that flagellation is essential for colonization. Harrington, Thompson-Carter, and Carter (1997) obtained evidence for recombination in the flagellin locus of *C. jejuni* influencing the flagellin gene-typing scheme. This genomic instability may contribute to the organism's ability to colonize a wide variety of hosts.

Flagella, however, are not the only factor involved in colonization of the GI tract. A 37-kDa outer membrane protein, designated CadF, has been described that facilitates the binding of *C. jejuni* to the glycoprotein fibronectin in the membrane of INT407 intestinal epithelial cells (Konkel et al., 1997). Oral administration to day-old chicks of a human clinical isolate of *C. jeuni* lacking the *cadF* gene resulted in the inability to recover the organism from the cecum (Ziprin et al., 1999). In contrast, the parental strain readily colonized the cecum of control chicks. These results indicated that disruption of the *cadF* gene via homologous recombination with a suicide vector renders *C. jejuni* incapable of colonizing the cecum of young poultry. Fibronectin-facilitated invasion of T84 colonic cells has been shown to occur preferentially at the basolateral surface of the mammalian cell (Monteville and Konkel, 2002).

A 28-kDa protein PEB1 (CBF1) has been found to be a common antigen located on the outer membrane of *C. jejuni* and *C. coli* and is the product of the *peb1A* gene (Pei and Blaser, 1993). PEB1 has been found to play a major role in adherence of *C. jejuni* to Hela cells, suggesting that it may be involved in *Campylobacter* colonization of the intestine (Fauchèr et al., 1989). The *peb1A* gene has been cloned and sequenced and found to exhibit significant homology to *Enterobacteriaceae* glutamine-binding protein, lysine/arginine/ornithine-binding protein, and histidine binding protein. A pair of primers (Table 10.6) were developed by Pei and Blaser (1993) that amplify a 702-bp sequence of the *peb1A* gene of *C. jejuni* strains but not from *C. coli*, *C. lari*, or *C. fetus* strains. Restriction digestion of the *peb1A* PCR amplicons from several *C. jeuni* strains yielded identical agarose banding patterns, indicating that *peb1A* is highly conserved in *C. jejuni* strains.

C. Lipopolysaccharide (LPS)

LPS is one of the main virulence factors of gram-negative bacteria. The LPS from *Campylobacter* spp. has been shown to have endotoxic properties (Branquinbo, Alviano, and Ricciardi, 1983; Naess and Hofstad, 1984), has been reported to be involved in adherence (McSweegan and Walker, 1986), and is thought to play a role in antigenic variation (Mills et al., 1992). The core oligosaccharide of the LPS in *C. jejuni* strains has been found to contain N-acetylneuraminic (sialic acid), which when linked to galactose resembles gangliosides in structure (Aspinall et al., 1992, 1993). This molecular similarity is thought to play a role in the autoimmune neuropathological disorders Guillain–Barré syndrome (GBS) and Miller–Fisher syndrome (Salloway et al., 1996; Schwerer et al., 1995). Expression of such ganglioside-like LPS in the outer membrane of *C. jejuni* results in a host immune response that cross-reacts with gangliosides of the peripheral nervous system (Ang et al., 2002). Genes *cst-II*, *cgtA*, and *cgtB* appear to be critical in *C. jejuni* ganglioside mimicry. In particular, *cst-II* encoding sialyltransferase has been shown to be involved in the addition of the terminal sialic acid residue that forms the GP_{1a} epitope. The cgtA gene product, $\beta1,4$-N-acetylgalactocyltransferase, and the *cdtB* gene product, $\beta1,3$-galactocytransferase, add the substrates for sialylation to the LPS backbone. Nachamkin et al. (2002) found the uniform presence of these genes in HS:19 serotype isolates which are strongly associated with GBS.

Rabbits immunized with gangliosides mimicking *C. jejuni* lipooligosaccharides (LOS) have been found to produce high titers of anti-LOS antibodies that immunologically cross-reacted with a panel of gangliosides (Moran, Annuk, and Prendergast, 2005). This is thought to disrupt the normal function of Na^+ and K^+ channels in human neurons and to thus interfere with nerve conduction.

The gene cluster (*wla*) involved in the synthesis of LPS by *C. jejuni* was cloned and sequenced by Fry et al. (1998). The first gene in this cluster encodes UDP-galactose-4-epimerase (GALE). The GALE epimerase encoded by the *galE* gene catalyzes the interconversion of UDP-galactose and UDP-glucose. UDP-galactose is used for the synthesis of carbohydrate polymers composed of galactose, inducing the galactose portion of LPS and exopolysaccharide. Deletion of a portion of the *galE* gene was found to result in the absence of galactose from the LPS core (Fry et al., 2000). The lipid A-core complex in the mutant no longer reacted to antiserum raised against the parental strain. The ability of the *galE*-deficient mutant to adhere to and invade human embryo intestinal epithelial INT407 cells was reduced 20- and 100-fold, respectively, and reflects a major role of LPS in infection. The ability of *C. jejuni* to invade INT407 cells was not blocked by chelation of extracellular Ca^{++} from host cells (Hu, Raybourne, and Kopecko, 2005). In contrast, invasion was markedly reduced by chelating host intracellular Ca^{++} and by blocking the release of Ca^{++} from intracellular host cell stores. Increasing INT407 intracellular free Ca^{++} levels stimulated cell invasion.

D. Capsule Formation

C. jejuni has been found to produce a capsular polysaccharide (CPS) that was stained with alcian blue and visualized by electron microscopy (Karlyshev, McCrossan, and Wren, 2001). DNA sequence data earlier indicated that the CPS was related to groups II and III CPS found among other pathogenic bacteria (Karlyshev et al., 2000). The authors concluded that the CPS may function to facilitate survival of the organism under various environmental conditions and to enhance pathogenicity.

E. Virulence Plasmid

C. jejuni has been found to harbor a plasmid that has been completely sequenced and found to be 37,468 nucleotides in length with an unusually low G + C content of 26% (Bacon et al., 2002). Seven genes were on the plasmid in a contiguous region of 8.9 kb that encoded orthologs of type IV secretion proteins. Type IV secretion systems are involved in bacterial DNA export, conjugation, transformation, protein secretion (Bacon et al., 2000; Christie and Vogel, 2000; Covacci et al., 1999), and in vitro invasion of INT407 enteric epithelial cells (Bacon et al., 2000). Seven additional pVir-encoded proteins exhibited significant similarities encoded by the plasticity zones of *Heliobacter pylori*. Five additional genes were found present that affect the in vitro invasion of INT407 cells, including one additional gene encoding a component of a type IV secretion system.

IV. ISOLATION OF CAMPYLOBACTERS FROM FOODS

Oxygen scavenging agents such as hemin and charcoal are frequently incorporated into culture media for growth and isolation of the organism, in addition to the use of a microaerobic atmosphere consisting of 5% O_2, 10% CO_2, and 85% N_2.

Liquid enrichments are often performed with continual gas bubbling to maintain the required microaerobic environment. Anaerobic jars with Campy paks, gas-generating envelopes, or pouches are also used to generate a suitable atmosphere with agar plate cultivation. Recovery of *C. jejuni* from foods has been found to be greater with enrichment broths with a constant flow of gas (presumably 5% O_2, 10% CO_2, and 85% N_2) through the broths than with an evacuation-replacement method (Heisick, Lanier, and Peeler, 1984).

A number of selective media have been developed for the isolation of *C. jejuni*. Doyle and Roman (1982) developed an enrichment broth to selectively recover small numbers of *C. jejuni* and *C. coli* in addition to nalidixic acid-resistant thermophilic campylobacters from foods (Table 10.4). After enrichment cultivation at 42°C for 16–18 h the cultures were plated onto Campy-*Brucella* agar plates (CBAP; Table 10.4). The methodology was able to recover as few as 0.3 to 1.0 campylobacters per gram of food from 25-g samples.

Wesley, Swaminathan, and Stadelman (1983) developed a selective enrichment medium designated ATB (Table 10.4) for isolation of very low numbers of *C. jejuni* from poultry products. Samples were enriched in ATB at 42°C for 48 h under an atmosphere of 5% O_2, 10% CO_2, and 85% N_2. Enrichment cultures were plated onto *Brucella* agar containing hematin, FBP supplement, and the antibiotics as listed in Table 10.4 for ATB with incubation at 42°C.

Strains of *C. jejuni* were found by Abram and Potter (1984) to grow in *Brucella* broth at 42°C with 0% to 1% NaCl. However, with 2% and 3% NaCl, the initial number of CFU declined rapidly.

Park and Sanders broth (PSB) has been reported to be superior to Preston broth (PB) for enrichment recovery of *Campylobacter jejuni* from chicken tissue and rinse solution (Tangvatcharin et al., 2005). The authors speculated that the difference may be due to the improved recovery of sublethaly injured cells in PSB, which consists of a rich basal medium supplemented with yeast extract, sodium citrate, and oxygen-quenching ingredients such as sodium pyruvate.

The direct agar plating of samples has been found to be consistently superior to the MPN selective enrichment technique for enumerating refrigerated *C. jejuni* (Ray and Johnson, 1984; Beuchat, 1985). This is thought to be most probably due to overgrowth of other organisms in the enrichment broths. Beuchat (1985) found that modified Butzler agar (MBA) was inferior to CBAP and blood-free *Campylobacter* medium (BFCM; Table 10.4). MBA supported colony development of *C. jejuni* poorly. BFCM not only supported good growth of *C. jejuni*, but also supported a larger number of other bacteria than did CBAP, which must be looked upon as a negative factor for MBA.

The U.S. Food and Drug Administration (FDA) *Bacteriological Analytical Manual* (Hunt, Abeyta, and Tran, 1998) recommends the pre-enrichment of food samples in Bolton broth (Table 10.4) incubated for 4 hr at 35 or 37°C followed by more selective enrichment by shifting to 42°C for 20 to 44 hr under microaerobic conditions throughout. Selective agar media (Table 10.4) are then streaked for isolation.

In a multilaboratory collaborative study Scotter, Humphrey, and Henle (1993) found that pre-enrichment of poultry skin samples with a gradual addition of

Table 10.4   Selective Media for Campylobacters

Campylobacter Culture Broths

Campylobacter broth of Park et al. (1981)

	Per L
Brucella broth	28 g
Vancomycin	8 mg
Trimethoprim	4 mg
Polymyxon B	8000 U

Campylobacter enrichment broth of Park et al. (1981) as modified by Lovett, Francis, and Hunt (1983)

	Per L
Brucella broth	28 g
Ferrus sulphate	0.25 g
Sodium metabisulfide	0.25 g
Sodium pyruvate	0.25 g
Vancomycin	15 mg
Trimethoprim	7.5 mg
Polymyxin B	5000 U

Campylobacter enrichment broth (CEB) of Martin et al. (1983)

	Per L
Brucella broth	28 g
5-fluorouracil	333 mg
Cephaperazone	32 mg
Trimethoprim	32 mg

Selective enrichment medium of Doyle and Roman (1982)[a]

Brucella broth	Per L	
Tryptone	10 g	After autoclaving, 7% lysed horse blood,
Peptamin	10 g	3.0g/L sodium succinate, 0.1 g/L cystine,
Dextrose	1 g	HCl, 15 mg/L of vancomycin, 5 mg/L of
Yeast extract	2.0 g	trimethoprim, 20,000 IU/L of polymyxin B,
NaCl	5.0 g	and 50 mg/L of cycloheximide are added.
Sodium bisulfite	0.1 g	Final pH 7.0

Bolton enrichment broth[a]

	Per L	
Meat peptone	10	After autoclaving in bottles with screw caps
Lactalbumin hydrolysate	5	and intact cap liners, add 50 ml of horse
Yeast extract	5	blood lysed by freezing and thawing twice.
NaCl	5	Before use add after filter sterilization: (1)
Haemin	0.01	4.0 ml/L of freshly prepared stock soln. of
Sodium pyruvate	0.5	Na cepoperazone in d.H_2O (0.5%); (2) 4.0
α-ketoglutamic acid	1.0	ml/L of tri-methoprim lactate stock soln. (20
Sodium metabisulfite	0.5	mg/L); (3) 4.0 ml/L of stock soln. of
Sodium carbonate	0.6	vancomycin in d.H_2O (0.5%), and 4.0 ml of
		cycloheximide stock soln. (1.25 g dissolved
		in 25 ml ethanol and made to 100 ml with
		d.H_2O). Final pH, 7.4

Continued

Table 10.4   Selective Media for Campylobacters (Continued)

Broth enrichment medium (BEM; Rogol et al., 1985)

	Per L	
Nutrient broth no. 2 (Oxoid)	13.0 g	After steam sterilization, 5 mg of
Agar	0.75 g	vancomycin, 2.5 mg of trimethoprim
Ferrous sulphate	0.5 g	lactate, and 1,250 IU of polymyxin B are
Sodium metabisulfate	0.5 g	added in addition to 50 ml of human
Sodium pyruvate	0.5 g	defibrinated blood. Distribute in 5-ml
Yeast extract	1.0 g	volumes to tubes. Final pH, 7.4
Bile salts no. 3 (Difco)	1.5 g	

Rosef broth (Rosef, 1981)

	Per L
Peptone (Difco, cat. no. 0118)	10 g
Lab lemco powder (Oxoid L29)	8 g
Yeast extract (Oxoid, cat. no. L21)	1 g
NaCl	5 g
Resazurin (0.025% solution)	16 ml
Vancomycin	10 mg
Trimethoprim lactate	5 mg
Polymyxin B	2000 IU

Preston broth (PB; Bolton and Robertson, 1982)

	Per L
Nutrient broth No. 2 (Oxoid CM67)	25 g
New Zealand agar	12 g
Saponin-lysed horse blood	50 ml
Polymyxin	5000 IU
Rifampin	10 mg
Trimethoprim	10 mg
Actidione	100 mg
Final pH 7.5	

Plates are incubated for 48 h at 43°C in an atmosphere of approximately 6% oxygen, 10% carbon dioxide, and 84% hydrogen (vol/vol).

Campylobacter enrichment broth (CEB; Christopher, Smith, and Vanderzant, 1982b)

	Per L
Brucella broth	28 g
Agar	1.5 g
Sodium pyruvate	0.5 g
Vancomycin	10 mg
Trimethoprin	5.0 mg
Polymyxin B sulphate	2500 IU
Amphotericin B	2.0 mg
Cephalothin	15 g

Table 10.4   Selective Media for Campylobacters (Continued)

Rosef broth (nonselective; Waage et al., 1999)

	Per L
Peptone	10.0 g
LabLemco (Oxoid)	8.0 g
Yeast extract	1.0 g
Sodium chloride	5.0 g
Rezasurin solution (0.025%, w/v)	1.6 ml

Selective agar media for isolation of *C. jejuni*

ATB selective enrichment medium (Wesley, Swaminathan, and Stadelman, 1983)

	Per L
Typtone	20 g
Yeast extract	2.5 g
NaCl	5.0 g
Agar	1.0 g
FBP supplement:	
FeSO$_4$	0.12 g
Sodium metabisulfite	0.25 g
Bicine	10 g

Hematin solution (6.25 ml/L) is prepared by dissolving 0.032 g of bovine hemin in 10 ml of 0.15 N NaOH and autoclaving for 30 min. Rifampin (25 mg/L), cefsulodin (6.25 mg/L), and polymyxin sulfate (20,000 IU/L) are added after autoclaving. Final pH 8.0

Blood-free Campylobacter agar (Bolton and Coates, 1983)

	Per L
Nutrient broth (Oxoid)	13 g
Charcoal	4 g
Ferrous sulphate	0.25 g
Sodium pyruvate	0.25 g
Agar	20 g

Butzler agar (Lauwers, De Boeck, and Butzler, 1978)

	Per L
Fluid thioglycollate medium	29.8 g
Defibrinated sheep blood	100 ml
Bacitracin	25,000 IU
Novobiocin	5 mg
Actidione	50 mg
Cephalothin	15 mg
Colistin[a] (20,070 U/mg)	10,000 U
Agar	30
Final pH 7.4	

Increasing the colistin level to 40,000 U with 72 hr incubation has been found to increase direct positive isolation (Patton et al., 1981).

Continued

Table 10.4   Selective Media for Campylobacters (*Continued*)

SK agar (Skirrow, 1977)

	Per L
Oxoid blood agar base No. 2	40 g
Lysed horse blood	70 ml
Vancomycin	10 mg
Polymyxin B	2500 IU
Trimethoprim	5 mg
Final pH 7.3–7.4	

Abeyta–Hunt–Bark agar (Hunt, Abeyta, and Tran, 1998)

	Per L
Heart infusion agar	40 g
Yeast extract	2 g
d.H$_2$O	950 ml
Final pH, 7.4	

Autoclave, cool, and add after filter sterilizing: (a) 6.4 ml of 0.5% stock soln. of cefoperazone to final conc. of 32 mg/ml, (b) 4.0 ml of 0.25% aqueous rifampicin to final conc. of 10 mg/ml; (c) 4.0 ml of 0.05% stock soln. of amphotericin B to final conc.; (d) 4 ml of FBP (prepared by dissolving 6.25 g sodium pyruvate in 20 ml of d.H$_2$O, adding 6.25 g ferrous sulphate and 6.25 g sodium metabisulfite made up to 100 ml with d.H$_2$O.

Modified Campylobacter blood-free selective agar base (CCDA; Hunt, Abeyta, and Tran, 1998)

	Per L
CCDA agar base (Oxoid)	45.5 g
Yeast extract	2.0 g
Final pH, 7.4	

Autoclave, cool, and add 6.4 ml of a 0.5% stock solution of sodium cepoperazone, 4.0 ml of 0.25% rifampicin stock soln., and 4.0 ml of 0.05% stock soln. of amphotericin B.

---

antibiotics suppressed competing organisms, resulting in an improved recovery of *Campylobacter*, as did a nonselective blood agar isolation medium used in combination with a membrane filtration technique.

Murphy, Carroll, and Jordan (2005) found that the extent to which *C. jejuni* survived acid stress at pH 4.5 was dependent on the identity of the broth used to culture and stress the cells. Survival following thermal stress of broth cultures at 55°C was also found dependent on the identity of the culture broth, as was tolerance to 21% oxygen. The authors concluded that studies involving stress using different media may not be comparable.

Until recently *C. jejuni* has been considered to be an obligate microaerophile, requiring oxygen concentrations of 3% to 15% and CO$_2$ concentrations of 3% to 5% for satisfactory growth, using amino acids as carbon and energy sources. The organism has a respiratory metabolism based on oxygen as a terminal electron acceptor. The partial reduction of oxygen to water during microbial respiration results in toxic reactive oxygen intermediates (ROI) such as the superoxide ion (O2$^-$) and H$_2$O$_2$. *C. jejuni* produces superoxide dismutase (SOD) and catalase. SOD catalyzes the conversion of superoxide to H$_2$O$_2$ and oxygen and catalase prevents the accumulation of H$_2$O$_2$. Protective components such as blood, charcoal, FeSO$_4$, metabisulfite, and pyruvate are thought to furnish protection of *C. jeuni* against such oxidative stress by acting as quenchers of ROI. Verhoeff-Bakkenes et al. (2008) grew *C. jejuni* in the presence of pyruvate (25 mmoles/L) in continuous cultures in brain–heart infusion broth (BHI) with a dissolved oxygen tension (DOT) a high as 90%, which is the maximum that can be achieved during aeration with compressed oxygen. Cultivation in the presence of pyruvate from 0.1% to 90% DOT yielded uniformly high growth yields. The authors attributed this high DOT growth-enabling effect of pyruvate to its ability to function as a quencher of ROI. This conclusion was based on the author's observation that catalase activity notably decreased with growth at 90% DOT in the presence of pyruvate.

V. SEROTYPING OF CAMPYLOBACTERS AND IMMUNOLOGICAL DETECTION

The "Penner" serotyping method is one of two commonly used techniques for serotyping *Campylobacter* isolates and is based on soluble heat stable (HS) antigens extracted from the cell walls, which are incorporated into a passive hemagglutination assay (PHA), because direct cellular agglutination is weak. The HS antigens are sometimes referred to as "Pen" and "O" to reflect the assumption they are lipopolysaccharides (LPS). In reality some are LPS and others are lipooligosaccharides (LPO). A number of the HS antigens are present in both *C. jejuni* and *C. coli* strains. The use of hippurate hydrolysis to distinguish these two species has allowed separate HS typing systems to be developed for both species. The systems presently include 42 serostrains of *C. jejuni* and 18 for *C. coli*. Penner serotyping of campylobacters is based on the fact that PHA is a highly sensitive technique for examining the distinctions among LPS and LOS components of cell walls. Heat-extracted LPS and LOS molecules adhere to the surface of mammalian erythrocytes (referred to as sensitization) and antibodies specific for individual LPS and LOS cause agglutination of the sensitized erythrocytes. Because the erythrocyte is much larger than the bacterial cell, detection of the antigen–antibody reaction is more sensitive with the PHA technique. A highly detailed review of the Penner typing system has been given by Moran and Penner (1999).

Lior et al. (1982) developed a serotyping scheme for isolates of *C. jejuni* based on slide agglutination of live bacteria with whole-cell antisera absorbed with homologous heated and heterologous unheated cross-reactive antigen. Among 815 isolates from human and nonhuman sources, 21 serogroups were recognized. Among isolates from all sources, eight serogroups (1, 2, 4, 5, 7, 8, 9, and 11) were encountered most frequently. Serogroups 1, 2, 4, 5, 9, and 11 were most common among human isolates. The majority of poultry and all of the swine isolates belonged to the same serogroups as the human isolates. The addition of DNase eliminated the problem encountered with mucoid slimy strains in the slide agglutination assay.

Serotyping has been found not to agree with random amplified polymorphic DNA (RAPD) typing. Aarts, van Lith, and Jacobs-Reitsma (1995) subjected 34 *C. jejuni* and *C. coli* strains from various livestock and darkling beetles from two

Dutch poultry farms to Penner serotyping and RAPD analysis using primer L1 of Jensen, Webster, and Straus (1993) and primer ERIC2 of Versalovic, Koeuth, and Lupski (1991). L1 is derived from the most conserved 23S sequence immediately following the spacer region between the 16S and 23S rRNA genes. ERIC2 is designed from the core inverted repeat of the enterobacterial repetitive intergeneric consensus sequence. Some strains with identical serotypes exhibited different PCR banding profiles and, conversely, strains with different serotypes produced identical RAPD banding profiles. The inconsistency between serotyping and RAPD banding profiles suggested that conventional typing methods should be used in combination with RAPD to reliably address the valid relationships among strains. The major 45-kDa outer membrane protein of *C. jejuni*, *C. coli*, *C. upsalienis*, and *C. lari* was used by Griffiths, Moreno, and Park (1992) as antigen for the production of species-specific antibodies in mice. A dot-ELISA was used in conjunction with nitrocellulose membranes. Species specificity was achieved by repetitive adsorption of cross-reacting antibodies to cross-reacting species and their removal from the antibody preparations.

Hoorfar et al. (1999) evaluated two automated ELISA assays, EIA-1 and EIA-2, for rapid detection of thermophilic campylobacters from cattle and swine feces using Preston broth enrichments. Overall sensitivities and specificities with cattle and swine samples with method EIA-1 were 95%, 88%, 71%, and 76%, respectively. With method EIA-2, overall sensitivities and specificities with cattle and swine samples were 84%, 69%, 32%, and 100%. The low level of specificity with method EIA-2 and fecal samples from cattle may have been due to the isolation and identification of only one colony per sample. Actual specificity may have been much higher.

VI. BACTERIOPHAGE TYPING OF *C. JEJUNI*

Connerton et al. (2004) undertook a longitudinal study of *C. jejuni* bacteriophages and their hosts in a broiler house identified as having a population of *Campylobacter*-specific bacteriophages. Cloacal and excreta samples were collected from three successive broiler flocks reared in the same barn. *C. jejuni* was isolated from each flock, whereas phages were isolated only from flocks 1 and 2 but not from flock 3. All *Campylobacter* isolates from flock 1 were indistinguishable by pulsed field gel electrophoresis (PFGE) and multilocus sequence typing (MLST), indicating that this *C. jejuni* type was maintained by flock 1 to flock 2 and was largely replaced in flock 3 by three genetically distinct *C. jejuni* types insensitive to the resident phage. These results indicated the obstacle of introducing phage into a flock to reduce the number of *Campylobacter* among poultry. The authors concluded that succession of phage resistance was due largely to introduction of new genotypes rather than the acquisition of phage resistance. Phage typing was not found to be useful for studying the relationships among these groups of strains, due presumably to the lysogenic state possibly influencing the susceptibility to the phages used in typing.

VII. MOLECULAR METHODS OF DETECTING AND TYPING CAMPYLOBACTERS

A. Genes Used

Genes used for molecular taxonomy of campylobacters are given in Table 10.5. Table 10.6 presents primers and probes for PCR, real-time PCR, and random amplified polymorphic DNA analysis involving these genes.

B. PCR

C. jejuni and *C. coli* have been found to possess two tandem genes encoding flagellar proteins designated *flaA* and *flaB*, which are 92% and 92.8% homologous in *C. coli* and *C. jeuni,* respectively (Nuijten et al., 1990). The overall identity of these flagellin genes between *C. coli* and *C. jejuni* is 73.6% to 82% (Alm, Guerry, and Trust, 1993). However, the central region of the *flaA* gene has been found to be highly polymorphic (Alm, Guerry, and Trust, 1993).

PCR primers have been designed that are specific for certain portions of the *flaA* and *flaB* gene sequences. Oyofo et al. (1992) developed a direct PCR assay specific for detection of only *C. jejuni* and *C. coli* without enrichment. The primers pg50 and pg3 (Table 10.6) amplified a 450-bp sequence spanning both *flaA* and *flaB* tandem genes that are highly conserved in both species. A 273-bp digoxygenin-labeled probe was used to assess the sensitivity of the assay via Southern blots. Starting with stool samples, the limit of sensitivity with the blot assay was 30 to 60

Table 10.5 Virulence, Toxin, and Other Genes of *C. jejuni* Used in PCR Assays for Detection of Thermophilic Campylobacters

Gene Designation	Description of Gene Product
cadF	Encodes CadF, a 37-kDa outer membrane surface protein that binds fibronectin of mammalian INT407 cells, mediating adhesion and colonization by *C. jejuni*.
ceuE	A 35-kDa lipoprotein component of a binding protein-dependent transport system for the siderophore enterochelin produced by *C. coli*.
pVirB11	A 35-kb virulence plasmid involved in adherence and invasion by *C. jejuni*.
pVir	A 37.5-bp virulence plasmid encoding proteins involved in invasion of INT407 cells by *C. jejuni*.
cdtA, cdtB, cdtC	The *cdt* gene cluster consists of three adjacent genes that encode proteins of 30, 29, and 2 kDa, respectively, involved in cytolethal distending toxin (CLDT) effect on Vero cells produced by *C. jejuni*.
flaA and *flaB*	Encode flagellin proteins involved in adhesion of *C. jejuni*.
HipO	Encodes the hippurate hydrolysis enzyme (hippuricase) of *C. jejuni*.
Asp	Encodes aspartokinase of *C. coli*.
VS1	Encodes an unnamed protein product unique to *C. jejuni*.
peb1A	Encodes CBF1, a 28-kDa protein located on the outer membrane involved in adherence to Hela cells by *C. jejuni* and *C. coli*.

Table 10.6   PCR Primers and DNA Probes

Primer or Probe	Sequence (5′ → 3′)[a]	Size of Amplified Sequence (bp)	Gene or DNA Target Sequence	References
Campylobacter spp. C1-F	GAT-GCT-TCA-GGG-ATG-GCG	1300 and 1700	*IaA*	Birkenhead et al. (1993)
C2-R	TTT-GTG-ATT-CTG-CTG-CTT-TAA-C			
Forward primer	GGA-TTT-CGT-ATT-AAC-ACA-AAT-GGT-GC	1728	*FlaA*	Nachamkin, Bohachick, and Patton (1993a)
Reverse primer	CTG-TAG-TAA-TCT-TAA-AAC-ATT-TG			
P2	ATC-ACC-AAG-AAT-ACC-CAT-TGC-G	603	16S rRNA	Purdy, Ash, and Patton (1996)
P4	TGC-TCT-GCA-GTT-GCA-GAG-AAC			
P2	ATC-ACC-AAG-AAT-ACC-CAT-TGC-G	531	16S rRNA	Purdy, Ash, and Fricker (1996)
P3	GTC-TCA-TTG-TAT-ATG-CCA-TTG-T			
P1.C5 431	AAA-GGA-TCC-GCG-TAT-TAA-CAC-AAA-TGT-TGC-AGC	1490	*FlaA*	Ayling et al. (1996)
P3.C5 433	GAT-TTG-TTA-TAG-CAG-TTT-CTG-CTA-TAT-CC			
P2.C5 432	AAA-GGA-TCC-GAG-GAT-AAA-CAC-CAA-CAT-CGG-T	1490	*FlaB*	Ayling et al. (1996)
P3.C5 43 P3.C5 433	GAT-TTG-TTA-TAG-CAG-TTT-CTG-CTA-TAT-CC			
RAPD 1290	GTG-GAT-GCG-A	—	—	Carvalho and Ruiz-Palacios (1996)
CF03 (sense)	GCT-CAA-AGT-GGT-TCT-TAT-GCN-ATG-G	340–380	*FlaA–FlaB*	Wegmüller, Lütty, and Candrian (1993)

Primer	Sequence	Size	Target	Reference
CF04 (antisense)	GCT-GCG-GAG-TTC-ATT-CTA-AGA-CC		Intergenic sequence	
CF03 (sense)	GCT-CAA-AGT-GGT-TCT-TAT-GCN-ATG-G	180–220	FlaA–FlaB Intergenic sequence	Wegmüller, Lütty, and Candrian (1993)
CF02 (antisense)	AAG-CAA-GAA-GTG-TTC-CAA-GTT-T			
C. coli cc-32-237(F)	ACT-CGG-ATG-TAA-AAT-ATA-CAA-ATT-CTA-CTC-TT	182	cadF	Englen and Fedorka-Cray (2002)
C. coli cc-30-rc682 (R)	TTT-TTC-TTC-AAA-GGC-TGG-ATT-GAT-ATC-TAC			
C. jejuni cj-30-560 (F)	AAA-GGA-AAA-AGC-TGT-AGA-AGA-AGT-TGC-TGA	560	colF	Englen and Fedorka-Cray (2002)
C. jejuni cj-30-rc682 (R)	TTT-TTC-TTG-AAA-AGT-TGG-ATT-TAT-AGT-AGT			
C. jejuni HIP400 F	GAA-GAG-GGT-TTG-GGT-GGT	735	hipO	Englen and Fedorka-Cray (2002)
C. jejuni HIP1134R	AGC-TAG-CTT-CGC-ATA-ATA-ACT-TG			Englen and Fedorka-Cray (2002) from Englen and Kelly (2000)
C. coli Col1 (F)	ATG-AAA-AAA-TAT-TTA-GTT-TTT-GCA	894	ceuE	Englen and Fedorka-Cray (2002)
C. coli Col2 (R)	ATT-TTA-TTA-TTT-GTA-GCA-GCG			Englen and Fedorka-Cray (2002) from Gonzalez et al. (1997)
Campylobacters C412F	GGA-TGA-CAC-TTT-TCG-GAG-C	815	16S rRNA	Inglis, Kellischuk, and Busz (2003) from Linton, Owen, and Stanley (1996b)
Campylobacters C1228R	CAT-TGT-AGC-ACG-TGT-GTC			

Continued

Table 10.6   PCR Primers and DNA Probes (Continued)

Primer or Probe	Sequence (5′ → 3′)[a]	Size of Amplified Sequence (bp)	Gene or DNA Target Sequence	References
C. coli (nested) CCceuEN3F	AAG-CGT-TGC-AAA-ACT-TTA-TGG	330	ceuE	Inglis, Kalischuk, and Busz (2003) from Inglis and Kalischuk (2003)
C. coli (nested) CCceuEN3R	CCT-TGT-GCG-CGT-TCT-TTA-TT			
C. jejuni (nested) CJmapAN3F	TGG-TGG-TTT-TGA-AGC-AAA-GA	413	mapA	Inglis, Kalischuk, and Busz (2003) from Inglis and Kalischuk (2003)
C. jejuni (nested) CJmapAN3R	GCT-TGG-TGC-GGA-TTG-TAA-A			
C. coli and C. jeuni MD16S1Upper	ATC-TAA-TGG-CTT-AAC-CAT-TAA-AC	857	16S rRNA	Denis et al. (1999)
C. coli and C. jejuni MD16S2 Lower	GGA-CGG-TAA-CTA-GTT-TAG-TAT-T			
C. coli COL3	AAT-TGA-AAA-TTG-CTC-CAA-CTA-TG	462	ceuE	Denis et al. (1999) from Gonzalez et al. (1997)
C. coli MDCOL2	TGA-TTT-TAT-TAT-TTG-TAG-CAG-CG			
C. jenuni MDmapA1Upper	CTA-TTT-TAT-TTT-TGA-GTG-CTT-GTG	589	mapA	Inglis, Kalischuk, and Busz (2003) from Denis et al. (1999)
C. jenuni MDmapA2Lower	GCT-TTA-TTT-GCC-ATT-TGT-TTT-ATT-A			
QCjmapANF	GGT-TTT-GAA-GCA-AAG-ATT-AAA-GG	94	mapA	Inglis and Kalischuk (2004)
QCjmapANR	AAG-CAA-TAC-CAG-TGT-CTA-AAG-TGC			

Name	Sequence	Target	Size (bp)	Reference
CAMP1F-B	Bio-GTT-AAG-AGT-CAC-AAG-CAA-GT	*C. jejuni* 16S/23S rRNA spacer	344	Grennan et al. (2001) from O'Sullivan et al. (2000)
B1-B	Bio-CYR-YTC-CCA-AGG-CAT-CCA-CC	*C. coli* 16S/23S rRNA spacer	503	
C. jejuni probe CJEJ7	AMI-GCT-TAG-TTG-AGA-CTA-AAT-CA	16S/23S rRNA spacer		
C. coli probe CCOL2	AMI-GAC-TTA-GTT-TAG-ATA-TTT-TTA-G	16S/23S rRNA spacer		
Campylobacter probe CAMP4	AMI-GGT-AAG-CTACTA-AGA-GCG	16S/23S rRNA spacer		
C. jejuni and *C. coli* JCF1(F)	TTA-GTA-TGA-GCG-ATG-AGG-GTG	*ceuE*	610	Hong et al. (2003)
C. jejuni and *C. coli* JCF1(R)	CTT-TTT-CCG-TGT-GTG-CCT-AC			
CeuE probe	ATC-ATT-TCT-GGA-CGC-CAA-AG-Bio			
Campylobacter P1 OT1118	AAT-TCT-AAT-ACG-ACT-CAC-TAT-AGG-GAG-AGT-GTG-ACT-GAT-CAT-CCT-CTC-A	16S rRNA	~200	Uyttendaele et al. (1994)
Campylobacter OT1547	GAC-AAC-AGT-TGG-AAA-CGA-CTG-CTA-ATA			
Probe OT1555	CTG-CTT-AAC-ACA-AGT-TGA-GTA-GG	16S rRNA	287	Lübeck et al. (2003) from Uyttendaele et al. (1994)
OT1559	CTG-CTT-AAC-ACA-AGT-TGA-GTA-GG			

Continued

Table 10.6   PCR Primers and DNA Probes (Continued)

Primer or Probe	Sequence (5′ → 3′)[a]	Size of Amplified Sequence (bp)	Gene or DNA Target Sequence	References
18-1rev	TTC-TGA-CGG-TAC-CTA-AGG-AA	—	16S rRNA	Lübeck et al. (2003)
Campylobacter FP	CTG-CTT-AAC-ACA-AGT-TGA-GTA-GG			Josefsen, Jacobsen, and Hoorfar (2004b) from Lübeck et al. (2003)
Campylobacter RP	TTC-CTT-AGG-TAC-CGT-CAG-AA			Josefsen, Jacobsen, and Hoorfar (2004b)
Target probe	FAM-TGT-CAT-CCT-CCA-CGC-GGC-GTT-GCT-GC-TAMRA			
Internal control probe	VIC-TTC-ATG-AG-ACA-CCT-GAG-TTG-A-TAMRA			
CA1	CCA-AAT-CGG-TTC-AAG-TTC-AAA-TCA-AAC	810–813	*FlaA* and *FlaB*	Rasmussen et al. (1996)
CA2	CCA-CTA-CCT-ACT-GAA-AAT-CCC-GAA-CC			
Probe	GGA-ACA-GGT-CTT-GGA-GCT-TTG-GC			
CHCU146F	GGG-ACA-ACA-CTT-AGA-AAT-GAG	878	16S rRNA	Lawson et al. (1997)
CU1024R	CAC-TTC-CGT-ATC-TCT-ACA-GA	1225 and 1375		
CH1371R	CCG-TGA-CAT-GGC-TGA-TTC-AC			
RAPD Gen 3-60-25	CCT-GTT-AGC-C	—	—	Madden, Moran, and Scates (1996a)
R1F	CGC-GGA-TCC-ATC-TAG-AAT-GTC-TTT-AAG-CAG-ACT-TAG-TTC-A	1480 and 1502	*flaA* and *flaB*	Thomas et al. (1997)

Primer	Sequence	Size	Target	Reference
R3R	GCG-CTG-GAT-CCT-CTA-GAT-CAA-ACG-TTT-ACT-TGA-GTA-ACA-GT			
Campyl. sp. C412F	GGA-TGA-CAC-TTT-TCG-GAG-C	816	16S rRNA	Linton, Owen, and Stanley (1996b)
C1288R	CAT-TGT-AGC-ACG-TGT-GTC		16S rRNA	
C. upsaliensis CHCU146F	GGG-ACA-ACA-CTT-AGA-AAT-GAG	878	16S rRNA	
CU1024R	CAC-TTC-CGT-ATC-TCT-ACA-GA		16S rRNA	
C. helveticus CHCU146F	GGG-ACA-ACA-CTT-AGA-AAT-GAG	1375	16S rRNA	
CH1371R	CCG-TGA-CAT-GCC-TGA-TTC-AC		16S rRNA	
C. fetus CFCH57F	GCA-AGT-CGA-ACG-GAG-TAT-TA	967	16S rRNA	
CF1054R	GCA-GCA-CCT-GTC-TCA-ACT		16S rRNA	
C. hyointestinalis CFCH57F	GCA-AGT-CGA-ACG-GAG-TAT-TA	1287	16S rRNA	
CH1344R	GCG-ATT-CCG-GCT-TCA-TGC-TC		16S rRNA	
C. lari CL594F	CAA-GTC-TCT-TGT-GAA-ATC-CAA-C	561	16S rRNA	
CL1155R	ATT-TAG-AGT-GCT-CAC-CCG-AAG		16S rRNA	
FLA4F	GGA-TTT-CGT-ATT-AAC-ACA-AAT-GGT-GC	1702	flaA	Broman et al. (2000)
FLA1728R	CTG-TAG-TAA-TCT-TAA-AAC-ATT-TTG			
C16SF	TAT-GGA-GAG-TTT-GAT-CCT-GGC-TCA-G	1464	16S rRNA	
B37	TAC-GGY-TAC-CTT-GTT-ACG-A			
RAPD	CAA-TCG-CCG-T	—	—	Hernández et al. (1996)

Continued

Table 10.6   PCR Primers and DNA Probes (Continued)

Primer or Probe	Sequence (5′ → 3′)[a]	Size of Amplified Sequence (bp)	Gene or DNA Target Sequence	References
130f	GGA-TTT-CCG-AAT-GGG-GCA-ACC-C	2646	23S rRNA	Iriarte and Owen (1996) from Lane (1991)
2747r	GTT-TCG-TGC-TTA-GAT-GCT-TTC			
CF1	GGA-GGA-TGA-CAC-TTT-TCG-GAG-CG	840	16S rRNA	Vanniasinkam, Lanser, and Barton (1999)
CR2	TCG-CGG-TAT-TGC-GTC-TCA-TTG-TAT-ATG-C			
CP3 probe	GGA-AGA-ATT-CTG-ACG-GTA-CCT-AAG			
JE	CCT-GCT-ACG-GTG-AAA-GTT-TTG	793	ceuE	Gonzalez et al. (1997)
JEJ2	GAT-TTT-TTG-TTT-TGT-GCT-GC			
JEJ3 probe	AGT-TAA-AAG-TAG-CTC-CAA-CTT-TA			
COL1	ATG-AAA-AAA-TAT-TTA-GTT-TTT-GCA	894	ceuE	Gonzalez et al. (1997)
COL2	ATT-TTA-TTA-TTT-GTA-GCA-GCG			
COL probe	AAT-TGA-AAA-TTG-CTC-CAA-CTA-TG			
THERM1	TAT-TCC-AAT-ACC-AAC-ATT-AGT	491	23S rRNA	Fermér and Engvall (1999) from Eyers et al. (1993)
THERM4	CTT-CGC-TAA-TGC-TAA-CCC			Fermér and Engvall (1999)
BO4263	AGA-ACA-ACA-CGC-GGA-CCT-ATA-TA	256	—	Jackson, Fox, and Jones (1996)
BO4264	CGA-TGC-ATC-CAG-GTA-ATG-TAT			
C.jejuni VS15	GAA-TGA-AAT-TTT-AGA-ATG-GGG	358	VS1	Occhialini et al. (1996)
C.jejuni VS16	GAT-ATG-TAT-GAT-TTT-ATC-CTG-C			

C. coli CSF	ATA-TTT-CCA-AGC-GCT-ACT-CCC-C	258	—	
C. coli CSR	CAG-GCA-GTG-TGA-TAG-TCA-TGG			
VS15	GAA-TGA-AAT-TTT-AGA-ATG-GGG	358	VS1	Yang et al. (2003) from Occihalini et al. (1996)
VS16	GAT-ATG-TAT-GAT-TTT-ATC-CTG-C			
TaqMan probe	[FAMJ-TTT-AAC-TTG-GCT-AAA-GGC-TAA-GGC-T-[TAMRA]			
CL1	ATT-GTA-TTC-TTG-GCG-TGG-CCC	402	—	Ng et al. (1997)
CR3	CCA-TCA-TCG-CTA-AGT-GCA-AC			
Forward	CTG-AAT-TTG-ATA-CCT-TAA-GTG-CAG-C	86	—	Nogva et al. (2000)
Reverse	AGG-CAC-GCC-TAA-ACC-TAT-AGC-T			
TaqMan probe	FAM-TCT-CCT-TGC-TCA-TCT-TTA-GGA-TAA-ATT-CTT-TCA-CA TAMRA			
pg50	ATG-GGA-TTT-CGT-ATT-AAC	450	flaA-flaB	Oyoto et al. (1992)
pg3	GAA-CTT-GAA-CCG-ATT-TG			
C. jejuni and pg-3	GAA-CTT-GAA-CCG-ATT-TG	460	flaA	Harmon, Ransom, and Wesley (1997) from Oyoto et al. (1992)
C. coli pg-50	ATG-GGA-TTT-CGT-ATT-AAC	160	—	
C. jejuni C-1	CAA-ATA-AAG-TTA-GAG-GTA-GAA-TGT			
C-4	GGA-TAA-GCA-CTA-GCT-AGC-TGA-T			

Continued

Table 10.6 PCR Primers and DNA Probes (Continued)

Primer or Probe	Sequence (5′ → 3′)[a]	Size of Amplified Sequence (bp)	Gene or DNA Target Sequence	References
C. jejuni AB-F	CTG-AAT-TTG-ATA-CCT-TAA-GTG-CAG	~550	—	Rudi et al. (2004) from Nogva et al. (2000)
C. jejuni AB-R	CTG-AAT-TTG-ATA-CCT-TAA-GTG-CAG-C			Rudi et al. (2004)
pg 50	ATG-GGA-TTT-CGT-ATT-AAC	1448	flaA	Alm, Guerry, and Trust (1993)
RAA19	GCA-CCY-TTA-AGW-GTR-GTT-ACA-CCT-GC			
pg50	ATG-GGA-TTT-CGT-ATT-AAC	1459	flaB	Alm, Guerry, and Trust (1993)
RAA9	AAG-GAT-TTA-AAA-TGG-GTT-TTA-GAA-TAA-ACA-CC			
C. jejuni 82F	TTG-GTA-TGG-CTA-TAG-GAA-CTC-TTA-TAG-CT	115	ORF-C	Sails et al. (2003)
C. jejuni 197R	CAC-ACC-TGA-AGT-ATG-AAG-TGG-TCT-AAG-T			
Probe CJTP2	FAM-TGG-CAT-ATC-CTA-ATT-TAA-ATT-ATT-TAC-CAG-GAC-TAMRA			
C. jejuni cj1	ATC-GGG-CTG-TTA-TGA-TGA-TA	265	—	Day et al. (1997)
C. jejuni cj4	CAT-ATC-CAG-AGC-CTC-TGG-AT			
C. jejuni	GCA-GAA-GGT-AAA-CTT-GAG-TCT-ATT	702	peb1A	Pei and Blaser (1993)
C. jejuni	TTA-TAA-ACC-CCA-TTT-TTT-CGC-TAA			
C. jejuni and C. coli VAT2	GTN-GCN-ACN-TGG-AAY-CTN-CAR-GG	494	cdtB	Eyigor et al. (1999) from Picket et al. (1996)
C. jejuni and C. coli VMI1	RTT-RAA-RTC-NCC-YAA-DAT-CAT-CC			

	Sequence	Size	Gene	Reference
C. jejuni and C. coli VAT2	GTN-GCN-ACN-TGG-AAY-CTN-CAR-GG	999	cdtA, cdtB, cdtC	Eyigor et al. (1999) from Picket et al. (1996)
C. jejuni and C. coli LPF-X	AAA-YTG-MAC-DTA-DCC-AAA-AGC			
C. jejuni CdtA$_1$	CGC-GTC-TAG-AAC-TAT-GGA-AAA-TGT-AAA-TCC-TTT-GGG-GCG-TTC-ATT-TGC	—	cdtaA	Hassane et al. (2001)
CdtA$_2$	GCG-GGT-CGA-CTT-TTC-ATG-GTA-CCT-CTC-CTT-GGC-G	—		
CdtB$_1$	CGC-GTC-TAG-AAA-TAT-ATG-GAA-AAT-TTT-AAT-GTT-GGC-ACT-TGG	—	cdtB	Hassane et al. (2001)
CtdB$_2$	GGC-GGT-CGA-CTG-TCC-TAA-AAT-TTT-CTA-AAA-TTT-ACT-GG	—		
CtdC$_1$	GCG-CTC-TAG-AAC-AAT-GGG-AGA-TTT-GAA-AGA-TTT-TAC-CGA-AAT	—	cdtC	Hassane et al. (2001)
CdtC$_2$	GGC-GGT-CGA-CCA-AGA-TAA-AAA-TCT-TAT-TCT-AAA-GGG-GTA-GC	—		
C. jejunji CadF-F2B	TTG-AAG-GTA-ATT-TAG-ATA-TG	~400	cadf	Poulsen et al. (2005) from Konkel et al. (1999)
CadF-R1B	CTA-ATA-CCT-AAA-GTT-GAA-AC			
C. jejuni LPS galE1	GCG-GTG-GTG-CAG-GTT-ATA-TAG-G	9.6 kb	LPS gene cluster	Knudsen et al. (2005) from Shi et al. (2002)
wlaH3	TCA-GTT-CTT-GCC-ATT-AAA-TTT-CTC			

Continued

Table 10.6 PCR Primers and DNA Probes (Continued)

Primer or Probe	Sequence (5' → 3')[a]	Size of Amplified Sequence (bp)	Gene or DNA Target Sequence	References
C. coli Cc-F	GTT-GGA-GCT-TAT-CTT-TTT-GCA-GAC-A	80	glyA	Jensen et al. (2005)
C. coli Cc-R	TGA-GGA-AAT-GGA-CTT-GGA-TGC-T			
C. coli probe	TET-TGC-TAC-AAC-AAG-TCC-AGC-AAT-GTG-TGC-A-TAMRA			
C. jejuni Cj-F	TAA-TGT-TCA-GCC-TAA-TTC-AGG-TTC-TC	135	glyA	Jensen et al. (2005)
C. jejuni Cj-R	GAA-GAA-CTT-ACT-TTT-GCA-CCA-TGA-GT			
C. jejuni probe	FAM-AAT-CAA-AGC-CGC-ATA-AAC-ACC-TTG-ATT-AGC-TAMRA			
C. lari Cl-F	CAG-GCT-TGG-TTG-TAG-CAG-GTG	96	glyA	Jensen et al. (2005)
C. lari Cl-R	ACC-CCT-AGT-CCA-TTC-CCT-TAT-GCT-CAT-GTT			
C. lari probe	TET-CAT-CCT-AGT-CCA-TTC-CCT-TAT-GCT-CAT-GTT-TAMRA			
C. upsaliensis Cu-F	TCG-TAG-CTG-GTG-AGC-ATC-CTA-G	65	glyA	Jensen et al. (2005)
C. upsaliensis Cu-R	GGT-TTT-GTG-TGT-GGT-TGA-GCT-T			
C. upsaliensis probe	FAM-CCT-TTC-CCT-CAC-GCA-CAC-ATC-G-TAMRA			
CJE195 probe	TAMRA-CCC-TAC-TCA-ACT-TGT	—	16S rRNA	Lehtola, Loaders, and Keevil (2005)
hipO-F156	AAT-AGG-AAA-AAC-AGG-CGT-TG	566	hipO	Fitch et al. (2005)

hypo-R721	GTC-CTG-CAT-TAA-AAG-CTC-CT		
C. jejuni iCJ-UP	CTT-AGA-TTT-ATT-TTT-ATC-TTT-AAC-T	349	Kahn and Edge (2007)
C. jejuni iCJ-DN	ACT-AAA-TGA-TTT-AGT-CTC-A	—	
C. lari ICL-UP	CTT-ACT-TTA-GGT-TTT-AAG-ACC	279	
C. lari ICL-DN	CAA-TAA-AAC-CTT-ACT-ATC-TC	—	
C. coli ICC-UP	GAA-GTA-TCA-ATC-TTA-AAA-AGA-TAA	72	
C. coli ICC-DN	AAA-TAT-ATA-CTT-GCT-TTA-GAT-T		
CJ-FIP	ACA-GCA-CCG-CCA-CCT-ATA-GTA-GAA-GCT-TTT-TTA-AAC-TAG-GGC	—	Yamazaki et al. (2008)
CJ-BIP	AGG-CAG-CAG-AAC-TTA-CGC-ATT-GAG-TTT-GAA-AAA-ACA-TTC-TAC-CTC-T	—	
CJ-F3	GCA-AGA-CAA-TAT-TAT-TGA-TCG-C	—	
CJ-B3	CTT-TCA-CAG-GCT-GCA-CTT	—	
CJ-LF	CTA-GCT-GCT-ACT-ACA-GAA-CAA-C	—	
CJ-LB	CAT-CAA-GCT-TCA-CAA-GGA-AA	—	
CC-F-P	AAG-AGA-TAA-ACA-CCA-TGA-TCC-CAG-TCA-TGA-ATG-AGC-TTA-CTT-TAG-C	—	Yamazaki et al. (2008)
CC-BIP	CCG-GCA-AAG-ACT-TAT-GAT-AAA-GCT-ACC-GCC-ATT-CCT-AAA-ACA-AG	—	
C-F3	TGG-GAG-CGT-TTT-TGA-TCT-CC-B3 AAT-CAA-ACT-CAC-CGC-CAT	—	
CC-LF	CCA-CTA-CAG-CAA-AGG-TGA-TG	—	
CC-LB	CCA-CGA-TAG-CCT-TTA-TGG-A	—	

[a] N = A + G + C + T, Y = C + T, R = A + G, M = A + C, W = A + T, D = T + G + A.

cells per PCR, which was tenfold more sensitive than visual observation of stained agarose gels.

Alm, Guerry, and Trust (1993) used a primer pair designated pg50 and RAA19 (Table 10.6) to specifically amplify a 1448-bp product of the *flaA* gene from strains of both *C. coli* and *C. jejuni*. The primer pair designated RAA9 and RAA19 (Table 10.6) was used to specifically amplify a 1459-bp product of the *flaB* gene from strains of both species.

Birkenhead et al. (1993) amplified portions of the *flaA* gene of *Campylobacter* spp. using the PCR. Primers were chosen that amplified 1.3-kb segments of the *flaA* gene in *C. jejuni* and *C. coli* (Table 10.6). The amplicon for *C. upsaliensis* was approximately 1.7 kb in size. Other species of *Campylobacter* failed to yield an amplicon.

Wegmüller, Lütty, and Candrian (1993) developed a direct PCR for detection of *C. jejuni* and *C. coli* based on the intergenic sequence residing between the flagellin genes *flaA* and *flaB* (common to both species) using a triple set of primers (Table 10.6) resulting in a seminested PCR. One microliter of amplified samples derived from an initial PCR with primers CF03 and CF04 was used in a second PCR with primers CF03 and CF02. The length of the amplicons varied because of variations in the length of the flagellin intergenic sequence among the various strains examined. The initial PCR resulted in amplicons of 340–380 bp. The second (seminested) PCR resulted in amplicons of 180–220 bp. The detection limit with water, milk, and soft cheese was 10 CFU per PCR reaction. A set of 93 dairy samples yielded 6.5% samples positive for the presence of *C. jejuni* or *C. coli*, and none were found with a conventional culture method.

Waage et al. (1999) developed a rapid and sensitive PCR assay for detection and species identification of small numbers of *C. jejuni* and *C. coli*, in addition to other campylobacters in water, sewage, and food samples. Water and sewage samples were filtered, and the filters enriched overnight in a nonselective medium (Rosef broth). A seminested PCR, based on specific amplification of the intergenic sequence between *Campylobacter* flagellin genes *flaA* and *flaB*, was performed. The primers used (Table 10.6) were those of Wegmüller, Lütty, and Candrian (1993). Primers CF03 and CF04 were for the primary PCR and amplified a sequence of 340 to 380 bp depending on the species, whereas primers CF03 and CF02 were used for the seminested PCR and amplified a 180- to 220-bp sequence, also depending on the species. *Campylobacter* species were identified on the basis of the presence of bands in the primary or seminested PCRs and on the size of the bands. The detection level was 3 to 5 CFU of *C. jejuni* per 100 ml of water samples. Dilution of enriched cultures 1:10 with sterile broth prior to the PCR was sometimes necessary to obtain positive results. Overnight enrichment of food samples allowed detection of 3 CFU per gram of food. Variable results were obtained with food samples without prior enrichment.

Kirk and Rowe (1994) developed a direct PCR assay for detection of *C. jejuni* and *C. coli* in water. Water samples (20 ml) were passed through polycarbonate filters (0.4 μ, 13 mm diam.). Each membrane filter was then placed in a 0.5 ml capacity PCR tube and 110 μl of sterile d.H_2O added, followed by sonication for 2 min. After sonication, each membrane was removed, and the samples subjected to six freeze–thaw cycles, followed by heating at 85°C to inactivate enzymes. The PCR mix of

200 µl contained the entire sonicated sample. The triple primer seminested PCR procedure of Wegmüller, Lütty, and Candrian (1993) was utilized, which amplifies sequences in the *flaA* and *flaB* gene sequences. Primers CF03 and CF04 (Table 10.6) were used for initial detection, yielding 340–380-bp amplicons. Primers CF02 and CF03 (Table 10.6) were used for confirmation, yielding amplicons of 180 to 220 bp. The sensitivity was 10–20 CFU/ml of water, which was reduced to 2 CFU/ml of water with 100 ml samples.

Purdy, Ash, and Fricker (1996) developed a seminested PCR procedure for the specific detection of campylobacters from a variety of water samples following over-night selective enrichment. The primers were designed from *Campylobacter* DNA encoding 16S rRNA genes. Primers P2 and P4 (Table 10.6) were used to amplify a 603-bp sequence. Seminested PCR reactions were performed by using 1 µl of PCR products from the first amplification reaction with primers P2 and P4 as template in a second PCR using primers P2 and P3 amplifying a 531-bp sequence. Sensitivity was 1 CFU per PCR. Strong amplicon bands of appropriate size were obtained with isolates of *C. jejuni* and *C. coli*, whereas weak bands were obtained with *C. fetus*, and no bands with *C. sputorum*.

Most PCR primers specific for strains of a given species are usually designed from gene sequencing data. Day, Pepper, and Joens (1997) developed a species-specific pair of PCR primers for *C. jejuni* strains by first generating banding pat-terns from a single *C. jejuni* strain using a random primer. DNA agarose bands were screened for species specificity by dot-blot hybridization with DNA from several *C. jejuni* strains, other species of *Campylobacter*, and various enteric species. A 486-bp DNA band that hybridized only to *C. jejuni* DNA was sequenced, and a pair of primers cj1 and cj4 (Table 10.6) were identified and synthesized that amplified a 265-bp product from *C. jejuni* isolates only. This same primer pair yielded an 800-bp product with a *C. coli* isolate that was readily distinguishable from the 265-bp product from *C. jejuni* strains.

Fermér and Engvall (1999) reported on the development of a sensitive PCR assay involving digestion of PCR products with *Alu*I and *Rsp*509I to identify the thermophilic *Campylobacter* species *C. jejuni*, *C. coli*, *C. lari*, and *C. upsaliensis*. The primers used, THERM1 and THERM 4 (Table 10.6), are based on the most variable part of the 23S rRNA gene, and amplify a 491-bp sequence from all four thermophilic species.

Occhialini et al. (1996) compared the efficiency of identification of *C. jejuni* and *C. coli* with two sets of PCR primers (VS15/VS16 for *C. jejuni* and CSF/CSR for *C. coli*), each specific for each species (Table 10.6), to that of the Api Campy (Biomérieux) test kit. Very good correlation was obtained. The two closely related species were readily distinguished by the two pairs of PCR primers. The Api Campy test kit was also highly effective. The authors did not indicate the identity of the target genes for PCR. The utilization of propionate was found to be a singularly unique character diagnostic for *C. coli* isolates, whereas hippuric acid utilization was diagnostic for *C. jejuni*.

A PCR-based method for rapid detection of foodborne thermotolerant campy-lobacters was evaluated through a collaborative trial format with 12 laboratories testing 12 seeded chicken and 12 seeded swine carcass rinse samples (Josefsen et al., 2004a). The participating laboratories purchased their own primers (Lübeck et al., 2003) OT1559 and 18-1rev (Table 10.6) and *Tth* DNA polymerase. The method yielded interlaboratory diagnostic sensitivity and specificity values with chicken samples of 96.7% and 100%, respectively, and with swine samples 94.2% and 83.3%, respectively.

Bang et al. (2004) subjected 117 *C. jejuni* isolates from Danish turkeys to PCR analysis for detection of seven virulence and toxin genes. All 117 isolates were positive for *fla*A, *cad*F, and *deu*E genes. 103 isolates were positive for the *dct* gene cluster, whereas 101, 102, and 110 isolates were positive for *cdt*A, *cdt*B, and the *cdt*C gene cluster, respectively. Only 39 of the isolates were positive for the *virB11* gene. Among the 117 isolates, 114 produced CLDT in Vero cell assays, 105 produced CLDT in colon 205 cell assays, and 109 produced CLDT in chicken embryo assays. The high prevalence of the seven virulence and toxin genes indicated that these putative pathogenic determinants are widespread among *C. jejuni* isolates from turkeys. Primer sequences used in this study are given in Table 10.6.

Mateo et al. (2005) compared the detection of campylobacters from retail poultry products using PCR and plating of poultry rinses. The primer pair MD16S1/MD16S2 (Table 10.6), as described by Denis et al. (1999), which amplify a sequence of the 16S rRNA gene of members of the genus *Campylobacter*, was used in addition to the respective primer pairs MDmapA1/MDmapA2 and COL3/MDCOL2 for amplifying sequences of the *mapa* gene of *C. jejuni* and the *ceuE* gene of *C. coli*, respectively (Table 10.6). All five cooked samples failed to yield campylobacters by plate cultivation. However, two of the five cooked samples were positive by PCR due presumably to the presence of DNA from dead cells. Direct plating of tissue rinses yielded 41/68 (60.2%) positive samples compared to direct PCR, which yielded 47/68 (69.1%). Enrichment plating yielded 47/68 (67.6%) positive samples compared to enriched PCR, which yielded 54/68 (79.4%) positive samples. 51/68 (75%) of the raw samples were proven to be positive for *C. jejuni* and 28/68 (41.2%) were proven to be positive for *C. coli*. It is interesting that no samples were positive for *C. coli* alone.

In contrast, El-Shibiny, Connerton, and Connerton (2005) found that *C. coli* was the predominant *Campylobacter* species isolated from both organic and free-range chickens. Only two *Campylobacter* positive birds (5%) from the free-range flock harbored both species. Campylobacters were isolated from the free-range flock much earlier than from the organic flock and were initially all *C. jejuni* from day 8 until replaced by *C. coli* at day 31 and returned toward the end of the rearing period. The appearance of either bacteriophages or the production of bacteriocins was associated with change in the levels of colonization and the predominant genotypes were determined by PFGE. *C. jejuni* counts in the ceca of chickens in the presence of bacteriophage were found to be lower by a factor of 1.4 logs compared to samples with no detectable bacteriophage (Atterbury et al. 2005). The authors speculated that phage-resistant mutants may be less virulent and less efficient colonizers than the wild type.

Jensen et al. (2005) developed a real-time PCR assay utilizing dual-labeled probes for the detection and identification of thermophilic *Campylobacter* spp. in swine fecal samples. The primers were based on the *glyA* gene, which encodes serine hydroxymethyltransferase of *C. jejuni, C. coli, C. lari,* and *C. upsaliensis.* This gene is highly conserved but exhibits sufficient sequence variation to allow differentiation of these four species with the use of separate pairs of primers. The probes were labeled with either 6-carboxyflorescein (FAM) or TET at the 5'-end and with the quencher tetramethylrhodamine (TAMRA) at the 3'-end (Table 10.6).

A microfabricated PCR chip has recently been developed for detection of *C. jejuni* (Poulsen et al., 2005). The chip has a 20-µl chamber volume and was able to detect 56 CFU/µl, resulting in a minimum detection level of 1120 CFU from pure cultures per PCR. The primers used were F2B and R1B (Table 10.6) from Konkel et al. (1999) that amplify a ~400-bp sequence of the *cadF* gene that encodes the 37-kDa membrane protein CadF (short for campylobacter adhesion to fibronectin). Advantages of the chip were rapid thermal transfer time (less than 5 s per cycle) resulting in 40 cycles being completed in 1.5 hr. Cell lysis to release DNA prior to PCR was found not to be necessary.

Rasmussen et al. (1996) developed a pair of PCR primers (Table 10.6) derived from the conserved regions of the flagellin gene sequences of *flaA* and *flaB* of *C. coli* to amplify gene sequences from *C. jejuni* and *C. coli.* The resulting amplicons were 810 and 813 bp, respectively. The PCR products were analyzed by hybridization to an internal probe (Table 10.6) immobilized onto the surface of microtiter wells. With pure cultures, the minimum detection limit for *C. coli* was strain-dependent and varied from 10 to 200 cells with agarose gels and from 5 to 20 cells with microtiter well assays. With *C. jejuni,* the minimum detection limit was less than 10 cells with agarose gels and less than 1 cell with microtiter wells. With poultry fecal enrichment cultures, the minimal detection limit with the microtiter well assay was 1–20 cells per PCR tube.

Linton, Owen, and Stanley (1996b) used 16S rRNA gene sequences to design PCR assays specific for the genus *Campylobacter* and for five *Campylobacter* species important in veterinary and human medicine. Primers C412F and C1288R amplify an 816-bp sequence from all *Campylobacter* species; primers CHCU146F and CU1024R (Table 10.6) amplify an 878-bp sequence only from *C. upsaliensis,* which is primarily associated with cats and dogs, causing occasional human enteric disease. Primers CHCU146F and CH1371R amplify a 1225-bp or 1375-bp amplicon (Table 10.6) only from *C. helveticus,* which occurs in cats and dogs and is not associated with human disease. *C. fetus* subsp. *fetus* is an agent of septic abortion in livestock animals, whereas *C. fetus* subsp. *venerealis* produces infertility in cattle. The primers CFCH57F and CF1054R (Table 10.6) amplify a sequence of 997 bp from both subspecies of *C. fetus* only. *C. hyointestinalis* is a pathogen of swine, which causes proliferative enteritis, and can also cause a nonbloody diarrhea in humans. Primers CFCH57F and CH1344R (Table 10.6) generate an amplicon of 1287 bp solely from C. *hyointestinalis. C. lari* occurs widely in avian species and in various domestic and wild animals and is considered to cause isolated cases of human gastroenteritis and bacteremia. The primers CL594F and CL1155R (Table 10.6) generate a 561-bp amplicon solely from *C. lari.* These primers should greatly facilitate the identification of non–*C. jejuni* and non–*C. coli* campylobacters from clinical and environmental sources.

Lawson et al. (1997) developed a PCR assay based on the 16S rRNA gene for the direct detection and differentiation of *C. upsaliensis* and *C. helveticus* in seeded human feces. An upstream primer CHCU146F (Table 10.6) that recognizes a region of the gene common only to *C. helveticus* and *C. upsaliensis* was used with two downstream primers. The downstream primer CU1024R (Table 10.6) is specific for *C. upsaliensis* and yields an 878-bp amplicon with the forward primer. The reverse primer CH1371R (Table 10.6) with CHCU146F produces a 1225- or 1375-bp amplicon specific for *C. helveticus*. Cells were lysed with guanidine isothiocyanate (GUSCN), and the resulting DNA was purified with polyvinyl pyrolidine to remove PCR inhibitors. The minimum level of detection was 34 to 250 CFU per PCR tube.

Ng et al. (1997) developed a PCR assay specific for *C. jejuni*. The primers CL1 and CR3 (Table 10.6) amplified a 402-bp sequence. Amplicons were confirmed by *Dde*I digestion which generated 166-, 118-, 105-, and 13-bp fragments. Overnight enrichment of poultry rinse samples spiked initially with as few as 10 CFU/ml yielding PCR amplicons. The assay was specific for *C. jejuni* and *C. jejuni* subsp. *doylei*.

Vanniasinkam, Lanser, and Barton (1999) reported on the use of PCR for the detection of *Campylobacter* spp. in human clinical fecal specimens following enrichment in BHI broth. PCR detected significantly more positive specimens than culture methodology. An 804-bp sequence of the 16S rRNA gene was amplified using primers designated CF1 and CR2 (Table 10.6). This primer pair was specific for all species of *Campylobacter*. The DNA extraction method used was rapid and efficient and consisted of treating 300 µl of BHI broth culture with lysozyme (2 mg/ml) for 10 min at 32°C, followed by the DNA extraction method of Saunders et al. (1990).

Eyigor et al. (1999) undertook a study to determine whether isolates of *C. jejuni* and *C. coli* from poultry carcasses carry CLDT. Hela cells were used for detection of CLDT. Only 1 out of 70 *C. jejuni* isolates failed to produce CLDT. All 35 *C. coli* isolates examined produced little or no CLDT even though the *cdtB* gene was present. The primer pair VAT2 and VMI1 (Table 10.6) of Picket et al. (1996) was used to amplify a 494-bp sequence of the *cdtB* gene. The primer pair VAT2 and LPF-X (Table 10.6) from Picket et al. (1996) was used to amplify a 999-bp sequence of the *cdt* gene cassette. *Eco*RI nuclease was used to restrict the 494-bp amplicon from the *C. jejuni* isolates that resulted in a single cut, yielding fragments of 0.35 and 0.15 kb. No restriction occurred with the 494-bp amplicon from the *C. coli* strains.

Denis et al. (1999) developed a multiplex PCR (m-PCR) assay with three sets of primers for simultaneous identification of *C. jejuni* and *C. coli*. The primers MD16S1 and MD16S2 amplified an 857-bp sequence of the 16S rRNA gene conserved in the genus *Campylobacter* (Table 10.6). The primers MDmapA1 and MDmapA2 amplified a 589-bp sequence of the mapA gene of *C. jejuni* (Table 10.6). The primers COL3 and MDCOL2 amplified a 462-bp sequence of the *ceuE* gene of *C. coli* (Table 10.6). When both species were present, three DNA bands were generated. When one species was present, two DNA bands were produced, one from the genus *Campylobacter* and the second from one of the two species. The efficiency of biochemical identification was found to be only 34%, compared to 100% efficiency with the m-PCR assay when applied to a total of 294 isolates comprised of 66 *C. jejuni* isolates and 228 *C. coli* isolates.

Gonzalez et al. (1997) developed a PCR assay for rapid identification of *C. jejuni* and *C. coli* based on the *ceuE* gene that encodes a protein involved in siderophore transport. A nucleotide sequence divergence of approximately 13% allowed the design of two species-specific PCR primer sets. The primers for *C. jejuni* JE and JEJ2 (Table 10.6) amplified a 793-bp sequence of the *ceuE* gene and the primers for *C. coli* COL1 and COL2 (Table 10.6) amplified an 894-bp sequence. The identity of amplicons was determined with the use of specific radio-labeled oligonucleotide probes and Southern hybridization dot blot analysis. The probes were designated JEJ3 and COL for *C. jejuni* and *C. coli*, respectively (Table 10.6).

A multiplex PCR assay for simultaneous detection of *C. jeuni* and *C. coli* was developed by Harmon, Ransom, and Wesley (1997). Two pairs of primers were used. Primer set I consisted of primers pg-3 and pg-50 (Table 10.6) from Oyofo et al. (1992) and amplified a 460-bp sequence derived from the *flaA* gene present in both *C. coli* and *C. jejuni*. Primer set II consisted of primers C-1 and C-4 (Table 10.6) and amplified a 160-bp sequence unique to *C. jejuni*. Strains of *C. jejuni* yielded both the 160- and 460-bp amplicons. *C. coli* strains yielded only the 460-bp amplicon.

Grennan et al. (2001) described the development of a PCR-ELISA for the detection of *Campylobacter* species and the discrimination of *C. jejuni* and *C. coli* in poultry tissue. The PCR assay targeted the 16S/23S ribosomal RNA intergenic spacer region conserved among *Campylobacter* species employing the biotinylated primers CAMP1F-B and B1-B (Table 10.6). The PCR reactions were performed on Malthus enrichment broths that were boiled for release of DNA that was directly incorporated into the PCRs. DNA oligonucleotide probes, designed from the adapter sequences of *C. jejuni*, *C. coli*, and other *Campylobacter* species (O'Sullivan et al., 2000) modified with an amine and a 12-T linker between the amino groups and the species-specific sequence, were immobilized onto Nucleolink wells (Nakg Nunc International, Roskilde, Denmark). The oligonucleotide probes CJEJ7, CCOL2, and CAMP4 (Table 10.6) were used for specific detection of *C. jejuni*, *C. coli*, and members of the genus *Campylobacter*, respectively. Biotinylated PCR products amplified from *Campylobacter* species were alkali denatured, hybridized to the immobilized probe on the surface of the wells, and detected using streptavidin-HRP conjugate and the chromogenic substrate 3,3',5,5'-tetramethylbenzidine (TMB). The limit of detection of the PCR-ELISA was 40–120 cells for *C. jejuni* and *C. coli* with their respective species-specific probes and four cells with the *Campylobacter* genus-specific probe.

Englen and Fedorka-Cray (2002) evaluated the selectivity of a newly developed multiplex commercial PCR assay (DuPont Qualicon) for simultaneous identification of *C. jenuni* and *C. coli* compared to previously established PCR procedures. The PCR commercial primer sequences (cc-32-237, cc-30-rc682, cj-30-560, and cj-30-rc682; Table 10.6) were designed from regions of the *cadF* virulence gene unique to *C. jejuni* and *C. coli*. Noncommercial PCR primers (HIP400F and HIP1134R; Table 10.6), specific for the hippuricase gene (*hipO*) of *C. jejuni* produced a 735-bp amplicon and the primers Col1F and Col2R (Table 10.6) produced a 894-bp amplicon and were used for the siderophore transport gene (*ceuE*) of *C. coli* (Table 10.6). We examined 133 *Campylobacter* isolates (*C. jejuni* and *C. coli*) from poultry. The commercial multiplex PCR yielded results identical to the noncommercial multiplex PCR system.

Inglis and Kalischuk (2003) reported on the development of PCR to directly detect and distinguish *Campylobacter* species without enrichment. Primers specific for 10 species of *Campylobacter* were used. A total of 24 fecal samples were obtained from eight milk cows (three samples from each cow on separate dates). The QI amp® DNA stool mini kit (Qiagen, Valencia, California) was used to extract and purify DNA from 200 mg of the fecal samples. With genus-specific primers, *Campylobacter* DNA was detected in 75% of bovine fecal samples, representing an increase in sensitivity of 8% compared to microbial isolation using four media. With nested primers, *C. jeuni* and *C. lanienae* were detected in 25% and 67% of the samples, respectively. In no instance was DNA from *C. coli*, *C. fetus*, or *C. hyointestinalis* detected.

Inglis, Kalischuk, and Busz (2003) made use of the PCR to survey campylobacters associated with feces from 382 beef cattle. An internal DNA control sequence was seeded into feces samples providing a smaller amplicon than that of the *Campylobacter* genus with the same primer pair C412F and C1228R (Table 10.6). Samples PCR positive for the genus *Campylobacter* were then subjected to a primary multiplex PCR targeting the *ceuE* gene of *C. coli* with the primers COL3 and MDCOL2 (Table 10.6) from Denis et al. (1999) and the *b* gene of *C. jejuni* with the primers MDmapA1Upper and MDmapA2Lower (Table 10.6). A secondary nested multiplex PCR was then performed targeting the *ceuE* gene of *C. coli* with the primers CCceuEN3F and CCceuEN3R (Table 10.6) and the *mapA* gene of *C. jejuni* with the primers CJmapAN3F and CJmapAN3R (Table 10.6). Primer pairs were used for detection of a total of five *Campylobacter* species known to be associated with cattle. Among the 382 samples, 305 (80.3%) were positive for the genus *Campylobacter*. The most frequently detected species was *C. lanienae* (49%). *C. jejuni* was detected in 38% of the samples and *C. hyointestinalis* and *C. coli* were detected in 8% and 0.5% of the samples, respectively. *C. fetus* DNA was not detected. With some samples, the genus primer set yielded positive results but species primers yielded negative results. Sequencing of the 16S rRNA amplicons indicated the presence of at least two undescribed species of *Campylobacter*.

Lübeck et al. (2003) reported on screening the effectiveness of 15 primers targeting the conserved 16S rRNA gene of *C. jeuni*, *C. coli*, and *C. lari* in addition to two primers targeting the conserved 23S rRNA gene for selectively detecting these three species. Only one primer pair, OT1559 and 18-1rev (Table 10.6) was found suitably selective. Several DNA polymerases were assessed for resistance to PCR inhibitors derived from poultry rinse samples. The DNA polymerase *Tth* was found not to be inhibited at a chicken carcass rinse concentration of 2% (vol/vol), unlike both *Taq* DNA polymerase and DyNAzyme.

Nogva et al. (2000) developed a real-time PCR (Rti-PCR) specific for *J. jejuni* involving the use of a TaqMan probe labeled at the 5′-end with 6-carboxyfloresceine (FAM) and at the 3′-end with 6-carboxy-N,N,N′,N′-tetramethylrhodamine (TAMRA) in conjunction with a pair of primers (Forward/Reverse; Table 10.6). The resulting 86-bp amplicon was derived from an unknown gene. Good correlation was obtained between CFU counts and the quantitative aspect of Rti-PCR (C_T values) using stressed cells. This was expected due to degradation of DNA in dead cells resulting from internal DNases. The limit of detection was linear over at least six log cycles.

Sails et al. (2003) developed an Rti-PCR assay for the quantitative detection of *C. jejuni* in foods after enrichment culture. The primers 82F and 197R (Table 10.6) targeted a 115-bp portion of the open reading frame C (ORF-C) sequence specific for *C. jejuni* and had a linear range of quantification over six orders of magnitude with a limit of detection of 12 genome equivalents (~12 cells). The 5'-end of the probe CJTP2 (Table 10.6) was labeled with FAM and the 3'-end with TAMRA. Of interest was the replacement of all the thymidine residues with 5-propyne-2'-deoxyuridine, which increases the melting point by 1°C per substitution. This allows shorter probes to be utilized while maintaining an optimum melting point of 66 to 70°C.

An Rti-PCR assay for quantitative detection of *C. jejuni* in poultry, milk, and environmental water samples without enrichment was developed by Yang et al. (2003). The entire assay was completed in 1 hr with a detection limit of approximately 6 to 15 CFU per PCR. Primers VS15 and VS16 (Table 10.6) based on a region of the *VS1* gene amplify a 358-bp sequence. A TaqMan probe (Table 10.6) was labeled at the 5'-end with FAM and at the 3'-end with TAMRA.

Josefsen, Jacobsen, and Hoorfar (2004b) developed an Rti-PCR assay for detection of *C. jejuni, C. coli,* and *C. lari.* The system involved enrichment of poultry tissue in Bolton Broth for 24 hr followed by a simple Chelex 100 resin-based sample treatment and thermal lysis of cells. Rti-PCR using a pair of primers designated FP and RP (Table 10.6) amplifying a sequence of the 16S rRNA gene was then used. A target *Campylobacter* probe labeled with an FAM reporter dye at the 5'-end and with TAMRA at the 3'-end as quencher dye was used in combination with an internal amplification control and a corresponding probe labeled with VIC (proprietary dye, Applied Biosystems, Foster City, California) at the 5'-end and TAMRA at the 3'-end (Table 10.6). A detection level of 1×10^3 and 2×10^3 CFU/ml of enrichment broths was achieved with the RotorGene and ABI-PRISM real-time thermal cycling units, respectively.

Inglis and Kalischuk (2004) reported on the development of a nested RTQ-PCR with the fluorescent dye SYBR green to directly quantify *C. jejuni* in cattle feces. The single copy *mapA* gene was selected, and a 589-bp sequence was amplifed with the use of the primary primers MDmapA1Upper and MDmapA2Lower (Table 10.6) from Denis et al. (1999). Nested PCR was performed with the use of primers QCjmapANF and QCjmapANR that ampified a 94-bp sequence of the *mapA* gene (Table 10.6). The QIamp DNA stool minikit (Qiagen) was used to extract DNA from 200 mg of feces. Nested PCR did not increase the specificity or sensitivity of *C. jejuni* quantification, and the limit of detection was 19 to 25 genome copies (~3×10^3 CFU/g of feces).

C. Randomly Amplified Polymorphic DNA (RAPD) Analysis

Wassenaar and Newell (2000) have briefly reviewed the methods used for genotyping of *Campylobacter* spp. indicating advantages and limitations of each of seven different molecular methods. It is presently recognized that the use of a single set of random primers in RAPD may not distinguish genomically different isolates and that the use of two or three random primers is often required (Levin, 2004).

Carvalho and Ruiz-Palacios (1996) subjected 88 isolates of *C. jejuni* from five countries to RAPD analysis with the random primer 1290 (Table 10.6). Results yielded four clusters indicating a tendency of the strains to form clusters according to their geographic origin. Isolates forming cluster 1 were from North America, clusters 2 and 3 were comprised mainly of isolates from Mexico, and isolates comprising cluster 4 were from bloody diarrhea in marmosets in Brazil.

Madden, Moran, and Scates (1996a) isolated 200 strains of *C. coli* from the swine of a single herd and subjected the isolates to RAPD analysis after screening 20 random sequence 10-mer primers. Primer Gen 3-60-25 (Table 10.6) was selected as best. Individual swine were found to be colonized by a single RAPD type. Isolates of *C. coli* and *C. jejuni* from different sources yielded clearly different RAPD patterns, whereas isolates from common sources exhibited identical or very similar patterns.

Madden, Moran, and Scates (1996b) subjected 18 isolates of *C. jejuni* from a single poultry processing plant to RAPD and PCR-restriction fragment length polymorphism (PCR-RFLP) analyses. PCR-RFLP typing was based on the method of Nachamkin, Bohachick, and Patton (1993a) involving the use of a primer pair (Table 10.6) that amplified a 1728-basepair sequence of the *flaA* gene. The resulting amplicon was then restricted with *Dde*1. PCR-RFLP yielded three types of *C. jejuni,* and RAPD yielded 10 types. RAPD typing was therefore far superior to PCR-RFLP for distinguishing between isolates. Several RAPD types were found to repeatedly occur over the period of a year from the poultry tissue derived from the single processor.

Bolton et al. (1996) undertook a multicenter collaborative study to compare established subtyping techniques for *C. jejuni* and included 16S rRNA RFLPs, PFGE, RAPD, PCR-RFLP, and multilocus enzyme electrophoresis (MLEE). RAPD yielded the greatest diversity of types. The use of more than one restriction nuclease with 16S ribotyping and PFGE increased discrimination.

Hong et al. (2003) reported on the use of the *Campylobacter ceuE* gene for designing a probe for use in PCR-ELISA assays for the rapid and direct detection of *C. jejuni* and *C. coli* from poultry swab samples without enrichment. The PCR primers JCF1(F) and JCF1(R) (Table 10.6) amplified a 610-bp sequence of the conserved *ceuE* gene of *C. jejuni* and *C. coli* encoding a lipoprotein component of a binding-protein-dependent transport system for the siderophore enterochelin. A digoxygenin (DIG) labeled *ceuE* probe (Table 10.6) was incorporated directly into the PCR reaction mix with a biotin molecule added to the 3'-end to prevent the probe from serving as a primer in the PCR. The bottom of an ELISA plate was coated with streptavidin and used to bind amplicon–probe complexes. Anti-DIG antibody-HRP conjugate was then used to colorimetrically detect the bound amplicons. The detection limit was 4×10^1 CFU/ml of carcass rinse water. The ELISA procedure increased the sensitivity of the conventional PCR by 10^2 to 10^3-fold.

1. Pulsed Field Gel Electrophoresis (PFGE)

Lior biotype II strains of *C. jejuni* produce extracellular DNAse, which can interfere with PFGE analysis of genomic DNA. Gibson, Sutherland, and Owen (1994) treated DNase positive strains of *C. jejuni* with 3.7% formaldehyde for 1 hr at room temperature after suspending the cells in 0.85% NaCl. The cells were then washed three times with saline and were finally suspended in saline prior to PFGE analysis. This procedure satisfactorily denatured the cellular DNase and allowed *Sma*I digestion patterns to be visualized in agarose gels.

On et al. (1998) subjected 36 Danish strains of *C. jejuni* to PFGE typing using the restriction nucleases *Sma*I, *Sal*I, *Ppn*I, and *Bam* HI. Human origin accounted for 20 strains, with 12 from a single outbreak of campylobacteriosis, 10 from poultry, 5 from cattle feces, and 1 water isolate. *Sma*I yielded 5–9 DNA bands per strain, *Sal*I yielded 5–7 bands, and *Bam*HI yielded 3–6; *Kpn*I yielded 10–16 bands and was therefore far more discriminatory. All 12 outbreak strains were indistinguishable by each of the restriction nucleases, whereas two strains yielding unique *Sma*I profiles were found to be distinct by other restriction nucleases. Among the 22 other strains assigned to one of six *Sma*I profile groups, 14 were not differentiated further with the other nucleases. However, eight isolates belonging to five of the six *Sma*I profile groups were shown to be distinct from other strains belonging to their respective *Sma*I profile group with other nucleases. Certain strains that yielded the same *Sma*I profile yielded distinguishable profiles with *Kpn*I. The authors concluded that PFGE analysis should be performed with the use of at least two restriction nucleases including *Kpn*I.

Nielsen et al. (2000) undertook a study comparing six subtyping methods for 90 *C. jeuni* isolates from poultry, cattle, and sporadic human clinical cases from a waterborne outbreak. The methods were evaluated on the basis of their abilities to identify isolates from one outbreak and discriminate between unrelated isolates and the agreement between methods in identifying clonal lines. All six methods identified the outbreak strain. RAPD and PFGE were the most discriminating methods, followed by *fla*-RFLP and RiboPrinting. Fla-DGGE and serotyping were least discriminating. None of the subtypes could be related to only one source. It is interesting that in two cases isolates from cattle and human patients were found identical according to all six methods.

Korolik, Moorthy, and Coloe (1995) analyzed the banding patterns of 120 poultry and 49 human isolates of *C. jejuni* and *C. coli* resulting from the application of the restriction endonucleases *Cla*I and *Eco*RV to genomic DNA. *Cla*I generated a multitude of DNA fragments with all strains from 9.5 to 3.5 kb. However, *Eco*RV generated a single unique DNA fragment of 3.0 kb present in all of the *C. jejuni* strains, which was absent from the *C. coli* strains. The resulting banding patterns were further examined by Southern blot analysis with a DNA probe pMO2005, which is able to distinguish between *C. jejuni* and *C. coli* isolates (Korolik, Krishnapillai, and Coloe, 1988). The probe hybridized with *Cla*I generated 18.5-kb genomic DNA fragment of *C. jejuni* from humans and with a 14.5- and 4.0-kb fragment of *C. jejuni* from poultry. With all *C. coli* strains, the pMO2005 probe hybridized strongly with a 9.0-kb *Cla*I generated fragment of *C. coli* strains. The DNA probe pMO2005 hybridized with 78% of the banding patterns derived from *C. jejuni* poultry isolates, yielding characteristic bands of 14.5 and 4.0 kb, and 22% of these strains yielded a single 18.5-kb fragment. In contrast, 71% of human *C. jejuni* isolates hybridized with the single 18.5-kb fragment and only 29% hybridized with the 14.5- and 4.0-kb fragments. These results suggested that only a small proportion of *C. jejuni* strains that colonize poultry may cause disease. These results also indicated that strains of *C. jejuni* and *C. coli* of poultry origin cannot be distinguished from those of human origin on the basis of restriction endonuclease (RE) banding patterns generated with *Cla*I and *Eco*RV.

Geilhausen, Koenen, and Mauff (1996) undertook a PFGE study of *Campylobacter* isolates from chicken breast meat from different retail sources in Cologne, Germany, over a period of 21 weeks. *Campylobacter* strains isolated from patients with intense enteric disease were compared to poultry strains isolated 1 week later from the same area using *Sma*I digests *of Campylobacter* DNA. The large number of banding patterns (commonly encountered with food and clinical isolates) and the lack of identities between poultry and human isolates suggested that a DNA shift may occur during passage from poultry to humans. One patient who had consumed almost uncooked poultry developed enteritis with excretion of isolates having a banding pattern totally identical to the isolates of the incriminated poultry meat.

Hageltorn and Berndtson (1996) subjected 20 isolates of *C. jejuni* to PFGE analysis using the restriction nucleases *Sma*I, *Sal*I, and *KSP*I. The isolates were from three suspected milkborne outbreaks in Sweden involving patients who consumed raw milk and the dairy cows that produced the suspected milk. Each of the outbreaks yielded different banding patterns, but within each outbreak, strains from patients and cow feces yielded identical patterns. A large flock of wild geese, visiting the cow's grazing fields the week before the outbreaks, was suspected as the source.

Steel et al. (1998) undertook the molecular typing of 110 *C. jejuni* and 31 *C. coli* isolates from human and environmental sources derived from several abattoirs in Ontario, Canada. The isolates were subjected to analysis by PFGE, fatty acid profile typing, serotyping, and biotyping. PFGE was found to be the most discriminating of the typing methods.

Nadeau, Messier, and Quessy (2003) undertook a comparative study of 173 *Campylobacter* isolates from poultry broilers and 24 from cases of human diarrhea in a region of Quebec, Canada. Most of the isolates were *C. jejuni* and the remainder *C. coli*. Adherence to and invasion of tissue culture cells in addition to cytotoxicity were used as criteria of virulence. All isolates adhered to and 63% invaded INT-407 cells, whereas only 13% were cytotoxic for CHO cells. The proportion of isolates exhibiting a high invasive potential or Vero cell cytotoxicity was significantly higher for human than for poultry isolates. PFGE typing of the isolates involved the use of the restriction nuclease *Kpn*I. The 197 *Campylobacter* isolates examined yielded 57 *Kpn*I genotypes. Isolates that clustered in a particular *Kpn*I genotype had similar invasion abilities and had the same CHO cell cytotoxicity status (cytotoxic or not). Nearly all strongly invasive isolates belonged to biotypes 1 and 2 of *C. jejuni*, whereas CHO cell-cytotoxic isolates were associated with *C. jejuni* biotypes 3 and 4. The data revealed an association of in vitro virulence properties with biotype, genotype, and host of origin.

The genomic stability of 12 *C. jejuni* strains after passage through newly hatched chicks' intestines was examined by Hänninen, Hakkinen, and Rautelin (1999). The chicks were completely free of campylobacters before implantation of the strains. Two of the 12 strains were found by PFGE (using formaldehyde-treated cells to denature DNase) to undergo alteration of genotype. This genetic instability was confirmed by ribotyping. In addition, one of these two strains also underwent changes from serotype 57 to serotype 27. The study suggested that during intestinal colonization, genomic rearrangement may occur.

Fitzgerald et al. (2001) used flagellin gene typing and *Sma*I-PFGE to determine the genetic diversity among 315 *C. jejuni* and *C. coli* isolates from cattle, sheep, and turkeys from farm environments and sporadic cases of human campylobacteriosis in the same geographic area of the United Kingdom. There were 48 *fla* types obtained and 71 different PFGE patterns were observed. There were 57 isolates from diverse hosts, times, and sources that had an identical combined PFGE/*fla* type. Molecular evidence was presented, suggesting a link between isolates in the farm environment with those causing disease in the community.

Hänninen et al. (2003) studied three waterborne outbreaks of gastroenteritis caused by *C. jejuni* in Finland, where in sparsely populated areas, groundwater is commonly used without disinfection. Large sample volumes (10–20 L) were required for detection of low numbers of *C. jejuni* in water samples. Isolates of *C. jejuni* were subjected to Penner serotyping and to genotyping with PFGE using the restriction nucleases *Sma*I and *Kpn*I. In outbreak 1, *C. jejuni* was isolated from a tapwater sample and was found to be of serotype Pen 12, the same as 10 isolates from 10 patients, and to also yield identical *Sma*I and *Kpn*I PFGE patterns as the 10 identical clinical isolates. The origin of outbreak 2 was indeterminate. With outbreak 3, *C. jejuni* was isolated from one well and *C. coli* from another well. A large dike running from a nearby swine and duck farm toward the wells was implicated. In August of 1998, an outbreak of campylobacteriosis due to *C. jejuni* occurred in a northern municipality of Finland having a population of 15,000. The total number of cases was estimated to be 2700 (Kuusi et al., 2005). The outbreak was found due to drinking nonchlorinated municipal tapwater. All outbreak isolates of *C. jejuni* were found to have the same respective *Sma*I and *Kpn*I-PFGE restriction patterns, clearly indicating that a highly dominant clone was responsible. Water samples were negative for campylobacters and coliforms. The authors concluded that the repair of water mains most probably resulted in contamination and that nonchlorinated groundwater systems may be susceptible to contamination by campylobacters.

Saito et al. (2005) determined the significance of poultry and bovine sources as reservoirs of human campylobacteriosis in Japan. *C. jejuni* was isolated from 52 of 73 retail poultry samples, 19 of 87 bovine bile samples, and 18 of 78 bovine fecal samples. A total of 159 human clinical isolates of *C. jejuni*, 68 poultry isolates, and 42 bovine isolates were subjected to serotype analysis. The most dominant serotype among human and poultry isolates was O:2; human isolates 35 (22%), and poultry isolates 14 (20.6%), and the most dominant serotype among bovine isolates was O:2; 10 isolates (23.8%). *Sma*I PFGE patterns of O:2 serotypes from human, poultry, and bovine sources revealed three major clusters, I, II, and III. Five PFGE patterns in cluster I were common to poultry and human isolates, whereas in cluster II one pattern was common to human and poultry isolates; one pattern was shared by human and bovine isolates. The serotype and PFGE data therefore indicated a possible link between sporadic human campylobacteriosis and *C. jejuni* isolates from retail poultry and bovine sources, suggesting that bovine sources serve as reservoirs for human infections in Japan, which has been observed in other countries.

Campylobacter isolates (48 chicken and 22 turkey) from retail poultry in Estonia were subjected to PFG typing by Praakle-Amin et al. (2007). The isolates were derived from 580 raw broiler chicken and 30 turkey meat samples. Among the isolates, 64 were *C. jejuni*, 4 *C. coli*, and 2 Campylobacter spp. *Sma*I and *Kpn*I PFGE yielded 39 and 34 PFGE types, respectively, revealing a high level of genetic diversity. The majority of the isolates sharing a similar PFGE genotype were found to originate from the same country.

2. Restriction Fragment Length Polymorphism (RFLP) Analysis

Jackson, Fox, and Jones (1996) developed a novel PCR assay for the detection and speciation of *C. jejuni*, *C. coli*, and *C. upsaliensis*. The primers BO4263 and BO4264 (Table 10.6) amplify a 256-bp sequence of a novel open reading frame adjacent to and downstream of a novel *C. jejuni* two-component regulator gene. The assay was based on the RFLP of PCR products digested with three restriction nucleases (*Alu*I, *Dde*I, and *Dra*I). The sensitivity of detection was 25 CFU/ml of water. The assay system was successfully applied to a water supply suspected of being contaminated and giving rise to *C. jejuni* infections due to the presence of viable but nonculturable (VBNC) cells of the organism, as reflected by negative selective enrichments. There were 25 liter samples employed and concentrated on membrane filters. It should, however, be emphasized that the inability to recover campylobacters from the selective enrichments may have been due to the presence of injured *Campylobacter* cells, sensitive to the inhibitory components of the selective medium used and not necessarily to the presence of true VBNC *Campylobacter* cells. Two isolates of *Campylobacter* from each of seven patients with campylobacteriosis were subjected to restriction endonuclease analysis of genomic DNA using the restriction nucleases *Bgl*II and *Eco*RV. Clonal homogeneity was revealed between both isolates from five of the seven patients. Infection with two different strains was found to have occurred in two of the seven patients. Restriction analysis in conjunction with Southern hybridization blot analysis proved to be a reliable technique to achieve precise identification of strains and to elucidate clonal heterogeneity among *Campylobacter* isolates from a single patient.

Molecular typing of *C. jejuni* isolates is usually performed with pure cultures. Waegel and Nachamkin (1996) applied PCR and RFLP typing to DNA extracted and purified from frozen human clinical fecal samples for confirmation of the presence of *C. jejuni* and typing. The *flaA* gene was amplified using the primers of Nachamkin, Bohachick, and Patton (1993a; Table 10.6) directed against a conserved region producing a 1728-bp amplicon. The resulting amplicons were then restricted with *Dde*I. This approach allowed typing without the requirement for isolation. The operational assumption was that each patient was infected with only one strain of *C. jejuni*. The limit of detection with purified DNA was about 1×10^4 *Campylobacter* genome equivalents per PCR.

Ayling et al. (1996) applied PCR-RFLP and PFGE typing to isolates of *C. jejuni* from 157 (10%) of the broiler houses in Great Britain. Three primers P1.C5 431, P2.C5 432, and P3.C5 433 (Table 10.6) were designed to amplify a 1490-bp fragment from either the *fla*A or the *fla*B genes that encode flagellins A and B. PCR-RFLP was performed by digesting the PCR product in separate reactions with the restriction nucleases *Dde*I and *Hinf*I. The method of Gibson, Sutherland, and Owen (1994) was used for PFGE analysis. A total of 75 (48%) out of the 157 broiler houses were positive for *Campylobacter*, with 97% of the isolates identified as *C. jejuni* and 3% as *C. coli*. A total of 54 different PCR-RFLP profile types were identified using both restriction nucleases. Isolates from 49 broiler houses showed only one profile type with nine houses, suggesting a single source of infection. Two profile types were found in 17 houses, and nine houses had more than two profile types, suggesting multiple sources of infection. PFGE yielded 27 profiles from the 140 strains investigated. In seven houses, isolates within a house yielded the same PFGE profile even though different PCR-RFLP profiles were obtained. In some cases, isolates belonging to an individual PCR-RFLP profile could be subdivided into three PFGE profiles. The authors concluded that PFGE provided little additional discriminatory power over that of PCR-RFLP and because it is more labor intense was less preferred than PCR-RFLP.

Using similar methodology, Koenraad et al. (1996) applied PCR-RFLP analysis of the flagellin genes of *Campylobacter* isolates from a sewage plant and the wastewater from a poultry abattoir using the restriction nuclease *Dde*I. Among a total of 182 isolates, 22 PCR-RFLP profiles emerged. Of the isolates, 52% were confined to only four profiles, suggesting that certain strains either have enhanced abilities to colonize host sources or have enhanced survival possibilities in the environment. Results also suggested that some strains may be confined to poultry only and may not occur in human waste.

Linton, Lawrence, and McGuiggan (1996a) made use of PCR-RFLP of the intergeneric region between the *fla*A and *fla*B genes of *C. jejuni*, *C. coli*, and *C. lari* for typing strains of each species. The primers CF03 and CF04 (Table 10.6) that amplify 340- to 380-bp of the intergeneric region as described by Wegmüller, Lütty, and Candrian (1993) were used. The resulting amplicons were restricted with *alu*I. Thrity-two strains of *C. jejuni* yielded five banding profiles, 25 strains of *C. coli* yielded four profiles, and four strains of *C. lari* failed to be restricted. The limited number of bands did not provide a high level of strain discrimination. The use of additional restriction nucleases would presumably have yielded larger numbers of banding profiles and enhanced discrimination. The observation that strains of *C. jejuni* and *C. coli* produced identical restriction profiles emphasizes the high degree of homology existing between the two species.

PCR-RFLP analysis has been found by Iriarte and Owen (1996) to be of limited value if derived from PCR amplification of 23S rRNA gene sequences due to extensive sequence conservation. These investigators applied PCR-RFLP analysis of 23S rRNA genes to 47 strains of *C. jejuni*. Seven different molecular profiles were detected by a combination of *Hpa*I, *Alu*I, and *Dde*I digest analysis. Most (83%) of the strains, including those with different Penner serotypes and from different hosts, had the same molecular profiles. The high level of conservation within the 23S rDNA sequences confirmed their value as species-specific targets in PCR identification but not for subtyping of *C. jejuni* isolates.

Investigation of a free-range broiler flock during the rearing period and at the slaughterhouse by PCR-RFLP of the flagellin (*fla*A) gene was undertaken by Rivoal et al. (1999). Restriction of the amplicons was with *Dde*I. Results indicated that poultry carcasses were contaminated with *Campylobacter* spp., which were previously present in the poultry feces. *Fla*A typing indicated that cross-contamination between batches from different flocks had occurred.

PCR-RFLP profiles derived from the flagellin genes (*fla*A and *fla*B) of *C. jejuni* were used by Thomas et al. (1997) to determine the genetic diversity among 287 PCR-positive isolates of *C. jejuni* from 60 poultry fecal samples from a commercial broiler farm. The primers R1F and R3R (Table 10.6) were used to generate amplicons that were then digested with the restriction nuclease *Alu*I. The resulting fragments yielded five RFLP profiles. Three of these profiles were dominant during the 3 weeks of isolation, and the other two profiles were detected at low frequency.

Rivoal et al. (1999) investigated a free-range broiler flock during the rearing period and at the slaughterhouse by PCR-RFLP of *Campylobacter* isolates. The primers pg50 and RAA19 (Table 10.6) of Alm, Guerry, and Trust (1993) amplifying a 1448-bp amplicon derived from the *fla*A gene were used. The resulting amplicons were then restricted with the nuclease *Dde*I and banding patterns compared. Results indicated that poultry carcasses were contaminated by strains of *Campylobacter* that were previously present in the poultry feces. The results further confirmed that contamination of the broilers during processing was due principally to the rupture of the GI tract during evisceration. *Fla*A typing by PCR-RFLP indicated that cross-contamination between batches from different flocks had occurred.

Knudsen et al. (2005) developed a genotyping method for isolates of *C. jejuni* based on amplification of a 9.6-kb sequence from the LPS gene cluster using the primer pair galE1/wlaH3 (Table 10.6) from Shi et al. (2002). The resulting amplicon was restricted with *Dde*I and *Rsa*I for generation of typing patterns. A low correlation between Penner serotyping and LPS genotyping of poultry isolates occurred. This was in contrast to a high correlation reported by Shi et al. (2002). The authors attributed this discrepancy to the use of reference strains of human origin and recent poultry isolates, whereas Shi et al. (2002) used all human isolates. The authors also hypothesized that genetic exchange of LPS in the GI tract of poultry may have occurred without necessarily altering the Penner serotype.

The use of PCR for detection of *Campylobacter* spp. in human clinical fecal samples has been claimed to be more reliable than direct culture on a selective medium (Vanniasinkam, Lanser, and Barton, 1999). A total of 493 samples from patients with enteritis were subjected to enrichment in brain–heart infusion (BHI) at 37°C and cell lysates subjected to PCR. The same samples were also cultured directly onto modified Butzler medium (Lauwers, De Boeck, and Butzler, 1978; Patton et al. 1981). The primers used, CF1 and CR2 (Table 10.6), amplified an 843-bp sequence of the 16S rRNA gene of all species of *Campylobacter.* A 24-mer DNA probe CP3 (Table 10.6) was used to confirm the identity of amplicons via Southern blot hybridization. Of the 493 clinical fecal samples, 24 were found to be PCR positive but only 13 were positive by direct selective agar isolation. In the same study, the PCR primers were able to detect *Campylobacter* spp. in 14 of 83 bovine fecal samples and in 12 of 109 unpasteurized milk samples.

D. Amplified Fragment Length Polymorphism (AFLP)

Duim et al. (1999) made use of AFLP, which is based on the selective amplification of restriction fragments of chromosomal DNA for genetic typing of *C. jejuni* and *C. coli* isolates from humans and poultry. The restriction nucleases *Hind*III and *Hha*I were used together to digest the genomic DNA. Digestion products were then ligated to *Hind*III and *Hha*I to restriction-site adapters. Preselected PCR primers were then used for amplification. This combination of restriction nucleases resulted in 40 to 65 agarose bands ranging from 50 to 500 bp in length with evenly distributed banding patterns. Up to 50 sharp DNA bands were clearly resolved for each strain. Unrelated *C. jejuni* strains produced heterogeneous patterns, whereas genetically related strains produced similar AFLP patterns.

Hänninen et al. (2001) compared the PFGE patterns, AFLP patterns, *Hae*III ribotypes, and heat-stable serotypes of 35 *C. jejuni* strains. The discriminatory powers of the first three methods were similar, and these methods assigned the strains into the same groups. The PFGE and AFLP patterns within a genotype were highly similar, indicating genetic relatedness. The same serotypes were found distributed among different genotypes, and different serotypes were identified within one genotype. The results indicated that common *C. jejuni* genotypes from genetic lineages colonized both humans and poultry.

1. In situ Colony Hybridization

Jensen et al. (2005) used the primer pair HIP400 F/HIP1134R (Table 10.6) from Englen and Fedorka-Cray (2002) to generate a 735-bp DIG-labeled probe derived from the *hipO* gene, which was applied to hip⁻ colonies of *C. jejuni*. Among a total of 89 positive PCR isolates of *C. jejuni* from swine fecal samples, 25 isolates (28%) were negative for hippuric acid hydrolysis, although the *hipO* gene was consistently present.

A high-affinity peptide nucleic acid (PNA) oligonucleotide probe was utilized by Lehtola, Loades, and Keevil (2005) for specifically detecting *C. coli, C. jejuni*, and *C. lari* using fluorescence in situ colony hybridization (FISH). The probe designated CJE195 (Table 10.6) was labeled with TAMRA at the 5′-end and was based on a 15-bp 16S rRNA sequence shared by these three organisms and distinct from other campylobacters and other bacteria. When the same probe was labeled with FAM as a DNA probe, it was unable to hybridize to the genomic DNA of the cells. The authors speculated that the reason for this was probably due to better permeability of the uncharged backbone of the PNA probe and its higher affinity to the Class 3-4 rRNA sites than DNA oligonucleotide probes.

E. Multilocus Sequence Typing (MLST)

Dingle et al. (2001) developed a multilocus sequencing system for *C. jejuni* that utilized the sequence variation present in seven housekeeping loci derived from genes encoding enzymes responsible for various aspects of intermediary metabolism. The sequence data derived from 194 strains indicated that *C. jejuni* is genetically diverse, with a weak clonal population structure, and that intra- and interspecies horizontal genetic exchange was common, including import of alleles from at least two other *Campylobacter* species, including *C. coli.*

Manning et al. (2003) subjected 266 *C. jejuni* isolates, mainly from veterinary sources, including cattle, sheep, poultry, swine, pets, and the environment, as well as human clinical isolates to MLST. Specific sequences from six genes encoding enzymes of intermediary metabolism were amplified by PCR for sequence comparison among all strains. The populations of veterinary and human isolates were found to overlap, suggesting that most veterinary sources could be considered reservoirs of pathogenic *C. jejuni.* There were some associations between source and sequence type complex, indicating that host or source adaptation may exist. The swine isolates formed a distinct MLST group, suggesting a potential swine-adapted clone of *C. jejuni.*

Colles et al. (2003) applied MLST to 112 *C. jejuni* isolates from poultry, cattle, sheep, starlings, and slurry. The MLST procedure consisted of amplifying by PCR-specific sequences from seven housekeeping genes encoding enzymes of intermediary metabolism. The nucleotide sequence of each amplicon was determined and compared among all isolates. A total of 30 genotypes emerged. All but two of these genotypes belonged to one of nine clonal complexes previously identified in isolates from human infections and retail food samples. There was some evidence for the association of certain clonal complexes with particular farm animals. Comparison with MLST data from 91 human clinical isolates showed small but significant genetic differentiation between farm and human clinical isolates.

MLST employing seven housekeeping genes was used by Fitch et al. (2005) to compare *C. jejuni* strains isolated from retail chicken products and humans with gastroenteritis in central Michigan. Sequence comparisons demonstrated overlapping diversity between chicken and human isolates. In examining the sequence variation in the short variable region (SVR) of the *flaA* gene, it was found that *C. jejuni* isolates from human clinical sources had a greater diversity of flagellin alleles in addition to a higher incidence of quinoline resistance than isolates from retail chicken products. The authors also examined the allelic variation in the hippuricase gene (*hipO*) by designing primers hipO-F156 and hipO-R721 (Table 10.6) to amplify a 566-bp sequence of the *hipO* gene. Sequencing of a 377-bp region of the resulting amplicon revealed 17 polymorphic sites and 14 distinct *hipO* alleles among 30 *C. jejuni* isolates with nucleotide variation comparable to that of the seven MLST loci. The authors concluded that retail chicken products can serve as reservoirs for *C. jejuni* that lead to human gastroenteritis and that additional selection of human pathogenic strains possibly occurs in the GI tract of the human host.

Although *C. coli* is responsible for fewer foodborne illnesses than *C. jejuni*, its impact is still substantial with more than 25,0000 cases of foodborne illness in the United Kingdom estimated to be caused by *C. coli* in the year 2000. Miller et al. (2006) described the MLST of 488 *C. coli* strains from four different animal sources (cattle, chickens, swine, and turkeys) collected over a 6-year period from different geographic locations in the United States. A total of 149 sequence types (Sts) were identified using seven housekeeping loci and allele endpoints. The majority of Sts and alleles were found to be host associated in that they were found primarily in a single food-animal source. Only 12/149 (8%) of Sts were found in multiple sources. The presence of host-associated *C. coli* MLST alleles has potential value for tracking sporadic and human clinical outbreaks to animal sources.

F. Loop-Mediated Isothermal Amplification (LAMP)

Yamazaki et al. (2008) developed a LAMP assay for the rapid and simple detection of *C. jejuni* and *C. coli*. The assay provides a specific AMP product for each of these two species detected by turbidity. Each assay correctly identified 65 *C. jejuni* and 45 *C. coli* strains, but not 75 non–*C. jejuni* and non–*C. coli* strains. The sensitivity of the LAMP assay for *C. jejuni* and *C. coli* in spiked stool samples was 5.6×10^3 CFU/g (1.4 CFU per amplification tube) and 4.8×10^3 CFU (1.2 CFU per amplification tube), respectively. When 90 stool specimens from patients with diarrhea were tested by LSMP and direct plating, the LAMP results showed 81.3% sensitivity and 96.6% specificity compared to isolation of *C. jejuni* and *C. coli* by direct plating. In addition, the LAMP assay requires less than 2 hr for detection of *C. jejuni* and *C. coli* in stool samples and the temperature for amplification is isothermal at 65°C.

G. Nucleic Acid Sequence Based Amplification (NASBA)

Uyttendaele et al. (1994) undertook detection of *C. jejuni* added to foods by using a combined selective enrichment and NASBA, which is distinct from PCR. A pair of oligonucleotide primers T1547 and OT1118 (Table 10.6) was chosen from the 16S rRNA sequence alignment of the *Campylobacter* species *C. jejuni*, *C. coli*, and *C. lari*. The amplification of the RNA sequence in NASBA was achieved through the concerted action of avian myeloblastosis virus reverse transcriptase, T7 RNA polymerase, and RNase H. Major advantages of NASBA over PCR are that NASBA is performed isothermally, precluding the use of a thermocycler, and no separate reverse transcription step is required for RNA amplification. A rapid nonradioactive "in solution" hybridization assay (ELGA) using an oligonucleotide probe (OT1559; Table 10.6) labeled at the 5′-end with horse radish peroxidase (HRP) was used to identify the NASBA products. Hybridization was at 60°C for 10 min. After hybridization, excess nonhybridized ELGA probes were separated from the homologous hybridized product by vertical gel electrophoresis on an acrylamide gel. The ELGA probes hybridized to the amplicons were then stained with an HRP substrate solution. Hybridization with the internal probe was found to occur only between the probe and amplicons derived from *C. jejuni, C. coli,* and *C. lari* and not with other species of *Campylobacter*. The NASBA-ELGA technique used is therefore not species-specific. The presence of high numbers of indigenous bacteria in meat was found to prevent the detection of *C. jejuni* via the NASBA-ELGA technique (Uyttendaele et al., 1995). Detection of *C. jejuni* was possible up to a ratio of indigenous bacteria to *C. jejuni* of 10,000:1 following enrichment. The limit of detection was less than 10 CFU/g of *C. jejuni* seeded into food samples.

H. Restriction Endonuclease Analysis

Jiménez et al. (1997) obtained two random isolates of *C. jejuni* from each of seven patients with campylobacteriosis. The chromosomal DNA was then extracted and purified from each isolate and subjected to restriction using the nucleases *Bgl*II and *Eco*RV. The resulting patterns of restricted fragments derived from agarose gel electrophoresis were then visually compared. Clonal homogeneity was revealed between both isolates from five of the seven patients. Infection with two different strains was found to have occurred in two of the seven patients. Restriction analysis was found to be highly reliable for determining clonal heterogeneity among *C. jejuni* isolates from a single patient.

VIII. THE VIABLE BUT NONCULTURABLE (VBNC) STATE OF CAMPYLOBACTERS

The main manifestations of sublethal injury to *C. jejuni* are an inability to grow at 42°C, increased sensitivity to antimicrobials, sensitivity to H_2O_2 and photochemical-induced toxic oxygen radicals (Mason, Humphrey, and Martin, 1996). The first two are associated with the VBNC state. The VBNC state is manifested when the organism is unable to grow at its normal permissible temperature of 42°C and requires incubation at a lower temperature (35°C) to achieve repair and overcome enhanced sensitivity to antimicrobial agents incorporated into selective culture media that uninjured cells tolerate.

Elevated temperature has been found to convert cells of *C. jejuni* to a VBNC state due to membrane damage, depending on the culture medium employed. Palumbo (1984) subjected cells of *C. jejuni* to heat injury in 0.1 M phosphate buffer at 46°C for up to 45 min. Little or no loss in CFU occurred when plating was performed onto *Brucella* agar supplemented with 0.25g/L each of Na metabisulfite, $FeSO_4$, and Na pyruvate. However, a pronounced decrease in CFU did occur with heating time when heated cells were plated onto brilliant green bile salts (BGBS) agar containing the same supplements. Cells were found to lose 260-nm absorbing material during heat injury, reflecting membrane injury. Repair of heated cells in supplemented *Brucella* broth to full resistance to the dye brilliant green and bile occurred in 4 hr at 37°C.

Preliminary findings by Rollins and Colwell (1986) found that VBNC *C. jejuni* formed in stream water at 4°C were culturable after animal passage. Saha, Saha, and Anyal (1991) reported that among 16 freeze–thaw-injured nonculturable strains of *C. jejuni* that had passed into coccoid forms, 7 were successfully reisolated following passage through the intestines of rats. Consecutive passage through rat intestines restored the original toxin production. These strains in the VBNC state were unable to develop colonies on Campy-BAP agar until passage through rats.

Mason, Humphrey, and Martin (1996) stored unseeded river water at 4°C for up to 21 days and periodically cultured campylobacters. The basal medium was nutrient broth containing 5% lysed horse blood and aerotolerant supplement (0.02% each of ferrous sulphate, sodium pyruvate, and sodium bisulfite). Exeter broth was used for selective enrichment and contains five antimicrobials including rifampicin and polymyxin B. With refrigerated river water, 80% of the samples were positive for *Campylobacter* when the five Exeter antimicrobials were added to Exeter enrichment broth with a delay of 8 hr followed by incubation for 48 hr at 37°C with plating onto Exeter agar. In contrast only 30% positive samples were obtained when all five antimicrobials were present at zero time in the enrichment broth. When aqueous suspensions of *C. jejuni* were stored at –20°C for 3 days and plated onto blood agar containing 10 µg/ml of rifampicin, a 9-log reduction in CFU occurred. In contrast, when rifampicin was deleted from Exeter agar, only a 4-log reduction in CFU occurred. The authors indicated that the rifampicin and polymyxin B in Exeter selective broth are most likely responsible for the enhanced inhibition of injured cells and that a recovery period of 8 hr at a permissible temperature in the absence of antimicrobials is required for maximum recovery.

Fearnley et al. (1996) suspended strains of *C. jejuni* in deionized water to a density of 10⁷CFU/ml followed by storage at 4°C. Culturability counts were obtained on 10% blood agar. Nonculturability was obtained on Exeter agar. The total viable count was obtained by staining with 5-cyano-2,3-ditolyl tetrazolium chloride (CTC) and the number of dead cells by staining with 4′,6-diamidino-2-phenylindole (DAPI). Over a 14-day period, a 6-log decline in CFU occurred and the stained viable count decreased by only 1 log. An 820-fold higher number of orally administered VBNC cells of a poultry isolate over that of the minimal colonization dose of the culturable form failed to colonize 1-day-old chicks.

Medema et al. (1992) found that only one of seven strains of *C. jejuni* exhibited a prolonged VBNC state induced by suspending cells at 25°C to a density of 10⁷/ml for 3 days in sterilized reservoir water. When day-old chicks were orally administered 1.8×10^5 VBNC cells, colonization failed to occur. In addition, inoculation of embryonated eggs with VBNC cells also resulted in failure to recover culturable cells. A modification of the method of Kogure, Simkidu, and Taga (1979) was used for enumeration of dead/viable cells (DVCs). The negative recovery results of VBNC cells, particularly from embryonated eggs, may have been due to a temperature of 25°C used for induction of the VBNC state in contrast to 4°C.

C. jejuni suspended in surface water at pH 6.0 to a density of 1×10^8 cells/ml and maintained at 4°C resulted in plate counts decreasing below 1 CFU/ml after 15 to 18 days and viable counts determined microscopically with CTC reduction remained around 1×10^6 cells/ml (Cappelier et al., 1999b). When cell suspensions maintained for 30 days at 4°C were diluted to 25 VBNC cells/ml and 1 ml injected into the yolk sac of 7-day-old embryonated eggs, culturable *C. jejuni* were successfully recovered from 77.5% to 87.5% of embryonated eggs. Dilution to below 0.003 culturable cells/ ml (equivalent to 25 VBNC cells) ensured that no culturable cells were injected.

Hald et al. (2001) examined the ability of a broiler isolate, a human clinical isolate, and a laboratory-adapted strain of *C. jejuni* to colonize day-old chicks following storage of the isolates in saline at 2 to 4°C. The ability of all three isolates to become intestinally established terminated beyond 9 days of saline storage, which was 3 to 4 weeks before the strains became nonculturable.

Cells of *C. jejuni* at a density of 1×10^7 CFU/ml in surface water at pH 6.0 and 4°C were found to enter into the VBNC state after 14 to 16 days, as reflected in the decrease in CFU to less than 1 per ml and the observation that about 100% of the cells contained CTC formazan crystals, indicating active metabolic activity (Tholozan et al., 1999). The VBNC forms were found to survive for at least 30 days at 4°C. Viability was determined with Columbia agar containing 5% lysed horse blood. In addition, the absence of culturable cells was confirmed by filtering 10 ml of cell suspensions onto membrane filters that were incubated on the surface of Columbia agar plates. No intracellular protein losses occurred after 30 days at 4°C; however, cellular water volume increased about 6.3-fold. After 15 days at 4°C, the ATP and ADP levels were below the detection level, whereas the internal AMP concentration varied from 0.2 to 0.4 nmol per mg of internal protein, representing a mean decrease of 35% for all three strains studied. Both the internal K content and membrane potential were significantly lower in VBNC cells than in culturable cells.

When strains of *C. jejuni* and *C. coli* were suspended at a density of 3.2×10^6 to 6.3×10^7 CFU/ml in Mueller–Hinton broth adjusted to a pH of 4.0 with formic acid and held at 37°C, the cells became nonculturable on blood agar after 2 hr (Chaveerach et al., 2003). The entry of the cells into the VBNC state was confirmed by staining with CTC/DAPI. Several of these strains that had entered the VBNC state were resuscitated by inoculating the yolk sacs and amniotic fluid of embryonated eggs. The CTC/DAPI technique was used to enumerate the DVCs.

Solid phase cytometry (SPC) in conjunction with fluorescent viability staining was found by Cools et al. (2005) to effectively detect VBNC cells of *C. jejuni* in drinking water. Seeded water samples were filtered through a black 25-mm polyester membrane of 0.40 μ porosity, and the retained cells stained using the ester carboxyfluorescein diacetate (ChemChrome V6). This ester is cleaved by intracellular esterases to yield green fluorescent carboxyfluorescein and therefore requires intracellular metabolic activity and membrane integrity. The number of green fluorescent bacterial cells was counted by an automatic laser scanning device that scanned the entire membrane in 3 min. The number of culturable cells decreased from the initial value of 10^8/ml to below detectable plate count levels in less than 50 days with seeded water samples held at 4°C. In contrast, the number of fluorescent bacteria remained at initial levels for at least 85 days. The authors attributed the discrepancy between these two results to the transition of culturable *C. jejuni* cells to the VBNC state.

Several studies published to date, however, suggest that the regrowth of cells in the VBNC state with respect to *Vibrio cholerae* (Ravel et al., 1993), *V. vulnificus* (Weichart, Oliver, and Kjelleberg, 1992), and various members of the *Enterobacteriaceae* (Bogosian, Morris, and O'Neil, 1998) may erroneously be due to the survival of a very small number of viable cells and that the VBNC state represents dead cells. A truly definitive study with *C. jejuni* to clarify this point has not as yet been undertaken.

IX. THE COCCOID FORM OF *C. JEJUNI*

It is well established that *C. jejuni* undergoes a morphological change from spiral-shaped rods to a coccoid form in the stationary growth phase or in response to unfavorable conditions (Moran and Upton, 1987a,b). Some investigators have considered the coccoid form to be dormant and in a nonculturable state, but metabolically active (Lázaro et al., 1999; Rollins and Colwell, 1986; Tholozan et al., 1999) and recoverable in suitable hosts (Cappelier et al., 1999a; Jones, Sutcliffe, and Curry, 1991; Stern et al., 1995; Talibart et al., 2000). However, other reports have indicated that the coccoid form is a nonviable degenerative form (Beumer, de Vries, and Rombouts, 1992; Boucher et al., 1994; Hazeleger et al., 1995; Medema et al., 1992; Moran and Upton, 1986, 1987b). Hudock, Borger, and Kaspar (2005) in a definitive study found that newly formed coccoid cells (1 to 3 days old) had a *Sma*I-digestion profile identical to that of spiral-shaped cells and that there was a progressive degradation of the DNA with continued incubation at 37°C resulting in a complete loss of the *Sma*I-digestion profile after 2 days and the concurrent appearance of DNA in supernatants of coccoid cells. In contrast, cells incubated at 4°C retained the spiral shape and their characteristic *Sma*I digestion profile for 8 weeks and released little DNA into the medium. The authors concluded that the coccoid form is a manifestation of cellular degradation of spiral-shaped cells and that possibly coccoid cells formed at low temperature are the most likely VBNC form of *C. jejuni*.

A fluorescent probe that efficiently hybridized to its homologous 16S rRNA sequence in young spiral-shaped cells in situ was found unable to bind to aged coccoid cells due presumably to lower rRNA content of the coccoid cells or to lower permeability of the cell wall (Lehtola, Loades, and Keevil, 2005).

X. IMMUNOMAGNETIC CAPTURE OF *C. JEJUNI*

Thermophilic campylobacters including *C. jejuni* were detected from poultry meat samples and milk by Lamoureux et al. (1997) with the use of immunomagnetic (IM) capture of cells with monoclonal antibody (mAb) against an outer membrane protein specific to thermophilic campylobacters. After enrichment, detection of the IM captured cells was achieved using two different DNA hybridization methods. In one, the captured cells were lysed by guanidine isothiocyanate (GITC) and the 23S rRNA was reacted with an rDNA probe immobilized on the surface of microtiter plate wells. Alternately, the IM captured cells were lysed by ultrasonication and the genomic DNA reacted with an RNA probe similarly immobilized. Detection of the RNA–DNA hybrids formed on the surface of the wells was performed using a monoclonal anti-RNA–DNA hybrid antibody followed by application of anti-mouse IgM–peroxidase conjugate and o-phenylenediamine.2HCl.

Lund et al. (2003) evaluated the use of PCR for detection of *Campylobacters* in poultry feces. DNA was isolated from poultry rectal swabs using immunomagnetic beads followed by PCR. The primers used were C412F and C1288R (Table 10.6) from Linton, Owen, and Stanley (1996b). Results could be obtained in less than 6 hr. The method was evaluated on 1282 samples and compared to conventional culture methodology with a resulting diagnostic specificity of 0.99 and with an agreement of 0.98. With both methods, the detection limit was 36 CFU/ml of fecal suspension. A DNA isolation kit was used to remove PCR inhibitors. Undiluted NA was found to contain PCR inhibitors. However, the addition of 1 mg/ml of BSA to the PCR mixtures reduced this inhibition.

Rudi et al. (2004) developed an immunomagnetic bead (IMB) procedure for *C. jejuni* derived from poultry feces. The IMB-captured cells were lysed, the DNA purified, and then incorporated into Rti-PCR assays. BSA (0.2%) was found to notably decrease PCR inhibition due to contaminants. The sensitivity of the assay was 2 to 25 CFU per Rti-PCR. The primers AB-F and AB-R (Table 10.6) were specific for *C. jejuni* and were used in conjunction with a probe (Table 10.6) labeled at the 5′-end with FAM and at the 3′-end with TAMRA previously described by Nogva et al. (2000).

REFERENCES

Aarts, H., van Lith, L., Jacobs-Reitsma, W. 1995. Discrepancy between Penner serotyping and polymerase chain reaction fingerprinting of *Campylobacter* isolates from poultry and other animal sources. *Lett. Appl. Microbiol.* 20:371–374.

Abram, D., Potter, N. 1984. Survival of *Campylobacter jejuni* at different temperatures in broth, beef, chicken and cod supplemented with sodium chloride. *J. Food Prot.* 47:795–800.

Adark, G., Meakins, S., Yip, H., Lopman, B., O'Brien, S. 2005. Disease risks from foods, England and Wales, 1996–2000. *Emerg. Infect. Dis.* 11:365–372.

Alm, R., Guerry, P., Trust, T. 1993. Distribution and polymorphism of the flagellin genes from isolates of *Campylobacter coli* and *Campylobacter jejuni*. *J. Bacteriol.* 175:3051–3057.

Alter, T., Gaull, F., Kasimir, S., Gürtler, M., Mielke, H., Linnebur, M., Fehlhaber, K. 2005. Prevalences and transmission routes of *Campylobacter* spp. strains within multiple pig farms. *Vet. Microbiol.* 108:251–261.

Ang, C., Laman, J., Willison, H., Wagner, E., Endtz, H., De Klerk, M., Tio-Gillen, A., van den Braak, N., Jacobs, B., Doorn, P. 2002. Structure of *Campylobacter jejuni* lipopolysaccharide determines antiganglioside specificity and clinical features of Guillain-Barre and Miller Fisher patients. *Infect. Immun.* 70:1202–1208.

Aspinall, G., McDonald, A., Raju, T., Pang, H., Mills, S., Kurjanczk, L., Penner, J. 1992. Serological diversity and chemical structures of *Campylobacter jejuni* low molecular weight lipopolysaccharides. *J. Bacteriol.* 174:1324–1332.

Aspinall, G., McDonald, A., Raju, T., Pang, H., Moran, A., Penner, J. 1993. Chemical structures of the core regions of *Campylobacter jejuni* serotypes O:1, O:4, O:23, and O:36 lipoplysaccharides. *Eur. J. Biochem.* 216:880.

Atterbury, R., Dilln, E., Swift, C., Connerton, P., Frost, J., Dodd, C., Rees, C., Connerton, I. 2005. Correlation of *Campylobacter* bacteriophage with reduced presence of hosts in broiler chicken ceca. *Appl. Environ. Microbiol.* 71:4885–4887.

Ayling, R., Johnson, L., Evans, S., Newell, D. 1996. PCR/RFLP and PFGE sub-typing of thermophilic *Campylobacter* isolates from poultry epidemiological investigation. In: *Campylobacters, Helicobacters, and Related Organisms. Proceedings of the Eighth International Workshop on Campylobacters, and Related Organisms*, July 10–13, 1995, Winchester, UK. D. Newell, J. Ketley, R. Feldman, Eds., Plenum Press: New York, pp. 181–185.

Bacon, D., Alm, R., Burr, D., Hu, L., Kopecko, D., Ewing, C., Trust, T., Guerry, P. 2000. Involvement in a plasmid in the virulence of *Campylobacter jejuni* 81-76. *Infect. Immun.* 68:4384–4390.

Bacon, D., Alm, R., Hu, L. Hickey, T., Ewing, C., Batchelor, R., Trust, T., Guerry, P. 2002. DNA sequence and mutational analyses of the pVir plasmid of *Campylobacter jejuni* 81-176. *Infect. Immun.* 70:6242–6250.

Bang, D., Borck, B., Nielsen, E., Scheutz, F., Pederson, K., Madsen, M. 2004. Detection of seven virulence and toxin genes of *Campylobacter jeuni* isolates from Danish turkeys by PCR and cytolethal distending toxin production of the isolates. *J. Food Prot.* 67:2171–2177.

Beery, J., Hugdahl, M., Doyle, M. 1988. Colonization of gastrointestinal tracts of chicks by *Campylobacter jejuni*. *Appl. Environ. Microbiol.* 54:2365–2370.

Beuchat, L. 1985. Efficacy of media and methods for detecting and enumerating *Campylobacter jejuni* in refrigerated chicken meat. *Appl. Environ. Microbiol.* 50:934–939.

Beumer, R., de Vries, J., Rombouts, F. 1992. *Campylobacter jejuni* non-culturable coccoid cells. *Int. J. Food Microbiol.* 15:153–163.

Bhaduri, S., Cotrell, B. 2004. Survival of cold-stressed *Campylobacter jejuni* on ground chicken and chicken skin during frozen storage. *Appl. Environ. Microbiol.* 70:7103–7109.

Birkenhead, D., Hawkey, P., Heritage, J., Gascoyne-Binzi, D., Kite P. 1993. PCR for the detection and typing of campylobacters. *Lett. Appl. Microbiol.* 17:235–237.

Blaser, M., Berkowitz, I., LaForce, F., Cravens, J., Relleer, L., Wang, W. 1979. *Campylobacter enteritis*: Clinical and epidemiological features. *Ann. Intern. Med.* 91:179–185.

Blaser, M., Hardesty, H., Powers, B., Wang, W. 1980. Survival of *Campylobacter fetus* subsp. *jenuni* in biological milieus. *J. Clin. Microbiol.* 127:309–313.

Blaser, M., Hardesty, H., Powers, B., Wang, W. 1986. Extra intestinal *Campylobacter jejuni* and *Campylobacter coli* infections: Host factors and strain characteristics. *J. Infect. Dis.* 153:552–559.

Bogosian, G., Morris, P., O'Neil, J. 1998. A mixed culture recovery method indicates that enteric bacteria do not enter the viable but nonculturable state. *Appl. Environ. Microbiol.* 64:1736–1742.

Bolton, F., Coates, D. 1983. Development of a blood-free *Campylobacter* medium: Screening tests on basal media and supplements, and the ability of selected supplements to facilitate aerotolerance. *J. Appl. Bacteriol.* 54:115–125.

Bolton, F., Fox, A., Gibson, J., Madden, R.K., Moore, J., Moran, L., Murphy, P., Owen, R., Pennington, T., Stanleyh, T., Thompson-Carter, F., Wareing, D., Wilson, T. 1996. A multi-centre study of methods for sub-typing *Campylobacter jejuni*. In: *Campylobacters, Helicobacters, and Related Organisms. Proceedings of the Eighth International Workshop on Campylobacters, and Related Organisms*, July 10–13, 1995, Winchester, UK. D. Newell, J. Ketley, R. Feldman, Eds., Plenum Press: New York, pp. 187–189.

Bolton, F., Holt, A., Hutchinson, D. 1985. Urease-positive campylobacters. *Lancet*. 323, no. 8370:1217–1218.

Bolton, F., Robertson, L. 1982. A selective medium for isolating *Campylobacter jejuni/coli*. *J. Clin. Pathol.* 35:462–467.

Boucher, S., Slater, B., Chamberlain, A., Adams, M. 1994. Production and viability of coccoid forms of *Campylobacter jejuni*. *J. Appl. Bacteriol.* 77:303–307.

Branquinbo, M., Alviano, C., Ricciardi, L. 1983. Chemical composition and biological action of lipopolysaccharide (LPS) of *Campylobacter fetus* ss. *Jeuni*. *Rev. Microbiol.* 14:90–96.

Broman, T., Bergström, S., Palmgren, H., McCafferty, D., Sellin, M., Olsen, B. 2000. Isolation and characterization of *Campylobacter jejuni* subsp. *jejuni* from macaroni penguins (*Eudyptes chrysolophus*) in the subantarctic region. *Appl. Environ. Microbiol.* 66:449–452.

Burgess, F., Little, C., Allen, G., Williamson, K., Mitchell, T. 2005. Prevalence of *Campylobacter, Salmonella,* and *Escherichia coli* on the external packaging of raw meat. *J. Food Prot.* 68:469–475.

Cabrita, J., Rodrigues, J., Bragança, F., Morgado, C., Pires, I., Gonçalves, A. 1992. Prevalence, biotypes, plasmid profile and antimicrobial resistance of *Campylobacter* isolated from wild and domestic animals from Northeast Portugal. *J. Appl. Bacteriol.* 73:279–285.

Cappelier, J., Magras, C., Louve, J., Federighi, M. 1999a. Recovery of viable but non-culturable *Campylobacter jejuni* cells in two animal models. *Food Microbol.* 16:375–383.

Cappelier, J., Minet, J., Colwell, R., Federight, M. 1999b. Recovery in embryonated eggs of viable but not culturable *Campylobacter jejuni* cells and maintenance of ability to adhere to HeLa cells after resuscitation. *Appl. Environ. Microbiol.* 65:5154–5157.

Carvalho, A., Ruiz-Palacios, G. 1996. Phylogenetic studies of *Campylobacter jejuni* using arbitrary primer–PCR fingerprinting. In: *Campylobacters, Helicobacters, and Related Organisms. Proceedings of the Eighth International Workshop on Campylobacters, and Related Organisms*, July 10–13, 1995, Winchester, UK. D. Newell, J. Ketley, R. Feldman, Eds., Plenum Press: New York, pp. 241–244.

Casanovas, L., de Simon, M., Ferrer, M., Arqués, J., Monzón, G. 1995. Intestinal carriage of campylobacters, salmonellas, yersinias, and listerias in pigeons in the city of Barcelona. *J. Appl. Bacteriol.* 78:11–13.

Castillo, A., Escartin, E. 1994. Survival of *Campylobacter jejuni* on sliced watermelon and papaya. *J. Food Prot.* 57:166–168.

Chaveerach, P., ter Huurne, A., Loipman, L., Knapen, F. 2003. Survival and resuscitiation of ten strains of *Campylobacter jejuni* and *Campylobacter coli* under acid conditions. *Appl. Environ. Microbiol.* 69:711–714.

Christie, P., Vogel, J. 2000. Bacterial type IV secretion: Conjugation systems adapted to deliver effector molecules to host cells. *Trends Microbiol.* 8:354–360.

Christopher, F., Smith, G., Vanderzant, C. 1982a. Effect of temperature and pH on the survival of *Campylobacter fetus*. *J. Food Prot.* 45:253–259.

Christopher, F., Smith, G., Vanderzant, C. 1982b. Examination of poultry giblets, raw milk and meat for *Campylobacter fetus* subsp. *jejuni*. *J. Food Prot.* 45:260–262.

Clark, A., Bueschkens, D. 1986. Survival and growth of *Campylobacter jejuni* in egg yolk and albumen. *J. Food Prot.* 49:135–141.

Colles, F., Jones, K., Harding, R., Maiden, M. 2003. Genetic diversity of *Campylobacter jejuni* isolates from farm animals and the farm environment. *Appl. Environ. Microbiol.* 69:7409–7413.

Connerton, I. 2004. Longitudinal study of *Campylobacter jejuni* bacteriophages and their hosts from broiler chickens. *Appl. Environ. Microbiol.* 70:3877–3883.

Cools, I., D'Haese, E., Uyttendaele, M., Storms, E., Nelis, H., Debevere, J. 2005. Solid phase cytometry as a tool to detect viable but non-culturable cells of *Campylobacter jejuni*. *J. Microbiol. Meth.* 63:107–114.

Covacci, A., Tedford, J., Del Giudice, G., Parsonnets, J., Rapouli, R. 1999. *Helicobacter pylori* virulence and genetic geography. *Science* 284:1328–1333.

Cui, S., Ge, B., Zheng, J., Meng, J. 2005. Prevalence and antimicrobial resistance of *Campylobacter* spp. and *Salmonella* serovars in organic chickens from Maryland retail sources. *Appl. Environ. Microbiol.* 71:4108–4111.

Day, W., Jr., Pepper, I., Joens, L. 1997. Use of an arbitrary primed PCR product in the development of a *Campylobacter jejuni*-specific PCR. *Appl. Environ. Microbiol.* 63:1019–1023.

Denis, M., Soumet, C., Rivoal, K., Ermel, G., Blivet, D., Salvat, G., Colin P. 1999. Development of an m-PCR assay for simultaneous identification of *Campylobacter jeuni* and *C. coli*. *Lett. App. Microbiol.* 29:406–410.

Devane, M., Nicol, C., Ball, A., Klena, J., Scholes, P., Hudson, J., Baker, M., Gilpin, B. 2000. The occurrence of *Campylobacter* subtypes in environmental reservoirs and potential transmission routes. *J. Appl. Microbiol.* 98:980–990.

Daikoku, T., Kaawaguchi, M., Takama, K., Suzuki, S. 1990. Partial purification and characterization of the enterotoxin produced by *Campylobacter jejuni*. *Infect. Immun.* 58:2414–2419.

Dingle, K., Colles, F., Wareing, D., Ure, R., Fox, A., Bolton, F., Bootsma, H., Willems, R., Urwin, R., Maiden, M. 2001. Multilocus sequence typing system for *Campylobacter jejuni*. *J. Clin. Microbiol.* 39:14–23.

Dlakić, M. 2000. Functionally unrelated signaling proteins contain a fold similar to Mg^{2+}-dependant endonucleases. *Trends Biochem. Sci.* 25:272–273.

Donta, S., Viner, J. 1975. Inhibition of the steroidogenic effects of cholera and heat-labile *Escherichia coli* enterotoxins by GM_1 ganglioside: Evidence for a similar receptor site for the two toxins. *Infect. Immun.* 11:982–985.

Doyle, M., Roman, D. 1982. Recovery of *Campylobacter jejuni* and *Campylobacter coli* from inoculated foods by selective enrichment. *Appl. Environ. Microbiol.* 43:1342–1353.

Duim, B., Wassenaar, T., Rigter, A., Wagenaar, J. 1999. High-resolution genotyping of *Campylobacter* strains isolated from poultry and humans with amplified fragment length polymorphism fingerprinting. *Appl. Environ. Microbiol.* 65:2369–2375.

El-Shibiny, A., Connerton, P., Connerton, I. 2005. Enumeration and diversity of campylobacters and bacteriophages isolated during the rearing cycles of free-range and organic chickens. *Appl. Environ. Microbiol.* 71:1259–1266.

Elwell, C., Dreyfus, L. 2000. DNaseI homologous residues in CdtB are critical for cytolelthal distending toxin-mediated cell cycle arrest. *Mol. Microbiol.* 37:952–963.

Englen, M., Fedorka-Cray, P. 2002. Evaluation of a commercial diagnostic PCR for the identification of *Campylobacter jejuni* and *Campylobacter coli*. *Lett. Appl. Microbiol.* 35:353–356.

Englen, M., Kelly, L. 2000. A rapid DNA isoation procedure for the identification of *Campylobacter jejuni* by the polymerase chain reaction. *Lett. Appl. Microbiol.* 31:421-426.

Everest, P., Cole, A., Knutton, S., Goossens, H., Butzler, J., Ketley, J., Williams, P. 1993. Roles of leukotriene B₄, prostoglandin E₂, and cyclic AMP in *Campylobacter jejuni*-induced intestinal fluid secretion. *Infect. Immun.* 61:4885–4887.

Eyers, M., Chapelle, S., van Camp, G., Goosens, H., de Wachter, R. 1993. Discrimination among thermophilic Campylobacter species by polymerase chain reaction amplification of 23S rRNA gene fragments. *J. Clin. Microbiol.* 31:3340–3343.

Eyigor, A., Dawson, K., Langlois, B., Pickett, C. 1999. Detection of cytolethal distending toxin activity and *cdt* genes in *Campylobacter* spp. isolated from chicken carcasses. *Appl. Environ. Microbiol.* 65:1501–1505.

Fauchèr, J., Kervella, M., Rosenau, A., Mohanna, K., Veron, M. 1989. Adhesion to HeLa cells of *Campylobacter jeuni* and *C. coli* outer membrane components. *Res. Microbiol.* 140:379–392.

Fearnley, C., Ayling, R., Cawthraw, S., Newell, D. 1996. The formation of viable but noncul-turable *C. jejuni* and their failure to colonize one-day-old chicks. In: *Campylobacters, Helicobacters, and Related Organisms. Proceedings of the Eighth International Workshop on Campylobacters, and Related Organisms*, July 10–13, 1995, Winchester, UK. D. Newell, J. Ketley, R. Feldman, Eds., Plenum Press: New York, pp. 101–104.

Fermér, C., Engvall, E. 1999. Specific PCR identification and differentiation of the thermo-philic campylobacters, *Campylobacter jejuni, C. coli, C. lari,* and *C. upsaliensis. J. Clin. Microbiol.* 37:3370–3373.

Fernández, H., Neteo, U., Fernandez, F., Pedra, M., Trabulsi, L. 1983. Culture supernatants of *Campylobacter jejuni* induce a secretory response in jejunal segments of adult rats. *Infect. Immun.* 40:429–431.

Fitch, B., Sachen, K., Wilder, S., Burg, M., Lacher, D., Khalife, W., Whittman, T., Young, V. 2005. Genetic diversity of *Campylobacter* sp. isolates from retail chicken products and humans with gastroenteritis in central Michigan. *J. Clin. Microbiol.* 43:4221–4224.

Fitzgerald, C., Stanley, K., Andrew, S., Jones, K. 2001. Use of pulsed-field gel elecrophoresis and flagellin gene typing in identifying clonal groups of *Campylobacter jejuni* and *Campylobacter coli* in farm and clinical environments. *Appl. Environ. Microbiol.* 67:1429–1436.

Florin, I., Antillon, F. 1992. Production of enterotoxin and cytotoxin in *Campylobacter jejuni* strains isolated in Costa Rica. *J. Med. Microbiol.* 37:22–29.

Fragoso, G., Peres, I., da Silva, V., Cabrit, J. 1996. A new toxin in *Campylobacter jejuni* and *Campylobacter coli*. In: *Campylobacters, Helicobacters, and Related Organisms. Proceedings of the Eighth International Workshop on Campylobacters, and Related Organisms*, July 10–13, 1995, Winchester, UK. D. Newell, J. Ketley, R. Feldman, Eds., Plenum Press: New York, pp. 599–605.

French, N., Barrigas, M., Brown, P., Ribiero, P., Williams, N., Leatherbarrow, H., Birtles, R., Bolton, E., Fearnhead, P., Fox, A. 2005. Spatial epidemiology and natural population structure of *Campylobacter jejuni* colonizing a farmland ecosystem. *Environ. Microbiol.* 7:1116–1126.

Friis, L., Pin, C., Peason, B., Wells, J. 2005. In vitro cell culture methods for investigating campylobacter invasion mechanisms. *J. Microbiol. Meth.* 61:145–160.

Fry, B., Feng, S., Chen, Y., Newell, D., Coloe, P., Korolik, V. 2000. *The galE* gene of *Campylobacter jejuni* is involved in lipopolysaccharide synthesis and virulence. *Infect. Immun.* 68:2594–2601.

Fry, B., Korolik, V., ten Brinke, J., Pennings, M., Zalm, R., Teunis, B., Coloe, P., van der Zeijst, B. 1998. The lipopolysacharide biosynthesis locus of *Campylobacter jejuni* 81116. *Microbiology* 144:2049–2061.

Geilhausen, B., Koenen, R., Mauff, G. 1996. Pulsed field electrophoresis in *Campylobacter* epidemiology. In: *Campylobacters, Helicobacters, and Related Organisms. Proceedings of the Eighth International Workshop on Campylobacters, and Related Organisms*, July 10–13, 1995, Winchester, UK. D. Newell, J. Ketley, R. Feldman, Eds., Plenum Press: New York, pp. 191–195.

Gibson, J., Sutherland, K., Owen, R. 1994. Inhibition of DNAse activity in PFGE analysis of DNA from *Campylobacter jejuni. Lett. Appl. Microbiol.* 19:357–358.

Gill, C., Harris, L., 1982. Contamination of red-meat carcasses by *Campylobacter fetus* subsp. *jejuni. Appl. Environ. Microbiol.* 43:977–980.

Gonzalez, I., Grant, K., Richardson, P., Park, S., Collins, M. 1997. Specific identification of the enteropathogens *Campylobacter jejuni* and *Campylobacter coli* by using a PCR test based on the *ceuE* gene encoding a putative virulence determinant. *J. Clin. Microbiol.* 35:759–763.

Grennan, B., O'Sullivan, N., Fallon, R., Carrol, C., Smith, T., Glennon, M., Jaher, M. 2001. PCR-ELISAs for the detection of *Campylobacter jejuni* and *Campylobacter coli* in poultry samples. *Biotechniques* 30:2001.

Griffiths, P., Moreno, G., Park, R. 1992. Differentiation between thermophilic *Campylobacter* species by species-specific antibodies. *J. Appl. Bacteriol.* 72:467–474.

Guerrant, R., Wanke, C., Pennie, R., Barrett, L., Lima, A., O'Brien, A. 1987. Production of a unique cytotoxin by *Campylobacter jejuni. Infect. Immun.* 55:2526–2530.

Hageltorn, M., Berndtson, E. 1996. Pulsed field gel electrophoresis: A useful epidemiological tool for comparing campylobacters in milkborne outbreaks in Sweden. In: *Campylobacters, Helicobacters, and Related Organisms. Proceedings of the Eighth International Workshop on Campylobacters, and Related Organisms*, July 10–13, 1995, Winchester, UK. D. Newell, J. Ketley, R. Feldman, Eds., Plenum Press: New York, pp. 457–459.

Hald, B., Knudsen, K., Lind, P., Madsen, M. 2001. Study of the infectivity of saline-stored *Campylobacter jejuni* for day-old chicks. *Appl. Environ. Microbiol.* 67:2388–2392.

Hänninen, M., Haajanen, H., Pummi, T., Wermendsen, K., Katila, M., Sarkkinen, H., Miettinen, I., Rautelin, H. 2003. Detection and typing of *Campylobacter jejuni* and *Campylobacter coli* and analysis of indicator organisms in three waterborne outbreaks in Finland. *Appl. Environ. Microbiol.* 69:1391–1396.

Hänninen, M., Hakkinen, M., Rautelin, H. 1999. Stability of related human and chicken *Campylobacter jejuni* genotypes after passage through chick intestine studied by pulsed-field gel electrophoresis. *Appl. Environ. Microbiol.* 65:2272–2275.

Hänninen, M., PerkoMakela, P., Rautelin, H., Dim, B., Wagenaar, J. 2001. Genomic related-ness within five common Finnish *Campylobacter jejuni* pulsed-field gel electrophoresis genotypes studied by amplified fragment length polymorphism, analysis, ribotyping, and serotyping. *Appl. Environ. Microbiol.* 67:1581–1586.

Harmon, K., Ransom, G., Wesley, I. 1997. Differentiation of *Campylobacter jejuni* and *Campylobacter coli* by polymerase chain reaction. *Mol. Cell. Probes.* 11:195–200.

Harrington, C., Thompson-Carter, M., Carter, P. 1997. Evidence for recombination in the flagellin locus of *Campylobacter jejuni*: implications for the flagellin gene typing scheme. *J. Clin. Microbiol.* 35:2386–2392.

Hassane, D., Lee, R., Mendenhall, M., Picket, C. 2001. Cytolethal distending toxin demonstrates genotoxic activity in a yeast model. *Infect. Immum.* 69:5752–5759.

Hazeleger, W., Janse, J., Koenraad, P.K., Beumer, R., Rombouts, F., Abee, T. 1995. Temperature-dependent membrane fatty acid and cell physiology changes in coccoid forms of *Campylobacter jejuni*. *Appl. Environ. Microbiol.* 61:2713–2719.

Hazeleger, W., Wouters, J., Rombuts, F., Abee, T. 1998. Physiological activity of *Campylobacter jejuni* far below the minimal growth temperature. *Appl. Environ. Microbiol.* 64:3917–3922.

Hébert, G., Hollis, D., Weaver, R., Lambert, M., Blaser, M., Moss, C. 1982. 30 years of campylobacters: Biochemical characteristics and a biotyping proposal for *Campylobacter jejuni*. *J. Clin. Microbiol.* 15:1065–1073.

Heisick, J., Lanier, J., Peeler, J. 1984. Comparison of enrichment methods and atmosphere modification procedures for isolating *Campylobacter jejuni* from foods. *Appl. Environ. Microbiol.* 48:1254–1255.

Hernández, J., Fayos, A., Alonso, J., Owen, R. 1996. Ribotypes and AP-PCR fingerprints of thermophilic campylobacters from marine recreational waters. *J. Appl. Bacteriol.* 80:157–164.

Hernandez, J., Owen, R., Costas, M., Lastovica, A. 1991. DNA-DNA hybridization and analysis of restriction endonclease and rRNA gene patterns of atypical (catalase-weak/negative) *Campylobacter jejuni* from paediatric blood and faecal cultures. *J. Appl. Bacteriol.* 70:71–80.

Hickey, T., Baquar, S., Bourgeois, L., Ewing, C., Guerry, P. 1999. *Campylobacter jejuni*-stimulated secretion of interleukin-8 by INT-407 cells. *Infect. Immun.* 67:88–93.

Hickey, T., McVeigh, A., Scott, D., Michielutti, R., Bixby, A., Carroll, S., Bourgeois, A., Guery, P. 2000. *Campylobacter jejuni* cytolethal distending toxin mediates release of interleukin-8 from intestinal epithelial cells. *Infect. Immun.* 68:6535–6541.

Hong, Y., Berrang, M., Tongrui, L., Hofacre, C., Sanchez, S., Wang, L., Maurer, J. 2003. Rapid detection of *Campylobacter coli*, *C. jeuni*, and *Salmonella enterica* on poultry carcasses by using PCR-enzyme-linked immunosorbent assay. *Appl. Environ. Microbiol.* 69:3492–3466.

Hoorfar, J., Nielsen, E., Stryhn, H., Anderson, S. 1999. Evaluation of two automated enzyme-immunoassays for detection of thermophilic campylobacters in faecal samples from cattle and swine. *J. Microbiol. Methods* 38:101–106.

Hu, L., Raybourne, R., Kopecko, D. 2005. Ca^{2+} release from host intracellular stores and related signal transduction during *Campylobacter jejuni* B81-176 signal transduction internalization into human intestinal cells. *Microbiology* 151:3097–3105.

Hudock, J., Borger, A., Kaspar, C. 2005. Temperature-dependent genome degradation in the coccoid form of *Campylobacter jejuni*. *Curr. Microbiol.* 50:110–113.

Hudson, J., Nicol, C., Wright, J., Whyte, R., Hasell, S. 1999. Seasonal variation of *Campylobacter* types from human cases, veterinary cases, raw chicken, milk and water. *J. Appl. Microbiol.* 87:115–124.

Hunt, J., Abeyta, C., Tran, T. 1998. *Campylobacter*. *FDA Bacteriological Analytical Manual*, 8th ed. (rev. A). AOAC International: Gaithersburg, MD, pp. 7.01–7.24.

Inglis, G., Kalischuk, L. 2003. Use of PCR for direct detection of *Campylobacter* species in bovine feces. *Appl. Environ. Microbiol.* 69:3435–3447.

Inglis, G., Kalischuk, L. 2004. Direct quantification of *Campylobacter jejuni* and *Campylobacter lanienae* in feces of cattle by real-time quantitative PCR. *Appl. Environ. Microbiol.* 70:2296–2306.

Inglis, G., Kalischuk, L., Busz, H. 2003. A survey of *Campylobacter* species shed in faeces of beef cattle using polymerase chain reaction. *Can. J. Microbiol.* 49:655–661.

Iriarte, P., Owen, R. 1996. PCR-RFLP analysis of the large subunit (23S) ribosomal RNA genes of *Campylobacter jejuni. Lett. Appl. Microbiol.* 23:163–166.

Jackson, C., Fox, A., Jones, D. 1996. A novel polymerase chain reaction assay for the detection and speciation of thermophilic *Campylobacter* spp. *J. Appl. Bacteriol.* 81:467–473.

Jensen, A., Anderson, M., Dalsgaard, A., Baggesen, D., Nielsen, E. 2005. Development of real-time PCR and hybridization methods for detection and identification of thermophilic *Campylobacter* spp. in pig faecal samples. *J. Appl. Microbiol.* 99:292–300.

Jensen, M., Webster, J., Straus, N. 1993. Rapid identification of bacteria on the basis of polymerase chain reaction-amplified ribosomal DNA spacer polymorphisms. *Appl. Environ. Microbiol.* 59:945–952.

Jiménez, A., Barros-Valázquez, J., Rodríquez, J., Villa, T. 1997. Restriction endonuclease analysis, DNA relatedness and phenotypic characterization of *Campylobacter jeuni* and *Camp. coli* isolates involved in food-borne disease. *J. Appl. Microbiol.* 82:713–721.

Johnson, W., Lior, H. 1984. Toxins produced by *Campylobacter jejuni* and *Campylobacter coli. Lancet* 323, no. 8370:229–230.

Johnson, W., Lior, H. 1986. Cytotoxic and cytotonic factors produced by *Campylobacter jejuni, Campylobacter coli,* and *Campylobacter laridis. J. Clin. Microbiol.* 24:275–281.

Johnson, W., Lior, H. 1988. A new heat-labile cytolethal distending toxin (CLDT) produced by *Campylobacter* spp. *Microbial Pathogen.* 4:115–126.

Jones, D., Sutcliffe, E., Curry, A. 1991. Recovery of viable but non-culturable *Campylobacter jejuni. J. Gen Microbiol.* 137:2477–2482.

Josefsen, M., Cook, N., Agostino, M., Hansen, F., Wagner, M., Demnerova, K., Heuvelink, A., Tassios, P., Lindmark, H., Kmet, V., Barbanera, M., Fach, P., Loncarevic, S., Hoorfar, J. 2004a. Validation of a PCR-based method for detection of food-borne thermotolerant campylobacters in a multicenter collaborative trial. *Appl. Environ. Microbiol.* 70:4379–4383.

Josefsen, M., Jacobsen, N., Hoorfar, J. 2004b. Enrichment followed by quantitative PCR broth for rapid detection and as a tool for quantitative assessment of food-borne thermotolerant campylobacters. *Appl. Environ. Microbiol.* 70:3588–3592.

Kahn, I., Edge, T. 2007. Development of a novel triplex PCR assay for the detection and differentiation of thermophilic species of *Campylobacter* using 16s-23S rDNA internal transcribed spacer (ITS) region. *J. Appl. Microbiol.* 103:2561–2569.

Kaneko, A., Matsuda, M., Miyajima, M., Moore, J., Murphy, P. 1999. Urease-positive thermophilic strains of *Campylobacter* isolated from seagulls (*Larus* sp.). *Lett. Appl. Microbiol.* 29:7–9.

Karlyshev, A., Linton, D., Gregson, N., Lastovica, A., Wren, B. 2000. Genetic and biochemical evidence of a *Campylobacter jejuni* capsular polysaccharide that accounts for Penner serotype specificity. *Mol. Microbiol.* 35:529–541.

Karlyshev, A., McCrossan, M., Wren, B. 2001. Demonstration of polysaccahride capsule in *Campylobacter jejuni* using electron microscopy. *Infect. Immun.* 69:5921–5924.

Kinde, H., Genigeorgis, C., Pappaioanou, M. 1983. Prevalence of *Campylobacter jejuni* in chicken wings. *Appl. Environ. Microbiol.* 45:1116–1118.

Kirk, R., Rowe, M. 1994. A PCR assay for the detection of *Campylobacter jejuni* and *Campylobacter coli* in water. *Lett. Appl. Microbiol.* 19:301–303.

Klipstein, F., Engert, R. 1984a. Properties of crude *Campylobacter jeuni* heat-labile enterotoxin. *Infect. Immun.* 45:314–319.

Klipstein, F., Engert, R. 1984b. Purification of *Campylobacter jejuni* enterotoxin. *Lancet* 323, no. 8386:1123–1124.

Klipstein, F., Engert, R. 1985. Immunological relationship of the B subunits of *Campylobacter jejuni* and *Escherichia coli* heat-labile enterotoxins. *Infect. Immun.* 48:629–633.

Knudsen, K., Bang, D., Nielsen, E., Madsen, M. 2005. Genotyping of *Campylobacter jejuni* strains from Danish broiler chickens by restriction fragment length polymorphism of the LPS gene cluster. *J. Appl. Microbiol.* 99:392–399.

Koenraad, P., Ayling, R., Hazeleger, W., Rombouts, F., Newell, D. 1996. Subtyping of *Campylobacter* isolates from sewage plants and waste water from a connective poultry abattoir using molecular techniques. In: *Campylobacters, Helicobacters, and Related Organisms. Proceedings of the Eighth International Workshop on Campylobacters, and Related Organisms*, July 10–13, 1995, Winchester, UK. D. Newell, J. Ketley, R. Feldman, Eds., Plenum Press: New York, pp. 197–201.

Kogure, K., Simkidu, U., Taga, N. 1979. A tentative direct microscopic method for counting living bacteria. *Can. J. Microbiol.* 2:415–420.

Konkel, M., Garvis, S., Tipton, S., Anderson, D., Cieplak, J. 1997. Identification and molecular cloning of a gene encoding a fibronectin-binding protein (CadF) from *Campylobacter jejuni. Mol. Microbiol.* 24:953–963.

Konkel, M., Gray, S., Kim, B., Garvis, S., Yoon, J. 1999. Identification of the enteropathogens *Campylobacter jejuni* and *Campylobacter coli* based on the *cad*F virulence gene and its product. *J. Clin. Microbiol.* 37:510–517.

Korolik, V., Krishnapillai, V., Coloe, P. 1988. A specific DNA probe for the identification of *Campylobacter jejuni. J. Gen. Microbiol.* 134:521–529.

Korolik, V., Moorthy, L., Coloe, P. 1995. Differentiation of *Campylobacter jejuni* and *Campylobacter coli* strains by using restriction endonuclease DNA profiles and DNA fragment polymorphism. *J. Clin. Microbiol.* 33:1136–1140.

Kuusi, M., Nuorti, J., Hänninen, M., Koskela, M., Jussila, V., Kela, E., Miettinen, I., Ruutu, P. 2005. A large outbreak of campylobacteriosis associated with a municipal water supply in Finland. *Epidemiol. Infect.* 133:593–601.

Lamoureux, M., MacKay, A., Messier, S., Fliss, I., Blais, B., Holley, R., Simard, R. 1997. Detection of *Campylobacter jejuni* in food and poultry viscera using immunomagnetic separation and microtiter hybridization. *J. Appl. Microbiol.* 83:641–651.

Lane, D. 1991. 16S/23S rRNA sequencing. In: *Nucleic Acid Techniques in Bacterial Systematics.* E. Stackenbrandt, M. Goodfellow, Eds., John Wiley: Chichester, pp. 115–175.

Lara-Tejero, M., Galán, J. 2000. A bacterial toxin that controls cell cycle progression as a deoxyribonuclease I-like protein. *Science* 290:354–357.

Lara-Tejero, M., Galán, J. 2001. CdtA, CdtB, and CdtC form a tripartite complex that is required for cytolethal distending toxin activity. *Infect. Immun.* 69:4385–4365.

Lauwers, S., De Boeck, M., Butzler, J. 1978. *Campylobacter* enteritis in Brussels. *Lancet* 311, no. 8064:604–605.

Lawson, A., Linton, D., Stanley, J., Owen, R. 1997. Polymerase chain reaction detection and speciation of *Campylobacter upsaliensis* and *C. helveticus* in human faeces and comparison with culture techniques. *J. Appl. Microbiol.* 83:375–380.

Lázaro, B., Cárcamo, J., Audícana, A., Perales, I., Fernández-Asorga, A. 1999. Viability and DNA maintenance in nonculturable spiral *Campylobacter jejuni* cells after long-term exposure to low temperature. *Appl. Environ. Microbiol.* 65:4677–4681.

Lehtola, M., Loades, C., Keevil, C. 2005. Advantages of peptide nucleic acid oligonucleotides for sensitive site directed 16S rRSNA fluorescence in situ hybridization (FISH) detection of *Campylobacter jejuni, Campylobacter coli* and *Campylobacter lari. J. Microbiol. Methods* 62:211–219.

Levin, R.E. 2004. The application of real-time PCR to food and agricultural systems. A review. *Food Biotechnol.* 18:97–133.

Linton, D., Lawrence, L., McGuiggan, J. 1996a. Differentiation within *Campylobacter jejuni* and *C. coli* by PCR-RFLP of the intergeneric region between the *flaA* and *flaB* genes. In: *Campylobacters, Helicobacters, and Related Organisms. Proceedings of the Eighth International Workshop on Campylobacters, and Related Organisms*, July 10–13, 1995, Winchester, UK. D. Newell, J. Ketley, R. Feldman, Eds., Plenum Press: New York, pp. 209–211.

Linton, D., Owen, R., Stanley, J. 1996b. Rapid identification by PCR of the genus *Campylobacter* and of five *Campylobacter* species enteropathogenic for man and animals. *Res. Microbiol.* 147:707–718.

Lior, H., Woodward, D., Edgar, J., Laroche, J., Gill, P. 1982. Serotyping of *Campylobacter jejuni* by slide agglutination based on heat-labile antigenic factors. *J. Clin. Microbiol.* 15:761–768.

Lovett, J., Francis, D., Hunt, J. 1983. Isolation of *Campylobacter jejuni* from raw milk. *Appl. Environ. Microbiol.* 46:459–462.

Lübeck P., Wolffs, P., On, S., Ahrens, P., Radstrom, P., Hoorfar, J. 2003. Toward an international standard for PCR-based detection of food-borne thermotolerant campylobacters: Assay development and analytical validation. *Appl. Environ. Microbiol.* 69:5664–5669.

Lund, M., Wedderkopp, A., Wainø, M., Nordentoft, S., Bang, D., Person, K., Madsen, M. 2003. Evaluation of PCR for detection of *Campylobacter* in a national broiler surveillance programme in Denmark. *J. Appl. Microbiol.* 94:929–935.

MacCallum, A., Haddock, G., Everest, P. 2005. *Campylobacter jejuni* activates mitogen-activated protein kinases in Caco-2 cell monolayers and in vitro infected colonic tissue. *Microbiol.* 151:2765–2772.

Madden, R., Moran, L., Scates, P. 1996a. Sub-typing of animal and human *Campylobacter* spp. using RAPD. *Lett. Appl. Microbiol.* 23:167–170.

Madden, R., Moran, L., Scates, P. 1996b. Frequency of occurrence of *Campylobacter* spp. in meats and their subsequent sub-typing using RAPD and PCR-RFLP. In: *Campylobacters, Helicobacters, and Related Organisms. Proceedings of the Eighth International Workshop on Campylobacters, and Related Organisms*, July 10–13, 1995, Winchester, UK. D. Newell, J. Ketley, R. Feldman, Eds., Plenum Press: New York, pp. 141–145.

Mahajan, S., Rodgers, F. 1990. Isolation, characterization, and host-cell properties of a cytotoxin from *Campylobacter jejuni*. *J. Clin. Microbiol.* 28:1314–1320.

Manning, G., Dowson, C., Bagnall, M., Ahmed, I., West, M., Newell, D. 2003. Multilocus sequence typing for comparison of veterinary and human isolates of *Campylobacter jejuni*. *Appl. Environ. Microbiol.* 69:6370–6379.

Martin, W., Patton, C., Morris, G., Potter, M., Puhr, N. 1983. Selective enrichment broth medium for isolation of *Campylobacter jejuni*. *J. Clin. Microbiol.* 17:853–855.

Mason, M., Humphrey, T., Martin, K. 1996. Isolation of sub-lethally injured campylobacters from water. In: *Campylobacters, Helicobacters, and Related Organisms. Proceedings of the Eighth International Workshop on Campylobacters, and Related Organisms*, July 10–13, 1995, Winchester, UK. D. Newell, J. Ketley, R. Feldman, Eds., Plenum Press: New York, pp. 129–133.

Mateo, E., Cárcamo, J., Urquijo, M., Perales, I., Fernández-Astorga, A. 2005. Evaluation of a PCR assay for the detection and identification of *Campylobacter jejuni* and *Campylobacter coli* in retail poultry products. *Res. Microbiol.* 156:568–574.

Matsuda, M., Kaneko, A., Fukuyama, M., Itoh, T., Shingaki, M., Inoue, M., Moore, J., Murphy, P., and Isida, Y. 1996. First finding of urease-positive thermophilic strains of *Campylobacter* in river water in the Far East, namely Japan, and their phenotypic and genotypic characterization. *J. Appl. Bacteriol.* 81:608–612.

McCardell, B., Madden, J., Lee, E. 1984. *Campylobacter jejuni* and *Campylobacter coli* production of a cytotonic toxin immunologically similar to cholera toxin. *J. Food. Prot.* 47:943–949.

McSweegan, E., Walker, R.S. 1986. Identification and characterization of two *Campylobacter jejuni* adhesins for cellular and mucous substrate. *Infect. Immun.* 53:141–148.

Medema, G., Scheets, F., Giessen, A., Havelaar, A. 1992. Lack of colonization of 1 day old chicks by viable, non-culturable *Campylobacter jejuni*. *J. Appl. Bacteriol.* 72:512–516.

Mégraud, F., Chevrier, D., Desplaces, N., Sedallian, A., Guesdon, J. 1988. Urease-positive thermophilic *Campylobacter* (*Campylobacter laridis* variant) isolated from an appendix and from human feces. *J. Clin. Microbiol.* 26:1050–1051.

Meinersmann, R., Helsel, L., Fields, P., Hiett, K. 1997. Discrimination of *Campylobacter jejuni* isolates by *fla* gene sequencing. *J. Clin. Microbiol.* 35:2810–2814.

Melby, K., Kvien, T., Glennås, A. 1996. *Campylobacter jejuni* as trigger of reactive arthritis. In: *Campylobacters, Helicobacters, and Related Organisms. Proceedings of the Eighth International Workshop on Campylobacters, and Related Organisms,* July 10–13, Winchester, UK. D. Newell, J. Ketley, R. Feldman, Eds., Plenum Press: New York, pp. 487–490.

Michaud, S., Menard, S., Arbeitt, R. 2004. Campylobacteriosis, eastern townships, *Quebec. Emerg. Infect. Dis.* 10:1844–1847.

Miller, W., Englen, M., Katharious, S., Wesley, I., Wang, G., Pittinger-Alley, L., Siletz, R., Muraoka, W., Fedorka-Cray, P., Mandrell, R. 2006. Identification of host-associated alleles by multilocus sequence typing of *Campylobacer coli* strains from food animals. *Microbiol.* 152:245–255.

Mills, S., Kuzniar, B., Shames, B., Kurjanczyk, L., Penner, J. 1992. Variation of the O antigen of *Campylobacter jejuni* in vivo. *J. Med. Microbiol.* 36:215–219.

Misawa, N., Blaser, M. 2000. Detection and characterization of autoagglutination activity by *Campylobacter jejuni*. *Infect. Immun.* 68:6168–6175.

Monteville, M., Konkel, M. 2002. Fibronectin-facilitated invasion of T84 eukaryotic cells by *Campylobacter jejuni* occurs preferentially at the basolateral cell surface. *Infect. Immun.* 70:6665–6671.

Moran, A., Annuk, H., Prendergast, M. 2005. Antibodies induced by gangliosides-mimicking *Campylobacter jejuni* lipooligosaccharides recognize epitopes at the nodes of Ranvier. *J. Neuroimmunol.* 165:179–185.

Moran, A., Penner, J. 1999. Serotyping of *Campylobacter jejuni* based on heat-stable antigens: Relevance, molecular basis and implications in pathogenesis. *J. Appl. Microbiol.* 86:361–377.

Moran, A., Upton, M. 1986. A comparative study of the rod and coccoidal forms of *Campylobacter jejuni* ATCC 29428. *J. Appl. Bacteriol.* 60:103–110.

Moran, A., Upton, M. 1987a. Effect of medium supplements, illumination and superoxide dismutase on the production of coccoid forms of *Campylobacter jejuni* ATCC 29428. *J. Appl. Bacteriol.* 62:43–51.

Moran, A., Upton, M. 1987b. Factors affecting production of coccoid forms by *Campylobacter jejuni* on solid media during incubation. *J. Appl. Bacteriol.* 62:527–537.

Murphy, C., Carroll, C., Jordan, K. 2005. The effect of different media on the survival and induction of stress responses by *Campylobacter jejuni*. *J. Microbiol. Methods* 62:161–166.

Nachamkin, I., Bohachick, K., Patton, C. 1993a. Flagellin gene typing of *Campylobacer jejuni* by restriction fragment length polymorphism analysis. *J. Clin. Microbiol.* 31:1531–1536.

Nachamkin, I., Liu, J., Ung, H., Moran, A., Prendergast, M., Sheikh, K. 2002. *Campylobacter jejuni* from patients with Guillain-Barre syndrome preferentially expresses a GD_{1a}-like epitope. *Infect. Immun.* 7:5290–5303.

Nachamkin, I., Yang, X., Stern, N. 1993b. Role of *Campylobacter jejuni* flagella as colonization factors for three-day-old chicks: Analysis with flagellar mutants. *Appl. Environ. Microbiol.* 59:1269–1273.

Nadeau, E., Messier, S., Quessy, S. 2003. Comparison of *Campylobacter* isolates from poultry and humans, association between in vitro virulence properties, biotypes, and pulsed-field gel electrophoresis clusters. *Appl. Environ. Microbiol.* 69:6316–6320.

Naess, V., Hofstad, T. 1984. Chemical composition and biological activity of lipopolysaccharide prepared from type strains of *Campylobacter jejuni* and *Campylobacter coli*. *Acta. Pathol. Microbiol. Immunol. Scand.* 92:217–222.

Ng, L., Kingombe, C., Yan, W., Taylor, D., Hiratsuka, K., Malik, N., Garcia, M. 1997. Specific detection and confirmation of *Campylobacter jejuni* by DNA hybridization and PCR. *Appl. Environ. Microbiol.* 63:4558–4563.

Nichols, G. 2005. Fly transmission of *Campylobacter. Emerg. Infect. Dis.* 11:361–364.

Nielsen, E., Engberg, J., Fussing, V., Peterson, L., Brogren C., On, S. 2000. Evaluation of phenotypic and genotypic methods for subtyping *Campylobacter jejuni* isolates from humans, poultry, and cattle. *J. Clin. Microbiol.* 38:3800–3810.

Nogva, H., Bergh, A., Holck, A., Rudi, K. 2000. Application of the 5′-nuclease PCR assay in evaluation and development of methods for quantitative detection of *Campylobacter jejuni. Appl. Environ. Microbiol.* 66:4029–4036.

Nuijten, P., van Asten, F., Gaastra, W., van der Zeijst, B. 1990. Structural and functional analysis of two *Campylobacter jejuni* flagellin genes. *J. Biol. Chem.* 265:17798–17804.

Occhialini, A., Stonnet, V., Hua, J., Camou, C., Guesden, J., Mégraud, F. 1996. Identification of strains of *Campylobacter jejuni* and *Campylobacter coli* by PCR and correlation with phenotypic characteristics. In: *Campylobacters, Helicobacters, and Related Organisms. Proceedings of the Eighth International Workshop on Campylobacters, and Related Organisms*, July 10–13, 1995, Winchester, UK. D. Newell, J. Ketley, R. Feldman, Eds., Plenum Press: New York, pp. 217–219.

Ohara, M., Oswald, E., Sugai, M. 2004. Cytolethal distending toxin: A bacterial bullet targeted to nucleus. *J. Biochem.* 136:409–413.

On, S., Nielsen, W., Engberg, J., Madsen, M. 1998. Validity of *smaI*-defined genotypes of *Campylobacter jejuni* examined by *SalI, EpnI*, and *BamHi* polymorphisms: Evidence of identical clones infecting humans, poultry, and cattle. *Epidemiol. Infect.* 20:231.

O'Sullivan, N., Fallon, R., Carroll, C., Sith, T., Maher, M. 2000. Detection and differentiation of *Campylobacter jejuni* and *Campylobacter coli* in broiler chicken samples using a PCRS/DNA probe membrane based colorimetric detection assay. *Mol. Cell. Probes.* 14:7–16.

Oyofo, B., Thornton, S., Burr, D., Trust, T., Pavlovski, O., Guerry, P. 1992. Specific detection of *Campylobacter jejuni* and *Campylobacter coli* by using polymerase chain reaction. *J. Clin. Microbiol.* 30:2613–2619.

Pacha, R., Clark, G., Williams, E. 1985. Occurrence of *Campylobacter jejuni* and *Giardia* species in Muskrat (*Indatra zibethica*). *Appl. Environ. Microbiol.* 50:177–178.

Palumbo, S. 1984. Heat injury and repair in *Campylobacter jejuni. Appl. Environ. Microbiol.* 48:477–480.

Park, C., Stankiewicz, Z., Lovett, J., and Hunt, J. 1981. Incidence of *Campylobacter jejuni* in fresh eviscerated whole market chickens. *Can. J. Microbiol.* 27:841–842.

Patrick, M., Christiansen, L., Wainø, M., Ethelberg, S., Madsen, H., Wegener, H. 2004. Effects of climate on incidence of *Campylobacter* spp. in humans and prevalence in broiler flocks in Denmark. *Appl. Environ. Microbiol.* 70:7474–7480.

Patton, C., Mitchell, S., Potter, M., Kaufman, A. 1981. Comparison of selective media for primary isolation of *Campylobacter fetus* subsp. *jejuni. J. Clin. Microbiol.* 13:326–330.

Pearson, A., Greenwood, M., Feltham, R., Healing, T., Donaldson, J., Jones, D., Colwell, R. 1996. Microbial ecology of *Campylobacter jejuni* in a United Kingdom chicken supply chain: Intermittent common source, vertical transmission, and amplification by flock propagation. *Appl. Environ. Microbiol.* 62:4614–4620.

Pearson, A., Greenwood, M., Feltham, R., Healing, T., Rollins, D., Shahamat, M., Donaldson, J., Colwell, R. 1993. Colonization of broiler chickens by waterborne *Campylobacter jejuni*. *Appl. Environ. Microbiol.* 59:987–996.

Pei, Z., Blaser, M. 1993. PEB1, the major cell-binding factor of *Campylobacter jejuni*, is a homolog of the binding component of gram-negative nutrient transport systems. *J. Biol. Chem.* 268:18717–18725.

Peterson, L., Nielsen, E., Engberg, J., On, S., Dietz, H. 2001. Comparison of genotypes and serotypes of *Campylobacter jejuni* isolated from Danish wild mammals and birds and from broiler flocks and humans. *Appl. Environ. Microbiol.* 67:3115–3121.

Peterson L., Wedderkopp, A. 2001. Evidence that certain clones of *Campylobacter jejuni* persist during successive broiler rotations. *Appl. Environ. Microbiol.* 67:2739–2745.

Picket, C., Pesci, E., Cottle, D., Russell, G., Erdem, A., Zeytin, H. 1996. Prevalence of cytolethal distending toxin production in *Campylobacter jejuni* and relatedness of *Campylobacter* sp. *cdtB* genes. *Infect. Immun.* 66:2070–2078.

Poulsen, C., El-Ali, J., Perch-Nielsen, I., Bang, D., Telleman, P., Wolff, A. 2005. Detection of a putative virulence *cad*F gene of *Campylobacter jejuni* obtained from different sources using a microfabricated PCR chip. *J. Rapid Meth. Autom. Microbiol.* 13:111–126.

Praakle-Amin, K., Roasto, M., Korkeala, H., Hänninen, M. 2007. PFGE genotyping and antimicrobial susceptibility of Campylobacter in retail poultry meat in Estonia. *Int. J. Food Microbiol.* 114:105–112.

Purdy, D., Ash, C., Fricker, C. 1996. Polymerase chain reaction assay for the detection of viable *Campylobacter* species from potable and untreated environmental water samples. In: *Proceedings of the Eighth International Workshop on Campylobacters, and Related Organisms*, July 10–13, 1995, Winchester, UK. D. Newell, J. Ketley, R. Feldman, Eds., Plenum Press: New York, pp. 147–153.

Rasmussen, H., Olsen, J., Jorgensen, K., Rasmussen, O. 1996. Detection of *Campylobacter jejuni* and *Camp. coli* in chicken faecal samples by PCR. *Lett. Appl. Microbiol.* 23:363–366.

Ravel, J., Hill, R., Knight, I., Dubois, C., Colwell, R. 1993. Recovery of *Vibrio cholerae* from the viable but non-culturable state. *Abstracts of the 93rd General Meeting of the American Society for Microbiology 1993.* American Society for Microbiology: Washington, DC, Abst. Q-6, p. 347.

Ray, B., Johnson, C. 1984. Sensitivity of cold-stressed *Campylobacter jejuni* to solid and liquid selective environments. *Food Microbiol.* 1:173–176.

Rivoal, K., Denis, M., Salvat, G., Colin, P., Ermel, G. 1999. Molecular characterization of the diversity of *Campylobacter* spp. isolates collected from a poultry slaughterhouse: analysis of cross-contamination. *Lett. Appl. Microbiol.* 29:370–374.

Robinson, D. 1981. Infective dose of *Campylobacter jejuni* in milk. *Br. Med. J.* 282:1584.

Rogol, M., Shpak, B., Rothman, D., Sechter, I. 1985. Enrichment medium for isolation of *Campylobacter jejuni-Campylobacter coli*. *Appl. Environ. Microbiol.* 50:125–126.

Rollins, D., Colwell, R. 1986. Viable but non-culturable stage of *C. jejuni* and its role in survival in the aquatic environment. *Appl. Environ. Microbiol.* 52:521–538.

Rosef, O. 1981. Isolation of *Campylobacter fetus* subsp. *jejuni* from the gallbladder of normal slaughter house pigs, using an enrichment procedure. *Acta Vet. Scand.* 22:149–151.

Rozynek, E., Dzierzanowska-Fangrat, K., Jozwiak, P., Popowski, J., Korsak, D., Dzierzanowska, D. 2005. Prevalence of potential virulence markers in Polish *Campylobacter jejuni* and *Campylobacter coli* isolates obtained from hospitalized children and from chicken carcasses. *J. Med. Microbiol.* 54:615–619.

Rudi, K., Hoidal, H., Katla, T., Johansen, B., Nordal, J., Jakobsen, K. 2004. Direct real-time PCR quantification of *Campylobacter jejuni* in chicken fecal and cecal samples by integrated cell concentration and DNA purification. *Appl. Environ. Microbiol.* 70:790–797.

Ruiz-Palacios, G., Torres, J., Torres, N., Escamilla, E., Ruis-Palacios, B., Tamayo, J. 1983. Cholera-like enterotoxin produced by *Campylobacter jejuni. Lancet* 322, no. 8344:250–253.

Saha, S., Saha, S., Anyal, S. 1991. Recovery of injured *Campylobacter jejuni* cells after animal passage. *Appl. Environ. Microbiol.* 57:3388–3389.

Sails, A., Fox, A., Bolton, F., Wareing, D., Greenway, D. 2003. A real-time PCR assay for the detection of *Campylobacter jejuni* in foods after enrichment culture. *Appl. Environ. Microbiol.* 69:1383–1390.

Saito, S., Yatsuyanagi, J., Harata, S., Ito, Y., Sinagawa, K., Suzuki, N., Amano, K., Enomoto, K. 2005. *Campylobacter jejuni* isolated from retail poultry meat, bovine feces and bile, and human diarrheal samples in Japan: Comparison of serotypes and genotypes. *FEMS Immunol. Med. Microbiol.* 45:311–319.

Salloway, S., Mermel, L., Seamans, M., Aspinall, G., Shin, J., Kurjanczyk, L., Penner, J. 1996. Miller-Fisher syndrome associated with *Campylobacter jejuni* bearing lipopolysaccharide molecules that mimic human gangliosides GD3. *Infect. Immun.* 64:2945–2949.

Saunders, N., Harrison, T., Haththotuwa, A., Kachwall, N., Talor, A. 1990. A method for typing strains of *Legionella pneumophila* serogroup 1 by analysis of restriction fragment polymorphisms. *J. Med Microbiol.* 31:45–55.

Schoeni, J., Wong, A. 1994. Inhibition of *Campylobacter jejuni* colonization in chicks by defined competitive exclusion bacteria. *Appl. Environ. Microbiol.* 60:1191–1197.

Schönberg-Norio, D., Takkinen, J., Hänninen, M., Katila, M., Kaukoranta, S., Mattila, L., Rautelin, H. 2004. Swimming and *Campylobacter* infections. *Emerg. Infect. Dis.* 10:1474–1477.

Schulze, F., Hänel, I., Borrmann, E. 1996. Detection of a cytolethal distending toxin in Campylobacters of human and animal origin. In: *Campylobacters, Heliobacters, and Related Organisms. Proceedings of the Eighth International Workshop on Campylobacters, and Related Organisms*, July 10–13, Winchester, UK. D. Newell, J. Ketley, R. Feldman, Eds., Plenum Press: New York, pp. 631–635.

Schwerer, B., Neisser, A., Polt, R., Bernheimer, H., Moran, A. 1995. Antibody cross-reactivities between gangliosides and lipopolysaccharides of *Campylobacter jejuni* serotypes associated with Guillain-Barré syndrome. *J. Endotox. Res.* 2:395–403.

Scotter, S., Humphrey, T., Henle, A. 1993. Methods for the detection of thermotolerant campylobacters in foods: results of an inter-laboratory study. *J. Appl. Bacteriol.* 74:155–163.

Shi, F., Chen, Y., Wassenaar, T., Woods, W., Coloe, P., Fry, B. 2002. Development and application of a new scheme for typing *Campylobacter jejuni* and *Campylobacter coli* by PCR-based restriction fragment length polymorphism analysis. *J. Clin. Microbiol.* 40:1791–1797.

Skirrow, M. 1977. *Campylobacter* enteritis: A 'new' disease. *Br. Med. J.* 2:9–11.

Skirrow, M. 1986. *Campylobacter* infections in man. In: *Medical Microbiology*, Vol. 4. C.S.F. Easman, Ed., Academic Press: New York, pp. 105–141.

Snelling, W., Matsuda, M., Moore, J., Dooley, J. 2005a. Under the microscope—*Campylobacter jejuni*. *Lett. Appl. Microbiol.* 41:297–302.

Snelling, W., McKenna, J., Lecky, D., Dooley, J. 2005b. Survival of *Campylobacter jejuni* in water borne protozoa. *Appl. Environ. Microbiol.* 71:5560–5571.

Steel, M., McNab, B., Fruhner, L., DeGrandis, S., Woodward, D., Odumeru, J. 1998. Epidemiological typing of *Campylobacter* isolates from meat processing plants by pulsed-field gel electrophoresis, fatty acid profile typing, serotping, and biotyping. *Appl. Environ. Microbiol.* 64:2346–2349.

Steinhauserova, I., Fojtikova, K., Klimes, J. 2000. The incidence and PCR detection of *Campylobacter upsaliensis* in dogs and cats. *Lett. Appl. Microbiol.* 31:209–212.

Stern, N., Jones, D., Wesley, I., Rollins, D. 1995. Colonization of chicks by non-culturable *Campylobacter* spp. *Lett. Appl. Microbiol.* 18:333–336.

Stern, N., Kazmi, S. 1989. *Campylobacer jejuni*. In: *Food-Borne Bacterial Pathogens*. M.P. Doyle, Ed., Marcel Decker: New York, pp. 71–110.

Talibart, R., Denis, M., Castillo A., Cappelier, J., Ermel, G. 2000. Survival and recovery of viable but nonculturable forms of *Campylobacter* in aqueous microcosm. *Int. J. Food Microbiol.* 55:263–267.

Tangvatcharin, P., Chanthachum, S., Kopaiboon, P., Inttasungkha, N., Griffiths, M. 2005. Comparison of methods for the isolation of thermotolerant *Campylobacter* from poultry. *J. Food Prot.* 68:616–620.

Tee, W., Anderson, B., Rosas, B., Dwyer, B. 1987. Atypical campylobacters associated with gastroenteritis. *J. Clin. Microbiol.* 25:1248–1252.

Tenover, F., Williams, S., Gordon, K., Plorde, J. 1985. Survey of plasmids and resistance factors in *Camplyobace jejuni* and *Camplyobacter coli*. *Antimicrob. Agents Chemother.* 277:37–41.

Tholozan, J., Cappelier, J., Tissier, J., Delattre, G., Federighi, M. 1999. Physiological characterization of viable-but-nonculturable *Campylobacter jejuni* cells. *Appl. Environ. Microbiol.* 65:1110–1116.

Thomas, C., Gibson, D., Hill, J., Mabey, M. 1999. *Campylobacter* epidemiology: An aquatic perspective. *J. Appl. Microbiol. Symp. Suppl.* 85:168S–177S.

Thomas, L., Long, K., Good, R., Panaccio, M., Widders, P. 1997. Genotypic diversity among *Campylobacter jejuni* isolates in a commercial broiler flock. *Appl. Environ. Microbiol.* 63:1874–1877.

Tu, Z., Eisner, W., Kreiswirth, B., Blaser, M. 2005. Genetic divergence of *Campylobacter fetus* strains of mammal and reptile origins. *J. Clin. Microbiol.* 43:3334–3340.

Uyttendaele, M., Schukkink, R., van Gemen, B., Debevere, J. 1994. Identification of *Campylobacter jejuni*, *Campylobacter coli* and *Campylobacter lari* by the nucleic acid amplification system NASBA®. *J. Appl. Bacteriol.* 77:694–701.

Uyttendaele, M., Schukkink, R., van Gemen, B., Debevere, J. 1995. Detection of *Campylobacter jejuni* added to foods by using a combined selective enrichment and nucleic acid sequence-based amplification (NASBA). *Appl. Environ. Microbiol.* 61:1341–1347.

Vandamme, P., Van Doorn, L., al Rashid, S., Quint, W., van der Plas, J., Chan, V., On, S. 1997. *Campylobacter hyoilei* Alberton et al. 1995 and *Campylobacter coli* Veron and Chatelain 1973 are subjective synonyms. *Int. J. Syst. Bacteriol.* 47:1055–1060.

Vanniasinkam, T., Lanser, J., Barton, M. 1999. PCR for the detection of *Campylobacter* spp. in clinical specimens. *Lett. Appl. Microbiol.* 28:52–56.

Varma, L., Pushpa, R., Subramanyam, K. 2005. Incidence of *Campylobacter jejuni* in milk. *Indian Vet. J.* 82:818–820.

Verhoeff-Bakkenes, L., Arends, A., Snoep, J., Zwietering, M., de Jonge, R. 2008. Pyruvate relieves the necessity of high induction levels of catalase and enables *Campylobacter jejuni* to grow under fully aerobic conditions. *Lett. Appl. Microbiol.* 46:377–382.

Versalovic, J., Koeuth, T., Lupski, R. 1991. Distribution of repetitive DNA sequences in eubacteria and application to fingerprinting of bacterial genomes. *Nucleic Acids Res.* 19:6823–6831.

Waage, A., Vardund, T., Lund, V., Kapperud, G. 1999. Detection of small numbers of *Campylobacter jejuni* and *Campylobacter coli* cells in environmental water, sewage, and food samples by a semi-nested PCR assay. *Appl. Environ. Microbiol.* 65:1636–1643.

Waegel, A., Nachamkin, I. 1996. Detection and molecular typing of *Campylobacter jejuni* in fecal samples by polymerase chain reaction. *Mol. Cell. Probes* 10:75–80.

Waldenström, J., Broman, T., Carlsson, I., Hasselquist, D., Achterberg, R., Wagenaar, J., Olsen, B. 2002. Prevalence of *Campylobacter jejuni*, *Campylobacter lari*, and *Campylobacter coli* in different ecological guilds and taxa of migrating birds. *Appl. Environ. Microbiol.* 68:5911–5917.

Walker, R., Caldwell, M., Lee, E., Guerry, P., Trust, T., Ruiz-Palacios, G. 1986. Pathophysiology of *Campylobacter* enteritis. *Microbiol. Rev.* 50:81–94.

Wallace, J., Stanley, K., Currie, J., Diggle, P., Jones, K. 1997. Seasonality of thermophilic *Campylobacter* populations in chickens. *J. Appl. Microbiol.* 82:219–224.

Wassenaar, T., Newell, D. 2000. Minireview: Genotying of *Campylobacter* spp. *Appl. Environ. Microbiol.* 66:1–9.

Wegmüller, B., Lütty, J., Candrian, U. 1993. Direct polymerase chain reaction detection of *Campylobacter jejuni* and *Campylobacter coli* in raw milk and dairy products. *Appl. Environ. Microbiol.* 59:2161–2165.

Weichart, D., Ollver, J., Kjelleberg, S. 1992. Low temperature induced non-culturability and killing of *Vibrio vulnificus*. *FEMS Microbiol. Lett.* 100:205–210.

Weijtens, M., Plas, J., Bijker, P., Urlings, H., Koster, D., Logtestijn, J., Veld, J. 1997. The transmission of *Campylobacter* in piggeries; an epidemiological study. *J. Appl. Microbiol.* 83:693–698.

Weijtens, M., Urlings, H., Van der Plas, J. 2000. Establishing a *Campylobacter*-free pig population through a top-down approach. *Lett. Appl. Microbiol.* 30:479–484.

Wesley, I., Wells, S., Harmon, K., Green, A., Schroeder-Tucker, L., Glover, M., Siddique, I. 2000. Fecal shedding of *Campylobacter* and *Arcobacter* spp. in dairy cattle. *Appl. Environ. Microbiol.* 66:1994–2000.

Wesley, R., Swaminathan, B., Stadelman, W. 1983. Isolation and enumeration of *Campylobacter jejuni* from poultry products by a selective enrichment medium. *Appl. Environ. Microbiol.* 46:1097–2102.

Whitehouse, C., Balbo, P., Pesci, E., Cottle, D., Mirabito, P., Pickett, C. 1998. *Campylobacter jejuni* cytolethal distending toxin causes a G2-phase cell cycle block. *Infect. Immun.* 66:1934–1940.

Woodall, C., Jones, M., Barrow, P., Hinds, J., Marsden, G., Kelly, D., Dorrell, N., Wren, B., Maskell, D. 2005. *Campylobacter jejuni* gene expression in the chick cecum: Evidence for adaptation to a low-oxygen environment. *Infect. Immun.* 73:5278–5285.

Yamazaki, W., Taguchi, M., Ishibashi, M., Kiazato, M., Nukina, M., Misawa, N., Inoue, K. 2008. Development and evaluation of a loop-mediated isothermal amplification assay fo rapid and simple detection of *Campylobacter jejuni* and *Campylobacter coli*. *J. Med. Microbiol.* 57:444–451.

Yang, C., Jiang, Y., Huanag, K., Zhu, C., Yin, Y. 2003. Application of real-time PCR for quantitative detection of *Campylobacter jejuni* in poultry, milk and environmental water. *FEMS Immunol. Med. Microbiol.* 38:265–271.

Yeen, P., Puthucheary, S., Pang, T. 1983. Demonstration of a cytotoxin from *Campylobacter jejuni*. *J. Clin. Pathol.* 36: 1237–1240.

Ziprin, R., Young, C., Sanker, L., Hume, M., Konkel, M. 1999. The absence of cecal colonization of chicks by a mutant of *Campylobacter jejuni* not expressing bacterial fibrobnectin-binding protein. *Avian Dis.* 43:586–589.

CHAPTER 11

Staphylococcus aureus

I. CHARACTERISTICS OF THE ORGANISM

Staphylococcus aureus is a gram-positive, facultatively anaerobic coccus occurring in packets of four cells when grown in broth. Staphylococcal food poisoning (SFP) results from ingestion of one or more preformed staphylococcal enterotoxins (SEs) in *S. aureus* contaminated foods with humans being the main reservoir for the organism. Typical symptoms, which are usually self-limiting, involve vomiting, abdominal cramps, and diarrhea. There are presently 20 recognized antigenically distinguishable SEs assigned a letter of the alphabet in the order of their discovery from SEA to SEU encoded by the corresponding genes *sea* to *seu*.

SEs have been divided into two groups. The members of group 1 are considered classic emetic toxins and are designated SEA, SEB, SEC_1, SEC_{bov}, SED, and SEE (Table 11.1) and cause about 95% of staphylococcal food poisoning (SFP) in humans. SEA, SEB, and SED are among the most frequently encountered SEs (Holmberg and Blake, 1984; Zhang, Iandola, and Sewart, 1998). Serological subtypes can occur, especially with SEC. Group 2 includes 15 additional newly found toxins SEG to SEU thought to be involved in the remaining 5% of SFP outbreaks (Table 11.1). There is no SEF, in that TSST-1 was mistakenly designated SEF and therefore there is no serotpe F (SEF) enterotoxin nor is there an SES or SET. SEs are referred to as superantigens in that they have the ability to stimulate a much higher percentage of T cells than conventional antigens. SEA is encoded by structural genes of bacteriophage (Betley and Mekalanos, 1985) whereas only the SED and SEJ toxins are plasmid encoded (Bayles and Iandola, 1989; Zhang, Iandola, and Stewart, 1998).

In addition to SFP, toxic shock syndrome is attributable to infection by *S. aureus* and production of the toxic shock syndrome toxin TSST, which is encoded by the gene *tst*. The human nasal cavity is thought to be a staging site for *S. aureus* strains harboring the *tst* gene for subsequent infection of other areas of the body (Warner and Onderdonk, 2004). The staphylococcal scalded skin syndrome (SSSS) of infants is due to unique strains of *S. aureus* that produce the exfoliative toxin EFT encoded by the *eft* (*eta*) gene. Many of the studies involved in the molecular characterization of *S. aureus* strains have focused on these toxin genes. The reader is alerted to the fact that some authors refer to staphylococcal enterotoxin A as being encoded by the *sea* gene, whereas others refer to the same gene as the *entA* gene.

II. MOLECULAR TECHNIQUES FOR DETECTION AND IDENTIFICATION OF *S. AUREUS*

A. PCR Used to Detect and Identify *S. aureus*

The thermonuclease of *S. aureus* is encoded by the *nuc* gene and is a phenotypic characteristic of *S. aureus* isolates; however, thermonucleases are also produced by other species of staphylococci encoded by genes of differing sequence. Brakstad, Aasbakk, and Maeland (1992) developed a species-specific PCR assay for *S. aureus* based on the *nuc* gene. Primers Nuc1/Nuc2 (Table 11.2) amplified a ~270-bp sequence of the *nuc* gene. The amplicons were detected in agarose gels or by a ^{32}P-labeled probe Nuc^{32}P-P (Table 11.2) following blotting and membrane hybridization. The primers recognized all 90 reference clinical isolates of *S. aureus*. Amplicons were not produced with 80 strains representing 16 other *Staphylococcus* species in addition to 20 strains of non-*Staphylococcus* species. It is interesting that some of the non-*S. aureus* staphylococci produced a thermostable nuclease but were PCR negative.

Chesneau, Allignet, and El Solh (1993) made use of the *nuc* gene for specific recognition of *S. aureus* isolates. The primers W-F/Y-R (Table 11.2) amplified a 450-bp sequence of the *nuc* gene.

A ^{32}P-labeled *nuc* probe was developed for hybridization to the DNA of lysed colonies blotted onto Hyband-N$^+$ membranes (Amersham). A total of 360 *Staphylococcus* isolates belonging to 28 staphylococcal species and consisting of 146 *S. aureus* isolates with the remainder being other species of staphylococci were tested. Among the 360 *Staphylococcus* isolates, 197 produced a thermonuclease (TNase). Among the 146 *S. aureus* isolates, 142 were TNase-positive and 55 non–*S. aureus* isolates were TNase-positive. However, the *nuc* primers yielded amplicons from seven *S. aureus* isolates that included six strong TNase producers and one isolate that failed to produce detectable TNase activity, whereas no amplicons were obtained with the DNA from five TNase-producing non–*S. aureus* staphylococcus isolates. The TNase probe hybridized to all 146 *S. aureus* isolates including the four that failed to produce detectable TNase activity. No hybridization signal occurred with any of the 214 remaining non–*S. aureus* isolates, including the 55 non–*S. aureus* TNase producers. The *nuc* PCR assay was therefore highly specific for *S. aureus* and was also able to detect the *nuc* gene in TNase-negative isolates of *S aureus*. The authors concluded that the labeled probe was ideally suited for blot recognition of multiple colonies of *S. aureus* on the surface of agar plates.

On the basis of hybridization assays with randomly selected clones from a *S. aureus* genomic library, Martineau et al. (1998) identified a chromosomal DNA fragment specific for *S. aureus* that detected all 82 *S. aureus* test isolates.

This 442-bp fragment was sequenced and used to design a pair of PCR primers Sa442-1/Sa442-2 (Table 11.2) that yielded a 108-bp amplicon from the genomic DNA of all 195 clinical isolates of *S. aureus* isolates examined. That these primers did not amplify DNA from an extensive list of other *Staphylococcus* species and other bacterial genera attests to their specificity. An additional pair of universal primers 16S rRNA-F/16S rRNA-R (Table 11.2) yielding a 241-bp amplicon from the 16S rRNA gene of all bacteria was used as an internal amplification control to prevent false-negative results.

Marcos et al. (1999) developed a species PCR assay for detection of *S. aureus* in whole milk. The primers FA1/RA2 (Table 11.2) amplified a 1153-bp sequence of the *aroA* gene that encodes 5-enolpyruvyl-shikimate-3-phosphate synthase. Out of a total of 103 isolates of *S. aureus*, 38 were from cows, 6 from ewes with mastitis, and 59 were strains from human outbreaks. All 103 isolates plus *S. aureus* control strains yielded the 1153-bp amplicon. No amplicons were obtained with other *Staphylococcus* species or other genera.

Ramesh et al. (2002) developed a multiplex PCR for the direct detection of *S. aureus* and *Yersinia enterocolitica* in spiked milk samples. The primers Sa-1/Sa-2 (Table 11.2) amplified a 482-bp sequence of the *nuc* gene. DNA extraction with a combination of organic solvents, detergents, and alkali allowed detection of DNA from 10 CFU/ml of milk for both organisms in monoplex PCR assays. However, when equal numbers were present in milk samples the detection limit for both organisms was 10^3 CFU/ml.

Hein et al. (2001) applied two different quantitative real-time (qRti-PCR) approaches to *S. aureus* isolates by targeting the *nuc* gene. The primers Nuc1/Nuc2 and dual-labeled probe (Table 11.2) used were from Brakstad, Aasbakk, and Maeland (1992). Quantification with SYBR green I using a pure culture resulted in 60 genomic targets/µl compared to six genomic targets/µl obtained with the dual-labeled fluorogenic probe. Application of the assay to seeded cheese of different types achieved detection sensitivities from 7.5×10^1 to 3.0×10^2 genomic targets/g depending on the cheese matrix.

Stepan et al. (2001) developed a PCR assay specific for *S. aureus*. The primers JIRS-2/JIRS-1 (Table 11.2) amplified an 826-bp species-specific genomic sequence derived from a 44-kb *Sma*I restriction fragment common to all *S. aureus* isolates examined. A total of 216 *S. aureus* strains yielded the 826-bp amplicon, whereas no amplicon was observed with 40 strains of other *Staphylococcus* species nor with 45 clinical strains of coagulase-negative staphylococci.

Staphylococcal coagulase, a major phenotypic determinant of *S. aureus*, is encoded by the *coa* gene, which exists in multiple allelic forms, in part because of the existence of gene variants within the 3′-end coding region. This region contains a series of 81-bp DNA sequences that differ both in the number of tandem repeats and the location of *Alu*I restriction sites among different isolates. Goh et al. (1992) utilized this observation to develop a novel typing method for *S. aureus* isolates based on the nested PCR amplification of the variable region of the *coa* gene followed by *Alu*I restriction and analysis of the resulting restriction fragments. The primers COAG1/COAG4 (Table 11.2) amplified a 1557-bp sequence of the *coa* gene. Nested primers COAG2/COAG3 (Table 11.2) were then used to amplify the variable tandem 3'-region of the *coa* gene.

The size of the PCR products ranged from about 400 to about 915 bp. *Alu*I restriction generated multiple DNA fragments with each strain with bands in multiples of 81 bp (81, 162, 243, 324, 405, and 486 bp).

A typing procedure for *S. aureus* was developed by Hookey, Richardson, and Cookson (1998) based on improved amplification of the *coa* gene and restriction fragment length polymorphism (RFLP) of the resulting amplicons. Primers coa-F/ coa-R (Table 11.2) were designed to avoid the variable regions of the *coa* gene and to encompass the entire 3' repeat elements. All *coa*-positive staphylococci produced a single PCR amplicon of either 875, 660, 603, or 547 bp. All 16 strains of epidemic methicillin-resistant *S. aureus* (EMRSA) yielded a 547-bp amplicon. *Coa* amplicons digested with *Alu*I and *Cfo*I resulted in 10 distinct RFLP patterns among 85 MRSA isolates and 10 propagating strains of methicillin-sensitive *S. aureus* (MSSA).

Yang et al. (2007) developed a PCR assay for the direct detection of *S. aureus* without enrichment in seeded whole milk, skim milk, and cheese. Cells of *S. aureus* seeded into these dairy products were separated with the use of a novel cell extraction system consisting of 1 ml of NH_3, 1 ml of ethanol, and 1 ml of petroleum ether added to 5 ml of milk or homogenized cheese. Centrifugation at 12,000 × g for 10 min yielded a pellet with nearly all of the target cells. The pellet was then subjected to cell lysis followed by DNA purification with chloroform, and DNA precipitation with ammonium acetate and ethanol. The primers Pri-1/Pri-2 (Table 11.2) amplified a 279-bp sequence of the *nuc* gene. The limit of sensitivity for whole milk and skim milk was 10 CFU/ml and with cheese was 55 CFU/g.

Thomas et al. (2007) developed a multiplex real-time PCR (mRti-PCR) assay for detection of *S. aureus* in blood cultures. Primers LTnucF-/LTnucR (Table 11.2) amplified an 87-bp sequence of the *nuc* gene encoding the thermonuclease of all *S. aureus* strains. Primers LTmecAF/LTmecAR (Table 11.2) amplified an 88-bp sequence of the *mecA* gene encoding methicillin resistance. The corresponding dual-labeled probes LTmecAHp2 and LTnucHP1 (Table 11.2) were both labeled at the 3'-end with the quencher BHQ1. The former was labeled with the fluorophore 5'-hexachloro-6-carboxy-forscein (HEX) and the latter with 5'-6-carboxy-fluorscein (FAM). A critical aspect of the assay involved DNA extraction and purification to remove PCR inhibitors. The QIAamp® DNA Mini Kit (Quiagen) failed to allow amplification; the NucliSens mini MAG (BioMerieux) effectively allowed amplification but required about 40 min for each extraction. The Magna Pure LC total nucleic acid isolation kit (Roche Diagnostics) resulted in no PCR inhibition.

Elizaquivel and Aznar (2008) developed a multiplex real-time PCR assay for simultaneous detection of *S. aureus*, *Escherichia coli* O157:H7, and *Salmonella* spp. on fresh minimally processed vegetables. The primers R465/F5 (Table 11.2) amplified a 97-bp sequence of the *nuc* gene of *S. aureus*. The dual-labeled probe NucP402 (Table 11.2) was labeled at the 5'-end with FAM and at the 3'-end with a black hole or nonfluorescent quencher (NFQ). With seeded fresh vegetables (spinach and lettuce) a level of detection sensitivity of 10^3 CFU/g was achieved.

B. Peptide Nucleic Acid Fluorescence *In Situ* Hybridization (PNA FISH) for Detection of *S. aureus*

Peptide nucleic acid (PNA) molecules are pseudopeptides that obey Watson–Crick base-pairing rules when hybridized to complementary nucleic acid targets. Owing to their uncharged neutral backbones, PNA probes exhibit favorable hybridization characteristics such as high specificities, strong affinities, and rapid kinetics, resulting in improved hybridization (Oliveira et al., 2002). In addition, the relatively hydrophobic character of PNA probes enables them to penetrate the hydrophobic cell wall of bacteria following preparation of a standard fluorescence in situ hybridization (FISH). FISH with PNA probes targeting 16S rRNA combines the unique functional properties of PNA probes with the advantage of greatly enhanced sensitivity derived from the large number (about 15,000) of 16S rRNA targets available per bacterial cell. Oliveira et al. (2002) developed a PNA FISH assay for direct detection of *S. aureus* from blood culture bottles. The PNA probe (Table 11.2) targeted a 16S rRNA sequence unique to *S. aureus* and exhibited 100% sensitivity and 96% specificity.

Riyaz-Ul-Hassan, Verma, and Qazi (2008) developed a PCR assay for detection of *S. aureus* in foods that targeted three genes. Primers cat-F/cat-R (Table 11.2) amplified a 641-bp sequence of the *cat* gene encoding catalase. Primers fem-F/fem-R (Table 11.2) amplified a 594-bp sequence of the *femA* gene encoding a factor essential for methicillin resistance. Primers fmhA1-F/fmhA1-R (Table 11.2) amplified a 685-bp sequence of the *fmhA* gene encoding a factor of unknown function. In addition to these three primer pairs, primers 16S rDNA-F/16s rDNA-R (Table 11.2) were used to initially identify all 45 inclusive strains of *S. aureus* and amplified a 350-bp sequence of the gene encoding the 16S rRNA. A total of 55 non–*S. aureus* cultures failed to yield any of the three amplicons. In contrast, 45 inclusive strains of *S. aureus* all yielded the correct amplicons; the detection limit was the DNA from 100 cells per 20 μl PCR. Among 150 raw milk samples 36 (24%) yielded the corresponding amplicons with each of the three primer pairs. The study clearly indicated that the *femA*, *smhA*, and *cat* genes are universally present in all strains of *S. aureus*. In addition, catalase-negative isolates such as *S. aureus* subsp. *anaerobius* were still detected by all three primer pairs including the *cat* primers due to the presence of the mutationally unexpressed catalase gene.

C. Conventional and Real-Time PCR (Rti-PCR) Detection of Enterotoxin Genes

Johnson et al. (1991) developed eight pairs of primers for PCR detection of the SE toxin genes *sea-see*, exfoliative toxins *eta* and *etb*, and toxic shock syndrome toxin tst-1 (Table 11.2). The primers were applied to 88 strains of *S. aureus* from clinical sources and foods whose toxigenicity was biologically and immunologically established. Concordance between phenotypic and genotypic identification of the toxins produced by the strains was 97.7%.

Mäntynen et al. (1997) developed an MPN-PCR nested quantitative PCR assay for *S. aureus* based on the enterotoxin *c1* gene from cells seeded into fresh cheese. Primers 1-F/2-R (Table 11.2) amplified an 801-bp sequence of the *entc1* gene. Primers 3-F/4-R (Table 11.2) amplified a 631-bp nested sequence. DNA extracted from the cheese was decimally diluted to yield three tube MPNs, which were then subjected to the nested PCR assay. The minimum level of detection was 20 CFU/g of cheese.

Monday and Bohach (1999) developed a single multiplex PCR (mPCR) assay for the simultaneous detection of the classic SE toxin genes *sea-see* in addition to the newly found SE toxin genes *seg, seh,* and *sei,* along with *tst.* Primers amplifying a 228-bp sequence of the 16srRNA gene of all *S. aureus* strains were used in the mPCR to confirm the identity of the *S. aureus* strains tested. Primers for each of the 11 genes are given in Table 11.2.

McLauchlin et al. (2000) developed a PCR-based procedure for the detection of SE genes *sea-see* and *seg-sei* together with *tst-1* in isolates of *S. aureus.* Primers for *sea-see* and tst-1 (Table 11.2) were from Johnson et al. (1991). The authors designed the primers for *seg-sei* (Table 11.2). A total of 129 isolates of *S. aureus* were selected, 39 of which were from 38 suspected staphylococcal food poisoning incidents. Thirty-two different genotypes were recognized. The presence of SE genes was associated with *S. aureus* strains reacting with phages in group II, and the *tst-1* gene with phages in group I. There was a 96% agreement between the PCR results for detection of *sea-sed* and *tst-1* as compared to a commercial immunoassay for detection of SEs from broth cultures. SE genes were readily detected by PCR from seeded mushroom soup and ham. However, detection was less successful with three types of cheese and with cream.

Mehrotra, Wang, and Johnson (2000) developed a multiplex PCR assay for detection of *Staphylococcal* enterotoxins *sea-see,* in addition to *tst, eta, etb,* and *mecA.* Detection of *femA* was used as an internal positive control. The multiplex PCR assay combined the primers for *sea-see* and *femA* in one set and those for *eta, etb, tst, mecA,* and *femA* in the other set (Table 11.2). Validation of the assay was performed with 176 human isolates of *S. aureus.*

Zschöck et al. (2000) examined 94 strains of *S. aureus* from cases of bovine mastitis in Germany for the presence of the *sea, seb, sec, sed, see,* and *tst* genes using the PCR. Primers sea1-F/sea2-R (Table 11.2) were from Tsen and Chen (1992), and the remaining primer pairs were from Johnson et al. (1991; Table 11.2). The most frequently occurring was *sec,* being present in 22 (23%) of the strains and 19 (20.2) of the strains possessed the *tst* gene. The *see* gene was not found in any of the 94 strains.

Larsen et al. (2000) developed primers (Table 11.2) for PCR detection of *S. aureus* enterotoxin genes *sea-sec* and *seh,* in addition to the genes *tst1, extA,* and *extB.* Primers 16S rDNA-F/16S rDNA-R (Table 11.2) amplified a 917-bp 16S rDNA sequence from all *S. aureus* isolates and were incorporated into the PCR assays as an amplification control to prevent false negative results. A total of 414 isolates of *S. aureus* from Danish cows with bovine mastitis among 45 dairy herds were examined for these genes. In addition, 100 *S. aureus* from Danish human carriers were also included in the study. Only one of the 414 isolates from bovine mastitis carried the genes for and produced in vitro SEC and TSST-1, whereas *sea, seb, sec, seh*, *tst*, and *extA* or a combination of these genes were observed in 47 of the 100 *S. aureus* isolates from human carriers. All PCR-positive isolates were also phenotypically positive for the corresponding SEs. The authors concluded that the lack of SEs among Danish isolates from bovine mastitis clearly demonstrated that the exotoxins included in the study are not of any importance to *S. aureus* in the pathogenesis of bovine mastitis.

The major SEC subtypes are SEC1, SEC2, and SEC3. Chen et al. (2001) developed four PCR primers that, when used in appropriate pairs, were capable of detecting the corresponding genes *sec1*, *sec2*, and *sec3*. Primers C1/C2 (Table 11.2) amplified a 234-bp sequence of the *sec* gene present in all SEC strains. Primers ENTC1/ENTCR (Table 11.2) amplified a 402-bp sequence of the *sec1* gene. Primers ENTC2/ENTCR (Table 11.2) amplified a 501-bp sequence of the *sec2* gene. Primers ENTC3/ENTCR (Table 11.2) amplified a 672-bp sequence of the *sec3* gene. Among a total of 39 strains of *S. aureus* from foodborne outbreaks in Central Taiwan from 1995–2000 the major SEC subtypes were SEC2 (6/39) and SEC3 (10/39).

Stephan et al. (2001) examined 34 strains of SE producing *S. aureus* from milk samples of 34 dairy cows suffering from mastitis from 34 different dairy farms in northeast Switzerland for phenotypic and genotypic characteristics involving the presence of the *sea-see* genes plus plus seven additional genes (Table 11.2) . Primers for *seb-see* and *tst*, in addition to *eta* and *etb* (Table 11.2), were from Johnson et al. (1991). Among the 34 isolates, 26 produced a single SE consisting of 23 SEC and three SED producers that were detected immunologically. Eight isolates produced both SEA and SED. Remarkably, two of the SEC formers were also positive for TSST-1 production. None of the 34 isolates carried the *sea*, *seb*, or *see* genes. PFGE analysis revealed 11 patterns. The authors concluded that a broad distribution of identical or closely related ET-producing *S. aureus* clones appeared to contribute to the bovine mastitis problem in northeast Switzerland.

Atanassova, Meindl, and Ring (2001) examined 135 samples consisting of raw pork, salted pork, and uncooked smoked ham from three suppliers for the production of SEs A–D and for the presence of the corresponding genes by the PCR. The primers for detection of the *sea-sed* genes (Table 11.2) were from Johnson et al. (1991). In 35 (25.9%) samples *S. aureus* was detected by cultivation, whereas 69 (51.1%) of the samples were PCR positive using the PCR primers Sa442-1/Sa442-2 (Table 11.2) from Martineau et al. (1998) that are specific for all isolates of *S. aureus*. Whereas 100% of the samples from supplier A contained *S. aureus*, the organism was isolated from 21% of the raw pork samples from supplier B and 75% from supplier C. Among the 135 total samples, 24 (17.8%) were PCR-positive for one or more of the SE genes, with 79.2% of the samples possessing only one SE gene. It is interesting that none of the 35 samples that were *S. aureus*–positive produced immunologically detectable enterotoxins.

Fueyo et al. (2001) analyzed *S. aureus* strains from human carriers (110 strains) and manually handled foods (114 strains) in Spain for the production of SEs (A, B, C, D, E, and J) by immunoassay and for the presence of the corresponding genes by the PCR using the primers of Martin, Gonzalez-Hevia, and Mendoza (2003; Table 11.2). Among the total of 224 strains, 62 (28%) were immunologically positive for one or more of the six SEs. Healthy human carriers were found to harbor SE-producing strains. None of the strains harbored the *see* gene. RAPD typing of the 62 S-positive strains with primers S and C (Table 11.2) yielded seven and five profiles, respectively. Combining results from both RAPD primers yielded 10 profiles. The most frequently occurring RAPD type included a great diversity of strains with varying SEs from both food and human samples. The observation that some SE-positive and some SE-negative strains generated identical RAPD profiles suggested that SE-positive strains do not belong to a specific genetic class.

Larsen, Aarestrup, and Jensen (2002) determined the geographic variation for the presence of *se* genes and the β-hemolysin gene in *S aureus* isolated from bovine mastitis in Europe and the United States. A total of 462 *S. aureus* isolates were examined for the presence of *sea–see, seh, tst, hlb*, and 128 isolates were examined for the *extA* and *extB* genes. The *se* primers were from Larsen et al. (2000; Table 11.2) and those for the β-hemolysin gene *hlb* were from Aarestrup et al. (1999). None of the 128 isolates harbored the exfoliative (EXT) toxins A or B. The total proportion of *se*-positive isolates varied from 1 (2%) among the Danish isolates to 32 (65%) of the Norwegian isolates. The most common toxin genes were *sec, tst-1*, and *sed*. In contrast to the geographic variation among SEs, 97% of the isolates were PCR positive for the β-hemolysin gene *hlb*. The authors concluded that the SEA toxins do not play a role in the pathogenesis of bovine *S. aureus* mastitis, but that the β-hemolysin may be an active virulence factor in bovine mastitis.

Letertre et al. (2003b) described the development of real-time PCR (Rti-PCR) triplex assays for detection of the *sea–sej* genes with primers and probes for each gene (Table 11.2). The triplex assays consisted of: *sea, see, seh; seb, sec, sed*; and *seg, sei, sej*. The fluorophores at the 5′-ends of the dual labeled probes were FAM, VIC, and TET, with MGB as the quencher at the 3′-end (Table 11.2). The assays correctly detected the *se* genes in all 68 reference strains of *S. aureus* with excellent agreement with field strains except for some strains harboring variant *se* genes. All 17 isolates other than *Staphylococcus* species and other genera were *se* negative.

Martin et al. (2004) examined 32 isolates of *S. aureus* from three food poisoning restaurant outbreaks in a single province of Spain for their genetic relatedness. The methods used consisted of PCR analysis for nine SE genes: *sea–sej* using the primers of Martin, Gonzalez-Hervia, and Mendoza (2003; Table 11.2) and *seg, she*, and *sei* using the primers of McLauchlin et al. (2000; Table 11.2), PFGE, RAPD with two random primers S and C (Table 11.2) from Williams et al. (1990), and restriction profile analysis with *Hind*III. The 32 isolates were differentiated into three non-*se* and 12 *se* strains, which were outbreak-specific, except for one that was represented in two of the outbreaks. In outbreak 1, the 16 food isolates had *sec, seg*, and *sei* genes and represented a single genomic clone. In outbreak 2, the four food isolates fell into three distinct genotypes. In outbreak 3, the five food isolates fell into four *seg–sei* strains generating identical RAPD patterns but different PFGE and plasmid profiles and one *sea* strain from two nasal carriers.

Chen, Chiou, and Tsen (2004) designed primer pairs for the detection of the newly found SE genes *seg, seh*, and *sei* from 55 human isolates of *S. aureus* shown to be PCR negative for the classic SE genes *sea–see*. All of the isolates were from the fecal specimens of patients suffering from food poisoning outbreaks in Taiwan. The

PCR primers for detection of the *sea–see* genes were from Johnson et al. (1991) and those for detection of *tst* were from Tsen, Chen, and Yu (1994). The primers SEG1/ SEG2 (Table 11.2) amplified a 583-bp sequence of the *seg* gene. The primers SEH1/ SEH2 (Table 11.2) amplified a 548-bp sequence of the *seh* gene. The primers SEI1/ SEI2 (Table 11.2) amplified a 789-bp sequence of the *sei* gene. Only eight of the 55 clinical isolates were found to harbor the *seg, seh,* or *sei* genes. Among 139 *S. aureus* isolates from foods only 62 (44.6%) possessed one or more of the *sea–seh* genes. *sea* strains accounted for the majority of the *se* genes. In addition, only 13 (9.3%) of the 139 food isolates possessed the *seg* or *sei* genes and did not have the classical *sea– see* genes. The authors concluded that the *seg, seh,* and *sei* genes may play only a minor role in staphylococcal food poisoning in Taiwan.

Sharma, Rees, and Dodd (2000) developed a unique single multiplex reaction (SR-mPCR) to detect the *sea–see* genes using one common forward universal and one gene-specific reverse primer for each individual toxin gene. Because SR-mPCR takes advantage of both conserved and unique regions of the toxin genes, this PCR reaction requires fewer sets of primers than conventional mPCR. Kwon et al. (2004) extended the use of SR-mPCR to other SEs and used the method to characterize toxin types from 141 *S. aureus* isolates from South Korea (Table 11.2). RAPD was also performed on the isolates with three random primers (Table 11.2). Among the 141 isolates, 23 were from retail pork, 42 from chicken carcasses or rinse water, and 37 and 39 from porcine and chicken abattoirs, respectively. Results indicated that all of the primers (Table 11.2) worked well among the 141 isolates except those for *seg* (Table 11.2), which did not generate consistent bands. Among the 141 isolates, 25 had only one SE gene, whereas two isolates had two SE genes. The most prevalent SE gene was *sei* (74%) followed by *sea* (30%) and *seh* (4%). RAPD revealed that the 27 toxigenic strains yielded 9, 8, and 10 distinct types with primers API, ERIC2, and AP7, respectively. The use of only primer AP7 simplified the typing. Clonal relatedness of toxigenic strains was found to be dependent on the origin of isolation. Toxin typing revealed that retail pork and porcine abattoirs carried more toxigenic *S. aureus* than chicken samples in South Korea.

Zschöck et al. (2005) subjected 104 randomly selected *S. aureus* strains from bovine mastitis in Germany to the PCR for detection of the *seg–sej* and *tst* genes. The *seg–sej* primers were from Monday and Bohach (1999), and the *tst* primers were from Johnson et al. (1991). Sixty-one (58.7%) of the isolates were positive for one or more of these genes. Among the 104 strains *seg, sei,* and *sej* were harbored by 36, 22, and 23 of the strains, respectively. None carried the *seh* gene. Twenty of the isolates harbored the *tst* gene. Among the 61 *se*-positive strains 14 (23%) were positive for both *sed* and *sej*. These 14 *sed/sej* positive strains were further characterized by PFGE, which resulted in six PFGE genotypes, nine strains of which exhibited two patterns with a high degree of relationship. These nine strains were more dissemi- nated than others due presumably to the sale and purchase of cattle between farms and cow-to-cow spread of the infectious strain.

Cremonesi et al. (2005) developed a multiplex PCR for the simultaneous detec- tion of *S. aureus* 23S rRNA, the *coa* and *nuc* genes, as well as the SE genes *sea–sej* and *sel*. The primers are listed in Table 11.2. The method was used to determine the presence of SE types for 93 *S. aureus* strains isolated from milk and dairy products in Italy. The mPCR assays were found to amplify some SE genes whose toxins were undetectable by immunoassay. The relative primer concentration was the most important factor in obtaining approximately equal yields of amplicons for the 11 genes in a single PCR reaction. The sensitivity of the optimized mPCR assay was similar to that of uniplex PCR and was equal to the DNA from 10 CFU/ml. Among the 93 isolates of *S. aureus*, only one strain showed a lack of correlation between the presence of the *coa* and *nuc* genes and their expression.

Omoe et al. (2005) described the development of a PCR system for detection of the five classic SE genes *sea–see* plus the newly found genes *seg–set* (Table 11.2) using four sets of multiplex PCR. The primers for *sea–see* were from Becker, Roth, and Peters (1998; Table 11.2). The primers for *seg–sei* were from Omoe et al. (2002; Table 11.2). The primers for *selj–selr* were developed by Omoe et al. (2005). Primers for *tst*, *femA*, and *femB*, were from Becker, Roth, and Peters (1998), Mehrotra, Wang, and Johnson (2000), and Perez-Roth et al. (2001), respectively (Table 11.2). Genotyping of 69 *S. aureus* food poisoning isolates and 97 *S. aureus* isolates from nasal swabs of healthy humans revealed 32 SE genotypes and showed that many *S. aureus* isolates harbor multiple toxin genes.

A set of 269 *S. aureus* isolates recovered from human nasal carriers and manually handled foods in a region of Spain were analyzed by Fueyo, Mendoza, and Martin (2005b) for *sea, seb, sec, sed, seg, sej, ser*, and *tst* toxin genes. Primers for the first five genes were from Martin et al. (2003). Primers Ser-1/Ser-2 amplified a 700-bp sequence of the *ser* gene (Table 11.2). Primers tst-1/tst-2 amplified a 481-bp sequence of the *tst* gene (Table 11.2). There were 57 isolates found to produce at least one of the four SEs, SEA–SED by immunoassay with 10 isolates producing only TSST-1 and 10 isolates producing both toxin types. The 77 toxigenic isolates yielded 36 PFGE profiles and 13 *Eco*RI plasmid profiles. Eight lineages were differentiated; six of them grouped both human and food isolates and two of these also included outbreak-implicated isolates.

Fueyo et al. (2005a) screened strains of *S. aureus* isolates from nasal carriers for the five classic SE genes plus 12 additional new SE genes by PCR (Table 11.2). PCR detection of *sea–see* and *sej* genes utilized the primers of Martin, Gonzalez-Hevia, and Mendoza (2003; Table 11.2). The remaining 11 new SE gene primers in addition to primers used for detection of the ET and TSST genes are presented in Table 11.2. A total of 86 isolates examined yielded 17 toxin genotypes, all of which were *eta, etb, etd,* and *sep* negative and generated 40 PFGE profiles. Correlations between classical SE toxins and PFGE lineages were established.

Da Silva, Carmo, and Da Silva (2005) subjected 36 strains of *S. aureus* from goat mastitis and 64 from bovine mastitis in Brazil to mPCR involving two sets of primers for detection of the SE genes *sea, seb*, and *sec*. The primers used (Table 11.2) were from Mehrotra, Wang, and Johnson (2000). The production of SEs was detected by immunoassays in all strains that amplified the corresponding genes. Among the total of 100 strains, 37 (37%) were found to harbor one or more of the SE genes. Among the bovine mastitis strains, four (6.3%) coamplified the *sea* and *seb* genes and two (3.1%) were positive for the *sec* gene. Among the goat mastitis strains, 31 (86%)

were positive for the *sec* gene and none were positive for the *sea* and *seb* genes. The authors concluded that *S. aureus* isolates from goat mastitis have a higher entero-toxigenic potential than those from bovine mastitis. Similar results were previously observed by Orden et al. (1992), who observed that 67% of goat-mastitits isolates and only 19% of bovine-mastitis isolates produced SEs.

Pinto, Chenoll, and Aznar (2005) subjected a total of 158 cultures (15 *S. aureus* reference strains, 12 other staphylococcus species, and 131 presumptive *S. aureus* isolates picked from Baird–Parker agar plates from a wide variety of foods) to phe-notypic API-Staph characterization and to PCR for detection of the *sea–see* and *nuc* genes. Primers SEA-5/SEA-6 for the *sea* gene were from Becker, Roth, and Peters (1998), those for the *seb, sed,* and *see* were from Johnson et al. (1991), and those for *nuc* were from Brakstad, Aasbakk, and Maeland (1992; Table 11.2). Primers SEC-5/SEC-6 (Table 11.2) for the *sec-1* gene were newly developed by the authors. RAPD profiles were also obtained with primers M13, T3, and T7 (Table 11.2). Among the 131 food isolates, 41 (31%) tested positive for one or more of the *se* genes. Among these, 14 (11%) were positive for *sea*, 22 (1.7%) for *see*, 1 (0.89%) for *sed*, and 3 (2.2%) for *sea* and *sec*. No amplicons from *seb* or *see* were obtained. Among the 131 food iso-lates 93 (71%) were DNase- and coagulase-positive and had >60% identity with the *S. aureus* API-Staph profiles and were considered to be strains of other *Staphylococcus* species. RAPD results indicated a great diversity of the *S. aureus* isolates from food with a high relationship between *sec*-positive strains, most of which grouped into three clusters in contrast to *sea* and *sea–sec*-positive strains that grouped into seven clusters. A high level of concordance between *S. aureus* and *nuc* PCR-positive strains (99%) corroborated the specificity of the *nuc* primers and the suitability of the *nuc* PCR for rapid identification of *S. aureus* in routine food analysis.

Bania et al. (2006) subjected 50 isolates of *S. aureus* from raw minced meat and sausages to the PCR detection of the classic ET genes *sea–see* using the primers of Sharma, Rees, and Dodd (2000; Table 11.2). In addition, the new ET genes *seg–seq* were detected with newly developed primers (Table 11.2) along with the *seu* gene using the primers of Letertre et al. (2003a; Table 11.2). Among the 50 isolates, 27 were found to be positive for one or more *se* genes. Only 9 of the 27 positive strains carried the *sea* and *see* genes. In 18 *sea–see* negative strains the presence of newly described *se* genes was detected. All *sea–see* positive strains simultaneously carried new *se* genes. The *seh* gene was the most frequently detected (14/50). Additional *se* genes present were *sei, seg, sep, sel,* and *sek.*

Many SFP outbreaks are suspected of being caused by newly described SEs or staphylococcal enterotoxin-like SELs. However, immunological assay kits cur-rently used for the detection of SEs can only detect the classic five SEAs (*sea–see*). Recently, dual priming oligonucleotides have been developed that involve individual primers with two separate primer segments joined by a polydeoxyinosine link. This structure allows one primer to anneal at two sites with its two segments resulting in enhanced multiplex PCRs. Hwang et al. (2007) developed such multiplex PCR assays for the rapid screening of the 18 genes that encode all of the presently recognized SEs (*sea–see* and *seg–sei*) and SE-like (SEL) toxins (*sej–ser* and *seu*) in addition to *tst*. The primers (Table 11.2) were designed to detect all 19 *se* genes in three sets of multiplex PCRs and involved individual primers with polydeoxyinosine links. The multiplex PCRs were applied to 143 *S. aureus* strains from pork and chicken meat in Korea. Almost 50% of the strains possessed at least one of the 19 *se* genes. The most frequently found genes were *seg, sem,* and *sen,* which were often found simultaneously in the same isolate. In those isolates, the *seo* or *seu* genes were frequently found together and the combination (*seg, sei, sem,* and *seo* or *seu*) was considered to be a part of the enterotoxin gene cluster. Among the classic *se* genes (*sea–see*) the *sea* gene was the gene most frequently detected. Newly described *se* genes (*seg–ser*) and *seu* were more frequently detected than the classic *se* genes.

The distribution of *se* genes in a total of 112 strains of *S. aureus* from bovine, goat, sheep, buffalo milk, and dairy products in Italy was examined by Morandi et al. (2007) using a multiplex PCR for detection of *sea, sec, sed, seg, seh, sei, sej,* and *sel.* The identity of the primer pairs used was not revealed. Among the 112 strains, 75 (67%) were found to harbor *se* genes, but only 52% produced detectable amounts of the classic SE toxins. The bovine isolates frequently harbored *sea, sed,* and *sej,* whereas *sec* and *sei* genes predominated in the goat and sheep strains.

Lawrynowicz-Paciorek et al. (2007) examined 71 isolates of *S. aureus* from foods and 30 from nasal carriers in Poland for the presence of the five classic enterotoxin genes (*se*) and the 13 newly found staphylococcal enterotoxin-like genes (*sel*) and by PCR. The primers for *sea–see, seo,* and *seu* were from Sharma, Rees, and Dodd (2000), Fueyo et al. (2005c), and Letertre et al. (2003a), respectively. Primers for *seg–sei, sek–sen,* and *sep* and *seq* were from Bania et al. (2006). Two sets of primers for *sej* were from Blaiotta et al. (2004) and Omoe et al. (2005), respectively. Among the total of 101 strains 89 (80%) were *se/sel* positive. It is interesting that the frequency of these genes was higher among nasal carrier strains (93%) than among food sample isolates (76%). Certain *se* and *sel* genes were found to coexist, which is consistent with the concept of movable genetic elements in *S. aureus* strains. Only three of all *sea–sed* positive strains had silent genes, and these three isolates were of food origin.

Chiang et al. (2008) subjected 147 *S. aureus* isolates from patients in Taiwan with staphylococcal enteritis to PCR analysis for detection of the *se* genes *sen, seo, sep, seq, ser,* and *seu* with newly designed primer pairs (Table 11.2) in addition to the classic *sea–see* toxin genes and the *tst* genes. Among these 147 strains, 135 (91.8%) were positive for one or more of the *se* genes. The most frequent classic *se* gene was *sea* (29.2%), followed by *seb* (19.7%), *sec* (6.8%), and *sed* (2.0%). Among the new *se* types, the most frequent were *sei* (29.9%) and *sep* (27.9%).

D. Loop-Mediated Isothermal Amplification (LAMP) Assay for *se* Toxin Genes

Goto et al. (2007) developed a LAMP assay for *sea, seb, sec,* and *sed* using five or six primers for each gene sequence (Table 11.2). The authors concluded that for the detection of the *se* genes in *S. aureus* the LAMP assay is more sensitive and more efficient than conventional PCR and similar to Rti-PCR except for a much shorter assay time with LAMP.

E. PCR Detection of the *tst* Gene in *S. aureus* Strains

Deurenberg et al. (2005) developed an RTi-PCR assay for rapid detection of potential TSST-1 producing *S. aureus* strains. The primers TSST-1-FP/TSST-1-RP in conjunction with the dual-labeled probe TSST-1-PR (Table 11.2) amplified an 88-bp sequence of the *tst* gene. The assay was applied to 51 community (CA) isolates, 36 hospital-acquired (HA) isolates, and 16 strains from patients with Wegener's granulomatosis (WG) in the Netherlands. Twelve (24%) of the CA isolates were positive for the *tst* gene. In contrast, only five (14%) of the HA isolates were *tst* positive. Four (25%) of the WG isolates were *tst* positive. The resulting 21 *tst* positive isolates were subjected to PFGE analysis and yielded six clonal groups. A majority of the TSST-1 producing isolates (9 out of 17) fell into one clonal group. These results suggested that *tst*-positive strains from a single clonal group became predominant in the community during the period 1999 to 2003.

F. Molecular Typing of *S. aureus* Isolates (RAPD, PFGE, RFLP, PCR-RFLP)

Numerous studies with *S. aureus* isolates have been undertaken for strain typing involving PFGE (the "gold standard"), RAPD, and RFLP. In addition, several genes, including those encoding coagulase (*coa*) and the response regulator gene (*sae*) for several exoproteins, have also been utilized. The typing method utilizing the *coa* gene is based on heterogeneity of a region containing 81-bp tandem repeats at the 3'-end of the coagulase gene, which can be variable among unrelated strains of *S. aureus*.

Schwartzkopf and Karch (1994) used the primers COAG2/COAG3 (Table 11.2) from Goh et al. (1992) to amplify a portion of the 3'-terminal region of the *coa* gene from 30 *S. aureus* strains. Twenty of the strains yielded a single amplicon of 654 bp, 735 bp, or 816 bp. Ten of the strains yielded two different sized amplicons with one being 400 bp and an additional band of 654 bp, 735 bp, or 897 bp. From the number and sizes of the *coa* bands from the 30 strains, six subgroups could be distinguished. Restriction of the resulting amplicons with *Alu*I yielded 22 RFLP patterns among the 30 strains. Strains with identical RFLP patterns were found to yield notably distinguishable amplicons on the basis of sequence analysis of the restriction fragments. These observations indicated that unrelated strains may share identical *Alu*I RFLP patterns and that the discriminatory power of *Alu*I RFLP typing of the *coa* gene is not great enough to allow its use as a sole typing method.

Giraudo et al. (1999) mapped the *sae* locus to the *Sma*I-D fragment of *S. aureus* genomic DNA by PFGE. Sequence analysis of the cloned fragment revealed the presence of two genes, designated *saeR* and *saeS*, encoding a response regulator and a histidine protein kinase, respectively, having a high level of homology to other bacterial two-component regulatory systems.

In a multicenter study involving seven laboratories, van Belkum et al. (1995) typed 59 strains of *S. aureus* and 1 of *S. intermedius* by RAPD using the three random primers AP1, AP7, and ERIC2 (Table 11.2). The 60 strains yielded 16 to 30 different genotypes, depending on the laboratory and the deliberate introduction of certain experimental variables. However, this did not hamper the epidemiologically correct clustering of related strains. Comparison of RAPD with PFGE indicated the existence of strains with constant PFGE types but variable RAPD types. The reverse consisting of constant RAPD types and variable PFGE types was also observed. The use of different DNA polymerases was found to be a major source of typing variability. The interlaboratory reproducibility of DNA-banding patterns and interlaboratory standardization were found to need improvement.

Van Leeuwen et al. (1996) applied RAPD and probes based on sequences of selected random generated amplicons to 243 *S. aureus* strains and a single isolate of *S. intermedius*. The random primers ERIC-1R, ERIC2, and AP1, AP7, and AP1026 (Table 11.2) were used for RAPD analysis. Five strain-specific probes in a five-digit typing system accurately distinguished epidemiologically related and unrelated strains of *S. aureus*.

Raimundo et al. (1999) subjected 151 *S. aureus* isolates from chronically infected cows in Australia to molecular typing based on polymorphism of the coagulase (*coa*) gene. The primers COAG2/COAG3 (Table 11.2) yielded amplicons ranging in size from 10 to 1100 bp. Following digestion with *Alu*I, *Cfo*I, and *Hae*III, the number of bands resulting from PCR-RFLP increased but PCR products of the same size yielded identical restriction profiles for all of the strains examined. In addition, there was perfect correlation between the restriction profiles generated by the three different enzymes. The application of *coa* typing by PCR-RFLP to 26 epidemiologically unrelated *S. aureus* strains established 13 types. The same 26 isolates produced 12 major PFGE types. By combining both typing schemes, 19 molecular types were defined among the 26 isolates. *coa* typing of the 151 *S. aureus* isolates from seven farms generated only six genotypes with 110 (73.3%) assigned to type 1 and 23 (15.2%) assigned to type 2. Nine of eleven cows with chronic infections showed evidence of the resistance of a single genotype for periods of up to 9 months.

Karahan and Çetinkaya (2007) subjected 161 isolates of *S. aureus* from cattle with subclinical mastitis from 32 locations in eastern and southeastern regions of Turkey to RFLP analysis of the coagulase (*coa*) gene. The primers Sau327/Sau1645 (Table 11.2) amplified a 1300-bp sequence of the 23S rRNA gene and were used to confirm the identity of isolates as *S. aureus*. The primers COAG2/COAG3 (Table 11.2) from Goh et al. (1992) amplified various sized basepair sequences of the *coa* gene ranging from 500 to 1400 bp. The presence of double-banded amplicons in 26 (16%) of the isolates indicated the possibility of human contamination by milking personnel in that human and bovine *S. aureus* represent two populations that rarely cross-infect (Lange et al., 1999; Larsen et al., 2000). Restriction of the amplicons with *Alu*I and *Hin*6I yielded 23 and 22 different restriction profiles, respectively. The most common profiles by either *Alu*I or *Hin*6I belonged to 55 isolates (34.2%) that had a single 950-bp *coa* PCR band.

There were 66 isolates of *S. aureus* obtained from milk samples of dairy cows in Brazil suffering from mastitis that were subjected to five different genotyping methods for comparison of their discriminatory power by Lange et al. (1999). Plasmid profiling produced 27 different profiles and consisted of one to five plasmids from 2 to 50 kbp. No plasmids were detectable in 31 (46.9%) of the isolates. PCR amplification of the 3'-end of the *coa* gene using primers COAG2/COAG3 (Table 11.2) of Goh et al. (1992) yielded a single band for each of the 66 *S. aureus* isolates. Seven different sized *coa* PCR products, which ranged from 580 to 1060 bp, were distinguished. *Alu*I digestion of the *coa* amplicons generated 14 different patterns with fragments of 580, 650, and 1060 bp. Amplification of the X region of the *spa* gene using primers SPAX1/SPAX2 (Table 11.2) from Frénay et al. (1996) yielded a single amplicon for each isolate with seven different sized amplicons from 170 to 340 bp. PCR amplification of the variable spacer region between the 16S-23S rRNA with primers RRNA1/RRNA2 (Table 11.2) from Cuny, Lause, and Witte (1996) yielded 11 different patterns with bands varying from 700 to 950 bp. Five isolates gave no bands. The total of 66 isolates displayed 33 different PFGE patterns consisting of 9–14 fragments varying in size from 50 to 550 bp. The most common PFGE pattern was detected in 10 isolates. PFGE yielded the largest number of genotypes and proved to be superior in its discriminatory power compared to the other methods of genotyping. However, comparative analysis of the results obtained from the three PCR methods used yielded 38 different genotype groups. The authors concluded that PFGE is far more time consuming, labor-intensive, and more costly than the PCR-based methods such as *coa*-and *spa*-PCR and rRNA spacer typing.

Annemüller, Lämmler, and Zschöck (1999) subjected 25 *S. aureus* isolates from bovine subclinical mastitis to several genotyping methods. The isolates were obtained from six different dairy farms at five locations in one region of Germany. PFGE revealed five patterns. Clones with different PFGE patterns were found within a single herd. Amplification of the intergeneric 16S-23S rRNA spacer region with primers 16S-23S-F/16S-23S-R (Table 11.2) yielded amplicons of 380, 400, 430, and 510 bp. Amplification of the IgG-binding region of the protein A gene *spa* with primers SPA-F/SPA-R (Table 11.2) revealed amplicons of 620 bp for 20 of the isolates, 280 bp for four of the isolates, and no band with one isolate. Amplification of the X-region of the protein A gene *spa* with primers X-SPA-F/X-SPA-R (Table 11.2) yielded amplicons of 120, 150, 170, 250, 260, and 300 bp, which corresponded closely to the PFGE patterns. Amplification of the *coa* gene with primers COA-F/COA-R (Table 11.2) yielded amplicons of 740, 800, and 990 bp. Restriction of the *coa* amplicons with *Alu*I yielded six *coa* types. These results with the *cao* gene are derived from the fact that the coagulase gene *coa* can contain a series of repetitive 81-bp sequences, which can differ in the number of tandem repeats and the location of *Alu*I restriction sites. The authors concluded that single widely distributed clones appeared to be responsible for cases of bovine subclinical mastitis found in this one region of Germany.

Zadoks et al. (2000) genotyped 38 bovine mammary *S. aureus* isolates from diverse clinical, temporal, and geographic origins by PFGE by means of binary typing using 15 strain-specific DNA profiles. Seven major PFGE types and four subtypes were identified, as were 16 binary types. For 28 bovine *S. aureus* isolates, detailed clinical observations in vivo were compared to strain typing results in vitro. Associations were found between distinct genotypes and the severity of disease, suggesting strain-specific bacterial virulence. The authors concluded that binary typing is robust and simple and is an excellent method for comparison of *S. aureus* strains.

PFGE and coagulase gene restriction profiles (CRP) were used by Chiou, Wei, and Yang (2000) to analyze 71 *S. aureus* isolates from nine foodborne disease outbreaks in Taiwan. The primers COAG-5-F/COAG-2-R (Table 11.2) were used to amplify a sequence of the *coa* gene yielding mostly amplicons of 560-bp length with some amplicons varying from 1 to 8 bp from this CRP 560-bp amplicon. When the 560-bp amplicons of 39 isolates were digested with *AluI* 11, CRP profiles emerged in contrast to 22 PFGE profiles. In addition, when the 560-bp amplicons from the 39 isolates were sequenced, extensive identity indicated that PFGE is far superior for molecular typing than CRP.

S. aureus coagulase-positive type VII strains have been the most frequently isolated from staphylococcal food poisoning in Tokyo, Japan. Shimizu et al. (2000) applied PFGE typing to 129 such strains. Three major and 33 subtypes emerged. Strains of the same subtypes were isolated from food poisoning cases in the same districts at time intervals of one to five years. A total of 225 *S. aureus* isolates from bovine udders, human skin, milking equipment, and bovine milk were subjected to PFGE and phage typing. A subset of 142 isolates was characterized by binary typing. Among the 225 PFGE isolates, 208 were successfully phage typed, resulting in 21 phage types. PFGE identified 24 main types and 17 subtypes. Among the 142 isolates binary typed, 138 were typeable and yielded 20 binary types. Typeability and overall concordance with epidemiological data were lower for binary typing than for phage indicating that binary typing is not a suitable replacement for PFGE but may be useful in combination with PFGE to refine strain differentiation.

Kondoh et al. (2002) examined 78 clinical isolates of MRSA by RAPD with the single primer ERIC2 (Table 11.2) and by PFGE. PFGE yielded nine genotypes and 28 subtypes. RAPD yielded three genotypes and 22 subtypes. When used in combination, RAPD and PFGE were found to provide more precise discrimination of strains than the use of a single typing method.

Omoe et al. (2002) investigated the distribution of SE genes *sea* to *sei* (Table 11.2) in 146 isolates of *S. aureus* from food poisoning outbreaks, healthy humans, cows with mastitis, and bovine raw milk using multiplex PCR. Among the 146 isolates, 113 (77.7%) were found to harbor one or more *se* genes. There were 14 *se* genotypes distinguished. *Seg* and *sei* were found to coexist in some strains. A sandwich ELISA assay showed that most *seh* haboring strains were able to produce a significant amount of SEH. However, most of the isolates harboring *seg* and about 60% of the isolates harboring *sei* did not produce a detectable level of SEH or SEI, whereas reverse transcription-PCR analysis using primers SEG, SEH, and SEI (Table 11.2) proved that the mRNAs of SEG and SEI were transcribed in strains harboring the *seg* and *sei* genes. The authors concluded that a quantitative assessment of SEG and SEI is necessary to clarify the relationship between these new SEs and food poisoning. This assessment should also be applied to the other new SEs because their contribution to food poisoning has not been clearly established.

Cespedes et al. (2002) subjected 35 nasal isolates of *S. aureus* from medical personnel and 17 isolates from nonmedical personnel to PFGE typing. Medical personnel were colonized with more antibiotic-resistant isolates than nonmedical personnel, and the PFGE strain profiles indicated that they tended to be more clonal in origin, suggesting that exposure to hospital isolates alters the colonization profile.

A total of 293 S. aureus isolates obtained from 127 bulk milk tank samples of goats and sheep from Switzerland were characterized by Scherrer et al. (2004) using phenotypic and genotypic traits. The primers COA-F/COA-R (Table 11.2) from Hookey, Richardson, and Cookson (1998) were used to generate amplicons from the coa gene. An mPCR assay described by Mehrotra, Wang, and Johnson (2000) was used to detect the sea–see genes (Table 11.2), and a second mPCR assay described by McLauchlin et al. (2000) was used to detect the seg–sei genes. The primers sej-F/sej-R (Table 11.2) from Monday and Bohach (1999) were used to detect the sej gene. Primers GTSSTR-1/GTSSTR-2 (Table 11.2) from Mehrotra, Wang, and Johnson (2000) were used to detect the tst gene. Among the 293 isolates, 193 (65.9%) were egg-yolk (lecithinase) negative and 15 (5.1%) were negative for the clumping factor or protein A determined by latex agglutination assay. For 285 (97.3%) of the isolates, PCR amplification indicated the presence of the coa gene with a total of five different sized amplicons (500, 580, 660, 740, and 820 bp) indicating genotypic variation. SE genes were detected by PCR in 191 (65.2%) of the isolates with 123 (43.0%) of the isolates positive for these genes, 31 (10.6%) being positive for the seg gene, 28 (9.6%) for the sea gene, 26 (8.9%) for the sej gene, 24 (8.2%) for the sei gene, 4 (1.4%) for the seb gene, and 4 (1.4%) for the sed gene. In addition, 126 isolates (43.0%) were positive for the tst gene. Coagulase gene restriction profile analysis of the 145 isolates harboring sea or sec genes revealed six different patterns using AluI and five different patterns using HaeIII. The authors concluded that remarkable differences in phenotypic traits between S. aureus originating from goats, sheep, and bovine milk were found and that the high prevalence of SE-producing strains may be of importance regarding dairy food hygiene, especially with raw milk products.

The presence of the SE genes seg–sej were studied among 100 human clinical isolates of S. aureus in Jordan by El-Huneidi, Bdour, and Mahasneh (2006). The four pairs of primers (Table 11.2) were from Rosec and Gigaud (2002). Thirty-nine of the isolates possessed one or more of the seg, seh, or sei genes, and none of the isolates possessed the sej gene. PCR-RFLP analysis of the aroA gene by restriction with TaqI revealed 39 of the 100 isolates were of the same PCR-RFLP genotype designated A and 11% of the isolates were of genotype B. Fifty percent of the isolates yielded a third banding pattern designated N.

Gilbert et al. (2006) made use of multiple-locus variable-number tandem repeats analysis (MLVA) to study the genotypic variation of S. aureus isolates from milk. A total of 96 strains isolated between 1961 and 2003 from the milk of 90 dairy cows belonging to 75 French herds were assayed. Among 17 primer pairs used to generate tandem repeat sequences, 9 were found useful for MLVA. The use of primer pairs, clfA, clfB, fnb, and SAV1078 (Table 11.2) enabled the recognition of 61 MLVA types. The authors concluded that MLVA has the advantage of being less technically demanding, less time consuming, and less costly compared to PFGE and multilocus sequence typing (MLST) and exhibits a high level of strain discrimination.

Aires-de-Sousa et al. (2007) characterized 30 isolates of S. aureus from cows, sheep, goats, and buffalo with subclinical mastitis from 18 small dairy herds and from colonization of ostriches from a breeding farm in Rio de Janiero, Brazil. PFGE typing, *spa* X region sequence typing with primers arcCF2/arcC-Da (Table 11.2), and MLST with primers arcCF2/arcC-Da amplifying a portion of the carbamate kinase gene (Table 11.2) revealed five clonal types. Detection of major clone A in 63% of the isolates in different herds among all animal species studied and in infection and colonization samples indicated its spread in Rio de Janiero and no host preference among the various animal species. Comparison with *S. aureus* from human origin suggested that all but one clone might be animal-specific, supporting the concept of species specificity as previously suggested by earlier studies.

Rall et al. (2007) used the PCR for detection of nine staphylococcal SET genes (*sea, seb, sec, sed, see, seg, seh, sei,* and *sej*). The primers for *sea–sed* were from Johnson et al. (1991), those for *see* were from Mehrotra, Wang, and Johnson (2000), those for *seg–sei* were from Omoe et al. (2002), and those for *sej* were from Nashev et al. (2004; Table 11.2). *S. aureus* was found in 38/54 (70.4%) of raw milk samples at levels up to 8.9×10^5 CFU/ml and in 19/106 (18%) of samples from pasteurized milk. Among the 57 strains studied, 39 (68.4%) were positive for one or more genes encoding the nine enterotoxins and 12 different genotypes were identified. The *sea* gene was the most frequent (16 strains, 41%).

Krawczyk et al. (2007) subjected 37 strains of *S. aureus* from patients with furunculosis in Poland to amplification of the DNA fragments surrounding rare *Xba*I restriction sites (ADSRRS-fingerprinting) and the PCR melting profiles (PCR MP) of the amplified fragments determined and compared to PFGE profiles. PFGE yielded 21 unique profiles. ADSRRS-fingerprinting resulted in each isolate yielding 25 to 30 *Xba*I genomic fragments from 2- to 1300-bp and 22 unique profiles. PCR MP yielded 22 types. The authors concluded that the ADSRRS and PCR MP methods had similar levels of discrimination compared to PFGE, are less costly, and are therefore suitable alternative methods. However, because terminal adapters must be ligated to the cohesive ends of the restriction fragments to allow their amplification by a suitable pair of common primers, both of the alternative techniques may be labor intensive, as is PFGE.

Peles et al. (2007) subjected 59 isolates of *S. aureus* from 14 bulk milk tanks from 20 dairy farms and nine isolates of *S. aureus* from milk of cows with mastitis in Hungary to PFGE analysis in addition to PCR detection of *sea–seg* and *tst* genes. The primers for the genes *sea–see* and *tst* were from Mehrotra, Wang, and Johnson (2000), primers for *seg–sei* were from McLauchlin et al. (2000), and primers for *sej* were from Monday and Bohach (1999; Table 11.2). PFGE revealed 22 distinct genotypes. Only one or two main PFGE types were observed on each farm, indicating the absence of genetic diversity within each farm. Sixteen (27.1%) of the *S. aureus* isolates were found positive for one or more *se* genes, with 15 carrying just one *se* gene and one strain carrying two genes (*seg* and *sec*). The most frequently detected *se* genes were *seb, sea,* and sec. None of the isolates harbored the *see, seh, sej,* or *tst* genes.

Santos et al. (2008) subjected 54 isolates of *S. aureus* from ovine mastitis infections to PCR-restriction fragment length polymorphism (RFLP) analysis of a partial *groEL* gene sequence. The primers H279A/H280A (Table 11.2) amplified a 550-bp sequence of the *groEL* gene. The restriction nuclease *Alu*I yielded one to four DNA bands derived from the 550-bp amplicons from the 54 isolates and clearly distinguished 11 species of *Staphylococcus* from one another including *S. aureus* except for the species *S. chromogenes*, *S. hyicus*, and *S. capitis* that yielded similar RFLP patterns to one another. For these species a double digestion with *Hind*II and *Pvu*II yielded two to four bands, which clearly distinguished these three species from one another.

Ruzickova et al. (2008) subjected 28 *S. aureus* SEH-positive strains isolated from food samples and animal specimens in 11 districts of the Czech Republic to genotype analysis in an attempt to determine whether any predominant clones prevailed. Genotype analysis consisted of PFGE, *spa* gene polymorphism analysis by sequencing the *spa* amplicons (Table 11.2), enterobacterial repetitive intergeneric consensus sequence-based PCR (ERIC2-PCR) fingerprinting, and prophage carriage detection. The primers spa-1095F/spa-1517R (Table 11.2) amplified a sequence of the *spa* gene. The primers ERIC2-F/ERIC2-R (Table 11.2) from Versalovic, Koeuth, and Lupski (1991) were used for ERIC-PCR typing. Among 879 food samples tested, 220 were positive for *S. aureus* with 131 of the 220 *S. aureus* strains SE positive. Among the SE-positive strains, *sei* and *seg* occurred most frequently (53% and 52%, respectively). The *seh* gene appeared alone in 16 strains isolated mostly from raw milk and raw meat. PFGE distinguished 20 profiles derived from a total of 28 *seh*-carrying strains. Sequencing of the *spa* amplicons yielded 10 *spa* types. The fact that 15 of the 28 *seh* strains were of one *spa* type suggested that these *seh*-positive strains have a related genetic background despite their classification into 11 different PFGE types. Ten lysogenic types were identified among the 28 *seh*-positive strains. The authors concluded that all of the *S. aureus* strains were of diverse genotypes with none prevailing among the 28 *seh*-positive isolates studied.

Reinoso et al. (2008) genotypically characterized 45 *S. aureus* strains isolated from humans, bovine subclinical mastitis, and food samples in Argentina by RAPD and PCR amplification of virulence genes. Resistance to various antibiotics was observed for human *S. aureus* isolates, was less pronounced for the bovine strains, and was not observed with strains from food samples. The strains were classified genotypically by RAPD and amplification of the genes encoding protein A (*spa*), coagulase (*coa*), clumping factor (*clfA*), the collagen adhesin domains A and B (*cnaA* and *cnaB*), capsular polysaccharides 5 and 8, the accessory gene regulator *agr* classes I, II, III, and the *S. aureus* gene regulator *sae*. Primers are listed in Table 11.2. RAPD analyses with primer rep (Table 11.2) and the different gene patterns revealed that the strains could be divided into seven groups, mostly matching with the origin of the isolates. The *agr* III gene occurred in both human and bovine isolates but not in food isolates. The *sae* gene occurred more frequently in human clinical and bovine isolates than in human nonclinical isolates and foods.

G. Multilocus Sequence Typing (MLST) of *S. aureus* Isolates

MLST is a highly discriminatory method of characterizing bacterial isolates on the basis of the sequences of ~450-bp internal fragments of seven housekeeping genes *arcC, aroE, glpF, gmK, pta, tpi,* and *yqiL* (Table 11.1). For each gene fragment, the different sequences are designated a distinct allele, and each isolate is defined by the

Table 11.1   Genes Used to Identify and Characterize Strains of *S. aureus*

Gene	Gene Product or Function
Classical Enterotoxins	
sea	Enterotoxin A
seb	Enterotoxin B
sec (*c1, c2, c3*)	Enterotoxins C1, C2, C3
sed	Enterotoxin D
see	Enterotoxin E
New Enterotoxins	
seg	Enterotoxin G
seh	Enterotoxin H
sei	Enterotoxin I
sej	Enterotoxin J
sek	Enterotoxin K
sel	Enterotoxin L
sem	Enterotoxin M
sen	Enterotoxin N
seo	Enterotoxin O
sep	Enterotoxin P
seq	Enterotoxin q
ser	Enterotoxin R
seu	Enterotoxin U
Other Toxins	
tst	Encodes toxic shock syndrome toxin (TSST)
eft	Encodes exfoliative toxin (EFT) resulting in scalded skin syndrome of infants
eta (*exta*)	Encodes exfoliative toxin (ETA) resulting in scalded skin syndrome of infants
etb (*extb*)	Encodes exfoliative toxin (ETB) resulting in scalded skin syndrome of infants
Genes Used for MLST	
arcC	Encodes carbamate kinase
aroA	Encodes 5-enolpyruvylshikimate-3-phosphate synthase
aroE	Encodes Shikimate dehydrogenase
glpF	Encodes glycerol kinase
gmK	Encodes guanylate kinase
pta	Encodes phosphate acetyltransferase
tpi	Encodes triosephosphate isomerase
yqiL	Encodes acetyl coenzyme A acetyltransferase
Other Genes	
femA	A genomic DNA sequence unique to all *S. aureus* isolates

Table 11.1   Genes Used to Identify and Characterize Strains of *S. aureus* (Continued)

Gene	Gene Product or Function
femB	A genomic DNA sequence unique to all *S. aureus* isolates
nuc	Encodes the unique thermonuclease of *S. aureus*
coa	Encodes the unique coagulase of *S. areus*
groEL	Encodes a 60-kDa polypeptide (known as GroEL, 60 kDa chaperonin, or HSP60 for heat shock protein 60)
Cpn60	Encodes a 60-kDa chaperonin of *S. aureus*
Hsp60	Encodes the 60-kDa heat shock protein of *S. aureus*
orfX	Encodes a putative glycoprotease
Sa442	Random genomic DNA fragment used to type strains of *S. aureus*
hol	Encodes the *S. aureus* bacteriophage protein holin involved in permeabilizing (hole production) in cell membrane and wall resulting in cell lysis and the release of mature phage particles
lys	Encodes *S. aureus* phage lytic enzyme
bap	Encodes a protein involved in biofilm formation
icaADBC	Gene cluster that mediates biofilm formation by encoding proteins involved in the synthesis of biofilm matrix polysaccharide composed of linear B-1-6-linked N-acetylglucoseamine residues
pnp	Encodes polynucleotide phosphorylase of *S. aureus*
aroA	Encodes 5-enolpyruvylshikimate-3-phosphate synthase
16S	rRNA Conserved sequences used to identify *S. aureus* isolates
23S	rRNA Conserved sequences used to identify *S. aureus* isolates
Genes Enhancing Pathogenesis	
spa (X-region)	X region of staphylococcal protein A
spa (IgG-region)	IgG binding region of staphylococcal protein A
clfA	Cell clumping factor
cap5	Capsular polysaccharide 5
cap8	Capsular polysaccharide 8
agrI	Accessory gene regulator class I that functions as an activator for expression of α- and β-hemolysin
agrII	Accessory gene regulator class II
agrIII	Accessory gene regulator class III
saeA	Exoprotein expression gene regulator which is part of the *sae* gene locus
seas	Encodes histidine protein kinase which is part of the *sae* gene locus
scn	Encodes the chemotaxis inhibitory protein (CHIPS)
chp	Encodes staphylococcal complement inhibitor
sak	Encodes the immune evasion protein staphylokinase
mecA	Encodes methicillin resistance
femA	Encodes a factor essential for methicillin resistance
femB	Encodes a factor essential for methicillin resistance
femhA	Encodes a factor of unknown function with respect to methicillin resistance
hlb	Encodes the β-hemolysin
erm	Encodes erythromycin resistance

Continued

Table 11.1   Genes Used to Identify and Characterize Strains of *S. aureus* (Continued)

Gene	Gene Product or Function
	Adhesins
cnaA	Encodes collagen adhesin domain A
cnaB	Encodes collagen adhesin domain B
fnbA	Encodes fibronectin binding protein A
fnbB	Encodes fibronectin binding protein B
fib	Encodes a fibrinogen binding protein
cflA	Encodes a fibrinogen binding protein
fbp	Encodes a fibrinogen binding protein
ebpS	Encodes an elastin-binding adhesin
map	Encodes a broad specificity adhesin

alleles derived from each of the seven used housekeeping genes. Because there are many possible alleles at each of the seven loci, a high level of strain discrimination results. Enright et al. (2000) subjected 155 isolates of *S. aureus* to MLST using pairs of primers (Table 11.2) for each of the seven housekeeping genes and then sequenced the resulting amplicons. A total of 53 allelic profiles were obtained. Pairs of isolates with the same MLST profile produced very similar PFGE patterns. The authors concluded that MLST provides an unambiguous method for assigning methicillin resistant and sensitive isolates to known clones or assigning them to novel clones.

Grundmann et al. (2002) examined 117 strains of *S. aureus* representative of the natural population of the organism carried by humans using MLST, PFGE, RAPD, and phage typing. MLST revealed 46 profiles, PFGE 57 profiles, and RAPD with a single unspecified random primer yielded 16 profiles. Among the 117 isolates, 28 were not typeable by 23 standard phages. The remaining isolates were grouped into 23 phage types. The overall correlation of phage typing with the DNA methods was low. The authors concluded that only MLST was able to define clonal complexes unambiguously. It is presently recognized that the use of two or three random primers is required to adequately type isolates by RAPD.

Peacock et al. (2002) compared MLST with PFGE using 52 *S. aureus* isolates associated with human carriage and 28 invasive disease isolates in a busy regional renal unit in Oxford, England. PFGE yielded 31 profiles and MLST yielded 28 profiles. The authors concluded that PFGE and MLST performed equally well in this study.

H. PCR-Immuno Assays for Detection of *S. aureus* Exotoxin Genes

Becker, Roth, and Peters (1998) developed two multiplex PCR enzyme immunoassays (PCR-EIA) for detection of *S. aureus* exotoxins. One set of primers and probes was designed to detect *sea* to *see* (Table 11.2). The second set of primers and probes was designed to detect exfoliative toxin genes *eta* and *etb* and the TSST gene *tst* (Table 11.2). Streptavidin-coated microtiter strips were incubated with an individual biotinylated probe. The strips were then incubated with denatured amplicons and treated with anti–double-stranded DNA mouse antibody. Anti-mouse antibody labeled with horseradish peroxidase (HRP) was then added and an HRP chromogen-substrate added for color development.

Gilligan et al. (2000) developed a PCR-ELISA assay for detection of SE genes *sea* and *seb*. The primers SEAU1/SEAL1 and the probe SEAP1 (Table 11.2) were used to amplify and detect a 396-bp sequence of the *sea* gene. The primers SEBU1/SEBL1 and the probe SEBP1 (Table 11.2) were used to amplify and detect a 109-bp sequence of the *seb* gene. E amplicons were labeled during amplification with digoxygenin-11-dUT (DIG). Amplicons were denatured in an alkaline solution and hybridized to the corresponding biotinylated probe in a neutralizing solution. The mixture was then transferred to streptavidin-coated microtiter plate strips and then processed with HRP-anti DIG antibody conjugate and color developed with HRP substrates H_2O_2 plus ABTS. The sensitivity and specificity of the assays were both 100%. An antigen-capture ELISA assay for toxin detection resulted in a much lower specificity of the SEB toxin than for the SEA toxin. The overall correlations between the SEA and SEB toxin ELISAs and their corresponding PCR-ELISA assays were 100% and 29.6%, respectively.

Aitichou et al. (2004) developed PCR-enzyme immunoassays (PCR-EIA) involving an electrochemical system of detection. Primers SEAU/SEAL and probe SEAP (Table 11.2) were used to detect *sea*. Primers SEBU/SEBL and probe SEBP (Table 11.2) were used to detect *seb*. Biotinylated amplicons were converted to single strands with lambda-exonuclease. A fluorescein-labeled probe was added. Fifty microliters were then transferred to wells of a streptavidin-coated microtiter plate, and after 10 min the plate was washed. HRP-anti-fluorscein antibody was then added, and after 10 min the plate was washed and an electrochemical reaction was initiated by adding 3,3′,5,5′-tetramethylbenzidine and H_2O_2. The current resulting from the enzyme substrate reaction was measured with an electrochemical reader using intermittent pulse amperometry. Similar sensitivity levels were obtained with colorimetric assays (12 genomic copies per PCR reaction). The sensitivity and specificity of the assays for both was 100% with reference isolates. The specificity was 96% for *sea* and 98% for *seb*.

I. Exfoliative Toxin (ET) Producing Strains of *S. aureus*

The ET toxin has been divided into two serotypes, A and B (ETA and ETB; Bailey, de Azavedo, and Arbuthnott, 1980). The *eta* gene encoding ETA is located on the chromosome of *S. aureus,* whereas the gene for ETB (*etb*) is plasmid encoded (Keyhani et al., 1975; O'Toole and Foster, 1986).

Hayakawa et al. (2001) reported on the isolation of three ETA-producing isolates from bovine milk. The levels of ET production by these three bovine isolates were 312- to 625-fold less than that of human isolates. The primers ETA-F/ETA-R (Table 11.2) were used to amplify a 741-bp sequence of the *eta* gene in the three bovine ETA-positive isolates and in those of three ETA-positive human isolates. The nucleotide sequences of the PCR products from the bovine and human isolates were found to be identical. The reason for the large difference in ETA levels produced between bovine and human isolates was not resolved.

Ruzickova et al. (2003) subjected 16 exfoliative toxin-producing strains of *S. aureus* isolated from maternity units of two distant hospitals in the Czech Republic to molecular typing. Genotyping methods used were PFGE, ribotyping, PCR ribotyping, and prophage carriage. Three strains secreted combined ETA and ETB, and the remaining strains produced ETA and enterotoxin c or TSST-1. A comparison of the various genomic profiles resulted in the identification of nine genotypes. The presence of one prevailing genotype was demonstrated in each hospital. Genomic profiling indicated that the skin disease "pemphigus neonatum" disseminated in both hospitals did not originate from a single ET-positive source or a common ancestor. The authors concluded that ET-positive *S. aureus* genotypes may have been present in the human population. A high genotypic similarity was found among the strains originating from the same hospital, but no relatedness was found between strains isolated from the two hospitals.

El Helali et al. (2005) reported that over a 3-month period 13 neonates developed staphylococcal scalded skin syndrome (SSSS) in a maternity unit in a French hospital, between 4 and 8 days after birth. The causative factor was an ancillary nurse assigned to postdelivery infant care who suffered from acute eczema of her hands and failed to use latex gloves when handling the infants. PFGE analysis with *Sma*I distinguished four different PFGE types among a total of 23 *S. aureus* isolates. *S. aureus* isolated from nine of the ten SSSS cases from colonized neonates, from two nasal swabs of staff members, and from the hands of the ancillary nurse suffering from acute eczema of the hands were of a single PFGE clonal type. Removal of the ancillary nurse and stringent infection control measures led to control of the epidemic. The authors concluded that tight surveillance of chronic dermatitis in healthcare workers is needed.

J. PCR Detection of Methicillin-Resistant *S. aureus* Strains

Methicillin resistance in strains of *S. aureus* (MRSA) is mediated by the acquisition of the *mecA* gene. The *mecA* gene encodes for an altered penicillin binding protein (PBP-2a) that has reduced affinity for β-lactam antibiotics and hence imparts resistance to the penicillin family of antibiotics. This gene product has made the treatment of MRSA-associated infections difficult because an increasing proportion of MRSA is also resistant to other antibiotics. In addition, there is an increasing proportion of MRSA that has reduced susceptibility to vancomycin, a major alternative antibiotic for *S. aureus* infections.

In the early 1990s, severe enteritis caused by methicillin-resistant *S. aureus* (MRSA) was prevalent in Japan, the incidence of which has since decreased. Okii et al. (2006) subjected 186 nonenteritis isolates and 12 enteritis isolates to phenotype and genotype analysis. Genotype analysis involved PFGE plus the use of primers for detection of *sea–sed* and *tst* genes. The 12 enteric isolates yielded four PFGE types. Only 7 of the 186 nonenteritis isolates had PFGE patterns indistinguishable from enteritis isolates. Eight of the twelve enteritis isolates possessed the *sea, sec,* and *tst* genes and produced high levels of SEA and TSST-1 but not SEC. The authors concluded that the disappearance of MRSA enteritis may have resulted from the decreased incidence of enteritis-causing clones and phenotypic changes.

Ünal et al. (1992) developed a PCR assay for detection of *mecA* in staphylococci including *S. aureus*. The methodology employed a rapid cell lysis procedure involving lysostaphin and proteinase K and nested PCR for detection of the *mecA* gene. Primers 1-F/2-R (Table 11.2) amplified a 1800-kb sequence of the *mec* gene. The nested primers 3-F/4-R (Table 11.2) amplified a 1108-bp sequence of the initial amplicon. In addition, primers 5-F/6-R (Table 11.2) were used to amplify a 900-bp sequence of the *femA* gene to confirm the identity of *S. aureus* and non–*S. aureus* strains. A total of 51 *mecA*-positive (17 *S. aureus* strains and 34 coagulase-negative staphylococci including *S. epidermidis*, *S. haemolyticus*, and *S. simulans*) were studied. All *S. aureus* strains regardless of the presence of the *mecA* gene were *femA*-positive. Among the 34 coagulase negative staphylococci (non–*S. aureus* isolates) none possessed the *femA* gene confirming their non–*S. aureus* identity. The *mecA* gene was detected in a number of coagulase-negative staphylococci (non–*S. aureus* species).

Tambic et al. (1997) subjected 17 strains of methicillin-resistant *S. aureus* (MRSA) from patients infected in an intensive care unit to RAPD with four random primers (Table 11.2) and PFGE analysis. Identical RAPD and PFGE profiles were found in 15 of the 17 isolates. The authors concluded that RAPD typing with multiple primers is useful for typing isolates nontypeable by phage typing and that the technique is easier and less time consuming than PFGE and does not require pulsed-field equipment.

Van Leeuwen et al. (1998) examined 103 strains of MRSA using binary typing, RAPD with primers AP-1, AP-7, and ERIC2 (Table 11.2), PFGE, and MecA/tn554 probe typing. Three different strain collections were analyzed comprised of locally, nationally, and internationally disseminated genotypes. MRSA strains recovered during an outbreak in a New York City hospital and Portuguese MRSA isolates all resembling the so-called Iberian clone were included in the local and national collections. The outbreak strains showed subclonal variation, whereas the Portuguese isolates displayed an increased number of genotypes. Among the epidemiologically unrelated MRSA strains, the different genotyping methods revealed a wide heterogeneity of types. The authors concluded that binary typing and RAPD are the typing methods of choice for discriminating between strains that have a recent common ancestor and have undergone as yet limited dissemination.

Lee (2003) examined 1913 specimens from cattle, pigs, and chickens in the Republic of Korea for the presence of MRSA. A total of 421 isolates was obtained, and of these 28 were resistant to methicillin at concentrations higher than 2 mg/ml. Isolates from 15 of the 28 specimens were found to be positive for the *mecA* gene using the PCR primers mecA1/mecA2 (Table 11.2) that amplified a 533-bp sequence of the *mecA* gene. Among the 15 *mecA*-positive MRSA isolates, 12 were from dairy cows and 3 were from chickens. All such MRSA isolates were also resistant to members of the penicillin family such as ampicillin, oxacillin, and penicillin. RAPD analysis with primers M13 and H12 (Table 11.2) indicated that six of these isolates from animals yielded identical patterns to certain human isolates. These observations suggested that animals were a possible source of human infections caused by consuming contaminated food products made from these animals.

Akpaka et al. (2007) found that among 1912 clinical isolates of *S. aureus* from three hospitals in the country of Trinidad and Tobago of the West Indies, 244 (12.8%) were resistant to methicillin (MRSA). Most of the MRSA isolates (96.7%) were from hospitalized patients from the three hospitals. PFGE analysis was striking and showed a similar banding pattern among all of the 244 MRSA isolates from the three hospitals. The authors concluded that the MRSA infections studied were hospital-acquired, inasmuch as over 96% of the isolates recovered were from patients who were hospitalized and the remaining 4% of the isolates were from outpatient clinics and could equally have been hospital-acquired infections because these patients had a past history of hospitalization.

Sabet et al. (2007) developed a triplex Rti-PCR assay for simultaneous detection of *mecA* (methicillin resistance), *ermA* (erythromycin resistance), and *fem* (*S. aureus* identification). The primers and dual-labeled probes used are given in Table 11.2. The assay was applied to 93 clinical *S. aureus* isolates from a hospital in Malaysia, of which 48 were MRSA and 45 methicillin-sensitive *S. aureus* (MSSA). The triplex Rti-PCR assay detected *mecA, erm*, and *femA* in all of the 48 MRSA isolates. All of the MRSA isolates contained the *femA* gene confirming their identity as *S. aureus*, and none were positive for *mecA* or *ermA*.

K. Enterotoxin Production by *S. intermedius*

S. intermedius, as is *S. aureus*, is coagulase-positive, is known to be involved in various infectious processes, and has been implicated in at least one staphylococcal food poisoning outbreak (Khambaty, Bennett, and Shah, 1994). Becker et al. (2001) examined the enterotoxigenic potential of 281 veterinary and 11 human isolates of *S. intermedius* by using a multiplex PCR DNA-enzyme immunoassay system (Becker, Roth, and Peters, 1998) targeting the enterotoxin (ET) genes *sea, seb, sec, sed,* and *see*. A total of 33 (11.3%) of the *S. intermedius* strains, including one human isolate, was found to possess the *sec* gene but possessed none of the other four ET genes. Immunoassay detected SEC toxin production by 30 (90.9%) of these 33 isolates. The authors concluded that an enterotoxigenic role of this organism in staphylococcal food poisoning via contamination of food products may be assumed.

Rajkovic et al. (2006) developed an Rti-PCR immunoquantitative sandwich PCR (iqPCR) method for detection of *S. aureus* enterotoxin B (SE) in pure cultures and foods (Figure 11.1). The assay consisted of immobilizing sheep polyclonal antibody against SEB on the surface of microwells. This was followed sequentially by the addition of samples containing SEB followed by sheep biotinylated polyclonal antibody to SEB. Streptavidin was incubated with biotinylated reporter DNA. The resulting streptavidin-reporter DNA complex was added to the wells followed by PCR to amplify the immobilized DNA. The double-stranded reporter DNA was 246 bp in length. The primers REPT-F/REPT-R (Table 11.2) amplified a 67-bp sequence of the reporter DNA with SYBR green as the Rti-PCR fluorophore.

III. THE ROLE OF *S. AUREUS* BACTERIOPHAGE (PHAGE) IN PATHOGENESIS

It is well established that *S. aureus* phages mediate the simultaneous double or triple lysogenic conversion of ETA encoded by the *sea* gene, staphylokinase

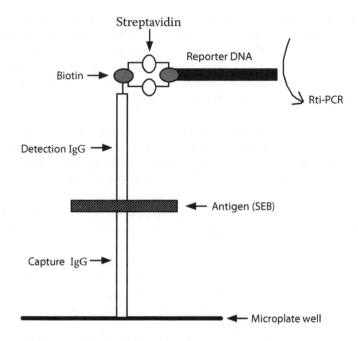

Figure 11.1 Adapted schematic diagram of the sandwich iqPCR of Rajkovic et al. (2006) using the same capture and detection antibody for detection of SEB.

encoded by the *sak* gene, and β-hemolysin encoded by the *hlb* gene (Coleman et al., 1989). Expression of β-hemolysin can be lost following lysogenization due to insertional inactivation of the *hlb* gene as a result of integration of phage NA sequences (Coleman et al., 1991). Van Wamel et al. (2006) found that the *scn* gene that encodes the chemotaxis inhibitory protein and the *chp* gene encoding *Staphylococcus* complement inhibitor are located on β-hemolysin–converting bacteriophages. These phages also carry the genes for the immune evasion molecules staphylokinase encoded by the *sak* gene and the *sea* and *sep* genes. These genes are located on an 8-kb region at the conserved 3′-end of β-hemolysin-negative converting phages, thereby forming an immune evasion cluster in human isolates of *S. aureus*.

Kumagai et al. (2007) found by PCR that in 13 of 43 non–β-hemolysin–producing bovine isolates of *S. aureus* there were two truncated β-hemolysin (*heb*) genes. One truncated *heb* gene was located beside the integrase (*int*) gene of phage origin. In addition, all 13 isolates with truncated *hlb* genes were found to harbor the holin (*hol*), lytic enzyme (*lgt*), and *sak* genes. *sea* and *sep* genes were also found in 5 and 2 of the 13 isolates, respectively. Among the 13 isolates, the *scn* and *chp* genes were detected in 13 and 4 of the isolates, respectively. These phage isolates induced by mitomycin C treatment of several of the β-hemolysin–negative isolates were unable to form plaques on the 13 isolates, suggesting lysogenic-mediated immunity. However, one of these phages did form turbid plaques on a β-hemolysin producing isolate, and the *sak* and *scn* genes were detected in three of five lysogenized isolates.

The authors concluded that their results suggested quadruple or quintuple conversion of *hlb, sak, sea* (or *sep*), *scn,* and *chp* genes by phages among non–β-hemolysin–producing bovine isolates of *S. aureus*. Primer pairs for detection of these genes are given in Table 11.2. Primers for detection of *sea* were from Monday and Bohach (1999; Table 11.2). The primers *hlb-1/hlb-2* were used to detect the normal untruncated *hlb* gene, whereas primers Hlb-1/Int-1 were used to detect the truncated *hlb* gene (Table 11.2).

Endo et al. (2003) isolated temperate phages from two ETA-positive bovine isolates of *S. aureus*. PCR analysis with primers Eta-1-F/Eta-2-R (Table 11.2) of the phage genomes indicated that the temperate phages carried the structural gene for the exfoliative toxin ETA. A bovine *eta*-carrying phage was able to lysogenize *eta*-negative bovine isolates of *S. aureus,* and the lysogenized *S. aureus* isolates had the ability to produce ETA. These results suggest the possibility of horizontal transmission of the *eta* gene by temperate phage among bovine isolates of *S. aureus* with potential transmission to humans through raw milk.

Devriese (1984) found that biotyping based on fibrinolysin, β-hemolysin, coagulase, and crystal violet agar reaction was able to differentiate *S. aureus* isolates from humans and animals into host-specific ecovars and biotypes that are not host-specific. Reinoso et al. (2004) subjected 40 strains of *S. aureus* from bovine milk and 40 strains from human clinical sources in Argentina to biotyping according to Devriese (1984). In addition, all of the strains were also RAPD typed using three random primers OLP6, OLP11, and OLP13 (Table 11.2). Eight different biotypes (three host-specific, human, bovine, and poultry, and five non–host-specific) were identified among the bovine and human isolates. Forty-two (52.5%) of the isolates belonged to a host-specific biotype. Among bovine isolates, 12 (30%) were of the bovine ecotype and 11 (27.5%) were of the human ecotype. The remaining bovine isolates (42.5%) were found to belong to non–host-specific biotypes. Among the human isolates, 17 (41.5%) belonged to the human ecovar. The remaining isolates from human hosts belonged to the poultry ecovar (5%) and non–host-specific biotypes (52.5%). It is interesting that no human isolates were of the bovine ecovar. RAPD profiles yielded amplicons ranging in size from 300 to 900 bp with primers OLP6, OLP11, and OLP13 yielding 47, 67, and 68 profiles, respectively. The combined RAPD banding profiles from all three primers yielded a total of 79 RAPD types. The genetic diversity among the bovine isolates (0.36) was relatively low compared to that of the human isolates (0.56).

Table 11.2 PCR Primers and DNA Probes

Primer or Probe	Sequence (5' → 3')	Size of Amplified Sequence (bp)	Gene or DNA Target Sequence	References
PNA probe	GCT-TCT-CGT-CCG-TTC	—	*16s rRNA*	Oliveira et al. (2002)
MLST arcCF2	CCT-TTA-TTT-GAT-TCA-CCA-GCG	—	—	Aires-de-Sousa et al. (2007)
MLST arcC-Da	AGG-TAT-CTG-CTT-CAA-TCA-GCG			Aires-de-Sousa et al. (2007) from Enright et al. (2000)
COAG-5-F	GGT-ATT-CGT-GAA-TAC-AAC-GAT-GGA	—	*coa*	Chiou, Wei, and Yang (2000)
COAG-2-R	AAA-GAA-AAC-CAC-TCA-CAT-CA			
SEA-3	CCT-TTG-GAA-ACG-GTA-AAA-ACG	127	*sea*	Omoe et al. (2002) from Becker, Roth, and Peters (1998)
SEA-4	TCT-GAA-CCT-TCC-CAT-CAA-AAA-C			
SEB-3	TCG-CAT-CAA-ACT-GAC-AAA-CG	277	*seb*	Omoe et al. (2002) from Becker, Roth, and Peters (1998)
SEB-4	GCA-GGT-ACT-CTA-TAA-GTG-CCT-GC			
SEC-3	CTC-AAG-AAC-TAG-ACA-TAA-AAG-CTA-GG	271	*sec*	Omoe et al. (2002) from Becker, Roth, and Peters (1998)
SEC-4	TCA-AAA-TCG-GAT-TAA-CAT-TAT-CC			
SED-3	CTA-GTT-TGG-TAA-TAT-CTC-CTT-TAA-ACG	319	*sed*	Omoe et al. (2002) from Becker, Roth, and Peters (1998)
SED-4	TTA-ATG-CTA-TAT-CTT-ATA-GGG-TAA-ACA-TC			

Continued

Table 11.2   PCR Primers and DNA Probes (Continued)

Primer or Probe	Sequence (5' → 3')	Size of Amplified Sequence (bp)	Gene or DNA Target Sequence	References
SEE-3	CAG-TAC-TAT-TAG-ATA-AAG-TTA-AAA-CAA-GC	178	see	Omoe et al. (2002) from Becker, Roth, and Peters (1998)
SEE-2	TAA-CTT-ACC-GTG-GAC-CCT-TC			
SEG-1	AAG-TAG-ACA-TTT-TTG-GCG-TTC-C	287	seg	Omoe et al. (2002)
SEG-2	AGA-ACC-ATC-AAA-CTC-GTA-TAG-C			
SEH-1	GTC-TAT-ATG-GAG-GTA-CAA-CAC-T	213	seh	Omoe et al. (2002)
SEH-2	GAC-CTT-TAC-TTA-TTT-CGC-TGT-C			
SEI-1	GGT-GAT-ATT-GGT-GTA-GGT-AAC	454	sei	Omoe et al. (2002)
SEI-2	ATC-CAT-ATT-CTT-TGC-CTT-TAC-CAG			
SEGF1	CCC-CGG-ATC-CCA-ACC-CGA-TCC-TAA-ATT-AGA-CGA-AC	722	seg	Omoe et al. (2002)
SEGR1	CCC-CGA-ATT-CTC-AGT-GAG-TAT-TAA-GAA-ATA-CTT-CC			
SEHF1	CCC-CGG-ATC-CGA-AGA-TTT-ACA-CGA-TAA-AAG-TGA-GTT	676	seh	Omoe et al. (2002)
SEHR1	CCC-CGA-ATT-CGA-TTA-TAC-TTT-TTT-CTT-AGT-ATA-TAG			
SEIF1	CCC-CGG-ATC-CCA-AGG-TGA-TAT-TGG-TGT-AGG-TAA-CT	677	sei	Omoe et al. (2002)
SEIR1	CCC-CGA-ATT-CTT-AGT-TAC-TAT-CTA-CAT-ATG-ATA-TTT-CG			
Sa-1	GAA-AGG-GCA-ATA-CGC-AAA-GA	482	nuc	Ramesh et al. (2002)

Primer	Sequence	Product size	Target gene	Reference
Sa-2	TAG-CCA-AGC-CTT-GAC-GAA-CT			
JIRS-2	AAA-AAC-ACT-TGT-CGA-TAT-GG	826	—	Stepan et al. (2001)
JIRS-1	GTT-TCA-ATA-CAT-CAA-CTG-C			
Eta-1-F	CTA-TTT-ACT-GTA-GGA-GCT-AG	—	*eta*	Endo et al. (2003)
Eta-2-R	ATT-TAT-TTG-ATG-CTC-TCT-AT			
TSST-1-FP	TCA-TCA-GCT-AAC-TCA-AAT-ACA-TGG-ATT	88	*tst*	Deurenberg et al. (2005)
TSST-1-RP	TGT-GGA-TCC-GTC-ATT-CAT-TGT-T			
TSST-1-PR	TAMRA-TCC-AAT-AAC-CAC-CCG-TTT-TAT-TAT-CGC-TTG-AA-FAM			
16S rDNA-F	TAG-ATG-GAT-CCG-CGC	917	*16S rDNA*	Larsen et al. (2000)
16S rDNA-R	CTT-AAT-GAT-GGC-AAC-TAA-GC			
ADE-F	ATT-TGT-GAA-AAA-AGG-CGG-AAT	304 and 431	*sea, sed, see*	Larsen et al. (2000)
ADE-R	GTG-ACA-CCA-CCG-TAC-GTA-CAA-GC			
BC-F	GGT-TTG-ATG-GAA-AAT-ATG-AA	590	*seb* and *sec*	Larsen et al. (2000)
BC-R	CAT-TAA-ATA-TTT-AGA-TTG-GTC-AAA			
H-F	AGA-AAT-CAA-GGT-GAT-AGT-GGC	441	*seh*	Larsen et al. (2000)
H-R	ATC-TAT-CTC-ATA-GTC-ATT-TTC-TCC			
TSST-1-F	TGA-ATT-TTT-TTA-TCG-TAA-GCC-C	468	*tst*	Larsen et al. (2000)
TSST-1-R	GTT-TTT-TAT-CGA-ACT-TTG-GCC			

Continued

Table 11.2   PCR Primers and DNA Probes (Continued)

Primer or Probe	Sequence (5' → 3')	Size of Amplified Sequence (bp)	Gene or DNA Target Sequence	References
EXT-F	GCA-TTA-ATC-AGA-TTA-AAA-CC	233 and 224	extA and extB	Larsen et al. (2000)
EXT-R	ATA-CCT-CAT-CCA-(GA)TT-TCC			
spa-1095F	AAG-ACG-ATC-CTT-CGG-TGA-GC	—	spa	Ruzickova et al. (2008)
spa-1517R	GCT-TTT-GCA-ATG-TCA-TTT-ACT-G			
RAPD ERIC2	AAG-TAA-GTG-ACT-GGG-GTG-AGC-G	—	—	Ruzickova et al. (2008) from Versalovic, Koeuth, and Lupski (1991)
SEA-3	CCT-TTG-GAA-ACG-GTT-AAA-ACG	127	sea	Becker, Roth, and Peters (1998)
SEA-4	TCT-GAA-CCT-TCC-CAT-CAA-AAA-C			
Probe SEA-7B	GGA-GTT-GGA-TCT-TCA-AGC-AAG-ACG			
SEB-1	TCG-CAT-CAA-ACT-GAC-AAA-CG	477	seb	Becker, Roth, and Peters (1998)
SEB-4	GCA-GGT-ACT-CTA-TAA-GTG-CCT-GC			
Probe	SEB-3B GAG-AAT-AGC-TTT-TGG-YAT-GAC-ATG			
SEC-3	CTC-AAG-AAC-TAG-ACA-TAA-AAG-CTA-GG	271	sec	Becker, Roth, and Peters (1998)
SEC-4	TCA-AAA-TCG-GAT-TAA-CAT-TAT-CC			
Probe SEC-5B	AAC-GGC-AAT-ACT-TTT-TGG-TAT-GAT			
SED-3	CTA-GTT-TGG-TAA-TAT-CTC-CTT-TAA-ACG	319	sed	Becker, Roth, and Peters (1998)
SED-4	TTA-ATG-CTA-TAT-CTT-ATA-GGG-TAA-ACA-TC			
Probe SED-5B	TAA-AGC-CAA-TGA-AAA-CAT-TGA-TTC-A			

Primer	Sequence	Size	Gene	Reference
SEE-3	CAG-TAC-CTA-TAG-ATA-AAG-TTA-AAA-CAA-GC	178	see	Becker, Roth, and Peters (1998)
SEE-2	TAA-CTT-ACC-GTG-GAC-CCT-TC			
Probe	SEE-5B CTT-TGG-CGG-TAA-GGT-GCA-AAG-AGG-C			
TST-3	AAG-CCC-TTT-GTT-GCT-TGC-G	445	tst	Becker, Roth, and Peters (1998)
TST-6	ATC-GAA-CTT-TGG-CCC-ATA-CTT-T			
Probe	TST-5B AAG-CCA-ACA-TAC-TAG-CGA-AGG-AAC			
ETA-3	CTA-GTG-CAT-TTG-TTA-TTC-AAG-ACG	119	eta	Becker, Roth, and Peters (1998)
ETA-4	TGC-ATT-GAC-ACC-ATA-GTA-CTT-ATT-C			
Probe ETA-5B	CCA-TGC-AAA-AGC-AGA-AGT-TTC-AGC			
ETB-3	ACG-GCT-ATA-TAC-ATT-CAA-TTC-AAT-G	262	etb	Becker, Roth, and Peters (1998)
ETB-4	AAA-GTT-ATT-CAT-TTA-ATG-CAC-TGT-CTC			
Probe	ETB-5B TAC-CAC-CTA-ATA-CCC-TAA-TAA-TCC-AA			
Sa442-1	AAT-CTT-TGT-CGG-TAC-ACG-ATA-TTC-ACG	108	—	Martineau et al. (1998)
Sa442-2	CGT-AAT-GAG-ATT-TCA-GTA-GAT-AAT-ACA-ACA			
16S rRNA-F	GGA-GGA-AGG-TGG-GGA-TGA-CG	241	16S rRNA	Martineau et al. (1998)
16S rRNA-R	ATG-GTG-TGA-CGG-GCG-GTG-TG			
se[adej]-F	AAA-GAT-TTG-GGA-AAA-AAG-TGT-GAA-TT	669, 668, 669	se[adej]	Fueyo, Mendoza, and Martin (2005b) from Martin, Gonzalez-Havia, and Mendoza (2003)
se[adej]-R	TT(G/T)-(C/T)(A/G)T-ATA-AAT-A(G/T)A-(A/T)RT-CAA-TAT-G			

Continued

Table 11.2 PCR Primers and DNA Probes (Continued)

Primer or Probe	Sequence (5′ → 3′)	Size of Amplified Sequence (bp)	Gene or DNA Target Sequence	References
se[bc]-F se[bc]-R	TAT-GAT-AAA-CAT-GTA-TCA-GCA-A (A/T)RT-CAA-TAT-G TTT-ATC-TCC-TGG-TGC-AGG-CAT-CAT	539 and 540	se[bc]	Fueyo, Mendoza, and Martin (2005a) from Martin, Gonzalez-Hevia, and Mendoza (2003)
Ser-1 Ser-2	AAA-CCA-GAT-CCA-AGG-CCT-GGA-G TCA-CAT-TGT-AGT-CAG-GTG-AAC-TT	700	ser	Fueyo, Mendoza, and Martin (2005b)
tst-1 tst-2	AGC-ATC-TAC-AAA-CGA-TAA-TAT-AAA-GG CAT-TGT-TAT-TTT-CCA-ATA-ACC-ACC-CG	481	tst	Fueyo, Mendoza, and Martin (2005b)
Staur4-F Staur6-R	ACG-GAG-TTA-CAA-AGG-ACG-AC AGC-TCA-GCC-TTA-ACG-AGT-AC	1240	S. aureus	Stephan et al. (2001) from Straub, Hertel, and Hammes (1999)
sea1-F sea2-R	AAA-GTC-CCG-ATC-AAT-TTA-TGG-CTA GTA-ATT-AAC-CGA-AGG-TTC-TGT-AGA	120	sea	Stephan et al. (2001) from Tsen and Chen (1992)
clfA-F clfA-R	GGC-TTC-AGT-GCT-TGT-AGG TTT-TCA-GGG-TCA-ATA-TAA-GC	980	clfA	Stephan et al. (2001)
spa-F spa-R	CAA-GCA-CCA-AAA-GAG-GAA CAC-CAG-GTT-TAA-CGA-CAT	100, 200, 280, 300	spa (X region)	Stephan et al. (2001) from Frénay et al. (1996)

Primer	Sequence	Product size	Target	Reference
IgG-F IgG-R	CAC-CTG-CTG-CAA-ATG-CTG-CG GGC-TTG-TTG-TTG-TCT-TCC-TC	750 and 920	*spa (IgG region)*	Stephan et al. (2001) from Seki et al. (1998)
coa-F coa-R	ATA-GAG-ATG-CTG-GTA-CAG-G GCT-TCC-GAT-TGT-TCG-ATG-C	580 and 660	*coa*	Stephan et al. (2001) from Hookey, Richardson, and Cookson (1998)
coa-F coa-R	ATA-GAG-ATG-CTG-GTA-CAG-G GCT-TCC-GAT-TGT-TCG-ATG-C	875, 660, 603, 547	*coa*	Hookey, Richardson, and Cookson (1998)
16S-23S-F 16S-23S-R	TCT-TCA-GAA-GAT-GCG-GAA-TA TAA-GTC-AAA-CGT-TAA-CAT-ACG	380, 400, 430, 510	*16S-23S sparer region*	Annmüller, Lämmler, and Zschöck (1999)
IgG-SPA-F IgG-SPA-R	CAA-AGA-TCA-ACA-AAG-CGC-C CGA-AGG-ATC-GTC-TTT-AAG-GC	28C and 620	*spa (IgG region)*	Annmüller, Lämmler, and Zschöck (1999)
COA-F COA-R	CGA-GAC-GAT-TCA-ACA-AG AAA-GAA-AAC-CAC-TCA-CAT-CA	740, 800, 990	*coa*	Annmüller, Lämmler, and Zschöck (1999) from van Belkum et al. (1997)
X-SPA-F	TGT-AAA-ACG-ACG-GCC-AGT-GCT-AAA-AAG-CTA-AAC-GAT-GC	120–300	*spa (X region)*	Annmüller, Lämmler, and Zschöck (1999) from Toshkova et al. (1997)
X-SPA-R	CAG-GAA-ACA-GCT-ATG-ACC-CCA-CCA-AAT-ACA-GTT-GTA-CC	—		

Continued

Table 11.2   PCR Primers and DNA Probes (Continued)

Primer or Probe	Sequence (5' → 3')	Size of Amplified Sequence (bp)	Gene or DNA Target Sequence	References
COAG2	CGA-GAC-CAA-GAT-TCA-ACA-AG	580–1060	coa	Lange et al. (1999) from Goh et al. (1992)
COAG3	AAA-GAA-AAC-CAC-TCA-CAT-CA			
SPAX1	CAA-GCA-CCA-AAA-GAG-GAA	170–340	spa (X region)	Frénay et al. (1996)
SPAX2	CAC-CAG-GTT-TAA-CGA-CAT			
RRNA1	TTG-TAC-ACA-CCG-CCC-GTC-A	700–950	16S-23S rRNA	Lange et al. (1999) from Cuny, Lause, and Witte (1996)
RRNA2	GGT-ACC-TTA-GAT-GTT-TCA-GTT-C			
SEAU1	ATG-GTA-GCG-AGA-AAA-GCG-AA	396	sea	Gilligan et al. (2000)
SEAL1	GCC-ATA-AAT-TGA-TCG-GCA-CT			
SEAP1	TGA-ATT-GCA-GGG-AAC-AGC-TTT-AG			Gilligan et al. (2000)
SEBU1	CAT-TAA-CCC-CTT-GTT-GCC-AT	109	seb	
SEBL1	ACA-AAT-CGT-TAA-AAA-CGG-CG			
SEBP1	CCA-ACT-TTA-GCT-GAA-ATT-GGG-G			Gilligan et al. (2000)
SEAU	Biotin-ATG-GTA-GCG-AGA-AAA-GCG-AA	396	sea	Aitichou et al. (2004) from Gilligan et al. (2000)
SEAL	GCC-ATA-AAT-TGA-TCG-GCA-CT			
SEAP	Fluorescein-CTA-AAG-CTG-TTC-CCT-GCA-ATT-CA			

Primer	Sequence	Size (bp)	Gene	Reference
SEBU SEBL	Biotin-TGT-ATG-TAT-GGT-GGT-GTA-AC ACA-AAT-TCG-TTA-AAA-ACG-GCG	—	seb	Aitichou et al. (2004) from Sharma, Rees, and Dodd (2000)
SEBP	Fluorescein-ATA-GTG-ACG-AGT-TAG-GTA			Aitichou et al. (2004) from Gilligan et al. (2000)
femA1 FemA2	AAA-AAA-GCA-CAT-AAC-AAG-CG GAT-AA-GAA-GAA-ACC-AGC-AG	134	femA	Omoe et al. (2005) from Becker, Roth, and Peters (1998)
femB1 FemB2	TTA-CAG-AGT-TAA-CTG-TTA-CC ATA-CAA-ATC-CAG-CAC-GCT-CT	651	femB	Omoe et al. (2005) from Pérez-Roth et al. (2001)
SEG1 SEG2	TGC-TAT-CGA-CAC-ACT-ACA-ACC CCA-GAT-TCA-AAT-GCA-GAA-CC	704	seg	Fueyo et al. (2005a) from McLauchlin et al. (2000)
SEH1 SEH2	CGA-AAG-CAG-AAG-ATT-TAC-ACG GAC-CTT-TAC-TTA-TTT-CGC-TGT-C	495	seh	Fueyo et al. (2005a) from McLauchlin et al. (2000)
SEI-1 SEI-II	CTC-AAG-GTG-ATA-TTG-GTG-TAG-G AAA-AAA-CTT-ACA-GGC-AGT-CCA-TCT-C	576	sei	Fueyo et al. (2005a) from Jarraud et al. (2002)
sek-F sek-R	ACC-GCT-CAA-GAG-ATT-GAT GAT-CGG-ATC-CTT-ATA-TCG-TTT-CTT-TAT-AAG-AA	278	sek	Fueyo et al. (2005a) from Yarwood et al. (2002)

Continued

Table 11.2 PCR Primers and DNA Probes (Continued)

Primer or Probe	Sequence (5′ → 3′)	Size of Amplified Sequence (bp)	Gene or DNA Target Sequence	References
sel-1-F	AAT-ATA-TAA-CTA-GTG-ATC-TAA-AGG-A	359	*sel*	Fueyo et al. (2005a) from
sel-2-R	TAT-GGA-ATA-CTA-CAC-ACC-CCT-TAT-A			Fueyo et al. (2005c)
sem-1-F	ATG-CTG-TAG-ATG-TAT-ATG-GTC	473	*sem*	Fueyo et al. (2005a) from
Sem-2-R	CGT-CCT-TAT-AAG-ATA-TTT-CTA-CAT-C			Fueyo et al. (2005c)
SEN-1	ATG-AGA-TTG-TTC-TAC-ATA-GCT-GCA-AT	680	*sen*	Fueyo et al. (2005a) from
SEN-2	AAC-TCT-GCT-CCC-ACT-GAA-C			Jarraud et al. (2002)
seo-F	TGT-AGT-GTA-AAC-AAT-GCA-TAT-GCA-AAT-G	722	*seo*	Fueyo et al. (2005a) from
seo-R	TTA-TGT-AAA-TAA-ATA-AAC-ATC-AAT-ATG-ATG-TC			Fueyo et al. (2005c)
sep-F	TTA-GAC-AAA-CCT-ATT-ATC-ATA-ATG-G	276	*sep*	Fueyo et al. (2005a) from
sep-R	TAT-TAT-CAT-GTA-ACG-TTA-CAC-CGC-C			Fueyo et al. (2005c)
seq-F	AAG-AGG-TAA-CTG-CTC-AAG	285	*seq*	Fueyo et al. (2005a) from
seq-R	GAT-CGG-ATC-CTT-ATT-CAG-TCT-TCT-CAT-ATG			Yarwood et al. (2002)
ser-1	AAA-CCA-GAT-CCA-AGG-CCT-GGA-G	700	*ser*	Fueyo et al. (2005a) from
ser-2	TCA-CAT-TGT-AGT-CAG-GTG-AAC-TT			Fueyo et al. (2005c)

PSE1	TGA-TAA-TTA-GTT-TTA-ACA-CTA-AAA-TGC-G	141	*seu*	Fueyo et al. (2005a) from Letertre et al. (2003a)
PSE4	CGT-CTA-ATT-GCC-ACG-TTA-TAT-CAG-T			
ET-1	CTA-TTT-ACT-GTA-GGA-GCT-AG	741	*eta*	Fueyo et al. (2005a) from Yamaguchi et al. (2002)
ET-2	ATT-TAT-TTG-ATG-GTC-TCT-AT			
ET-3	ATA-CAC-ACA-TTA-CGG-ATA-AT	629	*etb*	Fueyo et al. (2005a) from Yamaguchi et al. (2002)
ET-4	CAA-AGT-GTC-TCC-AAA-AGT-AT			
ET-14	AAC-TAT-CAT-GTA-TCA-AGG	376	*etd*	Fueyo et al. (2005a) from Yamaguchi et al. (2002)
ET-15	CAG-AAT-TTC-CCG-ACT-CAG			
tst-1	AGC-ATC-ACA-AAA-CGA-TAA-TAT-AAA-GG	481	*tst*	Fueyo et al. (2005c)
tst-2	CAT-TGT-TAT-TTT-CCA-ATA-ACC-ACC-CG			
1-F	ATG-AAT-AAG-AGT-CGA-TTT-ATT-TCA-T	801	*c1*	Mäntynen et al. (1997)
2-R	TTA-TCC-ATT-CTT-TGT-TGT-AAG-GTG-G			
3-F	ACA-CCC-AAC-GTA-TTA-GCA-GAG-AGC-C	631	*c1*	
4-R	CCT-GGT-GCA-GGC-ATC-ATA-TCA-TAC-C			

Continued

Table 11.2   PCR Primers and DNA Probes (Continued)

Primer or Probe	Sequence (5′ → 3′)	Size of Amplified Sequence (bp)	Gene or DNA Target Sequence	References
RAPD S	TCA-CGA-TGC-A	—	—	Fueyo et al. (2001) from Williams et al. (1990)
RAPD C	AGG-GAA-CGA-G	—	—	Fueyo et al. (2001) from Lin et al. (1996)
COAG1	ATA-CTC-AAC-CGA-CGA-CAC-CG	1557	coa	Goh et al. (1992)
COAG4	GAT-TTT-GGA-TGA-AGC-GGA-TT			
COAG2	CGA-GAC-CAA-GAT-TCA-ACA-AG	440	—	915
COAG3	AAA-GAA-AAC-CAC-TCA-CAT-CA			
Nuc1	GCG-ATT-GAT-GGT-GAT-ACG-GTT	~270	nuc	Brakstad, Aasbakk, and Maeland (1992)
Nuc2	AGC-CAA-GCC-TTG-ACG-AAC-TAA-AGC			
Nuc³²P-P	GGT-GTA-GAG-AAA-TAT-GGT-CCT-GAA-GCA-GCA-AGT-GCA			
SEG-F	GTT-AGA-GGA-GGT-TTT-ATG	198	seg	Bania et al. (2006)
SEG-R	TTC-CTT-CAA-CAG-GTG-GAG-A			
SEH-F	CAA-CTG-CTG-ATT-TAG-CTC-AG	173	seh	Bania et al. (2006)
SEH-R	CCC-AAA-CAT-TAG-CAC-CA			
SEI-F	GGC-CAC-TTT-ATC-AGG-ACA	328	sei	Bania et al. (2006)
SEI-R	AAC-TTA-CAG-GCA-GTC-CA			

Primer	Sequence	Size	Gene	Reference
SEJ-F	GTT-CTG-GTG-GTA-AAC-CA	131	*sej*	Bania et al. (2006)
SEJ-R	GCG-GAA-CAA-CAG-TTC-TGA			
SEK-F	GGA-GAA-AAG-GCA-ATG-AA	516	*sek*	Bania et al. (2006)
SEK-R	TAG-TGC-CGT-TAT-GTC-CA			
SEL-F	CGA-TGT-AGG-TCC-AGG-A	369	*sel*	Bania et al. (2006)
SEL-R	TTC-TTG-TGC-GGT-AAC-CA			
SEM-F	CAT-ATC-GCA-ACC-GCT-GA	148	*sem*	Bania et al. (2006)
SEM-R	TCA-GCT-GTT-ACT-GTC-GA			
SEN-F	GGC-AAT-TAG-ACG-AGT-CA	237	*sen*	Bania et al. (2006)
SEN-R	ATC-GTA-ACT-CCT-CCG-TA			
SEO-F	GTC-AAG-TGT-AGA-CCT-TA	288	*seo*	
SEO-R	TGT-CAG-GCA-GTA-TCC			
SEP-F	TCA-AAA-GAC-ACC-GCC-AA	396	*sep*	Bania et al. (2006)
SEP-R	ATT-GTC-CTT-GAG-CAC-CA			
SEQ-F	GGA-ATT-ACG-TTG-GCG-AA	530	*seq*	Bania et al. (2006)
SEQ-R	AAC-TCT-CTG-CTT-GAC-CA			
SEA-F	AAA-GAT-TTG-CGA-AAA-AAG-TCT-GAA-TT	106	*sea*	Letertre et al. (2003b)
SEA-R	CGT-GAC-TCT-CTT-TAT-TTT-CAG-TTT-TAG-C			
SEA-Probe	FAM-AGC-TTT-AGG-CAA-TCT-TA-MGB			

Continued

Table 11.2   PCR Primers and DNA Probes (Continued)

Primer or Probe	Sequence (5′ → 3′)	Size of Amplified Sequence (bp)	Gene or DNA Target Sequence	References
SEB-F	GAG-AAT-AGC-TTT-TGG-TAT-GAC-ATG-ATG	101	*seb*	Letertre et al. (2003b)
SEB-R	TTA-GAA-TCA-ACC-ATT-TTA-TTG-TCA-TTG			
SEB-Probe	FAM-ACC-AGG-AGA-TAA-ATT-MGB			
SEC1-F	GGC-AAT-ACT-TTT-TGG-TAT-GAT-ATG-ATG	101	*sec1*	Letertre et al. (2003b)
SEC1-R	TTA-GAA-TCA-ACC-GTT-TTA-TTG-TCG-TTG			
SEC1-Probe	VIC-ACC-AGG-CGA-TAA-GTT-MGB			
SEC3-F	GGC-AAT-ACT-TTT-TGG-TAT-GAT-ATG-ATG	101	*sec3*	Letertre et al. (2003b)
SEC3-R	TTA-GAA-TCA-ACC-GTT-TTA-TTG-TCG-TTG			
SEC3-Probe	VIC-ACC-AGG-CGA-TAA-GTT-MGB			
SED-F	TTG-TAT-AAT-AAT-GAT-ACT-CTC-GGA-GGA-AAA	73	*sed*	Letertre et al. (2003b)
SED-R	TAG-ACC-CAT-CAG-AAG-AAT-CAA-ACT-CTA-T			
SED-Probe	TET-ACA-GCG-CGG-AAA-A-MGB			
SEE-F	AAA-GAT-TTG-CGA-AAA-AAG-TCT-GAA-TT	106	*see*	Letertre et al. (2003b)
SEE-R	CAT-CAC-TCT-CTT-TGT-TTT-CAG-TTA-TAG-C			
SEE-Probe	VIC-TGC-TTT-AAG-CAA-TCT-TA-MGB			
SEG-F	GTT-ACR-ATT-CAG-GAA-CTA-GAT-TAC-AAA-GCA-A	74	*seg*	Letertre et al. (2003b)
SEG-R	CCA-TCA-AAC-TCG-TAT-AGC-TTT-TTT-TC			
SEG-Probe	FAM-ACA-CTG-GCT-CAC-TAA-A-MGB			

Primer	Sequence	Size	Gene	Reference
SEH-F	TGA-ATG-TCT-ATA-TGG-AGG-TAC-AAC-ACT-AAA-TA	78	*seh*	Letertre et al. (2003b)
SEH-R	ACC-CAA-ACA-TTA-GCA-CCA-ATC-AC			
SEH-Probe	TET-AAT-TGG-CAC-AGG-AAA-G-MGB			
SEI-F	TCA-CAG-ATA-AAA-ACC-TAC-CTA-TTG-CAA	69	*sei*	Letertre et al. (2003b)
SEI-R	CTG-AGA-TCA-AAT-CAT-TGG-TAC-CTG-TT			
SIB-Probe	VIC-TCA-ACT-CGA-ATT-TTC-MGB			
SEJ-F	GGA-TTG-ATA-GCA-TCA-GAA-CTG-TTG-TT	140	*sej*	Letertre et al. (2003b)
SEJ-R	ATC-AAA-GGT-ACT-AGG-GTT-GTA-TAA-ATT-ATA-TTG-T			
SEJ-Probe	TET-CTT-CAG-GCA-AGA-TAT-TA-MGB			
SEG1	GCT-ATC-GAC-ACA-CTA-CAA-CC	583	*seg*	Chen, Chiou, and Tsen (2004)
SEG2	CCA-AGT-GAT-TGT-CTA-TTG-TCG			
SEH1	CAC-ATC-ATA-TGC-GAA-AGC	548	*seh*	Chen, Chiou, and Tsen (2004)
SEH2	CGA-ATG-AGT-AAT-CTC-TAG-G			
SEI1	GAT-ACT-GGA-ACA-GGA-CAA-GC	789	*sei*	Chen, Chiou, and Tsen (2004)
SEI2	CTT-ACA-GGC-AGT-CCA-TCT-CC			
OLP6	GAG-GGA-AGA-G	—	—	Reinoso et al. (2004)
OLP11	ACG-ATG-AGC-C	—	—	
OLP13	ACC-GCC-TGC-T	—	—	

Continued

Table 11.2   PCR Primers and DNA Probes (Continued)

Primer or Probe	Sequence (5′ → 3′)	Size of Amplified Sequence (bp)	Gene or DNA Target Sequence	References
R465	FAM-TGC-ACT-ATA-TAC-TGT-TGG-ATC-TTC-AGA-A	97	*nuc*	Elizaquivel and Aznar (2008) from Alarcon, Vicedo, and Aznar (2006)
F5	CGC-TAC-TAG-TTG-CTT-AGT-GTT-AAC-TTT-AGT-TG			
NucP402	FAM-TGC-ATC-ACA-AAC-AGA-TAA-CGG-CGT-AAA-TAG-AAG-NFQ			
femA-F	ACT-GTG-ACG-ATG-AAT-GCG-ACA-A	132	*femA*	Sabet et al. (2007)
femA-R	ATG-TTG-TGG-TGT-TCT-TAT-ACC-AAA-TCC			
femA-P	Texas red-CGA-CAA-CTG-GCA-CAT-TGG-CTA-TCG-CTT-T-BHQ-2			
mecA-F	AAA-ACT-AGG-TGT-TGG-TGA-AGA-TAT-ACC	143	*mecA*	Sabet et al. (2007)
mecA-R	GAA-AGG-ATC-TGT-ACT-GGG-TTA-ATC-AG			
mecA-P	FAM-TTC-ACC-TTG-TCC-GTA-ACC-TGA-ATC-AGC-T-BHQ-1			
ermA-F	TCC-TTA-CTT-AAT-GAC-CGA-TGT-ACT-CT	147	*ermA*	Sabet et al. (2007)
ermA-R	TCT-TCG-CTT-TCG-CCA-CTT-TGA			
ermA-P	HEX-CAT-GAA-GCC-GAT-AAT-TTC-ACG-GTC-GCC-A-BHQ-1			
clfA-F	GCA-TTT-AAT-AAC-GGA-TCA-GG	—	*clfA*	Gilbert et al. (2006)
clfA-R	TGA-ATT-AGG-CGG-AAC-TAC-AT			

Primer	Sequence		Gene	Reference
clfB-F	GAT-GGT-GAT-TCA-GCA-GTA-AAT-CC	—	*clfB*	Gilbert et al. (2006)
clfB-R	GTT-GTT-TTG-ATA-CCT-TAT-TAG-AAT-G			
fnb-F	ATA-CAC-TTC-CAA-AAG-TAA-GCG-G	—	*fnb*	Gilbert et al. (2006)
fnb-R	ATT-TCA-ATA-ACA-GGT-GTT-ACT-ACT			
SAV1078-F	GTG-CAT-AAT-GGC-TTA-CGA-AT	—	*SAV1078*	Gilbert et al. (2006)
SAV1078-R	TGG-GAG-GAA-TTA-ATC-ATG-TC			
arcC-F	TTG-ATT-CAC-CAG-CGC-GTA-TTG-TC	—	*arcC*	Enright et al. (2000)
arcC-R	AGG-TAT-CTG-CTT-CAA-TCA-GCG			
aroE-F	ATC-GGA-AAT-CCT-ATT-TCA-CAT-TC	—	*aroE*	Enright et al. (2000)
aroE-R	GGT-GTT-GTA-TTA-ATA-ACG-ATA-TC			
glpF-F	CTA-GGA-ACT-GCA-ATC-TTA-ATC-C	—	*glpF*	Enright et al. (2000)
glpF-R	TGG-TAA-AAT-CGC-ATG-TCC-AAT-TC			
gmK-F	ATC-GTT-TTA-TCG-GGA-CCA-TC	—	*gmK*	Enright et al. (2000)
gmK-R	TCA-TTA-ACT-ACA-ACG-TAA-TCG-TA			
pta-F	GTT-AAA-ATC-GTA-TTA-CCT-GAA-GG	—	*pta*	Enright et al. (2000)
pta-F	GAC-CCT-TTT-GTT-GAA-AAG-CTT-AA			
tpi-F	TCG-TTC-ATT-CTG-AAC-GTC-GTG-AA	—	*tpi*	Enright et al. (2000)
tpi-R	TTT-GCA-CCT-TCT-AAC-AAT-TGT-AC			

Continued

Table 11.2   PCR Primers and DNA Probes (Continued)

Primer or Probe	Sequence (5′ → 3′)	Size of Amplified Sequence (bp)	Gene or DNA Target Sequence	References
yqiL-F	CAG-CAT-ACA-GGA-CAC-CTA-TTG-GC	—	yqiL	Enright et al. (2000)
yqiL-R	CGT-TGA-GGA-ATC-GAT-ACT-GGA-AC			
GSEAR-1	GGT-TAT-CAA-TGT-GCG-GGT-GG	102	sea	Mehrotra, Wang, and Johnson (2000)
GSEAR-2	CGG-CAC-TTT-TTT-CTC-TTC-GG			
GSEBR-1	GTA-TGG-TGG-TGT-AAC-TGA-GC	164	seb	Mehrotra, Wang, and Johnson (2000)
GSEBR-2	CCA-AAT-AGT-GAC-GAG-TTA-GG			
GSECR-1	AGA-TGA-AGT-AGT-TGA-TGT-GTA-TGG	451	sec	Mehrotra, Wang, and Johnson (2000)
GSCER-2	CAC-ACT-TTT-AGA-ATC-AAC-CG			
GSEDR-1	CCA-ATA-ATA-GGA-GAA-AAT-AAA-AG	278	sed	Mehrotra, Wang, and Johnson (2000)
GSDER-2	ATT-GGT-ATT-TTT-TTT-CGT-TC			
GSEER-1	AGG-TTT-TTT-CAC-AGG-TCA-TCC	209	see	Mehrotra, Wang, and Johnson (2000)
GSEER-2	CTT-TTT-TTT-CTT-CGG-TCA-ATC			
GFAMAR-1	AAA-AAA-GCA-CAT-AAC-AAG-CG	132	femA	Mehrotra, Wang, and Johnson (2000)
GFAMAR-2	GAT-AAA-GAA-GAA-ACC-AGC-AG			
GMECAR-1	ACT-GCT-ATC-CAC-CCT-CAA-AC	163	mecA	Mehrotra, Wang, and Johnson (2000)
GMECAR-2	CTG-GTG-AAG-TTG-TAA-TCT-GG			

GETAR-1 GETAR-2	GCA-GGT-GTT-GAT-TTA-GCA-TT AGA-TGT-CCC-TAT-TTT-TGC-TG	93	*eta*	Mehrotra, Wang, and Johnson (2000)
GETBR-1 GRTBR-2	ACA-AGC-AAA-AGA-ATA-CAG-CG GTT-TTT-GGC-TGC-TTC-TCT-TG	226	*etb*	Mehrotra, Wang, and Johnson (2000)
GTSSTR-1 GTSSTR-2	ACC-CCT-GTT-CCC-TTA-TCA-TC TTT-TCA-GTA-TTT-GTA-ACG-CC	326	*tst*	Mehrotra, Wang, and Johnson (2000)
sea-F sea-R	GCA-GGG-AAC-AGC-TTT-AGG-C GTT-CTG-TAG-AAG-TAT-GAA-ACA-CG	520	*sea*	Monday and Bohach (1999)
seb-sec-F seb-sec-R	ATG-TAA-TTT-TGA-TAT-TCG-CAG-TG TGC-AGG-CAT-CAT-ATC-ATA-CCA	643	*seb-sec*	Monday and Bohach (1999)
sec-F sec-R	CTT-GTA-TGT-ATG-GAG-GAA-TAA-CAA TGC-AGG-CAT-CAT-ATC-ATA-CCA	283	*sec*	Monday and Bohach (1999)
sed-F sed-R	GTG-GTG-AAA-TAG-ATA-GGA-GTG-C ATA-TGA-AGG-TGC-TCT-GTG-G	384	*sed*	Monday and Bohach (1999)
see-F see-R	TAC-CAA-TTA-ACT-TGT-GGA-TAG-AC CTC-TTT-GCA-CCT-TAC-GGC	170	*see*	Monday and Bohach (1999)
seg-F seg-R	TAC-CAA-TTA-ACT-TGT-GGA-TAG-AC CCA-AGT-GAT-TGT-CTA-TTG-TCG	327	*seg*	Monday and Bohach (1999)

Continued

Table 11.2   PCR Primers and DNA Probes (Continued)

Primer or Probe	Sequence (5′ → 3′)	Size of Amplified Sequence (bp)	Gene or DNA Target Sequence	References
seh-F	CAA-CTG-CTG-ATT-TAG-CTC-AG	360	seh	Monday and Bohach (1999)
seh-R	GTC-GAA-TGA-GTA-ATC-TCT-AGG			
sei-F	CAA-CTC-GAA-TTT-TCA-ACA-GGT-AC	465	sei	Monday and Bohach (1999)
sei-R	CAG-GCA-GTC-CAT-CTC-CTG			
sej-F	CAT-CAG-AAC-TGT-TGT-TCC-GCT-AG	142	sej	Monday and Bohach (1999)
sej-R	CTG-AAT-TTT-ACC-ATC-AAA-GGT-AC			
tst-F	GCT-TGC-GAC-AAC-TGC-TAC-AG	559	tst	Monday and Bohach (1999)
tst-R	TGG-ATC-CGT-CAT-TCA-TTG-TTA-A			
16S rRNA-F	GTA-GGT-GGC-AAG-CGT-TAT-CC	228	16S rRNA	Monday and Bohach (1999)
16S rRNA-R	CGC-ACA-TCA-GCG-TCA-G			
W-F	GGC-GTA-AAT-AGA-AGT-GGT-TCT-GAA-GAT-CCA	450	nuc	Chesneau, Allignet, and El Solh (2003)
Y-R	GAC-TAT-TAT-TGG-TTG-ATC-CAC-CTG			
SEC-5	GAA-CTA-GAC-ATA-AA-GCT-AGG	244	sec-1	Pinto, Chenoll, and Aznar (2005)
SEC-6	CAT-TCT-TTG-TTG-TAA-GGT-GG			
RAPD M13	GAA-ACA-GCT-ATG-ACC-ATG	—	—	
RAPD T7	AAT-ACG-ACT-CAC-TAT-AGG	—	—	Pinto, Chenoll, and Aznar (2005)
RAPD T3	ATT-AAC-CCT-CAC-TAA-AGG	—	—	

COAG2 COAG3	CGA-GAC-CAA-GAT-TCA-ACA-AG AAA-GAA-AAC-CAC-TCA-CAT-CA	180–1100	*coa*	Raimundo et al. (1999)
23S-F1200 23S-R1698	AGC-TGT-CGA-TTG-TCC-TTT-GG TCG-CTC-GCT-CAC-CTT-AGA-AT	499	*23S rRNA*	Cremonesi et al. (2005)
NUC-F166 NUC-R565	AGT-TCA-GCA-AAT-GCA-TCA-CA TAG-CCA-AGC-CTT-GAC-GAA-CT	400	*nuc*	Cremonesi et al. (2005)
COA-F2591 COA-R2794	CCG-CTT-CAA-CTT-CAG-CCT-AC TTA-GGT-GCT-ACA-GGG-GCA-AT	204	*coa*	Cremonesi et al. (2005)
SEA-F1170 SEA-R2794	TAA-GGA-GGT-GGT-GCC-TAT-GG CAT-CGA-AAC-CAG-CCA-AAG-TT	180	*sea*	Cremonesi et al. (2005)
SEC-F97 SEC-R467	ACC-AGA-CCC-TAT-GCC-AGA-TG TCC-CAT-TAT-CAA-AGT-GGT-TTC-C	371	*sec*	Cremonesi et al. (2005)
SED-F578 SED-R916	TCA-ATT-CAA-AAG-AAA-TGG-CTC-A TTT-TTC-CGC-GCT-GTA-TTT-TT	339	*sed*	Cremonesi et al. (2005)
SEG-F322 SEG-R753	CCA-CCT-GTT-GAA-GGA-AGA-GG TGC-AGA-ACC-ATC-AAA-CTC-GT	432	*seg*	Cremonesi et al. (2005)
SEH-F260 AEH-R722	TCA-CAT-CAT-ATG-CGA-AAG-CAG TCG-GAC-AAT-ATT-TTT-CTG-ATC-TTT	463	*seh*	Cremonesi et al. (2005)

Continued

Table 11.2   PCR Primers and DNA Probes (Continued)

Primer or Probe	Sequence (5′ → 3′)	Size of Amplified Sequence (bp)	Gene or DNA Target Sequence	References
SEI-F71	CTC-AAG-GTG-ATA-TTG-GTG-TAG-G	529	sei	Cremonesi et al. (2005)
SEI-R637	CAG-GCA-GTC-CAT-CTC-CTG-TA			
SEJ-F349	GGT-TTT-CAA-TGT-TCT-GGT-GGT	306	sej	Cremonesi et al. (2005)
SEJ-R654	AAC-CAA-CGG-TTC-TTT-TGA-GG			
SEL-F158	CAC-CAG-AAT-CAC-ACC-GCT-TA	240	sel	Cremonesi et al. (2005)
SEL-R397	CTG-TTT-GAT-GCT-TGC-CAT-TG			
ESG 1	ACG-CTC-CAC-CTG-TTG-AAG-G	400	seg	El-Huneidi, Bdour, and Mahasneh (2006) from Rosec and Gigaud (2002)
ESG 2	TGA-GCC-AGT-GTC-TTG-CTT-TG			
ESH 1	TCA-CAT-CAT-ATG-CGA-AAG-CAG	357	seh	El-Huneidi, Bdour, and Mahasneh (2006) from Rosec and Gigaud (2002)
ESH 2	TAG-CAC-CAA-TCA-CCC-TTT-CC			
ESI-1	TGG-AAC-AGG-ACA-AGC-TGA-AA	467	sei	El-Huneidi, Bdour, and Mahasneh (2006) from Rosec and Gigaud (2002)
ESI-2	TAA-AGT-GGC-CCC-TCC-ATA-CA			
ESJ 1	CAG-CGA-TAG-CAA-AAA-TGA-AAC-A	426	sej	El-Huneidi, Bdour, and Mahasneh (2006) from Rosec and Gigaud (2002)
ESJ 2	TCT-AGC-GGA-ACA-ACA-GTT-CTG-A			

Primer	Sequence	Size	Gene	Reference
Hlb-1-F Hlb-2-R	GTT-GCA-ACA-CTT-GCA-TTA-GC ACG-TAG-TAA-TAT-GGG-AAC-GCA	908	*hlb*	Kumagai et al. (2007)
Hlb-1-F Hlb-3-F	GTT-GCA-ACA-CTT-GCA-TTA-GC GTT-GCT-CGC-TTC-ATA-CCG-TA	915	*hlb*	Kumagai et al. (2007)
Int-1-F Hlb-2-R	TGA-AAT-CAG-CCT-GTA-GAG-TC ACG-TAG-TAA-TAT-GGG-AAC-GCA	1056	*int*	Kumagai et al. (2007)
Sak-1-F Sak-2-R	AAA-AGG-CGA-TGA-CGC-GCG-TT GCT-CTG-ATA-AAT-CTG-GGA-CA	326	*sak*	Kumagai et al. (2007)
Sak-1-F Sep-1-F	AAA-AGG-CGA-TGA-CGC-GCG-TT CTT-GGT-TCA-AAA-GAC-ACC-GC	201	*sep*	Kumagai et al. (2007)
lyt-1-F lyt-1-R	ACA-GTG-AAG-CAA-GAG-CAA-GG AGT-AGC-ACC-TAA-GCC-GTC-GA	566	*lyt*	Kumagai et al. (2007)
hol-1-F hol-2-R	GGG-TAG-CGA-TAT-TGT-CAG-CT AAA-GCT-TGG-TGG-CTA-TCA-CC	·84	*hol*	Kumagai et al. (2007) from van Wamel et al. (2006)
Scn-1 Scn-2	AGC-ACA-AGC-TTG-CCA-ACA-TCG TTA-ATA-TTT-ACT-TTT-TAG-TGC	258	*scn*	Kumagai et al. (2007) from van Wamel et al. (2006)

Continued

Table 11.2   PCR Primers and DNA Probes (Continued)

Primer or Probe	Sequence (5′ → 3′)	Size of Amplified Sequence (bp)	Gene or DNA Target Sequence	References
Chp-1	TTT-ACT-TTT-GAA-CCG-TTT-CCT-AC	—	chp	Kumagai et al. (2007) from van Wamel et al. (2006)
Chp-2	CGT-CCT-GAA-TTC-TTA-GTA-TGC-ATA-TTC-ATT-AG			
RAPD rep	TCG-CTC-AAA-ACA-ACG-ACA-C	—	—	Reinoso et al. (2008) from van Belkum et al. (1995)
spa	CAA-GCA-CCA-AAA-GAG-GAA	100–315	Spa (x-region)	Reinoso et al. (2008) from Frénay et al. (1996)
spa	CAC-CAG-GTT-TAA-CGA-CAT			
spa	CAC-CTG-CTG-CAA-ATG-CTG-CG	700 and ~900	spa (IgG–binding region)	Reinoso et al. (2008) from Seki et al. (1998)
spa	GGC-TTG-TTG-TTG-TCT-TCC-TC			
coa	ATA-GAG-ATG-CTG-GTA-CAG-G	400–1000	coa	Reinoso et al. (2008) from Hookey, Richardson, and Cookson (1998)
coa	GCT-TCC-GAT-TGT-TCG-ATG-C			
clfA	GGC-TTC-AGT-GCT-TGT-AGG	900–1000	clfA	Reinoso et al. (2008) from Stephan et al. (2001)
clfA	TTT-TCA-GGG-TCA-ATA-TAA-GC			

cna cna	ATA-TGA-ATT-CGA-GTA-TAA-GGA-AGG-GGT-T TTT-GGA-TCC-CTT-TTT-CAG-TAT-TAG-TAA-CCA	1700	*A domain*	Reinoso et al. (2008) from Switalski et al. (1993)
cna cna	AGT-GGT-TAC-TAA-TAC-TG CAG-GAT-AGT-TGG-TTT-A	1200	*B domain*	Reinoso et al. (2008) from Switalski et al. (1993)
cap 5 cap 5	ATG-ACG-ATG-AGG-ATA-GCG CTC-GGA-TAA-CAC-CTG-TTG-C	880	*cap 5*	Reinoso et al. (2008) from Moore and Lindsay (2001)
cap 8 cap 8	ATG-ACG-ATG-AGG-ATA-GCG CAC-CTA-ACA-TAA-GGC-AAG	1150	*cap 8*	Reinoso et al. (2008) from Moore and Lindsay (2001)
agr I agr I	CAC-TTA-TCA-TCA-AAG-AGC-C CCA-CTA-ATT-ATA-GCT-GG	350	*agr I*	Reinoso et al. (2008) from Moore and Lindsay (2001)
agr II agr II	GTA-GAG-CCG-TAT-TGA-TTC-C GTA-TTT-CAT-CTC-TTT-AAG-G	460	*agr II*	Reinoso et al. (2008) from Moore and Lindsay (2001)
agr III agr III	CTG-CAT-TTA-TTA-GTG-GAA-TAC-G GTT-TCA-TTT-CTT-TAA-GAG	550	*agr III*	Reinoso et al. (2008) from Moore and Lindsay (2001)

Continued

Table 11.2 PCR Primers and DNA Probes (Continued)

Primer or Probe	Sequence (5′ → 3′)	Size of Amplified Sequence (bp)	Gene or DNA Target Sequence	References
sae	TGT-GGG-GTT-CAG-GAA-TTG-TT	—	*sae*	Reinoso et al. (2008) from Giraudo et al. (1999)
sae	ATT-GAT-GAG-AAG-GAT-GCC-CA			
RAPD EPO17	TAC-ACC-CGT-CAA-CAT-TGA-GG	—	—	Tambic et al. (1997)
RAPD EP015	ACA-ACT-GCT-C	—	—	
RAPD EPOO7	AGC-ACG-CTG-TCA-ATC-ATG-TA	—	—	
RAPD KAY1	AGC-AGC-CTG-C	—	—	Tambic et al. (1997) from van Belkum et al. (1993)
UNIV-F	TAC-ATG-TAT-GGA-GGT-GCC-AC	—	—	Kwon et al. (2004)
sea-R	TTG-AAC-ACT-GTC-CTT-GAG-C	304	*sea*	
seb-R	ATA-GTG-ACG-AGT-AAG-GTA	165	*seb*	Kwon et al. (2004) from Sharma, Rees, and Dodd (2000)
sec-R	TTG-TAA-GGT-GGA-CTT-CTA-TC	379	*sec*	
sed-R	GGC-TTT-AGT-GTC-TAA-TGT	422	*sed*	
see-R	GCC-AAA-CCT-GTC-TGA-G	213	*see*	Kwon et al. (2004) from Sharma, Rees, and Dodd (2000)
seg-R	AAG-TCA-CAT-TGT-CTA-TTG-TCG	93	*seg*	
seh-R	GAA-TGA-GTA-ATG-TCT-AGG-AG	255	*seh*	
sei-R	GCA-GTC-CAT-GTC-CTG-TAT-AA	334	*sei*	
sej-R	AGT-TCA-CCT-GAC-TTC-AAC-G	531	*sej*	
RAPD AP1	GGT-TGG-GTG-AGA-ATT-GCA-CG	—	—	Kwon et al. (2004) from van Belkum et al. (1995)

Primer	Sequence	Size	Gene	Reference
RAPD AP7	GTG-GAT-GCG-A	—	—	Kwon et al. (2004) from van Leeuwen et al. (1998)
RAPD ERIC2	AAG-TAA-GTG-ACT-GGG-GTG-ACG-G	—	—	Kwon et al. (2004) from Kondoh et al. (2002)
16S rDNA-F 16S rDNA-R	AGA-GTT-TGA-TCC-TGG-CTC-AG CCC-ACT-GCT-GCC-TCC-CGT-AG	350	*16S rDNA*	Riyaz-Ul-Hassan, Verma, and Qazi (2008)
cat-F cat-R	TTC-GAA-GCC-ATT-GAA-AAA-GG ACA-TCA-TCC-GTT-ACG-CCT-TC	641	*cat*	Riyaz-Ul-Hassan, Verma, and Qazi (2008)
fem-F fem-R	GCA-AAC-TGT-TGG-CCA-CTA-TG TCA-TCA-CGA-TCA-GCA-AAA-GC	594	*femA*	Riyaz-Ul-Hassan, Verma, and Qazi (2008)
fmhA1-F fmhA1-R	CAA-TTC-AGA-GAG-GGA-TGG-GA ACT-CAC-GGA-ATT-TGA-AAC-CG	685	*fmhA*	Riyaz-Ul-Hassan, Verma, and Qazi (2008)
RAPD ERIC-1R	ATG-TAA-GCT-CCT-GGG-GAT-TCA-C	—	—	van Leeuwen et al. (1996) from van Belkum et al. (1994)
RAPD AP-1026	TAC-ATT-CGA-GGA-CCC-CTA-AGT-G	—	—	van Leeuwen et al. (1996) from Versalovic, Koeuth, and Lupski (1991)
LTmecAF LTmecAR LTmecAHp2	AAA-GAA-CCT-CTG-CTC-AAC-AAG-T TGT-TAT-TTA-ACC-CAA-TCA-TTG-CTG-TT HEX-CCA-GAT-TAC-AAC-TTC-ACC-AGG-TTC-AAC-T-BHQ1	88	*mec*	Thomas et al. (2007)

Continued

Table 11.2 PCR Primers and DNA Probes (Continued)

Primer or Probe	Sequence (5′ → 3′)	Size of Amplified Sequence (bp)	Gene or DNA Target Sequence	References
LTnucF	AAA-TTA-CAT-AAA-GAA-CCT-GCG-ACA	87	nuc	Thomas et al. (2007)
LTnucR	GAA-TGT-CAT-TGG-TGG-ACC-TTT-GTA			
LTnucHP1	FAM-AAT-TTA-ACC-GTA-TCA-CCA-TCA-ATC-GCT-TT-BHQ1			
C1	AAC-ATT-AGT-GAT-AAA-AAA-CTG-AAA	234	sec	Chen et al. (2001)
C2	TTG-TAA-GTT-CCC-ATT-ATC-AAA-GTG			
ENTC1	ACA-GAG-TTA-TTA-AAT-GAA-GG	402	sec1	Chen et al. (2001)
ENTCR	ATC-ATA-CCA-AAA-AGT-ATT-GC			
ENTC2	GTA-TCA-GCA-ACT-AAA-GTT-AT	501	sec2	Chen et al. (2001)
ENTCR	ATC-ATA-CCA-AAA-AGT-ATT-GC			
ENTC3	AC-AGA-TTA-TTT-ATT-TCA-CGT	672	sec3	Chen et al. (2001)
ENTCR	ATC-ATA-CCA-AAA-AGT-ATT-GC			
FA1	AAG-GGC-GAA-ATA-GAA-GTG-CCG-GGC	1153	aroA	Marcos et al. (1999)
RA2	CAC-AAG-CAA-CTG-CAA-GCA-T			
SEA FIP	GAT-CCA-ACT-CCT-GAA-CAG-TTA-CAA-TAC-AGT-ACC-TTT-GGA-AAC-G	—	sea	Marcos et al. (1999)
SEA BIP	CTG-ATG-TTT-TTG-ATG-GGA-AGG-TTC-CCG-AAG-GTT-CTG-TAG-AAG-T	—	sea	

SEA F3	TCA-ATT-TAT-GGC-TAG-ACG-GT	—	*sea*	Marcos et al. (1999)
SEA B3	CTT-GAG-CAC-CAA-ATA-AAT-CG	—	*sea*	
SEA LB	AGA-GGG-GAT-TAA-TCG-TGT-TTC-A	—	*sea*	
SEB FIP	CAC-CAA-ATA-GTG-ACG-AGT-TAG-GTA-AGA-CGT-ACA-AAC-TAA-GAA-AAA-GG	—	*seb*	Marcos et al. (1999)
SEB BIP	ACT-CTA-TGA-ATT-AAA-CAA-CTC-GCC-TTG-TCA-TAC-CAA-AAG-CTA-TTC-TCA	—	*seb*	
SEB F3	GTT-CGG-GTA-TTT-GAA-GAT-GG	—	*seb*	
SEB B3	TTG-GTC-AAA-TTT-ATC-TCC-TGG	—	*seb*	
SEB LF	TCT-AAT-TCT-TGA-GCA-GTC-A	—	*seb*	
SEB LB	ATG-AAA-CGG-GAT-ATA-TTA-AAT-TTA-T	—	*seb*	
SEC FIP	TCC-TTC-ATG-TTT-TGT-TAT-TCC-TCC-AAG-ATA-ATG-TAG-GTA-AAG-TTA-CAG-GT	—	*sec*	Marcos et al. (1999)
SEC BIP	ACC-ACT-TTG-ATA-ATG-GGA-ACT-TAC-ATT-TGC-ACT-TCA-AAA-GAA-ATT-GTG	—	*sec*	
SEC F3	TGT-AAA-CTG-CTA-TTT-TTC-ATC-CA	—	*sec*	
SEC B3	CTT-TTA-TGT-CTA-GTT-CTT-GAG-CT	—	*sec*	
SEC LB	TGT-ACT-TAT-AAG-AGT-TTA-TGA-AAA-T	—	*sec*	
SED FIP	CGC-TGT-ATT-TTT-CCT-CCG-AGA-GTG-CGC-TAT-TTG-CAA-AAG-GAT	—	*sed*	Marcos et al. (1999)
SED BIP	AGA-GTT-TGA-TTC-TTC-TGA-TGG-GTC-TTA-TTC-GTA-ATT-GTT-TTT-CGG-GAA	—	*sed*	
SED F3	ACA-AGA-ATT-AGA-TGC-ACA-AGC	—	*sed*	
SED B3	TGA-AGG-TGC-TCT-GTG-GAT	—	*sed*	
SED LB	TGA-TTT-ATT-TGA-TGT-TAA-GGG-TG	—	*sed*	

Continued

Table 11.2   PCR Primers and DNA Probes (Continued)

Primer or Probe	Sequence (5′ → 3′)	Size of Amplified Sequence (bp)	Gene or DNA Target Sequence	References
1-F	GTT-GTA-GTT-GTC-GGG-TTT-GG	1800	mecA	Ünal et al. (1992)
2-R	CCA-CCC-AAT-TTG-TCT-GCC-AGT-TTC			
3-F	GAC-CGA-AAC-AAT-GTG-GAA-TTG-GCC	1108	mecA	Ünal et al. (1992)
4-R	CAC-CTT-GTC-CGT-AAC-CTG-AAT-CAG-C			
5-F	CGA-GGT-CAT-TGC-AGC-TTG-CTT-AC	900	femA	Ünal et al. (1992)
6-R	CTA-GAC-CAG-CAT-CTT-CAG-C			
mecA1	AAA-ATC-GAT-GGT-AAA-GGT-TGG-C	533	mecA	Lee (2003)
mecA2	AGT-TCT-GCA-GTA-CCG-GAT-TTG-C			
RAPD M13	GAG-GGT-GGC-GGT-TCT	—	—	Lee (2003)
RAPD H12	ACG-CGC-ATG-T	—	—	
ETA-F	CTA-TTT-ACT-GTA-GGA-GCT-AG	741	eta	Hayakawa et al. (2001)
ETA-R	ATT-TAT-TTG-ATG-CTC-TGT-AT			
nuc-probe	FAM-CCA-CAT-CTC-TTT-ATA-CCA-GGA-CTT-CGT-TCA-CGT-TAMRA	—	—	Hein et al. (2001) from Brakstad, Aasbakk, and Maeland (1992)
SEN1	CTT-CTT-GTT-GGA-CAC-CAT-CTT	135	sen	Chiang et al. (2008)
SEN2	GAA-ATA-AAT-GTG-TAG-GCT-T			

Primer	Sequence	Size	Gene	Reference
SEO1	AAA-TTC-AGC-AGA-TAT-TCC-AT	172	seo	Chiang et al. (2008)
SEO2	TTT-GTG-TAA-GAA-GTC-AAG-TGT-AG			
SEP1	ATC-ATA-ACC-AAC-CGA-ATC-AC	148	sep	Chiang et al. (2008)
SEP2	AGA-AGT-AAC-TGT-TCA-GGA-GCT-A			
SEQ1	TCA-GGT-CTT-TGT-AAT-ACA-AAA	359	seq	Chiang et al. (2008)
SEQ2	TCT-GCT-TGA-CCA-GTT-CCG-GT			
SER1	AGA-TGT-GTT-TGG-AAT-ACC-CTA-T	123	ser	Chiang et al. (2008)
SER2	CTA-TCA-GCT-GTG-GAG-TGC-AT			
SEU1	ATT-TGC-TTT-TAT-CTT-CAT	167	seu	Chiang et al. (2008)
SEU2	GGA-CTT-TAA-TGT-TTG-TTT-CTG-GAT			
SEJ-1	CAT-CAG-AAC-TGT-TGT-TCC-GCT-AG	142	sej	Rall et al. (2007) from Nashev et al. (2004)
SEJ-2	CTG-AAT-TTT-TAC-CAT-CAA-AGG-TAC			
H279A	GAI-III-GCI-GGI-GA(TC)-GGI-ACI-AC	550	grpEL	Santos et al. (2008) from Goh et al. (1997)
H280A	(TC)(TG)I-(TC)(TG)I-TCI-CC(AG)-AAI-CCI-GGI-GC(TC)-TT			
Pri-1	GCG-ATT-GAT-GGT-GAT-ACG-GTT	279	nuc	Yang et al. (2007)
Pri-2	AGC-CAA-GCC-TTG-ACG-AAC-TAA-AGC			

Continued

Table 11.2 PCR Primers and DNA Probes (Continued)

Primer or Probe	Sequence (5′ → 3′)	Size of Amplified Sequence (bp)	Gene or DNA Target Sequence	References
Sau327	GGA-CGA-CAT-TAG-ACG-AAT-CA	~1300	23S rRNA	Karahan and Çetinkaya (2007) from Riffon et al. (2001)
Sau1645	CGG-GCA-CCT-ATT-TTC-TAT-CT			
SEA-F	ATG-GTT-ATC-AAT-GTG-GGG-GTG-III-IIC-CAA-ACA-AAA-C	344	sea	Hwang et al. (2007)
SEA-R	TGA-ATA-CTG-TCC-TTG-AGC-ACC-AII-III-ATC-GTA-ATT-AAC			
SEB-F	TGG-TAT-GAC-ATG-ATG-CCT-GCA-CIII-III-GAT-AAA-TTT-GAC	196	seb	Hwang et al. (2007)
SEB-R	AGG-TAC-TCT-ATA-AGT-GCC-TGC-CTI-III-IAC-TAA-CTC-TT			
SEC-F	GAT-GAA-GTA-GTT-GAT-GTG-TAT-GGA-TCI-III-IAC-TAT-GTA-AAC	399	sec	Hwang et al. (2007)
SEC-R	AGA-TTG-GTC-AAA-CTT-ATC-GCC-TGG-III-IIG-CAT-CAT-ATC			
SED-F	CTG-AAT-TAA-GTA-GTA-CCG-CGC-TII-III-ATA-TGA-AAC	451	sed	Hwang et al. (2007)
SED-R	TCC-TTT-TGC-AAA-TAG-CGC-CTT-GII-III-GCA-TCT-AAT-TC			
SEE-F	CGG-GGG-TGT-AAC-ATT-ACA-TGA-TII-III-CCG-ATT-GAC-C	286	see	Hwang et al. (2007)
SEE-R	CCC-TTG-AGC-ATC-AAA-CAA-ATC-ATA-AII-III-CGT-GGA-CCC-TTC			

SEG-F	ATA-GAC-TGA-ATA-AGT-TAG-AGG-AGG-TII-III-GAA-GAA-ATT-ATC	594	*seg*	Hwang et al. (2007)
SEG-R	TTA-GTG-AGC-CAG-TGT-CTT-GCI-III-IAA-TCT-AGT-TC			
SEH-F	CAT-TCA-CAT-CAT-ATG-CGA-AAG-CAG-III-IIT-TAC-ACG	218	*seh*	Hwang et al. (2007)
SEH-R	CTT-CTG-AGC-TAA-ATC-AGC-AGT-TGC-III-IIT-TAC-TCT-C			
SEI-F	AGG-CGT-CAC-AGA-TAA-AAA-CCT-ACC-III-IIC-AAA-TCA-ACT-C	154	*sei*	Hwang et al. (2007)
SEI-R	ACA-AGG-ACC-ATT-ATA-ATC-AAT-GCC-III-IIT-ATC-CAG-TTT-C			
SEJ-F	TGT-ATG-GTG-GAG-TAA-CAC-TGC-ATG-III-IIA-ATC-AAC-TTT-ATG	102	*sej*	Hwang et al. (2007)
SEJ-R	CTA-GCG-GAA-CAA-CAG-TTC-TGA-TGC-III-IIA-TCC-ATA-AAT			
SEK-F	GTG-TCT-CTA-ATA-ATG-CCA-GCG-CTI-III-ICG-ATA-TAG-G	282	*sek*	Hwang et al. (2007)
SEK-R	CGT-TAG-TAG-CTG-TGA-CTC-CAC-CII-III-TGT-ATT-TAG			
SEL-F	ATT-CAC-CAG-AAT-CAC-ACC-GCT-III-IIT-ACT-CGT-A	469	*sel*	Hwang et al. (2007)
SEL-R	GTG-TAA-AAT-AAA-TCA-TAC-GAG-III-IIA-GAA-CCA-TCA-TTC			
SEM-F	CGC-AAC-CGC-TGA-TGT-CGG-III-IIT-GAA-TCT-TAG-G	572	*sem*	Hwang et al. (2007)
SEM-R	CAG-CTT-GTC-CTG-TTC-CAG-TAT-CII-III-AGT-CAT-AAG			

Continued

Table 11.2   PCR Primers and DNA Probes (Continued)

Primer or Probe	Sequence (5′ → 3′)	Size of Amplified Sequence (bp)	Gene or DNA Target Sequence	References
SEN-F	TCA-GCT-TAT-ACG-GAG-GAG-TTA-CGI-III-ITG-ATG-GAA-ATC	103	*sen*	Hwang et al. (2007)
SEN-R	AAC-CTT-CTT-CTT-GGA-CAC-CAT-CII-III-ATA-CAT-TAA-CGC			
SEO-F	GTG-GAA-TTT-AGC-TCA-TCA-GCG-ATT-TCI-III-IAA-TTT-CTA-GG	116	*seo*	Hwang et al. (2007)
SEO-R	GTA-CAG-GCA-GTA-TCC-ACT-TGA-TGC-III-IIA-TGA-CAA-TGT-GC			
SEP-F	ATC-ATA-ACC-AAC-CGA-ATC-ACC-AGI-III-IGG-GTG-AAA-CTC	547	*sep*	Hwang et al. (2007)
SEP-R	GTC-TGA-ATT-GCA-GGG-AAC-TGC-III-IIG-CAA-TCT-TAG			
SEQ-F	GGT-GGA-ATT-ACG-TTG-GCG-AAT-CAI-III-ITA-GAT-AAA-CC	330	*seq*	Hwang et al. (2007)
SEQ-R	CTC-TGC-TTG-ACC-AGT-TCC-GGT-GII-III-CAA-ATC-GTA-TG			
SER-F	TTC-AGT-AAG-TGC-TAA-ACC-AGA-TCC-III-IIC-TGG-AGA-ATT-G	368	*ser*	Hwang et al. (2007)
SER-R	CTG-TGG-AGT-GCA-TTG-TAA-CGC-CII-III-ATA-TGC-AAA-CTC-C			
SEU-F	ATG-GCT-CTA-AAA-TTG-ATG-GTT-CTA-III-IIT-TAA-AAA-CAG	410	*seu*	Hwang et al. (2007)
SEU-R	GCC-AGA-CTC-ATA-AGG-CGA-ACT-AII-III-TTC-ATA-TAA-A			

Primer	Sequence	Size (bp)	Gene	Reference
TST-F	GTT-GCT-TGC-GAC-AAC-TGC-TAC-AGI-III-IAC-CCC-TGT-TC	209	*tst*	Hwang et al. (2007)
TST-R	TCA-AGC-TGA-TGC-TGC-CAT-CTG-TGI-III-ITA-TAC-GCA-TAG			
Fem-F	ACA-GCT-AAA-GAG-TTT-GGT-GCC-TII-III-GAT-AGC-ATG-C	723	*fem*	Hwang et al. (2007)
Fem-R	TTC-ATC-AAA-GTT-GAT-ATA-CGC-TAA-AGG-TII-III-CAC-ACG-GTC			
SEG1-F	TGC-TAT-CGA-CAC-ACT-ACA-ACC	704	*seg*	McLauchlin et al. (2000)
SEG2-R	CCA-GAT-TCA-AAT-GCA-GAA-CC			
SEH1-F	CGA-AAG-CAG-AAG-ATT-TAC-ACG	495	*seh*	McLauchlin et al. (2000)
SEH2-R	GAC-CTT-TAC-TTA-TTT-CGC-TGT-C			
SEI1-F	GAC-AAC-AAA-ACT-GTC-GAA-ACT-G	630	*sei*	McLauchlin et al. (2000)
SEI2-R	CCA-TAT-TCT-TTG-CCT-TTA-CCA-G			
Sea1-F	TTG-GAA-ACG-GTT-AAA-ACG-AA	~20	*sea*	Johnson et al. (1991)
Sea2-R	GAA-CCT-TCC-CAT-CAA-AAA-CA			
seb1-F	TCG-CAT-CAA-ACT-GAC-AAA-CG	478	*seb*	Johnson et al. (1991)
seb2-R	GCA-GGT-ACT-CTA-TAA-GTG-CC			
sec1-F	GAC-ATA-AAA-GCT-AGG-AAT-TT	257	*sec*	Johnson et al. (1991)
sec2-R	AAA-TCG-GAT-TAA-CAT-TAT-CC			

Continued

Table 11.2 PCR Primers and DNA Probes (Continued)

Primer or Probe	Sequence (5′ → 3′)	Size of Amplified Sequence (bp)	Gene or DNA Target Sequence	References
sed1-F sed2-R	CTA-GTT-TGG-TAA-TAT-CTC-CT TAA-TGC-TAT-ATC-TTA-TAG-GG	312	*sed*	Johnson et al. (1991)
see1-F see2-R	TAG-ATA-AAG-TTA-AAA-CAA-GC TAA-CTT-ACC-GTG-GAC-CCT-TC	170	*see*	Johnson et al. (1991)
TSST-1-F TSST-2-R	ATG-GCA-GCA-TCA-GCT-TGA-TA TTT-CCA-ATA-ACC-ACC-CGT-TT	350	*tst*	Johnson et al. (1991)
eta1-F eta2-R	CTA-GTG-CAT-TTG-TTA-TTC-AA TGC-ATT-GAC-ACC-ATA-GTA-CT	119	*eta*	Johnson et al. (1991)
etb1-F etb2-R	ACG-GCT-ATA-TAC-ATT-CAA-TT TCC-ATC-GAT-AAT-ATA-CCT-AA	200	*etb*	Johnson et al. (1991)
REPT-F REPT-R	AAG-CCT-TGC-AGG-ACA-TCT-TCA GCC-GCC-AGT-GTG-ATG-GAT-AT	67	*seb*	Rajkovic et al. (2006)

REFERENCES

Aarestrup, F., Larsen, H., Erikson, N., Elsberg, C., Jensen, N. 1999. Frequency of α- and β-hemolysin in *Staphylococcus aureus* of bovine and human origin. *APMIS*. 107:425-430.

Aires-de-Sousa, M., Parente, C., Vieira-da-Motta, O., Bonna, I., Silva, D., Lencastre, H. 2007. Characterization of *Staphylococcus aureus* isolates from buffalo, bovine, ovine, and caprine milk samples collected in Rio de Janeiro state, Brazil. *Appl. Environ. Microbiol.* 73:3845-3849.

Aitichou, M., Henkins, R., Sultana, A., Ulrich, R., Ibrahim, M. 2004. Detection of *Staphylococcus aureus* enterotoxin A and B genes with PCR-EIA and hand-held electrochemical sensor. *Mol. Cell. Probes* 18:373-377.

Akpaka, P., Kissoon, S., Rutherford, C., Swanston, W., Jayaratne, P. 2007. Molecular epidemiology of methicillin-resistant *Staphylococcus aureus* isolates from original hospitals in Trinidad and Tobago. *Int. J. Infect. Dis.* 11:544-548.

Alarcon, B., Vicedo, B., Aznar, R. 2006. PCR-based procedures for the detection and quantification of *Staphylococcus aureus* and their application in food. *J. Appl. Microbiol.* 100:352-364.

Annemüller, C., Lämmler, C., Zschöck, M. 1999. Genotyping of *Staphylococcus aureus* isolated from bovine mastitis. *Vet. Microbiol.* 69:217-224.

Atanassova, V., Meindl, A., Ring, C. 2001. Prevalence of *Staphylococcus aureus* and staphylococcal enterotoxins in raw pork and uncooked smoked ham—A comparison of classical culturing detection and RFLP-PCR. *Int. J. Food Microbiol.* 68:105-113.

Bailey, C., de Azavedo, J., Arbuthnott, J. 1980. A comparative study of two serotypes of epidermolytic toxin from *Staphylococccus aureus*. *Biochim. Biophs. Acta* 624:111-120.

Bania, J., Dabrowska, A., Bystron, J., Korzekwa, K., Chrzanowska, J., Molenda, J. 2006. Distribution of newly described enterotoxin-like genes in *Staphylococcus aureus* from food. *Int. J. Food Microbiol.* 108:36-41.

Bayles, K., Iandola, J. 1989. Genetic and molecular analyses of the gene encoding staphylococcal enterotoxin D. *J. Bacteriol.* 171:4799-4806.

Becker, K., Keller, B., von Eiff, C., Bruck, M., Lubritz, G., Etienne, J., Peters, G. 2001. Enterotoxigenic potential of *Staphylococcus intermedius*. *Appl. Environ. Microbiol.* 67:5551-5557.

Becker, K., Roth, R., Peters, G. 1998. Rapid and specific detection of toxigenic *Staphylococcus aureus*: Use of two multiplex PCR enzyme immunoassays for amplification and hybridization of staphylococcal enterotoxin genes, exfoliative toxin genes, and toxic shock syndrome toxin 1. *J. Clin. Microbiol.* 36:2548-2553.

Betley, M., Mekalanos, J. 1985. Staphylococcal enterotoxin A is encoded by phage. *Science* 229:185-187.

Blaiotta, G., Ercolini, D., Pennacchia, C., Fusco, V., Casaburi, A., Pepe, O., Villain, F. 2004. PCR detection of staphylococcal enterotoxin genes in *Staphylococcus* spp. strains isolated from meat and dairy products. Evidence for new variants of seG and seI in *S. aureus* AB-8802. *J. Appl. Microbiol.* 97:719-730.

Brakstad, O., Aasbakk, K., Maeland, J. 1992. Detection of *Staphylococcus aureus* by polymerase chain reaction amplification of the *nuc* gene. *J. Clin. Microbiol.* 30:1654-1660.

Cespedes, C., Miller, M., Quagliarello, B., Vavagiakis, P., Klein, S., Lowy, F. 2002. Differences between *Staphylococcus aureus* isolates from medical and nonmedical hospital personnel. *J. Clin. Microbiol.* 40:2594-2597.

Chen, T., Chiou, C., Tsen, H. 2004. Use of novel PCR primers specific to the genes of staphylococcal enterotoxin G, H, I for the survey of *Staphylococcus aureus* strains isolated from food-poisoning cases and food samples in Taiwan. *Int. J. Food Microbiol.* 92:189–197.

Chen, T., Hsiao, M., Ciou, C., Tsen, H. 2001. Development of PCR primers for the investigation of C1, C2, and C3 enterotoxin types of *Staphylococcus aureu*s strains isolated from food-borne outbreaks. *Int. J. Food Microbiol.* 71:63–70.

Chesneau, O., Allignet, J., El Solh, N. 1993. Thermonuclease gene as a target nucleotide sequence for specific recognition of *Staphylococcus aureus*. *Mol. Cell. Probes* 7:301–310.

Chiang, Y., Liao, W., Fan, C., Pai, W., Ciou, C., Tsen, H. 2008. PCR detection of staphylococcal enterotoxins (SEs) N, O, P, Q, R, U and survey of SE types in *Staphylococcus aureus* isolates from food-poisoning cases in Taiwan. *Int. J. Food Microbiol.* 121:66–73.

Chiou, C., Wei, H., Yang, L. 2000. Comparison of pulsed-field gel electrophoresis and coagulase gene restriction profile techniques in the molecular typing of *Staphylococcus aureus*. *J. Clin. Microbiol.* 38:2186–2190.

Coleman, D., Knights, J., Russell, R., Shanley, D., Birkbeck, T., Dugan, G., Charles, I. 1991. Insertional inactivation of the *Staphylococcus aureus* beta-toxin by bacteriohage phi 13 occurs by site- and orientation-specific integration of the phi 13 genome. *Mol. Microbiol.* 5:933–939.

Coleman, D., Sullivan, D., Russell, R., Arbuthnott, J., Carey, B., Pomeroy, H. 1989. *Staphylococcus aureus* bacteriophages mediating the simultaneous lysogenic conversion of beta-lysin, staphylokinase and enterotoxin A: Molecular mechanism of triple conversion. *J. Gen. Microbiol.* 15:1679–1697.

Cremonesi, P., Luzzana, M., Brasca, M., Morandi, S., Lodi, R., Vimercati, C., Agnellini, D., Caramenti, G., Moroni, P., Castilioni, B. 2005. Development of a multiplex PCR assay for the identification of *Staphylococcus aureus* enterotoxigenic strains isolated from milk and diary products. *Mol. Cell. Probes* 19:299–305.

Cuny, C., Lause, H., Witte, W. 1996. Discrimination of *S. aureus* strains by PCR for rRNA gene spacer size polymorphism and comparison to *SmaI* macrorestriction patterns. *Zbl. Bakt.* 283:466–476.

Da Silva, E., Carmo, L., Da Silva, N. 2005. Detection of enterotoxins A, B, and C genes in *Staphylococcus aureus* from goat and bovine mastitis in Brazilian dairy herds. *Vet. Microbiol.* 106:103–107.

Deurenberg, R., Nieuwenhuis, R., Driessen, C., London, N., Stassen, F., van Tiel, F., Stobberingh, E., Vink, C. 2005. The prevalence of the *Staphylococcus tst* gene among community- and hospital-acquired strains and isolates from Wegener's Granulomatosis patients. *FEMS Microbiol. Lett.* 245:185–189.

Devriese, L. 1984. A simplified system for biotyping *Staphylococcus aureus* strains isolated from different animal species. *J. Appl. Bacteriol.* 56:215–220.

El Helali, N., Carbonne, A., Naas, T., Kerneis, S., Fresco, O., Giovangrandi, Y., Fortineau, N., Normann P., Astagneau, P. 2005. Nosocomial outbreak of staphylococcal scalded skin syndrome in neonates: Epidemiological investigation and control. *J. Hosp. Infect.* 61:130–138.

El-Huneidi, W., Bdour, S., Mahasneh, A. 2006. Detection of enterotoxin genes *seg, seh, sei,* and *sej* and of a novel *aroA* genotype in Jordanian clinical isolates of *Staphylococcus aureus*. *Diag. Microbiol. Infect. Dis.* 56:127–132.

Elizaquivel, P., Aznar, R. 2008. A multiplex RTi-PCR reaction for simultaneous detection of *Escherichia coli* O157:H7, *Salmonella* spp. and *Staphylococcus aureus* on fresh minimally processed vegetables. *Food Microbiol.* 25:705–713.

Endo, Y., Yamada, T., Matsunaga, K., Haakawa, Y., Kaidoh, T., Takeuchi, S. 2003. Phage conversion of exfoliative toxin A in *Staphylococcus aureus* isolates from cows with mastitis. *Vet. Microbiol.* 96:81–90.

Enright, M., Day, N., Davies, C., Peacock, S., Spratt, B. 2000. Multilocus sequence typing for characterization of methicillin resistance and methicillin-susceptible clones of *Staphylococcus aureus*. *J. Clin. Microbiol.* 38:1008–1015.

Frénay, H., Bunschoten, A., Scouls, L., van Leeuwen, W., Vandenbrouke-Grauls, C., Verhoef, J., Mooi, F. 1996. Molecular typing of methicillin-resistant *Staphylococccus aureus* strains on the basis of protein A gene polymorphism. *Eur. J. Clin. Microbiol. Infect. Dis.* 15:60–64.

Fueyo, J., Martin, M., Gonzalez-Hevia, M., Mendoza, M. 2001. Enterotoxin production and DNA fingerprinting in staphylococcus isolates from human and food samples. Relations between genetic types and enterotoxins. *Int. J. Food Microbiol.* 67:139–145.

Fueyo, J., Mendoza, M., Alverez, M., Martin, M. 2005a. Relationships between toxin gene content and genetic background in nasal isolates of *Staphylococcus aureus* from Asturias, Spain. *FEMS Microbiol. Lett.* 243:447–454.

Fueyo, J., Mendoza, M., Martin M. 2005b. Enterotoxins and toxic shock syndrome toxin in *Staphylococcus* recovered from human nasal carriers and manually handled foods: Epidemiological and genetic findings. *Microb. Infect.* 7:187–194.

Fueyo, J., Mendoza, M., Rodio, M., Muniz, J., Alvaz, M., Martin, M., 2005c. Cytotoxin and pyrogenic toxin superantigen gene profiles of *Staphylococcus aureus* associated with subclinical mastitis in dairy cows and relationships with macrorestriction genomic profiles. *J. Clin. Microbiol.* 43:1278–1284.

Gilbert, F., Fromageau, A., Gelineau, L., Poutrel, B. 2006. Differentiation of bovine *Staphylococcus aureus* isolates by use of polymorphic tandem repeat typing. *Vet. Microbiol.* 117:297–303.

Gilligan, K., Shipley, M., Stiles, B., Hadfield, T., Ibrahim, S. 2000. Identification of *Staphylococcus aureus* enterotoxins A and B genes by PCR-ELISA. *Mol. Cell. Probes* 14:71–78.

Giraudo, A., Calzolari, A., Cataldi, A., Bogni, C., Nagel, R. 1999. The *sae* locus of *Staphylococcus aureus* encodes a two-component regulatory system. *FEMS Lett.* 177:15–22.

Goh, S., Byrne, S., Zhang, J., Chow, A. 1992. Molecular typing of *Staphylococcus aureus* on the basis of coagulase gene polymorphisms. *J. Clin. Microbiol.* 30:1642–1645.

Goh, S., Santucci, Z., Kloos, W., Faltyn, M., George, C., Driedger, D., Hemmingsen, S. 1997. Identification of *Staphylococcus* species and subspecies by the chaperonin 60 gene identification method and reverse checkerboard hybridization. *J. Clin. Microbiol.* 35:3116–3121.

Goto, M., Hayashidani, H., Takatori, K., Hara-Kudo, Y. 2007. Rapid detection of enterotoxigenic *Staphylococcus aureu*s harbouring genes for four classical enterotoxins, SEA, SEB, SEC and SED, by loop-mediated isothermal amplification assay. *Lett. Appl. Microbiol.* 45:100–107.

Grundmann, H., Hori, S., Enright, M., Webster, C., Tami, A., Feil, J., Pitt, T. 2002. Determining the genetic structure of the natural population of *Staphylococcus aureus*: a comparison of multilocus sequence typing with pulse-field gel electrophoresis, randomly amplified polymorphic DNA analysis, and phage typing. *J. Clin. Microbiol.* 40:4544–4546

Hayakawa, Y., Hasimoto, N., Imaizumi, K., Kaidoh, T., Takeuchi, S. 2001. Genetic analysis of exfoliative toxin A-producing *Staphylococcus aureus* isolated from mastitic cow's milk. *Vet. Microbiol.* 78:39–48.

Hein, I., Lehner, A., Rieck, P., Klein, K., Brandle, E., Wagner, M. 2001. Comparison of different approaches to quantify *Staphylococcus aureus* by real-time quantitative PCR and application of this technique for examination of cheese. *Appl. Environ. Microbiol.* 67:3122–3126.

Holmberg, S., Blake, P. 1984. Staphylococcal food poisoning in the United States. *J. Am. Med. Assoc.* 251:487–489.

Hookey, J., Richardson, J., Cookson, B. 1998. Molecular typing of *Staphylococcus aureus* based on PCR restriction fragment length polymorphism and DNA sequence analysis of the coagulase gene. *J. Clin. Microbiol.* 36:1083–1089.

Hwang, S., Kim, S., Jang, E., Kwon, N., Par, Y., Koo, H., Jung, W., Kim, W., Park, Y. 2007. Novel multiplex PCR for the detection of the *Staphylococcus aureus* superantigen and its application to raw meat isolates in Korea. *Int. J. Food Microbiol.* 117:99–105.

Jarraud, S., Mugel, C., Thioulouse, J., Lina, G., Meugnier, H., Forey, F., Nesme, X., Etienne, J., Vanenesch, F. 2002. Relationships between *Staphylococcus aureus* genetic background virulence facors, *agr* groups (alleles), and human disease. *Infect. Immun.* 70:631–641.

Johnson, W., Tyler, S., Ewan, E., Ashton, F., Pollard, D., Roze, K. 1991. Detection of genes for enterotoxins, exfoliative toxins, and toxic shock syndrome toxin 1 in *Staphylococcus aureus* by the polymerase chain reaction. *J. Clin. Microbiol.* 29:426–430.

Karahan, M., Çetinkaya, B. 2007. Coagulase gene polymorphisms detected by PCR in *Staphylococcus aureus* isolated from subclinical bovine mastitis in Turkey. *Vet. J.* 174:428–431.

Keyhani, M., Rogolsky, M., Wiley, B., Glasgow, L. 1975. Chromosomal synthesis of staphylococcal exfoliative toxin. *Infect. Immun.* 12:193–197.

Khambaty, F., Bennett, R., Shah, D. 1994, Application of pulsed-field gel electrophoresis to the epidemiological characterization of *Staphylococcus intermedius* implicated in a food-related outbreak. *Epidemiol. Infect.* 113:75–81.

Kondoh, K., Furuya, D., Yagihashi, A., Uehara, N., Nakamura, M., Kobayashi, D., Tsuji, N., Watanabe, N. 2002. Comparison of arbitrary primed-polymerase chain reaction and pulse-field gel electrophoresis for characterizing methicillin-resistant *Staphylococcus aureus*. *Lett. Appl. Microbiol.* 35:62–67.

Krawczyk, B., Leibner, J., Baranska-Rybak, W., Samet, A., Nowicki, R., Kur, J. 2007. ADSRRS-fingerprinting and PCR MP techniques for studies of intraspecies genetic relatedness in *Staphylococcus aureus*. *J. Microbiol. Methods* 71:114–122.

Kumagai, R., Nakaani, K., Ikeya, N., Kito, Y., Kaidoh, T., Takeuchi, T. 2007. Quadruple or quintuple conversion of *hlb, sak, sea* (or *sep*), *scn,* and *chp* genes by bacteriophages in non-β-hemolysin-producing bovine isolates of *Staphylococcus aureus*. *Vet. Microbiol.* 122:190–195.

Kwon, N., Kim, S., Park, K., Nae, W., Kim, J., Lim, J., Ahn, J., Lyoo, K., Kim, J., Jung, W., Noh, K., Bohach, G., Park, Y. 2004. Application of extended single-reaction multiplex polymerase chain reaction for toxin typing of *Staphylococcus aureus* isolates in South Korea. *Int. J. Food Microbiol.* 97:137–145.

Lange, C., Cardoso, M., Senczek, M., Schwarz, S. 1999. Molecular subtyping of *Staphylococcus aureus* isolates from cases of bovine mastitis in Brazil. *Vet. Microbiol.* 67:127–141.

Larsen, H., Aarestrup, F., Jensen, N. 2002. Geographical variation in the presence of genes encoding superantigenic exotoxins and β-hemolysin among *Staphylococcus aureus* isolated from bovine mastitis in Europe and USA. *Vet. Microbiol.* 85:61–67.

Larsen, H., Huda, A., Erikson, N., Jensen, N. 2000. Differences between Danish bovine and human *Staphylococcus aureus* isolates in possession of superantigens. *Vet. Microbiol.* 76:153–162.

Lawrynowicz-Paciorek, M., Kochman, M., Piekarska, P., Grochowska, A., Windyga, B. 2007. The distribution of enterotoxin and enterotoxin-like genes in *Staphylococcus aureus* strains isolated from nasal carriers and food samples. *Int. J. Food Microbiol.* 117:319–323.

Lee, J. 2003. Methicillin (oxacillin)-resistant Staphylococcus strains isolated from major food animals and their potential transmission to humans. *Appl. Environ. Microbiol.* 69:6489–6494.

Letertre, C., Perelle, S., Dillasser, F., Fach, P. 2003a. Identification of a new putative enterotoxin SEU encoded by the *egc* cluster of *Staphylococcus aureus*. *J. Appl. Microbiol.* 95:38–43.

Letertre, C., Perelle, S., Dillasser, F., Fach, P. 2003b. A strategy based on 5'-nuclease multiplex PCR to detect enterotoxin genes *sea* to *sej* of *Staphylococcus aureus*. *Mol. Cell. Probes.* 17:227–235.

Lin, A., Usera, M., Barret, T., Goldsby, R. 1996. Application of random amplified polymorphic DNA analysis to differentiate strains of *Staphylococcus enteritidis*. *J. Clin. Microbiol.* 34:870–876.

Mäntynen, V., Niemelia, S., Kaijalainen, K., Pirhonen, T., Lindstrom, K. 1997. MPN-PCR— Quantification method for *Staphylococcal* enterotoxin c1 gene from fresh cheese. *Int. J. Food Microbiol.* 36:135–143.

Marcos, J., Soriano, A., Salazar, M., Moral, H., Ramos, S., Smeltzer, M., Carasco, G. 1999. Rapid identification and typing of *Staphylococcus aureus* by PCR-restriction fragment length polymorphism and analysis of the *aroA* gene. *J. Clin. Microbiol.* 37:5760–5764.

Martin, M., Fueyo, J., Gonzalez-Hevia, M., Mendoza, M., 2004. Genetic procedures for identification of enterotoxigenic strains of *Staphylococcus aureus* from three food poisoning outbreaks. *Int. J. Food Microbiol.* 94:279–286.

Martin, M., Gonzalez-Hevia, M., Mendoza, M. 2003. Usefulness of a two-step PCR procedure for detection and identification of enterotoxigenic staphylococci of bacterial isolates and food samples. *Food Microbiol.* 20:605–610.

Martineau, F., Picard, F., Roy, P., Quellette, M., Bergeron, M. 1998. Species-specific and ubiquitous-DNA based assays for rapid identification of *Staphylococcus aureus*. *J. Clin. Microbiol.* 36:618–623.

McLauchlin, J., Narayanan, G., Mitani, V., O'Neill, G. 2000. The detection of enterotoxins and toxic shock syndrome toxin genes in *Staphylococcus aureus* by polymerase chain reaction. *J. Food Prot.* 63:479–488.

Mehrotra, M., Wang, G., Johnson, W. 2000. Multiplex PCR for detection for *Staphylococcus aureus* enterotoxins, exfoliative toxins, toxic shock syndrome toxin 1, and methicillin resistance. *J. Clin. Microbiol.* 38:1032–1035.

Monday, S., Bohach, G. 1999. Use of multiplex PCR to detect classical and newly described pyrogenic toxin genes in *Staphylococcus* isolates. *J. Clin. Microbiol.* 37:3411–3414.

Moore, P., Lindsay, J. 2001. Genetic variation among hospital isolates of methicilllin sensitive *Staphylococcus aureus*: Evidence for horizontal transfer of virulence genes. *J. Clin. Microbiol.* 39:2760–2767.

Morandi, S., Brasca, M., Lodi, R., Cremonesi, P., Castiglionii, B. 2007. Detection of classical enterotoxins and identification of enterotoxin genes in *Staphylococcus aureus* from milk and dairy products. *Vet. Microbiol.* 124:66–72.

Nashev, D., Toshkova, K., Isrina, S., Salaisa, S., Hassan, A., Lämmler, C., Zschöck, M. 2004. Distribution of virulence genes of *Staphylococcus aureus* isolated from stable nasal carriers. *FEMS Microbiol. Lett.* 233:45–52.

Okii, K., Hiyama, E., Takesue, Y., Kodaira, M., Sueda, T., Yokoyama, T. 2006. Molecular epidemiology of enteritis-causing methicillin-resistant *Staphylococcus aureus*. *J. Hosp. Infect.* 62:37–43.

Oliveira, K., Procop, G., Wilson, D., Cull, J., Stender, H. 2002. Rapid identification of *Staphylococcus aureus* directly from blood cultures by fluorscence in situ hybridization with peptide nucleic acid probes. *J. Clin. Microbiol.* 40:247–251.

Omoe, K., Hu, D., Takahasi-Omoe, H., Nakame, A., Shinagawa, K. 2005. Comprehensive analysis of classical and newly described staphylococcal superantigenic toxin genes in *Staphylococcus aureus* isolates. *FEMS Microbiol. Lett.* 246:191–198.

Omoe, K., Ishikawa, M., Shimoda, Y., Hu, D., Ueda, S., Shinagawa, K. 2002. Detection of *seg, seh*, and *sei* genes in *Staphylococcus aureus* isolates and determination of the enterotoxin productivities of *S. aureus* isolates harboring *seg, seh*, or *sei* genes. *J. Clin. Microbiol.* 40:857–862.

Orden, J., Goyache, J., Hernández, J., Domenech, A., Suárez, G., Gémez-Lucia, E. 1992. Detection of enterotoxins and TSST-1 secreted by *Staphylococcus aureus* isolated from ruminant mastitis—Comparison of ELISA and immunoblot. *J. Appl. Bacteriol.* 72:486–489.

O'Toole, P., Foster, T. 1986. Epidermolytic toxin serotpe B of *Staphylococcus aureus* is plasmid-encoded. *FEMS Microbiol. Lett.* 36:311–314.

Peacock, S., de Silva, G., Justice, A., Cowland, A., Moore, C., Winearls, C., Day, N. 2002. Comparison of multilocus sequence typing and pulsed-field gel electrophoresis as tools for typing *Staphylococcus aureus* isolates in a microepidemiological setting. *J. Clin. Microbiol.* 40:3764–3770.

Peles, F., Wagner, M., Varga, L., Hein, I., Rieck, P., Gutseer, K., Kereszturi, P., Kardos, G., Turcsanyi, I., Beri, B., Szabo, A. 2007. Characterization of *Staphylococcus aureus* strains isolated from bovine milk in Hungary. *Int. J. Food Microbiol.* 118:186–193.

Perez-Roth, E., Claverie-Martin, F., Villar, J., Mendez-Alverez, S. 2001. Multiplex PCR for simultaneous identification of *Staphylococcus aureus* and detection of methicillin and mupirocin resistance. *J. Clin. Microbiol.* 39:4037–4041.

Pinto, B., Chenoll, E., Aznar, R. 2005. Identification of typing of food-borne *Staphylococcus aureus* by PCR-based techniques. *Syst. Appl. Microbiol.* 28:340–352.

Raimundo, O., Deighton, M., Capstick, J., Gerraty, N. 1999. Molecular typing of *Staphylococcus aureus* of bovine origin by polymorphisms of the coagulase gene. *Vet. Microbiol.* 66:275–284.

Rajkovic, A., Moualij, B., Uyttendaele, M., Brolet, P., Zorzi, W., Heinen, E., Foubert, E., Debevere, J. 2006. Immunoquantitative real-time PCR for detection and quantification of *Staphylococcus aureus* enterotoxin B in foods. *Appl. Environ. Microbiol.* 72:6593–6599.

Rall, V., Vieira, F., Rall, R., Vieitis, R., Fernandez, A., Caneias, J., Cardoso, K., Araújo, J., Jr. 2007. PCR detection of staphylococcal enterotoxin genes in *Staphylococcus aureus* strains isolated from raw and pasteurized milk. *Vet. Microbiol.* 132:408–413.

Ramesh, A., Padmapriya, P., Chandrashekar, A., Varadaraj, M. 2002. Application of a convenient DNA extraction method and multiplex PCR for the direct detection of *Staphylococcus aureus* and *Yersinia enterocolitica* in milk samples. *Mol. Cell. Probes* 16:307–314.

Reinoso, E., Bettra, S., Frigero, C., Direnzo, M., Calzolari, A., Bogni, C. 2004. RAPD-PCR analysis of *Staphylococcus aureus* strains isolated from bovine and human hosts. *Microbiol. Res.* 159:245–255.

Reinoso, E., El-Sayed, A., Lämler, C., Bogni, C., Zschöck, M. 2008. Genotyping of *Staphylococcus aureus* isolated from humans, bovine subclinical mastitis and food samples in Argentina. *Microbiol. Res.* 163:314–322.

Riffon, R., Sayasith, K., Khalil, H., Dubruil, P., Drolet, M., Lagace, J. 2001. Development of a rapid and sensitive test for identification of major pathogens in bovine mastitis by PCR. *J. Clin. Microbiol.* 39:2584–2589.

Riyaz-Ul-Hassan, S., Verma, V., Qazi, G. 2008. Evaluation of three different molecular markers for the detection of *Staphylococcus aureus* by polymerase chain reaction. *Food Microbiol.* 25:452–459.

Rosec, J., Gigaud, O. 2002. Staphylococcal enterotoxin genes of classical and new types detected by PCR in France. *Int. J. Food Microbiol.* 77:61–70.

Ruzickova, V., Karpiskova, R., Pantucek, R., Pospisilova, M., Cernikova, P., Doskar, J. 2008. Genotype analysis of enterotoxin H-positive *Staphylococcus aureus* strains isolated from food samples in the Czech Republic. *Int. J. Food Microbiol.* 121:60–65.

Ruzickova, V., Pantucek, R., Petras, P., Doskar, J., Sedlacek, I., Rosypal, S. 2003. Molecular typing of exfoliative toxin-producing *Staphylococcus aureus* strains involved in epidermolytic infections. *Int. J. Med. Microbiol.* 292:541–545.

Sabet, N., Subramaniam, G., Navaraatnam, P., Sekaran, S. 2007. Detection of *mecA* and *ermA* genes and simultaneous identification of *Staphylococus aureus* using triplex real-time PCR from Malaysian *S. aureus* strain collections. *Antimicrob. Agents.* 29:582–585.

Santos, O., Baros, E., Aparecida, M., Brito, V., Bastos, M., Santos, K., Giambiagi-deMarval, M. 2008. Identification of coagulase-negative staphylococci from bovine mastitis using RFLP-PCR of the *groeEL* gene. *Vet. Microbiol.* 130:134–140.

Scherrer, D., Corti, S., Muehlherr, J., Zweifel, C., Stephan, R. 2004. Phenotypic and genotypic characteristics of *Staphylococcus aureus* isolates from raw bulk-tank milk samples of goats and sheep. *Vet. Microbiol.* 101:101–107.

Schwartzkopf, A., Karch, H. 1994. Genetic variation in *Staphylococcus aureus* coagulase genes: potential and limits for use as epidemiological marker. *J. Clin. Microbiol.* 32:2407–2412.

Seki, K., Sakurada, J., Seong, K., Murai, M., Tachi, H., Ishii, H., Masuda, S. 1998. Occurrence of coagulase serotype among *Staphylococcus aureus* strains isolated from healthy individuals—Special reference to correlation with size of protein-A gene. *Microbiol. Immunol.* 43:407–409.

Sharma, N., Rees, C., Dodd, C. 2000. Development of a single-reaction multiplex PCR toxin typing assay for *Staphylococcus aureus* strains. *Appl. Environ. Microbiol.* 66:1347–1353.

Shimizu, A., Fujita, M., Igarashi, H., Takagi, M., Nagase, N., Sasaki, A., Kawano, J. 2000. Characterization of *Staphylococcus aureus* coagulase type VII isolates from staphylococcal food poisoning outbreaks (1980–1995) in Tokyo, Japan, by pulsed field gel electrophoresis. *J. Clin. Microbiol.* 38:3746–3749.

Stepan, J., Pantucek, R., Ruzickova, V., Rosypal, S., Hajek, V., Doskar, J. 2001. Identification of *Staphylococcus aureus* based on PCR amplification of species specific genomic 826 bp sequence derived from a common 44-kb SmaI restriction fragment. *Mol. Cell. Probes* 15:249–257.

Stephan, R., Annemüller, C., Hassan, A., Lammler, C. 2001. Characterization of enterotoxigenic *Staphylococcus aureus* strains isolated from bovine mastitis in north-east Switzerland. *Vet. Microbiol.* 78:373–382.

Straub, J., Hertel, C., Hammes, W. 1999. A 23S rDNA-targeted polymerase chain reaction-based system for detection of *Staphylococcus aureus* in meat starter cultures and dairy products. *J. Food Prot.* 62:1150–1156.

Switalski, L., Patti, J., Butcher, W., Gristina, A., Speziale, P., Höök, M. 1993. Collagen receptor on *Staphylococcus* strains isolated from patients with septic arthritis mediates adhesion to cartilage. *Mol. Microbiol.* 7:99–107.

Tambic, A., Poer, E., Talsania, H., Antony, R., French, G. 1997. Analysis of an outbreak of non-phage-typeable methicillin resistant *Staphylococcus aureus* by using a randomly amplified polymorphic DNA assay. *J. Clin. Microbiol.* 35:3092–3097.

Thomas, L., Giding, H., Ginn, A., Olma, T., Irdell, J. 2007. Development of a real-time *Staphylococcus aureus* and MRSA (SAM-) PCR for routine blood culture. *J. Microbiol. Methods* 68:296–302.

Toshkova, K., Savov, E., Soedarmanto, I., Lämmler, C., Chankova, D., van Belkum, A., Verbruch, H., van Leeuwen, W. 1997. Typing of *Staphylococcus aureus* isolated from nasal carriers. *Zbl. Bakt.* 286:547–559.

Tsen, H., Chen, T., 1992. Use of the polymerase chain reaction for the specific detection of type A, D, E enterotoxigenic *S. aureus* in food. *Appl. Microbiol. Biotechnol.* 37:685–690.

Tsen, H., Chen, T., Yu, G. 1994. Application of polymerase chain reaction (PCRS) for the specific detection of enterotoxigenic *Staphylococcus aureus* in various food samples. *J. Food Drug Anal.* 2:217–224.

Ünal, S., Hoskins, J., Flokowitsch, J., Wu, C., Preston, D., Skatrud, P. 1992. Detection of methicillin-resistant staphylococci by using the polymerase chain reaction. *J. Clin. Microbiol.* 30:1685–1691.

van Belkum, A., Bax, R., Peerbooms, P., Goessens, W., Leeuwen, N., Quint, W. 1993. Comparison of phage typing and DNA fingerprinting by polymerase chain reaction for discrimination of methicillin-resistant *Staphylococcus aureus* strains. *J. Clin. Microbiol.* 31:798–803.

van Belkum, A., Bax, R., van der Straaten, P., Quint, W., Veringa, E. 1994. PCR fingerprinting for epidemiological studies of *Staphylococcus aureus*. *J. Microbiol. Methods* 20:235–347.

van Belkum, A., Erikson, H., Sijmonds, M., van Leeuwen, W., van den Bergh, M., Kluytmans, J., Espersen, F., Verbrugh, H. 1997. Coagulase and protein A polymorphisms do not contribute to persistence of nasal colonization by *Staphylococcus aureus*. *J. Med. Microbiol.* 46:222–232.

van Belkum, A., Kluytmans, J., van Leewen, W., Bax, R., Quint, W., Peters, E., Fluit, A., Vandenbroucke-Grauls, C., van den Brule, A., Koeleman, H., Melchers, W., Meis, J., Elaichouni, A., Vaneecoutt, M., Moonens, F., Mais, N., Struelens, M., Tenover, F., Verbrugh, H. 1995. Multicenter evaluation of arbitrarily primed PCR for typing of *Staphylococcus aureus* strains. *J. Clin. Microbiol.* 33:1537–1547.

van Leeuwen, W., Sijmons, M., Sluijs, J., Berbrugh, H., van Belkum, A. 1996. On the nature and use of randomly amplified DNA from *Staphylococus aureus*. *J. Clin. Microbiol.* 34:2770–2777.

van Leeuwen, W., van Belkum, A., Kreiswirth, B., Verbrugh, H. 1998. Genetic diversification of methicillin-resistant *Staphylococcus aureus* as a function of prolonged geographic dissemination and as measured by binary typing and other genotyping methods. *Res. Microbiol.* 149:497–507.

van Wamel, W., Rooijakkers, S., Ruyken, M., van Kessel, K., Srijp, J. 2006. The innate immunomodulators staphylococcal complement inhibitor and chemotaxis inhibitor protein of *Staphylococcus aureus* are located on β-hemolysin-converting bacteriophages. *J. Bacteriol.* 188:1310–1315.

Versalovic, J., Koeuth, T., Lupski, J. 1991. Distribution of repetitive DNA sequences in eubacteria and application to fingerprinting of bacterial genomes. *Nucleic Acids Res.* 19:6823–6831.

Warner, J., Onderdonk. A. 2004. Diversity of toxic shock syndrome toxin 1-postive *Staphylococcus aureus* isolates. *Appl. Environ. Microbiol.* 70:6931–6935.

Williams, J., Kubelin, A., Livak, K., Rafalski, J., Tingy, S. 1990. DNA polymorphisms amplified by arbitrary primers are useful as genetic markers. *Nucleic Acids Res.* 18:6531–6535.

Yamaguchi, T., Nishifuji, K., Sassaki, M., Fudaba, Y., Aepfelbacher, M., Takata, T., Ohara, M., Komatsuzawa, H., Magai, M., Sugai, M. 2002. Identification of the *Staphylococcus aureus* etd pathogenicity island which encodes a novel exfoliative toxin, ETD, and EDIN-B. *Infect. Immun.* 70:5835–5845.

Yang, Y., Xu-dong, S., Yao-wu, Y., Chun-yu, K., Ying-jun, L., Xiao-ying, Z. 2007. Detection of *Staphylococcus aureus* in dairy products by polymerase chain reaction assay. *Agric. Sci. China* 6:857–862.

Yarwood, J., McCormick, J., Paustian, M., Orwin, P., Kapur, V., Schlievert, P. 2002. Characterization and expression analysis of *Staphylococcus aureus* pathogenicity island 3. *J. Biol. Chem.* 277:13138–13147.

Zadoks, R., van Leeuwen, W., Barkema, H., Sampimon, O., Verbrugh, H., Schukken, Y., van Belkum, A. 2000. Application of pulsed-field gel electrophoresis and binary typing as tools in veterinary clinical microbiology and molecular epidemiologic analysis of bovine and human staphylococcus isolates. *J. Clin. Microbiol.* 38:1931–1939.

Zhang, S., Iandola, J., Sewart, G. 1998. The enterotoxin D plasmid of *Staphylococcus aureus* encodes a second enterotoxin determinant (sej). *FEMS Microbiol. Lett.* 168:227–233.

Zschöck, M., Botzler, D., Blocher, S., Sommerhauser, J., Hamann, H. 2000. Detection of genes for enterotoxins (ent) and toxic shock syndrome toxin-1 (tst) in mammary isolates of *Staphylococcus aureus* by polymerase-chain-reaction. *Int. Dairy J.* 10:569–574.

Zschöck, M., Klopper, B., Wolter, W., Hamann, H., Lämmler, C. 2005. Pattern of enterotoxin genes *seg, seh, sei* and *sej* positive *Staphylococcus aureus* isolated from bovine mastitis. *Vet. Microbiol.* 108:243–249.

CHAPTER 12

Listeria monocytogenes

I. CHARACTERISTICS OF THE ORGANISM

Listeria monocytogenes is a peritrichously flagellated, gram-positive, intracellular bacterial pathogen that is capable of growth at refrigerator temperatures. The organism has a wide host range including mammals, birds, fish, and crustacea (Davis et al., 1973) and is widely distributed, being found in plant, soil, and surface water samples (Weiss and Seeliger, 1975). The organism has been isolated from a wide variety of foods, including milk, cheese, coleslaw, frankfurters, beef, poultry, and fish (Lovett, 1989) and is capable of growth at refrigerator temperatures as low as 1°C (Seeliger and Jones, 1986). Most healthy adults infected with *L. monocytogenes* experience only mild flu-like symptoms (Lovett, 1989). Listeriosis, however, is an infectious disease, which is characterized by monocytosis, growth of the organism in macrophages, septicemia, and the formation of multiple focal abscesses in the viscera. Infection of pregnant women may lead to invasion of the fetus, resulting in stillbirth or abortion (Lovett, 1989). The commonest form of listeriosis is meningitis, which develops predominantly in newborns and the aged, resulting in approximately 70% mortality if untreated (Seeliger and Finger, 1976; Killinger, 1970).

Among the seven species of *Listeria*, only *L. monocytogenes* and *L. ivanovii* are pathogens for humans and animals. Among the virulence factors that contribute to the pathogenicity of the organism and its ability to enter, survive, and grow within mammalian cells is the SH-activated α hemolysin *(listeriolysin O)*. The *listeriolysin O* gene *(hlyA) from L. monocytogenes* has been sequenced and was shown to be absent in other *Listeria* species (Mengaud et al., 1988). Although two other *Listeria* species, *L. ivanovii* and *L. seeligeri,* are also hemolytic, the listeriolysin O gene was not detected in their genetic sequence (Leimeister-Wächter and Chakraborty, 1989). The high level of species specificity of this gene has allowed it to be used as a valuable target for detection of *L. monocytogenes* by the PCR.

The European community directive on milk and milk-based products specifies zero tolerance of *L. monocytogenes* for soft cheeses and the absence of the organism in 1 g of other products (Nogva et al., 2000). Great Britain's provisional guidelines for some ready-to-eat foods establishes four quality groups based on the number of *L. monocytogenes*: not detected in 25 g is satisfactory; <10^2/25 g is fairly satisfactory; 10^2 to 10^3 is unsatisfactory; and numbers >10^3 make the product unacceptable. The U.S. Department of Agriculture specifies a zero tolerance for *L. monocytogenes* in ready-to-eat meat products and other foods. The International Commission on Microbiological Specification for Foods has concluded that if this organism does not exceed 100 organisms/g of food at the point of consumption, the food is considered acceptable for individuals who are not at risk. A rapid quantitative method for detection of the organism in foods is therefore of value, in addition to the utility of such a method in determining the infectious dose during outbreaks of listeriosis.

A. Gene Sequences Used in PCR Assays for *L. monocytogenes*

Several genes coding for various aspects of virulence in *L. monocytogenes* have been identified and used for PCR detection of the organism. The *actA* gene product ActA is a surface protein responsible for actin polymerization into an elongated filament required for intracellular bacterial propulsion and cell-to-cell invasion (Cossart and Kocks, 1994; Domann et al., 1992; Kocks et al., 1992). The *hly* gene (Mengaud et al., 1988) codes for listeriolysin O, a protein having pore-forming activity, and is involved in lysis of the phagocyte vacuole allowing cells of *L. monocytogenes* to escape from the vacuole. Activity is usually determined by the lytic activity on erythrocytes, hence the designation hemolysin. The *hly* (*hlyA*) gene is also present in *L. seeligeri* and *L. ivanovii*, in addition to *L. monocytogenes* (Leimeister-Wächter and Chakraborty, 1989). However, sequence heterogeneity among the *hly* determinants of these species has allowed primers specific for *L. monocytogenes* to be developed (Deneer and Boychuk, 1991; Rossen et al., 1991). The *iaP* gene codes for an invasion-associated protein p60 (Köhler et al., 1990). The *inlA* gene product internalin A is a novel surface protein conferring invasiveness to human enterocytes (Gaillard et al., 1991). The *lmaA* gene product is a 21-kDa polypeptide antigen termed *L. monocytogenes* antigen (lmaA), which is capable of eliciting a specific delayed-type hypersensitivity response in *Listeria*-immune mice (Göhmann et al., 1990) and which has also been referred to as the DTH-18 factor (Notermans et al., 1989) encoded by the *Dth-18* gene (Wernars et al., 1991). The *lmaA* gene is present in both *L. monocytogenes* and *L. ivanovii*, is absent in all other species of *Listeria*, and is expressed only by *L. monocytogenes* strains (Göhmann et al., 1990). It therefore lacks PCR species specificity for *L. monocytogenes*.

The *flaA* gene codes for flagellin (Dons, Rasmussen, and Olsen, 1992) and is specific for all seven species of the genus *Listeria*. The *plcB* gene encodes a phospholipase C (Geoffroy et al., 1991). The product of the *prfA* gene, PrfA, regulates the expression of a cascade of virulence factors including listeriolysin O (Leimeister-Wächter et al., 1990; Mengaud et al., 1991). In addition, PCR primers based on 16S rRNA sequences have been designed that are specific for the genus *Listeria* (Border et al., 1990) and others that are specific for the species *L. monocytogenes* (Wang, Cao, and Johnson, 1992).

B. PCR Identification of *L. monocytogenes* and Other Members of the Genus *Listeria*

The polymerase chain reaction (PCR), as applied to the detection of *Listeria monocytogenes* in foods, encompasses a variety of cell extraction, cell lysis, and DNA purification techniques. These techniques have been applied to foods directly and to enrichment broths with wide variations in sensitivity reported by numerous investigators. The selection of the specific gene for amplification of a selected sequence is also important in that some genes associated with pathogenicity have been found to be present in certain of the other species of *Listeria*.

Border et al. (1990) used a set of primers designated U1/LI1 (Table 12.1) for detection of any *Listeria* species based on a 16S rRNA sequence that is highly conserved only in species of *Listeria* with a resulting 938-bp amplification product. They also confirmed that the LM1/LM2 set of primers of Mengaud et al. (1988) for amplification of a 702-bp sequence of the listeriolysin O gene were species-specific for *L. monocytogenes*.

Deneer and Boychuk (1991) developed a two-phase amplification PCR detection assay specific for *L. monocytogenes* using a set of primers (Table 12.1) derived from a 174-bp sequence of the *hlyA* gene in 100 μl PCR reaction volumes. The use of stringent reaction conditions restricted the formation of a PCR-amplified gene product solely from *L. monocytogenes*. Cell suspensions were first treated with lysozyme and then with SDS-proteinase K, followed by extraction and purification of the DNA using two different protocols, both of which successfully yielded the appropriate amplified DNA product. A faint DNA band was observed in EB-agarose gels with 35 amplification cycles after starting with 542 bacterial cells. When 10 μl of amplified product were subjected to an additional 35 amplification cycles with fresh PCR reagents, the minimum level of detection was reduced to 5 to 50 cells.

Wernars et al. (1992) made use of the *prfA* gene to develop a highly specific PCR for recognition of pathogenic *L. monocytogenes* strains. The primers used, prfA-A/prfA-B (Table 12.1), flank a 1060-bp sequence encompassing the entire *prfA* gene, were found to be specific for all virulent strains of *L. monocytogenes* tested, and readily distinguished them from an avirulent strain of *L. monocytogenes* and representatives of the other six *Listeria* species. Cell lysis consisted of suspending a portion of a colony in 100 μl of PCR buffer, followed by heating for 5 min at 95°C.

Wiedman et al. (1997), using PCR-restriction fragment length polymorphism (PCR-RFLP), identified 8 *hly*, 11 *inlA,* and 2 *actA* alleles among strains of *L. monocytogenes* (see Table 12.1 for primers used).

Three genetic lineages have been identified for *L. monocytogenes* using various subtyping procedures. Lineage I contains serotpes 1/2b, 3b, 3c, and 4b. Lineage II contains serotypes 1/2a, 1/2c, and 3a. Lineage III contains serotypes 4a and 4c. Lineage I isolates include major epidemic clones of *L. monocytogenes* that have caused a large number of cases of human listeriosis. Lineage II isolates are derived mostly from foods and the environment. Lineage III strains are isolated mostly from animals.

Norton et al. (2001a) subjected 117 isolates of *L. monocytogenes* from a smoked-fish processing plant and 275 human clinical isolates to molecular characterization.

Table 12.1 PCR Primer Pairs Used for Detection of *L. monocytogenes* and *Listeria spp.*[a]

Primer	Sequence (5′ → 3′)[b]	Size of Amplified Sequence (bp)	Gene	Reference
U1	CAG-CMG-CCG-CGG-TAA-TWC[b]	938	16S rRNA	Lane et al. (1985)
L1	CTC-CAT-AAA-GGT-GAC-CCT			Stackenbrandt and Curiale (1988)
LM1	CCT-AAG-ACG-CCA-ATC-GAA	702	*hly*	Mengaud et al. (1988)
LM2	AAG-CGC-TTG-CAA-CTG-CTC			
hly-F	TGC-GTT-TCA-TCT-TTA-GAA-GC	1193 and 812	*hly*	Wiedman et al. (1997)
hly-B$_n$	GTC-GAT-GAT-TTG-AAC-TTC-ATC-TTT			
actA-F	TAG-CGT-ATC-ACG-AGG-AGG	1993 and 1888	*actA*	Wiedman et al. (1997)
actA-R	TTT-TGA-ATT-TCA-TAT-CAT-TCA-CC			
actA-946F	TAA-AAG-TGC-AGG-GTT-ATT-G[c]	539	*actA*	Wiedman et al. (1997)
actA-1834R	GGA-TTA-CTG-GTA-GGC-TCG-G[c]			
inlA-F	CAG-GCA-GCT-ACA-ATT-ACA-CA	2341	*inlA*	Wiedman et al. (1997)
inlA-R	ATA-TAG-TCC-GAA-AAC-CAC-ATC-T			
prfA-A	CTG-TTG-GAG-CTC-TTC-TTG-GTG-AAG-CAA-TCG	1060 TAA-CTC	*prfA*	Wernars et al. (1992)
prfA-B	AGC-AAC-CTC-GGT-ACC-ATA-TAC-			
(1) dth-18	CCG-GGA-GCT-GCT-AAA-GCG-GT	326	*Dth-18 (lmaA)*	Wernars et al. (1991)
(2) dth-18	GCC-AAA-CCA-CCG-AAA-AGA-CC			

Primer	Sequence	Size	Gene	Reference
hly-1	CTA-ATC-AAG-ACA-ATA-AAA-TC	521	*hly*	Fluit et al. (1993)
hly-2	GTT-AGT-TCT-ACA-TCA-CCT-GA			
hly-F	CGG-AGG-TTC-CGC-AAA-AGA-TG	234	*hly*	Furrer et al. (1991)
hly-R	CCT-CCA-GAG-TGA-TCG-ATG-TT			
iap-F	ACA-AGC-TGC-ACC-TGT-TGC-AG	131	*iaP*	
iap-R	TGA-CAG-CGT-GTG-TAG-TAG-CA			
hlyA-a	ATT-GCG-AAA-TTT-GGT-ACA-GC	234	*hly*	Johnson et al. (1992)
hlyA-b	ACT-TGA-GAT-ATA-TGC-AGG-AG			
lmaA-a	AAC-AAG-GTC-TAA-CTG-TAA-AC	257	*lmaA*	
lmaA-b	ACT-ATA-GTC-AGC-TAC-AAT-TG			
prfA-F	CCCC-AAG-TAG-CAG-GAC-ATG-CTA-A	571	*prfA*	Cooray et al. (1994)
prfA-R	GGT-ATC-ACA-AAG-CTC-ACG-AG			
hlyA-F	CAC-TCA-GCA-TTG-ATT-CG	276	*hly*	Cooray et al. (1994)
hlyA-R	ATT-TTC-CCT-TCA-CTG-ATT-CG			
plcB-F	GCA-AGT-GTT-CTA-GTC-TTT-CCG-G	795	*plcB*	Cooray et al. (1994)
plcB-R	ACC-TGC-CAA-AGT-TTG-CTG-TGA			
hlyA-234	CAT-CGA-CGG-CAA-CCT-CGG-AGA	417	*hly*	Fitter, Heuzenroeder, and Thomas (1992)
hlyA-319	ATC-AAT-TAC-CGT-TCT-CCA-CCA-TTC			

Continued

Table 12.1   PCR Primer Pairs Used for Detection of *L. monocytogenes* and *Listeria spp.*[a] (Continued)

Primer	Sequence (5′ → 3′)	Size of Amplified Sequence (bp)	Gene	Reference
flaA-F	AGC-TCT-TAG-CTC-CAT-GAG-TT	420	*flaA*	Gray and Kroll (1995) from Dons, Rasmussen, and Olsen (1992)
flaA-R	AGT-AGC-AGC-ACC-TGT-AGC-AGT			
hly-LL1-F	GAC-ATT-CAA-GTT-GTG-AA	LL1/LL4: 560	*hly*	Thomas et al. (1991)
hly-LL3-F	ATT-GCG-AAA-TTT-GGT-AC	LL1/LL6: 299		
hly-LL4-R	CGC-CAC-ACT-TGA-GAT-AT	LL3/LL4: 240		
hly-LL5-F	AAC-CTA-TCC-AGG-TGC-TC	LL5/LL4: 520		
hly-LL6-R	CTG-TAA-GCC-ATT-TCG-TC	LL5/LL6: 267		
L-1-F	CAC-GTG-CTA-CAA-TGG-ATA-G	70	16S rRNA	Wang, Cao, and Johnson (1992)
L-2-R	AGA-ATA-GTT-TTA-TGG-GAT-TAG			
a1-hlyA	CCT-AAG-ACG-CCA-ATC-GAA-AAG-AAA	858	*hly*	Norton et al. (2001)
β1-hlyA	TAG-TTC-TAC-ATC-ACC-TGA-GAC-AGA			Norton et al. (2001) from Bsat and Batt (1993)
C1-*prfA*	TCT-CCG-AGC-AAC-CTC-GGA-ACC	1052	*prfA*	Dickinson, Kroll, and Grant (1995)
S6-*prfA*	TGG-ATT-GAC-AAA-ATG-GAA-CA			
Lis-1-hlyA	GCA-TCT-GCA-TTC-AAT-AAA-GA	174	*hly*	Deneer and Boychuk (1991)
Lis-2-hlya	TGT-CAC-TGC-ATC-TCC-GTG-GT			
LM2	CCT-TTG-ACC-ACT-CTG-GAG-ACA-GAG-C	553	16S rRNA	Lantz et al. (1994)

	Sequence	Size	Gene	Reference
ru8	AAG-GAG-GTG-ATC-CA[G/A] CCG-CA[G/C]-[G/C]TT-C	275	16S rRNA	
LM1	GGA-GCT-AAT-CCC-ATA-AAA-CTA			
ru8	AAG-GAG-GTG-ATC-CA[G/A] CCG-CA[G/C]-[G/C]TT-C			Herman, De Block, and Moermans (1995)
LM$_1$	CCT-AAG-ACG-CCA-ATC-GAA	701	*hly*	
LM$_2$	AAG-CGC-TTG-CAA-CTG-CTC			
LL5	AAC-CTA-TCC-AGG-TGC-TC	267	*hly*	
LL6	CTG-TAA-GCC-ATT-TCG-TC			
LF	CAA-ACG-TTA-ACA-ACG-CAG-TA	750	*hly*	Bansal (1996)
LR	TCC-AGA-GTG-ATC-GAT-GTT-AA			
PCRGO	GAA-TGT-AAA-CTT-CGG-CGC-AAT-CAG	388	*hly*	Bohnert et al. (1992)
PCRDO	GCC-GTC-GAT-GAT-TTG-AAC-TTC-ATC			
αn	AAA-GAA-AAT-TCA-ATT-TCA-TCC-ATG	987	*hly*	Bsat and Batt (1993)
βn	GTC-GAT-GAT-TTG-AAC-TTC-ATC-TTT			
α1	CCT-AAG-ACG-CCA-ATC-GAA-AAG-AAA	858	*hly*	
β1	TAG-TTC-TAC-ATC-ACC-TGA-GAC-AGA			
Forward	TGC-AAG-TCC-TAA-GAC-GCC-A	113	*hlyA*	Nogva et al. (2000)
Reverse	CAC-TGC-ATC-TCC-GTG-GTA-TAC-TAA			
Probe	FAM-CGA-TTT-CAT-CCG-CGT-GTT-TCT-TTT-CG-TAMRA			
23S-MF	AGG-ATT-TTG-GCT-TAG-AAG	894	23s rDNA	Soejima et al. (2008) from Hong et al. (2004)
23S-MR	CAC-TAC-CCC-GAC-AAG-GAA-T			

Continued

Table 12.1   PCR Primer Pairs Used for Detection of *L. monocytogenes* and *Listeria spp.*ᵃ (Continued)

Primer	Sequence (5′ → 3′)	Size of Amplified Sequence (bp)	Gene	Reference
ORF2372-F	AGA-TAG-CCT-GAT-GCG-AGT-TTT	595	ORF2372	Zhang and Knabel (2005)
ORF2372-R	ACG-TTT-TAC-GAT-CTC-CAC-CTG			
inlB-F	CAT-GGG-AGA-GTA-ACC-CAA-CC	500	inlB	
inlB-R	GCG-GTA-ACC-CCT-TTG-TCA-TA			
inlC-F	CCC-ACA-ATC-AAA-TAA-GTG-ACC-TT	400	inlC	
inlC-R	CTG-GGT-CTT-TGA-CAG-TAT-TTG-TT			
lmo0171-F	TTG-CAA-TCC-GGG-AGA-TTA-T	200	lmo0171	
lmo0171-R	GTA-ACT-ACC-GCG-CCA-GAT-TT			
P1	ACC-AGG-ATT-TTG-GCT-TAG-AAG	~900	23s rDNA	Hong et al. (2004)
P2	CAC-TTA-CCC-CGA-CAA-GGA-AT			
No. 11 Probe	TGC-TCT-ATT-AGG-GTG-CAA-GCC-CGA-GA			
ECI-F	AAT-AGA-AAT-AAG-CGG-AAG-TGT	303	LMOf2365_2798	Chen and Knabel (2007)
ECI-R	TTA-TTT-CCT-GTC-GGC-TTA-G			
ECII-F	ATT-ATG-CCA-AGT-GGT-TAC-GGA	889	inla	Chen and Knabel (2007)
ECII-R	ATC-TGT-TTG-CGA-GAC-CGT-GTC			
ECIII-F	TTG-CTA-ATT-CTG-ATG-CGT-TGG	497	LMOF6854_2463.4	
ECIII-R	GCG-CTA-GGG-AAT-AGT-AAA-GG			
Serotype 4b-F	AGT-GGA-CAA-TTG-ATT-GGT-GAA	597	ORF2110	

Primer	Sequence	Size (bp)	Target	Reference
Serotype 4B-R	CAT-CCA-TCC-CTT-ACT-TTG-GAC			
Serotype 1/2a-F	GAG-TAA-TTA-TGG-CGC-AAC-ATC	724	lmo0737	
Serotype 1/2a-R	CCA-ATC-GCG-TGA-ATA-TCG-G			
L.mono.-F	TGT-CCA-GTT-CCA-TTT-TTA-ACT	420	lmo2234	
L.mono.-R	TTG-TTG-TTC-TGC-TGT-ACG-A			
Listeria spp.-F	ATG-AAT-ATG-AAA-AAA-GCA-AC	145C–1600	*iap*	
Listeria spp.-R	TTA-TAC-GCG-ACC-GAA-GCC-AAC			
inlA-F	ACG-AGT-AAC-GGG-ACA-AAT-GC	800	*inlA*	Liu et al. (2007)
inlA-R	CCC-GAC-AGT-GGT-GCT-AGA-TT			
inlC-F	AAT-TCC-CAC-AGG-ACA-CAA-CC	517	*inlC*	
inlC-R	CGG-GAA-TGC-AAT-TTT-TCA-CTA			
inlJ-F	TGT-AAC-CCC-GCT-TAC-ACA-GTT	238	*inlJ*	
inlJ-R	AGC-GGC-TTG-GCA-GTC-TAA-TA			
LM4-F	CAG-TTG-CAA-GCG-CTT-GGA-GT	446	*hly*	Jinneman and Hill (2001)
LM5-R	CCT-CCA-GAG-TGA-TCG-ATG-TT		*hly*	
LMA-F	AAG-CCG-TAA-TTT-ACG-GTG-AC		*hly*	
LMB-R	GTA-AGT-CTC-CGA-GGT-TGC-AA		*hly*	
LMC-R	GAA-CTC-CTG-GTG-TTT-CTC-AA		*hly*	
LMD-F	CAC-CAG-GAG-TTC-CCA-TTG-AC		*hly*	

Continued

Table 12.1 PCR Primer Pairs Used for Detection of *L. monocytogenes* and *Listeria spp.*[a] (Continued)

Primer	Sequence (5′ → 3′)	Size of Amplified Sequence (bp)	Gene	Reference
LMrt3F	CAA-AGC-GAG-AAT-GTG-GCT-ATA-AAT-GA	—	*actA*	Oravcová, Kuchta, and Kaclíková (2007)
LMrt3R	TAA-TTT-CCG-CTG-CGC-TAT-CCG			
ListP	ACC-CTG-GAT-GAC-GAC-GCT-CCA-CT			
D1-F	CGA-TAT-TTT-ATC-TAC-TTT-GTC-A	214	Div. I or III	Borucki and Call (2003)
D1-R	TTG-CTC-CAA-AGC-AGG-GCA-T			
D2-F	GCG-GAG-AAA-GCT-ATC-GCA	140	Div. I	Borucki and Call (2003)
D1-R	TTG-TTC-AAA-CAT-AGG-GCT-A			
FlaA-F	TTA-CTA-GAT-CAA-ACT-GCT-CC	538	Serotypes 1/2a and 3a	Borucki and Call (2003)
FlaA-R	AAG-AAA-AGC-CCC-TCG-TCC			
GLT-F	AAA-GTG-AGT-TCT-TAC-GAG-ATT-T	483	Serotypes 1/2b and 3b	Borucki and Call (2003)
GLT-R	AAT-TAG-GAA-ATC-GAC-CTT-CT			

[a] F denotes forward primer, R denotes reverse primer.
[b] M denotes A or C; W denotes A or T.
[c] Sequence is internal to that flanked by actA-F and actA-R primers.

Ribotyping and allelic typing of the *actA* and *hlyA* DNA sequences by PCR-RFLP differentiated 23 subtypes and allowed classification of the isolates into three genetic lineages. PCR primers for amplifying the *actA* and *hlyA* gene sequences were those of Wiedman et al. (1997) in Table 12.1. Among the human clinical isolates 69.1% were classified as lineage I compared to 36.8% of the industrial isolates. The remaining industrial isolates (63.2%) were allocated to lineage II, which contained only human sporadic isolates. A mouse cell culture plaque assay indicated that the median plaque size with lineage I industrial isolates was 11% larger than lineage II isolates. Isolates in lineage I were found to be more likely to cause human disease than isolates in lineages II and III. Serotype 1/2 strains are the most frequent isolates from foods and are in lineage II. Serotype 4b strains, which are most frequently associated with human listeriosis, are not among the common food isolates and are in lineage I. Serotype 1/2c is the serotype most frequently isolated from meat products, is rarely associated with human disease (Farber and Peterkin, 1991), and is in lineage I. The authors hypothesized that lineage I strains may have greater human pathogenic potential compared to lineage II.

Jinneman and Hill (2001) developed a unique mismatch amplification mutation PCR assay (MAMA) to rapidly screen and characterize *L. monocytogenes* isolates with regard to lineage genotpes based on the listeriolysin (*hly*) gene. Six primers were developed consisting of a standard primer pair LM4-F/LM5-R (Table 12.1) that amplified a 446-bp sequence of the *hly* gene of all *L. monocytogenes* isolates and four mismatch primers that were designed so that when paired with the standard primer of the reverse complementary sequence a predicted amplicon of unique size was produced for each lineage group. Primers LMA-F/LM5-R and LMD-F/LM5-R (Table 12.1) amplified a 247-bp sequence and a 139-bp sequence, respectively, from only lineage I strains. Primers LM4-F/LMC-R (Table 12.1) amplified a 319-bp sequence from only lineage II strains. Primers LM4-F/LMB-R (Table 12.1) amplified a 268-bp sequence from only lineage III strains. Among a total of 97 isolates, only one could not be allocated to one of the three lineages. The method is faster than other techniques such as RFLP oligonucleotide hybridization or sequencing that have been used for the allocation of strains to the three lineage groups and is highly specific.

Aznar and Alarcón (2002) undertook an extensive examination of nine sets of primers for detection of *L. monocytogenes* and found that the primer pair LM1/LM2 developed by Mengaud et al. (1988) and derived from the *hlyA* gene sequence (Table 12.1) was superior to the other eight primer pairs in terms of specificity for *L. monocytogenes*.

Gray and Kroll (1995) used the primers flaA-F/flaA-R (Table 12.1) from Dons, Rasmussen, and Olsen (1992) to amplify a 420-bp sequence of the *flaA* gene for detection of all seven species of the genus *Listeria*. No other bacterial isolates were detected among 17 additional genera comprising 20 gram-positive and gram-negative species.

Loessner, Schneider, and Scherer (1995) developed a unique cell lysing system by cloning a *Listeria* bacteriophage-encoded lysin that was overexpressed in a strain of *Escherichia coli*. The lysin was effective with both *L. monocytogenes* and *L. ivanovii* and may also be effective with other *Listeria* species. This enzyme appears to offer the advantage of a rapid single-step lysing method for PCR detection and confirmation of *L. monocytogenes* without inhibition of amplification.

Makino, Okada, and Maruyama (1995) developed a unique DNA extraction procedure for soft cheese samples utilizing a high concentration of NaI as a chaotropic agent for removal of PCR inhibitors. A sample (0.5 g) of soft cheese was homogenized with water seeded with *L. monocytogenes* and 0.5 ml of water added. A portion (200 μl) of the aqueous phase was mixed with 400 μl of lysis buffer (0.5% N-laurylsarcosine, 50 mM Tris-HCl, and 25 mM EDTA, pH 8.0). After vortexing, the sample was centrifuged at 15,000 rpm for 5 min. The pellet was suspended in 200 μl of lysis buffer containing glycogen (0.03 mg/μl). And 4 μl of proteinase K (2 mg/ml) added and incubated for 1 hr at 37°C. NaI solution (300 ml) consisting of 6 M NaI in 50 mM Tris-HCl, 25 mM EDTA, pH 8.0 was added followed by the addition of 500 ml of isopropanol. The sample was centrifuged at 15,000 rpm for 5 min the pellet was washed with 35% isopropanol, partially dried, and dissolved in 20 μl of water. The primers LA1/LB1 (Table 12.1) amplified a 625-bp sequence of an unspecified gene. The primers, however, also resulted in 625-bp amplicons from *L. innocua*, *L. welshimeri*, *L. ivanovii*, and *L. murrayi*, and therefore lacked species specificity for *L. monocytogenes*. The method did allow the direct PCR detection of 1×10^3 CFU/g of soft cheese sample. An increased sample size would presumably allow lower numbers of CFU to be detected.

Molecular typing studies (Katharion, 2003; Chen, Zhang, and Knabel, 2007) have identified four major epidemic clones of *L. monocytogenes* (ICI, ECII, ECII, and ECIV). Among these epidemic clones, ECI, a serotype 4b cluster, has been implicated in a number of major outbreaks in different countries. ECII isolates have been associated with hot dog and turkey deli meat outbreaks in the United States. ECIII isolates are serotype 1/2a associated with hot dog and turkey deli meat in the United States. Chen and Knabel (2007) developed a multiplex PCR (mPCR) for simultaneous detection of all members of the genus *Listeria*, *L. monoctogenes*, and major serotypes and epidemic clones of *L. monocytogenes*. Primers ECI-F/ECI-R (Table 12.1) amplified a 303-bp sequence derived from ECI epidemic clones. Primers ECII-F/ECII-R (Table 12.1) amplified an 889-bp sequence derived from ECII epidemic clones. Primers ECIII-F/ECIII-R (Table 12.1) amplified a 497-bp sequence of a putative helicase-like gene derived from ECIII epidemic clones. Primers Serotype 4b-F/Serotype 4b-R (Table 12.1) amplified a 597-bp sequence from a gene encoding a putative secreted protein derived from all serotype 4b strains. Primers Serotype 1/2a-F/Serotype 1/2a-R (Table 12.1) amplified a 724-bp sequence from all 1/2a serotypes. Primers *L. mono.*-F/*L. mono.*-R (Table 12.1) amplified a 420-bp sequence from all *L. monocytogenes* strains. Primers *Listeria* spp.-F/*Listeria* spp.-R (Table 12.1) amplified a 1450–1600-bp sequence of the *iap* gene derived from all species of *Listeria*.

C. Direct PCR Detection of *L. monocytogenes* in Foods without Enrichment Cultivation

Bessesen et al. (1990) used the PCR procedure of Saiki et al. (1988) involving a pair of 24-mer primers for PCR amplification of a 606-bp segment of the *hlyA* gene for direct detection of *L. monocytogenes* in milk and spinal fluid. The limit of detection was 1×10^4 CFU per PCR reaction with cells lysed by boiling in d.H_2O for 2 min. The treatment of cells with 1.0% SDS and 1 mg of lysozyme per milliliter was less effective due presumably to inhibition of *Taq* polymerase by one or both of these agents.

Wernars et al. (1991) used a pair of primers (Table 12.1) to amplify a 326-bp sequence of the *dth-18* gene for direct detection of *L. monocytogenes* in soft cheeses. A single strain, *L. monocytogenes* 1/2a ScottA was used with a strain of *L. grayii* as negative control. Pure cell suspensions in 100 μl of lysis buffer (1.0% w/v Triton X-100) were heated for 5 min at the optimum temperature of 40°C, centrifuged, and 10 μl of supernatant used in 100 μl PCRs. The detection limit of cell suspensions was between 1 and 10 CFU per assay. Cheese extracts were prepared by homogenizing 0.5 g of cheese in 0.4 ml of homogenization buffer (25 mmol/l Tris/HCl pH 7.5, 50 mmol/l glucose). One hundred milliliters of a cell suspension was added, followed by 250 μl of 5× lysis buffer. The mixture was then heated at 40°C and centrifuged, and the DNA in 0.5 ml of the aqueous phase was then purified by ethanol precipitation. Discernible agarose bands failed to occur following PCR due to inhibition of *Taq* polymerase. The authors then made use of three additional purification procedures using 0.5 ml of the aqueous phase extracts. Dialysis failed to allow amplification. Phenol extraction yielded an amplified band but notably reduced the sensitivity of detection about 100-fold due presumably to loss of DNA. The use of Qiagen-5™ DNA affinity columns (Qiagen, Inc., Valencia, CA) resulted in the detection of 1×10^3 CFU/0.5 g with two cheese samples. One sample of cheese required 1×10^6 CFU/0.5 g for detection of an amplified agarose band, while with other soft cheese samples, CFU levels of $10^8/0.5$ g failed to yield a discernible PCR product.

Johnson et al. (1992) used a set of 20-mer primers for detecting a 234-bp sequence of the *hlyA* gene and a set of 20-mer primers for detecting a 257-bp sequence of the *lmaA* gene (Table 12.1). A strain of *L. innocua* was found negative for the presence of the *hlyA* gene but positive for the presence of the *lmaA* gene. The limits of sensitivity were 10 pg of DNA for *hlyA* and 1 pg for *lmaA* equivalent to ~10^5 and ~10^4 CFU, respectively (Bej et al., 1990). Among 53 strains of *L. monocytogenes*, 17 lacked the *lmaA* gene. Seven of the 53 strains were of serotype 4c and were negative for both the *hlyA* and *lmaA* genes and were of food origin; eight serotype 4c strains of nonhuman origin were positive for the *hlyA* gene but negative for the *lmaA* gene.

Powell et al. (1994) found that fresh milk inhibited the PCR. The addition of trypsin inhibitors completely eliminated the inhibition, suggesting that the enzyme plasmin in milk was responsible by degrading *Taq* polymerase.

Dickinson, Kroll, and Grant (1995) described two methods for the extraction of DNA from bacterial cells in foods for direct PCR. The first method consisted of homogenizing 10 g of food with 90 ml of quarter-strength Ringer's solution followed by centrifuging 1.0 ml of the homogenate at high speed and suspending the resulting pellet in a rapid lysis buffer (100 mmolar Tris-HCl at pH 8.5, 5 mmolar EDTA, 0.2% SDAS, 200 mmolar NaCl, and 200 μg/ml of Proteinase K) followed by incubation at 55°C for 2 hr and precipitation of the DNA with isopropanol. This procedure failed to yield amplified PCR products with *L. monocytogenes* due presumably to inefficient cell lysis. The second method involved a modification of the cell lysis procedure of

Pitcher, Saundear, and Owen (1989) and consisted of treating 1.0 ml of homogenate with 20 μl of toluene followed by centrifugation at high speed. The resulting pellet was then suspended in 0.25 ml of mutanolysin solution (20 units of enzyme in 50 mmolar Tris-HCl at pH 6.5) followed by incubation at 37°C for 30 min. Cell lysing reagent (0.5 ml; 5 mmolar guanidium thiocyanate, 100 mmolar EDTA, and 0.5% sarkosyl) was then added and followed by incubation at 50°C for 1 hr. DNA was then precipitated using the procedure of Pitcher, Saundear, and Owen (1989). The precipitated DNA was then suspended in 50–100 ml Tris-EDTA buffer. Some 2.5 μl representing 1/20 to 1/40 of the total DNA sample was then added to PCR reactions totaling 25 μl. This direct procedure without enrichment resulted in minimal levels of detection of 10^4, 10^3–10^4, and 10^3 CFU per gram of Camembert cheese, raw chicken, and coleslaw, respectively. The addition of 10 μl of DNA samples to PCR reactions would have expectedly resulted in tenfold greater levels of sensitivity.

D. PCR Detection of *L. monocytogenes* in Foods Following Enrichment Cultivation

Thomas et al. (1991) developed a protocol for identification of *L. monocytogenes* from skimmed milk and ground beef involving enrichment of 10 ml of seeded milk samples in 90 ml of *Listeria* enrichment broth (LEB) and 25 g of ground beef samples in 225 ml of LEB. A 0.1 ml volume of lysis buffer (10 mM EDTA, 100 mM Tris at pH 8.0 containing 7.5 mg/ml of lysozyme) and 750 units/ml of mutanolysin were then added to enrichment cultures followed by incubation at 37°C for 15 min. DNA was precipitated from samples by adding 50 ml of 3 M sodium acetate and 1.0 ml of ice-cold abs. ethanol. Three forward and two reverse primers, all of 17 bps, were used in a total of five combinations for amplification of variable sequences of the *hly* gene (Table 12.1). Formation of a PCR product was completely inhibited by 10 μg of herring sperm DNA per 100 μl reaction volumes. All of the PCR products derived from the five primer combinations were of approximately equal intensity and were specific for *L. monocytogenes*. No amplified products were obtained from 47 gram-positive and gram-negative species representing 17 additional bacterial genera. The minimum number of CFU detected from enrichment broths was 2.5/ml of milk and 1.2/g of beef.

Rossen et al. (1991) developed a PCR procedure for *L. monocytogenes* following a 40-hr enrichment at 37°C of food samples diluted 1:10 in LEB using a set of primers designated LM14 and LM16 flanking the *hlyA* gene. The primer pair was highly specific for *L. monocytogenes* but the primer sequences were not described. The detection limit with soft cheese following enrichment was 2×10^5 CFU ml corresponding to 100 CFU/PCR reaction. In contrast, the detection limit with a pure culture was 1×10^4 CFU/ml corresponding to 5 CFU/PCR reaction. Nonhomologous DNA derived from 10^8 CFU of *L. innocua* was found to be inhibitory to the PCR. Rossen et al. (1992) used the same set of primers LM14 and LM16 to determine the relative inhibition of the PCR by several foods and by an extensive list of components of culture media and DNA extraction reagents using 100 μl PCR reaction volumes. Detergents, lysozyme, NaOH, alcohols, EDTA, and EGTA exhibited inhibition. Soft brie cheese was notably inhibitory when as little as 1 µl of homogenate (1:10 dilution in 0.9% NaCl) was added to PCR reaction mixtures. In contrast, 50 µl of boiled ham homogenate and 25 µl of salami homogenate exhibited no inhibition. Chicken salad was completely inhibitory with 10 µl of homogenate but not with 5 µl. The PCR was tolerant toward relatively high levels of oil, salt, carbohydrate, and amino acids. When 0.1 mg of casein hydrolysate was added, no inhibition occurred, whereas with 1.0 mg, inhibition was observed. With ovalbumin, 0.1 mg resulted in no inhibition whereas 0.25 mg was inhibitory. It is interesting that 0.4 mg of unrelated DNA caused smearing of the gel and the absence of visible PCR products. The level of inhibition by selective media depended on the volume added to the PCR reaction. The PCR was notably sensitive to bile salts, acriflavine, and ferric ammonium citrate, which are present in certain selective enrichment media.

Fitter, Heuzenroeder, and Thomas (1992) homogenized 1:10 dilutions of food samples in a selective enrichment broth (*Listeria* enrichment broth, LEB; or University of Vermont broth 1, UVM-1). Following incubation at 37°C for 18 hr, food particles from 10 ml of enrichment broths were removed by centrifugation at 1000 g for 2 min and bacterial cells were pelleted by centrifugation at 2000 g for 10 min, washed twice with saline, resuspended in 5 ml of d.H$_2$O, and then lysed by heating in a microwave oven for 2 min. A set of primers coding for a 417-bp sequence of the *hlyA* gene was then used for PCR (Table 12.1). Identity of the PCR product was confirmed by *Hind*III digestion of the amplified DNA that generated a 162-bp segment and a 255-bp segment as anticipated. Detection limits of 10–100 CFU/g were obtained following selective enrichment of seeded chicken skin and soft cheese incorporating 39 µl of lysed samples into 50 µl PCR. Detection limits of 25–250 CFU were obtained with dilutions of pure enrichment cultures on the basis of the above protocol.

Bohnert et al. (1992) used a set of two 24-mer primers designated PCRGO and PCRDO (Table 12.1) amplifying a 388-pb sequence of the *hly* gene for the species-specific detection of *L. monocytogenes* from 180 foods (milk, meat, ice cream, sausage, and chicken). Following primary and secondary enrichment totaling 48 hr, 1.0-ml samples of secondary enrichments were sequentially subjected to treatment with lysozyme and Proteinase K to facilitate thermal lysis of cells. Some samples were found to be positive as a result of PCR following enrichment that were negative using conventional selective isolation. The limit of sensitivity was 10^2 CFU/ml of secondary enrichment broth with 10 CFU/25 g of food detectable.

Wang, Cao, and Johnson (1992) developed a species-specific PCR for detection of *L. monocytogenes* in a variety of foods based on the use of a primer pair (Table 12.1) for amplification of a 70-bp sequence derived from 16S rRNA (Wang, Cao, and Johnson, 1991). The detection limit with pure cultures was 2 CFU per 25 µl PCR reaction. There were 25 g of food products homogenized with 225 ml of PBS, 0.3 ml of Triton X-100, and 2×10^8 CFU of *L. monocytogenes* and then filtered through Whatman no. 4 filter paper. The filtrate was centrifuged at 8000 g for 10 min, and 35 µl of PBS used to resuspend the pellet followed by centrifugation at 10,000 g for 10 min. PBS (1.0 ml) was then used to resuspend the pellet, which was centrifuged again at 16,000 g for 3 min. The pellet was then washed three times with PBS and once with water and resuspended in 100 µl of 1.0% Triton X-100 corresponding to the original 2×10^8 CFU added to the 25 g samples. Decimal dilutions with Triton X-100 were then made followed by heating at 100°C for 5 min to release DNA and 2 μl of dilutions added to 23 μl of PCR reaction mixtures.

The detection limit with foods was 4 CFU per PCR reaction. This observation, however, was based on the addition of 2×10^8 per 25 g sample and a subsequent 1×10^5 dilution that diluted out *Taq* polymerase inhibitors derived from the tissue. When 25-g samples were inoculated with 4 to 10 CFU, negative results were obtained, which were attributable to the loss of CFU during filtration, washing, or centrifugation. However, because their final preparation consisted of suspending pellets in 50 μl of 1.0% Triton X, and incorporating 2 μl into PCR reactions, a minimum inoculation level of 100 CFU/25 g would have been required for detection. When 0.5 ml of filtered food samples were inoculated with 1×10^4 CFU/ml and the samples concentrated to 50 μl and 2 μl (equivalent to 400 CFU) incorporated into PCR without dilution of CFU, amplified products were obtained with samples of chicken breasts, turkey frankfurters, chicken nuggets, fermented sausages, yogurt, and chicken drumsticks, but not with a soft cheese sample. This problem was overcome by overnight selective enrichment. The authors observed that when 1 μl of UVM Broth was added to 24 μl of PCR reaction mixture, inhibition of the PCR did not occur but fluorescent materials from the medium appeared in agarose gels, which could be eliminated by washing the cells to remove the medium prior to PCR. The authors made the additional observation that false-negative results occurred when more than 10^5 or 10^6 CFU of *L. monocytogenes* in 1-μl volumes were added to the 23-μl PCR mixture.

Norton et al. (2001) made use of the commercial BAX PCR system for detecting *L. monocytogenes* in three smoked fish processing plants. A total of 531 samples including raw fish, smoked fish, and environmental samples yielded 95 (17.9) positive samples. Using selective enrichment in LEB at 30°C for 48 hr followed by isolation on Oxford agar yielded 85 (16.0%) positive samples. A set of primers amplifying an 858-bp sequence of the *hlyA* gene was used for identification of cultural isolates of *L. monocytogenes*. The BAX PCR system had a sensitivity [(no. of BAX positive samples/total no. of positive samples) × 100] of 91.8%, an accuracy [(no. of BAX positive samples/total no. of samples) + (no. of BAX negative samples/total no. of samples) × 100] of 95.5%, and a specificity [(no. of BAX negative samples/total no. of negative samples) × 100] of 96.2%. Ribotyping indicated that specific ribotype strains exhibited long-term persistence and were part of the resident microflora of these plants.

Fluit et al. (1993) circumvented the problem of *Taq* inhibition by soft cheese components encountered by Wernars et al. (1991) by using a selective enrichment followed by magnetic immunocapture prior to the PCR. Magnetic particles were coated with a monoclonal antibody (mAB) specific for members of the genus *Listeria*. Primer set A (Table 12.1), specific for the *dth* gene (Wernars et al., 1991) and amplifying a 326-bp sequence, detected strains representing 13 of 14 *L. monocytogenes* serotypes but failed to detect serotype 4a strains. A second set of primers (Table 12.1) derived from a sequence of the listeriolysin O gene yielding a 521-bp amplified product detected all 78 β-hemolytic strains of *L. monocytogenes* representing all known serotypes of the organism. A nonhemolytic strain of *L. monocytogenes* failed to yield an amplified DNA band. Samples of cheese (25 g) were homogenized with 225 ml of LEB and incubated for 24 h at 30°C (enrichment I). From enrichment I, 0.1 ml was transferred into 10 ml of Fraser broth and incubated for 24 hr at 30°C (enrichment II). After secondary enrichment in Fraser broth, immunologic capture was performed with 100 ml of enrichment broth followed by PCR. Cell lysis was achieved by heating the beads in 20 μl of 0.05% SDS for 5 min at 100°C, and then adding 2.5 μl of 20% Nonidet P-40. Enrichment I resulted in detection of 40 CFU/25 g, whereas enrichment I followed by enrichment II allowed the detection of 1 CFU/g of cheese.

Bansal (1996) applied the PCR to detection of *L. monocytogenes* in foods by pelleting 10 ml of secondary enrichment broths and lysing the cells at 96°C for 12–15 min. The preparations were centrifuged, and the resulting supernatants used in PCR reactions. The primers LF and LR (Table 12.1) were selected to amplify a 750-bp region of the *hlyA* gene and were shown to be specific for the species *L. monocytogenes*. A *Listeria* internal positive control (LIPC) fragment was also incorporated into the PCRs. The LIPC is a cloned DNA sequence constructed by deleting a 225 fragment from the 750-bp *hlyA* sequence amplified by the LF and LR primers so as to result in a 525-bp amplified product distinguishable in size from the target sequence of 750 bp. The number of *L. monocytogenes* cells present in the sample can be estimated semiquantitatively by comparing the intensity of the target band and the LIPC band. The LF and LR primers were also used in an mPCR with the *Listeria* spp. specific set of primers U1 (Lane et al., 1985) and LT1 (Stackenbrandt and Curiale, 1988) described by Border et al. (1990; Table 12.1), yielding a 938-bp PCR product for confirmation of positive and negative assays.

Duffy et al. (1999) ground and then homogenized 25 g of chicken and beef with 225 ml of buffered peptone water (BPW). After incubation for 18–24 hr at 30°C, polycarbonate membranes (25 mm diameter, 0.6 μ porosity) were attached to glass slides using 1.0% molten agar. Membranes were then immersed into the enrichment cultures for 15 min to allow cellular adhesion. The membranes were then placed into a 1.5-ml microcentrifuge tube containing 200 μl of lysis buffer (2.0% Triton X-100, 1.0% SDS, 100 mM NaCl, 10 mM Tris pH 8.0, and 1 mM EDTA). A portion (200 μl) of phenol:chloroform:isoamyl alcohol (25:24:1) was then added, vortexed for 30 s, 0.3 g of glass beads (425–600 μ) added, and vortexed for an additional 2 min. This procedure extracted the DNA and dissolved the polycarbonate membranes. The DNA in 200 μl was then precipitated by adding 20 μl of 3 M sodium acetate and 600 ml of abs. ethanol. The primer set of Thomas et al. (1991) LL5/LL4, specific for a 520-bp sequence of the *hly* gene was then used for PCR detection of *L. monocytogenes*. Selective enrichment offered no significant advantage over nonselective enrichment. The limit of detection of a pure culture of *L. monocytogenes* in buffered peptone water was 1×10^3 CFU/ml and that following enrichment of beef and chicken samples was 1×10^4/ml.

E. Nested PCR

Bsat and Batt (1993) developed a nested PCR assay in combination with a dot-blot assay for detection of *L. monocytogenes*. The lysis of cells derived from 1.0 ml of broth cultures was achieved by first washing the cell pellet with PBS, resuspending the cells in 100 µl of PBS, freezing with dry ice, thawing, and boiling for 10 min. The lysate (2.5 µl) was then incorporated directly into a 25-µl PCR. With inoculated aluminum metal surfaces, cotton swab tips were placed into 1.0 ml of TSB and incubated overnight at 37°C in 1.5-ml microcentrifuge tubes with agitation. The swab tips were removed, the cells pelleted, resuspended in 100 ml of water, frozen in liquid nitrogen, and then boiled for 10 min. The first pair of primers αn and βn amplified a 987-bp fragment of the *hlyA* gene (Table 12.1). The second set of primers nested within the first PCR product α1 and β1 amplified an 858-bp fragment (Table 12.1). The first pair of primers αn and βn was used at a concentration of 46 nm each. The second pair of primers α1 and β1 was used at a concentration of 460 nm each in the same reaction volume as the first set of primers with a single PCR amplification of 30 thermal cycles. The lowest detectable cell number by PCR and EB-stained agarose gels was 2.5×10^4 CFU per PCR. With a digoxygenin-11-dUTP labeled dot-blot probe, the minimum level of detection with crude cell lysates without enrichment was between 2 and 25 CFU per PCR. With samples derived from inoculated aluminum surfaces, the minimum level of detection following enrichment and with the use of the dot-blot detection assay was 5 to 25 CFU/25 cm². The minimum level of sensitivity of the PCR assay alone with EB-stained agarose gels can most probably be notably decreased by incorporating 10 µl of sample DNA into a 50-µl PCR reaction volume. In addition, an initial PCR amplification with the first set of primers followed by incorporating 10 µl of the initially amplified product into a second PCR reaction vial with the internally nested second set of primers should also result in a further decrease in the minimal level of detection. This latter enhancement would be derived from preventing both sets of primers from having to compete for the *taq* polymerase simultaneously.

Lantz et al. (1994) developed an aqueous two-phase system involving 8% polyethylene glycol (PEG) 4000 and 11% dextran 40 for use in the direct PCR detection of *L. monocytogenes* in soft cheese. Soft cheese (25 g) was stomached with 225 ml of saline and 0.1 ml of homogenized sample then added to the two-phase system of 2.5 g final weight. Vigorous agitation and settling resulted in cells of *L. monoctogenes* partitioning primarily into the bottom dextran phase and PCR inhibitors partitioned into the top PEG phase. A set of two 16S rRNA based primers (Table 12.1) were then used for 50-µl nested PCRs containing 10-µl DNA samples using two thermal cycling sequences. After the first set of 30 thermal sequences involving primers LM2 and ru8, the amplified product was diluted tenfold and 10 µl added to a fresh PCR reaction mixture containing primers LM1 and ru8 followed by 30 thermal cycles. The minimum level of detection was 1×10^5 CFU/g of cheese. A loss of 0.2 to 0.4 orders of magnitude resulted presumably from some cells partitioning into the interphase.

Herman, De Block, and Moermans (1995) described a direct two-stage PCR amplification procedure without enrichment cultivation for detecting *L. monocytogenes* in milk. The procedure involved nested PCR with primers derived from a portion of the listeriolysin O (*hly*) gene sequence (Table 12.1). In the first PCR, the primers LM_1 and LM_2 were used to amplify a 701-bp sequence of the *hlyA* gene in a 50 µl PCR containing 10 µl of extracted DNA. For the second PCR, 5 µl of the first

PCR amplified product were used with primers LL5 and LL6 to amplify a 267-bp internal sequence (Table 12.1). After the first PCR, an average of 10^4 to 10^5 could be detected in 1 and 25 ml of raw milk. With the two-stage nested PCR, 1 CFU could be detected in 1 and 25 ml of raw milk. The background flora of up to 2×10^5 CFU/ml of milk did not affect the sensitivity. The statistical sensitivity of the method was calculated to be between 10 and 5 CFU per 25 ml of milk. This high level of sensitivity can be considered to be dependent on the efficiency of DNA capture associated with DNA extraction and purification, on the purity of the DNA sample, and on the second PCR amplification with the nested set of primers.

Simon, Gray, and Cook (1996) described four different protocols for extraction of total DNA from cold-smoked salmon for PCR detection of *L. monocytogenes* without enrichment. Each of the protocols used proteinase K to facilitate cell lysis and the precipitation of DNA with hexadecyl trimethylammonium bromide (CTAB) as described by Murray and Thompson (1980). Tissue samples (5 g) were homogenized with 45 ml of buffer and then subjected to DNA extraction. A nested PCR detection protocol was used.

The first PCR amplification incorporated 5 µl of extracted DNA in 50 µl PCR volumes and used primers PRFA and PRFB, which were directed against nucleotides 181 to 207 and 1462 1482 of the *prfA* gene sequence (Wernars et al., 1992) amplifying a 1060-bp product. The second PCR amplification incorporated 2 µl of amplified product from the first PCR into a second 50-µl reaction and employed primers LIP1 and LIP2, which were directed against nucleotides 634 to 654 and 886 to 907 of the *prfA* gene sequence, amplifying a 274-bp product. With protocol 1 involving CTAB extraction, the PCR was completely inhibited, due presumably to the presence of phenolic compounds in the DNA sample derived from the smoked tissue. It is interesting that when Tween 20 was incorporated into the PCR at a level of 2.5%, 6×10^4 CFU of *L. monocytogenes* per gram of tissue could then be detected with protocol 1. Protocol 2 involving CTAB extraction with ether separation resulted in detection of 94 CFU/g of tissue. Protocol 3, using CTAB extraction with silica column purification in place of ether separation resulted in the detection of 216 CFU/g of tissue. Protocol 4, which incorporated CTAB extraction with filter membrane separation to remove particles greater than 0.2 µ after cell lysis followed by silica column purification, allowed detection of 0.8×10-CFU/g of tissue.

F. Multiplex PCR (mPCR)

Furrer et al. (1991) made use of an mPCR procedure consisting of two 20-bp primer sets (Table 12.1), one coding for a 234-bp sequence of the *hlyA* gene, and a second coding for a 131-bp sequence thought to be of the *hlyB* gene and later shown by Köhler et al. (1990) to be a sequence of the *iap* gene for identification of *L. monocytogenes* isolates from cooked sausages. The detection limit with pure cultures with the hlyA primers was 10 cells. Commercial cooked sausage samples (10 g) were homogenized in 90 ml of LEB, incubated at 30°C for 24 hr, and then 0.1 ml was transferred to 10 ml of LEB II incubated at 30°C for 24 hr. One loop was then streaked onto PALCAM agar.

Typical colonies were then subcultured to blood agar plates and single colonies picked for metabolic tests for confirmation of the genus *Listeria*. Cells from growth on blood agar plates were suspended in 4 ml of PCR reaction buffer (~10^9 cells/ml) and samples containing ~5×10^5 cells/ml were then lysed by first treating with 1 mg/ ml of lysozyme at room temperature for 15 min followed by the addition of 200 µg/ ml of Proteinase K, incubation at 60°C for 30 min, and then boiling for 10 min. Six of 50 cooked sausage samples were confirmed positive for the genus *Listeria* with the use of metabolic tests. Only three of the samples were confirmed to be *L. mono-cytogenes* by PCR. The hlyA primer set detected the serotypes 1/2a, 1/2b, 1/2c, 4b, and 4a. The iap primer set detected the first four serotypes but failed to detect the rare 4a serotype.

With milk samples, various numbers of *L. monocytogenes* were added to 10 ml of milk and the cells recovered by centrifugation at 3000 g for 10 min, washed five times with 1 ml of PCR reaction buffer with intermittent centrifugation, and subjected to lysis with lysozyme/Proteinase K and boiling in a volume of 50 µl. The entire sample was then subjected to PCR using only the hlyA primers in a reaction volume of 100 µl. The direct detection limit in milk was 10 cells/10 ml.

Cooray et al. (1994) developed an mPCR assay for detection of *L. monocyto-genes* in seeded milk using three pairs of primers simultaneously. One primer pair amplified a 571-bp sequence of the *prfA* gene, a second primer pair amplified a 276-bp sequence of the *hlyA* gene, and a third primer pair amplified a 795-bp sequence of the *plcB* gene (Table 12.1). The authors noted that boiling for 5 min at 100°C was superior to heating at 95°C for releasing DNA from cells. The presence of 0.05% Tween 20 enhanced DNA release by boiling. The limit of detection was 4 CFU per PCR reaction (4×10^3 CFU/ml of milk) when 1 µl of lysed cell preparations was incorporated into PCR reactions without pre-enrichment. The incorporation of all three primer pairs in a single mPCR reaction resulted in the requirement for an increase in *Taq* polymerase to three units per reaction. Partial inhibition of *Taq* poly-merase by components in milk was eliminated by washing the bacterial pellet three times with PBS with intermittent centrifugation at 5000 g for 5 min.

Niederhauser et al. (1992) used the same two sets of primers as Furrer et al. (1991) for mPCR detection of the *hlyA* and *iap* genes simultaneously. The *hlyA* gene was detected in all 13 serotypes of *L. monocytogenes* examined, whereas the *iap* gene was absent in the rare 4a and 4c serotypes. A sequence of two selective enrich-ments was used followed by the lysis procedure of Furrer et al. (1991). The detection limit with pure cultures was 250 cells per PCR reaction equating to 1×10^4 cells/ml of enrichment broth. With the use of a centrifugation step to achieve a fivefold con-centration of cells in the enrichment broth prior to lysis and PCR, the detection limit was 2000 cells/ml of enrichment broth. The application of the procedure to various meat and dairy products resulted in a detection limit of less than 10 CFU/10 g with both primary and secondary enrichment except for raw meat, where 4 CFU/10 g was not detected by primary enrichment alone. A total of 330 naturally contaminated food samples yielded 20 positive samples. High levels of bacteria in some second-ary enrichments led to false-negative results due to inhibition of *Taq* polymerase. Dilution of secondary enrichments by a factor of 100 eliminated this problem.

Zhang and Knabel (2005) developed an mPCR procedure for identifying *L. monocytogenes* serotypes 1/2a and 4b, which are considered responsible for the majority of cases of human listeriosis worldwide. The primers ORF2372-F/ORF2371-R (Table 12.1) amplified a 595-bp sequence specific for serotype 4b. Primers inlB-F/inlB-R (Table12.1) amplified a 500-bp sequence of the internalin B (*inlB*) gene. Primers inlC-F/inlC-R (Table 12.1) amplified a 400-bp sequence of the internalin C (*inlC*) gene. Primers lmo0171-F/lmo0171-R (Table 12.1) amplified a 200-bp sequence derived from a gene encoding an internalin-type protein from serotypes 1/2a and 1/2c. Three agarose gel patterns emerged. Pattern I was specific for serotype 4b and yielded three bands of 595 bp (ORF2372), 500 bp (inlB), and 400 bp (inlC). Pattern II was specific for serotypes 1/2a and 1/2c with three bands of 500 bp (inlB), 400 bp (inlC), and 200 bp (lmo0171). Pattern II had only two bands (500-bp inlB and 400-bp inlC) and was common to all *L. monocytogenes* serotypes other than 1/2a, 1/2c, and 4b.

The species *L. monocytogenes* encompasses a diversity of strains with varied virulence and pathogenicity. Many strains of *L. monocytogenes* are notably virulent, whereas others are avirulent. Liu et al. (2007) developed an mPCR assay for confirming the identity of *L. monocytogenes* isolates and their mammalian virulence. The assay focused on the simultaneous amplification of sequences from three genes. The *InlA* gene encodes internalin A, which is species-specific and is present in all isolates of *L. monocytogenes* but in no other species of *Listeria*. The *inlC* gene encodes internalin C that contributes to the postintestinal stages of listeriosis. The *inlJ* gene encodes a novel internalin that facilitates the passage of *L. monocytogenes* through the intestinal barrier and is also involved in subsequent stages of infection. Although the *inlJ* gene is found in numerous virulent strains, certain unique lineage IIIB strains that are notably virulent for mice have been found to lack this gene (Liu et al., 2006). Primers inlA-F/inlA-R (Table 12.1) amplified an 800-bp sequence of the *inlA* gene. Primers inlC-F/inlC-R (Table 12.1) amplified a 517-bp sequence of the *inlC* gene. Primers inlJ-F/inlJ-R amplified a 338-bp sequence of the *inlJ* gene. The species identity of all 36 strains of *L. monocytogenes* was confirmed with the *inlA* primers. The virulence of these strains was confirmed with the *inlC* and *inlJ* primer pairs. Virulent strains able to cause mouse mortality following intraperitoneal injection were consistently detected with the *inlC* or *inlJ* primers. Nonpathogenic strains showing no mouse virulence were negative with these primer pairs.

G. Real-Time PCR (Rti-PCR)

Nogva et al. (2000) developed an Rti-PCR assay for quantitative detection of *L. monocytogenes* in water, skim milk, and unpasteurized whole milk utilizing TaqMan methodology (Table 12.1). The assay amplified a 113-bp sequence of the listeriolysin O gene (*hlyA*) and was positive for all 65 *L. monocytogenes* isolates tested and negative for all other *Listeria* strains (16 isolates representing five species). The detection limit was 6 to 60 CFU/PCR reaction with quantitative linearity over at least seven log cycles. The procedure could be completed within 3 hr. Nonspecific bacteria-binding magnetic beads (BB beads, Genpoint AS, Oslo, Norway) were used to capture and concentrate the target organism; they were found superior to other magnetic beads.

Ethidium bromide monoazide (EMA) is a DNA intercollating agent that with visible light activation cross-links the two strands of DNA so as to prevent PCR amplification. EMA penetrates only membrane-damaged cells. This has allowed the selective Rti-PCR amplification of DNA from viable bacterial pathogens including *L. monocytogenes* but not from dead cells (Nogva et al., 2003). Their results, utilizing the same 113-bp amplicon of Nogva et al. (2000), however, indicated that with heat-killed cells of *L. monocytogenes,* complete inhibition of amplification by EMA was not achieved.

Rudi et al. (2005) used EMA Rti-PCR to distinguish viable from dead cells on gouda-like cheeses. The Rti-PCR procedure of Nogva et al. (2000) was used. Nonselective enrichment for 16 hr allowed detection of 10 CFU/g of *L.monocytogenes* initially seeded onto cheese samples.

Oravcová, Kuchta, and Kacliková (2007) developed an Rti-PCR assay for the detection of *L. monocytogenes* in foods. A 24-hr incubation in half-Frazer Broth followed by a 6-hr subculture in Frazer broth was used. The primers LMrt3F/LMrt3R (Table 12.1) amplified an unspecified base-pair sequence length of the *actA* gene. A dual-labeled probe listP with FAM at the 5′-end and TAMRA at the 3′-end was used for detection of amplification. The limit of detection with seeded samples of cheese, fish, and salami was 10 CFU/25 g.

Soejima et al. (2008) found that 10 mg/ml of EMA plus intense visible light randomly cleaved chromosomal DNA from heat-killed cells of *L. monocytogenes* but not from viable cells. With primers 23S-MF/23S-MR (Table 12.1) the amplicon was large (894 bp) and amplification was completely suppressed by EMA plus intense visible light. With the 113-bp amplicon (Table 12.1) from Nogva et al. (2000) DNA amplification was not completely suppressed by EMA plus light only. The treatment of heat-killed cells with topoisomerase poisons following exposure of cells to EMA plus light completely prevented amplification of a 113-bp sequence of the *hly* gene but did not prevent amplification of DNA from live cells. Topoisomerase poisons penetrate heat-killed cells but not live cells and function by accelerating the forward breakage of DNA by topoisomerase (which survives the heat treatment) and inhibits its DNA reunion activity. This resulted in the cleavage of genomic DNA below 113 bp so as to prevent such a small sequence from being amplified.

H. Application of Random Amplified Polymorphic DNA (RAPD) Analysis to *L. monocytogenes* Isolates

The application of RAPD analysis to *L. monocytogenes* was evaluated as part of a World Health Organization (WHO) multicenter study (Wernars et al., 1996). Six laboratories were requested to use a standard protocol derived from that of Mazurier and Wernars (1992) for RAPD analysis of 80 strains of *L. monocytogenes*. Among these strains were 22 groups of epidemiologically linked isolates and 11 pairs of duplicate strains. Using three different 10-mer primers (Table 12.2), the median reproducibility derived from the 11 pairs of duplicate strains was 86.5% (range 0–100%).

Failure in reproducibility was mainly due to results obtained with one particular primer, HLWL74, that was considered to be more sensitive to experimental variables than the other two primers used. The overall correlation between the results from the different participating laboratories ranged from 32% to 85%. One set of duplicate strains was found to be associated with numerous problems of profile reproducibility, indicating that RAPD analysis of some strains is more problematic than others with respect to reproducibility. A low correlation between some laboratories was due to a lack of discriminating capacity with certain of the primers that can be compensated for by using other primers of greater discriminating capacity. These observations indicated that despite the use of a standardized protocol, the reproducibility of the RAPD analysis can vary between different laboratories. The authors stressed that a standardized protocol is essential to reproducible results and that variations in the source of *Taq* polymerase and nonhomogeneity of the temperature heating-block, especially during the annealing step, can result in poor reproducibility.

Lawrence, Harvey, and Gilmour (1993) made use of a 10-mer primer designated OPM-01 (Table 12.2) to produce RAPD profiles with 91 strains of *L. monocytogenes* from raw milk, food, veterinary, medical, and food-environmental sources. Cells were cultured overnight in brain–heart infusion broth, washed in saline, centrifuged, resuspended in distilled water, and heated at 100°C for 10 min. A 100-fold dilution in distilled water was then made and 2.5 μl used for amplification in 50 μl PCR reaction volumes. Reproducibility was enhanced by annealing at low stringency (30°C) for 2 min and by the introduction of a 1-min ramp time between annealing and extension temperatures. This resulted in increased efficiency of PCR amplification due to stabilization of the primer during annealing and was found to allow resolution of DNA fragments across a broader molecular weight range with increased reproducibility. The resulting profiles contained 1 to 10 bands. A total of 33 RAPD profiles were obtained, with specific profiles representing each source. With food strains, one RAPD profile was more common than others, suggesting this to be a common type associated with foods.

Farber and Addison (1994) subjected 52 strains of *L. monocytogenes* to RAPD analysis using three random primers. Cells were grown overnight in tryptic soy broth containing 0.6% yeast extract (TSB-YE). Cell pellets were ground for 30 s with a disposable micropestle. One hundred microliters of DNA extraction buffer (200 mmolar Tris-HCl at pH 7.5, 250 mmolar NaCl, 25 mmolar EDTA, and 0.5% SDS) were added followed by 400 μl of extraction buffer. After centrifugation, the supernatant was precipitated with cold isopropanol and dissolved in TE buffer. The three primers used (Table 12.2) were selected as the best among 200 screened. Primer UBC155 yielded 23 profiles and was found to provide a banding pattern that was the most clearly interpreted. Primers UBC156 and UBC157 generated 19 and 26 reproducible profiles, respectively, but were less discerning because of the faintness of some bands. Most banding patterns consisted of between 5 and 15 distinct bands. When only one primer was used, strains from different serotypes were occasionally found to produce identical banding profiles, which were distinguishable with the use of the other two primers. A total of 31 composite profiles were obtained using all three primers with the 52 isolates.

Table 12.2   Primers Used for RAPD Analysis of *L. monocytogenes* Isolates

Primer Designation	Sequence (5′ → 3′)	Number of Strains Examined	Number of Banding Patterns	Reference
OPM-01	GTT-GGT-GGC-T	91	33	Lawrence, Harvey, and Gilmour (1993)
OPM-01	GTT-GGT-GGC-T	289	18	Lawrence and Gilmour (1995)
HLWL74	ACG-TAT-CTG-C	51	21	Mazurier and Wernars (1992)
HLWL74	ACG-TAT-CTG-C	104	38	Mazurier et al. (1992)
HR4	AGT-GCG-AGC-AGC-CAG-GTC-A	48	20	Niederhauser et al. (1994)
—	ACC-GCC-TGC-T	18	10	Black et al. (1995)
UBC 155	CTG-GCG-GCT-G	115	11	Destro, Leitao, and Farber (1996)
UBC 127	ATC-TGG-CAG-C	115	16	Destro, Leitao, and Farber (1996)
UBC155	CTG-GCG-GCT-G	148	16	Vogel et al. (2001) from Farber and Addison (1994)
HLWL85	ACA-ACT-GCT-C	148	16	Vogel et al. (2001) from Wernars et al. (1996)
DAF4	CGG-CAG-CGC-C	148	16	Vogel et al. (2001) from Wiedman-al-Ahmad et al. (1994)
OPM-01	GTT-GGT-GGC-T	148	16	Vogel et al. (2001) from Lawrence, Harvey, and Gilmour (1993)
HLW74	ACG-TAT-CTG-C	79	—	Kerr et al. (1995) from Mazurier and Wernars (1992)
LURP I	TTT-CTT-CAG-GTC-TTC-GC	79	—	Kerr et al. (1995) from Mazurier and Wernars (1992)
LURP2	GGT-TTG-CTT-GAG-CAA-GC	79	—	Kerr et al. (1995) from Mazurier and Wernars (1992)
UBC155	CTG-GCG-GCT-G	52	23	Farber and Addison (1994)

UBC156	GCC-TGG-TTG-C	52	19	Farber and Addison (1994)
UBC127	ATC-TGG-CAG-C	52	26	Farber and Addison (1994)
1-Universal	TTA-TGT-AAA-ACG-ACG-GCC-AGT	—	—	Levett et al. (1993) from Welsh and McClelland (1990)
2	ATC-TGC-AGC-TGA-ACG-GTC-TGG	—	—	
3	CAG-ATT-TCA-TGC-CAC-GTC-GTT-CC	—	—	
4	GGG-CGT-TGT-CGG-TGT-TCA-TG	—	—	
5	ACA-GGT-CCA-ACA-AAA-GCT-GG	—	—	
6	AAC-AGC-ACT-CTG-TTC-AGG-C	—	—	
1-Universal	TTA-TGT-AAA-ACG-ACG-GCC-AGT	95	7	O'Donoghue et al. (1995) from Welsh and McClelland (1990)
2	ATC-TGC-AGC-TGA-ACG-GTC-TGG	95	5	
3	CAG-ATT-TCA-TGC-CAC-GTC-GTT-CC	95	6	
4	GGG-CGT-TGT-CGG-TGT-TCA-TG	95	6	
5	ACA-GGT-CCA-ACA-AAA-GCT-GG	95	8	
6	AAC-AGC-ACT-CTG-TTC-AGG-C	95	9	
1-Universal	TTA-TGT-AAA-ACG-ACG-GCC-AGT	—	—	MacGowan et al. (1993) from Welsh and McClelland (1990)
2	GGG-CGT-TGT-CGG-TGT-TCA-TG	—	—	
3	ACA-GGT-CCA-ACA-AAA-GCT-GG	—	—	
PB1	GGA-ACT-GCT-A	100	25	Boerlin et al. (1995)
PB4	AAG-GAT-CAG-C	100	22	
HLWL74	ACG-TAT-CTG-C	100	22	Boerlin et al. (1995) from Mazurier and Wernars (1992)

Continued

Table 12.2   Primers Used for RAPD Analysis of *L. monocytogenes* Isolates (Continued)

Primer Designation	Sequence (5′ → 3′)	Number of Strains Examined	Number of Banding Patterns	Reference
PJ108	GCT-TAT-TCT-TGA-CAT-CCA	51	29	Louie et al. (1996)
PJ118	TGT-TCG-TGC-TGT-TTC-TG	51	31	
HLWL74	ACG-TAT-CTG-C	287	13	Giovannacci et al. (1999) from Mazurier and Wernars (1992)
PB4	AAG-GAT-CAG-C	287	1	Giovannacci et al. (1999) from Boerlin et al. (1995)
UBC127	ATC-TGG-CAG-C	287	12	Giovannacci et al. (1999) from Farber and Addison (1994)
Lis5	GCT-GGA-GTC-A	287	5	
Lis11	AGC-CAG-GTC-A	287	6	
OPL-8	AGC-AGG-TGG-A	3	3	Czajka et al. (1993)
OPL-13	ACC-GCC-TGC-T	3	3	
OPL-20	TGG-TGG-ACC-A	3	3	
HLWL85	ACA-ACT—GCT-C	429	55	Vogel et al. (2001) from Wernars et al. (1996)
M13	GTT-GTA-AAA-CGA-CGG-CCA-GT	61	—	Franciosa et al. (2001)
HLWL74	ACG-TAT-CTG-C	80	—	Wernars et al. (1996) from Mazurier and Wernars (1992)
HLWL82	CGG-CCT-CTG-C	80	—	
HLWL85	ACA-ACT—GCT-C	80	—	Wernars et al. (1996) from Wernars et al. (1996)

Lawrence and Gilmour (1995) isolated a total of 289 strains of *L. monocytogenes* from a single poultry-processing plant and resulting poultry products over a 6-month period and subjected them to RAPD analysis using primer OPM-01 (Table 12.2). Eighteen RAPD profiles (A to R) were identified with 184 (64%) displaying a single RAPD profile (A). This genotype was widespread on poultry contact surfaces, floors, and drains. This was the only genotype that persisted throughout the entire 6-month period. This genotype and a second (B) were the only genotypes found in both raw and cooked poultry-processing environments. *L. monocytogenes* strains isolated from the cooked processing environment up to 1 year later (17 strains) contained only RAPD types A and B, reflecting the potential for cross-contamination by persistent strains. Multilocus enzyme electrophoresis provided no further differentiation within RAPD types B through R, reflecting the high level of discrimination provided by RAPD analysis.

Mazurier and Wernars (1992) used a 10-mer random primer HLWL74 (Table 12.2) for the establishment of RAPD profiles of isolates representing all seven species of *Listeria*, including 51 strains of *L. monocytogenes*. To ensure uniformity of cellular DNA, the cells from overnight brain–heart infusion broth cultures were washed and resuspended in saline to an absorbance at 600 nm of 1.5 ($7.5 \pm 0.5 \times 10^6$ CFU/ml) and 5 μl used for PCR amplification without prior cell lysis. A total of 29 different banding profiles were obtained. Among these, 21 patterns were found exclusively among the 51 strains of *L. monocytogenes*. The remaining eight profiles were derived from the additional six species of *Listeria*. RAPD analysis was found to discriminate between isolates of different species and also between isolates of a given species. Serotypes of *L. monocytogenes* were not found to be consistently correlated with RAPD banding patterns.

Mazurier et al. (1992) subjected 104 strains of *L. monocytogenes* to RAPD typing using the single 10-mer primer HLWL74 (Table 12.2). A total of 38 RAPD profiles was obtained. The correlation between RAPD typing and phage typing for 53 human outbreak strains was 98%.

Levett et al. (1993) used six separate random primers (Table 12.2) ranging from 19 to 23 mer to confirm a relapsed infection by a single strain of *L. monocytogenes*. DNA was extracted by boiling cell suspensions for 3–5 min. The number of bands obtained with each primer varied from two to five.

MacGowan et al. (1993) subjected representative isolates of all seven species of *Listeria* to RAPD analysis using three separate primers (Table 12.2) ranging in length from 19 to 21 mer. Several colonies were suspended in 15–20 μl of d.H_2O, which was boiled for 3–5 min and then centrifuged for 20 s. The supernatant was then used for RAPD analysis. Five isolates of *L. innocua* and four isolates of *L. selligeri* were all distinguishable from one another. The four isolates of *L. ivanovii* tested, although distinguishable from other *Listeria* spp., were not differentiated. Fourteen neonatal cross-infection sets of *L. monocytogenes* isolates, shown to be indistinguishable by serotyping and phage typing, were examined with the three primers. With one primer, three of the sets were shown to consist of closely related but distinguishable strains. In the other 11 cases, each set of strains was indistinguishable with all three primers. Smaller 10-mer primers may have yielded higher levels of discrimination.

Niederhauser et al. (1994) evaluated 11 different primers for RAPD analysis of isolates representing *Listeria* spp. and found that only one HR4, a notably long 19-mer primer (Table 12.2), generated reproducible and specific profiles for members of the genus *Listeria*. To ensure the generation of clear and strong RAPD profiles, the isolates were grown on tryptic soy agar plates for a minimum of 24 hr. The cells were harvested with 2.5 ml of PCR buffer. A tenfold dilution of the cell suspension in PCR buffer was boiled for 15 min and centrifuged at 12,000 g, and the absorbance of the supernatant at 260 nm was adjusted to 0.7. The diluted lysates (10 µl) were then used directly for RAPD analysis. A total of 48 strains of *L. monocytogenes* from food and clinical sources yielded 20 RAPD profiles with the HR4 primer.

Boerlin et al. (1995) typed 100 strains of *L. monocytogenes* by RAPD with three different 10-mer random primers (Table 12.2) and compared the results to those obtained by serotyping, ribotyping, multilocus enzyme electrophoresis, restriction enzyme analysis, and phage typing. Preliminary studies with untreated cells indicated that the RAPD profiles obtained were dependent on the state of growth of the cells. A portion (2.5 ml) of overnight BHI broth cultures was therefore centrifuged and the pellets washed in 1 ml d.H_2O. The cell pellets were resuspended in d.H_2O to an absorbance at 600 nm of 1.3. The cell suspensions were then subjected to four freeze–thaw cycles (1 min frozen in an alcohol–dry ice bath and thawed for 1 min at 98°C). Portions (5 µl) of the disrupted cell suspensions were then added to 50 µl PCR mixtures. This procedure for releasing DNA was found to enhance the reproducibility of RAPD profiles. The RAPD profiles appeared to be stable during epidemics over periods of several years. Discrimination by RAPD typing using two or three primers was found to be superior to the other five typing methods.

O'Donoghue et al. (1995) typed 25 serogroup 1/2 strains (19 strains of serovar 1/2a and six of serovar 1/2b) and 70 serovar 4b strains of *L. monocytogenes* from human clinical sources using RAPD and the six random primers used by Levett et al. (1993; Table 12.2). DNA was extracted from single colonies by boiling as described by MacGowan et al. (1993). All 70 serovar 4b strains gave the same major banding patterns with each of the six primers. With the use of minor bands, the 4b isolates could be differentiated and yielded 59 different banding patterns. In contrast, the 25 serogroup 1/2 isolates yielded 12 different profile groups. The discrimination index for RAPD major band analysis was 0.452 and for RAPD major and minor band analysis was 0.974. Serovar 1/2 isolates were found to be more heterogeneous than 4b isolates because they were more easily distinguished on the basis of major band analysis alone. The relatively low number of banding profiles (5–9) obtained with each primer can be attributed to the relatively high number of nucleotides per primer (19–23).

Black et al. (1995) used a novel capillary thermal cycler with a 10-mer random primer (Table 12.2) to generate RAPD profiles with strains of *Listeria* species involving less than 60 min for 30 thermal cycles. The cells from four large colonies were suspended in 100 µl of d.H_2O to yield a heavy cell suspension, which was then heated at 95°C for 5 min to achieve cell lysis. One microliter of the supernatant resulting from centrifugation at 8000 g for 1 min was then used for amplification. Representative strains from each of the seven species of *Listeria* yielded distinguishable profiles. Eighteen strains of *L. monocytogenes* yielded 10 profiles with similar profiles observed with serotypes 1/2a, 4b, 4d, and 7. Increased surface binding of $MgCl_2$ and surface denaturation of *Taq* polymerase due to the increase in the surface area to volume ratio with capillary tubes was overcome by increasing the concentration of $MgCl_2$ to 3.2 mMol and adding 0.25 mg/ml of bovine serum albumin (BSA) to PCR reaction mixtures. The addition of BSA allowed the reduction of *Taq* polymerase by 50%.

Destro, Leitao, and Farber (1996) subjected 115 strains of *L. monocytogenes* collected from different areas of a shrimp processing plant in Brazil over a 5-month period to RAPD and pulsed field gel electrophoresis (PFGE) analysis. For RAPD analysis, isolates were grown overnight at 37°C in TSB containing 0.6% yeast extract. The cells were washed in saline and resuspended in distilled water to an absorbance at 600 nm of 1.8 (~10^7 cells/ml). One microliter of the cell suspension was then added to the PCR mix and heated at 96°C for 6 min, followed by the addition of *Taq* polymerase and subjected to 35 cycles of amplification. Two random primers were used for RAPD analysis designated UBC 155 and UBC 127 (Table 12.2) that generated 11 and 16 different RAPD profiles, respectively. The authors found that the concentration of *Taq* polymerase greatly influenced reproducibility of RAPD profiles and used 0.85 units per PCR reaction in presumably 50-μl reaction volumes. The use of composite profiles derived from both RAPD and PFGE resulted in an increase in strain discrimination. Strains from the processing environment fell into unique composite profile groups, whereas strains from both processing water and utensils showed another composite profile group. Isolates from fresh shrimp of one profile group were found in different areas of the processing line. This same profile group was also present on food handlers from the processing and packaging areas of the plant. In addition to RAPD analysis, PFGE was performed on all 115 isolates with the use of two restriction endonucleases *Sma*I and *Apa*I, which yielded 13 and 15 restriction endonuclease digestion profiles (REDP), respectively. When the DNA was digested with *Sma*I the number of bands per profile ranged from 16 to 25, and when digested with *Apa*I, 11 to 17 number bands per profile resulted. Use of the two restriction nucleases with PFGE increased the discriminatory ability. Some overlap among serogroups resulted when only RAPD profiles were used. However, when composite RAPD–PFGE profiles were compared, no overlap occurred, clearly indicating that certain strains are capable of colonizing plant surfaces.

Vogel et al. (2001a) compared the RAPD profiles of 148 isolates of *L. monocytogenes* from vacuum-packed cold-smoked salmon derived from 10 different Danish smokehouses. A total of 16 different RAPD profiles was obtained using four separate primers. The grouping of all 148 strains was exactly the same with each of the four primers (Table 12.2) used. Isolates, which were indistinguishable using a single primer, were on no occasion found to be dissimilar with the other three primers. The authors noted that the same RAPD types were found in products produced after 6 and 8 months for two of the smokehouses, indicating long-term establishment of specific strains in smokehouses. A subset of 20 strains typed by PFGE using the restriction endonuclease *Apa*I resulted in only one strain allocated to a different group as compared to grouping by RAPD. Different RAPD types dominated products from different smokehouses. Some identical RAPD types were isolated from several smokehouses. Each smokehouse carried its own specific RAPD type, suggesting a possible persistence of closely related strains of *L. monocytogenes*. The application of only heat (Mazurier and Wernars, 1992) to lyse the cells from different cell preparations of the same isolates yielded RAPD profiles with differences in band intensities and numbers of bands. The authors therefore used the DYNAL DIRECT System I for cell lysis and DNA purification.

Kerr et al. (1995) subjected 79 strains of *L. monocytogenes* derived from human clinical sources, food, and the hands of food handler personnel to RAPD analysis. One 10-mer and two 17-mer random primers were used (Table 12.2). Cells were harvested from the entire surface of blood agar plates and washed with saline, and 0.5 ml of DNA extraction buffer (0.1 M NaOH, 1 M NaCl, and 0.3% SDS) was added to the cell pellet followed by heating at 95°C for 15 min. The released DNA was then phenol-chloroform (1:1 v/v) purified, precipitated with ethanol, washed with ether, resuspended in d.H$_2$O, and stored at 4°C. Epidemiologically related strains previously shown to be indistinguishable by phage typing yielded identical RAPD profiles. Strains isolated from the hands of three workers in a retail food establishment showed the presence of a single predominant *L. monocytogenes* RAPD type. Several strains, found to be indistinguishable with one primer, were found to be dissimilar when the two additional primers were used. Significant differences in reproducibility of banding patterns occurred with template DNA produced by direct heating of cells at 100°C for 10 min. This problem was eliminated by the use of phenol-chloroform extraction of DNA. Such extracted DNA could be stored for up to 5 months at 4°C without alteration in profile. The authors concluded that using a single primer may fail to distinguish between strains, whereas the use of at least three primers resulted in a level of discrimination that allowed satisfactory use of just RAPD for valid comparison of epidemiologically linked strains.

Cao et al. (2005a) subjected 99 randomly selected isolates of *Listeria monocytogenes* from several processing environment locations in a shrimp processing plant obtained during a 5-month sampling period to randomly amplified polymorphic DNA (RAPD) analysis with the use of four primers UBC155, PB1, PB4, and HLWL74 (Table 12.2). Preliminary studies indicated that the number of DNA bands and their intensity differed greatly with respect to the commercial source of the *Taq* polymerase used with individual isolates. There were 18 composite RAPD types discerned with the use of the four primers. Among these 18 composite RAPD types, type 1 was comprised of 14 indistinguishable isolates, and type 9 was comprised of 49 indistinguishable isolates. These results indicated that the shrimp processing plant was dominated by these two RAPD types that comprised 63.6% of the 99 randomly selected isolates.

Cao et al. (2005b) sampled fresh fish fillets over a 24-month period from two major supermarket retail outlets in Amherst, Massachusetts, designated A and B for the incidence of *Listeria monocytogenes* and numbers of the organism present per 100 g of tissue. There were 15 species of fish represented, and 74 samples out of a total of 320 were confirmed by PCR as yielding *L. monocytogenes*. From retail source A, a total of 171 samples yielded 59 (34.5%) that were positive for the presence of *L. monocytogenes*. In contrast, from retail source B, a total of 149 samples yielded 15 (10.0%) that were positive. A total of 221 strains of *L. monocytogenes* were derived from the MPN cultures, 164 from retail source A and 57 from retail source B. All 221 strains were subjected to RAPD analysis using three random primers. Primer PB1 (Table 12.2) yielded 21 RAPD profiles, primer PB4 (Table 12.2) yielded 19 profiles, and primer HLWL74 (Table 12.2) yielded 26 profiles. A total of 55 composite profiles was identified by combining the profiles derived from the three primers. Source A yielded 50 composite RAPD profiles, whereas source B yielded only 10 composite profiles. In addition, 27 of the 55 composite profiles were derived from individual isolates, and RAPD types 11 and 18 included 49 and 27 isolates, respectively. Fish from retail source A clearly harbored far more RAPD types than did source B. The results clearly indicated that two major retail sources in close geographic proximity can vary considerably with respect to the incidence and numbers of *L. monocytogenes* present on the fish tissue. It was not possible to determine whether the processors furnishing fish to retail outlet A or the supermarket itself was responsible for the notably higher incidence and numbers of *L. monocytogenes* on fish from retail source A compared to fish from retail source B.

I. Application of Pulsed Field Gel Electrophoresis (PFGE) to *L. monocytogenes* Isolates

Listeria monocytogenes strains of serovar 4b are the most frequently encountered among cases of listeriosis (Buchrieser et al., 1992; Wagner and Allerberger, 2003). Brosch, Buchrieser, and Rocourt (1991) analyzed 42 strains of the 4b serovar using PFGE and the three restriction endonucleases *Apa*I, *Sma*I, and *Not*I. There were 16 profiles obtained with *Apa*I and *Sma*I and 7 with *Not*I (Table 12.3). The number of bands ranged from 15 to 18 with *Apa*I, 6 to 15 with *Sma*I, and 3 to 7 with *Not*I. PFGE distinguished between strains that were indistinguishable by serotyping, ribotyping, and phage typing.

Buchrieser, Brosch, and Rocourt (1991) applied PFGE to the analysis of 35 *L. monocytogenes* of serovars 1/2 and 3. The number of bands from *Api*I, *Sma*I, and *Not*I was 13–18, 25, and 3–7, respectively, and the number of profiles was 17, 18, and 15, respectively (Table 12.3). The combination of profiles from all three enzymes yielded 24 PFGE types within the 35 strains. No correlation was observed between the restriction profiles and the serovars.

Brosch et al. (1996) subjected 80 strains of *L. monocytogenes* to a WHO multicenter international PFGE typing study involving four different laboratories. The endonucleases *Apa*I and *Sma*I were used in all four laboratories and yielded 23 to 28 and 21 to 24 PFGE profiles, respectively (Table 12.3). *Asc*I was used in one laboratory and yielded 21 profiles. The combination of *Apa*I, *Sma*I, or *Asc*I profiles yielded 25 to 33 types. Agreement of typing data among the four laboratories ranged from 79% to 90%. On average 84% of 11 duplicate (22 cultures) were identified by all four laboratories. Reproducibility in each laboratory was extremely high. Serovars of the 80 strains could be predicted from the RAPD profiles.

Table 12.3   Restriction Nucleases used with Strains of *L. monocytogenes* for PFGE Analysis

Restriction Nuclease	Number of Strains Examined	Number of Digestion Profiles	Reference
*Sma*I	115	13	Destro, Leitao, and Farber (1996)
*Apa*I	115	15	Destro, Leitao, and Farber (1996)
*Apa*I	148	—	Vogel et al. (2001)
*Apa*I	287	17	Giovannacci et al. (1999)
*Apa*I	51	22	Louie et al. (1996)
*Sma*I	51	26	Louie et al. (1996)
*Apa*I	34	—	Brett, Short, and McLauchlin (1998)
*Sma*I	34	—	Brett, Short, and McLauchlin (1998)
*Apa*I	41	6	Miettinen, Björkroth, and Kkorkeala (1999)
*Asc*I	41	8	Miettinen, Björkroth, and Kkorkeala (1999)
*Sma*I	41	7	Miettinen, Björkroth, and Kkorkeala (1999)
*Apa*I	18	8	Unnerstad et al. (1996)
*Asc*I	18	8	Unnerstad et al. (1996)
*Sma*I	18	9	Unnerstad et al. (1996)
*Apa*I	42	16	Brosh, Buchrieser, and Rocourt (1991)
*Sma*I	42	16	Brosh, Buchrieser, and Rocourt (1991)
*Not*I	42	7	Brosh, Buchrieser, and Rocourt (1991)
*Asc*I	303	9	Autio et al. (1999)
*Sma*I	303	6	Autio et al. (1999)
*Apa*I	75	—	Buchrieser et al. (1992)
*Sma*I	75	—	Buchrieser et al. (1992)
*Not*I	75	—	Buchrieser et al. (1992)
*Asc*I	45	32	Harvey and Gilmour (2001)
*Asc*I	7	7	Howard, Harsono, and Luchansky (1992)
*Sma*I	7	—	Howard, Harsono, and Luchansky (1992)
*Apa*I	7	—	Howard, Harsono, and Luchansky (1992)
*Not*I	7	—	Howard, Harsono, and Luchansky (1992)
*Apa*I	35	17	Buchrieser, Brosch, and Rocourt (1991)
*Sma*I	35	18	Buchrieser, Brosch, and Rocourt (1991)
*Not*I	35	15	Buchrieser, Brosch, and Rocourt (1991)

Table 12.3   Restriction Nucleases used with Strains of *L. monocytogenes* for
PFGE Analysis (*Continued*)

Restriction Nuclease	Number of Strains Examined	Number of Digestion Profiles	Reference
AscI	176	63	Brosch, Chen, and Luchansky (1994)
ApaI	176	72	Brosch, Chen, and Luchansky (1994)
ApaI	80	23–28	Brosch et al. (1996)
SmaI	80	21–24	Brosch et al. (1996)
AscI	80	21	Brosch et al. (1996)
ApaI	153	30	Vela et al. (2001)
SmaI	153	28	Vela et al. (2001)
AscI	53	30	Wagner and Allerberger (2003)
ApaI	53	—	Wagner and Allerberger (2003)
SmaI	53	—	Wagner and Allerberger (2003)
SmaI	247	—	Ojeniyi et al. (1996)
ApaI	247	—	Ojeniyi et al. (1996)
ApaI	27	23	Nakama et al. (1998)
AscI	27	25	Nakama et al. (1998)
SmaI	27	25	Nakama et al. (1998)
Sse8387I	27	21	Nakama et al. (1998)

There are relatively few reports of listeriosis derived from seafood (FAO, 1999), however, one such outbreak involving two cases that occurred in New Zealand in 1992 was studied in detail by Brett, Short, and McLauchlin (1998). PFGE profiles derived from the two restriction endonucleases *Apa*I and *Sma*I (Table 12.3) indicated that the isolates from both patients were identical to those obtained from refrigerated mussels of a specific brand still possessed by one patient. Isolates from refrigerated retail packets of the same brand and the processing environment from which they were derived yielded isolates of the same PFGE profile as that of the two patients. This strain was found to persist in the processing environment from 1990 to 1993.

Miettinen, Björkroth, and Kkorkeala (1999) subjected 41 isolates of *L. monocytogenes* from an ice cream plant to PFGE analysis using the three restriction endonucleases *Apa*I, *Asc*I, and *Sma*I. *Asc*I resulted in the highest level of discrimination among the three endonucleases and produced eight profiles, whereas *Apa*I and *Sma*I produced six and seven, respectively (Table 12.3). On the basis of one-band differences, 12 different PFGE types were distinguished using profiles from all three enzymes combined. The dominant PFGE type was found to have persisted in the ice cream plant for 7 years. Improved cleaning and disinfection of the packaging machine combined with structural changes to its conveyer belt to facilitate cleaning eliminated *L. monocytogenes* from the plant.

Strains of *L. monocytogenes* serovar 3b were repeatedly isolated from the processing environment in a Scandinavian dairy and from cheeses it produced from

1988 to 1995 by Unnersstad et al. (1996). Ten of these isolates were analyzed using PFGE with the restriction endonucleases *Apa*I, *Sma*I, and *Asc*I (Table 12.3) and yielded identical profiles with the respective enzymes. An additional eight strains of the same serovar from other sources yielded unique combinations of the three restriction enzyme profiles. The results indicated that a single clone of *L. monocytogenes* can persist in a dairy plant for at least 7 years.

Nakama et al. (1998) used PFGE analysis to trace *L. monocytogenes* contamination in shredded cheese products and the two shredded cheese processing plants from which they were produced. The restriction endonucleases *Apa*I, *Asc*I, *Ama*I, and *Sse8387*I yielded 23, 25, 25, and 21 PFGE types, respectively, with 27 unrelated reference strains from various sources (Table 12.3). Profiles generated with *Asc*I or *Sse8387*I were easier to interpret. The combined profiles with all four endonucleases resulted in the 27 reference strains yielding 27 different genotypes. Strains (6/6) isolated from different lots of cheese processed in plant A over a 2-month period and a strain from a drain displayed an identical genotype. In plant B, three of three strains isolated over a 10-day period from different lots of cheese and three of three strains from the processing equipment and environment exhibited identical PFGE profiles. The authors noted interestingly that the strains isolated from the cheese and processing environments in the two plants had indistinguishable PFGE profiles with all four endonucleases. The authors speculated that because shredded cheese in Japan is usually processed from imported semihard cheeses and that these imported cheeses entering both plants may have been of common origin, they therefore might be contaminated with the identical clonal strains.

Brosch, Chen, and Luchansky (1994) analyzed 176 *L. monocytogenes* strains using PFGE. The endonucleases *Asc*I and *Apa*I resulted in 63 and 72 profiles, respectively (Table 12.3). Statistical analysis of the profiles obtained with *Asc*I revealed two distinct genomic divisions of *L. monocytogenes* that also correlated with the flagella (H) antigen type: division I contained serovar 1/2a, 1/2c, 3a, and 3c strains; and division II contained serovar 1/2b, 3b, 4b, 4d, and 4e strains. Both of these major divisions could be further subgrouped into serovar clusters using *Apa*I-generated profiles.

L. monocytogenes is not considered a natural contaminant of fish (FAO, 1999; Autio et al., 1999). Contamination of fishery products is considered to be a result of processing contamination. The processing of cold-smoked rainbow trout does not inactivate *L. monocytogenes* (Autio et al., 1999). In addition, most such products are vacuum packaged and consumed without cooking, which can pose a potential listeriosis threat. Only one sample of rainbow trout among a total of 60 was found positive before processing in a cold-smoked processing plant (Autio et al., 1999). None of 49 fillets sampled were positive. The frequency of fish contaminated with *L. monocytogenes* was found to clearly rise after brining, and the most contaminated processing sites were the brining and postbrining areas. A total of 303 isolates of *L. monocytogenes* from the raw fish, processing environment, and final product were characterized by Autio et al. (1999) using PFGE. *Asc*I and *Sma*I yielded nine and six profiles, respectively (Table 12.3), and in combination resulted in a total of nine types. The predominating types of the final product were associated with brining and slicing. The use of hot steam, hot air, and hot water was effective in eliminating the organism from the plant and final product.

Buchrieser et al. (1992) applied PFGE using *Apa*I, *Sma*I, and *Not*I to analyze 75 *L. monocytogenes* strains of serovar 4b isolated during six major and eight smaller listeriosis outbreaks. Twenty genomic varieties or types were identified. Thirteen of fourteen strains isolated during major epidemics in Switzerland, the United States, and Denmark from 1983 to 1987 yielded indistinguishable PFGE profiles. In contrast, strains responsible for other outbreaks in Canada, the United States, France, New Zealand, and Austria from 1969 to 1989 yielded individually unique PFGE profiles. From these results it was hypothesized that certain clones of *L. monocytogenes* are more capable of causing globally occurring epidemics than others. A major factor may involve the ability of certain 4b strains to form biofilms with enhanced surface adhesion properties resulting in their persistence in processing environments (Norwood and Gilmour, 1999).

Vela et al. (2001) applied PFGE analysis to 153 strains of *L. monocytogenes* from different sources (72 from sheep, 12 from cattle, 18 from feedstuffs, and 51 from humans). *Apa*I yielded 12–17 bands with 30 profiles, and *Sma*I yielded 11–18 bands with 28 profiles. Composite profiling derived from both endonucleases yielded 55 PFGE types indicating considerable genetic diversity among the 153 strains. In most cases, clinical strains from different animals of the same flock had identical PFGE types. Strains with PFGE types identical to those of clinical strains were isolated from silage, potatoes, and maize stalks.

Wagner and Allerberger (2003) typed 41 human clinical strains of *L. monocytogenes* isolated from listeriosis cases in Austria between the years 1997 and 2000 using PFGE. The resulting PFGE profiles were compared to those from nine reference strains isolated from seven outbreaks in Europe and the United States and three fecal strains from healthy human carriers. *Asc*I, *Apa*I, and *Sma*I yielded 37 combined profile types for the 41 Austrian clinical isolates. The PFGE profiles suggested that a similar genetic background was present in strains derived from global outbreaks of listeriosis and isolates associated with sporadic listeriosis in Austria.

J. Comparison of RAPD, PFGE, and Other Molecular Methods for Typing *L. monocytogenes* Isolates

Czajka et al. (1993) found that although some distinction could be made among strains of *L. monocytogenes* from their 16S rDNA sequence, a far greater discrimination within the species was achieved with RAPD profiles from chromosomal DNA. The use of three random 10-mer primers OPL-8, OPL-13, and OPL-20 (Table 12.2) in a notably limited study allowed differentiation between various serotypes with the same 16S rDNA sequences. The number of RAPD bands obtained with the three primers ranged from one to three. Identical banding patterns, however, were obtained with six strains of serovar 4b. Cell lysis in TEN buffer (10 mM tris-HCl, pH 7.6; 1 mM EDTA, 10 mM NaCl) was achieved with lysozyme (10 mg/ml) at 37°C for 30 min followed by treatment with SDS.

Louie et al. (1996) typed 51 clinical isolates of *L. monocytogenes* (15 isolates from two outbreaks and 36 epidemiologically unrelated isolates) using serotyping, ribotyping (RT), PFGE, and RAPD. Serotyping failed to distinguish between related and unrelated strains. Restriction with *Eco*RI and *Pva*II gave 16 and 23 RT patterns, respectively, with 8–15 bands per isolate. Restriction with *Apa*I or *Sma*I yielded 22 and 26 PFGE profiles, respectively. *Apa*I profiles were found easier to interpret, with 10 to 15 bands each, whereas *Sma*I profiles had 15 to 20 bands each. RAPD with two different primers PJ108 (18-mer) and PJ118 (17-mer; Table 12.2) yielded 29 and 31 profiles, respectively, and 6 to 10 bands per isolate. Among the three molecular techniques evaluated, RT was the least discriminatory. The abilities of RAPD and PFGE to differentiate strains were comparable, with close numerical values for the discriminatory index of each. The use of at least two independent random primers with RAPD was recommended for enhanced discrimination.

Ojeniyi et al. (1996) isolated *L. monocytogenes* from 111/236 (4.7%) cecal samples from parent poultry flocks in Denmark, providing broilers to seven abattoirs investigated. Cecal samples from 2078 broilers representing 90 randomly selected broiler flocks were negative for *L. monocytogenes*. A total of 3080 samples from the seven abattoirs including poultry processing line samples and final products yielded 0.3% to 18.7% positive samples from the individual abattoirs. PFGE analysis of 247 abattoir isolates with *Sma*I and *Apa*I resulted in 27 PFGE types and ribotyping 25 types. Combined profiles from both typing methods yielded 62 types. The number of combined PFGE types obtained with *Sma*I and *Apa*I present in each abattoir varied from 1 to 12, and the number of ribotypes varied from 1 to 11. Both serotyping and phage typing were of limited epidemiological value in that only half of the *L. monocytogenes* isolates were typeable with phage, whereas serotyping lacked discrimination, which others have also reported (Brosch, Buchrieser, and Rocourt, 1991; MacGowan et al., 1993; Boerlin et al., 1995; Louie et al., 1996; Nakama et al., 1998). The abattoirs appeared to be contaminated with only a few main clonal types of *L. monocytogenes* found present on the processing line and in ready-to-eat products. The study indicated that *L. monocytogenes* may be introduced into the abattoirs by the live animals but that it is unlikely they contribute significantly to the total numbers of *L. monocytogenes* in the abattoirs. The results also indicated that the main source of contamination of the processed poultry and poultry products originated from improper cleaning and sanitization of the processing lines and abattoir processing environment.

Giovannacci et al. (1999) examined 287 isolates of *L. monocytogenes* from five French pork slaughtering and cutting plants, involving samples from live swine and cut pork. RAPD was performed with five different 10-mer primers (Table 12.2). PFGE was performed using the restriction endonuclease *Apa*I (Table 12.3). Results obtained from RAPD and PFGE were closely related, with 17 RAPD and 17 PFGE types distinguished. Dominance of a single strain type occurred in addition to other very closely related types over a 1-year period in the environment of two plants, even after cleaning and disinfection procedures.

The routes of contamination of two Danish cold-smoked salmon processing plants by *L. monocytogenes* were investigated by Vogel et al. (2001b) by analyzing

3585 samples from products (1995 to 1999) and processing environments (1998 and 1999). The level of product contamination in plant I varied from 31% to 85% and no *L. monocytogenes* was found on raw fish. In plant II, the levels of both raw fish and product contamination varied from 0% to 25%. A total of 429 isolates of *L. monocytogenes* were subjected to RAPD analysis with a single 10-mer primer HLWL85 (Table 12.1) and 55 different profiles resulted. DNA for RAPD was prepared with Dynabeads DYNAL DIRECT system 1. The RAPD types detected on the product were identical to types found on the processing equipment and the processing environment, suggesting that contamination in both plants was from the processing environment and not from the raw fish. In plant I, the same predominant RAPD type was found over a 4-year period. In plant II, which had a lower prevalence of *L. monocytogenes*, no RAPD type persisted over long periods of time. Persistent strains (125) were also typed by PFGE and amplified fragment length polymorphism (AFLP) analysis, which confirmed the results obtained by RAPD profiling. The authors concluded that persistent strains may be avoided by vigorous cleaning and sanitation.

Franciosa et al. (2001) subjected 29 unrelated strains of *L. monocytogenes*, 16 strains from a recent outbreak of invasive listeriosis, and 16 strains from two outbreaks of noninvasive listeriosis occurring in Italy to three PCR-based typing techniques. Among the three techniques, analysis by infrequent-restriction-site PCR (IRS-PCR) exhibited the highest level of discrimination among strains from the invasive outbreak, in that three different clusters of strains were identified compared to two clusters resulting from both RT and RAPD. Strains among the two outbreaks of noninvasive listeriosis were practically identical with all three techniques. Only IRS-PCR clearly discriminated between strains derived from noninvasive listeriosis and those from invasive listeriosis. RAPD was performed with only a single 18-mer primer M13 (Table 12.2) derived from the core sequence of bacteriophage M13. Low-resolution bands were consistently obtained with a high background and resulted in a low degree of discrimination with RAPD. The use of two or three 10-mer primers may have resulted in a higher level of discrimination with RAPD.

Random amplified polymorphic DNA and pulsed-field gel electrophoresis analyses have been found to be powerful molecular methods for differentiating isolates of a given bacterial species. When applied to *Listeria monocytogenes*, both methods have been found highly effective in tracking isolates involved in foodborne outbreaks of listeriosis and in identifying routes of contamination in food processing plants. Between the two methods, PFGE is considered somewhat superior in discriminatory power. However, the use of two or more independent random primers with RAPD is considered to result in a level of discrimination equal to that of PFGE. When results from both methods are combined, a maximum level of discrimination that exceeds that obtained with both methods independently can be achieved. Individually, both methods far exceed the discriminatory power of serotyping and phage typing of *L. monocytogenes* strains in that serotypes 1/2a, 1/2b, and 4b, represent over 90% of all human isolates, and phage typing at times has allowed typing of no more than about 50% of isolates. In addition, both RAPD and PFGE on occasion have been found to be superior to ribotyping, multilocus enzyme electrophoresis, and restriction enzyme analysis of *L. monocytogenes* isolates.

Harvey and Gilmour (2001) analyzed 14 sporadic and 9 recurrent isolates of *L. monocytogenes* from raw milk and 10 sporadic and 12 recurrent isolates from non-dairy foods using PFGE and multilocus enzyme electrophoresis (MEE). *Asc*I band patterns were simpler and more clearly resolved compared to those generated with *Apa*I and were therefore used solely for typing, which resulted in 32 PFGE profiles. The grouping of strains by PFGE and MEE were in broad agreement. Plasmid carriage and resistance to cadmium were found to occur more frequently in recurrent than in sporadic strains and may possibly be associated with persistence in food and food processing environments.

K. Microarray Technology

Hong et al. (2004) developed an oligonucleotide array on a nylon membrane for detection of 10 foodborne pathogens including *L. monocytogenes*. PCR template DNA was derived from single colonies. The universal primers P1/P2 (Table 12.1) were used to amplify a ~900-bp sequence of bacterial 23S rDNA of each species. The P2 primer was labeled at the 5′-end with digoxygenin (DIG) for color development. DNA probes were first applied to a specific location of the membrane for each organism and the membranes washed and dried. Probe no. 11 (Table 12.1) was used for detection of *L. monocytogenes*. Five microliters of PCR DIG-labeled product from each specific organism was then added to a hybridization solution in which the membrane was immersed and hybridization allowed to occur for 4 hr at 50°C. The membranes were then processed for color development using ELISA-based anti-DiG antibody methodology. 23S rRNA gene amplification products for the 10 species yielded high levels of sensitivity and specificity.

L. PCR Determination of Serotype Divisions

The serotypes of *L. monocytogenes* are considered to comprise three divisions (I, II, and III). Borucki and Call (2003) developed serological-division-specific primers for serological identification of *L. monocytogenes* isolates. The primers D1-F/D1-R (Table 12.1) amplified a 214-bp sequence of anion transport protein produced by serological divisions I and II. Primers D2-F/D2-R (Table 12.1) amplified a 140-bp gene sequence present solely in serological division II isolates. Primers FlaA-F/FlaA-R (Table 12.1) amplified a 38-bp sequence of the *flaA* gene, which encodes the *L. monocytogenes* flagella protein of the division II serotype. Primers GLT-F/GLT-R (Table 12.1) amplified a 485-bp sequence of the 1/2b serotype-specific DNA region flanking the *gltA-gltB* cassette described by Lei et al. (2001). D1 and D2 primer pairs were used in an mPCR to distinguish division II serotypes from divisions I and II. Strains identified as belonging to division I or II were further subtyped by using primers GLT-F/GLT-R to differentiate serotypes 4 and 1/2b (Table 12.1). Strains identified as serotype 4 were further serotyped for identification of division II strains using the MAMA-C primers (Table 12.2) of Jinneman and Hill (2001). If strains tested positive with these primers then the serogroup was identified as 4a/c. Serovar 4 strains that tested negative with these primers were considered serotype 4b.

REFERENCES

Autio, T., Hielm, S., Miettinen, M., Sjöberg, A., Aarnisalo, K., Björkroth, J., Mattila-Sandholm, T., Korkeala, H. 1999. Sources of *Listeria monocytogenes* contamination in a cold-smoked rainbow trout processing plant detected by pulsed-field gel electrophoresis typing. *Appl. Environ. Microbiol.* 65:150–155.

Aznar, R., Alarcón, B. 2002. On the specificity of PCR detection of *Listeria monocytogenes* in food: A comparison of published primers. *Syst. Appl. Microbiol.* 25:109–119.

Bansal, N. 1996. Development of a polymerase chain reaction assay for the detection of *Listeria monocytogenes* in foods. *Lett. Appl. Microbiol.* 22:353–356.

Bej, A., Steffan, R., DiCesare, J., Haff, L., Atlas, R.M. 1990. Detection of coliform bacteria in water by polymerase chain reaction and gene probes. *Appl. Environ. Microbiol.* 56:307–314.

Bessesen, M., Luo, Q., Rotbart, H., Blaser, M., Ellison, III, R. 1990. Detection of *Listeria monocytogenes* by using the polymerase chain reaction. *Appl. Environ. Microbiol.* 56:2930–2932.

Black, S., Gray, D., Fenlon, D., Kroll, R. 1995. Rapid RAPD analysis for distinguishing *Listeria* species and *Listeria monocytogenes* serotypes using a capillary air thermal cycler. *Lett. Appl. Microbiol.* 20:188–190.

Boerlin, P., Bannerman, E., Ischer, F., Rocurt, J., Bille, J. 1995. Typing of *Listeria monocytogenes*: A comparison of random amplification of polymorphic DNA with 5 other methods. *Res. Microbiol.* 146:35–49.

Bohnert, M., Dilasser, F., Dalet, C., Mengaud, J., Cossart, P. 1992. Use of specific oligonucleotides for direct enumeration of *Listeria monocytogenes* in food samples by colony hybridization and rapid detection by PCR. *Res. Microbiol.* 143:271–280.

Border, P., Howard, J., Plastow, G., Siggens, K. 1990. Detection of *Listeria* species and *Listeria monocytogenes*. *Lett. Appl. Microbiol.* 11:158–162.

Borucki, M., Call, D. 2003. *Listeria monocytogenes* serotype identification by PCR. *J. Clin. Microbiol.* 41:5537–5540.

Brett, M., Short, P., McLauchlin, J. 1998. A small outbreak of listeriosis associated with smoked mussels. *Int. J. Food Microbiol.* 43:223–229.

Brosch, R., Brett, M., Catimel, B., Luchansky, J., Ojeniyi, B., Roacourt, J. 1996. Genomic fingerprinting of 80 strains from the WHO multicenter international typing study of *Listeria monocytogenes* via pulsed-field gel electrophoresis (PFGE). *Int. J. Food Microbiol.* 32:343–355.

Brosch, R., Buchrieser, C., Rocourt, J. 1991. Subtyping of *Listeria monocytogenes* serovar 4b by use of low-frequency-cleavage restriction endonucleases and pulsed-field gel electrophoresis. *Res. Microbiol.* 142:667–675.

Brosch, R., Chen, J., Luchansky, J. 1994. Pulsed-field fingerprinting of listeriae: Identification of genomic divisions for *Listeria monocytogenes* and their correlation with serovar. *Appl. Environ. Microbiol.* 60:2584–2592.

Bsat, N., Batt, C., 1993. A combined modified reverse dot-blot and nested PCR assay for the specific non-radioactive detection of *Listeria monocytogenes*. *Mol. Cell Probes* 7:199–207.

Buchrieser, C., Brosch, R., Catimel, B., Rocourt, J. 1992. Pulsed-field gel electrophoresis applied for comparing *Listeria monocytogenes* strains involved in outbreaks. *Can. J. Microbiol.* 39:395–401.

Buchrieser, C., Brosch, R., Rocourt, J. 1991. Use of pulsed field gel electrophoresis to compare large DNA-restriction fragments of *Listeria monocytogenes* strains belonging to serogroups 1/2 and 3. *Int. J. Food Microbiol.* 14:297–304.

Cao, J., Cronin, C., McLandsborough, L., Levin, R. 2005a. Effects of primers and *Taq* polymerase on randomly amplified polymorphic DNA analysis for typing *Listeria monocytogenes* fom the environment of a shrimp processing plant. *Food Biotechnol.* 19:217–226.

Cao, J., Witkowski, R., Lu, H., Abolmaaty, A., Lu, S., and Levin, R.E. 2005b. Detection, enumeration, and RAPD analysis of *Listeria monocytogenes* isolates in fish derived from retail sources in Western Massachusetts. *Food Biotechnol.* 19:145–160.

Chen, Y., Knabel, S. 2007. Multiplex PCR for simultaneous detection of bacteria of the genus *Listeria, Listeria monocytogenes,* and major serotypes and epidemic clones of *L. monocytogenes. Appl. Environ. Microbiol.* 73:6299–6304.

Chen, Y., Zhang, W., Knabel, S. 2007. Multi-virulence-locus sequence typing identifies single nucleotide polymorphism which differentiates epidemic clones and outbreak strains of *Listeria monocytogenes. J. Clin. Microbiol.* 45:835–846.

Cooray, K., Nishibori, T., Xiong, H., Matsuyama, T., Fujita, M., Mitsuyama, M. 1994. Detection of multiple virulence-associated genes of *Listeria monocytogenes* by PCR in artificially contaminated milk samples. *Appl. Environ. Microbiol.* 60:3023–3026.

Cossart, P., Kocks, C. 1994. The actin-based motility of the intracellular pathogen *Listeria monocytogenes. Mol. Microbiol.* 13:395–402.

Czajka, J., Bsat, N., Piani, M., Russ, W., Sultana, K., Wiedman, M., Whitaker, R., Batt, C. 1993. Differentiation of *Listeria monocytogenes* and *Listeria innocua* by 16S rRNA genes and intraspecies discrimination of *Listeria monocytogenes* strains by random amplified polymorphic DNA polymorphisms. *Appl. Environ. Microbiol.* 59:304–308.

Davis, B., Dulbecco, R., Eisen, H., Ginsberg, H., Wood, B., and McCarty, M. 1973. *Listeria monocytogenes.* In: *Microbiology.* New York: Harper and Row, pp. 946–948.

Deneer, H., Boychuk, I. 1991. Species specific detection of *Listeria monocytogenes* by DNA amplification. *Appl. Environ. Microbiol.* 57:606–609.

Destro, M., Leitao, M., Farber, J. 1996. Use of molecular typing methods to trace the dissemination of *Listeria monoctogenes* in a shrimp processing plant. *Appl. Environ. Microbol.* 62:705–711.

Dickinson, J., Kroll, R., Grant, K. 1995. The direct application of the polymerase chain reaction to DNA extracted from foods. *Lett. Appl. Microbiol.* 20:212–216.

Domann, E., Wehland, J., Rohde, M., Pistor, S., Hartl, M., Goebel, W., Leimester-Wächter, M., Wuenscher, M., Chakraborty, T. 1992. A novel bacterial gene in *Listeria monocytogenes* required for host cell microfilament interaction with homology to the proline-rich region of vinculin. *EMBO J.* 11:1981–1990.

Dons, I., Rasmussen, O., Olsen, J. 1992. Cloning and characterization of a gene encoding flagellin of *Listeria monocytogenes. Mol. Microbiol.* 6:2919–2929.

Duffy, G., Cloak, O., Sheridan, J., Blair, I., McDowell, D. 1999. The development of a combined surface adhesion and polymerase chain reaction technique in the rapid detection of *Listeria monocytogenes* in meat and poultry. *Int. J. Food Microbiol.* 49:151–159.

FAO. 1999. FAO Fisheries Report no. 604. Report of the FAO expert consultation on the trade impact of *Listeria* in fish products. Amherst, MA, 17–20 May 1999. United Nations: Rome, Italy.

Farber, J., Addison, C. 1994. Rapid typing for distinguishing species and strains in the genus *Listeria. J. Appl. Bacteriol.* 77:242–250.

Farber, J., Peterkin, P. 1991. *Listeria monocytogenes*, a food-borne pathogen. *Microbiol. Rev.* 55:476–511.

Fitter, S., Heuzenroeder, M., Thomas, C. 1992. A combined PCR and selective enrichment method for rapid detection of Listeria monocytogenes. *J. Appl. Microbiol.* 73:53–59.

Fluit, A., Torensma, R., Visser, M., Aarsman C., Poppelier, M., Keller, B., Klapwijk, P., Verhoef, J. 1993. Detection of *Listeria monocytogenes* in cheese with the magnetic immuno-polymerase chain reaction assay. *Appl. Environ. Microbiol.* 59:1289–1293.

Franciosa, G., Tartara, S., Wedell-Neergaard, C., Aureli, P. 2001. Characterization of *Listeria monocytogenes* strains involved in invasive and noninvasive listeriosis outbreaks by PCR-based fingerprinting techniques. *Appl. Environ. Microbiol.* 67:1793–1799.

Furrer, B., Candrian, U., Hoefelein, C., Leuthy, J. 1991. Detection and identification of *Listeria monocytogenes* in cooked sausage products and in milk by in vitro amplification of hae-molysin gene fragments. *J. Appl. Bacteriol.* 70:372–379.

Gaillard, J., Berche, P., Frehel, C., Gouin, E., Cossart, P. 1991. Entry of *L. monocytogenes* into cells is mediated by internalin, a repeat protein reminiscent of surface antigens from Gram-positive cocci. *Cell* 65:1127–1141.

Geoffroy, C., Raveneau, J., Beretti, J., Lecroisey, A., Vazquez-Boland, J., Alouf, J., Berche, P. 1991. Purification and characterization of an extracellular 29-kilodalton phospholipase C from *Listeria monocytogenes. Infect. Immun.* 59:2382–2388.

Giovannacci, I., Ragimbeau, C., Queguiner, S., Salvat, G., Vendeuvre, J., Carlier, V., Ernel, G. 1999. *Listeria monocytogenes* in pork slaughtering and cutting plants: Use of RAPD, PFGE and PCR-REA for tracing and molecular epidemiology. *Int. J. Food Microbiol.* 53:127–140.

Göhmann, S., Leimester-Wächter, M., Schlitz, E., Goebel, W., Charaborty, T. 1990. Characterization of a *Listeria monocytogenese*-specific protein capable of inducing delayed hypersensitivity in *Listeria*-immune mice. *Mol. Microbiol.* 4:1091–1099.

Gray, D., Kroll, R. 1995. Polymerase chain reaction amplification of the flaA gene for the rapid identification of *Listeria* spp. *Lett. Appl. Microbiol.* 20:65–68.

Harvey, J., Gilmour, A. 2001. Characterization of recurrent and sporadic *Listeria monocy-togenes* isoates from raw milk and nondairy foods by pulsed-field gel electrophoresis, monocin typing, plasmid profiling, and cadmium and antibiotic resistance determina-tion. *Appl. Environ. Microbiol.* 67:840–847.

Herman, L., De Block, J., Moermans, R. 1995. Direct detection of *Listeria monocytogenes* in 25 milliliters of raw milk by a two-step PCR with nested primers. *Appl. Environ. Microbiol.* 61:817–819.

Hong, B., Jiang, L., Hu, Y., Fang, D., Guo, H. 2004. Application of oligonucleotide array technology for the rapid detection of pathogenic bacteria of foodborne infections. *J. Microbiol. Methods* 58:403–411.

Howard, P., Harsono, K., Luchansky, J. 1992. Differentiation of *Listeria monocytogenes, Listeria innocua, Listeria ivanovii,* and *Listeria seeligeri* by pulsed-field gel electropho-resis. *Appl. Environ. Microbiol.* 58:709–712.

Jinneman, K., Hill, W. 2001. *Listeria monocytogenes* lineage group classification by MAMA-PCR of the listeriolysin gene. *Curr. Microbiol.* 43:129–133.

Johnson, W., Tyler, S., Ewan, E., Ashton, F., Wang, G., Rozee, K. 1992. Detection of genes coding for listeriolysin and *Listeria monocytogenes* antigen (lmaA) in *Listeria* spp. by the polymerase chain reaction. *Microbial Pathol.* 12:79–86.

Katharion, S. 2003. Foodborne outbreaks of listeriosis and epidemic associated lineages of *Listeria monocytogenes*. In: *Microbial Food Safety in Animal Agriculture*. M.E. Torrence, R.E. Isaacson, Eds., Iowa State University Press: Ames, pp. 243–256.

Kerr, K., Kite, P., Heritage, J., Hawkey, P., 1995. Typing of epidemiologically associated environmental and clinical strains of *Listeria monocytogenes* by random amplification of polymorphic DNA. *J. Food Prot.* 58:609–613.

Killinger, A.H. 1970. *Listeria monocytogenes*, In: *Manual of Clinical Microbiology*. J.E. Blair, E.H. Lennette, J.P. Truant, Eds., American Society for Microbiology: Bethesda, MD, pp. 95–105.

Kocks, C., Gouin, E., Tabouret, M., Berche, P., Ohayon, H., Cossart, P. 1992. *Listeria monocytogenes* induced actin assembly requires the ActA gene product, a surface protein. *Cell* 68:521–531.

Köhler, S., Leimeister-Wächter, M., Chakraborty, T., Lottspeich, F., Goebel, W. 1990. The gene coding for protein p60 of *Listeria monocytogenes* and its use as a specific probe for *Listeria monocytogenes*. *Infect. Immun.* 58:1943–1950.

Lane, D., Pace, B., Olsen, G., Stahl, D., Sogin, M., Pace, N. 1985. Rapid determination of 16S ribosomal RNA sequences for phylogenetic analyses. *Proc. Natl. Acad. Sci. USA*, 82:6955–6959.

Lantz, P., Tjerneld, F., Borch, E., Hahn-Hägerdal, B., Rådström P. 1994. Enhanced sensitivity in PCR detection of Listeria monocytogenes in soft cheese through use of an aqueous two-phase system as a sample preparation method. *Appl. Environ. Microbiol.* 60:3416–3418.

Lawrence, L., Gilmour, A. 1995. Characterization of *Listeria monocytogenes* isolated from poultry products and from poultry-processing environment by random amplification of polymorphic DNA and multilocus enzyme electrophoresis. *Appl. Environ. Microbiol.* 61:2139–2144.

Lawrence, L., Harvey, J., Gilmour, A. 1993. Development of a random amplification of polymorphic DNA typing method for *Listeria monocytogenes*. *Appl. Environ. Microbiol.* 59:3117–3119.

Lei, X., Fiedler, F., Lan, Z., Katharious, S. 2001. A novel serotype-specific gene cassette (*gltA-gltB*) is required for expression of teichoic acid-associated surface antigens in *Listeria monocytogenes* of serotype 4b. *J. Bacteriol.* 183:1133–1139.

Leimeister-Wächter, M., Chakraborty, T. 1989. Detection of listeriolysin, the thiol-dependant hemolysin in *Listeria monocytogenes, Listeria ivanovii*, and *Listeria seeligeri*. *Infect. Immun.* 57:2350–2357.

Leimeister-Wächter, M., Haffner, C., Domann, E., Goebel, W., Chakraborty, T. 1990. Identification of a gene that positively regulates expression of listeriolysin, the major virulence factor of Listeria monocytogenes. *Proc. Natl. Acad. Sci. USA* 87:8336–8340.

Levett, P., Bennett, P., O'Donoghue, K., Bowker, K., Reeves, D., MacGowan, A. 1993. Relapsed infection due to *Listeria monocytogenes* confirmed by random amplified polymorphic DNA (RAPD) analysis. *J. Infect.* 27:205–207.

Liu, D., Lawrence, M., Austin, F., Ainsworth, A. 2007. A multiplex PCR for species- and virulence-specific determination of *Listeria monocytogenes*. *J. Microbiol. Meth.* 71:133–140.

Liu, D., Lawrence, M., Wiedman, M., Gorski, M., Mandrell, R., Ainsworth, A., Austin, F. 2006. Listeria monocytogenes subgroups IIA, IIIB, and IIC delineate genetically distinct populations with varied pathogenic potential. *J. Clin. Microbiol.* 44:4229–4233.

Loessner, M., Schneider, A., Scherer, S. 1995. A new procedure for efficient recovery of DNA, RNA, and proteins from *Listeria* cells by rapid lysis with a recombinant bacteriophage endolysin. *Appl. Environ. Microbiol.* 61:1150–1152.

Louie, M., Jayaratne, P., Luchsinger, I., Devenish, J., Yao, J., Schlech, W., Simor, A. 1996. Comparison of ribotyping arbitrary primed PCSR, and pulsed-field gel electrophoresis for molecular typing of *Listeria monocytogenes. J. Clin. Microbiol.* 34:15–19.

Lovett, J. 1989. *Listeria monocytogenes.* In: *Foodborne Bacterial Pathogens.* M.P. Doyle, Ed., Marcel Dekker: New York, pp. 283–310.

MacGowan, A., O'Donaghue, K., Nicholls, S., Bowker, K., McLauchlin, J., Bennett, P., Reeves, D.S. 1993. Typing of *Listeria* sp. by random amplified polymorphic DNA (RAPD) analysis. *J. Med. Microbiol.* 38:322–327.

Makino, S., Okada, Y., Maruyama, T. 1995. A new method for direct detection of *Listeria monocytogenes* from foods by PCR. *Appl. Environ. Microbiol.* 61:3745–3747.

Mazurier, S., Audurier, A., Marquet-Van der Mee, N., Notermans, S., Wernars, K. 1992. Comparative study of randomly amplified polymorphic DNA analysis and conventional phage typing for epidemiological studies of *Listeria monocytogenes* isolates. *Res. Microbiol.* 143:507–512.

Mazurier, S., Wernars, K. 1992. Typing of *Listeria* strains by random amplification of polymorphic DNA. *Res. Microbiol.* 143:499–505.

Mengaud, J., Dramsi, S., Gouin, E., Vasquez-Boland, J., Milon, G., Cossart, P. 1991. Pleiotropic control of *Listeria monocytogenes* virulence factors by a gene which is autoregulated. *Mol. Microbiol.* 5:2273–2283.

Mengaud, J., Vicente, M., Chenevert, J., Pereira, J., Geoffrey, C., Gicquel-Sanzey, B., Baquero, F., Perez-Diaz, J., Cossart, P. 1988. Expression in *Escherichia coli* and sequence analysis of the listeriolysin O determinant of *Listeria monocytogenes. Inf. Immun.* 56:766–772.

Miettinen, M., Björkroth, K., Kkorkeala, H. 1999. Characterization of *Listeria monocytogenes* from an ice cream plant by serotyping and pulsed-field gel electrophoresis. *Int. J. Food Microbiol.* 46:187–192.

Murray M., Thompson, W. 1980. Rapid isolation of high molecular weight plant DNA. *Nucleic Acids Res.* 8:4321–4325.

Nakama, A., Matsuda, M., Itoh, T., Kaneuchi, C. 1998. Molecular typing of *Listeria monocytogenes* isolated in Japan by pulsed-field gel electrophoresis. *J. Vet. Med. Sci.* 60:749–752.

Niederhauser, C., Candrian, U., Höfelein, C., Jermini, M., Bühler, H., Lüthy, J. 1992. Use of the polymerase chain reaction for detection of *Listeria monocytogenes* in food. *Appl. Environ. Microbiol.* 58:1564–1568.

Niederhauser, C., Höfelein, C., Alimann, M., Burkhalater, P., Lüthy, J., Candrian, U. 1994. Random amplification of polymorphic bacterial DNA: Evaluation of 11 oligonucleotides and application to food contaminated with *Listeria monocytogenes. J. Appl. Bacteriol.* 77:574–582.

Nogva, H., Drømtorp, S., Nissen, H., Rudi, K. 2003. Ethidium monoazide for DNA-based differentiation of viable and dead bacteria by 5′-nuclease PCR. *BioTechniques.* 34:804–813.

Nogva, H., Rudi, K., Naterstad, K., Holck, A., Lillehaug, D. 2000. Application of 5′-nuclease PCR for quantitative detection of *Listeria monocytogenes* in pure cultures, water, skim milk, and unpasteurized whole milk. *Appl. Environ. Microbiol.* 66:4266–4271.

Norton, D., Scarlet, J., Horton, K., Sue, D., Thimothe, J., Boor, K., Wiedmann, M. 2001a. Characterization and pathogenic potential of *Listeria monocytogenes* isolates from the smoked fish industry. *Appl. Environ. Microbiol.* 67:646–653.

Norton, D., McCamey, M., Gall, K., Scarlett, J., Boor, K., Wiedman, M. 2001b. Molecular studies on the ecology of *Listeria monocytogenes* in the smoked fish processing industry. *Appl. Environ. Microbiol.* 67:198–205.

Norwood, D., Gilmour, A. 1999. Adherence of *Listeria monocytogenes* strains to stainless steel coupons. *J. Appl. Microbiol.* 86:576–582.

Notermans, S., Chakraborty, T., Leimeister-Wachter, M., Dufrenne, J., Heuvelman, C., Maas, H., Jansen, W., Wernars, K., Guinee, P. 1989. A specific gene probe for detection of biotyped and serotyped *Listeria* strains. *Appl. Environ. Microbiol.* 55:902–906.

Ojeniyi, B., Wegener, H., Jensen, N., Bisgaard, M. 1996. *Listeria monoctogenes* in poultry and poultry products: Epidemiological investigations in seven Danish abattoirs. *J. Appl. Bacteriol.* 80:395–401.

O'Donoghue, K., Bowker, K., McLauchlin, J., Reeves, D., Bennett, P., MacGowan, A., 1995. Typing of *Listeria monocytogenes* by random amplified polymorphic DNA (RAPD) analysis. *Int. J. Food Microbiol.* 27:245–252.

Oravcová, K., Kuchta, T., Kacliková, E. 2007. A novel real-time PCR-based method for the detection of *Listeria monocytogenes* in food. *Lett. Appl. Microbiol.* 45:568–573.

Pitcher, D., Saundear, M., Owen, R. 1989. Rapid extraction of bacterial genomic DNA with guanidium thiocyanate. *Lett. Appl. Microbiol.* 8:151–156.

Powell, H., Gooding, C., Garrett, S., Lund, B., McKee, R. 1994. Proteinase inhibition of the detection of *Listeria monocytogenes* in milk using the polymerase chain reaction. *Lett. Appl. Microbiol.* 18:59–61.

Rossen, L., Holmstrøm, K., Olsen, J., Rasmussen, O. 1991. A rapid polymerase chain reaction (PCR)-based assay for identification of *Listeria monocytogenes* in food samples. *Int. J. Food Microbiol.* 14:145–152.

Rossen, L., Nørskov, P., Holmstrøm, K., Rasmussen, O. 1992. Inhibition of PCR by components of food samples, microbial diagnostic assays and DNA-extraction solutions. *Int. J. Food Microbiol.* 17:37–45.

Rudi, K., Naterstad, K., Dromtorp, S., Holo, H. 2005. Detection of viable and dead *Listeria monocytogenes* on gouda-like cheeses by real-time PCR. *Lett. Appl. Microbiol.* 40:301–306.

Saiki, R., Gelfand, D., Stoffel, S., Scharf, S., Higuchi, R., Horn, G., Mullis, K., Erlich, H. 1988. Primer-directed enzymatic amplification of DNA with a thermostable DNA polymerase. *Science* 239:487–491.

Seeliger, H., Finger, H. 1976. Listeriosis. In: *Infectious Diseases of the Fetus and Newborn Infant.* J.S. Remington, J.O. Klein, Eds., W. B. Saunders: Philadelphia, p. 333.

Seeliger, H., Jones, D. 1986. Genus *Listeria.* In: *Bergy's Manual of Systematic Bacteriology,* Vol. 2. P. Sneath, N. Mair, M. Sharpe, Eds., Williams & Wilkins: Baltimore, MD, pp. 1335–1345.

Simon, M., Gray, D., Cook, N. 1996. DNA extraction and PCR methods for the detection of *Listeria monocytogenes* in cold-smoked salmon. *Appl. Environ. Microbiol.* 62:822–824.

Soejima, T., Iida, K., Qin, T., Taniae, H., Seki, M., Oshida, S. 2008. Method to detect only live bacteria during PCR amplification. *J. Clin. Microbiol.* 46:2305–2313.

Stackenbrandt, E., Curiale, M. 1988. Detection of *Listeria.* European Patent Application 88308820.5.

Thomas, E., King, R., Burchak, J., Gannon, V. 1991. Sensitive and specific detection of *Listeria monocytogenes* in milk and ground beef with the polymerase chain reaction. *Appl. Environ. Microbiol.* 57:2576–2580.

Unnerstad, H., Bannerman, E., Bille, J., Danielsson-Tham, M., Waak, E., Tham, W. 1996. Prolonged contamination of a dairy with *Listeria monocytogenes*. *Neth. Milk Dairy J.* 50:493–499.

Vela, A.J., Fernandez-Garayzabal, J., Vazquez, J., Latre, M., Blanco, M., Moreno, M., De La Fuente, L., Marco, J., Franco, C., Cepeda, A., Rodriguez-Moure, A., Suarez, G., Doninguez, L. 2001. Molecular typing by pulsed-field gel electrophoresis of Spanish animal and human *Listeria monocytogenes* isolates. *Appl. Environ. Microbiol.* 67:5840–5843.

Vogel, B., Jørgensen, L., Ojeniyi, B., Huss, H., Gram, L. 2001a. Diversity of *Listeria monocytogenes* isolates from cold-smoked salmon produced in different smokehouses as assessed by random amplified polymorphic DNS analysis. *Int. J. Food Microbiol.* 65:83–92.

Vogel, B., Huss, H., Ojeniyi, B., Ahrens, P., Gram, L. 2001b. Elucidation of *Listeria monocytogenes* contamination routes in cold-smoked salmon processing plants detected by DNA-based typing methods. *Appl. Environ. Microbiol.* 67:2586–2595.

Wagner, M., Allerberger, F. 2003. Characterization of *Listeria monocytogenes* recovered from 41 cases of sporadic listeriosis in Austria by serotyping and pulsed-field gel electrophoresis. *FEMS Immunol. Med. Microbiol.* 35:227–234.

Wang, R., Cao, W., Johnson, M. 1991. Development of a 16S rRNA-based oligomer probe specific for *Listeria monocytogenes*. *Appl. Environ. Microbiol.* 57:3666–3670.

Wang, R., Cao, W., Johnson, M. 1992. 16SA rRNA-based probes and polymerase chain reaction method to detect *Listeria monocytogenes* cells added to foods. *Appl. Environ. Microbiol.* 58:2827–2831.

Weiss, J., Seeliger, H. 1975. Incidence of *Listeria monocytogenes* in nature. *Appl. Microbiol.* 30:29–32.

Welsh, J., McClelland, M. 1990. Fingerprinting genomes using PCR with arbitrary primers. *Nucleic Acids Res.* 18:7213–7218.

Wernars, K., Boerlin, P., Audurier, A., Russell, E., Curtis, G., Herman, L., van der Mee-Marquet, N. 1996. The WHO multicenter study on *Listeria monoctogenes* subtyping: Random amplification of polymorphic DNA (RAPD). *Int. J. Food Microbiol.* 32:325–341.

Wernars, K., Heuvelman, C., Chakraborty, T., Notermans, S. 1991. Evaluation of the polymerase chain reaction for detection of *Listeria monocytogenes* in soft cheese. *J. Appl. Bacteriol.* 70:121–126.

Wernars, K., Heuvelman, K., Notermans, S., Domann, E., Leimeister-Wächter M., Chakraborty, T. 1992. Suitability of the *prfA* gene, which encodes a regulator of virulence genes in *Listeria monocytogenes*, in the identification of pathogenic *Listeria* spp. *Appl. Environ. Microbiol.* 58:765–768.

Wiedman, M., Bruce, J., Keating, C., Johnson, A., Mcdonough, P., Batt, C., 1997. Ribotypes and virulence gene polymorphisms suggest three distinct *Listeria monocytogenes* lineages with differences in pathogenic potential. *Inf. Immun.* 65:2707–2716.

Wiedmann-al-Ahmad, M., Tichy, H., Schön, G. 1994. Characterization of *Acinetobacter* type strains and isolates obtained from wastewater treatment plants by PCR fingerprinting. *Appl. Environ. Microbiol.* 60:4066–4071.

Zhang, W., Knabel, S. 2005. Multiplex PCR assay simplifies serotyping and sequence typing of *Listeria monocytogenes* associated with human outbreaks. *J. Food Prot.* 68:1907–1910.

CHAPTER 13

Clostridium botulinum

I. CHARACTERISTICS OF THE ORGANISM

C. botulinum is a gram-positive, obligately anaerobic, spore-forming rod of which there are seven types, A–G, based on serological distinction of the respective neurotoxins produced. Human botulism is caused by types A, B, E, and, though rarely, type F. Types C and D cause botulism in animals. Type G is not associated with neurotoxicity in humans or animals. All isolates of *C. botulinum* can be placed into one of four groups based on physiological differences: group I, all type A strains and proteolytic B and F types; group II, all type E strains and nonproteolytic B and F strains; group III, C and D type strains; and group IV, G type strains. 16S and 23S rRNA gene sequence studies (Hutson et al., 1993a,b, 1994) have confirmed this grouping and have documented a high level of relatedness among strains within each group and little relatedness between members of the different groups (Hatheway, 1993).

In addition to *C. botulinum* producing these botulinum neurotoxins (BoNTs), isolates of *C. butyricum* and *C. barati* have also been found to harbor certain of these *bont* genes resulting in human botulism.

II. RELATIONSHIP BETWEEN BOTULISM AND SEAFOOD

In recent years an average of 450 botulism outbreaks have been annually reported in the international literature, 12% of the outbreaks being caused by type E (Hatheway, 1995). Coastal marine environments usually exhibit serotype E as predominant, although certain marine sediments have been found to contain serotype B predominantly. Although isolates of types B, E, and F are known to be psychrotolerant, isolates of type E are truly psychrotrophic and exhibit the ability to grow in seafood tissue under refrigerated conditions (~4°C). Type E has been the most frequent cause of botulism derived from seafood.

There is a well-established history of salted fish causing type E botulism. Uneviscerated, salt-cured fish have been implicated in a number of additional botulism outbreaks (Badhey et al., 1986; Kotev et al., 1987; Telzak et al., 1990). The intestines of uneviscerated salted fish are thought to result in a low-salt environment, allowing spores of *C. botulism* to germinate, grow, and produce toxin. Weber, Hibbs, and Sarswish (1993) reported on a massive outbreak of type E botulism associated with the consumption of traditional salted fish in Cairo. Low levels of type E toxin are known to result in primarily gastrointestinal (GI) symptoms. Sobel, Malavet, and John (2007) reported on an outbreak of clinically mild botulism type E illness among five individuals resulting in predominantly GI symptoms consisting of nausea, vomiting, abdominal pain, dry mouth, shortness of breath, and, in one individual, diplopia (double vision). Fresh uneviscerated whitefish with salt had been placed in a sealed Ziploc bag and stored for less than 1 month at ambient temperature prior to consumption. Remnant fish tested positive for botulinum type E toxin.

Commercially produced vacuum-packaged hot-smoked fish is presently considered one of the most important botulism food vehicles. Hot-smoked Canadian whitefish was reported by Korkeala et al. (1998) to be the cause of a single family outbreak of type E botulism in 1997. The fish was smoked only 5 days before consumption, indicating that toxin production had been rapid and that there had been marked temperature abuse during storage or transport of the fish. Type E toxin was confirmed by toxin neutralization and the mouse bioassay and by PCR.

In the Baltic Sea area, where nonproteolytic group II *C. botulinum* is known to predominate, a particularly high prevalence of type E has been reported (Hielm et al. 1998c; Hyytiä, Hielm, and Korkeala, 1998). Hyytiä, Hielm, and Korkeala (1998) described contamination levels of type E in Finland of 10–40% in raw fish and fish intestines, with the highest prevalence being Baltic herring, and 4–14% in fish roe. In addition, 30% of German raw fish have been found to contain type E spores (Hyytiä-Trees et al., 1999).

The prevalence of *C. botulinum* types A, B, E, and F in river lamprey caught in Finnish rivers was determined by Merivirta et al. (2006) using a quantitative PCR-MPN analysis. The multiplex PCR assay and primers (Table 13.1) of Lindström et al. (2001) were utilized. Among 67 raw whole lampreys one (1.5%) was positive for the *bont/E* gene with an estimated *C. botulinum* spore count of 100 spores/kg. Two type E strains were isolated from the positive sample and confirmed as different genotypes by PFGE using *Sma*I and *Xho*I. The authors concluded that vacuum packaging with refrigerated storage may constitute a safety hazard in processed lamprey from the Baltic Sea area and recommended a storage temperature of 3°C or below for such products.

III. INFANT BOTULISM

A. Historical Aspects and Incidence

The first clinical case of infant botulism was described in 1976 (Midura and Arnon, 1976; Pickett et al., 1976). Infant botulism results from ingestion of spores of *C. botulinum* or other neurotoxin producing clostridia. Inmature intestinal flora are thought to allow the spores to germinate, leading to growth of the organism in the intestine and production of BoNT. Children between the ages of 2 weeks and 6 months are most susceptible (Arnon, Dumas, and Chin, 1981), with types A and B being the most frequent cause. In most cases the vehicle for infection is unknown.

Barash, Tang, and Arnon (2005) reported on a 38-hr newborn infant in California stricken in late 2003 with what was ultimately diagnosed as infant botulism. Three rounds of hemodialysis over a 12-day period resulted in complete recovery. A stool specimen yielded colonies that when subjected to biochemical characterization, 16S rRNA sequencing, and mouse bioassay identified the responsible organism as *C. barati* producing BoNT type F toxin. Epidemiological investigation suggested that the patient's exposure to *C. barati* may have occurred at the birthing hospital in the immediate perinatal period. As of 2005, a total of five cases of infant botulism due to *C. barati* producing type BoNT/F had been reported globally (Barash, Tang, and Arnon, 2005).

Johnson et al. (2005) reported on the sixth case of infant botulism to occur in the United Kingdom in 2001. The case was caused by a type B strain of *C. botulinum*. *C. botulinum* type B isolated from the infant's feces and from an opened container of infant milk formula yielded identical PFGE profiles with the restriction nucleases *Nru*I, *Sac*I, *Sma*I, and *Xho*I; however, the strains differed in cell morphology and level of toxin produced in culture. In addition, a strain of *C. botulinum* type B was isolated from an unopened container of infant milk formula, which yielded a distinctly different PFGE profile from the former isolates. Inasmuch as the source of the *C. botulinum* type B isolate from the opened container could not be determined with certainty, the authors concluded that the source of the causative organism was uncertain. The reader is reminded of the possibility that passage of the organism through the intestine of the infant may have influenced its morphology and toxin production, which was about fivefold higher in the strain isolated from the infant's fecal sample than from the opened container.

B. Implication of Honey in Infant Botulism

Honey is the only foodstuff to date that has been implicated as the cause of infant botulism (Midura, 1996). Studies have shown that more than 20% of stricken children had ingested honey prior to the onset of botulism (Arnon et al., 1979; Chin et al., 1979; Morris et al., 1983). Midura et al. (1979) reported levels of 10^3–10^4 spores of *C. botulinum*/kg of honey samples associated with cases of infant botulism.

To develop a maximally sensitive method of detecting *C. botulinum* using enrichment and the PCR three different methods for preparing samples of honey were examined by Nevas et al. (2002): (1) PCR detection after diluting 25 g of honey 1:10 and centrifuging at 9000 × g for 30 min and then using the resulting pellet for enrichment in 10 ml of trypticase peptone glucose yeast (TPGY) broth; (2) dilution of 25 g of honey with 25 ml of distilled water followed by heating at 65°C for 30 min and then transfer of the entire sample to 450 ml of TPGY broth; (3) the supernatant of method 1 was subjected to filtration through a 0.5-m membrane filter and the entire filter transferred to 10 ml of TPGY broth. Incubation of all enrichments was for 5 days at 30°C. The third method involving membrane filtration was found to be the most sensitive and allowed the detection of 2.5 spores/25 g sample. A total of 190 honey samples from 29 countries was assayed for the presence of *C. botulinum*. The primers (Table 13.1) of Franciosa, Ferreira, and Hatheway (1994) were employed for amplification and detection of sequences derived from the *bont/A* and *bont/B genes*. The PCR yielded 20 (10.5%) positive samples (17 A and 12 B types).

Nevas et al. (2005a) applied the mPCR assay of Lindström et al. (2001) for detection of BoNT genes *A, B, E,* and *F* (Table 13.1) in honey from Denmark, Norway, and Sweden. Positive samples were 29/112 (26%), 12/122 (10%), and 1/61 (2%), respectively. The Danish samples yielded 1 A and 18 B type strains with the PCR; the Norwegian samples yielded 74 type E and 1 type F; and the Swedish samples yielded 1 type E. PFGE with *Sac*II yielded eight different patterns among the total of 42 positive strains.

Zhou, Sugiyama, and Johnson (1993) reported achieving transduction of the *bont/E* gene from a toxigenic *C. butyricum* strain (from infant botulism) to a nontoxigenic *C. botulinum* strain. The growth of the nontoxigenic *C. botulinum* strain in the presence of a phage induced from the type E toxin-producing strain of *C. butyricum* failed to result in transduction. However, when the nontoxigenic *C. botulism* strain was grown in broth in the presence of a nontoxigenic *C. butyricum* "helper" strain and the induced phage, transduction of the *bont/E* gene was achieved. The gene transfer appeared to be a transduction by a defective phage made infective by the helper strain.

IV. MOLECULAR TECHNIQUES APPLIED TO *C. BOTULINUM*

A. PCR Detection of *C. botulinum*

C. botulinum type C is found naturally in the sediments of lakes and marshes and is the primary cause of botulism in wild waterfowl. Williamson et al. (1999) developed seminested PCR assays for detection of type C strains of *C. botulinum* during avian botulism outbreaks from wetland sediments from areas in California and North Dakota. Among 18 site samples, 16 were positive for the presence of *C. botulinum* type C. The primers ToxC-384F/ToxC-850R/ToxC-625F (Table 13.1) yielded a 226-bp amplicon, as did the primers ToxC-625F/ToxC-1049R/ToxC850R (Table 13.1).

A PCR assay was developed by Campbell, Collins, and East (1993) for the specific detection of the BoNT genes *B, E,* and *F* of *C. botulinum*. Degenerative primers BoNT1/BoNT2 (Table 13.1) amplified a specific fragment of approximately 1100-bp from strains of *C. botulinum* types A, B, E, F, and G in addition to neurotoxin-producing strains of *C. barati* type F and *C. butyricum*. DNA probes (Table 13.1) labeled with fluorescein-dUTP were developed for detection of the specific neurotoxin-type strains. An additional probe (Table 13.1) was designed for the detection of the *bont/F* gene of *C. barati,* which differed in sequence from the *bont/F* genes of both proteolytic and nonproteolytic strains of *C. botulinum* type F.

Szabo, Pemberton, and Desmarchelier (1993) developed PCR assays for detection of BoNT genes *A, B, C, D,* and *E* using the primer pairs listed in Table 13.1.

Among five methods examined for DNA extraction the method of choice involved suspending spores or vegetative cells in 200 ml of 1 M NaOH (pH 10.0) and heating at 60°C for 5 min followed by centrifugation at 8000 × g for 5 min. Purification of released DNA with a commercial DNA binding matrix was recommended. The sensitivity of detection using broth cultures was the DNA from approximately three cells per PCR reaction with types A, B, and E.

Fach et al. (1995) developed a PCR assay using a degenerate primer pair P260/P261 to specifically amplify a 260-bp fragment from *C. botulinum* type A, B, E, F, and G strains. In addition, five individual digoxygenin (DIG) labeled probes (Table 13.1) allowed identification of each toxin type by hybridization of the probes to the PCR products on a nylon membrane. The level of sensitivity of detection after 18 hr of enrichment in tryptone yeast extract glucose broth (TYG) was 10 CFU/g of seeded food.

Franciosa, Ferreira, and Hatheway (1994) studied the effectiveness of the PCR in detecting type A, B, and E *bont* genes among 209 strains of *C. botulinum* and 29 strains of other *Clostridium* species. Individual specific primer pairs were used for types A and E, yielding amplicons of 2278 and 762 bp, respectively (Table 13.1), and three pairs of primers were used for detection of type B yielding amplicons of 1284, 1150, and 881 bp, respectively (Table 13.1), in addition to a pair of primers yielding the entire *bont/E* gene as an amplicon of 3874 bp (Table 13.2). A universal bacterial primer pair (Table 13.1) was also used to yield a 763-bp amplicon derived from the 16S rRNA, which served as a positive amplification control. Both type A and B genes were detected in 43 type A toxin-producing strains, only one of which could be shown to produce type B toxin with the mouse bioassay, the type B toxin being produced in greater quantity than the type A toxin. The type B gene was also detected in two strains of *C. subterminale*, which were found to be nontoxigenic with the mouse bioassy.

Hielm et al. (1996) developed PCR methodology for the MPN-PCR detection and enumeration of BoNT types A, B, E, and F in fish and sediment samples. The general BoNT primers BoNT1/BoNT2 (Table 13.1) of Campbell, Collins, and East (1993) were used for detection of all BoNT types and yielded a 1184-bp amplicon from all types except A, which yielded an amplicon of 1038 bp. The 16S rRNA universal bacterial primers BAC-3/BAC-4 (Table 13.1) from Franciosa, Ferreira, and Hatheway (1994) yielding a 760-bp amplicon were incorporated into each PCR reaction to ensure against PCR inhibition and false-negative results. The primers NKB-1/NKB-5, B1-a/B2-d, GF-1/GF-3, and 48F/50R (Table 13.1) yielded amplicons of 2278, 1284, 762, and 1138 bp for types A, B, E, and F, respectively. Rainbow trout were seeded with spores of *C. botulinum* type E at 10^2 to 10^6 spores/kg of tissue in addition to the inoculation of fish intestines. Each sample was subjected to a 5-day enrichment in TPGY broth followed by transfer of 0.5 ml into 10 ml with overnight incubation prior to PCR reactions. Washed vegetative cells from such enrichment broth cultures were boiled for 10 min and 1 ml incorporated into PCR reactions. All seeded samples were detected as positive. Among ten sediment samples tested, eight (80%) were positive for *C. botulimum* type E spores, with spore counts of *C. botulinum* type E ranging from 95 to 2710/kg of sample.

Most toxigenic strains of *C. botulinum* produce a single antigenically distinct toxin (A–G). Rare isolates, however, have been found to produce two BoNTs in different proportions. It is interesting that all such isolates share the ability to produce either BoNT/A or BoNT/B. Franciosa, Ferriera, and Hathaway (1997) reported on a new strain of *C. botulinum* producing both BoNT/A and BoNT/B from a canned macrobiotic food (whole-meal flour, soybean sauce, herbs, seaweed, and water) of commercial origin suspected of causing a fatal case of foodborne botulism. Remnants of the canned food product were lethal to mice, and *C. botulinum* colonies were isolated and confirmed by PCR to harbor both *bont/A* and *bont/B* genes using the primer pairs (Table 13.1) of Franciosa, Ferriera, and Hatheway (1994) that yielded amplicons of ~2300 and 1300 bp, respectively. The strain produced ten times the level of BoNT/A than BoNT/B and was classified as *C. botulinum* subtype Ab (A/B ratio, 10:1). The authors discussed the possibility that the existence of subtypes may have been previously underestimated because of the difficulty in detecting the minor toxin, which is masked by the major one in mouse bioassays.

Dahlenborg, Borch, and Radsröm (2001) developed a sensitive combined selection and enrichment PCR procedure for detection of *C. botulinum* types B, E, and F in fecal samples from slaughtered pigs. Two enrichment PCR assays using the DNA polymerase *rTth* were utilized. One assay was specific for the *bont/B* gene and the other was specific for the *bont/E* and *bont/F* genes. Sample preparation prior to PCSR involved heat treatment of feces homogenates at 70°C for 10 min, enrichment in TPGY broth at 30°C for 18 hr, and DNA extraction using a commercial DNA extraction kit. The primers fB/rB (Table 13.1) amplified a 480-bp sequence of the *bont/B* gene. Internally nested primers fBn/rBn (Table 13.1) amplified a 220-bp sequence derived from the initial 480-bp *bont/B* amplicon. The primers fEF/rEF (Table 13.1) amplified a 630-bp sequence from both the *bont/E* and *bont/F* genes. The internally nested primers fEFn/rEFn amplified a 200-bp sequence derived from the initial 630-bp amplicons from the *bont/E* and *bont/F* genes. The detection limits were 10 spores/g of fecal sample for nonproteolytic type B and 3.0×10^3 spores/g of fecal sample for type E and nonproteolytic type F.

Lindström et al. (2001) described the development of a multiplex PCR (mPCR) for detection of the *bont A, B, E*, and *F* genes in food and fecal samples. The primer pairs (Table 13.2) yielded amplicons of 782, 205, 389, and 543 bp, respectively, and were therefore easily resolved and detected in agarose gels. With a two-step enrichment the detection limit of the mPCR with seeded food and fecal samples was at least 1 spore/10 g of sample.

Alsallami and Kotlowski (2001) developed improved primer pairs for detection of the *bont/B* and *bont/E* genes. The detection limit was increased from 1 to 0.1 ng of DNA by increasing the annealing temperature from 50°C to 62°C. The primers BoTE1/BoTE2 (Table 13.1) amplified a 307-bp sequence of the *bont/E* gene.

Kimura et al. (2001) developed an Rti-PCR assay for quantifying *C. botulinum* type E in modified-atmosphere packaged fish samples (jack mackerel). The primers BE1430F/BE1709R (Table 13.1) amplified a 269-bp sequence of the *bont/E* gene. The dual-labeled probe BE1571FP (Table 13.1) was labeled at the 5′-end with 6-carboylfluorescein (FAM) and at the 3′-end with TAMRA. The quantifiable range was 10^2 to

10^8 CFU/g, which allowed detection much earlier than toxin could be detected with the mouse bioassay.

Fach et al. (2002) determined the prevalence of *C. botulinum* types A, B, E, and F in 214 fresh fish and environmental samples collected in Northern France with the use of a PCR-ELISA assay. The primers CB1/P261 and CB/P260 (Table 13.1) allowed simultaneous detection of all four BoNT genes A, B, E, and F. Primer pairs CB1/CBA, CB1/CBB, CB1/CBE, and CB1/CBF (Table 13.1) were used to generate amplicons derived from the individual BoNT genes A, B, E, and of 527, 523, 506, and 591 bp, respectively. After amplification, the amplicons were alkali denatured and detected in a sandwich hybridization assay using microwells coated with streptavidin, a 5′-biotin-labeled probe designated CB, and a 3′-DIG labeled probe designated P260 (Table 13.1). The biotin-labeled probe was added to the streptavidin-coated wells, the amplified DNA added next, followed by the DIG-labeled probe and then HRP-labeled anti-DIG IgG added and processed for color development with appropriate HRP substrates (TMB plus H_2O_2). An internal amplification control (IC) consisting of a recombinant pMO*Blue* plasmid DNA, with primer binding regions flanking a DNA sequence of the chloramphenicol-resistance gene (*Cm*) was used in conjunction with the 5′-biotin-labeled probe CatCap and the 3′-DIG-labeled probe CatRev (Table 13.1). The incidence of *C. botulinum* in fresh marine fish and sediments was 16.6% and 4%, respectively. Among the 214 samples, 31 were BoNT-positive with the PCR-ELISA assay, whereas only 24 of these 31 samples were positive with the mouse bioassay. The *bont/F* gene was absent from all samples.

Sharkey, Markos, and Haylock (2004) developed a competitive reverse transcription PCR assay (cRT-PCR) to quantify toxin-encoding mRNA production by a type E strain in media with either sorbic acid or sodium nitrite. The primers mRNA-EF/mRNA-ER (Table 13.1) amplified a 250-bp sequence of the BoNT/E mRNA. A 10-fold reduction in toxin mRNA production and a 25-fold reduction in the proportion of mRNA to total RNA were estimated when either 1 mg/ml of sorbic acid or 100 mg/ml of sodium nitrite was added to the medium at pH 7.0.

Strains of *C. botulinum* are known to cause various external abscesses and infections. Akbulut, Grant, and McLauchlin (2005) developed Rti-PCR assays for detection of the neurotoxin gene of *C. tetani* and *C. botulinum* neurotoxin genes *bont/A*, B, and E in material from wound infections. The primers TQ A1/TQ A2, TQ B1/TQ B2, TQ E1/TQ E2 (Table 13.1) yielded amplicons of 107, 117, and 139 bp from BoNT genes *A, B,* and *E*, respectively. Dual-labeled probes A(FAM), B(VIC), and E(FAM; Table 13.1) were labeled at the 5′-end with the indicated fluorophore and at the 3′-end with the black hole quencher (MGB).

Artin et al. (2007) reported on the first case of a wound infection caused by type E *C. botulinum*. The patient was an injecting heroin user with a previous history of recurrent skin and soft tissue abscesses and presented with intense symptoms of botulinum toxicity. Samples of an abdominal wall abscess were cultured and subjected to Rti-PCR with FRET probes and an internal amplification control. The primers fToxe/rToxe (Table 13.1) amplified a specific sequence of the *bont/E* gene. The FRET probes ToxE-Fluo and ToxE-Red (Table 13.1) were used to detect amplification and to confirm the identity of the type E infecting strain.

Lindström et al. (2001) described the development of a multiplex PCR (mPCR) for detection of *C. botulinum* types A, B, E, and F in food and fecal material. The primer pairs (Table 13.1) yielded amplicons of 782, 205, 389, and 543 bp, respectively, and were therefore easily resolved and detected in agarose gels. Sensitivity of the mPCR for types A, E, and F was the DNA from 10^2 cells and for type B was the DNA from 10 cells per mPCR reaction. With a two-step enrichment, the detection limit in food and fecal samples was one spore per 10 g sample or less.

Akbulut, Grant, and McLauchlin (2005) developed Rti-PCR assays for detection of *bont* genes A, B, and E in clinical samples. The primers TQ A1/TQ A2, TQ B1/TQ B2, and TQ E1/TQ E2 (Table 13.1) yielded amplicons of 107, 117, and 139 bp from the *bont* genes A, B, and E, respectively. Dual-labeled probes A(FAM), B(VIC), and E(FAM; Table 13.1) were labeled at the 5'-end with the indicated fluorophore and at the 3'-end with the black hole quencher MGB.

Myllykoski et al. (2006) examined a number of different protocols for PCR detection of *C. botulinum* types A, B, E, and F in pig intestinal samples and found that a double enrichment was optimum, involving initial enrichment in TPGY at 30 and 37°C for 3 days followed by transfer of 1 ml enrichment to 10 ml of TPGY with 16-hr incubation. The cells from 1 ml of the second enrichment were washed, heated at 99°C to rupture the cells, and DNA purified with a commercial kit. The mPCR primers (Table 13.1) and protocol of Lindström et al. (2001) were then used for amplification of target DNA.

Fenicia et al. (2007) developed a SYBR green RTi-PCR assay for detection of *C. botulinum* type A. Four different methods were used for the extraction and purification of DNA from broth cultures. Among the four extraction methods, boiling was significantly less efficient than Chelex 100, DNeasy tissue kit, and Instagene matrix DNA. Chelex 100 was selected because it was the least expensive. The primers IOA1a-F/IOA2a-R (Table 13.1) amplified a 101-bp sequence of the *bont/A* gene. To eliminate false-negative results, an internal amplification control (IAC) that yielded a 197-bp amplicon was synthesized and included in each amplification mixture. The detection limit was less than 60 copies of *C. botulinum* DNA.

The nontoxic nonhemagglutinin (NTNH) is a conserved component of the toxin complex in all seven BoNT types (A–G). East and Collins (1994) demonstrated that the gene encoding NTNH is present in strains of *C. botulinum* that produce BoNTs and is absent from strains that are nontoxic. Raphael and Andreadis (2007) developed an Rti-PCR assay that utilized a locked nucleic acid probe to target the NTNH gene to detect bacterial strains harboring the neurotoxin gene cluster. A dual-labeled probe NTNH (Table 13.1) labeled at the 5'-end with 6-FAM and with a black hole quencher (BHQ) at the 3'-end was used. Among all toxin types in a multiple sequence alignment of the NTNH gene, 25 out of 28 nucleotides in the complementary probe sequence were identical. In order to permit mismatches within the probe site, locked nucleic acid bases (LNA) at four positions were used (Table 13.1), which bracketed the complementary bases to the probe that showed degeneracy among strains in order to increase specificity for the conserved nucleotides. A locked nucleic acid

(LNA) is a modified DNA nucleotide. The deoxyribose moiety of an LNA nucleotide is modified with an extra bridge connecting the 2' and 4' carbons. The bridge "locks" the deoxyribose in the 3'-end structural conformation, which is often found in the A-form of DNA. The locked base conformation has an increased affinity for complementary DNA nucleotides and enhances base stacking and backbone preorganization. This significantly increases the thermal stability (melting temperature) of oligonucleotides and increases the T_m value. Because the regions flanking the probe site contained a high level of nucleotide diversity among serotypes, a total of seven primers (Table 13.1) consisting of four forward and three reverse primers were used in a multiplex PCR. Genomic DNA was prepared using a commercial kit. Stools spiked with 1 CFU/ml and enriched for 24 hr in Chopped Meat Glucose Starch Medium yielded positive PCR results. Without enrichment, spiked stool samples seeded with 10^2–10^3 CFU/ml could be detected by Rti-PCR with a 1:10 dilution of stool.

B. Molecular Typing of *Clostridium botulinum* Strains

The genetic structure of neurotoxin genes in a strain of *C. botulinum* producing both type A and B neurotoxins (type Ab) was investigated by Fujinaga et al. (1995). Analyses by PCR using type-specific primers corresponding to the coding regions for N-terminals of light chain of type A and B neurotoxins and Southern hybridization of total DNA were undertaken. Primers A1/A2 and B1/B2 (Table 13.1) amplified light chain regions of *bont/A* and *bont/B*. Primers A3/A4 and B3/B4 (Table 13.1) amplified heavy-chain regions of *bont/A* and *bont/B*. Results indicated that the type Ab strain carried at least one copy each of the type A and B neurotoxin genes.

Scheinert et al. (1996) developed a PCR assay for differentiating a number of bacterial genera and species including distinguishing among *C. botulinum* serotypes A, B, E, and F based on the length and sequence variability of the spacer region between 16S and 23S rRNA. A single pair of primers was used for clostridia with primer A at the 3'-end of 16S rRNA and primer B near the 5'-end of 23S rRNA (Table 13.1; Figure 13.1). Resulting amplicons for various organisms varied from 280 to 1300 bp with some species yielding one characteristic band and others yielding two or three bands of different size. *C. botulinum* types A, B, E, and F yielded the following characteristic bands: A, 300 bp; B, 250 and 300 bp; E, 250 and 290 bp; and F, 300 and 430 bp.

Lin and Johnson (1995) subjected four strains of type A *C. botulinum* to PFGE typing. Among 15 restriction enzymes tested, *Mlu*I, *Rsr*II, *Sma*I, *Nru*I, *Ksp*I, *Nae*I, and *Xho*I generated satisfactory patterns with various type A strains. The three restriction nucleases *Mlu*I, *Ksp*I, and *Sma*I were selected for final PFGE analysis of the four type A strains, all of which yielded different PFGE patterns. Genes encoding proteins involved in the toxigenicity of *C. botulinum*, including neurotoxin A, hemagglutinin A, and genes for temperate phages, as well as various transposon Tn*916* insertion sites in *C. botulinum* strain 62A, were mapped by PFGE and gene probes. The genes encoding neurotoxin A and hemagglutinin A-1 were located on the same fragment with each of five restriction nuclease digests, indicating their probable physical linkage.

Hielm et al. (1998a) applied PFGE to 21 strains of *C. botulinum* nonproteolytic group II. DNA degradation problems caused by extracellular DNases were overcome by fixing of cells with 4.0% formaldehyde prior to cell lysis. Among 15 restriction nucleases *Apa*I, *Mlu*I, *Nru*I, *Sma*I, and *Xho*I yielded the most revealing patterns, allowing strain differentiation. Twenty of the strains yielded 13 PFGE types, with one strain consistently untypeable.

The distribution of *C. botulinum* serotypes A, B, E, and F in Finnish trout farms was assessed using the PCR by Hielm et al. (1998b). The PCR primers from Hielm et al. (1996) including those for *bont/E* (Table 13.1) were used. A total of 333 samples were tested with neurotoxin gene-specific PCR assays. *C. botulinum* type E was found in 68% of farm sediment samples, in 15% of fish intestinal samples, and in 5% of the fish skin samples. No other serotypes were found. The average spore count in sediments, fish intestines, and skin were 2×10^3, 1.7×10^2, and 3×10^2 per kg, respectively. PFGE with *Sma*I of 42 Finnish isolates plus 12 North American reference strains generated 28 PFGE profiles indicating extensive genetic diversity.

The genetic diversity of 92 type E *C. botulinum* strains was assessed by Hyytiä et al. (1999b). There were 67 from Finnish seafood and fishery origin, 15 were from German farmed fish, and 10 from North American seafoods. PFGE performed with *Sma*I-*Xma*I resulted in 75 typeable strains that yielded 33 profiles. PFGE performed with *Xho*I allowed 91 strains to be typed yielding 51 profiles. All 92 strains were typeable with RAPD primers OPJ-6 and OPJ-13 (Table 13.1), which yielded 27 and 19 banding patterns, respectively. The frequent occurrence of small fragments and faint bands made RAPD interpretation difficult. A high level of genetic diversity among the isolates was observed regardless of their source, presumably because of the absence of strong evolutionary selection factors.

Wang et al. (2000) analyzed type E botulinum toxin producing strains isolated from botulinum cases or soil specimens in Italy and China, using sequencing of the *bont/E* gene, RAPD (Table 13.1), PFGE, and Southern blot hybridization for the *bont/E* gene. The deduced amino acid sequences of the BoNT/Es of 11 *C. butyricus* isolates from China were identical and exhibited 95.0% and 96.9% identity with those of the Italian BoNT/E strains of *C. butyricum* BL6340 and *C. botulinum* type E, respectively. The results indicated that BoNT/E-producing *C. butyricum* is clonally distributed globally.

All type A strains of *C. botulinum* can be subdivided into three groups A, AB, and Ab depending on expression of the accompanying B toxin gene (Franciosa, Ferreira, and Hatheway, 2004). Franciosa et al. (2004) developed PCR methodology for identifying *C. botulinum types* A, AB, and A(B) based on the combination of the *ha33* and *p47* genes and the RFLP analysis of the BoNT/A and BoNT/B genes using the restriction endonuclease *Xba*I. The *ha33* gene of *C. botulinum* is found in A1 strains, the *p47* gene is found in A2 strains, and both *ha33* and *p47* genes are found in Ab strains, where B is expressed along with A (Franciosa et al., 1997), and AB strains (where B is unexpressed). The primers P47-A/P47-B amplified a 1236-bp sequence of the *p47* gene (Table 13.1). Primers H1/H2 amplified a 774-bp sequence of the *ha33* gene (Table 13.1). The primers NKB-1/NKB-2 (Table 13.1) were used to amplify a 2278-bp sequence of the *bont/A* gene (Table 13.1). The primers B1-a/B2-d were used to amplify a 1284-bp sequence of the *bont/B* gene. PCR differentiation of the BoNT/A gene clusters were found to correlate well with strain genotyping using PFGE and RAPD. On the basis of their results, the authors allocated Type A strains to the A1 gene cluster (*ha33* gene present and *p47* gene absent), type Ab strains to the A2 gene cluster (*ha33* gene absent and *p47* gene present), and type A(B) and Ab to the A3 gene cluster (both *ha33* and *p47* genes present).

PFGE was applied by Nevas et al. (2005b) to 55 *C. botulinum* strains from diverse sources. All strains were proteolytic with 19 of type A, 28 of type B, 3 of type F, and 5 of type A/B. *Sac*II, *Sma*I, and *Xho*I were found to produce the most convenient number of bands with clear macrorestriction patterns. With the majority of strains PFGE allowed discrimination between individual strains of types A and B. The different toxin types were discriminated at an 86% similarity level with *Xho*I. *Sac*II produced the highest number of PFGE patterns (33) as a single enzyme, with *Sma*I and *Xho*I individually producing 29 and 32 patterns, respectively. The authors concluded that *Sac*II could be used as a single-restriction nuclease with PFGE for epidemiological purposes. When the patterns of all three restriction nucleases were combined, a total of 37 patterns was discerned, and the patterns derived from *Sac*II and *Xho*I when combined yielded the largest number of patterns (38) with the highest discriminatory index of 0.952. Contrary to intuitive insight, increasing the number of restriction nucleases from two to three does not always increase the number of combined patterns.

Leclair et al. (2006) undertook a comparative typing study involving PFGE, RAPD, and automated ribotyping of *C. botulinum* type E strains derived from clinical and food sources associated with four botulinum outbreaks that occurred in the Canadian Arctic. All type E strains previously untypeable by PFGE, even with the use of a formaldehyde fixation step, could be typed by the addition of 50 mM thiourea to the electrophoresis running buffer. Digestion with *Sma*I and *Xho*I followed by PFGE was used to link food and clinical isolates from the four different type E botulinum outbreaks and to differentiate them from among 31 recently isolated Arctic environmental group II *C. botulinum* strains. *Sma*I PFGE typing yielded 18 profiles and *Xho*I PFGE typing yielded 23 profiles. Strain differentiation was unsuccessful with the automated ribotyping system, which yielded only two profiles. RAPD analysis of the group II strains was not consistently reproducible with primers OPJ-6 and OPJ-13 (Table 13.1). Primer OPJ-13 did, however, yield 28 profiles.

Paul et al. (2007) undertook a detailed study of the flagella of *C. botulinum* and found that multiple flagellin proteins are produced. The flagellin genes *flaA1* and *flaA2* were found to encode the major structural flagellins of *C. botulinum*. Primers cbotflaF/cbotflaR (Table 13.1) were used to amplify 827-bp sequences from both the *flaA1* and *flaA2* genes. Colony PCR and sequencing of *fla1/fla2* variable regions classified 80 environmental and clinical strains into group I and group II and clustered isolates into 12 flagellar types. Flagellum type was distinct from neurotoxin type. The *flaA1/flaA2* gene locus was found to be present in all 80 *C. botulinum* isolates. A second flagellum gene *flaB* was found to encode a large type E–specific flagellin protein.

Johnson et al. (2008) elucidated the neurotoxin cluster gene sequence and gene arrangements for strains of *C. botulinum* encoding BoNT subtypes A3, A4, and a unique A1-producing strain (HA- OrX⁺ A1). A total of 51 primer pairs was used for the comprehensive comparison of the individual neurotoxin cluster genes and *bont* genes among the three different subtypes.

Macdonald et al. (2008) used ten variable-number tandem-repeat (VNTR) regions identified within the complete genomic sequence of *C. botulinum* strain ATCC 3502 to characterize 59 *C. botulinum* serotype A strains involving subtypes A1 to A4. Ten pairs of VNTR primers (Table 13.1) amplified fragments within all of the type A strains and were less successful with strains of other serotypes. Composite multilocus VNTR analysis of the 59 BoNT (A1–A4) strains plus three bivalent B strains identified 38 different genotypes. Among BoNTA1 and BoNTA1(B) strains, 30 genotypes were identified, demonstrating discrimination below the subtype level. The authors concluded that the 10 VNTR primer pairs furnished a highly discriminating tool to distinguish BoNTA1 strains for epidemiological studies.

C. Molecular Methods for Differentiating between Group I and Group II Strains

C. botulinum groups I and II are known to differ phenotypically. The optimum growth temperature for group I strains is 35–40°C (Smith and Sugiyama, 1988), and the minimum growth temperature is 10°C (Hatheway, 1993). The optimum growth temperature of group II strains is 30°C with growth occurring as low as 3°C (Eklund, Wieler, and Poysky, 1967; Graham, Mason, and Peck, 1997). *C. Botulinum* group I (proteolytic) and group II (nonproteolytic) are genetically and physiologically distinguishable with both groups being involved in human botulism. As a result of differences in spore heat resistance and growth characteristics, the two groups are considered to pose different risks involving food and the environment.

RAPD and repetitive element sequence-based PCR (rep-PCR) were used by Hyytiä et al. (1999a) to characterize 15 group I and 21 group II strains of *C. botulinum*. RAPD primer OPJ6 from a commercial random primer kit yielded the most discriminatory patterns among four arbitrary primers and distinguished Group II *C. botulinum* serotypes at the strain level. Group I strains were mainly discriminated at the serotype level. The discriminatory power of rep-PCR was found to be inferior to that of RAPD. The degenerate rep-PCR primers REP1R-Dt/REP2R-Dt (Table 13.1) generated group I- and group II-specific fragments, and the random primer OPJ6 produced a serotype E–specific fragment of 1300 bp. Both PD and rep-PCR yielded a typeability of 100%.

Amplified fragment length polymorphism (AFLP) analysis was applied by Keto-Timonen, Nevas, and Korkeala (2005) to 33 group I and 37 group II *C. botulinum* isolates. The restriction nuclease combination *Hind*III and *Hpy*CH4IV with primers having one selective nucleotide apiece (Hind-C and Hpy-A) were utilized. AFLP clearly differentiated between *C. botulinum* groups I and II. All strains studied were typeable by APLP and a total of 42 AFLP types were distinguished.

An enzyme complex FldA(I)BL (phenyllactate dehydrogenase) plays a key role in the use of phenylalanine (Phe) for ATP synthesis among Group I *C. botulinum* strains. In addition, Phe is also required in culture media by Group I *C. botulinum* strains in amounts exceeding biosynthesis requirements and at levels notably higher than those required by group II isolates (Whitmer and Johnson, 1988). Hence, the mechanism involved in the metabolism of Phe for generation of ATP can be expected to be present in group I *C. botulinum* strains but not in group II isolates. Based on this a priori reasoning, Dahlsten et al. (2008) developed a PCR assay for distinguishing group I from group II isolates of *C. botulinum*. The assay is based on the *fldB* gene, which is associated with phenylalanine metabolism in proteolytic clostridia and employs an internal amplification control targeted to certain conserved regions of 16S rRNA in groups I and II of *C. botulinum*. The primers CBEPfldB-F/CBEPfldB-FR (Table 13.1) amplified a 552-bp sequence of the *fldB* gene present only in group I isolates. The primers CBEDac-F/CBEDac-R amplified a 761-bp sequence of the conserved regions of 16S rRNA as an internal amplification control. Among 36 group I strains, all yielded the 552- and 761-bp amplicons. In contrast, among 24 group II isolates all yielded only the 761-bp control amplicon. The assay provides a substantial improvement in discriminating between groups I and II *C. botulinum* which has been traditionally based on a time-consuming culture method.

D. Molecular Methods for Detection of BoNTs

1. Reverse Transcription PCR for Indirect Measurement of BoNT Production

The standard method for quantitatively determining BoNT production is the mouse bioassay, where results are expressed in mouse lethality units. McGrath, Dooley, and Haylock (2002) developed a competitive reverse-transcription PCR (RT-PCR) for indirectly quantifying BoNT production as a replacement assay for the mouse bioassay. The method was found to accurately measure the level of toxin encoding mRNA in *C. botulinum* cells. The basis of the competitive RT-PCR assay is coamplification of the target DNA and the control DNA. This is possible if the target and the control DNA have almost the same DNA sequence and identical primer binding sites. Both targets compete for the common primers and reagents in the same PCR reaction tube. Quantification is then accomplished by quantitatively comparing the PCR signal of the specific target template with the PCVR signal obtained with known concentrations of the competitor DNA. The two amplicons are distinguished by constructing a competitor template having the same sequence as the cDNA but containing a deletion or an insertion so as to result in an amplicon of distinguishable size in an agarose gel. The point at which the concentrations of the control and target amplicons are of equal fluorescence intensity (equivalence point) is then used for quantification of the cDNA and calibrated to mouse units. A type E strain was grown in two different media: brain–heart infusion broth (BHI) and type E broth. Type E broth is known to produce high levels of type E toxin by the mouse bioassay. At the stationary growth phase, the target RNA from the type E broth culture corresponded to the control RNA at a dilution of 1:2500. However, the target RNA from a BHI broth culture corresponded to the control RNA at a dilution of 1:250,000 thereby reflecting a 100-fold difference in the amount of toxin produced in the two culture broths when competitive RT-PCR was used. With the mouse bioassay, the Type E broth produced a level of toxin lethal to mice at a 1:100 dilution but not beyond, whereas the BHI broth produced a level of toxin lethal to mice at a 1:1 dilution but not beyond, revealing a 100-fold difference in the relative amount of toxin produced by the two broths, agreeing with the RT-PCR assay.

Lövenklev et al. (2004) developed a quantitative reverse transcription-PCR (qRT-PCR) to monitor the relative expression of the *C. botulinum* type B toxin. cDNA was obtained with reverse transcriptase, and the levels of BoNT/B mRNA in five type B strains were accurately monitored by applying primers fBn/rBn (Table 13.1), which were specific for the *bont/B* gene and primers f16S/r16S (Table 13.1) for a reference gene encoding a portion of the 16S rRNA. Corresponding dual-labeled probes (Table 13.1) were used for Rti-PCR. The levels of *bont/B* mRNA were correlated with the levels of extracellular type B toxin using an ELISA assay and the mouse bioasssay.

2. Immunopolymerase Chain Reaction Assays for Ultrasensitive BoNT Detection

Chao et al. (2004) reported on unique and extremely sensitive immuno-PCR assays for detection of BoNT type A. Both an indirect immuno-PCR and an indirect sandwich immuno-PCR assay were developed. Biotinylated reporter DNA was generated by amplification of commercially available target DNA pCMS-EGFP with a 5′-biotinylated forward primer pCMS-EGFP2449-2469 and a nonbiotinylated reverse primer pCMS-EGFP3168-3148. The indirect immuno-PCR involved first coating the wells of polystyrene microtiter plates with BoNT/A followed by the addition of blocking agent. Mouse monoclonal IgG was then added followed by goat-antimouse IgG labeled with biotin. Streptavidin was then added followed by the addition of 5′-labeled biotinylated reporter DNA and PCR performed with unlabeled primers pCMS-EGFP2455-2475/pCMS-EGFP3168-3148 directed to the bound reporter DNA. The indirect sandwich immuno-PCR assay involved initially binding mouse monoclonal IgG to microtiter plate wells. BoNT/A was then added to the wells followed by blocking agent. Rabbit polyclonal IgG to BoNT/A was then added followed by the addition of goat anti-rabbit IgG labeled with biotinylated reporter DNA. Streptavidin was then added followed by the addition of biotinylated reporter DNA and PCR performed with primers pCMS-EGF2455-2475/pCMS-EGFP3168-3148. To avoid confusion, the reader is reminded that the assays are not designed to detect the *bont/A* gene, but are instead designed to amplify greatly the bound level of reporter DNA, which is quantitatively dependent on the initial amount of type A toxin bound to the wells. The detection limit for BoNT/A was 50 g for both immuno-PCR assays, which is 10^5-fold lower than conventional ELISA assays.

Table 13.1 PCR Primers and DNA Probes for Clostridium botulinum

Primer or Probe	Sequence (5′ → 3′)	Sequence (bp)	Gene	Reference
BoTE1	GTG-AGT-TAT-TTT-TTG-TGG-CTT-CCG-AGA	307	*bont/E*	Alsallami and Kottowski (2001)
BoTE2	TTA-TTT-TCA-CCT-TCG-GGC-ACT-TTC-TG			
fToxe	TGA-ATC-AGC-ACC-TGG-A	325	*bont/E*	Artin et al. (2007)
rToxe	GGT-TAG-CTT-CAG-TAG-TAA-A			
ToxE-Fluo	TGT-CAA-TAA-ACC-TGT-GCA-AGC-AGC-FI	—		
ToxE-Red	LC640-TAT-TTG-TA(A/G)-GCT-GGA-TAC-AAC-AAG-T(A/G)-T-TAG-TAG-AT			
mRNA-EF	AGC-AAA-TAG-AAA-ATG-AAC	250	*BoNT/E* mRNA	Sharkey, Markos, and Haylock (2004)
mRNA-ER	GAA-TAC-TAT-TAT-TTA-GGG-TA			
RAPD OPJ-6	TCG-TTC-CGC-A	—	—	Leclair et al. (2006)
RAPD OPJ-13	CCA-CAC-TAC-C	—	—	
BE1430F	GTG-AAT-CAG-CAC-CTG-GAC-TTT-CAG	269	*bont/E*	Kimura et al. (2001)
BE1709R	GCT-GCT-TGC-ACA-GGT-TTA-TTG-A			
BE1571FP	6-FAM-ATG-CAC-AGA-AAG-TGC-CCG-AAG-GTG-A-TAMRA			
RAPD Primer 1	GGT-GCG-GGA-A	—	—	Wang et al. (2000)
RAPD Primer 2	CCC-GTC-AGC-A	—	—	

Continued

Table 13.1   PCR Primers and DNA Probes for Clostridium botulinum (Continued)

Primer or Probe	Sequence (5' → 3')	Sequence (bp)	Gene	Reference
cbotflaF	CGC-GGG-GAT-CCA-TGA-TAA-TTA-ATC-ACA-ATT-TAA-ATG	827	flaA1/flaA2	Paul et al. (2007)
cbotflaR	CGC-GGG-GAT-CCC-TTA-ATA-ATT-GAA-GAA-CTC-CTT-GTC			
BoNT1	TAT-(A/G)TA-GGA-TC(T/C)-TGC-TTT-AAA-TAT-A(G/A)(G/T)-(A/T)A(A/T)-TGA	1100	bont A, B, E, F, G	Campbell, Collins, and East (1993)
BoNT2	TTA-GT(T/A)-ATA-GTT-ACA-AAA-ATC-CA(T/C)-(T/C)T(A/G)-TTT-ATA-TA			
Con.BoNT	TA-ATT-C(A/G)G-G(A/C)T-GGA-AA(G/A)-TAT-C		bont A, B, E, F, G	Campbell, Collins, and East (1993)
Cbot.B	TGC-TTT-TGA-GAT-TGC-AGG-A(G/T)C-CAG		bont/B	
Cbot.E	GAG-CTT-ACA-AAT-AAA-TAT-GAT-ATT-AAG-C		bont/E	
Cbot.F	AGA-CTT-GAA-TCT-GAA-TAT-A(A/G)T-ATC-AAT-A		bont/F	
Cbar.F	AGA-CTT-AGA-GCT-GAA-TAT-AAT-ATC-TAT-AG		barF	
CBMLA1	AGC-TAC-GGA-GGC-AGC-TAT-GTT	782	bont/A	Lindström et al. (2001)
CBMLA2	GGT-ATT-TGG-AAA-GCT-GAA-AAG-G			
CBMLB1	CAG-GAG-AAG-TGG-AGC-GAA-AA	205	bont/B	Lindström et al. (2001)
CBMLB2	CTT-GCG-CCT-TTG-TTT-TCT-TG			
CBMLE1	CCA-AGA-TTT-TCA-TCC-GCC-TA	389	bont/E	Lindström et al. (2001)
CBMLE2	GCT-ATT-GAT-CCA-AAA-CGG-TGA			

Primer	Sequence	Size	Target	Reference
CBMLF1	CGG-CTT-CAT-TAG-AGA-ACG-GA	543	*bont/F*	Lindström et al. (2001)
CBMLF2	TAA-CTC-CCC-TAG-CCC-CGT-AT			
NKB-1	GAT-ACA-TTT-ACA-AAT-CCT-GAA-GGA-GA	2278	*bont/A*	Franciosa, Ferreira, and Hatheway (1994)
NKB-5	AAC-CGT-TTA-ACA-CCA-TAA-GGG-ATC-ATA-GAA			
B1-a	GAT-GGA-ACC-ATT-TGC-TAG	1284	*bont/B*	Franciosa, Ferreira, and Hatheway (1994)
B2-d	AAC-ATC-AAT-ACA-TAT-TCC-TGG			
B3	CCA-GGA-ATA-TGT-ATT-GAT-GTT	1450	*bont/B*	Franciosa, Ferreira, and Hatheway (1994)
B4	AAA-TCA-AGG-AAC-ACA-CTA			
B5	TGG-ATA-AGA-ATA-CCT-AAA-TAT-AAG	881	*bont/B*	Franciosa, Ferreira, and Hatheway (1994)
B6	AAG-CAA-CTG-ACA-ACT-ATA-TGT			
JF-B1	ATG-CCA-GTT-ATA-AAT-TTT-AAT-TAT	3873	*bont/B*	Franciosa, Ferreira, and Hatheway (1994)
JF-B2	TTC-AGT-CCT-CCC-TTC-ATC-TTT-AGG			
GF-1	AAA-AGT-CAT-ATC-TAT-GGA-TA	762	*bont/E*	Franciosa, Ferreira, and Hatheway (1994)
GF-3	GTG-TTA-TAG-TAT-ACA-TTG-TAG-TAA-TCC			
BAC-3	ACG-GCC-CAG-ACT-CCT-ACG-GGA-GGC	763	16S rRNA	Franciosa, Ferreira, and Hatheway (1994)
BAC-4	GGG--TTG-CGC-TCG-TTG-CGG-CAC-TTA			
48F	GCT-TCA-TTA-GAG-AAC-GGA-AGC-AGT-GCT	1138	*bont/ F*	Hielm et al. (1996) from Ferreira et al. (1994)
50R	GTG-GCG-CCT-TTG-TAC-CTT-TTC-TAG-G			

Continued

Table 13.1   PCR Primers and DNA Probes for Clostridium botulinum (Continued)

Primer or Probe	Sequence (5′ → 3′)	Sequence (bp)	Gene	Reference
BAC-3	ACG-GCG-CAG-ACT-CCT-ACG-GGA-GGC	763	16S rRNA	Hielm et al. (1996) from Franciosa, Ferreira, and Hatheway (1994)
BAC-4	GGG-TTG-CGC-TCG-TTG-CGG-CAC-TTA			
A1	TAT-GGA-ATA-GCA-ATT-AAT-CC	457	bont/A	Szabo, Pemberton, and Desmarchelier (1993)
A2	GTG-TAA-TTT-ACC-TTA-GGT-AC			
B1	GAT-GGA-ACC-ACC-ATT-TGC-C(T/A)A-AG	1284	bont/B	Szabo, Pemberton, and Desmarchelier (1993)
B2	(T/A)AC-ATC-(T/A)AT-CA(T/A)-ATT-CC(TA)-GG			
B3	AGA-TAG-ACG-TGT-TCC-ACT-CG	727	bon/tB	Szabo, Pemberton, and Desmarchelier (1993)
B4	CTG-CTA-TAT-TAG-TTT-CTG			
C1	ATA-TAC-TCC-GGT-TAC-GGC-G	998	bont/C	Szabo, Pemberton, and Desmarchelier (1993)
C2	CCT-GGA-TAA-CCA-CGT-TCC-C			
D1	TAA-GTA-AAC-CGC-CCA-GAC-C	403	bont/D	Szabo, Pemberton, and Desmarchelier (1993)
D2	TAG-TAT-AGA-TAA-TGT-TCC-A			
E1	TAT-ATA-TTA-AAC-CAG-GCG-G	745	bont/E	Szabo, Pemberton, and Desmarchelier (1993)
E2	TAG-AGA-AAT-ATT-GGA-ACT-G			

P260	C(C/A)(G/A)-(G/A)(T/A)(T/A)-(T/A)TT-TAA-A(G/A)G-A(G/T)T-TTT-GGG-G	260	*bont A, B, E, F, G*	Fach et al. (1995)
P261	(T/G)AT-ATA-(C/T)(C/A)(C/T)-(G/A)AT-C(A/T)T-(T/C)(C/A)T-TTC-TAA-C	—		
Pr265	GGG-CCT-AGA-GGT-AGC-GTA-A	—	*bont/A*	
Pr266	CAC-GTA-GCA-AAT-ATA-ATC-AAA-A	—	*bont/B*	
Pr270	CTA-CTT-TAA-GCA-TTA-ATA-ATA-TA	—	*bont/E*	
Pr272	TTA-TAA-GAA-AAA-ATG-GTC-CTA-TAG	—	*bont/F*	
Pr280	GGG-AAA-CTG-CAC-CAC-GTA-CAA	—	*bont/F*	
IOA1a-F	GGG-CCT-AGA-GGT-AGC-GTA-(A/G)TG	101	*bont/A*	Fenicia et al. (2007)
IOA2a-R	TCT-T(C/T)A-TTT-CCA-GAA-GCA-TAT-TTT			
B1	GAT-GGA-ACC-ACC-ATT-TGC-(T/A)AG	1018	*bont/B*	Szabo, Pemberton, and Desmarchelier (1993)
B2	(T/A)AC-ATC-(T/A)AT-ACA-(T/A)AT-TCC-TGG			
B3	TAT-TAT-AAG-GCT-TTC-AAA-ATA-ACA-GAT-AGA-ATT-TGG-ATA			
ToxC-384-F	AAA-CCT-CCT-CGA-GTT-ACA-AGC-C	226	*bont/A*	Williamson, Rocke, and Aiken (1999)
TocC-850-R	GAA-AAT-CTA-CCC-TCT-CCT-ACA-TCA			
ToxC-625-F	CTA-GAC-AAG-GTA-ACA-ACT-GGG-TTA			
ToxC-625-F	CTA-GAC-AAG-GTA-ACA-ACT-GGG-TTA	226	*bont/A*	Williamson, Rocke, and Aiken (1999)
ToxC-1049-R	AAT-AAG-GTC-TAT-AGT-TGG-ACC-TCC			
Toxc-850-R	GAA-AAT-CTA-CCC-TCT-CCT-ACA-TCA			

Continued

Table 13.1   PCR Primers and DNA Probes for Clostridium botulinum (Continued)

Primer or Probe	Sequence (5′ → 3′)	Sequence (bp)	Gene	Reference
F1	GAT-TTA-AGT-GAA-AAA-TTA-TTT-AAT-ATA-T	—	*ntnh*	Raphael and Andreadis (2007)
F2	CCA-CTA-AAT-GAT-TTA-AAT-GAA	—	*ntnh*	
F3	TGA-TGA-AAT-ACC-TAA-TAG-TAT-GTT-AAA-T	—	*ntnh*	
CD2F	GAC-ATA-TCA-GAT-AGT-TTA-TTG-GGA	—	*ntnh*	
R1	TTT-AGC-CAT-ACA-AAT-TAA-ATC	—	*ntnh*	
R2	ACT--AGC-CAT-ACA-AAT-TAG-ATC	—	*ntnh*	
R3	TAT-TAA-ACT-TTC-TTG-AGC-TA	—	*ntnh*	
NTNH410	FAM-ATC-AAT-GGT-GGA-CAC-AAT-ATT-ATA-GTC-A-BHQ[a]	—	*ntnh*	
TQ A1	CGA-AAT-GGT-TAT-GGC-TCT-ACT-CAA	107	*bont/A*	Akbulut, Grant, and McLauchlin (2005)
TQ A2	TTG-CCT-GCA-CCT-AAA-AGA-GGA-T			
A (FAM)	FAM-ACT-TCA-AGT-GAC-TCC-TCA-AAA-CCA-AAT-GTA-AAA-TCT-G-MGB			
TQ B1	GAT-TAT-AAA-TGG-TAT-ACC-TTA-TCT-TGG-AGA-TAG-A	117	*bont/B*	Akbulut, Grant, and McLauchlin (2005)
TQ B2	CGC-TCC-ACT-TCT-CCT-GGA-TTA-C			
B (VIC)	VIC-TGT-TCC-ACT-CGA-AGA-GTT-TAA-CAC-AAA-CAT-TGC-MGB			
TQ E1	AAT-ATT-GTT-TCT-GTA-AAA-GGC-ATA-AGG-AA	139	*bont/E*	Akbulut, Grant, and McLauchlin (2005)
TQ E2	AAG-TTA-CTG-TAT-CGT-CAA-TTT-CTT-TAG-GAG			
E (FAM)	FAM-TTC-TCG-GAA-GCC-ACA-AAA-AAT-AAC-TCA-CCA-T-MGB			

	Sequence	Size	Target	Reference
CB1 A(A/T)	A-ATA-ATT-C(A/G)G-GAT-GGA-AA(A/G)-TAT-C	—	*bont A, B, E, F*	Fach et al. (2002)
P261	(G/T)(A/G)T-AT(A/G)-(C/T)(A/T/C)(A/T/C)-(A/G)AT-C(G/A/T)T-(C/T)(A/T/C)T-TTC-TAA-C			
CB	Biot-TAT-ATA-AAT-A(A/G)A-TGG-ATT-TTT-GTA-ACT-AT			
P260	C(A/C)(A/G)-(A/G)((A/T)(A/T)-(A/T)TT-TAA-A(A/G)(A/G)-A(C/T)T-TTT-GGG-G-Dig			
CBA	CAC-TAC-GCT-ACC-TCT-AGG-CCC	527	*bont/A*	
CBB	GAA-TTT-TGA-TTA-TAT-TTG-CTA-CGT-G	523	*bont/B*	
CBE	GCT-TCT-TAT-ATT-AAT-GCT-TAA-AGT-AG	506	*bont/E*	
CBF	CTA-TAG-GAC-CAT-TTT-TTC-TTA-TAA-TGA-C	591	*bont/F*	
CatCap	Biot-TCG-CAA-GAT-GTG-GCG-TGT-TAC-GGT-GAA-AAC	—	—	
CatRev	CAA-GGC-GAC-AAG-GTG-CTG-ATG-CCG-CTG-GCG-Dig	—	—	
P47-A	ACT-TAT-GGT-TGG-GAT-ATT-GTT-T	1236	*p47*	Franciosa et al. (2004)
P47-B	TAT-ATT-CTC-ACC-TTC-TTT-AAT-TTG-C			
H1	AAA-TTG-TTA-CCA-TCT-CCT-GTA-AGG-C	774	*h33*	Franciosa et al. (2004) from East et al. (1996)
H2	AAT-TAA-ATA-CTT-GAA-TAG-CAG-TTC-CGT			
A	AAG-TCG-TAA-CAA-A(A/G)C		16S–23S spacer	Scheinert et al. (1996)
B	CT(A/G)-GTG-CCA-AGG-CAT-TCA-CC			

Continued

Table 13.1   PCR Primers and DNA Probes for *Clostridium botulinum* (Continued)

Primer or Probe	Sequence (5' → 3')	Sequence (bp)	Gene	Reference
fB	TGG-ATA-TTT-TTC-AGA-TCC-AGC-CTT-G	480	*bont/B*	Dahlenborg, Borch, and Radström (2001)
rB	TGG-TAA-GGA-ATC-ACT-AAA-ATA-AGA-AGC			
fBn	AAA-GTA-GAT-GAT-TTA-CCA-ATT-GTA	220	*bont/B*	
rBn	GTT-AGG-ATC-TGA-TAT-GCA-AAC-TA			
fEF	CAA(A/G)AT-ATG-ATT-CTA-ATG-G(A/T)A-CAA-GTG-A	630	*bont E, F*	
rEF	TG(CT)-AAA-GC(C/T)-TGA-TAC-ATT-TG(C/T)-TCT-TTT-C			
fEFn	CAG-CA(C/T)-TAT-TT(A/G)-TA(A/G)-(A/G)(C/T)T-GGA-TA	200	*bont E, F*	
rEFn	TCT-AA(C/T)-AAA-ATA-CC(C/T)-(A/G)C(C/T)-CCT-A			
REP1R-Dt	III-(A/G/C/T)CG-(A/G/C/T)CG-(A/G/C/T)CA-TC(A/G/C/T)-GGC	—	BoNT types I and II	Hyytiä et al. (1999b)
REP2R-Dt	(A/G/C/T)CG-(A/G/C/T)CT-TAT-C(A/G/C/T)G-GCC-TAC	—		
1-F	GCA-ATA-AGA-ATA-AAT-GTT-TCG	526–631	*bont/A*	Macdonald et al. (2008)
1-R	TAC-TGG-CTG-TTA-CAA-GAA-TGT			
2-F	GAG-GAC-GAG-GGT-AAC-TTA-GT	327–618	*bont/A*	Macdonald et al. (2008)
2-R	GGT-ACT-TAT-GAT-TTC-CGG-TA			
3-F	GTT-ATA-TTT-GCT-GTA-TTC-TGT-T	367–517	*bont/A*	Macdonald et al. (2008)
3-R	CCA-TTT-ATT-TAT-GGG-ATA-TCT			
4-F	GAT-GCG-GAT-ATA-GAG-CTT-T	374–401	*bont/A*	Macdonald et al. (2008)
4-R	GTA-GAA-TAT-AGT-TTG-CAC-CCT			

Primer	Sequence	Position	Gene	Reference
5-F	CAT-ATA-AGC-CAC-AAC-AAG-AAT	300–318	*bont/A*	Macdonald et al. (2008)
5-R	CTT-GAT-AAA-CCT-TTC-CTT-TG			
6-F	GAG-GTG-TAG-TTA-TGA-GAG-ATG-G	359–500	*bont/A*	Macdonald et al. (2008)
6-R	CTT-TCA-TAT-GCT-TCT-CTT-TCA			
7-F	CAA-ATC-CGT-TAA-AGA-ATG-TAT	505–545	*bont/A*	Macdonald et al. (2008)
7-R	CCA-TTG-CTT-TAG-AAA-CTG-AT			
8-F	CTA-CTC-AAA-CAA-TAG-ACA-GGC	421–457	*bont/A*	Macdonald et al. (2008)
8-R	GTA-TTC-CTG-CTA-ATG-TTC-AAG			
9-F	GTG-CTA-ATA-CAG-AAC-CTG-CTA	370–424	*bont/A*	Macdonald et al. (2008)
9-R	GAA-CTA-ATA-CAG-AAC-CTG-CTA			
10-F	GGT-GCT-TTT-TCT-GAT-ACA-TTG	268–436	*bont/A*	Macdonald et al. (2008)
10-R	GCG-CTG-AAG-TAG-TAA-AGT-CTA-A			
A1	TGC-AGG-ACA-AAT-GCA-ACC-AGT	283	*bont/A*	Fujinaga et al. (1995)
A2	TCC-ACC-CCA-AAA-TGG-TAT-TCC			
A3	TTA-CAG-TGG-CCT-TTC-TCC-CC	2297	*bont/A*	Fujinaga et al. (1995)
A4	TCT-ATA-TTA-GGC-ATA-AGT-TC			
B1	CCT-CCA-TTT-GCG-AGA-GGT-ACG	315	*bont/B*	Fujinaga et al. (1995)
B2	CTC-TTC-GAG-TGG-AAC-ACG-TCT			

Continued

Table 13.1 PCR Primers and DNA Probes for Clostridium botulinum (Continued)

Primer or Probe	Sequence (5′ → 3′)	Sequence (bp)	Gene	Reference
B3	TTA-TTC-AGT-CCA-CCC-TTC-AT	2243	bont/B	Fujinaga et al. (1995)
B4	TCT-GAG-AGT-ATA-AAT-ATT-GA			
fBn	AAG-GTA-GAT-GAT-TTA-CCA-ATT-GTA	216	bont/B	Lövenklev et al. (2004) Dahlenborg, Borch, and Radstrom (2001)
rBn	GT-AGG-ATC-TGA-TAT-GCA-AAC-TA			
cntB probe	FAM-ACC-T̲T̲G-TTA-AGT-CTA-TCA-ACT-ATC-CC[b]			
f16S	GTG-TCG-TGA-GAT-GTT-GGG-TTA-A	207	16S rRNA	Lövenklev et al. (2004) Dahlenborg, Borch, and Radstrom (2001)
r16S	TAG-CTC-CAC-CTC-GCG-GTA-TT			
16S probe	FAM-TCC-CGC-A̲A̲C-GAG-CGC-AAC-CCT-T[b]			

[a] Underline indicates location of locked nucleic acid bases.
[b] Double underline indicates location of internal dark quencher.

REFERENCES

Akbulut, D., Grant, K., McLauchlin, J. 2005. Improvement in laboratory diagnosis of wound botulism and tetanus among injecting illicit-drug users by use of real-time PCR assays for neurotoxin gene fragments. *J. Clin. Microbiol.* 43:4342–4348.

Alsallami, A., Kotlowski, R. 2001. Selection of primers for specific detection of *Clostridium botulinum* types B and E neurotoxin genes using PCR method. *Int. J. Food Microbiol.* 69:247–253.

Arnon, S., Dumas, K., Chin, J. 1981. Infant botulism: Epidemiology and relation to sudden infant death syndrome. *Epidemiol. Rev.* 3:45–66.

Arnon, S., Midura, T., Damus, K., Thompson, B., Woods, R., Chin, J. 1979. Honey and other environmental risk factors for infant botulism. *J. Pediat.* 94:331–336.

Artin, I., Bjorkman, P., Cronqvist, J., Radström, P., Ost, E. 2007. First case of type E wound botulism diagnosed using real-time PCR. *J. Clin. Microbiol.* 45:3589–3594.

Badhey, H., Cleri, D., D'Amaato, R., Vernaleo, R., Veinni, V., Hochstein, L. 1986. Two fatal cases of type E adult food-borne botulism with early symptoms and terminal neurological signs. *J. Clin. Microbiol.* 23:616–618.

Barash, J., Tang, T., Arnon, S. 2005. First case of infant botulism caused by *Clostridium baratii* type F in California. *J. Clin. Microbiol.* 43:4280-4282.

Campbell, K., Collins, M., East, A. 1993. Gene probes for identification of the botulinal neurotoxin gene and specific identification of neurotoxin types B, E, and F. *J. Clin. Microbiol.* 31:2255–2262.

Chao, H., Wang, Y., Tang, S., Liu, H. 2004. A highly sensitive immuno-polymerase chain reaction assay for *Clostridium botulinum* neurotoxin type A. *Toxicon* 43:27–34.

Chin, J., Arnon, S., Midura, T. 1979. Food and environmental aspects of infant botulism in California. *Reviews of Infectious Diseases* 1:693–696.

Dahlenborg, M., Borch, E., Radsröm, P. 2001. Development of a combined selection and enrichment PCR procedure for *Clostridium botulinum* types B, E, and F and its use to determine prevalence in fecal samples from slaughtered pigs. *Appl. Environ. Microbiol.* 67:4781–4788.

Dahlsten, E., Korkeala, H., Somervuo, P., Lindström, M., 2008. PCR assay for differentiating between Group I (proteolytic) and Group II (nonproteoytic) strains of *Clostridium botulinum*. *Int. J. Food Microbiol.* 124:108–111.

East, A., Bhandari, M., Stacey J., Campbell, K., Collins, M. 1996. Organization and phylogenetic interrelationships of genes encoding components of the botulinum complex in proteolytic *Clostridium botulinum* Types A, B, and F: Evidence of chimeric sequences in the gene encoding the nontoxic nonhemagglutinin component. *Int. J. Syst. Bacteriol.* 46:1105–1112.

East, A., Collins, M. 1994. Conserved structure of genes encoding components of botulinum neurotoxin complex M and the sequence of the gene coding for the nontoxic component in nonproteolytic *Clostridium botulinum* type F. *Curr. Microbiol.* 29:69–77.

Eklund, M., Wieler, D., Poysky, F. 1967. Outgrowth and toxin production of nonproteolytic type B *Clostridium botulinum* at 3.3 to 5.6°C. *J. Bacteriol.* 93:1461–1462.

Fach, P., Gibert, M., Griffais, R., Guillou, J., Popoff, M. 1995. PCR and gene probe identification of botulinum neurotoxin A-, B-, E-, F-, and G-producing *Clostridium* spp. and evaluation in food samples. *Appl. Environ. Microbiol.* 61:389–392.

Fenicia, L., Anniballi, F., De Medicin, D., Delibato, E., Aureli, P. 2007. SYBR green real-time PCR method to detect *Clostridium botulinum* type A. *Appl. Environ. Microbiol.* 73:2891–2896.

Fach, P., Perelle, S., Dilasser, F., Grout, J., Dargaignaratz, C., Botella, L., Gourreau, J., Carlin, F., Popoff, M., Broussolle, V. 2002. Detection by PCR-enzyme-linked immunosorbent assay of *Clostridium botulinum* in fish and environmental samples from a coastal area in Northern France. *Appl. Environ. Microbiol.* 68:5870–5876.

Ferriera, J., Hamdy, M., McCay, S., Hemphill, M., Nameer, K., Baumstark, B. 1994. Detection of *Clostridium botulinum* type F using the polymerase chain reaction. *Mol. Cell Probe.* 8:365–373.

Franciosa, G., Fenicia, L., Pourshaban, M., Aureli, P. 1997. Recovery of a strain of *Clostridium botulinum* producing both neurotoxin A and neurotoxin B from canned macrobiotic food. *Appl. Environ. Microbiol.* 63:1148–1150.

Franciosa, G., Ferreira, J., Hatheway, C. 1994. Detection of type A, B, and E botulinum neurotoxin genes in *Clostridium botulinum* and other *Clostridium* species by PCR: Evidence of unexpressed type B toxin genes in type A toxigenic organisms. *J. Clin. Microbiol.* 32:1911–1917.

Franciosa, G., Florida, F., Maugliani, A., Aurelo, P. 2004. Differentiation of the gene clusters encoding botulinum neurotoxin type A complexes in *Clostridium botulinum* type A, Ab, and A(B) strains. *Appl. Environ. Microbiol.* 70:7192–7199.

Fujinaga, Y., Takeshi, K., Inoue, K., Fujita, R., Ohyama T., Moriishi, K., Oguma, K. 1995. Type A and B neurotoxin gene in a *Clostridium botulinum* type AB strain. *Biochem. Biophys. Res. Commun.* 213:737–745.

Graham, S., Mason, D., Peck, M. 1997. Effect of pH and NaCl on growth from spores of non-proteolytic *Clostridium botulinum* at chill temperatures. *Lett. Appl. Microbiol.* 24:95–100.

Hatheway, C. 1993. *Clostridium botulinum* and other clostridia that produce botulism neurotoxin. In: *Clostridium botulinum: Ecology and Control in Foods*. A.H.W. Hauschild and K.L. Dodds, Eds., Marcel Dekker: New York, pp. 3–20.

Hatheway, C. 1995. Botulism: The present status of the disease. *Curr. Top. Microbiol.* 195:55–75.

Hielm, S., Björkroth, J., Hyytiä, E., Korkeala, H. 1998a. Genomic analysis of *Clostridium botulinum* group II by pulsed-field gel electrophoresis. *Appl. Environ. Microbiol.* 64:703–708.

Hielm, S., Björkroth, J., Hyytiä, E., Korkeala, H. 1998b. Prevalence of *Clostridium botulinum* in Finnish trout farms: Pulsed-field gel electrophoresis typing reveals extensive genetic diversity among type E isolates. *Appl. Environ. Microbiol.* 64:4161–4167.

Hielm, S., Hyytiä, E., Andersin, A., Korkeala, H. 1998c. A high prevalence of *Clostridium botulinum* type E in Finnish freshwater and Baltic Sea sediment samples. *J. Appl. Microbiol.* 84:133–137.

Hielm, S., Hyytiä, E., Ridell, J., Korkala, H. 1996. Detection of *Clostridium botulinum* in fish and environmental samples using polymerase chain reaction. *Int. J. Food Microbiol.* 31:357–365.

Hutson, R., Collins, M., East, A., Thompson, D. 1994. Nucleotide sequence of the gene coding for non-proteolytic *Clostridium botulinum* type B neurotoxin: Comparison with other clostridial neurotoxins. *Curr. Microbiol.* 28:1–1–110.

Hutson R., Thompson, D., Collins, M., 1993a. Genetic interrelationships of saccharolytic *Clostridium botulinum* types B, E and F and related clostridia as revealed by small-subunit rRNA gene sequences. *FEMS Microbiol. Lett.* 108:103–110.

Hutson, R., Thompson, D., Lawson, R., Schocken-Itturino, R., Bottger, E., Collins, M. 1993b. Genetic interrelationships of proteolytic *Clostridium botulinum* types A, B, and F and other members of the *Clostridium botulinum* complex as revealed by small-subunit rRNA gene sequences. *Anton. Leeuwen.* 64:273–283.

Hyytiä, E., Björkroth, J., Hielm, S., Korkeala, H. 1999a. Characterization of *Clostridium botulinum* groups I and II by randomly amplified polymorphic DNA analysis and repetitive element sequence-based PCR. *Int J. Food Microbiol.* 48:179–189.

Hyytiä, E., Hielm, S., Björkroth, J., Korkeala, H. 1999b. Biodiversity of *Clostridium botulinum* type E strains isolated from fish and fishery products. *Appl. Environ. Microbiol.* 65:2057–2064.

Hyytiä, E., Hielm, S., Korkeala, H. 1998. Pevalence of *Clostridium botulinum* type E in Finnish fish and fishery products. *Epidemiol. Infect.* 120:245–250.

Hyytiä-Trees, E., Lindström, M., Schalch, B., Stolle, A., Korkeala, H. 1999. *Clostridium botulinum* type E in Bavarian fish. *Arch. Für Lebensmittelhyg.* 50:79–82.

Johnson, E., Teppk, W., Bradshaw, M., Gilbert, R., Cook, P., McIntosh, E. 2005. Characterization of *Clostridium botulinum* strains associated with an infant botulism case in the United Kingdom. *J. Clin. Microbiol.* 43:2602–2607.

Johnson, M., Lin, G., Raphael, B., Andreadis, J., Johnson, E. 2008. Analysis of neurotoxin cluster genes in *Clostridium botulinum* strains producing botulinum neurotoxin serotype A subtypes. *Appl. Environ. Microbiol.* 74:2778–2786.

Keto-Timonen, R., Nevas, M., Korkeala, H. 2005. Efficient DNA fingerprinting of *Clostridium botulinum* types A, B, E, and F by amplified fragment length polymorphism analysis. *Appl. Environ. Microbiol.* 71:1148–1154.

Kimura, B., Kawasaki, S., Nakano, H., Fujii, T. 2001. Rapid, quantitative PCR monitoring of growth of *Clostridium botulinum* type E in modified-atmosphere-packaged fish. *Appl. Environ. Microbiol.* 67:206–216.

Korkeala, H., Stengel, G., Hyytiä, E., Vogelsang, B., Bohl, A., Wihlman, H., Pakkala, P., Hielm, S. 1998. Type E botulism associated with vacuum-packaged hot-smoked white-fish. *Int. J. Food Microbiol.* 43:1–5.

Kotev, S., Leventhal, A., Bashary, A., Zahavi, H., Cohen, A. 1987. International outbreak of type E botulism associated with ungutted, salted whitefish. *Morb. Mortal. Wkly. Rep.* 36:812–813.

Leclair, D., Pagotto, F., Farber, J., Cadieux, B., Austin, J. 2006. Comparison of DNA finger-printing methods for use in investigation of type E botulism outbreaks in the Canadian Arctic. *J. Clin. Microbiol.* 44:1635–1644.

Lin, W., Johnson, E. 1995. Genome analysis of *Clostridium botulinum* type A by pulsed-field gel electrophoresis. *Appl. Environ. Microbol.* 61:4441–4447.

Lindström, M., Keto, R., Markkula, A., Nevas, M., Hielm, S., Korkeala, H. 2001. Multiplex PCR assay for detection and identification of *Clostridium botulinum* types A, B, E, and F in food and fecal material. *Appl. Environ. Microbiol.* 67:5694–5699.

Lövenklev, M., Holst, E., Borch, E., Rådstrom, P. 2004. Relative neurotoxin gene expression in *Closridium botulinum* type B, determined using quantitative reverse transcription-PCR. *Appl. Environ. Microbiol.* 70:2919–2927.

Macdonald, T., Helma, C., Ticknor, L., Jackson, P., Okinaka, R., Smith, L., Smith, T., Hill, K. 2008. Differentiation of *Clostridium botulinum* serotype A strains by multiple-locus variable-number tandem-repeat analysis. *Appl. Environ. Microbiol.* 74:875–882.

McGrath, S., Dooley, J., Haylock, R. 2000. Quantification of *Clostridium botulinum* toxin gene expression by competitive reverse transcription-PCR. *Appl. Environ. Microbiol.* 66:1423–1428.

Merivirta, L., Lindström, M., Jörkroth, K., Korkeala, H. 2006. The prevalence of *Clostridium botulinum* in European river lamprey (*Lampetra fluviatilis*) in Finland. *Int. J. Food Microbiol.* 109:234–237.

Midura, T. 1996. Update: Infant botulism. *Clin. Microbiol. Rev.* 9:119–125.

Midura, T., Arnon, S. 1976. Infant botulism: Identification of *Clostridium botulinum* and its toxin in feces. *Lancet* 308, no. 7992:934–936.

Midura, T., Snowden, S., Wood, R., Arnon, S. 1979. Isolation of *Clostridium botulinum* from honey. *J. Clin. Microbiol.* 9:282–283.

Morris, J., Snyder, J., Wilson, R., Feldman, R. 1983. Infant botulism in the United States: an epidemiological study of the cases occurring outside California. *Am. J. P. H.* 73:1385–1388.

Myllykoski, J., Nevas, M., Lindström, M., Korkeala, H. 2006. The detection and prevalence of *Clostridium botulinum* in pig intestinal samples. *Int. J. Food Microbiol.* 110:172–177.

Nevas, M., Hielm, S., Lindström, M., Horn, H., Koivulehto, K., Korkeala, H. 2002. High prevalence of *Clostridium botulinum* types A and B in honey samples detected by polymerase chain reaction. *Int. J. Food Microbiol.* 72:45–52.

Nevas, M., Lindström, M., Hautamäki, K., Puoskari, S., Korkeala, H. 2005a. Prevalence and diversity of *Clostriium botulinum* types A, B, E, and F in honey produced in the Nordic countries. *Int. J. Food Microbiol.* 105:145–151.

Nevas, M., Lindström, M., Hielm, S., Björkroth, K., Peck, M., Korkeala, H. 2005b. Diversity of proteolytic *Clostridium botulinum* strains, determined by a pulsed-field gel electrophoresis approach. *Appl. Environ. Microbiol.* 71:1311–1317.

Paul C., Twine, S., Tam, K., Mullen J., Kelly, J., Austin, J., Logan, S. 2007. Flagellin diversity in *Clostridium botulinum* groups I and II: A new strategy for strain identification. *Appl. Environ. Microbiol.* 73:2963–2975.

Pickett, J., Berg, B., Chaplin, E., Brunstetter-Schafer, M. 1976. Syndrome of botulism in infancy: Clinical and electrophysiologic study. *New Engl. J. Med.* 295:770–772.

Raphael, B., Andreadis, J. 2007. Real-time PCR detection of the nontoxic nonhemagglutinn gene as a rapid screening method for bacterial isolates harboring the botulinum neurotoxin (A-G) gene complex. *J. Microbiol. Methods* 71:343–346.

Scheinert, P., Krausse, R., Ullmann, U., Söller, R., Krupp, G. 1996. Molecular differentiation of bacteria by PCR amplification of the 16S-23s rRNA spacer. *J. Microbiol. Methods* 26:103–117.

Sharkey, F., Markos, S., Haylock, R. 2004. Quantification of toxin-encoding mRNA from *Clostridium botulinum* type E in media containing sorbic acid or sodium nitrite by competitive RT-PCR. *FEMS Microbiol. Lett.* 232:139–144.

Smith, L., Sugiyama, H. 1988. Cultural and serological characteristics. In: *Botulism. The Organism, Its toxins, the Disease*. L. Smith, H. Sugiyama, Eds., Charles C. Thomas: Springfield, IL, pp. 23–37.

Sobel, J., Malavet, M., John, S. 2007. Outbreak of clinically mild botulism type E illness from home-salted fish in patients presenting with predominantly gastrointestinal symptoms. *Clin. Infect. Dis.* 45:e14–16.

Szabo, E., Pemberton, J., Desmarchelier, P. 1993. Detection of the genes encoding botulinum neurotoxin types A to E by the polymerase chain reaction. *Appl. Environ. Microbiol.* 59:3011–3020.

Telzak, E., Bell, E., Kautter, D., Crowell, L., Budnick, L., Morse, D., Schulz, S. 1990. An international outbreak of type E botulism due to uneviscerated fish. *J. Infect. Dis.* 161:340–342.

Wang, X., Maegawa, T., Karasawa, T., Kozaki, S., Tsukamoto, K., Gyobu, Y., Yamakawa, K., Oguma, K., Sakaguchi, Y., Nakamura, S. 2000. Genetic analysis of type E botulinum toxin-producing *Clostridium butyricum* strains. *Appl. Environ. Microbiol.* 66:4992–4997.

Weber, J., Hibbs, R., Sarswish, A. 1993. A massive outbreak of type E botulism associated with traditional salted fish in Cairo. *J. Infect. Dis.* 167:451–454.

Whitmer, M., Johnson, E. 1988. Development of improved defined media for *Clostridium botulinum* serotypes A, B and E. *Appl. Environ. Microbiol.* 54:753–759.

Williamson, J., Rocke, T., Aiken, J. 1999. In situ detection of the *Clostridium botulinum* type C_1 toxin gene in wetland sediments with a nested PCR assay. *Appl. Environ. Microbiol.* 65:3240–3243.

Zhou, Y., Sugiyama, H., Johnson, E. 1993. Transfer of neurotoxigenicity from *Clostridium butyricum* to a nontoxigenic *Clostridium botulinum* type E-like strain. *Appl. Environ. Microbiol.* 59:3825–3831.

CHAPTER 14

Clostridium perfringens

I. CHARACTERISTICS OF THE ORGANISM

Clostridium perfringens is a gram-positive, spore-forming, nonmotile, obligately anaerobic, encapsulated rod. Notable differences among strains of *C. perfringens* include the extent of β-hemolysis and the heat resistance of spores. "Heat-sensitive" food poisoning strains are less common than "heat-resistant" strains, the heat sensitivity of *C. perfringens* spores being defined as not surviving 100°C for 60 min.

The organism produces several extracellular proteins, some of which are toxins and others enzymes. These include the *C. perfringens* enterotoxin (CPE) in addition to a lethal necrotizing hemolytic lecithinase plus several additional lethal necrotizing toxins, several hemolysins, collagenase, protease, hyaluronidase, and DNAse (Smith and Williams, 1984). Four toxins, alpha, beta, epsilon, and iota, are used for classifying all strains into five types (A–E) depending on which combination of these four toxins a given strain produces (Table 14.1). Most cases of food poisoning due to *C. perfringens* are due to type A, which produces only the alpha toxin, which is distinct from CPE (*cpe*) that is also produced. CPE is not a true exotoxin in that cells do not secrete the toxin. The toxin is released when cells lyse following spore formation and acts on epithelial cells causing diarrhea and loss of water and ions. The alpha toxin (*cpa*) is considered the main lethal toxin of *C. perfringens* and is a multifunctional phospholipase causing hydrolysis of membrane phospholipids resulting in cell lysis. The beta toxin (*cpb*) is a highly trypsin-sensitive protein causing mucosal necrosis resulting in central nervous system symptoms in domestic animals. The epsilon toxin (*etx*) is a potent toxin responsible for lethal enterotoxemia in livestock. The iota toxin (*cpi*) increases vascular permeability and is dermonecrotic. An additional toxin beta2 (*cpb2*) is associated with enteric diseases in piglets and horses.

Food poisoning due to *C. perfringens* usually occurs 8–24 hr after ingestion of food containing large numbers of vegetative cells. Diarrhea and severe abdominal pain usually occur, with nausea less common and fever and vomiting unusual. A more severe illness from foodborne type C is a necrotizing hemorrhagic jejunitis

Table 14.1 Typing of *C. perfringens* Strains Based on Toxins Produced

	Toxin Produced			
Type	Alpha	Beta	Epsilon	Iota
A	+	−	−	−
B	+	+	+	−
C	+	+	−	−
D	+	−	+	−
E	+	−	−	+

reported from Papua, New Guinea. In addition, *C. perfringens* is the classic cause of gas gangrene in extraintestinal wound infections.

II. MOLECULAR METHODS FOR DETECTION OF *C. PERFRINGENS*

A. PCR

Wang et al. (1994) developed a 16S rDNA-based PCR assay for the rapid and specific detection of *C. perfringens* on poultry. *C. perfringens* was inoculated onto the skin of slaughtered poultry. The skin and attached tissue were then rinsed with vigorous shaking using saline. The rinses were then passed through filter paper on a Buchner funnel and the filtrates were then centrifuged at $10,000 \times g$ for 10 min. The pellets were incubated in PRAS BHI broth at 37°C for 17 hr. Enrichments (0.2 ml) were diluted in PBS to 1.2 ml, centrifuged, washed, and resuspended in 1% Triton X-100 and then boiled to release DNA that was not purified. The primers CP.1/CP.2 (Table 14.2) amplified a 279-bp sequence of the 16S rDNA. Only strains of *C. perfringens* yielded the expected amplicon, and 12 other species of *Clostridium* failed to undergo DNA amplification with the primer pair. When the skin of 100 g samples of poultry was inoculated with 20 cells, positive results were obtained.

Fach and Popoff (1997) developed a duplex PCR assay for the detection of *C. perfringens* in food and biological samples and for identification of enterotoxigenic strains. Primers PL3/PL7 amplified a 283-bp sequence of the phospholipase *cpa* gene (Table 14.2). Primers P145/P146 (Table 14.2) amplified a 426-bp sequence of the *cpe* gene. When the assay was applied to 24 strains of *C. perfringens* from a food poisoning outbreak, all were found to harbor the *cpa* gene. Seven of the 24 isolates were PCR positive for the *cpe* gene. Without enrichment, the detection level in stools and foods was 10^5 CFU/g. With overnight enrichment at 37°C, 10 CFU/g of seeded foods were detected.

Miwa et al. (1998) made use of a combined MPN and nested PCR after culturing meat samples for *C. perfringens*. Primers CPE-1/CPE-2 (Table 14.2) were used for the initial PCR. Primers CPE-3/CPE-4 (Table 14.2) were used for the internally nested PCR. The results obtained by this method were closely correlated with CFU counts obtained by plating. Application of the method to randomly selected meat samples yielded *C. perfringens* in 16% of beef and 84% of poultry samples out of

Table 14.2 PCR Primers and DNA Probes for *Clostridium perfringens*

Primer, Probe, or Adapter	Sequence (5' → 3')	Sequence (bp)	Gene	Reference
CP.1	AAA-GAT-GGC-ATC-ATT-CAA-C	279	16S rDNA	Wang et al. (1994)
CP2	AAA-CCC-CTT-CTA-TTA-CTG-CCA-T			
Probe Cperf191	GCT-CCT-TTG-GTT-GAA-TGA-TG	—	16S rDNA	Fallini et al. (2006)
RAPD HLWL74	ACG-TAT-CTG-C	—	—	Leflon-Guibout et al. (1997)
RAPD MHN1	ACG-TCT-ATG-C	—	—	
RAPD R108	GTA-TTG-CCC-T	—	—	
RAPD A3	TGG-ACC-CTG-C	—	—	
CPE4F	TTA-GAA-CAG-TCC-TTA-GGT-GAT-GGA-G	1300	cpe	Miyamoto, Wen, and McClane (2004)
IS1470R1.3	CTT-CTT-GAT-TAC-AAG-ACT-CCA-GAA-GAG			
3F	GAT-AAA-GGA-GAT-GGT-TGG-ATA-TTA-GG	~600	cpe	Miyamoto, Wen, and McClane (2004)
4R	GAG-TCC-AAG-GGT-ATG-AGT-TAG-AAG			
CPE4F	TTA-GAA-CAG-TCC-TTA-GGT-GAT-GGA-G	1600	cpe	Miyamoto, Wen, and McClane (2004)
IS1470	CTT-TGT-GTA-CAC-AGC-TTC-GCC-AAT-GTC-			
CPE4F	TTA-GAA-CAG-TCC-TTA-GGT-GAT-GGA-G	~800	cpe	Miyamoto, Wen, and McClane (2004)
IS1151R0.8	ATC-AAA-ATA-TGT-TCT-TAA-AGT-ACG-TTC			

Continued

Table 14.2 PCR Primers and DNA Probes for Clostridium perfringens (Continued)

Primer, Probe, or Adapter	Sequence (5′ → 3′)	Sequence (bp)	Gene	Reference
Clper-F	GCA-TGA-GTC-ATA-GTT-GGG-ATG-ATT	—	cpa	Shannon et al. (2007)
Clper-R	CCT-GCT-GTT-CCT-TTT-TGA-GAG-TTA-G			
Clper-PR	FAM-TGC-AGC-AAA-GGT-AAC-TT-BHQ			
CPE-F	TAA-CAA-TTT-AAA-TCC-AAT-GG	933	cpe	Lukinmaa, Takknen, and Siitonen (2002)
CPE-R	ATT-GAA-TAA-GGG-TAA-TTT-CC			
PL3	AAG-TTA-CCT-TTG-CTG-CAT-AAT-CCC	283	cpa	Fach and Popoff (1997)
PL7	ATA-GAT-ACT-CCA-TAT-CAT-CCT-GCT			
P145	GAA-AGA-TCT-GTA-TCT-ACA-ACT-GCT-GGT-CC	426	cpe	Fach and Popoff (1997)
P146	GCT-GGC-TAA-GAT-TCT-ATA-TTT-TTG-TCC-AGT			
CPEPS	TGT-AGA-ATA-TGG-ATT-TGG-AAT	363	cpe	Baez and Juneja (1995)
CPENS	AGC-TGG-GTT-TGA-GTT-TAA-TGC			
ETP-K-F	TCC-AAT-GGT-GTT-CGA-AAA-TG	821	cpe	Nakamura et al. (2004)
ETP-K-R1	CAT-CAC-CTA-AGG-ACT-GTT-C			
dcm	GTA-TTG-TAA-TCC-AGG-TAG-CAG	2954	cpe	
uapC	GCT-GCC-CTA-ATC-CTA-AAT-GCA-G	4134 and 3278	cpe	
Alpha-F	TGC-ACT-ATT-TTG-GAG-ATA-TAG-ATA-C	—	cpa	Gurjar et al. (2008)
Alpha-R	CTG-CTG-TGT-TTA-TTT-TAT-ACT-GTT-C			
Alpha-Pr	FAM-TCC-TGC-TAA-TGT-TAC-TGC-CGT-TGA-TAMRA			

Continued

Primer	Sequence	Gene	Size	Reference
Beta2-F	TAA-CAC-CAT-CAT-TTA-GAA-CTC-AAG	*cpb2*	—	Gurjar et al. (2008)
Beta2-R	CTA-TCA-GAA-TAT-GTT-TGT-GGA-TAA-AC			
Beta2-Pr	HEX-TGC-TTG-TGC-TAG-TTC-ATC-ATC-CCA-TAMRA			
Beta-F	ATT-TCA-TTA-GTT-ATA-GTT-AGT-TCA-C	*cpb*	—	Gurjar et al. (2008)
Beta-R	TTA-TAG-TAG-TAG-TTT-TGC-CTA-TAT-C			
Beta-Pr	HEX-AAC-GGA-TGC-CTA-TTA-TCA-CCA-ACT-TAMRA			
Epsilon-F	TTA-ACT-AAT-GAT-ACT-CAA-CAA-GAA-C	*etx*	—	Gurjar et al. (2008)
Epsilon-R	GTT-TCA-TTA-AAA-GGA-ACA-GTA-AAC			
Epsilon-Pr	FAM-TGC-TTG-TAT-CGA-AGT-TCC-CAC-AGT-TAMRA			
Iota-F	CAA-GAT-GGA-TTT-AAG-GAT-GTT-TC	*cpi*	—	Gurjar et al. (2008)
Iota-R	TTT-TGG-TAA-TTT-CAA-ATG-TAT-AAG-TAG			
Iota-Pr	FAM-TTC-CAT-CGC-CAT-TAC-CTG-GTT-CAT-TAMRA			
CPE-F	AAC-TAT-AGG-AGA-ACA-AAA-TAC-AAT-AG	*cpe*	—	Gurjar et al. (2008)
CPE-R	TGC-ATA-AAC-CTT-ATA-ATA-TAC-ATA-TTC			
CPE-Pr	HEX-TCT-GTA-TCT-ACA-ACT-GCT-GGT-CCA-TAMRA			
CPA-F	GCT-AAT-GTT-ACT-GCC-GTT-GA	*cpa*	324	Lin and Labbe (2003)
CPA-R	CCT-CTG-ATA-CAT-CGT-GTA-AG			
CPE-F	GGA-GAT-GGT-TGG-ATA-TTA-GG	*cpe*	233	Lin and Labbe (2003)
CPE-R	GGA-CCA-GCA-GTT-GTA-GAT-A			

Table 14.2 PCR Primers and DNA Probes for *Clostridium perfringens* (*Continued*)

Primer, Probe, or Adapter	Sequence (5′ → 3′)	Sequence (bp)	Gene	Reference
CPA-F	TTC-TAT-CTT-GGA-GAG-GCT-ATG-GCA-CAC-TAT-TTT-GG	319	*cpa*	Songer and Meer (1996)
CPA-R	AGT-AAA-GTT-ACC-TTT-GCT-GCT-GCA-ATA-ATC-CC			
CPB-F	TAC-TAT-AAC-TAG-AAA-TAA-GAC-ATC-AGA-TGG-C	554	*cpb*	Songer and Meer (1996)
CPB-R	CCA-CGA-GTA-GTT-TCT-GTA-AAT-TTT-GTA-TCG-C			
ETX-F	TAC-TCA-TAC-TGT-GGG-AAC-TCC-GAT-ACA-AGC	402	*etx*	Songer and Meer (1996)
ETX-R	CTC-ATC-TCC-CAT-AAC-TGC-ACT-ATA-ATT-TCC			
CPI-F	ACT-ACT-CTC-AGA-CAA-GAC-AG·	446	*cpi*	Songer and Meer (1996)
CPI-R	CTT-TCC-TTC-TAT-TAC-TAT-ACG			
CPE-F	TGT-TAA-TAC-TTT-AAG-GAT-ATG-TAT-CC	935	*cpe*	Songer and Meer (1996)
CPE-R	TCC-ATC-ACC-TAA-GGA-CTG			
PT1	TCT-AGA-ATA-TGG-ATT-TGG-AAT	364	*cpe*	Singh, Bhilegaonkar, and Agarwal (2005) from Saito, Matsumoto, and Funabashi (1992)
PT2	AGC-TGG-TTT-GAG-TTT-AAT-G			
Alpha 1	GCT-ACA-TTC-TAT-CTT-GGA-GA	407	*cpa*	Moller and Ahrens (1996)
Alpha 2	TCC-AAC-TGA-TGG-ATC-ATT-AC			
Beta 1	AAA-CGG-ATG-CCT-ATT-ATC-AC	514	*cpb*	Moller and Ahrens (1996)

Primer	Sequence	Size	Gene	Reference
Beta 2	AGA-TGA-TTC-AGC-ATA-TTC-GC			
Epsilon 1	CGG-TGA-TAT-CCA-TCT-ATT-CA	289	*etx*	Moller and Ahrens (1996)
Epsilon 2	TTC-ATT-GAT-GGT-TCT-CCA-TC			
Iota 1	AAC-TAC-TCT-CAG-ACA-AGA-CA	320	*cpi*	Moller and Ahrens (1996)
Iota 2	TGT-ATA-AGT-AGT-GGT-GTA-GG			
CPE 1	ATC-CAA-TGG-TGT-TCG-AAA-AT	499	*cpe*	Moller and Ahrens (1996)
CPE 2	ATT-TCC-TAA-GCT-ATC-TGC-AG			
Int. std. B3	GCA-TAT-TCG-CAA-ATC-GAG-TAT-CTA-ACC-GGA	341	*cpb*	Moller and Ahrens (1996)
Beta-F	CAA-AAG-GAG-GTT-TTT-TTA-TGA-AG	1040	*cpb*	Yamagishi et al. (1997)
Beta-R	AAA-TAT-AAA-AGT-CTA-AAT-AGC			
Epsilon-F	TAC-TCA-TAC-TGT-GGG-AAC-TTC	403	*etx*	Yamagishi et al. (1997)
Epsilon-R	CTC-ATC-TCC-CAT-AAC-TGC-ACT-ATA			
Iota-F	GAT-CAA-TTT-AAT-GCT-AG	414	*cpi*	Yamagishi et al. (1997)
Iota-R	TCC-TTT-AAT-TAT-TTC-AGA-TGT			
Alpha -F	TGC-TAA-TGT-TAC-TGC-CGT-TGA-TAG	247	*cpa*	Stagnitta, Micalizzi, and Guzmán (2002) from Uzal et al. (1997)
Alpha-R	ATA-ATC-CCA-ATC-ATC-CCA-ACT-ATG			
Beta-F	AGG-AGG-TTT-TTT-TAT-GAA-G	1025	*cpb*	Stagnitta, Micalizzi, and Guzmán (2002) from Uzal et al. (1997)
Beta-R	TCT-AAA-TAG-CTG-TTA-CTT-TGT			

Continued

Table 14.2   PCR Primers and DNA Probes for *Clostridium perfringens* (Continued)

Primer, Probe, or Adapter	Sequence (5′ → 3′)	Sequence (bp)	Gene	Reference
Epsilon-F Epsilon-R	TAC-TCA-TAC-TGT-GGG-AAC-CTT-CGA-TAC-AAG-C CTC-ATC-TCC-CAT-AAC-TGC-ACT-ATA-ATT-TCC	403	*etx*	Stagnitta, Micalizzi, and Guzmán (2002) from Uzal et al. (1997)
Iota-F Iota-R	TTT-TAA-CTA-GTT-CAT-TTC-CTA-GTT-A TTT-TTG-TAT-TCT-TTT-TCT-CTA-GAT-T	298	*cpi*	Stagnitta, Micalizzi, and Guzmán (2002) from Uzal et al. (1997)
CPE-F CPE-R	TGT-TAA-TAC-TTT-AG-GAT-ATG TCC-ATC-ACC-TAA-GCA-CTG	935	*cpe*	Stagnitta, Micalizzi, and Guzmán (2002) from Uzal et al. (1997)
Alpha CPA5L Alpha CPA5R	AGT-CTA-CGC-TTG-GGA-TGG-AA TTT-CCT-GG-TTG-TCC-ATT-TC	900	*cpa*	Baums et al. (2004)
Beta CPBL Beta CPBR	TCC-TTT-CTT-GAG-GGA-GGA-TAA-A TGA-ACC-TCC-TAT-TTT-GTA-TCC-CA	611	*cpb*	Baums et al. (2004)
Epsilon CPE TXL Epsilon CPE TXR	TGG-GAA-CTT-CGA-TAC-AAG-CA TTA-ACT-CAT-CTC-CCA-TAA-CTG-CAC	396	*etx*	Baums et al. (2004)
Iota CPIL Iota CPIR	AAA-CGC-ATT-AAA-GCT-CAC-ACC CTG-CAT-AAC-CTG-GAA-TGG-CT	293	*cpi*	Baums et al. (2004)

Continued

Beta2 CPB2L Beta 2 CPB2R	CAA-GCA-ATT-GGG-GGA-GTT-TA GCA-GAA-TCA-GGA-TTT-TGA-CCA	200	*cpb2*	Baums et al. (2004)
CPE CPEL CPE CPER	GGG-GAA-CCC-TCA-GTA-GTT-TCA ACC-AGC-TGG-ATT-TGA-GTT-TAA-TG	506	*cpe*	Baums et al. (2004)
CPA-L CPA-R	AAG-ATT-TGT-AAG-GCG-CTT ATT-TCC-TGA-AAT-CCA-CTC	1167	*cpa*	Miserez et al. (1998)
CPB-L CPB-R	AGG-AGG-TTT-TTT-TAT-GAA-G TCT-AAA-TAG-CTG-TTA-CTT-TGT-G	1025	*cpb*	Miserez et al. (1998)
CPE-L CPE-R	AAG-TTT-AGC-AAT-CGC-ATC TAT-TCC-TGG-TGC-CTT-AAT	961	*etx*	Miserez et al. (1998)
PLC-F PLC-R PLC-P1 PLC-P2	GGC-TGG-GGT-ATC-AAC-T AGT-AGA-ACC-TAA-TTG-AAG-CTC ACG-CTT-GGG-ATG-GAA-AAA-TTG-ATG-G-FITC 640RED-CAG-GAA-CTC-ATG-CTA-TGA-TTG-TAA-CTC-AAG-G-PO$_4$	187	*cpa*	Dela Cruz et al. (2006)
CPE-F CPE-R CPF-P1 CPE-P2	CTA-CAA-CTG-CTG-GTC-C AGT-TAG-AAG-AAC-GCC-AAT-C GTT-TAT-GCA-ACT-TAT-AGA-AAG-TAT-CAA-GCT-ATT-AGA-AT-FITC 640RED-CTC-ATG-GTA-ATA-TCT-CTG-ATG-ATG-GAT-CA-PO$_4$	336	*cpe*	Dela Cruz et al. (2006)

Table 14.2   PCR Primers and DNA Probes for *Clostridium perfringens* (Continued)

Primer, Probe, or Adapter	Sequence (5' → 3')	Sequence (bp)	Gene	Reference
CPerf165F	CGC-ATA-ACG-TTG-AAA-GAT-GG	105	16S rDNA	Wise and Siragusa (2005)
CPerf269R	CCT-TGG-TAG-GCC-GTT-ACC-C			
CPerf187F	FAM-TCA-TCA-TTC-AAC-CAA-AGG-AGC-AAT-CC-TAMRA			
Alpha-L	AAG-AAC-TAG-TAG-CTT-ACA-TAT-CAA-CTA-GTG-GTG	124	*cpa*	Albini et al. (2008)
Alpha-R	TTT-CCT-GGG-TTG-TCC-ATT-TCC			
Alpha-S	VIC-TTG-GAA-TCA-AAA-CAA-AGG-ATG-GAA-AAA-CTC-AAG-TAMRA			
Beta-L	TGG-AGC-GTG-AAA-GAA-ACT-GTT-ATT-A	85	*cpb*	Albini et al. (2008)
Beta-R	GGT-ATC-AAA-AGC-TAG-CCT-GGA-ATA-GA			
Beta-S	FAM-CTT-AAT-TGG-AAT-GGT-GCT-AAC-TGG-GTA-GGA-CAA-TAMRA			
Beta2-L	TAT-TTG-AAA-GTT-TAC-TGT-AAT-TTT-TAT-GTT-TTC-A	127	*cpb2*	Albini et al. (2008)
Beta2-R	CCA-TTA-CCT-TTC-TAT-AAG-CGT-CGA-TT			
Beta2-S	CY5-TGC-ACT-TGC-TTT-CAT-TGG-ACT-TAT-TGC-TCC-BHQ2			
Epsilon-L	TTT-GAT-AAG-GTT-ACT-ATA-AAT-CCA-CAA-GGA	121	*etx*	Albini et al. (2008)
Epsilon-R	AGA-GAG-CTT-TTC-CAA-CAT-AAA-CAT-CTT-C			
Epsilon-S	Cy5-TAA-TCC-TAA-AGT-TGA-ATT-AGA-TGG-AGA-ACC-A-BHQ2			

Primer	Sequence	Size	Gene	Reference
Iota-L	GCA-TTA-AAG-CTC-ACA-CCT-ATT-CCA	85	*cpi*	Albini et al. (2008)
Iota-R	GAG-ATG-TGA-GAG-TTA-ATC-CAA-ATT-CTT-G			
Iota-S	FAM-CTA-ACT-TAA-TTG-TAT-ATA-GAA-GGT-CTG-GTC-C-TAMRA			
CPE-L	AGC-TGC-TGC-TAC-AGA-AAG-ATT-AAA-TTT	88	*cpe*	Albini et al. (2008)
CPE-R	TGA-GTT-AGA-AGA-ACG-CCA-ATC-ATA-TAA			
CPE-S	FAM-ACT-GAT-GCA-TTA-AAC-TCA-AAT-CCA-GCT-TAMRA			
MET-1.5F	CTC-AGA-GTT-AGG-AGC-TAG-CCC-AAC-CC	~3300	*cpe*	Wen, Miyamoto, and McClane (2003)
CPE-up	CCT-AAT-ATC-CAA-CCA-TCT-CC			
CPE-4.5F	CAG-TCC-TTA-GGT-GAT-GGA	~2100	*cpe*	Wen, Miyamoto, and McClane (2003)
IS1470F	AAC-TAA-ATA-GGC-CTA-TAA-ATA-CC			
CPA-F	TGC-ATG-AGC-TTC-AAT-TAG-GT	400	*cpa*	Heikinheimo and Korkeala (2005)
CPA-R	TTA-GTT-TTG-CAA-CCT-GCT-GT			
CPB-F	GCG-AAT-ATG-CTG-AAT-CAT-CTA	196	*cpb*	Heikinheimo and Korkeala (2005) from Meer and Songer (1997)
CPB-R	GCA-GGA-ACA-TTA-GTA-TAT-CTT-C			
ETX-F	GCG-GTG-ATA-TCC-ATC-TAT-TC	655	*etx*	Heikinheimo and Korkeala (2005) from Meer and Songer (1997)
ETX-R	CA-CTT-ACT-TGT-CCT-ACT-AAC			

Continued

Table 14.2 PCR Primers and DNA Probes for *Clostridium perfringens* (Continued)

Primer, Probe, or Adapter	Sequence (5′ → 3′)	Sequence (bp)	Gene	Reference
CPI-F	ACT-ACT-CTC-AGA-CAA-GAC-AG	446	*cpi*	Heikinheimo and Korkeala (2005) from Meer and Songer (1997)
CPI-R	CTT-TCC-TTC-TAT-TAC-TAT-ACG			
CPE-F	GGA-GAT-GGT-TGG-ATA-TTA-GG	233	*cpe*	Heikinheimo and Korkeala (2005) from Meer and Songer (1997)
CPE-R	GGA-CCA-GCA-GTT-GTA-GAT-A			
CPA-F	AAG-ATT-TGT-AAG-GCG-CTT	1167	*cpa*	Aschfalk and Müller (2002)
CPA-R	ATT-TCC-TGA-CCA-AAT-CTC			
CPB-F	CGG-ATG-CCT-ATT-ATC-ACC-AA	861	*cpb*	Aschfalk and Müller (2002) from Saint-Joanis, Garnier, and Cole (1989)
CPB-R	ACC-CAG-TTA-GCA-CCA-TTC-CA			
ETX-F	AAG-TTT-AGC-AAT-CGC-ATC	961	*etx*	Aschfalk and Müller (2002) from Aschfalk and Müller (2001)
ETX-R	TAT-TCC-TGG-TGC-CTT-AAT			
CPI-F	TTT-TAA-CTA-GTT-CAT-TTC-CTA-GTT	298	*cpi*	Aschfalk and Müller (2002) from Hunter et al. (1993)
CPI-R	TTT-TTG-TAT-TCT-TTT-TCT-CTA-GGA-TT			
CPB2-F	GAA-AGG-TAA-TGG-AGA-ATT-ATC-TTA-ATG-C	573	*cpb2*	Aschfalk and Müller (2002) from Perelle et al. (1993)
CPB2-R	GCA-GAA-TCA-GGA-TTT-TGA-CCA-TAT-ACC			

CPE-F	TAA-CAA-TTT-AAA-TCC-AAT-GG	933	cpe	Aschfalk and Müller (2002) from Van Damme-Jongsten, Haagsma, and Notermans (1990)
CPE-R	ATT-GAA-TAA-GGG-TAA-TTT-CC			
A1	TGG-TTG-GAT-ATT-AGG-GGA-ACC	235	cpe	Ridell et al. (1998)
A2	TTC-ATT-TGG-ACC-AGC-AGT-TG			
A3	TGT-AGA-ATA-TGG-ATT-TGG-AAT	363	cpe	Ridell et al. (1998)
A4	AGC-TGG-ATT-TGA-GTT-TAA-TG			
Alpha-F	GTT-GAT-AGC-GCA-GGA-CAT-GTT-AAG	402	cpa	Yoo et al. (1997)
Alpha-R	CAT-GTA-GTC-ATC-TGT-TCC-AGC-ATC			
Beta-F	ACT-ATA-CAG-ACA-GAT-CAT-TCA-ACC	236	cpb	Yoo et al. (1997)
Beta-R	TTA-GGA-GCA-GTT-AGA-ACT-ACA-GAC			
Epsilon-F	ACT-GCA-ACT-ACT-ACT-CAT-ACT-GTC	541	etx	Yoo et al. (1997)
Epsilon-R	CTG-GTG-CCT-TAA-TAG-AAA-GAC-TCC			
Iota-F	GCG-ATG-AAA-AGC-CTA-CAC-CAC-TAC	317	cpi	Yoo et al. (1997)
Iota-R	GGT-ATA-TCC-TCC-ACG-CAT-ATA-GTC			
CPE-F	GGA-GAT-GGT-TGG-ATA-TTA-GG	233	cpe	Gholamiandekhordi et al. (2006) from Herholz et al. (1999)
CPE-R	GAC-AGG-GGC-ATA-CCC-ATA-TA			

Continued

Table 14.2 PCR Primers and DNA Probes for Clostridium perfringens (Continued)

Primer, Probe, or Adapter	Sequence (5′ → 3′)	Sequence (bp)	Gene	Reference
CP6-F	GTA-AAG-ATG-ATT-GCT-ATT-TAG-AGA-TAA	507	—	Sawires and Songer (2005)
CP6-R	TAA-GGT-ATC-ATC-AAA-ATC-CAC-TCC-AGG			
CPB13-F	AAG-GAA-GAT-GCT-ACT-CAA-GAT-G	262	nagK	Sawires and Songer (2005)
CPB13-R	CAT-TCT-CTT-TCA-TTC-TCT-GTA-A			
CP16-F	AAA-GTT-CCA-GGT-AAA-ATA-AGA-G	433	parB	Sawires and Songer (2005)
CP16-R	CAT-TCT-CTT-TCA-TTC-TCT-GTA-A			
CP19-F	CTC-AAT-CCT-AAC-AAT-ATG-TGC-TGA-CTA	481	—	Sawires and Songer (2005)
CP19-R	GTA-GCA-GCA-ATA-AAA-CCA-ACC-TAA-A			
CP42-F	GAT-GGC-CCA-AGA-AAC-AGA-AC	—	—	Sawires and Songer (2005)
CP42-R	GCT-GGG-AAT-AAA-GGG-TTT-GA			
ADH1	5′-ACG-GTA-TGC-GAC-AG	—	—	McLauchlin et al. (2000)
ADH2	3′-GAG-TGC-CAT-ACG-CTG-TCT-CGA	—	—	McLauchlin et al. (2000)
HI-X	GGT-ATG-CGA-CAG-AGC-TTG	200–1500	—	
HindIII	5′-CTC-GTA-GAC-TGC-GTA-CC	—	—	Keto-Timonen et al. (2006)
HindIII	3′-CTG-ACG-CAT-GGT-CGA	—	—	

Primer	Sequence	Range		Reference
HpyCH41V	5'-GAC-GAT-GAG-TCC-TGA-C		—	Keto-Timonen et al. (2006)
HpyCH41V	3'-TA-CTC-AGG-ACT-GGC			
Hind-O	GAC-TGC-GTA-CCA-GCT-T		—	Keto-Timonen et al. (2006)
Hpy-O	CGA-TGA-GTC-CTG-ACC-GT			
Hind-C	GAC-TGC-GTA-CCA-GCT-TC	75–450		Keto-Timonen et al. (2006)
Hpy-A	CGA-TGA-GTC-CTG-ACC-GTA			
APMLVA-01bR	GTT-TCT-TCA-AAA-CTT-GGT-ATG-CTC-TTT-GG	428–440	—	Chalmers et al. (2008)
LMLVA-01bL	GAA-GGG-GCA-GGA-AAT-AAG-GA			
ADMLVA-01eR	GTT-TCT-TGG-GGT-TAA-ACT-TGA-TTT-TCT-GTG	424–494	—	Chalmers et al. (2008)
LMLVA-01eL	ATG-GGG-TAA-GGG-GAT-CCA-A			
ADMLVA-07R	GTT-TCT-TTC-TTC-TGT-TTC-AAA-ATT-AGT-TTC-AT	127–139	—	Chalmers et al. (2008)
LMLVA-07L	TTC-AAA-GCC-AGG-TAC-ATT-CAA			
ADMLVA-17R	GTT-TCT-TTT-GTG-GAC-AAT-CTG-TAG-GAG-GA	141–183	—	Chalmers et al. (2008)
LMLVA-17L	TTA-ATG-TCC-ATA-TTC-CGC-CA			
ADMLVA-23bR	GTT-TCT-TAA-ATA-ATT-CAG-GCT-ATG-AGA-AGA-AAA-A	280–298	—	Chalmers et al. (2008)
LMLVA-23bL	GCC-TCT-CTC-CTT-GAG-CAA-AA			
ADMLVA-24dR	GTT-TCT-TGA-ATT-ACT-TTT-CTA-CAA-TTT-ATT-TTC-ATA-A	132–201	—	Chalmers et al. (2008)
LMLVA-24dL	TAT-AGT-GTG-GTT-AAA-ATA-TTA-AAT-TAT-AGT-T			

Continued

Table 14.2 PCR Primers and DNA Probes for *Clostridium perfringens* (Continued)

Primer, Probe, or Adapter	Sequence (5' → 3')	Sequence (bp)	Gene	Reference
CPE-1	ACA-GGT-ACC-TTT-AGC-CAA-TC	—	cpe	Miwa et al. (1998)
CPE-2	ATT-CTT-TCT-GTA-GCA-GCA-GC	—		
CPE-3	GGA-TTT-GGA-ATA-ACT-ATA-GG	~150	cpe	
CPE-4	CTG-CAG-ATG-TTT-TAC-TAA-GC	—		
CPA-F	GCT-AAT-GTT-ACT-GCC-GTT-GA	324	cpa	Erol et al. (2008) from Meer and Songer (1997)
CPA-R	CCT-CTG-ATA-CAT-CGT-GTA-AG			
CPB-F	GCG-AAT-ATG-CTG-AAT-CAT-CTA	196	cpb	Erol et al. (2008) from Meer and Songer (1997)
CPB-R	GCA-GGA-ACA-TTA-GTA-TAT-CTT-C			
ETX-F	GCG-GTG-ATA-TCC-ATC-TAT-TC	655	etx	Erol et al. (2008) from Meer and Songer (1997)
ETX-R	CCA-CTT-ACT-TGT-CCT-ACT-AAC			
CPI-F	ACT-ACT-CTC-AGA-CAA-GAC-AG	446	cpi	Erol et al. (2008) from Meer and Songer (1997)
CPI-R	CTT-TCC-TTC-TAT-TAC-TAT-ACG			
CPE-F	GGA-GAT-GGT-TGG-ATA-TTA-GG	233	cpe	Erol et al. (2008) from Meer and Songer (1997)
CPE-R	GGA-CCA-GCA-GTT-GTA-GAT-A			
CPB2-F	AGA-TTT-TAA-ATA-TGA-TCC-TAA-CC	567	cpb2	Erol et al. (2008) from Meer and Songer (1997)
CPB2-R	CAA-TAC-CCT-TCA-CCA-AAT-ACT-C			

Primer	Sequence	Gene	Size	Reference
Beta-F	TAA-ATG-ATA-TAG-GTA-AAA-CTA-CTA-CTA	*cpb*	—	Li, Sayeed, and McClane (2007)
Beta-R	CTA-AAT-AGC-TGT-TAC-TT-GTG-AGT-AAG			
Epsilon-F	TTT-TAA-CTT-GGG-TTT-TGT-CG	*etx*	—	Li, Sayeed, and McClane (2007)
Epsilon-R	GGT-TCT-CCA-TCT-AAT-TCA-AC			
Iota-F	ACA-TCC-TCT-GTA-AGT-AAT-CG	*cpi*	—	Li, Sayeed, and McClane (2007)
Iota-R	GGT-TCT-CCA-TCT-AAT-TCA-AC			
P1-F	ACC-AGG-ATT-TTG-GCT-TAG-AAG	23S rDNA	~900	Hong et al. (2004)
P2-R	CAC-TTA-CCC-CGA-CAA-GGA-AT			
C. perf. Probe	CAG-TAG-CGA-GAT-GTG-GGT-GA			
CPE554-F2	CTG-CAG-ATA-GCT-TAG-GAA-ATA-TTG-ATC-A	*cpe*	112	Fukushima et al. (2007)
CPE665-R2	GCA-GCT-AAA-TCA-AGG-ATT-TCT-TTT-TCT			
CPE1	GAT-AAA-GGA-GAT-GGT-TGG-AA	s *cpe*	328	Schoepe et al. (1998)
CPE2	GAT-CCA-TCA-TCA-GAG-ATA-TT			
CPE3	CAG-GTA-CCT-TTA-GCC-AAT-CA	*cpe*	420	Schoepe et al. (1998)
CPE4	CTT-TCT-GTA-GCA-GCA-GCAA			
CP224	AGG-AAC-TCA-TGC-TAT-GAT-TGT-AAC-TCA-AGG	*plc*	775	Schoepe et al. (1998)
CP9721I	ACC-ACT-AGT-TGA-TAT-GTA-AGC-TAC-TAG			

50 samples of each. However, *cpe* positive strains were found among only 2% of the beef and 12% of the poultry samples. No *C. perfringens* was found in pork.

Wise and Siragusa (2005) developed a quantitative real-time PCR (QRti-PCR) assay for detection and enumeration of *C. perfringens* in the cecum and ileum of broiler chickens The primers CPerf165F/CPerf269R in conjunction with dual-labeled fluorogenic probe CPerf187F (Table 14.2) amplified a 105-bp sequence of the 16S rDNA specific for *C. perfringens*. DNA from fecal samples was purified with a commercial fecal DNA kit. The limit of detection with ileal samples was 10^2 CFU/g of ileal material, but was only about 10^4 CFU/g of cecal samples due to an unknown PCR inhibitor in the DNA from cecal samples.

Dela Cruz et al. (2006) developed real-time PCR (Rti-PCR) assays for the *cpa* and *cpe* genes using fluorescence resonance energy transfer (FRET) probes. The primers PLC-F/PLC-R (Table 14.2) amplified a 187-bp sequence of the *cpa* gene in conjunction with the FRET probes PLC-P1/PLC-P2 (Table 14.2). The primers CPE-F/CPE-R (Table 14.2) amplified a 336-bp sequence of the *cpe* gene in conjunction with the FRET probes CPE-P1/CPE-P2 (Table 14.2). With stool samples, 2.0 g of stool were homogenized with 18.0 ml of sterile water. Stool homogenates (0.9 ml) were then seeded with 1×10^4 CFU. Samples were then centrifuged for 1 min at 14,000 rpm and the cell pellets resuspended in 200 µl of lysozyme buffer and incubated at 37°C for 5 min. DNA was then extracted with a commercial kit. A similar procedure was used for sausage samples except that 10 g was homogenized with 90 ml of water. The sensitivity of the assays was two genomic copies per PCR reaction.

Shannon et al. (2007) developed QRti-PCR assays for 13 pathogens and indicator bacteria including *C. perfringens* for detection at five stages of municipal wastewater treatment. The primers Clper-F/Clper-R (Table 14.2) amplified a sequence of the alpha toxin gene *cpa* and were used in conjunction with the dual-labeled probe Clper-PR (Table 14.2).

Fukushima et al. (2007) applied buoyant density gradient centrifugation for rapid separation and concentration following filtration and differential centrifugation prior to QRti-PCR. The methodology resulted in a 250-fold sample concentration and allowed Rti-PCR detection of *C. perfringens* at 3×10^3 CFU/g. The primers CPE554-F2/CPE665-R2 (Table 14.2) amplified a 112-bp sequence of the *cpe* gene of *C. perfringens*. SYBR green was used as the Rti-PCR fluorophore.

Gurjar et al. (2008) developed Rti-PCR assays for the alpha, beta, epsilon, iota, CTE, and beta2 toxin genes of *C. perfringens* using duplex protocols. The primer pairs and dual-labeled probes are listed in Table 14.2. Among 241 *C. perfringens* isolates from the feces of 307 lactating cattle from seven dairy farms in the state of Pennsylvania, *cpa*, *cpb*, *cpi*, *etx*, *cpb2*, and *cpe* were detected in 68 (28.2%), 6 (2.5%), 6 (2.5%), 4 (1.6%), 164 (68%), and 11 (4.5%), respectively. The study clearly indicated that strains harboring the *cpb2* gene were widely prevalent in lactating cattle.

B. Molecular Probes

Baez and Juneja (1995) developed a 364-bp PCR generated probe for detection and enumeration of *C. perfringens* in raw beef. The primers CPEPS/CPENS

(Table 14.2) amplified a 364-bp sequence of the *cpe* gene and were labeled with digoxygenin (DIG). Ground beef (20 g) was seeded with a pure culture of *C. perfringens* and 20 ml of phosphate buffered saline (PBS) added and then stomached. The meat homogenates (10 ml) were then centrifuged at 500 rpm for 5 min and 1.0 ml of the supernatants filtered onto cellulose nitrate membranes along with 5 ml of sterile 0.1% peptone water to uniformly distribute the cells on the filter membrane surfaces. The membranes were then transferred onto the surface of tryptone-sulfite-cycloserine (TSC) agar plates, which were incubated anaerobically at 37°C for 18 hr. The membranes were then placed on chromatography paper and denatured with denaturing solution (0.5 M NaOH plus 1.5 M NaCl) and transferred to a second piece of chromatography paper and saturated with neutralization solution (1.0 M Tris-HCl plus 1.5 M NaCl, pH 8.0). The membranes were then heated in a dry oven at 80°C for 2 hr or in a microwave oven to fix the DNA and then hybridized for 2 hr at 65°C with the DIG-labeled probe (20 ng/ml). Anti-DIG IgG conjugated to alkaline phosphatase was added followed by the substrates nitroblue tetrazolium chloride and 5-bromo-4-chloro-3-indolyl-phosphate (NBT-BCIP) to develop a blue precipitate at the colony sites. The level of detection was less than 10 CFU/g of meat.

Heikinheimo, Lindström, and Korkeala (2004) developed a hydrophobic grid membrane filter-colony hybridization method (HGMF-CH) for the enumeration and isolation of *cpe*-gene carrying *C. perfringens* spores from feces. Fecal samples were diluted tenfold in 0.1% peptone water and heated at 75°C for 20 min to kill vegetative cells and to enhance spore germination. A total of 0.1 g per sample was filtered using 10 HGMFs in a membrane filter system. After filtration, the HGMFs were placed on TSC plates and incubated 36 to 42 hr at 37°C. The colonies were replicated by placing a nylon membrane disk onto the HGMF and the nylon replicas transferred to dry blotting paper and hybridized with a 425-bp *cpe* probe labeled with digoxygenin (DIG) and color developed with a chromogenic system. The *cpe* primers of Miwa et al. (1996) were used to produce the DIG-labeled probe. *Cpe*-positive *C. perfringens* could be isolated using this method if the ratio of *cpe*-positive *C. perfringens* spores to total *C. perfringens* was 6×10^{-5} or higher.

Hong et al. (2004) developed an oligonucleotide array system for identification of 14 species of pathogenic bacteria. A pair of universal primers P1-F/P2-F (Table 14.2) was designed to amplify a ~900-bp sequence of the 23S rDNA of all the target organisms. The P2-R primer was labeled with DIG. Each of the 14 species probes including that of No. 18 for *C. perfringens* was spot immobilized onto a nylon membrane grid. Denatured ~900-bp DIG-labeled amplicons were then hybridized to the membrane spots and color developed with anti-DIG IgG and NBT-BCIP.

C. perfringens and *C. difficile* are potentially associated with gastrointestinal infections and allergy in infants. Fallini et al. (2006) developed two 16S rDNA probes, Cdif198 for *C. difficile* and Cperf191 (Table 14.2) for *C. perfringens*. These probes were used to assess the composition of the intestinal flora of 33 healthy infants from 1.5 to 18.5 months of age. Homogenized fecal samples were fixed with 4% paraformaldehyde; the cells were then washed and treated with lysozyme and then subjected to fluorescent in situ hybridization with the probes combined with flow cytometry (FISH-FC). Both organisms were detected by FISH-FT in all 33 unseeded infant fecal samples.

III. MOLECULAR TYPING OF ISOLATES

A. Molecular Organization of the CPE Gene

Only about 1% to 5% of all *C. perfringens* type A isolates produce the *C. perfringens* enterotoxin CPE. Type A North American and European isolates carry the *cpe* gene on either the chromosome or large plasmids. Food poisoning isolates carry the *cpe* gene on the chromosome, whereas nonfoodborne GI disease isolates carry the *cpe* gene on a plasmid. The chromosomal *cpe* locus organization is highly conserved, but the plasmidborne *cpe* locus can have at least two organizational arrangements: either an IS*1470*-like or an IS*1151* sequence can be immediately downstream of the plasmid *cpe* gene (Miyamoto et al., 2002).

Brynestad, Synstad, and Granum (1997) found that an IS*200*-like element, IS*1469*, is almost always upstream of *cpe*. A new insertion element was identified, IS*1470*, a member of the IS*30* family, which is found both up- and downstream of *cpe* in the type A strain NCTC 8239. PCR results confirmed that this configuration was conserved in type A human food poisoning strains. The enterotoxin gene was on a 6.3-kb transposon, which, in addition to the two flanking copies of IS*1470*, included IS*1469* and two 1-kb stretches, one on each side of *cpe*, with no open reading frames (Figure 14.1).

Sparks et al. (2001) examined 34 *cpe*-positive *C. perfringens* isolates from North American cases of food poisoning (6 isolates) and antibiotic-associated diarrhea (AAD) (28 isolates). All 34 isolates were type A. Restriction fragment length polymorphism (RFLP) utilizing *Nru*I and a DIG-labeled *cpe* probe with Southern blotting and PFGE genotyping utilizing *Ceu*I and the same probe showed that the North American AAD isolates all had a plasmid-borne *cpe* gene. In contrast, the North American food poisoning isolates all carried a chromosomalborne *cpe* gene.

Wen, Miyamoto, and McClane (2003) developed a duplex PCR assay to determine whether the *cpe* gene is chromosomal- or plasmid-borne in isolates of *C. perfringens*. The primers MET-1.5F/CPE-up (Table 14.2) amplified a ~3.3-kb product from an apparently conserved DNA region between the upstream sequences and the plasmid-borne *cpe* gene in typeA isolates (Figure 14.2). The primers CPE-4.5F/IS1470F amplified a ~2.1-kb product from an apparently conserved DNA region between the chromosomal *cpe* gene and downstream IS*1470* sequences in type A isolates (Figure 14.2).

Figure 14.1   Schematic representation of *cpe* gene and flanking regions. The *uapC* gene encodes purine permease, *cpe* encodes enterotoxin, and *nadC* encodes quinolate phosphoribosyl transferase. (Adapted from Brynestad, S., Synstad, B., Granum, P., *Microbiol.* 143:2109–2115, 1997.)

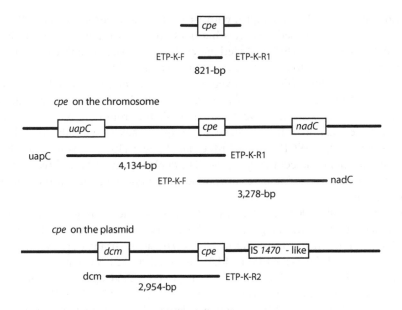

Figure 14.2 Schematic drawing of strategy used for PCR identification of the location of the *cpe* gene. (Redrawn from Nakamura, M. et al., *Int. J. Med. Microbiol.* 143:2109–2115, 2004. With permission from Elsevier.)

Miyamoto, Wen, and McClane (2004) developed a multiplex PCR for distinguishing these three organizational forms of the *cpe* gene. Primer pair (1) CPE4F/IS1470R1.3 (Table 14.2) amplified a 1300-bp sequence from the chromosomal *cpe* locus and was based on the observation that an IS*1470* sequence is present ~1 kb downstream from the chromosomal *cpe* gene. A second pair of primers (2) 3F/4R (Table 14.2) amplified a 600-bp sequence of the conserved *cpe* gene sequence of type A isolates. A third primer pair (3) CPE4F/IS1470 amplified a ~1600-bp sequence from type A isolates carrying an IS*1470*-like sequence downstream of the plasmid *cpe* locus. Primer pair (4) CPE4F/IS1151R0.8 (Table 14.2) amplified a ~800-bp sequence of the plasmidborne *cpe* locus. The six different primers were combined in a multiplex PCR to distinguish the three organizational forms of the *cpe* gene. The multiplex PCR effectively utilized cell lysates derived from single colonies without DNA purification and was used to prove that type A isolates carrying a plasmidborne *cpe* gene were responsible for four Japanese food poisoning outbreaks due to *C. perfringens,* in contrast to type A foodborne outbreak isolates in North America and Europe where the *cpe* gene is chromosomal.

Tanaka et al. (2003) described an outbreak of food poisoning in a nursing home in Japan involving 90 individuals who became ill. *C. perfringens* was cultured from 90/1–7 (84%) stool samples. PFGE Southern blot analysis revealed that the *cpe* gene was plasmid located. The principle of the technique is based on the fact that *C. perfringens* chromosomal DNA is too large to enter a pulsed-field gel without any restriction enzyme digestion. In addition, because the restriction nuclease I–*Ceu*I sites are located exclusively on the *C. perfringens* chromosome and not on the plasmid, digestion of DNA samples with I-*Ceu*I produces chromosomal DNA fragments that can enter pulsed-field gels but does not affect the migration of plasmid DNA. With the use of a 639-bp labeled *cpe* probe, the causative outbreak strain of *C. perfringens* was found to harbor the *cpe* gene on a plasmid.

Wen and McClane (2004) surveyed 887 raw meat and fish samples and found 13 that contained type A isolates of *C. perfringens* harboring the *cpe* gene chromosomally. These type A isolates were indistinguishable from food poisoning isolates in terms of the location of the *cpe* gene and in spores having a notably high level of heat resistance. The mPCR assay of Meer and Songer (1997) was used to identify the toxin type and to detect the *cpe* gene. The duplex assay of Wen, Miyamoto, and McClane (2003) was used to determine whether the *cpe* gene was chromosomal, or plasmid-borne.

Nakamura et al. (2004) developed PCR assays for determining the chromosomal or plasmid location of the *cpe* gene in food poisoning outbreak strains of *C. perfringens*. A total of 31 clinical and nonclinical *C. perfringens* strains were examined. The primers ETP-K-F/ETP-K-R1 (Table 14.2) amplified an 821-bp sequence of the *cpe* gene. Primers dcm/ETP-K-R1 (Table 14.2) amplified a 2954-bp sequence of the plasmid-located *cpe* gene plus a flanking sequence. Primers uapC/ETP-K-R1 (Table 14.2) amplified a 4134-bp and a 3278-bp sequence derived from upstream and downstream sequences flanking the *cpe* gene and containing the *cpe* sequence (Figure 14.2). The *cpe* genes of nine heat-sensitive (100°C for 10 min) strains isolated from three food poisoning outbreaks were located on a plasmid, whereas those of six heat-resistant strains from other food poisoning outbreaks were located on the chromosome.

Sarker et al. (2000) examined strains of *C. perfringens* in an attempt to explain why isolates harboring the *cpe* gene chromosomally are associated with food poisoning in contrast to nonfoodborne human GI infectious strains that carry a plasmid-borne *cpe* gene. They reasoned that because *C. perfringens* food poisoning usually involves cooked meat products that chromosomal *cpe* isolates are strongly associated with food poisoning because (1) they are more heat resistant than plasmid *cpe* isolates, (2) heat induces loss of the *cpe* plasmid, or (3) heating induces migration of the plasmid *cpe* gene to the chromosome. Vegetative cells of chromosomal *cpe* isolates were found to exhibit twofold higher decimal reduction values (*D* values) at 55°C than vegetative cells of plasmid-borne *cpe* isolates. In addition, the spores of chromosomal *cpe* isolates had on average an approximately 60-fold higher *D* value at 100°C than the spores of plasmid-borne *cpe* isolates. All survivors of heating were found to retain their *cpe* gene in the original plasmid or chromosome location. The authors concluded that chromosomal *cpe* isolates are strongly associated with food poisoning because their cells and spores possess a higher degree of heat resistance compared to those of plasmid-borne *cpe* strains, which presumably enhances survival in incompletely cooked or inadequately warmed foods.

In a subsequent study, Li and McClane (2006) found that chromosomal *cpe* isolates exhibited about an eightfold higher *D* value at 4°C and a threefold higher *D* value at –20°C than vegetative cells of plasmid-borne *cpe* isolates. In addition, *C. perfringens* type A isolates carrying chromosomal *cpe* also grew significantly faster than plasmid-borne *cpe* isolates at 25°C, 37°C, and 43°C. Chromosomal *cpe* isolates also grew at higher maximum (53.3°C versus 50.4°C) and lower minimum (12°C versus 18.9°C) temperatures than plasmid *cpe* isolates. Collectively, these results suggest that chromosomal-borne *cpe* isolates are commonly involved in food poisoning because of their greater resistance to low as well as high temperatures for both growth and survival compared to plasmid-borne *cpe* isolates.

Raju and Sarker (2005) cured a *cpe*-bearing plasmid from a nonfoodborne gastrointestinal isolate of *C. perfringens* and compared the heat resistance levels of wild-type, *cpe* knockout, and *cpe* plasmid-cured strains. Results indicated that the high level of heat resistance of vegetative cells and spores of a wild-type chromosomal *cpe* strain were similar to a *cpe* knockout mutant of the wild-type strain. This observation indicated that the chromosomal location of the *cpe* gene has no role mediating high-level heat resistance of spores and vegetative cells. When the heat resistance of vegetative cells and spores of a *cpe* plasmid-bearing strain and its plasmid-cured derivative was compared, the low level of heat resistance of both strains was similar. These results suggested that the *cpe* plasmid does not carry a gene that might confer heat sensitivity to strains harboring the *cpe* gene on a plasmid.

Li, Sayeed, and McClane (2007) investigated whether type A isolates of *C. perfringens* carrying a chromosomal *cpe* gene are present in two potential reservoirs, soil and home kitchen surfaces. No *C. perfringens* were recovered from kitchen surfaces (300 samples from 30 homes). Among a total of 502 soil samples, 343 (68.3%) were positive for vegetative cells of *C. perfringens* and 377 (75.1%) were positive for spores of *C. perfringens*. Ninety-nine percent of all isolates were type A. Primers for mPCR genotyping (Table 14.2) were those of Meer and Songer (1997). About 7% of the 502 soil samples contained *C. perfringens* carrying the *cpe* gene. The *cpe*-positive soil isolates were all type A, with their *cpe* genes on *cpe* plasmids. No chromosomal *cpe* isolates were detected. Several type C, D, and E isolates were obtained from soil samples, contradicting earlier studies. The results suggested that neither soil nor home kitchen surfaces represent major reservoirs for type A isolates with chromosomal *cpe* that cause food poisoning, although soil does appear to be a reservoir for *cpe*-positive isolates causing nonfoodborne gastrointestinal infections. When isolates were identified by the mPCR as belonging to type C, D, or E, their genotypes were confirmed using a second PCR. The primers Beta-F/Beta-R (Table 14.2) amplified a specific sequence of the *cpb* gene. The primers Epsilon-F/Epsilon-R (Table 14.2) amplified a specific sequence of the *etx* gene. The primers Iota-F/Iota-R (Table 14.2) amplified a sequence of the *cpi* gene. The mPCR assay and primers (Table 14.2) of Miyamoto, Wen, and McClane (2004) were used to determine the chromosomal and plasmid location of *cpe* genes.

Lahti et al. (2008) examined 53 type A *C. perfringens* isolates from 26 patients and 27 foods associated with 11 Finnish and 13 German food poisoning outbreaks from 1984 to 2007. Among the 53 isolates, 5 were negative by PCR for the presence of the *cpe* gene. Among the remaining 48 isolates, 36 (75%) carried the chromosomal IS*1470*-*cpe* gene, 8 (16.6%) carried the IS*1470*-like-*cpe* gene, and 3 (6.3%) carried the plasmid-borne IS*1151*-*cpe* gene. The finding that 11 of 48 strains (23%) harbored the *cpe* gene on a plasmid differs from the usual assumption that chromosomal *cpe* strains are responsible for almost all *C. perfringens* food poisoning outbreaks.

B. Genotyping

Moller and Ahrens (1996) developed PCR assays for detection of each of the four genes used for toxin typing of *C. perfringens* isolates plus the *cte* gene. An internal standard for use in coamplification with the beta toxin gene served as an amplification control. The internal standard was constructed to test whether amplification of the beta toxin gene (*cpb*) from intestinal mucosa was inhibited by introduction of a degenerate primer B3 that together with primer B1 amplified a part of the amplicon produced with primers Beta 1/Beta 2 yielding a product with 3′- and 5′-ends identical to the original Beta 1/Beta 2 generated amplicon but smaller. Primers are listed in Table 14.2. Among 95 Danish *C. perfringens* isolates including 37 from piglets suffering from necrotizing enteritis, none harbored the *cpe* gene. Among a total of 59 piglet isolates, 26 were positive with both the ELISA for the CPB toxin and for PCR detection of the *cpb* gene. These results clearly indicated that the *C. perfringens* enterotoxin (CPE) is not involved in diarrhea in certain animal species from the geographic area sampled.

Songer and Meer (1996) developed PCR assays involving primer pairs (Table 14.2) for detection of the *C. perfringens* genes for the alpha toxin (*cpa*), beta toxin, (*cpb*), epsilon toxin (*etx*), iota toxin (*cpi*), and enterotoxin (*cpe*) allowing classification into genotypes A (positive for *cpa*), B (positive for *cpa, cpb,* and *etx*), C (positive for *cpa* and *cpb*), D (positive for *cpa* and *etx*), or E (positive for *cpa* and *cpi*).

Yoo et al. (1997) developed a multiplex PCR assay for detection of the four toxins used in placing strains of *C. perfringens* into one of the five toxin types (A–E). The primer pairs used are listed in Table 14.2. Among farm animals exhibiting diarrhea, enterotoxemia, and necrotic enteritis only type A was isolated from calves and chickens, and type C (two of 14 isolates) in addition to type A was isolated from piglets.

Yamagishi et al. (1997) developed PCR assays for the detection of the three toxin genes alpha (*cpb*), epsilon (*etx*), and iota (*cpi*), which allowed allocation of *C. perfringens* strains to the five toxin types (A–E). Primers used are listed in Table 14.2. Strains lacking all three of these genes were assumed to be type A.

Miserez et al. (1998) developed PCR assays for detection of alpha, beta, and epsilon toxin genes in type D *C. perfringens* isolates from sheep and goats. The primers CPA-L/CPA-R (Table 14.2) amplified a 1167-bp sequence of the *cpa* gene. Primers CPB-L/CPB-R amplified a 1025-bp sequence of the *cpb* gene (Table 14.2). The primers CPE-L/CPE-R (Table 14.2) amplified a 961-bp sequence of the epsilon (*etx*) gene. Eighteen healthy animals yielded 13 *C. perfringens* isolates, all of which possessed the *cpa* gene and only 2 of which possessed the *etx* gene. In contrast, 52 animals suffering from enterotoxemia yielded 52 isolates of *C. perfringens*, all of which possessed the *cpa* and *etx* genes. None of the isolates harbored the *cpb* gene.

Schoepe et al. (1998) developed a multiplex PCR assay for detection of the *cpe* gene of *C. perfringens*. Primers CPE1/CPE2 (Table 14.2) amplified a 329-bp sequence of the *cpe* gene. Primers CPE3/CPE4 (Table 14.2) amplified a 421-bp sequence of the *cpe*. The primers CP224/CP9721I (Table 14.2) amplified a 775-bp sequence of the *cpa* gene present in all *C. perfringens* isolates and served as an amplification control in a multiplex PCR assay with the CPE3/CPE4 primers. The detection limit of the assay without enrichment was 10^5 CFU/g of meat.

Collie and McClane (1998) used RFLP and PFGE analyses to compare the genotypes of 43 *cpe*-positive *C. perfringens* isolates from diverse sources. All 11 North American and European food poisoning isolates were found to carry a chromosomal *cpe* gene, whereas all nonfoodborne human GI disease isolates and most veterinary isolates were found to carry their *cpe* gene on a plasmid.

Klaasen et al. (1999) made use of PCR assays to detect the presence of the *cta*, *ctb*, *etx*, and *cpb2* toxin genes in piglets with diarrhea or necrotic enteritis. Primer pairs used for *cpa*, *cpb*, and *etx* (Table 14.2) were from Miserez et al. (1998), those for *cpb2* (Table 14.2) were from Perelle et al. (1993), and those for *cpe* (Table 14.2) were from Van Damme-Jongsten, Haagsma, and Notermans (1990). The study involved 37 isolates from Holland and 51 isolates from Switzerland. All isolates possessed the *cpa* gene and were negative for *etx* as well as *cpe*. *Cpb2* was present in 28 of the 37 Dutch isolates and in 34 of the Swiss isolates. The predominance of the *cpb2* gene supports the concept of a causal relationship of CPB2 toxin-producing strains with digestive tract disease in piglets.

Ridell et al. (1998) examined a total of 71 *C. perfringens* isolates for the presence of the *cpe* gene using the PCR, 28 from food and 43 from diarrheal feces associated with 36 separate food poisoning outbreaks in Germany between 1984 and 1992. Primers A1/A2 amplified a 235-bp sequence of the *cpe* gene, and primers A3/A4 amplified a 363-bp sequence of the *cpe* gene (Table 14.2). Among the 28 food isolates, 24 (86%) were positive for the presence of the *cpe* gene. Among the 43 fecal isolates from clinical food poisoning cases, 38 (88%) were positive for the *cpe* gene. PFGE analysis with *Sma*I and *Apa*I indicated that multiple *cpe* positive clones were frequently present within one outbreak.

Garmory et al. (2000) examined 96 isolates of *C. perfringens* from 24 foals, 27 lambs, 33 piglets, and 12 calves with diarrhea. The isolates were genotyped using the primers (Table 14.2) for *cpa*, *cpb*, *etx*, *cpi*, *cpb2*, and *cpe* from Meer and Songer (1997). Most of the disease isolates were type A. The *cpe* gene was present in 4% of the isolates. The *cpb2* gene was present in 50% of the isolates; however, the prevalence of this gene varied among animal isolates. Of the 33 piglet isolates, 27 (82%) were positive for the *cpb2* gene compared with none of the 7 piglet control isolates, strongly suggesting that this gene is significantly associated with diarrhea in piglets. It is interesting that the prevalence of the *cpb2* gene was similar for both diseased and the 10 control foal isolates (50%). With lambs, the *cpb2* gene was present in 7 (33%) of the 27 disease isolates and in 8 (67%) of the 12 lamb control isolates.

Stagnitta, Micalizzi, and Guzmán, (2002) examined 515 meat samples (315 fresh sausages, 100 hamburgers, and 100 samples of minced meat) from San Luis, Argentina for the presence of *C. perfringens*. A total of 126 *C. perfringens* isolates were obtained with nine (7.1%) producing enterotoxin (CPE). Among the 126 isolates, 123 (97.2%) were type A, 2 (1.6%) were type C, and 1 (0.79%) was type E. All enterotoxigenic strains were type A. PCR primers used for strain typing are listed in Table 14.2.

Aschfalk and Müller (2002) cultured 95 fecal samples from Atlantic cod caught along the northern Norwegian coast for the occurrence of *C. perfringens*. The PCR was used for detecting the toxin genes *cpa, cpb, etx, cpi, cpb2*, and *cpe*. Primer pairs are listed in Table 14.2. Among 97 samples, 37 (38.9%) yielded *C. perfringens*. All isolates were type A. The isolates from two cod harbored the *cpb2* gene. Strains harboring the *cpb, etx, cpi*, and *cpe* genes were not found. The possible origin of such strains from sewage pollution of coastal waters was suggested by the authors.

Augustynowicz, Gzyl, and Slusarczyk (2002) developed a duplex PCR assay involving two sets of primers to detect *C. perfringens* phospholipase C (*cpa*) and enterotoxin (*cpe*) genes, which was applied to a collection of 64 predominantly food poisoning related *C. perfringens* isolates. The primers used (Table 14.2) were those of Fach and Popoff (1997). Among the 64 isolates, 26 were *cpe* and *cpa* positive and produced enterotoxin (CPE), 16 were *cpe* and *cpa* positive but did not produce CPE, and 21 were *cpa* positive and *cpe* negative.

Lin and Labbe (2003) examined the incidence of *C. perfringens* in 131 food samples from retail sources in the northeastern United States. There were 40 isolates obtained from each of 40 positive foods and the presence of the *cpe* and *cpa* genes was determined by PCR with primers listed in Table 14.2. None of the isolates harbored the *cpe* gene, whereas all possessed the *cpa* gene. PFGE with *Apa*I revealed that about 5% of the isolates were closely related and the remaining isolates unrelated. The authors concluded that their results demonstrated the rarity of the *cpe* gene in strains derived from retail foods.

Beta2 toxin, encoded by the *cpb2* gene, has been implicated in the pathogenesis of porcine, equine, and bovine enteritis by type A *C. perfringens*. By incorporating primers to *cpb2* into a multiplex genotyping PCR, Bueschel et al. (2003) screened 3270 field isolates of *C. perfringens*. Among these, 1216 (37.2%) were positive for the *cpb2* gene. The majority of isolates from cases of porcine enteritis were positive for *cpb2* (>85%), and this was even more true for *C. perfringens* isolated from cases of porcine neonatal enteritis (91.8%). In contrast, isolates from normal pigs harbored *cpb2* in only 11.1% of cases. The correlation between enteritis in other animal species and the presence of *cpb2* was not so strong. *cpb2* was found in 21.4% of *C. perfringens* isolates from cattle with enteritis, and in 47.3% of isolates from calves with enteritis or abomastitis. In addition, enterotoxigenic type D and E strains almost always carried *cpb2*. In *cpb2*-positive isolates of porcine origin, the production of CPB2 toxin was detected with 96.9% of the isolates. However, in *cpb2*-postive isolates from other animal species, only 50% expressed the CPB2 toxin. The high incidence of CPB2-positivity among strains from neonatal pigs with enteritis and the high correlation of genotype with phenotype supports the contention that the CPB2 toxin plays a role in the pathogenesis of these infections.

Baums et al. (2004) reported on the development of a multiplex PCR for detecting *cpa, cpb, etx, cpi, cpe*, and *cpb2* genes from heat-lysed cell suspensions of *C. perfringens*. The assay was notably efficient for toxin genotyping isolates. The primers used are listed in Table 14.2.

Al-Khaldi et al. (2004) developed a microarray-based method for detection of the six *C. perfringens* toxin genes *cpa, cpb, cpb2, etx, cpi*, and *cpe*. Three individual oligonucleotide probes, complementary to the unique sequences of each toxin gene, were designed and immobilized on the surface of aldehyde-coated glass slides. This oligoprobe redundancy was used to increase the confidence of identification. mPCR was used to simultaneously amplify DNA target regions of all six genes. Single-stranded DNA (ssDNA) samples for microarray analysis were prepared by bringing about a primer extension of previously generated amplicons with *Taq* polymerase in the presence of a mixture of all six reverse primers. The resulting ssDNAs were fluorescently labeled with Cy3 by chemical modification of guanine bases. The presence of toxin genes in *C. perfringens* isolates was established by hybridization of the fluorescently labeled ssDNAs representing different isolates to the microarray gene-specific oligonucleotide probes.

Singh, Bhilegaonkar, and Agarwal (2005) screened 211 meat samples of different animals (70 of buffalo, 70 of goat, and 71 of poultry) for the presence of *C. perfringens*. Goat meat yielded 64 (91.4%), poultry meat 50 (70.4%), and buffalo meat 46 (65.7) positive samples. A total of 114 of these isolates (buffalo-32, goat-37, and poultry-45) were screened for the presence of the *cpe* gene by PCR using the primers PT1/PT2 (Table 14.2) that amplified a 364-bp sequence of the *cpe* gene. Buffalo, goat, and poultry isolates yielded 9.3%, 32.4%, and 15.5%, respectively, possessing the *cpe* gene.

Heikinheimo and Korkeala (2005) developed a multiplex PCR assay for detection of the *cpa, cpb, etx, cpi,* and *cpe* genes from *C. perfringens* broiler chickens. The primer pairs used are listed in Table 14.2. All 118 isolates were of type A. None harbored the *cpb, etx, cpi,* or *cpe* genes.

Albini et al. (2008) reported on the development and validation of three multiplex Rti-PCR assays utilizing dual-labeled probes for the detection of the *C. perfringens* *cpa, cpb, etx, cpi,* and *cpe* genes. Primers and probes are listed in Table 14.2. One mRti-PCR assay consisted of primers and probes for detection of the *cpa, cpb,* and *cpb2* genes. A second mRti-PCR assay consisted of primers and probes for detection of the *cpa, etx,* and *cpe* genes. A third mRti-PCR assay consisted of primers and probes for detection of the *cpa* and *cpi* genes. Because the *cpa* gene is present in all strains of *C. perfringens* and is diagnostic for the species it was used as a positive amplification control in all three mRti-PCRs. The individual primer pairs in monoplex Rti-PCRS assays were used to quantify the copy numbers of plasmid-borne toxin genes in relation to the chromosomally located *cpa* gene.

Erol et al. (2008) determined the presence of toxin genes in 22 *C. perfringens* isolates from turkey meat samples by PCR. The *cpa, cpb, cpb2, etx, cpi,* and *cpe* toxin genes were detected by mPCR using the primer pairs listed in Table 14.2. All 22 turkey meat isolates were found to harbor the *cpa* gene, but none of the isolates harbored any of the other five toxin genes.

Miki et al. (2008) cultured 200 raw meat samples from retail sources in Japan for *C. perfringens*. A total of 142 samples (71%) were found contaminated with *C. perfringens,* and 212 strains were isolated. Multiplex PCR (Meer and Songer, 1997) determined that all of the isolates were of type A. Four percent of the samples were contaminated with *cpe*-positive *C. perfringens*, with most of these isolates bearing the *cpe* gene on a plasmid.

C. Random Amplified Polymorphic DNA (RAPD) Typing

Leflon-Guibout et al. (1997) undertook the typing of *C. perfringens* strains by use of RAPD in comparison to zymotyping. A total of 13 RAPD primers was screened for suitability and four (HLWL74, MHN1, R108, and A3; Table 14.2) were used for typing 28 strains of *C. perfringens* previously found to be genotypically distinct with pulsed field gel electrophoresis (PFGE). In addition, 24 further strains were included that had previously been typed by zymotyping, yielding 23 zymotypes for 27/28 typeable strains. Four epidemiologically linked strains yielded identical RAPD patterns. Zymotyping with a discriminatory index of 0.99 offered a higher level of discrimination than RAPD with a discriminatory value of 0.97. However, zymotyping was not 100% in that 1/28 strains could not be zymotyped, whereas RAPD was capable of typing all 52 strains examined. The authors concluded that RAPD is more rapid, less fastidious, and less labor intense than zymotyping, and is quite suitable for typing *C. perfringens* isolates.

D. Pulsed Field Gel Electrophoresis (PFGE) Typing

Maslanka et al. (1999) subjected 62 food poisoning isolates of *C. perfringens* to PFGE using *Sma*I, which was preferred over five other restriction nucleases because it produced 11 to 13 well-distributed bands of 40 of ~1100 bp, allowing good discrimination between isolates. A total of 17 distinct PFGE patterns were obtained with the 62 isolates from seven outbreaks. In general, multiple isolates from a single individual had identical PFGE patterns. Isolates from different individuals within an outbreak had similar, if not identical, patterns. Epidemiologically unrelated isolates had unique patterns.

Lukinmaa, Takknen, and Siitonen (2002) examined 47 *C. perfringens* isolates from nine foodborne outbreaks that occurred from 1984 to 1999 in Finland with the use of the PCR for detection of the *cpe* gene, reversed passive latex agglutination for detection of CPE production, and PFGE typing using the restriction nucleases *Sma*I and *Apa*I. The primers CPE-F/CPE-R (Table 14.2) amplified a 935-bp sequence of the *cpe* gene. Thirty-three of the 47 *C. perfringens* strains possessed the *cpe* gene. CPE production was detected with 31 strains. One *cpe*-positive isolate was CPE negative. A total of nine outbreak clusters was involved in the study. All six isolates in outbreak cluster I isolated in 1984 were *cpe* negative, as were all four isolates from outbreak cluster V isolated in 1994. All strains carrying the *cpe* gene in five outbreak clusters (II, IV, VI, VII, and IX) had an identical PFGE subtype within each cluster. The authors concluded that the *cpe*-negative strains were probably just members of the normal intestinal flora, as were all the strains in outbreak clusters I and V. These observations indicated that *C. perfringens* was not the cause of the outbreaks in 1984 and 1994 and that a viral agent may have been responsible inasmuch as no other bacterial pathogens were isolated from these outbreaks.

Nakamura et al. (2003) reported on an outbreak of food poisoning that occurred in Osaka, Japan, in 2001. Enrichment cultivation of fecal specimens was performed by inoculating liquid thioglycollate medium with fecal material, which was then heated at 75°C for 20 min and incubated anaerobically overnight. Isolation was on CW agar with kanamycin and egg yolk. Fecal specimens from 81 patients yielded 53 cpe-positive isolates of C. perfringens after enrichment and direct plating yielded only 13 cpe-positive isolates. Among the recovered 53 isolates, 36 had indistinguishable PFGE patterns with SmaI and the same serotype.

Nauerby, Pederson, and Madsen (2003) analyzed a total of 279 isolates of C. perfringens collected from 25 poultry farms using PFGE with SmaI. There were 208 isolates from 46 chickens suffering from severe necrotic enteritis (NE), 48 from healthy poultry, 19 from poultry suffering from cholangio-hepatitis (CH), and an additional 4 isolates from nondescript ill poultry. PFGE analysis yielded 25 different PFGE patterns with the 48 isolates from healthy poultry. Some of the healthy birds yielded up to 5 isolates of C. perfringens differing in PFGE type. In contrast, flocks suffering from NE or CH carried only one or two PFGE clones. All isolates were of toxin type A derived from PCR detection of toxin genes.

Waters et al. (2003) studied 29 cpb2 positive isolates of C. perfringens derived from pigs with GI infections. PFGE analysis with SmaI and MluI indicated that all 29 isolates harbored the cpb2 gene on a plasmid. The results indicated a significant association between CPB2 producing isolates of C. perfringens and diarrhea in piglets.

Gholamiandekhordi et al. (2006) genotypically and phenotypically characterized C. perfringens isolates from poultry flocks of different health status. A total of 27 isolates was obtained from 23 broiler flocks without clinical problems using cloacal swabs. Intestinal and liver samples of animals suffering from necrotic enteritis yielded 36 isolates from 10 flocks. The PCR was used for detection of the cpa, cpb, cpb2, etx, cpi, and cpe toxin genes. Primer pairs for cpa, cpb, etx, and cpi were from Yoo et al. (1997; Table 14.2). Primers for cpb2 (Table 14.2) were from Perelle et al. (1993). Primers for cpe (Table 14.2) were from Herholz et al. (1999). All isolates were of toxin type A. PFGE with SmaI indicated a high level of genetic diversity among isolates from different flocks with a total of 35 PFGE types differentiated. Isolates derived from diseased flocks were mostly of the same PFGE type, but each flock harbored a different clone. Isolates from 5 of the 35 PFGE types carried the cpb2 gene, encoding the beta2 necrotizing toxin, and isolates from 2 out of the 35 PFGE types harbored the cpe gene encoding the enterotoxin.

E. Amplified Fragment Length Polymorphism (AFLP)

McLauchlin et al. (2000) subjected 35 C. perfringens isolates from patients and foods implicated in seven outbreaks of suspected food poisoning by C. perfringens together with five isolates from unrelated incidents to serotyping and amplified fragment length polymorphism (AFLP). Restriction of genomic DNA was accomplished with HindIII. Ligation of adapter ADH1/ADH2 (Table 14.2) was with T4 DNA ligase. The single primer HI-X (Table 14.2) was used for PCR. AFLP was found to be highly reproducible and yielded 16 profiles (each unique to the 12 incidents). The authors concluded that AFLP can provide a rapid, sensitive, and reproducible method for the typing of *C. perfringens* isolates for outbreak investigations.

Engström et al. (2003) subtyped 21 *C. perfringens* isolates from 10 different poultry farms in Sweden using AFLP and PFGE. In a second study, 32 isolates of *C. perfringens* type A from three broiler farms with healthy flocks were subtyped by PFGE using *Sma*I and *Apa*I. All 53 isolates analyzed by PCR belonged to type A with the gene coding for alpha toxin uniformly present. Two isolates possessed the *cpb2* gene as well, but none had the other toxin genes. Both AFLP and PFGE differentiated 21 strains into 10 different subtypes. This differentiation correlated with the origin of the isolates. Only isolates from poultry of one farm demonstrated more than one subtype of *C. perfringens*. The subtyping of poultry from healthy flocks showed that each bird carried two to three different subtypes. AFLP and PFGE were found equally suitable for subtyping *C. perfringens* isolates.

Keto-Timonen et al. (2006) applied AFLP analysis to 37 strains of *C. perfringens* and obtained 29 different AFLP types. Genomic DNA was digested with *Hind*III. The restriction site-specific *Hind*III adapter and the HpyCH41V adapter (Table 14.2) were ligated with T4 DNA ligase. The digested and ligated DNA samples were then amplified by preselection PCR using the primers Hind-O/Hpy-O. An aliquot of the diluted amplicons was then subjected to selective amplification using the Hind-C primer labeled at the 5'-end with FAM and the Hpy-A primer (Table 14.2). AFLP proved to be highly reproducible, easy to perform, and relatively rapid.

F. Ribotyping

Schalch et al. (1997) characterized 34 *C. perfringens* isolates from 10 food poisoning outbreaks from 1984 to 1991 by ribotyping. A total of 12 different ribopatterns were generated by *Eco*RI digestion. In eight food poisoning outbreaks all of the ribotypes for each food and stool isolate were identical with respect to each outbreak group.

Schalch et al. (1999) examined 155 isolates of *C. perfringens* using ribotyping, plasmid profiling, and PFGE. Isolates were from 10 food poisoning outbreaks (34 isolates), food and fecal samples (24 isolates), and meat and fish pastes (121 isolates). For outbreaks 1, 2, and 5 all the *C. perfringens* isolates involved in the outbreaks had identical plasmid profiles with respect to their outbreak group. A total of 12 distinct ribotype patterns were discerned among 34 food poisoning isolates. PFGE could distinguish *C. perfringens* isolates with identical ribopatterns. With 111 *C. perfringens* isolates from minced meat, 107 distinctly different ribopatterns were discerned. Ribopatterns and PFGE patterns were more easily interpreted than plasmid profiling.

Kilic, Schalch, and Stolle (2002) subjected 111 *C. perfringens* isolates from commercially produced ground meat to ribotyping, which yielded 107 distinctly different ribopatterns. In only four cases did two isolates show an identical ribopattern. The discriminatory index was 0.99, reflecting a notably high level of discrimination.

Schalch et al. (2003) reported on a 1998 outbreak of acute gastroenteritis involving 21 inhabitants of a German nursing home: after consuming minced beef heart two residents died. A total of 17 strains of *C. perfringens* were isolated from patients' stools, autopsy material, and food. A majority of the strains were untypeable by PFGE with *Sma*I presumably due to DNase activity. Ribotyping distinguished four different groups.

G. Multiple-Locus Variable Number Tandem Repeat Loci (VNTR)

Sawires and Songer (2005) developed a VNTR assay for distinguishing strains of *C. perfringens* isolates. Five VNTR loci were characterized, four of which are contained within protein-encoding genes. The primers CP6-F/CP6-R (Table 14.2) amplified a repeat locus of 507 bp mapping to the sequence encoding the collagenlike protein. The most common allele was ~570 bp. The primers CPB13-F/CPP13-R (Table 14.2) amplified a ~262-bp allele mapping to *nagK*, which is involved in hyaluronidase production. The primers CP16-F/CP16-R (Table 14.2) amplified a repeat locus of a 433-bp sequence located within the open reading frame of *parB*. The most common allele was ~1058 bp. The primers CP19-F/CP19-R (Table 14.2) amplified a 481-bp repeat locus that mapped to the ferrous ion transport protein B gene. The null allele was the most common. The primers CP42-F/CP42-R amplified a repeat locus mapping to the riboflavin synthesis beta subunit and to a hypothetical protein. The most common allele was ~200 bp. A total of 112 *C. perfringens* isolates was subject to the MLVA typing system, and each isolate was assigned an MLVA genotype. A numerical index of discrimination for the five VNTR loci was 0.995.

Chalmers et al. (2008) applied MLVA to 54 epidemically unrelated isolates of *C. perfringens* from 11 different animal species in addition to 27 *C. perfringens* isolates previously subjected to PFGE typing. Six loci in noncoding regions of the chromosomal DNA were utilized involving six pairs of primers (Table 14.2). There were 35 unique MLVA types obtained from the 54 epidemically unrelated isolates. Sequencing confirmed that all six loci included tandem repeats and that the size variation of amplicons was the result of changes in repeat numbers. The number of alleles among the 54 strains varied from two to eight. With the exception of one PFGE type, all isolates belonging to the same PFGE major type were the same MLVA type. However, subtypes within a PFGE major type could not be differentiated with MLVA, demonstrating that the level of discrimination achieved by MLVA was lower than with PFGE when subtypes were employed. Isolates previously untypeable by PFGE because of the presence of nucleases and resulting DNA degradation were, however, successfully typed with MLVA.

REFERENCES

Albini, S., Broard, I., Jaussi, A., Wollschlaeger, N., Frey, J., Miserez, R., Abril, C. 2008. Real-time multiplex PCR assays for reliable detection of *Clostridium perfringens* toxin genes in animal isolates. *Vet. Microbiol.* 127:179–185.

Al-Khaldi, A., Meyers, K., Rasooly, A., Chizhikov, V. 2004. Genotyping of *Clostridium perfringens* toxins using multiple oligonucleotide microarray hybridization. *Mol. Cell. Probes* 18:359–367.

Aschfalk, A., Müller, W. 2001. *Clostridium perfringens* toxin types in hooded seals in the Greenland Sea, determined by PCR and ELISA. *J. Vet. Med. Ser. B* 48:765–769.

Aschfalk, A., Müller, W. 2002. *Clostridium perfringens* toxin types from wild-caught Atlantic cod (*Gadus morhua* L.), determined by PCR and ELISA. *Can. J. Microbiol.* 48:35–368.

Augustynowicz, E., Gzyl, A., Slusarczyk, J. 2002. Detection of enterotoxigenic *Clostridium perfringens* with a duplex PCR. *J. Med. Microbiol.* 51:169–172.

Baez, L., Juneja, V. 1995. Nonradioactive colony hybridization assay for detection and enumeration of enterotoxigenic *Clostridium perfringens* in raw beef. *Appl. Environ. Microbiol.* 61:807–810.

Baums, C., Schotte, U., Amtsberg, G., Goethe, R. 2004. Diagnostic multiplex PCR for toxin genotyping of *Clostridium perfringens* isolates. *Vet. Microbiol.* 100:11–16.

Brynestad, S., Synstad, B., Granum, P. 1997. The *Clostridium perfringens* enterotoxin gene is on a transposable element in type A human food poisoning strains. *Microbiol.* 143:2109–2115.

Bueschel, D., Jost, B., Billington, S., Trinh, H., Songer, J. 2003. Prevalence of *cpb2*, encoding beta2 toxin, in *Clostridium perfringens* field isolates: Correlation of genotype with phenotype. *Vet. Microbiol.* 94:121–129.

Chalmers, G., Martin, S., Prescott, J., Boerlin, P. 2008. Typing of *Clostridium perfringens* by multiple-locus variable number tandem repeats analysis. *Vet. Microbiol.* 128:126–135.

Collie, R., McClane, B. 1998. Evidence that the enterotoxin gene can be episomal in *Clostridium perfringens* isolates associated with non-food-borne human gatrointestinal diseases. *J. Clin. Microbiol.* 36:30–36.

Dela Cruz, W., Gozum, M., Lineberry, S., Stassen, S., Daughtry, M., Stassen, N., Jones, M., Johnson, O.J. 2006. Rapid detection of enterotoxigenic *Clostridium perfringens* by real-time fluorescence resonance energy transfer PCR. *J. Food Prot.* 69:1347–1353.

Engström, B., Fermér, C., Lindberg, A., Saarinn, E., Båverud, V., Gunnarsson, A. 2003. Molecular typing of isolates of *Clostridium perfringens* from healthy and diseased poultry. *Vet. Microbiol.* 94:225–235.

Erol, I., Goncuoglu, M., Ayaz, N., Ormanci, B., Hildebrandt, G. 2008. Molecular typing of *Clostridium perfringens* isolated from turkey meat by multiplex PCR. *Lett. Appl. Microbiol.* 47:31–34.

Fach, P., Popoff, M. 1997. Detection of enterotoxigenic *Clostridium perfringens* in food and fecal samples with a duplex PCR and the slide latex agglutination test. *Appl. Environ. Microbiol.* 63:4232–4236.

Fallini, M., Rigottier-Gois, L., Aguilera, M., Bridonneau, C., Collignon, A., Edwards, C., Corthier, G., Doré, J. 2006. *Clostridium difficile* and *Clostridium perfringens* species detected in infant faecal microbiota using 16S rDNA targeted probes. *J. Microbiol. Methods* 67:150–161.

Fukushima, H., Katsube, K., Hata, Y., Kishi, R., Fujiwara, S. 2007. Rapid separation and concentration of food-borne pathogens in food samples prior to quantification by viable-cell counting and real-time PCR. *Appl. Environ. Microbiol.* 73:92–100.

Garmory, H., Chanter, N., French, N., Bueschel, D., Songer, J., Titball, R. 2000. Occurrence of *Clostridium perfringens* β2-toxin amongst animals, determined using genotyping and subtyping PCR assays. *Epidemiol. Infect.* 124:61–67.

Gholamiandekhordi, A., Ducatelle, R., Heyndrickx, M., Haesebrouck, F., Immerseel, F. 2006. Molecular and phenotypical characterization of *Clostridium perfringens* isolates from poultry flocks with different disease status. *Vet. Microbiol.* 113:143–152.

Gurjar, A., Hegde, N., Love, B., Jayarao, B. 2008. Real-time multiplex PCR assay for rapid detection and toxin typing of *Clostridium perfringens* toxin producing strains in fecses of dairy cattle. *Mol. Cell Probes* 22:90–95.

Heikinheimo, A., Korkeala, H. 2005. Multiplex PCR assay for toxinotyping *Clostridium perfringens* isolates obtained from Finnish broiler chickens. *Lett. Appl. Microbiol.* 40:407–411.

Heikinheimo, A., Lindström, M., Korkeala, H. 2004. Enumeration and isolation of *cpe*-positive *Clostridium perfringens* spores from feces. *J. Clin. Microbiol.* 42:3992–3997.

Henderson, I., Duggleby, C., Turnbull, P. 1994. Differentiation of *Bacillus anthracis* from other *Bacillus cereus* group bacteria with the PCR. *Int. J. Syst. Bacteriol.* 44:99–105.

Herholz, C., Miserez, R., Nicolet, J., Frey, J., Popoff, M., Gilbert M., Gerber, H., Straub, R. 1999. Prevalence of beta2-toxigenic *Clostridium perfringens* in horses with intestinal disorders. *J. Clin. Microbiol.* 37:358–361.

Hong, B., Kiang, L., Hu, Y., Fang, D., Guo, H. 2004. Application of oligonucleotide array technology for the rapid detection of pathogenic bacteria of foodborne infections. *J. Microbiol. Methods* 58:403–411.

Hunter, S., Brown, J., Oyston, P., Sakurae, J., Titball, R. 1993. Molecular genetic analysis of beta-toxin of *Clostridium perfringens* reveals sequence homology with alpha-toxin, gamma-toxin and leukocidin of *Staphylococcus aureus*. *Infect. Immun.* 61:3958–3965.

Keto-Timonen, R., Heikinheimo, A., Ecrola, E., Korkala, II. 2006. Identification of *Clostridium* species and DNA fingerprinting of *Clostridium perfringens* by amplified fragment length polymorphism analysis. *J. Clin. Microbiol.* 44:4057–4065.

Kilic, U., Schalch, B., Stolle, A. 2002. Ribotyping of *Clostridium perfringens* from industrially produced ground meat. *Lett. Appl. Microbiol.* 34:238–243.

Klaasen, H., Molkenboer, M., Bakker, J., Miserez, R., Häni, H., Frey, J., Popoff, M., van den Bosch, J. 1999. Detection of the β2 toxin gene of *Clostridium perfringens* in diarrhoeic piglets in the Netherlands and in Switzerland. *FEMA Immunol. Med. Microbiol.* 24:325–332.

Kokai-Kun, J., Songer, J., Czeczulin, J., Chen F., McClane, B. 1994. Comparison of western immunoblots and gene detection assay for identification of potentially enterotoxigenic isolates of *Clostridium perfringens*. *J. Clin. Microbiol.* 32:2533–2539.

Lahti, P., Heikinheimo, A., Johansson, T., Kokeala, H. 2008. *Clostridium perfringens* type A strains carrying a plasmid-borne enterotoxin gene (genotype Is*1151-cpe* or IS*1470*-like-*cpe*) as a common cause of food poisoning. *J. Clin. Microbiol.* 46:371–373.

Leflon-Guibout, V., Pons, J., Heym, B., Nicolas-Chanoine, M. 1997. Typing of *Clostridium perfringens* strains by use of random amplified polymorphic DNA (RAPD) system in comparison with zymotyping. *Anaerobe* 3:245–250.

Li, J., McClane, B. 2006. Further comparison of temperature effects on growth and survival of *Clostridium perfringens* type A isolates carrying a chromosomal or plasmid-borne enterotoxin gene. *Appl. Environ. Microbiol.* 72:4561–4568.

Li, J., Sayeed, S., McClane, B. 2007. Prevalence of enterotoxigenic *Clostridium perfringens* isolates in Pittsburgh (Pennsylvania) area soils and home kitchens. *Appl. Environ. Microbiol.* 73:7218–7224.

Lin, Y., Labbe, R. 2003. Enterotoxigenicity and genetic relatedness of *Clostridium perfringens* isolates from retail foods in the United States. *Appl. Environ. Microbiol.* 69:1642–1646.

Lukinmaa, S., Takknen, E., Siitonen, S. 2002. Molecular epidemiology of *Clostridium perfringens* related to food-borne outbreaks of disease in Finland from 1984 to 1999. *Appl. Environ. Microbiol.* 68:3744–3749.

Maslanka, S., Kerr, J., Willsiams, G., Barbaree, J., Carson, L., Miller, J., Swaminathan, B. 1999. Molecular subtyping of *Clostridium perfringens* by pulsed-field gel electrophoresis to facilitate food-borne disease outbreak investigations. *J. Clin. Microbiol.* 37:2209–2214.

McLauchlin, J., Ripabelli, G., Brett, M., Threlfall, E. 2000. Amplified fragment length polymorphism (AFLP) analysis of *Clostridium perfringens* for epidemiological typing. *Int. J. Food Microbiol.* 56:21–28.

Meer, R., Songer, G. 1997. Multiplex polymerase chain reaction assay for genotyping *Clostridium perfringens*. *Am. J. Vet. Res.* 58:702–705.

Miki, Y., Miyamoto, K., Kaneco-Hirano, I., Fujiuchi, K., Akimoto, S. 2008. Prevalence and characterization of enterotoxin gene-carrying *Clostridium perfringens* isolates from retail meat products in Japan. *Appl. Environ. Microbiol.* 74:5366–5372.

Miserez, R., Frey, J., Buogo, C., Capaul, A., Burnens, A., Nicolet, J. 1998. Detection of α- and ε-toxigenic *Clostridium perfringens* type D in sheep and goats using a DNA amplification technique. *Lett. Appl. Microbiol.* 26:382–386.

Miwa, N., Nishina, T., Kubo, S., Atsumi, M., Honda, H. 1998. Amount of enterotoxigenic *Clostridium perfringens* in meat detected by nested PCR. *Int. J. Food Microbiol.* 42:195–200.

Miwa, N., Nishina, T., Kubo, S., Jujikura, K. 1996. Nested polymerase chain reaction for detection of low levels of enterotoxigenic *Clostridium perfringens* in animal feces and meat. *J. Vet. Med. Sci.* 58:197–203.

Miyamoto, K., Chakrabarti, G., Morino, Y., McClane, B. 2002. Organization of the plasmid *cpe* locus of *Clostridium perfringens* type A isolates. *Infect. Immun.* 70:4261–4272.

Miyamoto, K., Wen, Q., McClane, B. 2004. Multiplex PCR genotyping assay that distinguishes between isolates of *Clostridium perfringens* type A carrying a chromosomal enterotoxin gene (*cpe* locus, a plasmid *cpe* locus with an IS*1470*-like sequence, or a plasmid *cpe* locus with an IS*1151* sequence. *J. Clin. Microbiol.* 42:1552–1558.

Moller, K., Ahrens, P. 1996. Comparison of toxicity neutralization-, ELISA- and PCR tests for typing of *Clostridium perfringens* and detection of enterotoxin gene by PCR. *Anaerobe* 2:103–110.

Nakamura, H., Ogasawara, J., Monma, C., Hase, A., Suzuki, H., Kai, A., Haruki, K., Nishikawa, Y. 2003. Usefulness of a combination of pulse-field gel electrophoresis and enrichment culture in a laboratory investigation of a foodborne outbreak due to *Clostridium perfringens*. *Diag. Microbiol. Infect. Dis.* 47:471–475.

Nakamura, M., Kato, A., Tanaka, D., Gyobu, Y., Higaki, S., Karasawa, T., Yamagishi, T. 2004. PCR identification of the plasmid-borne enterotoxin gene (*cpe*) in *Clostridium perfringens* strains from food poisoning outbreaks. *Int. J. Med. Microbiol.* 294:261–265.

Nauerby, B., Pederson, K., Madsen, M. 2003. Analysis by pulsed-field gel electrophoresis of the genetic diversity among *Clostridium perfringens* isolates from chickens. *Vet. Microbiol.* 94:257–266.

Nilsson, J., Svensson, B., Ekelund, K., Christiansson, A. 1998. A RAPD-PCR method for large-scale typing of *Bacillus cereus*. *Lett. Appl. Microbiol.* 27:168–172.

Perelle, S., Gilbert, M., Boquet, P., Popoff, R. 1993. Characterization of *Clostridium perfringens* ioto-toxin genes and expression in *Escherichia coli. Infect. Immun.* 61:5147–5156.

Raju, D., Sarker, M. 2005. Comparison of the levels of heat resistance of wild-type, *cpe* knockout, and *cpe* plasmid-cured *Clostridium perfringens* type A strains. *Appl. Environ. Microbiol.* 71:7618–7620.

Ridell, J., Björkroth, J., Eisgruber, H., Schalch, B., Stolle, A., Korkeala, H. 1998. Prevalence of the enterotoxin gene and clonality of *Clostridium perfringens* strains associated with food-poisoning outbreaks. *J. Food Prot.* 61:240–243.

Saito, M., Matsumoto, M., Funabashi, M. 1992. Detection of *Clostridium perfringens* enterotoxin gene by the PCR amplification procedure. *Int. J. Food Microbiol.* 17:47–55.

Sarker, M., Shivers, R., Sparks, S., Juneja, V., McClane, B. 2000. Comparative experiments to examine the effects of heating on vegetative cells and spores of *Clostridium perfringens* isolates carrying plasmid genes versus chromosomal enterotoxin genes. *Appl. Environ. Microbiol.* 66:3234–3240.

Sawires, Y., Songer, J. 2005. Multiple-locus variable-number tandem repeat analysis for strain typing of *Clostridium perfringens. Anaerobe* 11:262–272.

Schalch, B., Bader, L., Schau, H., Bergmann, R., Rometsch, A., Maydl, G., Kebler, S. 2003. Molecular typing of *Clostridium perfringens* from a food-borne disease outbreak in a nursing home: Ribotyping versus pulsed-field gel electrophoresis. *J. Clin. Microbiol.* 41:892–895.

Schalch, B., Bjorkroth, J., Eisgruber, H., Korkeala, H., Stolle, A. 1997. Ribotyping for strain characterization of *Clostridium perfringens* isolates from food poisoning cases and outbreaks. *Appl. Environ. Microbiol.* 63:3992–3994.

Schalch, B., Sperner, B., Eisgruber, H., Stolle, A. 1999. Molecular methods for the analysis of *Clostridium perfringens* relevant for food hygiene. *FEMA Immunol. Med. Microbiol.* 24:281–286.

Schoepe, H., Potschka, H., Schlapp, T., Fiedler, J., Schau, H., Baljer, G. 1998. Controlled multiplex PCR of enterotoxigenic *Clostridium perfringens* strains in food samples. *Mol. Cell. Probes* 12:359–365.

Shannon, K., Lee, D., Trevors, J., Beaudette, L. 2007. Application of real-time quantitative PCR for the detection of selected bacterial pathogens during municipal waste water treatment. *Sci. Total Environ.* 382:121–129.

Singh, R., Bhilegaonkar, K., Agarwal, R. 2005. Studies on occurrence and characterization of *Clostridium perfringens* from select meats. *J. Food Safety* 25:146–156.

Smith, L., Williams, B. 1984. *Pathogenic Anaerobic Bacteria.* Charles C. Thomas: Springfield, IL, pp. 101–134.

Songer, J., Meer, R. 1996. Genotyping of *Clostridium perfringens* by polymerase chain reaction is a useful adjunct to diagnosis of clostridial enteric disease in animals. *Anaerobe* 2:197–203.

Sparks, S., Carman, R., Sarker, M., McClane, B. 2001. Genotyping of enterotoxigenic *Clostridium perfringens* fecal isolates associated with antibiotic-associated diarrhea and food poisoning in North America. *J. Clin. Microbiol.* 39:883–888.

Stagnitta, P., Micalizzi, B., Guzmán, S. 2002. Prevalence of enterotoxigenic *Clostridium perfringens* in meats in San Luis, Argentina. *Anaerobe* 8:253–258.

Tanaka, D., Isobe, J., Hosorogi, S., Kimata, K., Shimizu, M., Katori, K., Gyobu, Y., Nagai, Y., Yamagishi, T., Karasawa, T., Nakamura, S. 2003. An outbreak of food-borne gastroenteritis caused by *Clostridium perfringens* carrying the *cpe* gene on a plasmid. *Jpn. J. Infect. Dis.* 56:137–139.

Uzal, F., Plumb, J., Blackall, L., Kelly, W. 1997. PCR detection of *Clostridium perfringens* producing different toxins in faeces of goats. *Lett. Appl. Microbiol.* 25:339–344.

Van Damme-Jongsten, M., Haagsma, M., Notermans, S. 1990. Testing strains of Clostridium perfringens type A isolated from diarrhoeic piglets for the presence of the enterotoxin gene. *Vet. Rec.* 126:191–192.

Wang, R., Cao, W., Franklin, W., Campbell, W., Cerniogdlia, C. 1994. A 16S rDNA-based PCR method for rapid and specific detection of *Clostridium perfringens* in food. *Mol. Cell. Probes* 8:131–138.

Waters, M., Savoie, A., Garmory, H., Bueschel, D., Popoff, M., Songer, J., Titball, R., McClane, B., Sarker, M. 2003. Genotyping and phenotyping of beta2-toxigenic *Clostridium perfringens* fecal isolates associated with gastrointestinal diseases in piglets. *J. Clin. Microbiol.* 41:3584–3591.

Wen, Q., McClane, B. 2004. Detection of enterotoxigenic *Clostridium perfringens* type A isolates in American retail foods. *Appl. Environ. Microbiol.* 70:2685–2691.

Wen, Q., Miyamoto, K., McClane, B. 2003. Development of a duplex PCR genotyping assay for distinguishing *Clostridium perfringens* type A isolates carrying chromosomal enterotoxin (*cpe*) genes from those carrying plasmid-borne enterotoxin (*cpe*) genes. *J. Clin. Microbiol.* 41:1494–1498.

Wise, M., Siragusa, G. 2005. Quantitative detection of *Clostridium perfringens* in the broiler fowl gastrointestinal tract by real-time PCR. *Appl. Environ. Microbiol.* 71:3911–3916.

Yamagishi, T., Sugitani, K., Tanishima, K., Nanamura, S. 1997. Polymerase chain reaction test for differentiation of five toxin types of *Clostridium perfringens*. *Microbiol. Immunol.* 41:295–299.

Yoo, H., Lee, S., Park, K., Park, Y. 1997. Molecular typing and epidemiological survey of prevalence of *Clostridium perfringens* types by multiplex PCR. *J. Clin. Microbiol.* 35:228–232.

CHAPTER **15**

Bacillus cereus

I. CHARACTERISTICS OF THE ORGANISM

Bacillus cereus is a peritrichously flagellated, gram-positive, hemolytic, facultatively anaerobic rod-shaped bacterium producing endospores. Several distinct clinical forms of food poisoning can result from *B. cereus*. Food poisoning outbreaks due to *B. cereus* can result in profuse watery diarrhea ("diarrheal syndrome") as a result of large numbers of spores germinating in the intestine and undergoing vegetative growth and toxin production. Alternatively, the emetic form ("emetic syndrome") can occur and is associated exclusively with starchy foods such as cooked rice and is characterized by rapid onset within 1 to 5 hr resulting in nausea, vomiting, and malaise, which in some cases is followed by diarrhea with different toxins being involved.

The emetic form usually results from ingestion of foods in which extensive growth of the organism with toxin production has occurred prior to ingestion of the contaminated food. The emetic toxin is named cereulide and is a 12 amino acid peptide in a ring configuration that is nonantigenic. Cereulide is a product of nonribosomal peptide synthesis mediated by a peptide synthetase (Toh et al., 2004) borne on a plasmid (Hoton et al., 2005). The inability of emetic outbreak strains to hydrolyze starch has been found to correlate with cereulide production (Shinagawa, 1993; Agata, Ohta, and Mori, 1996). This inability to hydrolyze starch by emetic outbreak strains is considered to constrain the growth of *B. cereus* in starchy foods, resulting in the induction of cereulide. Thus, starch hydrolyzing strains, recognized as diarrheal, are considered unlikely to produce cereulide in starchy foods. *B. cereus* is a member of the *B. cereus* group consisting of six species of bacilli that includes, in addition to *B. cereus*, *B. anthracis*, *B. thuringiensis*, *B. mycoides*, *B. pseudomycoides*, and *B. weihenstephanensis*, all of which are considered closely related to *B. cereus*.

Because four species of the *B. cereus* group, *B. cereus*, *B. thuringiensis*, *B. anthracis*, and *B. mycoides*, share many phenotypic properties, some workers have questioned the taxonomic status of these separate species (Carlson and Kolosto, 1993; Carlson, Caugent, and Kolsto, 1994; Helgason et al., 1998). DNA–DNA hybridization studies on the first three of these bacilli have indicated that these organisms share relatively high levels of chromosomal base sequence similarity (Somerville and Jones, 1972; Drobniewski, 1993). In addition, the primary sequence of the 16S *rRNA* gene of these *Bacillus* species exhibits very high levels (>99%) of sequence similarity (Ash et al., 1991a,b). This sequence similarity among certain members of the *B. cereus* group explains the presence of many common toxin genes among these *Bacillus* species, which tends to complicate the use of molecular techniques for specifically distinguishing *B. cereus*.

II. SELECTIVE ISOLATION OF *B. CEREUS*

A. Selective Agar Media

Standard isolation and enumeration of *B. cereus* from food, environmental, and clinical samples involves the use of two commonly used conventional selective plating media, Polymyxin–egg yolk mannitol-bromothymol blue (PEMBA) agar and mannitol–egg yolk–polymyxin (MYP) agar. These two standard plating media differ in their base composition but are based on the same detection principles: (1) characteristic colony appearance, (2) egg yolk hydrolysis (phospholipase activity) resulting in a precipitation zone around suspected colonies, and (3) the inability of *B. cereus* group strains to produce acids from mannitol. Typical colonies of *B. cereus* grown on PEMB and MYP are surrounded by a zone of precipitate and a peacock-blue and pink color, respectively.

A test panel of 100 *B. cereus* from different sources and with different toxigenic potentials (40 food, 40 foodborne outbreaks, and 20 clinical isolates) was used by Fricker, Reissbrodt, and Ehling-Schulz (2008) in assessing the relative effectiveness of PEMBA, MYP, and two newly developed chromogenic selective agar media CBC and BCM for *B. cereus*. CBC contains 5-bromo-4-chloro-3-indolyl-β-glucopyranoside that is cleaved by β-D-glucosidase and results in white colonies with a blue-green center. BCM contains 5-bromo-4-chloro-3-indoxyl myoinositol-1-phoshate, is cleaved by phosphateidylinositol phospholipase C, and imparts a homogeneous blue-turquoise color to *B. cereus* colonies that are sometimes surrounded by a blue halo.

The growth of each of the 100 test isolates on each of these four media was assessed as "typical," "weak," or "atypical." In addition, the primers PlcR-F1/PlcR-R2 and PlcR-F2/PlcR-R1 (Table 15.2) were used for amplification of sequences derived from the *plcr* gene encoding the phospholipase C regulator from selected strains. In general, identification, isolation, and enumeration of *B. cereus* from foods was easier on the chromogenic plating media than on the standard plating media because *B. cereus* was frequently overgrown by background flora from foods. Both chromogenic media contain the additional inhibitors trimethoprim and cycloheximide that inhibit potential background flora more effectively than polymyxin B. BCM and CBC are therefore more selective for *B. cereus* than PEMB and MYP. However, a significant diversity of *B. cereus* colony appearance on CBC indicated its relative lack of suitability. On PEMBA 35% of the isolates exhibited either a weak or atypical colony appearance and 72% did not exhibit the expected blue color. In addition,

29% of the isolates showed unusual colony morphology on either PEMBA or MYP. The best overall performance was obtained with BCM, which resulted in no weak reactions, and only six strains exhibited atypical colonies.

III. THE EMETIC TOXIN CEREULIDE

A. Synthesis of Cereulide and Location of the Responsible Peptide Synthetase

Cereulide is a cyclic peptide with a molecular mass of 1.2 kDa having the structure [D-O-Leu-D-Ala-D-O-Val-D-Val]$_3$ (Arnesen, Fagerlund, and Granum, 2008). Toh et al. (2004) presented experimental evidence that the emetic toxin cereulide is a product of nonribosomal peptide synthesis mediated by a multidomain enzymatic complex referred to as a nonribosomal peptide synthetase (NRPS). The primers BEF/BER amplified a ~850-bp sequence of a putative *nrps* gene also designated *ces*. A gene sequence search indicated that *B. anthracis, B. halodurans*, and *B. subtilis* showed no significant sequence similarity to the *nrps* gene. However, similarities in gene sequence were found for the *nrps* gene in other *Bacillus* species. All 30 emetic cereulide-producing isolates were PCR positive. It is interesting that several diarrheal isolates were also PCR positive and produced cereulide.

Hoton et al. (2005) undertook plasmid curing studies of emetic strains of *B. cereus* producing cereulide. Plasmid loss was accompanied by a loss of cereulide production. The study documented the extrachromosomal location of the genetic determinants for cereulide formation to be located on a plasmid designated pCERE01.

B. Phenotypic Characteristics of Emetic Toxin Producing Strains

In addition to not degrading, starch emetic strains have been reported to have a common ribotype and a common RAPD-PCR pattern (Ehling-Schulz et al. (2005a) and are also unable to ferment salicin (Ehling-Schulz et al., 2005a; Raevuori, Kiutamo, and Kiskanen, 1977).

Carlin et al. (2006) phenotypically characterized 17 emetic toxin producing strains of *B. cereus*, 40 diarrheal foodborne strains, and 43 food-environment strains. The emetic strains showed a shift toward higher temperatures in growth limits, regardless of origin. None of the emetic toxin-producing strains were able to grow below 10°C. In contrast, 9 (11%) out of the 83 nonemetic strains were able to grow at 4°C and 41 (49%) at 7°C. All emetic strains were able to grow at 48°C, but only 32/83 (39%) of the nonemetic strains grew at 48°C. Spores from the emetic strains showed on average a higher heat resistance at 90°C compared to nonemetic strains.

C. RAPD Typing of Emetic Strains of *B. cereus*

A collection of 886 *B. cereus* isolates derived from 10 different dairies and dairy farms was screened for the presence of emetic toxin producing strains by Svensson et

al. (2006) using phenotypic traits, RAPD analysis, and a boar sperm motility inhibition test (Andersson et al., 2004). In addition, the emetic toxin (cereulide) in culture
extracts was measured by LC ion trap MS. Among the 886 isolates of *B. cereus*
tested, 84 presumptively emetic isolates were identified. Among 262 environmental
samples from one dairy farm, only 4 presumptively emetic strains were found with
all originating from soil. Among feed, grass, or feces, no isolates with the emetic
phenotype were found. There were 80 isolates of the emetic phenotype found from
samples taken during housing conditions on two farms but none from other farms.
All isolates found positive in the rapid sperm motility test including two reference
strains displayed a single RAPD pattern with primer bceP (Table 15.2). In contrast,
nonemetic isolates exhibited considerable variation in their RAPD patterns. A total
of 61% of the isolates with the "emetic" RAPD pattern were inhibitory to sperm
cells. Nontoxic strains with the emetic RAPD pattern were considered to be genetically related to the emetic toxin-producing strains.

IV. ENTEROTOXINS OF *B. CEREUS*

A. Major Enterotoxins of *B. cereus*

The major enterotoxins produced by *B. cereus* fall into three categories: hemolytic
(HBL), nonhemolytic (NHE), and CytK, a monomeric protein exhibiting necrotic
and hemolytic activity that is highly toxic to epithelial cells. HBL is encoded by the
hbl gene complex consisting of the genes *hblA*, *hblD*, and *hblC* (Table 15.1). NHE
is encoded by the *nhe* gene complex consisting of *nheA*, *nheB*, and *nheC* genes
(Table 15.1). Maximum cytotoxic activity of NHE against vero cells is dependent on
the presence of all three NHE components, which have been found to form pores in
planar lipid bilayers and to cause ATP and LDH release from Caco-2 and Vero cell
monolayers in addition to hemolysis of human erythrocytes (Fagerlund et al., 2008)
although the designation NHE refers to "nonhemolytic." CytK is encoded by the
cytK gene. Additional toxins are presented in Table 15.1. The genes encoding these
cytotoxins associated with diarrheal disease are generally chromosomally located
and are present in *B. cereus*, *B. thuringiensis*, and *B. anthracis* (Arnesen, Fagerlund,
and Granum, 2008).

V. MOLECULAR METHODS FOR DETECTION
AND CONFIRMATION OF *B. CEREUS*

A. PCR

1. PCR Detection of Emetic Strains of B. cereus

Fricker et al. (2007) developed a novel real-time PCRS (Rti-PCRS) assay
for detection of emetic strains of *B. cereus*. The primers Ces-TM-F/Ces-TM-R

Table 15.1 Genes Used to Identify and Characterize Isolates of the *B. cereus* and Members of the *B. cereus* Group

Gene	Gene Product or Function
Hemolytic Enterotoxin	
hbl	Hemolytic enterotoxin HBL (BL). Possesses hemolytic cytotoxic, dermonecrotic, and vascular permeability activities. Consists of a complex of three proteins (B, L, and L2) organized as a single operon nencoded by the *hblA, hblC*, and *hblD* genes, respectively
hblA	Cell binding component of BL designated HblB
hblC	Cell lytic component of HBL designated HblL$_2$
hblD	Cell lytic component of HBL designated HblL$_1$
hblD/hblA	Amplified sequence encompassing *hbld* and *hbla* genes
Nonhemolytic Enterotoxin	
Nhe	Nonhemolytic enterotoxin NHE. A complex of three proteins (NheA, NheB, and NHeC) Encoded by the *nheA, nheB*, and *nheC* genes
nheA	Cell lytic element of NHE designated NheA
nheB	Cell lytic element of NHE designated NheB
nhec	An NHE protein element of NHE of unknown function designated *NheC*
nheB/nheC	Amplified sequence encompassing *nheB* and *nheC* genes
Emetic Toxin	
ces	A 12 amino acid peptide in a ring configuration designated cereulide that is nonantigenic. Its synthesis is mediated by a peptide synthetase encoded by the cereulide synthase gene (*ces*), which is part of the cereulide gene cluster encoded by *nrps*
nrps	Nonribosomal peptide synthetase cereulide gene cluster
Phospholipases	
Plc	Phosphatidylcholine-specific phospholipase C designated PC-PLC
Piplc	Phosphatidylinositol-specific phospholipase C designated PI-PLC
Sph	Sphingomyelinase
cerAB	The cereolysen AB gene (*cerAB*) encodes phospholipase C (*cerA*) and sphingomyelinase (*cerB*), which constitutes abiologically functional two-component cytolysin, which acts synergistically in lysing human erythrocytes
Other Toxins	
bceT	A monomeric protein exhibiting diarrheal toxigenicity
intFM	An enerotoxin designated FEntFM
cytK	A monomeric protein exhibiting necrotic and hemolytic activity that is highly toxic to epithelial cells designated CytK
entFM	Cytotoxic to Vero cells and designated enterotoxin FM
Other Genes	
Cs	Cell surface component designated CS
spoIIIAB	Associated with sporulation stage designated SPOIIIAB

Continued

Table 15.1   Genes Used to Identify and Characterize Isolates of the *B. cereus* and Members of the *B. cereus* Group (Continued)

Gene	Gene Product or Function
gyr	DNA gyrase
cry	Parasporal crystalline protein insecticide produced only by *B. thuringiensis*
Housekeeping Genes	
Adk	Adenylate cyclase
ccpA	Catabolic control protein A
ftsA	Cell division protein
glpT	Glycerol-3-phoshate permease
pyrE	Orotate phosphoribosyl transfrase
recF	DNA replication and repair protein
SucC	Succinyl coenzyme A synthetase

(Table 15.2) amplified a 103-bp sequence of the cereulide synthetase gene. The dual-labeled probe Ces-TM-P (Table 15.2) was labeled at the 5′-end with FAM and with TAMRA at the 3′-end. An internal amplification control was also incorporated into the Rti-PCR assays using primers IAC-F/IAC-R (Table 15.2) that amplified a 118-bp sequence of a commercially available plasmid pUC19 employing a dual-labeled probe IAC-P with the 5′-end labeled with HEX and the 3′-end labeled with TAMRA. An Rti-PCR assay was also developed that was mediated by SYBR green utilizing primers Ces-SYBR-F/Ces-SYBR-R (Table 15.2) that amplified a 176-bp sequence of the cereulide synthetase gene. Both Rti-PCR assays were assessed using 16 non-emetic strains of *B. cereus* and 23 emetic strains of *B. cereus*, in addition to 25 isolates representing nine other species of bacilli and 36 isolates representing seven non-*Bacillus* genera. Both Rti-PCR assays yielded positive results for all 23 emetic strains of *B. cereus* and negative results for the remaining 77 isolates, including the 16 nonemetic *B. cereus* isolates.

Nakano et al. (2004) developed a PCR assay for the detection of emetic *B. cereus* strains that was based on the sequence of a unique RAPD fragment associated only with cereulide-producing strains. The primers RE234-F/RE234-R (Table 15.2) amplified a 234-bp sequence of this RAPD fragment. In addition, the universal Eubacteria specific primers prbaC1/prbaC2 (Table 15.2) were used to amplify a sequence of 296–300 bp as an amplification control. Among 12 emetic strains of *B. cereus,* all yielded the 234-bp amplicon. Among 18 nonemetic strains of *B. cereus,* one yielded the 234-bp amplicon. The authors speculated that this strain may have harbored a cryptic gene responsible for cereulide production. Representative isolates of other *Bacillus* species did not yield the amplicon.

Ehling-Schulz, Fricker, and Scherer (2004) developed a pair of degenerate primers targeting highly conserved motifs of known nonribosomal peptide synthetase (NRPS) for the specific detection of emetic *B. cereus* strains. The primers EM1F/EM1R (Table 15.2) were designed to amplify a 635-bp sequence of the *nrps* gene. The nested primers HpaIF/EM1R were used to generate a 260-bp digoxygenin (DIG) labeled probe internal to the 635-bp amplicon for Southern blot analysis.

The universal bacterial primers 8-26-56/1511-1493 (Table 15.2) amplified a 1600-bp sequence of 16S rRNA and were used as an internal amplification control. The primers EM1F/EM1R resulted in two distinct amplicons of 700–800 bp from emetic *B. cereus* strains and only one band (700 bp) from nonemetic *B. cereus* strains. Sequence analysis of the 800-bp fragment revealed no significant homologies to nonribosomal peptide synthetases. The 635-bp amplicon was highly specific for only emetic strains of *B. cereus*. Other species of the *B. cereus* group failed to yield both the 635- and 800-bp amplicons.

Altayar and Sutherland (2006) determined the incidence of emetic toxin producing *B. cereus* in 271 soil, animal feces, and selected vegetable produce samples. A total of 177 (45.8%) of the samples were positive for *B. cereus,* with 177 *B. cereus* isolates recovered at 7°C. Only three strains recovered at 30°C were positive for emetic toxin production, and only one strain recovered at 7°C was positive for emetic toxin production. All *B. cereus* isolates came from washed or unwashed potato skins. The primers BcAPR1/BcFF2 (Table 15.2) were previously shown by Francis et al. (1998) to yield a 284-bp band with mesophilic and psychrotrophic *B. cereus* isolates, whereas only primers BcAPR1/BcAPF1 (Table 15.2) yielded a band of 160 bp. The single psychrotrophic emetic strain was confirmed as psychrotrophic by production of a 160-bp amplicon with the corresponding primers.

McIntyre et al. (2008) recharacterized *Bacillus* food poisoning strains for 39 outbreaks and identified *B. cereus* in 23 outbreaks, *B. thuringiensis* in 4, *B. mycoides* in 1, and mixed strains of bacilli in 11 outbreaks. A pair of primers amplifying a 270-bp sequence of the *cry1* or *cry2* genes was used to distinguish *B. thuringiensis* from *B. cereus* strains in addition to electron microscopy to confirm the presence of the parasporal protein crystal in *B. thuringiensis* strains. The primers EM1F/EM1R (Table 15.2) from Ehling-Shultz, Fricker, and Scherer (2004) were used to amplify a 635-bp sequence of the *nrps* (emetic gene cluster) to verify emetic strains.

2. PCR Detection and Confirmation of B. cereus and Members of the B. cereus Group

A PCR assay was developed by Schraft and Griffiths (1995) to facilitate detection and identification of *B. cereus* in foods. The primers Pf/Pr (Table 15.2) amplified a 389-bp sequence of the *cereolysin AB* gene (*cerAB*) encoding phospholipase C. The primers Pf/Cr (Table 15.2) amplified a 1461-bp sequence encompassing the coupled sequences for phospholipase C (*cerA*) and the sphingomyelinase (*cerB*) of *B. cereus,* which constitutes a biologically functional two-component cytolysin that acts synergistically in lysing human erythrocytes. Because the activity of these enzymes includes hydrolysis of lecithin, these primers were expected to amplify DNA sequences from members of the *B. cereus* group. With both primer sets, all but one of 39 strains of *B. cereus* yielded the expected amplicons. Both primer sets also yielded amplicons of the expected sizes with isolates of *B. mycoides* and *B. thuringiensis* and therefore lacked species specificity. Other lecithinase-positive *Bacilli* not members of the *B. cereus* group did not yield the amplicons.

Table 15.2   PCR Primers and DNA Probes for *B. cereus*

Primer, Probe, or Adapter	Sequence (5′ → 3′)	Sequence (bp)	Gene	Reference
HD2 F	GTA-AAT-TAI-GAT-GAI-CAA-TTT-C	1091	*hbl*	Ehling-Schulz et al. (2006)
HA4 R	AGA-ATA-GGC-ATT-CAT-AGA-TT			
NA2 F	AAG-CIG-CTC-TTC-GIA-TTC	766	*nhe*	Ehling-Schulz et al. (2006)
NB1 R	ITI-GTT-GAA-ATA-AGC-TGT-GG			
CKF2	ACA-GAT-ATC-GGI-CAA-AAT-GC	421	*cytK*	Ehling-Schulz et al. (2006)
CKR5	CAA-GTI-ACT-TGA-CCI-GTT-GC			
CesF1	GGT-GAC-ACA-TTA-TCA-TAT-AAG-GTG	1271	*ces*	Ehling-Schulz et al. (2006)
CesR2	GTA-AGC-GAA-CCT-GTC-TGT-AAC-AAC-A			
RAPD PM13	GAG-GGT-GGC-TCT	—	—	Ehling-Schulz et al. (2005a) from Henderson et al. (1996)
RAPD-1	CCG-AGT-CCA	—	—	Ehling-Schulz et al. (2005a) from Nilsson et al. (1998)
RAPD-2	CCG-GCG-GCG	—	—	Ehling-Schulz et al. (2005a) from Nilsson et al. (1998)
8-26	AGA-GTT-TGA-TCC-TGG-CTC-A	1600	16s rDNA	Ehling-Schulz et al. (2005a) from Stackbrandt and Liesack (1992)
1511-1493	CGG-CTA-CCT-TGT-TAC-GAC			

Primer	Sequence	Gene	Size	Reference
16St-R2	GTT-GTA-CAC-ACC-GCC-CGT-C	*16S-23s spacer*	—	Ehling-Schulz et al. (2005a) from Gürtler and Stanisich (1996)
23St-R10	GC-TTT-CCC-TCA-CGG-TAC-TG			
CS-F2	GTT-AA(A/T)-GTG-CCA-ACT-AGG-TGG	*cs*	—	Ehling-Schulz et al. (2005a)
CS-R2	CGT-TTA-CAG-ATA-TCA-TTC-CAG-C			
Spo2F	CGA-CGA-GGA-TAA-CCC-AAT-TTG-C	*spoIIIAB*	—	Ehling-Schulz et al. (2005a)
Spo2R	CAG-TGA-GAG-ACC-GAG-GCA-AC			
CesF1	GGT-GAC-ACA-TTA-TCA-TAT-AAG-GTG	*ces*	—	Ehling-Schulz et al. (2005a) from Ehling-Schulz et al. (2005b)
CesR2	GTA-AGC-GAA-CCT-GTC-TGT-AAC-AAC-A			
HC F	GAT-AC(T/C)-AAT-GTG-GCA-ACT-GC	*hblC*	740	Ehling-Schulz et al. (2005a) from Guinebretiere, Broussolle, and Nguyen-The (2002)
HC R	TTG-AGA-CTG-CTC-G(T/C)T-AGT-TG			
HD F	ACC-GGT-AAC-ACT-ATT-CAT-GC	*hblD*	829	Ehling-Schulz et al. (2005a) from Guinebretiere, Broussolle, and Nguyen-The (2002)
HD R	GAG-TCC-ATA-TGC-TTA-GAT-GC			
LIF(for)	CGC-TCA-AGA-ACA-AAA-AGT-AGG	*hblD/hblA*	—	Ehling-Schulz et al. (2005a)
4R (rev)	CTC-CTT-GTA-AAT-CTG-TAA-TCC-CT			
HA F	AAG-CAA-TGG-AAT-ACA-ATG-GG	*hblA*	1154	Ehling-Schulz et al. (2005a) from Guinebretiere, Broussolle, and Nguyen-The (2002)
HA R	AGA-ATC-TAA-ATC-ATG-CCA-CTG-C			

Continued

Table 15.2   PCR Primers and DNA Probes for *B. cereus* (Continued)

Primer, Probe, or Adapter	Sequence (5′ → 3′)	Sequence (bp)	Gene	Reference
HA-F HB R	AAG-CAA-TGG-AAT-ACA-ATG-GG AAT-ATG-TCC-CAG-TAC-ACC-CG	2684	*hblB*	Ehling-Schulz et al. (2005a) from Guinebretiere, Broussolle, and Nguyen-The (2002)
NA F NA R	GTT-AGG-ATC-ACA-ATC-ACC-GC ACG-AAT-GTA-ATT-TGA-GTC-GC	755	*nheA*	Ehling-Schulz et al. (2005a) from Guinebretiere, Broussolle, and Nguyen-The (2002)
NB F NB R	TTT-AGT-AGT-GGA-TCT-GTA-CGC TTA-ATG-TTC-GTT-AAT-CCT-GC	743	*nheB*	Ehling-Schulz et al. (2005a) from Guinebretiere, Broussolle, and Nguyen-The (2002)
517 (for) 1142 (rev)	CGG-TTC-ATC-TGT-TGC-GAC-AGC GCA-TAT-GAA-TCC-ATT-GCA-AA	—	*nheB/nheC*	Ehling-Schulz et al. (2005a)
NC F NC R	TGG-ATT-CCA-AGA-TGT-AAC-G ATT-ACG-ACT-TCT-GCT-TGT-GC	683	*nheC*	Ehling-Schulz et al. (2005a) from Guinebretiere, Broussolle, and Nguyen-The (2002)
Fc (for) Rc (rev)	GTA-ACT-TTC-ATT-GAT-GAT-CC GAA-TAC-ATA-AAT-AAT-TGG-TTT-CC	505	*cytK*	Ehling-Schulz et al. (2005a) from Stenfors et al. (2002)
Adk-F Adk-R	CAG-CTA-TGA-AGG-CTG-AAA-CTG CTA-AGC-CTC-CGA-TGA-GAA-CC	450	*adk*	Ehling-Schulz et al. (2005a) from Helgason et al. (2004)

Primer	Sequence	Size	Gene	Reference
ccpA-F ccpA-R	GTT-TAG-GAT-ACC-GCC-CAA-ATG TGT-AAC-TTC-TTC-GCG-CTT-CC	418	ccpA	Ehling-Schulz et al. (2005a) from Helgason et al. (2004)
fts-F fts-R	TCT-TGA-CAT-CGG-TAC-ATC-CA GCC-TGT-AAT-AAG-TGT-ACC-TTC-CA	401	ftsA	Ehling-Schulz et al. (2005a) from Helgason et al. (2004)
glpT-F glpT-R	TGC-GGC-TGG-ATG-AGT-GA AAG-TAA-GAG-CAA-GGA-AGA	330	glpT	Ehling-Schulz et al. (2005a) from Helgason et al. (2004)
pyrE-F pyrE-R	TCG-CAT-CGC-ATT-TAT-TAG-AA CCT-GCT-TCA-AGC-TCG-TAT-G	404	pyrE	Ehling-Schulz et al. (2005a) from Helgason et al. (2004)
recF-F recf-R	GCG-ATG-GCG-AAA-TCT-CAT-AG CAA-ATC-CAT-TGA-TTC-TGA-TAC-ATC	470	recF	Ehling-Schulz et al. (2005a) from Helgason et al. (2004)
sucC-F sucC-R	GGC-GGA-ACA-GAA-ATT-GAA-GA TCA-CAC-TTC-ATA-ATG-CCA-CCA	504	sucC	Ehling-Schulz et al. (2005a) from Helgason et al. (2004)
G1 L1	GAA-GTC-GTA-ACA-AGG CAA-GGC-ATC-CAC-CGT	—	16S-23S spacer	Hansen and Hendriksen (2001) from Jensen, Webster, and Straus (1993)

Continued

Table 15.2 PCR Primers and DNA Probes for *B. cereus* (Continued)

Primer, Probe, or Adapter	Sequence (5′ → 3′)	Sequence (bp)	Gene	Reference
BCET1	CGT-ATC-GGT-CGT-TCA-CTC-GG	—	*bceT*	Hansen and Hendriksen (2001) from Agata et al. (1995)
BCET2	AGC-TTG-GAG-CGG-AGC-AGA-CT	—	*bceT*	
BCET3	GTT-GAT-TTT-CCG-TAG-CCT-GGG	—	*bceT*	
BCET4	TTT-CTT-TCC-CGC-TTG-CCT-TT	—	*bceT*	
BCET5	TTA-CAT-TAC-CAG-GAC-GTG-CTT	—	*bceT*	
BCET6	TGT-TTG-TGA-TTG-TAA-TTC-AGG	—	*bceT*	
hblA1	GTG-CAG-ATG-TTG-ATG-CCG-AT	320	*hblA*	Hansen and Hendriksen (2001) from Heinrichs et al. (1993)
hblA2	ATG-CCA-CTG-CGT-GGA-CAT-AT			
L1A	AAT-CAA-GAG-CTG-TCA-CGA-AT	430	*hblD*	Hansen and Hendriksen (2001) from Ryan et al. (1997)
L1B	CAC-CAA-TTG-ACC-ATG-CTA-AT			
L2A	AAT-GGT-CAT-CGG-AAC-TCT-AT	750	*hblC*	Hansen and Hendriksen (2001) from Ryan et al. (1997)
L2B	CTC-GCT-GTT-CTG-CTG-TTA-AT			
nheA344	TAC-GCT-AAG-GAG-GGG-CA	500	*nheA*	Hansen and Hendriksen (2001) from Granum, O'Sullivan, and Lund (1999)
nheA843	GTT-TTT-ATT-GCT-TCA-TCG-GCT			

Primer	Gene	Size	Reference	Sequence
nheB 1500 S	*nheB*	770	Hansen and Hendriksen (2001) from Granum, O'Sullivan, and Lund (1999)	CTA-TCA-GCA-CTT-ATG-GCA-G
nheB2269 S				ACT-CCT-AGC-GGT-GTT-CC
nheC2820 S	*nheC*	581	Hansen and Hendriksen (2001) from Granum, O'Sullivan, and Lund (1999)	CGG-TAG-TGA-TTG-CTG-GG
nheC3401 A				CAG-CAT-TCG-TAC-TTG-CCA-A
nheA 7173 F	*nheA*	560	Stenfors and Granum (2001)	GCT-CTA-TGA-ACT-AGC-AGG-AAA-C
nheA 7174 R				GCT-ACT-TAC-TTG-ATC-TTC-ACC-G
BEF	*nrps* (ces)	~350	Toh et al. (2004)	ACT-TAG-ATG-ATG-CAA-GAC-TG
BER				TTC-ATA-GGA-TTG-ACG-AAT-TTT
883-F	*hblA*	883	Prüß et al. (1999)	GCT-AAT-GTA-GTT-GTT-TCA-CCT-GTA-GCA-AC
883-R				AAT-CAT-GCC-ACT-GCG-TGG-ACA-TAT-AA
1F	*hblA* Int	622	Veld et al. (2001)	ACG-AAC-AAT-GGA-GAT-ACG-GC
2R				TTG-GTA-GAC-CCA-AAA-TAG-CAC-C
3F	*hblB*	232	Veld et al. (2001)	ATA-ACT-ATT-AAT-GGA-AAT-ACA
4R				CTC-CTT-GTA-AAT-CTG-TAA-TCC-CT
F-20	*hblD* (L_1)	810	Veld et al. (2001)	ATA-TTC-ACC-TTA-ATC-AAG-AGC-TGT-CAC-G
R-35				CCA-GTA-AAT-CTG-TAT-AAT-TTG-CGC-CC

Continued

Table 15.2   PCR Primers and DNA Probes for *B. cereus* (Continued)

Primer, Probe, or Adapter	Sequence (5′ → 3′)	Sequence (bp)	Gene	Reference
F-6	TAT-CAA-TAC-TCT-CGC-AAC-ACC-AAT-CG	977	*hblC (L2)*	Veld et al. (2001)
R-1	GTT-TCT-CTA-AAT-CAT-CTA-AAT-ATG-CTC-GC			
F	TTA-CAT-TAC-CAG-GAC-GTG-CTT	428	*bceT*	Veld et al. (2001)
R	TGT-TTG-TGA-TTG-TAA-TTC-AGG			
RE234-F	AAC-GTC-GGT-ATG-ATT-TTA-GG	234	*RAPD* fragment	Nakano et al. (2004)
RE234-R	CTC-TTC-TGC-TCT-CTA-TTT-ATG-TC			
prbaC1	ACT-ACG-TGC-CAG-CAG-C	296–300	*16S rDNA*	Nakano et al. (2004) from Rupf, Merte, and Eschrich (1999)
prbaC2	GGA-CTA-CCA-GGG-TAT-CTA-ATC-C			
FHBIC	CCT-ATC-AAT-ACT-CTC-GCA-A	695	*hblC*	Ngamwongsatit et al. (2008)
RHBIC	TTT-CCT-TTG-TTA-TAC-GCT-GC			
FHD	GAA-ACA-GGG-TCT-CAT-ATT-CT	1018	*hblD*	Ngamwongsatit et al. (2008)
RHD2	CTG-CAT-CTT-TAT-GAA-TAT-CA			
FHBIA	GCA-AAA-TCT-ATG-AAT-GCC-TA	884	*hblA*	Ngamwongsatit et al. (2008)
RHBIA	GCA-TCT-GTT-CGT-AAT-GTT-TT			
F2NHeA	TAA-GGA-GGG-GCA-AAC-AGA-AG	759	*nheA*	Ngamwongsatit et al. (2008)
RNHeA	TCA-ATG-CGA-AGA-GCT-GCT-TC			

Primer	Sequence	Product	Gene	Reference
F2NHeB RNHeB	CAA-GCT-CCA-GTT-CAT-GCG-G GAT-CCC-ATT-GTG-TAC-CAT-TG	935	*nheB*	Ngamwongsatit et al. (2008)
FNHeC R2NHeC	ACA-TCC-TTT-TGC-AGC-AGA-AC CCA-CCA-GCA-ATG-ACC-ATA-TC	618	*nheC*	Ngamwongsatit et al. (2008)
GCytK R2cytK	CGA-CGT-CAC-AAG-TTG-TAA-CA CGT-GTG-TAA-ATA-CCC-CAG-TT	565	*cytK*	Ngamwongsatit et al. (2008)
FEntFM REntFM	GTT-CGT-TCA-GGT-GCT-GGT-AC AGC-TGG-GCC-TGT-ACG-TAC-TT	486	*entFM*	Ngamwongsatit et al. (2008)
entFM-F evtFM-R	AAA-GAA-ATT-GGA-CAA-ACT-CAA-ACT-CA GTA-TGT-AGC-TGG-GCC-TGT-ACG-T	609	*entFM*	Sergeev et al. (2006)
piple-F piple-R	ATA-TTT-ATC-ACA-TTT-GTA-TTT-GCT-TTA-CAT-GA CTT-CTT-GAT-ACA-ATA-ATG-GTG-ACC-ACT	509	*piplc*	Sergeev et al. (2006)
nheA-F nheA-R	TTT-CTA-TCG-GTA-CTT-TAA-GTA-ATG-AAA-TTG-TA AAC-TGT-TTA-ATG-TAC-TTC-AAC-GTT-GTA-AAC	405	*nheA*	Sergeev et al. (2006)
nheB-F nheB-R	TTA-TAA-AGT-AAT-GGC-TCT-ATC-AGC-ACT TAC-TGC-ACC-ACC-GAT-AAT-TGC-AA	750	*nheB*	Sergeev et al. (2006)
nheC-F nheC-R	GTT-CAG-TTG-TGA-GCA-GGA-GCT-T AAA-CTA-TTT-GTA-TCT-TTC-GCC-ATT-CTA-T	620	*nheC*	Sergeev et al. (2006)

Continued

Table 15.2 PCR Primers and DNA Probes for *B. cereus* (Continued)

Primer, Probe, or Adapter	Sequence (5′ → 3′)	Sequence (bp)	Gene	Reference
hblA-F hblA-R	CGA-CGC-TAT-TAA-CTA-TTA-CAA-CTG-CTA GTA-ACA-GCA-TGT-GCC-CTT-GCA	265	*hblA*	Sergeev et al. (2006)
hblC-F hblC-R	TAT-AAC-AAA-GGA-AAA-GAA-ATT-AAC-AAC-TCT-A CAT-GAC-TAT-TCT-CCT-TCT-TTC-GCT-AA	641	*hblC*	Sergeev et al. (2006)
hblD-F hblD-R	TGC-ACA-AGA-AAC-GAC-CGC-TCA ATA-ATT-TGC-GCC-CAT-TGT-ATT-CCA-T	987	*hblD*	Sergeev et al. (2006)
spg-F spg-R	GAA-GAA-AGA-ATA-YCC-AAA-TCA-AAA-CAG-CA ARG-RCT-ATC-AGG-GAA-ATT-ATA-TTT-TGC-A	543	*spg*	Sergeev et al. (2006)
plc-F plc-R	CAC-TTG-TAA-AAC-AAG-ATC-GAG-TTG-CA TAC-ACC-TTT-TAG-CAA-TTT-ACC-TTT-CAC-GT	727	*plc*	Sergeev et al. (2006)
CytK-F CytK-R	GTA-ACA-GAT-ATC-GGK-CAA-AAT-GCA TGT-TAT-ATC-CRT-TAA-AGA-ATA-CGT-TCC-A	527	*CytK*	Sergeev et al. (2006)
beeT-F beeT-R	CCA-AGA-TTT-ATG-AAA-GAG-TTA-GTT-TCA-ACA CAT-CCA-TAA-TGA-CTG-ATT-GTA-AGA-ATG-T	720	*beeT*	Sergeev et al. (2006)
pa-F pa-R	CCA-GAC-CGT-GAC-AAT-GAT-G CAA-GTT-CTT-TCC-CCT-GCT-A	508	*pa*	Sergeev et al. (2006)

Primers	Sequences	Size	Gene	Reference
cya-F cya-R	GCG-ATG-AAA-ACA-ACG-AAG-TA TCG-TCT-TTG-TCG-CCA-CTA-TC	720	*cya*	Sergeev et al. (2006)
lef-F lef-R	GGT-GCG-GAT-TTA-GTT-GAT-TC CGC-TTC-ATT-TGT-TCT-CCC	851	*lef*	Sergeev et al. (2006)
ba-F ba-R	CAT-GGA-ATC-AAG-CTG-CAA-ATT-ATA-AAG CTA-AAT-TGT-CAT-TTG-GCA-AAT-TGA-AAT-TC	1017	*ba*	Sergeev et al. (2006)
S-S-Bc-200-a-S-18 S-S-Bc-470-a-A-18	TCG-AAA-TTG-AAA-GGC-GGC GGT-GCC-AGC-TTA-TTC-AAC	—	*16S rDNA*	Choo et al. (2007) from Hansen, Leser, and Hendrikson (2001)
BcF 16S2	GGA-TTA-AGA-GCT-TGC-TCT-TAT AAG-GCC-TAT-CTC-TAG-GGT-TT	—	*16S rDNA*	Choo et al. (2007) from Chen and Tsen (2002)
BC1 BC2r	ATT-GGT-GAC-ACC-GAT-CAA-ACA TCA-TAC-GTA-TGG-ATG-TTA-TTC	—	*gyrB*	Choo et al. (2007) from Yamada et al. (1999)
BCFW1 BCrevnew	GTT-TCT-GGT-GGT-TTA-CAT-GG TTT-TGA-GCG-ATT-TAA-ATG-C	374	*gyrB*	Choo et al. (2007) from Manzano et al. (2003a)
K3 K5	GCT-GTG-ACA-CGA-AGG-ATA-TAG-GCA-C AGG-ACC-AGG-ATT-TAC-AGG-AG	1600–1700	*cry*	Choo et al. (2007) from Kuo and Chak (1996)

Continued

Table 15.2　PCR Primers and DNA Probes for *B. cereus* (Continued)

Primer, Probe, or Adapter	Sequence (5′ → 3′)	Sequence (bp)	Gene	Reference
RAPD OPG 16	AGC-GTC-CTC-C	—	—	Choo et al. (2007)
Ces-SYBR-F	CAC-GCC-GAA-AGT-GAT-TAT-ACC-AA	176	*ces*	Fricker et al. (2007)
Ces-SYBR-R	CAC-GAT-AA-ACC-ACC-ACT-GAG-ATA-GTG			
Ces-TM-F	CGC-CGA-AAG-TGA-TTA-TAC-CAA	103	*ces*	Fricker et al. (2007)
Ces-TM-R	TAT-GCC-CCG-TTC-TCA-AAC-TG			
Ces-TM-P	FAM-GGG-AAA-ATA-ACG-AGA-AAT-GCA-TAMRA			
IAC-F	GCA-GCC-ACT-GGT-AAC-AGG-AT	118	—	Fricker et al. (2007)
IAC-R	GCA-GAG-CGC-AGA-TAC-CAA-AT			
IAC-P	HEX-AGA-GCG-AGG-TAT-GTA-GGC-GG-TAMRA			
EM1F	GAC-AAG-AGA-AAT-TTT-CTA-CGA-GCA-AGT-ACA-AT	635	*nrps*	Ehling-Schulz, Fricker, and Scherer (2004)
EM1R	GCA-GCC-TTC-CAA-TTA-CTC-CTT-CTG-CCA-CAG-T			
HpaIF	GCC-AGA-AGA-TGC-AAT-GAT-TCC-AGT-ATG			
HblA1	GCT-AAT-GTA-GTT-TCA-CCT-GTA-GCA-AC	874	*hblA*	Hsieh et al. (1999) from Mäntynen and Lindström (1998)
HblA2	AAT-CAT-GCC-ACT-GCG-TGG-ACA-TAT-AA			
Ph1	CGT-GCC-GAT-TTA-ATT-GGG-GC	558	*sph*	Hsieh et al. (1999)
Ph2	CAA-TGT-TTT-AAA-CAT-GGA-TGC-G			

Primer	Gene	Size	Reference	Sequence
ETF	*bceT*	428	Hsieh et al. (1999) from Agata et al. (1995)	TTA-CAT-TAC-CAG-GAC-GTG-CTT
ETR				TGT-TTG-TGA-TTG-TAA-TTC-AGG
ENTA	*entFM*	1269	Hsieh et al. (1999) Asano et al. (1997)	ATG-AAA-AAA-GTA-ATT-TGC-AGG
ENTB				TTA-GTA-TGC-TTT-TGT-GTA-ACC
HC F	*hblC*	740	Guinebretiere, Broussolle, and Nguyen-The (2002)	GAT-AC(T/C)-AAT-GTG-GCA-ACT-GC
HC R				TTG-AGA-CTG-CTC-G(T/C)T-AGT-TG
HD F	*hblD*	829	Guinebretiere, Broussolle, and Nguyen-The (2002)	ACC-GGT-AAC-ACT-ATT-CAT-GC
HD R				GAG-TCC-ATA-TGC-TTA-GAT-GC
HA F	*hblA*	1154	Guinebretiere, Broussolle, and Nguyen-The (2002)	AAG-CAA-TGG-AAT-ACA-ATG-GG
HA R				AGA-ATC-TAA-ATC-ATG-CCA-CTG-C
HA-F	*hblB*	2684	Guinebretiere, Broussolle, and Nguyen-The (2002)	AAG-CAA-TGG-AAT-ACA-ATG-GG
HA R				AAT-ATG-TCC-CAG-TAC-ACC-CG
NA F	*nheA*	755	Guinebretiere, Broussolle, and Nguyen-The (2002)	GTT-AGG-ATC-ACA-ATC-ACC-GC
NA R				ACG-AAT-GTA-ATT-TGA-GTC-GC
NB F	*nheB*	743	Guinebretiere, Broussolle, and Nguyen-The (2002)	TTT-AGT-AGT-GGA-TCT-GTA-CGC
NB R				TTA-ATG-TTC-GTT-AAT-CCT-GC

Continued

Table 15.2 PCR Primers and DNA Probes for *B. cereus* (Continued)

Primer, Probe, or Adapter	Sequence (5' → 3')	Sequence (bp)	Gene	Reference
NC F	TGG-ATT-CCA-AGA-TGT-AAC-G	683	*nheC*	Guinebretiere, Broussolle, and Nguyen-The (2002)
NC R	ATT-ACG-ACT-TCT-GCT-TGT-GC			
BCET1 F	CGT-ATC-GGT-GGT-TCA-CTC-GG	661	*bceT*	Guinebretiere, Broussolle, and Nguyen-The (2002)
BCET3 R	GTT-GAT-TTT-CCG-TAG-CCT-GGG			
BCET1 F	GGT-ATC-GGT-CGT-TCA-CTC-GG	924	*bceT*	Guinebretiere, Broussolle, and Nguyen-The (2002)
BCET4 R	TTT-CTT-TCC-CGC-TTG-CCT-TT			
CK F	ACA-GAT-ATC-GG(G/T)-CAA-AAT-GC	809	*cytK*	Guinebretiere, Broussolle, and Nguyen-The (2002)
CK R	GAA-CTG-(G/C)(/T)A-ACT-GGG-TTG-GA			
Pf	GAG-TTA-GAG-AAC-GGT-ATT-TAT-GCT-GC	389	*Plc*	Schraft and Griffiths (1995)
Pr	CTA-CTG-CCG-CTC-CAT-GAA-TCC			
Pf	GAG-TTA-GAG-AAC-GGT-ATT-TAT-GCT-GC	1461	*Plc-Sph*	Schraft and Griffiths (1995)
Cr	GCA-TCC-CAA-GTC-GCT-GTA-TGT-CCA-G			
HblA1	GCT-AAT-GTA-GTT-TCA-CCT-GTA-GCA-AC	874	*hblA*	Mäntynen and Lindström (1998)
HblA2	AAT-CAT-GCC-ACT-GCG-TGG-ACA-TAT-AA			
BceT1	GAA-TTC-CTA-AAC-TTG-CAC-CAT-CTC-G	—	*bcet*	Mäntynen and Lindström (1998)
BecT2	CTG-CGT-AAT-CGT-GAA-TGT-AGT-CAA-T			

Primer	Sequence	Size	Target	Reference
Cer-F	CAT-GCG-GCA-AAC-TTT-ACG-AAC-CT	639	cerAB	Kim, Czajka, and Batt (2000)
Cer-R	TAA-TCT-GCC-CCG-AAT-AAA-T			
CerTAQ-1	FAM-CGT-AAT-CTT-GTT TTC- GCA-ACT-ACT-GC[a]	—		
CerTAQ-2	FAM-(A/C)GT-AAT-CTT-TGT-TTC-GC(ACG)-ACT-ACT-GC[b]	—		
RAPD Bce-P	CCG-AGT-CCA	—	—	Svensson et al. (2006)
A-F	GGA-GAG-TTA-GAT-CTT-GGC-TCA	280	16S-23S rRNA spacer	Haque and Russell (2005)
B-R	CCA-GTG-TGG-CCG-GTC-GCC-CTC			
X	CGT-GCC-AGC-CGC-GGT-AAT-AT		16S-23S rRNA spacer	Haque and Russell (2005)
Y	TTG-ACG-TCA-TCC-CCA-CCT-TCC-T			
X	CGT-GCC-AGC-CGC-GGT-AAT-AT	600	16S-23S rRNA spacer	Haque and Russell (2005)
D	CCG-GGT-TTC-CCC-ATT-CGG			
PlcR-F1	GCA-CGC-AGA-AAA-ATT-AGG-AAG-TG	820	plcR	Fricker et al. (2008)
PlcR-R1	ATG-G(A/T)-TC(C/T)-TAA-TAT-ATC			
PlcR-F2	GCA-CGC-AGA-AAA-ATT-AGG-AAG-TG	889	plcR	Fricker et al. (2008)
PlcR-R2	CTA-ATA-TAT-C(A/G)A-AAA-AGA-AG(C/T)-(A/T)(A/T)G-C			

Continued

Table 15.2 PCR Primers and DNA Probes for *B. cereus* (Continued)

Primer, Probe, or Adapter	Sequence (5′ → 3′)	Sequence (bp)	Gene	Reference
RAPD PM13	GAG-GGT-GGC-GGC-TCT	—	—	Guinebretiere and Nguyen-The (2003)
DAF And2	CCG-GCG-GCG	—	—	
HA-F1 H1-R1	ATT-AAT-ACA-GGG-GAT-GGA-GAA-ACT-T TGA-TCC-TAA-TAC-TTC-TTC-TAG-ACG-CTT	237	*hblA*	Yang et al. (2005)
HC-F1 HC-R1	CCT-ATC-AAT-ACT-CTC-GCA-ACA-CCA-AT TTT-TCT-TGA-GTC-ATA-GCC-ATT-TCT	386	*hblC*	Yang et al. (2005)
HD-F1 HD-R1	AGA-TGC-TAC-AAG-ACT-TCA-AAG-GGA-AAC-TAT TGA-TTA-GCA-CGA-TCT-GCT-TTC-ATA-CTT	436	*hblD*	Yang et al. (2005)
NA-F1 NA-R1	ATT-ACA-GGG-TTA-TTG-GTT-ACA-GCA-GT AAT-CTT-GCT-TAC-TCT-CTT-GGA-TGC-T	475	*nheA*	Yang et al. (2005)
NB-F1 NB-R1	GTG-CAG-CAG-CTG-TAG-GCG-GT ATG-TTT-TTC-CAG-CTA-TCT-TTC-GCA-AT	328	*nheB*	Yang et al. (2005)
NC-F1 NC-R1	GCG-GAT-ATT-GTA-AAG-AAT-CAA-AAT-GAG-GT TTT-CCA-GCT-ATC-TTT-CGC-TGT-ATG-TAA-AT	558	*nheC*	Yang et al. (2005)

Continued

Primer	Sequence	Size (bp)	Gene	Reference
FM-F2	CAA-AGA-CTT-CGT-AAC-AAA-AGG-TGG-T	290	*entFM*	Yang et al. (2005)
FM-R2	TGT-TTA-CTC-CGC-CTT-TTA-CAA-ACT-T			
BceT-F1	AGT-TTG-GAG-CGG-AGC-AGA-CTA-TGT	701	*bceT*	Yang et al. (2005)
BceT-R1	GTA-TTT-CTT-TCC-CGC-TTG-CCT-TTT			
Cyt-F1	ATC-GGG-CAA-AAT-GCA-AAA-ACA-CAT	300	*cytK*	Yang et al. (2005)
Cyt-R1	ACC-CAG-TTT-GCA-GTT-AAT-GT			
Cyt-F2	ATC-GGT-CAA-AAT-GCA-AAA-ACA-CAT	300	*cytK*	Yang et al. (2005)
Cyt-R2	ACC-CAG-TTA-CCA-GTT-CCG-AAT-GT			
EM-F1	AGC-TTG-GAG-CGG-AGC-AGA-CTA-TGT	535	*nrps*	Yang et al. (2005)
EM-R1	GTA-TTT-CTT-TCC-CGC-TTG-CCT-TTT			
ITS-F1	AAT-TTG-TAT-GGG-CCT-ATA-GCT-CAG-CT	185	16S-23S rRNA spacer	Yang et al. (2005)
ITS-R1	TTT-AAA-ATA-GCT-TTT-TGG-TGG-AGC-CT			
HBLAup	GCT-AAT-GTA-GTT-TCA-CCA-GTA-ACA-AC	374	*hblA*	Liu et al. (2007)
HBLAlp	AAT-CAT-GCC-ACT-GCG-TGG-ACA-TAT-AA			
Capture probe HBLcp	Thiol-TCA-GTA-ATG-TTT-TAA-TGA-ACA-ACA-TAA-CT			
Detection probe HBLAdp	TTT-GCA-AGT-GAA-ATT-GAA-CAA-ACG-ATA-CA-Biotin			

Table 15.2 PCR Primers and DNA Probes for *B. cereus* (Continued)

Primer, Probe, or Adapter	Sequence (5′ → 3′)	Sequence (bp)	Gene	Reference
HBLCup	TAA-TGT-TTT-AAT-GAA-CAA-CAT-AAC-T	747	*hblB*	Liu et al. (2007)
HBLClp	ATA-TCC-ATG-CCT-TCC-TGT-TGA-GTT-T			
Capture probe HBLCcp	Thiol-TCA-GTA-ATG-TTT-TAA-TGA-ACA-ACA-TAA-CT			
Detection probe HBLCdp	GTA-TGA-CCA-GAC-AGA-AAG-GAT-AAG-GAC-TA-Biotin			
HBLDup	GCA-CAA-GAA-ACG-ACC-GCT-CAA-GAA-C	989	*hblD*	Liu et al. (2007)
HBLDlp	GTA-TAA-TTT-GCG-CCC-ATT-GTA-TTC-C			
Capture probe HBLDcp	Thiol-CAC-CTT-CAT-GCA-TTT-GCA-CAA-GAA-ACG-AC			
Detection probe HBLDdp	GCT-CAA-GAA-CAA-AAA-GTA-GGC-AAT-TGG-TA-Biotun			
BCETup	GGC-GTT-TTT-TTA-TTA-GAG-AGG-A	835	*bceT*	Liu et al. (2007)
BCETlp	GTT-GAT-TTT-CCG-TAG-CCT-GGG-CTT-G			
NHEAup	ACG-AAT-CAT-CAA-AAG-TTT-GCA-AAG-GG	310	*nheA*	Liu et al. (2007)
NHEAlp	CGG-CTT- TAA-TTG-ATA-AGT-TGT-TAC-T			
Capture probe NHEAcp	Thiol-ATT-CAC-GAA-TCA-TCA-AAA-GTT-TGC-AAA-GG			
Detection probe NHEAdp	ATG-TAC-GAG-AAT-GGA-TTG-ATG-AAT-ACT-AA-Biotin			
NHEBup	ATG-TAT-CGT-CTG-TTG-ATG-CGG-CTT-T	317	*nheb*	Liu et al. (2007)
NHEBlp	GTT-TTG-CGT-ATC-CGA-AGT-CAT-TTT-A			

Continued

Capture probe NHEBcp	Thiol-TTC-TAT-GTA-TCG-TCT-GTT-GAT-GCG-GCT-TT			
Detection probe NHEBdp	AAG-GGA-AAG-TAA-TTC-AGC-ACC-AAG-ACA-TA-Biotin			
NHECup NHECIp	CTG-GGG-TGG-CAA-CGA-GTA-ACG-CAT-T ATC-CGC-TTT-TAA-TTT-TCC-GCT-ATC-C	413	*nheC*	Liu et al. (2007)
Capture probe NHECcp	Thiol-CAT-TCT-GGG-GTG-GCA-ACG-AGT-AAC-GCA-TT			
Detection probe NHECdp	CTT-TAC-ATC-CTT-TTG-CAG-CAG-AAC-AACT-AC-Biotin			
CYTKup CYTKIp	GAT-ATC-GGG-CAA-AAT-GCA-AAA-ACA-C TGT-TTC-CAA-CCC-AGT-TTG-CAG-TTC-C	811	*cytK*	Liu et al. (2007)
Capture probe CYTKcp	Thiol-ATT-CTG-ACT-ACC-AAC-TTC-GTC-CAG-GCT-TC			
Detection probe CYTKdp	GGA-ACT-GCA-AAC-TGG-GTT-GGA-AAC-ATA-AT-Biotin			
NRPSup NRPSIp	GTT-ACC-GAC-GTT-AGA-GTT-GCC-GAC-A CGT-ATT-CAC-AAA-CAT-CCC-GAT-TAA-A	274	*nrps*	Liu et al. (2007)
Capture probe NRPScp	Thiol-ATA-GGT-TAC-CGA-CGT-TAG-AGT-TGC-CGA-CA			
Detection probe NRPSdp	ACA-GAC-AAC-GTC-CAC-TTT-TGA-AAA-CTC-TA-Biotin			
Cry1A-1	ACA-GTT-TCC-CAA-TTA-ACA-AGA-GAA-ATT-TAT-ACG-AAC-CCA-G	632	*cry1A*	Zhou et al. (2008)
CryA-2	TGT-GAT-GAA-GGA-ATT-ATA-TTA-AAT-TC			

Table 15.2   PCR Primers and DNA Probes for *B. cereus* (Continued)

Primer, Probe, or Adapter	Sequence (5′ → 3′)	Sequence (bp)	Gene	Reference
BCFW1-F	GTT-TCT-GGT-GGT-TTA-CAT-GG	326	*gyrB*	Jensen et al. (2005) from Manzano et al. (2003b)
BCFW1-R	CAA-CGT-ATG-ATT-TAA-TTC-CAC-C			
BcAPR1	CTT-(C/T)TT-GGC-CTT-CTT-CTA-A	284	—	Altayar and Sutherland (2006) from Francis et al. (1998)
BcFF2	GAG-ATT-TAA-ATG-AGC-YGY-AAÅ			
BcAPR1	CTT-(C/T)TT-GGC-CTT-CTT-CTA-A	160	—	Altayar and Sutherland (2006) from Francis et al. (1998)
BcAPF1	GAG-GAA-ATA-ATT-ATG-ACA-GTT			
CK F	ACA-GAT-ATC-GG(G/T)-CAA-AAT-GC	809	*cytK*	Aragon-Alegro et al. (2008) from Guinebretiere et al. (2006)
CK R	TCC-AAC-CCA-GTT-(A/T)(G/C)C-AGT-TC			
CK1 F	CAA-TTC-CAG-GGG-CAA-GTG-TC	426	*cytK-1*	Aragon-Alegro et al. (2008) from Guinebretiere et al (2006)
CK1 R	CCT-CGT-GCA-TCT-GTT-TCA-TGA-G			
CK2 F	CAA-TCC-CTG-GCG-CTA-GTG-CA	535	*cytK-2*	Aragon-Alegro et al. (2008) from Guinebretiere et al (2006)
CK2 R	GTG-IAG-CCT-GGA-CGA-AGT-TGG			
K3-CRYSTAL	GCT-GTG-ACA-CGA-AGG-TAG-CCA-C	1635	*crry*	Ankolekar, Rahmati, and Labbe (2008) from Kuo et al. (2007)
K5-CRYSTAL	ACG-ACC-AGG-ATT-TAC-AGG-AGG			

Primer	Sequence		Size	Gene	Reference
FHA₂	GGA-GAT-ACG-GCT-CTT-TCT		325	*hblA*	Thaenthanee, Wong, and Panbangred (2005)
RHA₂	CTC-CAT-CCC-CTG-TAT-TAA				
FHC	GAA-TGG-TCA-TCG-GAA-CTC-TA		672	*hblC*	Thaenthanee, Wong, and Panbangred (2005)
RHC	CAT-CAG-GTC-ATA-CTC-TTG-TGT				
FHD	GAA-ACA-GGG-TCT-CAT-ATT-C		487	*hblD*	Thaenthanee, Wong, and Panbangred (2005)
RHD	CTT-GAA-GTT-GTG-GAA-TCG-TT				
FHC	GAA-TGG-TCA-TCG-GAA-CTC-TA		1768	*hblC–hblD*	Thaenthanee, Wong, and Panbangred (2005)
RHD	CTT-GAA-GTT-GTG-GAA-TCG-TT				
FHC	GAA-TGG-TCA-TCG-GAA-CTC-TA		2815	*hblC–hblA*	Thaenthanee, Wong, and Panbangred (2005)
RHA₂	CTC-CAT-CCC-CTG-TAT-TAA				
FHD	GAA-ACA-GGG-TCT-CAT-ATT-C		1534	*hblD–hblA*	Thaenthanee, Wong, and Panbangred (2005)
RHA₂	CTC-CAT-CCC-CTG-TAT-TAA				
CytKF	GAT-AAT-ATG-ACA-ATG-TCT-TTA-AA		⁻011	*cytK*	Bartoszewicz, Hansen, and Swiecicka (2008) from Swiecicka and Mahillon (2006)
CytKR	GGA-GAG-AAA-CCG-CTA-TTT-GT				

a Detection probe NHEAdp.
b T indicates nucleotide labeled with TAMRA.

Kim, Czajka, and Batt (2000) developed an Rti-PCR assay for specific detection of *B. cereus* based on the sequence of the *cereolysen AB* gene (*cerAB*) that encodes phospholipase C and sphingomyelinase. The primers Cer-F/Cer-R (Table 15.2) amplified a 639-bp sequence of the *cerAB* gene. The dual-labeled probe CerTAQ-1 (Table 15.2) was labeled with the reporter dye FAM at the 5'-end and with the quencher dye TAMRA at nucleotide 13 from the 3'-end. A total of 51 out of 72 *B. cereus* strains tested positive with the CerTAQ-1 probe, and only one out of five *B. thuringiensis* strains tested positive. A second similarly dual-labeled probe CerTAQ-2 (Table 15.2) resulted in 35 out of 39 *B. cereus* strains testing positive (including 10 of 14 previously negative strains). A PCR assay using CerTAQ-1 was able to detect approximately 58 *B. cereus* CFU/g of seeded nonfat dry milk. A luminescent spectrometer was used to measure fluorescence.

Choo et al. (2007) cultured 140 samples of dried red pepper purchased in Korea for the presence of *B. cereus*. A multiplex PCR (mPCR) assay was also developed for the rapid confirmation of *B. cereus*. The primers S-S-Bc-200-a-S-18/S-S-Bc-470-a-A-18 (Table 15.2) amplified a sequence of the 16S rDNA of *B. cereus* group members. The primers BcF/16S2 (Table 15.2) amplified a sequence of the 16S rDNA of *B. cereus, B. thuringiensis*, and *B. mycoides*. The primers BC1/BC2r amplified a sequence of the DNA gyrase B gene of *B. cereus, B. thuringiensis*, and *B. mycoides*. The primers BCFW1/BCrevnew (Table 15.2) amplified a 374-bp sequence of the *gyrB* gene from all *B. cereus* and *B. thuringiensis* strains examined but not from *B. mycoides*. The primers K3/K5 (Table 15.2) amplified a 1600–1700-bp sequence of the *cry* gene highly specific for *B. thuringiensis*. Therefore, in an mPCR assay, strains that produced the 374-bp amplicon without the 1600–1700-bp amplicon were identified as *B. cereus*, and strains yielding the 1600–1700-bp amplicon regardless of generation of the 374-bp amplicon were identified as *B. thuringiensis*. *B. cereus* was found in 118 (84.3%) of the samples with a mean level of 1.9×10^4/g of the dried red peppers. RAPD with primer OPG 16 (Table 15.2) yielded 142 RAPD types, which allowed discrimination of the isolates.

VI. MOLECULAR TYPING OF *B. CEREUS* ISOLATES

A. Genotyping of *B. cereus* Isolates

Mäntynen and Lindström (1998) developed a rapid PCR assay for detection of enterotoxic *B. cereus*. The primers HblA1/HblA2 (Table 15.2) amplified an 874-bp sequence of the B component of the hemolysin *hblA* gene (Table 15.2). Primers Bcet1/BceT2 (Table 15.2) amplified a sequence of the *B. cereus* enterotoxin T *bcet* gene. Among 51 strains of *B. cereus*, 26 were found to harbor the *hblA* gene. The hemolysin *hblA* gene was detected in both enterotoxigenic and nonenterotoxigenic strains. In addition, several other species also harbored the *hblA* gene. The *bcet* gene was found absent from the enterotoxigenic strains examined and was therefore too rare to be used for assessing the enterotoxigenic potential of *B. cereus* strains. These results are in contrast to those of Agata et al. (1995), who found that all 10 strains of *B.*

cereus examined harbored the *bcet* gene. Restriction fragment length polymorphism (RFLP) analysis was performed on the *hblA* amplicons by digesting with *Rsa*I, *Taq*I, and *Xba*I and indicated that the *hblA* amplicons were slightly heterogeneous.

Hsieh et al. (1999) determined the enterotoxigenic profiles of 26 strains of *B. cereus*, 2 strains *of B. anthracis*, 3 strains of *B. mycoides*, 9 strains of *B. thuringiensis*, 18 strains of other *Bacillus* spp., and several non-*Bacillus* spp. In addition, 28 *B. cereus* strains from food samples and 30 other *B. cereus* strains from samples associated with food poisoning outbreaks were also examined. The primers HblA1/HblA2 (Table 15.2) amplified an 874-bp sequence of the *hblA* gene encoding the BC hemolysin. The primers Ph1/Ph2 (Table 15.2) amplified a 558-bp sequence of the *sph* gene that encodes sphingomyelinase. The primers ETF/ETR (Table 15.2) amplified a 428-bp sequence of the *B. cereus* enterotoxin T encoded by the *bcet* gene. Primers ENTA/ENTB amplified a 1269-bp sequence of the *entFM* gene that encodes the enterotoxin FM. These PCR primers are specific to all *B. cereus* group strains used. Only 26 of the 84 *B. cereus* isolates examined harbored the *hblA* gene. Among the 28 foodborne strains only four harbored the *hblA* gene, whereas among the 30 food poisoning outbreak strains seven harbored the *hbla* gene, indicating no significant difference in the distribution of this gene among both groups of isolates.

In addition to the 26 *B. cereus* strains harboring the *hblA* gene, eight of nine *B. thuringiensis* and two of three *B. mycoides* strains also harbored this gene and produced the encoded toxin. The *bcet* gene was found present in 41 of 84 *B. cereus* isolates, two of three *B. mycoides* strains, and eight of nine *B. thuringiensis* strains. The *entFM* gene was found present in 78 of 84 *B. cereus* strains, one of three *B. mycoides* strains, and in eight of nine *B. thuringiensis* strains. In addition, 27 of 28 foodborne isolates of *B. cereus* and all 30 outbreak-associated strains of *B. cereus* harbored the *entFM* gene. The *sph* gene was found present in all the *B. cereus* group isolates tested and absent in *Bacillus* spp. other than those of the *B. cereus* group. The Ph1/Ph2 primers were therefore found suitable for the specific detection of members of the *B. cereus* group.

Prüb et al. (1999) determined the presence of the *hblA* gene among strains representing all species of the *B. cereus* group except *B. anthracis*. The primers 883-F/883-R (Table 15.2) amplified an 883-bp sequence of the *hblA* gene. The results indicated that the HBL enterotoxin is broadly distributed among strains of the *B. cereus* group and is associated neither with a specific species nor with a specific environment.

Hansen and Hendriksen (2001) assessed the presence of the toxin genes *bceT, hblA, hblD, hblC, beheA, bheB,* and *bheC* in 22 *B. cereus* and 41 *B. thuringiensis* strains in addition to using a pair of 16S-23S rDNA spacer primers for confirming the identity of the isolates as members of the *B. cereus* group. Six primers (1, 2, 3, 4, 5, and 6) were used in five combinations (1/3, 1/4, 2/3, 2/4, and 5/6) for detection of *bcet* sequences. The primers are listed in Table 15.2. PCR analysis of the 16S-23S rDNA spacer region revealed identical patterns for all 63 strains examined and was considered as an additional argument for considering *B. cereus* and *B. thuringiensis* as one taxonomic group. At least one gene of the two protein complexes HBL and NHE was detected in all of the *B. thuringiensis* strains. Six of the *B. cereus* strains were devoid of all three HBL genes, three lacked at least two of the three NHE genes, and one lacked all three NHE genes. The use of five pairs of primers for detection of the *bcet* gene indicated that the *bcet* gene is widely distributed among both species and that the gene varies in sequence among different strains.

In't Veld et al. (2001) assessed 86 strains of *B. cereus* from milk for the presence of the three components of the HBL enterotoxin complex [*hblA, hblB, hblD*(L_1) *hblC* (L_2)] and enterotoxin-T (*bceT*). The primers are listed in Table 15.2. The *hblD* gene was detected in 56 (65%) of the isolates, and the *hblC* gene in 53 (62%) of the 86 strains. Only nine strains contained the *hblB* gene. All three HBL genes were found combined in 47 (55%) of the isolates. Only three strains contained just the *hbld* gene. The enterotoxin gene *bceT* was detected in 53 (62%) of the strains. In order to cause food poisoning all three components of the HBL complex (B, D, and C) must be present (Beecher et al., 1995).

Guinebretiere, Broussolle, and Nguyen-The (2002) compared the enterotoxigenic profiles of 51 *B. cereus* isolates from foods and 37 *B. cereus* isolates from food poisonings. A total of 10 primer pairs (Table 15.2) was used to amplify sequences from 10 toxin genes. The *cytK* gene and association of the *hbl-nhe-cytK* enterotoxin genes were more frequent among diarrheal strains, occurring at a frequency of 73% and 33%, respectively. Unlike diarrheal strains, foodborne strains showed frequent *nhe* and *hbl* gene polymorphism and were often low toxin producers.

Haque and Russell (2005) phenotypically and genotypically characterized seven strains of *B. cereus* from Bangladeshi rice. The strains were identified as *B. cereus* based on the API 50CHB test system and 16S RDNA analysis. They were distinguishable from one another on the basis of substrate-utilization patterns using the API system. RFLP analysis of the spacer region between the 16S and 23S rRNA genes using one of three sets of PCR primers (X/D; Table 15.2) and resulting amplicons digested with a mixture of *Dde*1, *Aau3a*1, and *Hha*1 yielded two groups, whereas sequence analysis of the variable 16S rRNA gene gave four different groups. Despite their association with rice, none of the strains produced the emetic toxin.

Yang et al. (2005) developed an mPCR for the detection of five different enterotoxins and the emetic toxin cereulide involving the simultaneous incorporation of 11 primer pairs (Table 15.2) into the mPCR. An additional pair of primers ITS-F1/ITS-R1 (Table 15.2) was incorporated that amplified a 185-bp sequence of the 16S–23S rRNA internal spacer region for use as an amplification control. The assay was applied to 94 food poisoning strains of *B. cereus*, 51 food-related strains of *B. cereus*, and 17 strains of *B. mycoides*. A total of 10 toxigenic patterns for all 162 tested strains was obtained. All of the *B. cereus* strains harbored at least one toxin gene. More than 70% of the *B. mycoides* strains harbored no known toxin genes. The presence of genes *hblA, hblC,* and *hblD* were highly associated with each other. These three *hbl* genes were detected in 46.8%, 29.4%, and 5.9% of the 162 isolates, respectively. All of the *B. cereus* strains carried the genes *nhea, nheB, nheC,* and *entFM,* as did five of the *B. mycoides* strains. The occurrence of the *bceT* gene was 21.3%, 11.8%, and 5.9% among these three populations, respectively. In addition, 66.0%, 48.0%, and 5.9% of the three populations, respectively, harbored the *cytK* gene. Only 3.9% of food-related *B. cereus* strains were of the emetic type. It is interesting that no statistical difference in the distribution of these toxin genes was found between food poisoning and food-associated strains of *B. cereus*.

Thaenthanee, Wong, and Panbangred (2005) assessed the presence of the *hblA,* *hblD,* and *hblC* genes in 339 *B. cereus* strains isolated in Thailand with primers listed in Table 15.2. All three *hbl* genes were detected in 222 (65.5%) of the isolates. Two, one, or no *hbl* genes were detected in 3 (0.9%), 61 (1.8%), and 108 (31.8%) of the strains, respectively. Among the 222 strains in which all three *hbl* genes were detected, 210 (61.9%) displayed discontinuous hemolysis, where hemolysis begins several millimeters distant from the colony forming a ring of lysis, which is characteristic of HBL producers, whereas 12 (3.5%) showed continuous hemolysis on sheep blood agar. Among strains in which none of the *hbl* genes were detected, 97 (28.6%) displayed continuous hemolysis and 11 (3.2%) did not show hemolysis activity. *hblC* was present in five of six strains where only one *hbl* gene was detected. The *Hpa*II restriction profiles of PCR fragments amplified from the hblC-A region in these five strains using primers FHC/RHA$_2$ (Table 15.2) displayed heterogeneous patterns. The authors concluded that HBL-encoding genes are widely distributed among *B. cereus* isolates in Thailand and there is a high degree of heterogeneity in both the genes and encoded proteins.

Ehling-Schulz et al. (2006) developed a multiplex PCR assay for detection of the *hbl, nhe, cytK,* and *ces* genes of *B. cereus.* The primers HD2 F/HA4 R (Table 15.2) amplified a 1091-bp sequence of the *hbl* gene. The primers NA2 F/NB1 R (Table 15.2) amplified a 766-bp sequence of the *nhe* gene. The primers CKF2/CKR5 (Table 15.2) amplified a 421-bp sequence of the *cytK* gene. The primers CesF1/CesR2 (Table 15.2) amplified a 1271-bp sequence of the *ces* gene. Sequencing revealed high levels of polymorphism of the enterotoxin genes that may have resulted in false PCR-negative results previously observed for the enterotoxins NHE and HBL. Observed point mutations in toxin gene sequences were taken into account when primers were designed and inosine was inserted at variable positions. The newly designed primers allowed the amplification of enterotoxin gene sequences from strains that were previously PCR negative but that were detected as positive for enterotoxin genes by Southern blot analysis.

The forward and reverse primers were located in two different genes of the corresponding operon, targeting two toxin genes in a single reaction. The forward primer, designed for the detection of the *nhe* complex, was located in *nhea* and the reverse primer was located in *nheB.* Primers for *cytK* were directed at highly conserved regions of the toxin gene so as to detect both forms of *cytK* (*cytK1* and *cytK2*) in a single reaction. A total of 60 *B. cereus* strains from different food and clinical diagnostic laboratories plus 80 dairy silo *B. cereus* group isolates were genotyped by the mPCR assay. The population of the silo tanks was dominated by strains with the toxin profile "C" (*nhe$^+$, hbl$^+$, cytK$^-$, ces$^-$*) and "F" (*nhe$^+$, hbl$^-$, cytK$^-$, ces$^-$*). The prevalence of the toxin gene profile "C" strains was much lower from diagnostic laboratories than "F" strains. The incidence of emetic strains was generally low and emetic strains carrying I was rare, although emetic strains were found in all environments examined.

Svensson et al. (2007) assayed 396 isolates of the *B. cereus* group from dairy farms, dairy silo tanks, and pasteurized milk production lines for the presence of the toxin genes *nheA, hblc,* and *cytK* by PCR. The primers nheA 7173 F/nheA 7174

R (Table 15.2) amplified a 560-bp sequence of the *nheA* gene. Primers hblC L2A/ hblc L2B (Table 15.2) from Hansen and Hendriksen (2001) amplified a 750-bp sequence of the *hblC* gene. Primers CKF2/CKR5 (Table 15.2) from Ehling-Schulz et al. (2006) amplified a 421-bp sequence of the *cytK* gene. Highly toxigenic strains producing NHE and HBL were less common among dairy isolates compared to farm and silo isolates. No producer of high levels of both toxins was found among 156 psychrotrophic dairy isolates. Only 3% of the psychrotrophic isolates were high producers of NHEA, although detectable enterotoxin production by psychrotrophic strains of *B. cereus* was found to occur by Van Netten et al. (1990) in a variety of food initially seeded with 10 CFU/g after 24 days at 4°C, after about 12 days at 7°C, and within 24 hr at 17°C.

A total of 490 stool sample specimens, consisting of 325 from diarrheal patients and 165 from nondiarrheal healthy individuals, was cultured for the presence of *B. cereus* by Al-Khatib et al. (2007). *B. cereus* was found in 31 (9.5%) of diarrheal samples compared to 3.0 (1.8%) for nondiarrheal samples. The hemolytic enterotoxin HBL genes *hblA*, *hblC*, and *hblC* were detected by the PCR using the primers (Table 15.2) of Guinebretiere, Broussolle, and Nguyen-The (2002) in 58%, 58%, and 68% of the isolates, respectively. The nonhemolytic enterotoxin NHE genes *nheA*, *nheB*, and *nheC* of the *B. cereus* isolates were detected by the PCR using the primers (Table 15.2) of Hansen and Hendriksen (2001) more frequently occurring in 71%, 84%, and 90% of the isolates, respectively. Among the 31 *B. cereus* isolates from diarrheal samples 15 (48%) possessed all six enterotoxin genes and four isolates possessed only a single enterotoxin gene (*hblD* or *nheC*). Among the three *B. cereus* isolates from healthy individuals, two possessed all six enterotoxin genes and one possessed only the *hblD* gene. The study suggested that *B. cereus* isolates harboring one or more enterotoxin genes can be a potential cause of diarrhea.

Among 433 honey samples collected in Argentina, López and Alippi (2007) found that 114 (27%) yielded *B. cereus* isolates and 60 (14%) yielded other species of *Bacillus*. Primers U1/U2 from Ash et al. (1993) were used for PCR amplification of the 16S rRNA and yielded a ~1100-bp amplicon from *Bacillus* species and closely related genera. Amplicons were restricted with *Hae*II, *Hinf*I, *Alu*I, *Cfo*I, *Taq*I, *Msp*I, or *Rsa*I for restriction fragment length polymorphism analysis. Each of these endonucleases generated only two RFLP patterns designated A and B for each enzyme. Repetitive DNA PCR (rep-PCRW) was performed with primers BOX, REP, and ERIC as described by Versalovic et al. (1994). A dendrogram showing phenotypic similarities among the isolates failed to correlate with a dendrogram obtained from rep-PCR analysis, indicating a high degree of phenotypic and genotypic diversity among the isolates.

Ngamwongsatit et al. (2008) designed eight new pairs of PCR primers (Table 15.2) for multiplex detection of eight toxin genes (*hblC*, *hblD*, *hblA*, *nheA nheB*, *nheeC*, *cytK*, and *entFM*) in strains of *B. cereus* (121 from food and 290 from soil) and *B. thuringiensis* (205). All isolates were allocated to one of four genotype groups. All eight genes were detected in group I. Group II lacked *hblCDA*, Group III lacked *cytK*. Group IV lacked both *hblCDA* and *cyt K*. Among the 121 food isolates of *B. cereus*, 60, 35, 7, and 19 were allocated to groups I, II, III, and IV, respectively. Among the 290 soil isolates of *B. cereus*, 194, 76, 10, and 10 and were allocated to groups I, II,

III, and IV, respectively. HbL, Nhe, and CytK are considered the primary virulence factors in *B. cereus* diarrhea.

Zhou et al. (2008) cultured a total of 54 samples of pasteurized full-fat milk packaged in cartons from chain supermarkets in Wuhan, China. Among 102 isolated *B. cereus*–like isolates, 92 were *B. thuringiensis*, and one was *B. mycoides*. The incidence of *B. cereus* in milk samples was 71.4% and 33.3% in spring and autumn samples, respectively, and the average count among the positive samples was 11.7 MPN/ml. PCR assays utilizing the primers of Hansen and Hendriksen (2001) in Table 15.2 revealed that the enterotoxin genes *hblA, hblC, hblD, nheA, nheB*, and *nheC* occurred in *B. cereus* isolates with frequencies of 37%, 66.3%, 71.1%, 71.7%, 62.0%, and 71.7%, respectively. Nine *B. thuringiensis* isolates were also identified from six pasteurized milk samples, and most of them harbored the six enterotoxic genes and the insecticidal toxin *cry1A* gene. A single *B. mycoides* isolate harbored the *nheA* and *nheC* genes.

Bartoszewicz, Hansen, and Swiecicka (2008) determined the level of milk contamination by *B. cereus*. There were 44 samples collected from a dairy farm and two independent dairies in northeastern Poland. A total of 680 *B. cereus* isolates was recovered. Based on spore counts, their highest level in milk was found during the spring and summer months. All strains were analyzed by the PCR for the presence of the *hblA, hblD, hblC, nheA, nheB*, and *nheC* genes using the primers of Hansen and Henrikson (2001; Table 15.2) in addition to the cereulide cluster gene (*nrps*) using the primers of Ehling-Schulz, Fricker, and Scherer (2004; Table 15.2) and the *cytdK* gene with the CytKF/CytKR primers (Table 15.2) of Swiecicka and Mahillon (2006). Among 227 randomly selected *B. cereus/B. weikenstaphanensis* isolates, 178 (78%) were positive for the *nheA* gene, 123 (54%) for the *hblA* gene, and 129 (57%) for the *cytK* gene. Similar results were obtained with 20 *B. thuringiensis* isolates with 17 (85%), 11 (55%), and 12 (60%) positive for *nehA, hblA*, and *cytK*, respectively. Among 25 *B. mycoides/B. pseudomycoides* isolates only 8 (30%) were positive for the *nheA*, 18 (70%) were positive for *hblA*, and none harbored the *cytK* gene. Among 272 isolates surveyed, only one *B. cereus/B. weikenstaphanensis* harbored the *ces* gene. Repetitive element sequence polymorphism PCR (rep-PCRS) was used for chromosomal comparison of strains. Isolates having identical rep-PCRS patterns were analyzed for their clonality by PFGE using *Sma*I. Isolates of *B. cereus/B. weikenstaphanensis* (155) showed high levels of genetic divergence with rep-PCRS and yielded 104 different rep-PCR profiles. A total of 29 isolates of *B. cereus/B. weikenstaphanensis* were genetically identical according to rep-PCR and also yielded identical PFGE patterns.

Ankoledar, Rahmati, and Labbe (2008) assessed 178 samples of raw rice from retail food sources for the presence of *B. cereus* spores. A total of 94 (52.8%) of the rice samples were positive for *Bacillus* species with an average of 32.6 CFU/g for *B. cereus*. Among the 94 isolates, 83 were identified as *B. cereus* and 11 were identified as *B. thuringiensis*. The presence of the gene *ces,* the genes for the HBL and NHE complexes, and the *cry* gene was assessed by the PCR using the primers (Table 15.2) of Ehling-Schulz et al. (2005a; 2006), Guinebretiere, Broussolle, and Nguyen-The (2002), and Kuo et al. (2007). The *ces* gene was not found in any of the isolates. By contrast, 47 (56.6%) of the *B. cereus* isolates harbored the *hblA* and *hblD* genes and 74 (89.1%) possessed the *nheA* and *nheB* genes. With the use of commercial assay kits 44 (53.0%) of the 83 *B. cereus* isolates produced both NHE and HBL enterototoxins. Protein toxin crystals were detected in all 11 *B. thuringiensis* isolates, and the *cry* gene was detected by PCRS in 10 of the isolates. None of the 94 total *Bacillus* isolates harbored the *ces* gene.

B. Multilocus Sequence Typing (MLST) of *B. cereus* Isolates

Ehling-Schulz et al. (2005a) subjected a total of 90 isolates of *B. cereus* of diverse geographical origins to RAPD typing with primers RAPD-1, RAPD-2, and PM13 (Table 15.1). In addition, the strains were also subjected to multilocus sequence typing (MLST) targeting seven housekeeping genes (*adk, ccpA, ftsA, glpT, pyre, recF,* and *succ*; Tables 15.1 and 15.2). The DNA of isolates was also subjected to 16S-23S rDNA spacer region 16S rDNA, and 23S RNA fragment sequence analysis. Primers are listed in Table 15.2. In addition, a total of 13 primer pairs (Table 15.2) was used for genotyping. M13 RAPD analysis of emetic isolates yielded two bands of 650 and 260 bp that were not present in nonemetic isolates. Sequencing of the seven housekeeping genes derived from emetic strains yielded the same sequence type. A total of 22 emetic isolates carried the *ces* gene for cereulide synthetase and produced the emetic toxin cereulide. None of the emetic isolates produced HBL, nor did any of them carry the *hbl* genes, whereas HBL production was frequently observed among diarrheal isolates.

From all typing methods used, the authors concluded that the emetic isolates of *B. cereus* constitute a distinct genotypic cluster that was unable to degrade starch or ferment salicin, did not possess the genes encoding hemolysin BL, and showed only weak or no hemolysis. In contrast, hemolytic-enterotoxin-producing strains showed a high degree of heterogeneity. These data were considered as evidence for a clonal population structure of cereulide-producing emetic *B. cereus* strains within a weakly clonal background population structure of the species and that the cereulide synthetase gene may have been recently acquired.

C. RAPD Typing

Guinebretiere and Nguyen-The (2003) typed 134 *B. cereus* strains from zucchini, milk, proteins, starch, and soil from which the zucchinis were grown using RAPD typing with primers M13 and And2 (Table 15.2). Sixteen combined patterns from the two RAPD primers resulted and indicated that the soil was the main source of contamination for the zucchinis.

D. Ribotyping

Pirttijärvi et al. (1999) characterized 64 isolates of the *B. cereus* group from industrial and environmental sources, 32 from food poisoning incidents, and 7 reference strains. Ribotyping using the restriction nucleases *Eco*RI and *Pvu*II yielded 40 ribotypes among the 93 isolates. Eleven strains from the 32 food-poisoning isolates produced the emetic toxin as determined by the boar spermatozoa assay. These emetic toxin–producing strains possessed closely similar ribotypes that were rare among strains of other origins. The emetic strains did not hydrolyze starch and did not produce hemolysin BL.

E. PCR Restriction Fragment Length Polymorphism (PCR-RFLP)

Jensen et al. (2005) examined the possibility of differentiating members of the *B. cereus* group utilizing PCR-RFLP. The study involved 12 *B. cereus*, 25 *B. thuringiensis*, 25 *B. mycoides*, and 2 *B. anthracis* pure cultures. Each strain was examined with the G1-F/G1-R primers (Table 15.2) of Jensen, Webster, and Straus (1993), yielding amplicons derived from the 16s-23S rRNA spacer region to confirm that each strain was a member of the *B. cereus* group. All of the isolates yielded the characteristic four-band pattern characteristic of the *B. cereus* group (Daffonchio et al., 1998). The primers BCFW1/BCRW1 (Table 15.2) were used to amplify a 326-bp sequence of the *gyrB* gene from each strain. The resulting amplicons were subsequently digested with Sau3A1. The results obtained suggested that only the *B. mycoides* isolates generated species-specific fragments following PCR-RED. The study, however, failed to distinguish strains of *B. cereus* from strains of *B. thuringiensis* using the *gyrB* restriction patterns. These results were consistent with those of Chen and Tsen (2002), who found that it was difficult to discriminate *B. cereus* from *B. thuringiensis* using the *gyrB*-based DNA method. However, Manzano et al. (2003) claimed that it was possible to differentiate among *B. cereus*, *B. thuringiensis*, and *B. mycoides* using the same PCR-RFLP methodology as Jensen et al. (2005).

F. Microarrays

Sergeev et al. (2006) developed a microarray analysis assay system for detection of 16 virulence genes associated in the *B. cereus* group. The method involved an initial mPCR followed by identification of the amplicons by hybridization to an oligonucleotide microarray containing the specific virulence genes. A total of 12 gene sequences for *B. cereus* and the *B. cereus* group were amplified in addition to four genes (*lef, pagA, cap*, and *cyaA*) specific for *B. anthracis* (Table 15.2). Four to ten DNA probes for each gene amplicon were immobilized onto coated glass slides. Single-stranded amplicons were labeled with the florescent dye Cy-5 and then hybridized to the immobilized probes followed by sequential washing with 6× SSC buffer containing 0.2% Tween 20, 2× SSC buffer, 1× SSC buffer, and air flow dried. The microarray assay system was applied to strains of *B. cereus*, *B. thuringiensis*, *B. subtilis*, *B. amyloliquefaciens*, and *B. anthracis*. No clear distinction was obtained with strains of the first three species. Strains of *B. amyloliquefaciens* lacked all 16 virulence genes. The microarray system failed to detect the presence of the *lef, pagA, cap*, and *cyaA* genes, which confirmed their specificity for *B. anthracis*.

Liu et al. (2007) developed an electrochemical microarray assay system for detection of seven known genes involved in diarrhea and the emetic *ces* gene using a 16-position electrical microarray chip. PCR primers were designed for generation of amplicons for each gene (Table 15.2) to confirm the presence of the genes in several positive control strains used to develop the methodology. The *bcet* gene was excluded because of its absence from the two strains used for the development studies. Capture probes for each amplicon were labeled with a thiol group at the 5'-end. Detection probes were labeled with biotin at the 3'-end. Capture probes were spotted onto all 16 microelectrode positions using a piezo nanodisperser. Denatured target DNA obtained from sonication of cell suspensions was applied to the immobilized capture probes in the presence of detection probes. Avidin alkaline phosphatase conjugate (EXT-ALP) was then added to label the detection probe. After washing, an electrochemical inactive enzyme substrate 4-amino phenyl phosphate (pAPP) was added. The enzyme ALP converts the substrate pAPP to the electrochemical active product pAP that is oxidized to quinoneimine (QI) at the anode. QI is then reduced back to pAP at the cathode resulting in a redox recycling driven current in the nanoampere range followed by a portable computer and appropriate software program. The entire process is presented in Figure 15.1.

The electrochemical microarray assay required 30 min. However, a problem encountered with the microarray assay system involved the application of crude cell sonicates derived from 5×10^8 CFU, which failed to generate a signal, even after removal of

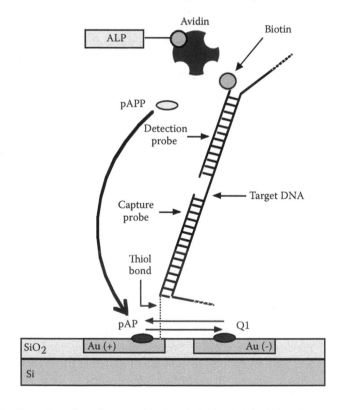

Figure 15.1   Operation of a microarray local and electrochemical signal generation at electrodes. (Redrawn from Liu et al., 2007. With permission from Elsevier.)

cellular debris by centrifugation. It was assumed that chemical contaminants in crude sonicates were trapped and deposited either on the probes, hampering hybridization, or the electrode surface hindering redox cycling. The final optimized protocol for sample preparation involved sonication for 5 min to rupture the cells and release DNA, extraction four times with phenol:chloroform:isoamyl alcohol (25:24:1), and sonication for an additional 10 min to reduce the size of DNA segments for optimum hybridization.

Aragon-Alegro et al. (2008) assessed 155 *B. cereus* strains from Brazilian food production for the production of HBL and NHE via immunoassays. Strains were also assessed for the presence of the HBL and NHE gene complexes, *ctdK, cytK-1, cytdK-2*, and *ces*, using the PCR. The *hbl* and *nhe* primers (Table 15.2) were from Guinebretiere et al. (2002). *ces* primers (Table 15.2) were from Ehling-Schulz et al. (2005b). Primers for *ctyK* (Table 15.2) were from Guinebretiere et al. (2006). HBL was detected in 105 (67.7%) of the strains and NHE in 154 (99.4%) of the strains. All the strains harbored at least one gene of the NHE complex. All strains exhibited toxigenic capacity and were considered to represent a potential risk for consumers.

REFERENCES

Agata, N., Ohta, M., Arakawa, Y., Mori, M. 1995. The *bceT* gene of *Bacillus cereus* encodes an enterotoxic protein. *Microbiology* 141:983–988.

Agata, N., Ohta, M., Mori, M. 1996. Production of an emetic toxin, cereulide, is associated with a specific class of *Bacillus cereus. Curr. Microbiol.* 33:67–69.

Al-Khatib, M., Khyami-Horani, H., Badran, E., Shehabi, A. 2007. Incidence and characterization of diarrheal enterotoxins of fecal *Bacillus cereus* isolates associated with diarrhea. *Diag. Microbiol. Infect. Dis.* 59:383–387.

Altayar, M., Sutherland, A. 2006. *Bacillus cereus* is common in the environment but emetic toxin producing isolates are rare. *J. Appl. Microbiol.* 100:7–14.

Andersson, M., Jääskeläinen, E., Shahen, R., Pirhonen, T., Wijnands, L., Salkinoja-Salonen, M. 2004. Sperm bioassay for rapid detection of cereulide producing *Bacillus cereus* in food and related environments. *Int. J. Food Microbiol.* 94:175–183.

Ankoledar, C., Rahmati,T., Labbe, R. 2008. Detection of toxigenic *Bacillus cereus* and *Bacillus thuringiensis* spores in U.S. rice. *Int. J. Food Microbiol.* 128:460–466.

Aragon-Alegro, L., Palcich, G., Lopes, G., Ribiero, V., Landgdraf, M., Destro, M. 2008. Enterotoxigenic and genetic profiles of *Bacillus cereus* strains of food origin in Brazil. *J. Food Prot.* 71:2115–2118.

Arnesen, L., Fagerlund, A., Granum, P. 2008. From soil to gut: *Bacillus cereus* and its food poisoning toxins. *FEMS Microbiol. Rev.* 32:579–606.

Asano, S., Nukumizu, Y., Bando, H., Hzuka, T., Yamamoto, T. 1997. Cloning of novel enterotoxin genes from *Bacillus cereus* and *Bacillus thuringiensis. Appl. Environ Microbiol.* 63:1054–1057.

Ash, C., Farrow, J., Dorsch, M., Stackebrandt, E., Collins, M. 1991a. Comparative analysis of *Bacillus anthracis, Bacillus cereus*, and related species on the basis of reverse transcriptase sequencing of 16S rRNA. *Int. J. Syst. Bacteriol.* 41:343–346.

Ash, C., Farrow, J., Dorsch, M., Wallbanks, S., Collins, M. 1991b. Phylogenetic heterogeneity of the genus *Bacillus* revealed by comparative analysis of small-subunit-ribosomal RNA sequences. *Lett. Appl. Microbol.* 13:202–206.

Ash, C., Priesdt, F., Collins, M. 1993. Molecular identification of rRNA group 3 bacilli (Ash, Farrow, Wallbanks and Collins) using a PCR probe test. Proposal for the creation of a new genus *Paenibacilli. Ant. Van Leeuwen. J. Microbiol.* 64:253–260.

Bartoszewicz, M., Hansen, B., Swiecicka, I. 2008. The members of the *Bacillus cereus* group are commonly present contaminants of fresh and heat-treated milk. *Food Microbiol.* 25:588–596.

Beecher, D., Schoeni, J., Wong, A. 1995. Enterotoxin activity of hemolysin from *Bacillus cereus. Immunol.* 63:4423–4428.

Beecher, D., Wong, A. 1994. Improved purification and characterization of hemolysin BL, a hemolytic dermonecrotic vascular permeability factor from *Bacillus cereus. Infect. Immun.* 62:980–986.

Carlin, F., Fricker, M., Pielaat, A., Heisterkamp, S., Shaheen, R., Salonen, M., Svensson, B., Nguyn-The, C., Ehling-Schulz, M. 2006. Emetic toxin-producing strains of *Bacillus cereus* show distinct characteristics within the *Bacillus cereus* group. *Int. J. Food Microbiol.* 109:132–138.

Carlson, C., Caugent, D., Kolsto, A. 1994. Genotypic diversity among *Bacillus cereus* and *Bacillus thuringiensis* strains. *Appl. Environ. Microbiol.* 60:1719–1725.

Carlson, C., Kolsto, A. 1993. A complete physical map of a *Bacillus thuringiensis* chromosome. *J. Bacteriol.* 175:1053–1060.

Chen, M., Tsen, H. 2002. Discrimination of *Bacillus cereus* and *Bacillus thuringiensis* with 16S rRNA and *gyrB* gene-based PCR primers and sequencing of their annealing sites. *J. Appl. Microbiol.* 92:912–919.

Choo, E., Jang, S., Kim, K., Lee, K., Heu, S., Syu, S. 2007. Prevalence and genetic diversity of *Bacillus cereus* in dried red pepper in Korea. *J. Food Prot.* 70:917–922.

Daffonchio, D., Borin, S., Consolandi, A., Mora, D., Manachini, P., Sorlini, C. 1998. 16S-23S rRNA internal transcribed spacers as molecular markers for the species of the 16S rRNA group I of the genus *Bacillus. FEMS Microbiol. Lett.* 163:229–239.

Drobniewski, F. 1993. *Bacillus cereus* and related species. *Clin. Microbiol. Rev.* 6:324–338.

Ehling-Schulz, M., Fricker, M., Scherer, S. 2004. Identification of emetic toxin producing *Bacillus cereus* strains by a novel molecular assay. *FEMS Microbiol. Lett.* 232:189–195.

Ehling-Schulz, M., Guinebretiere, M., Monthan, A., Berge, O., Fricker, M., Svensson, B. 2006. Toxin profiling of enterotoxic and emetic *Bacillus cereus. FEMS Microbiol. Lett.* 260:232–240.

Ehling-Schulz, M., Svensson, B., Guinebretiere, M., Lindback, T., Andersson, M., Schulz, A., Fricker, M., Christianson, A., Granum, P., Martlbauer, E., Nguyen-The, C., Salkinoja-Salonen, M., Scherer, S. 2005a. Emetic toxin formation of *Bacillus cereus* is restricted to a single evolutionary lineage of closely related strains. *Microbiology* 151:183–197.

Ehling-Schulz, M., Vukov, M., Schulz, N., Shaheen, A., Andersson, M., Märtibauer, E., Scherer, S. 2005b. Identification and partial characterization of the nonribosomal peptide synthase gene responsible for cereulide production in emetic *Bacillus cereus. Appl. Environ. Microbiol.* 71:105–113.

Fagerlund, A., Lindbäck, T., Storset, A., Granum, P., Hardy S. 2008. *Bacillus cereus* NHe is a pore-forming toxin with structural and functional properties similar to the ClyA (Hlye, Shea) family of haemolysins, able to induce osmotic lysis in epithelia. *Microbiology* 154:693–704.

Francis, K., Mayrs, R., von Stetten, F., Stewart, G., Scherer, S. 1998. Discrimination of psychrotrophic and mesophilic strains of the *Bacillus cereus* group by PCR targeting of major cold shock protein genes *Appl. Microbiol.* 64:3525–3529.

Fricker, M., Messelhäuber, M., Busch, U., Scherer, S., Ehling-Schulz, M. 2007. Diagnostic real-time PCR assays for the detection of emetic *Bacillus cereus* strains in foods and recent food-bone outbreaks. *Appl. Environ. Microbiol.* 73:1892–1898.

Fricker, M., Reissbrodt, R., Ehling-Shulz, M. 2008. Evaluation of standard and new chromogenic selective plating media for isolation and identification of *Bacillus cereus*. *Int. J. Food Microbiol.* 121:27–34.

Granum, P., O'Sullivan, K., Lund, T. 1999. The sequence of the non-hemolytic enterotoxin operon from *Bacillus cereus*. *FEMS Microbiol. Lett.* 177:225–229.

Guinebretiere, M., Broussolle, V., Nguyen-The, C. 2002. Enterotoxigenic profiles of food-poisoning and food-borne *Bacillus cereus* strains. *J. Clin. Microbiol.* 40:3053–3056.

Guinebretiere, M., Fagerlund, H., Granum, P., Nguyen-The, C. 2006. Rapid discrimination of *cytK-1* and *cytK-2* genes in *Bacillus cereus* strains by a novel duplex PCR system. *FEMS Microbiol. Lett.* 259:74–80.

Guinebretiere, M., Nguyen-The, C. 2003. Sources of *Bacillus cereus* contamination in a pasteurized zucchini purée processing line, differentiated by two PCR-based methods. *FEMS Microbiol. Ecol.* 43:207–215.

Gürtlier, V., Stanisich, V. 1996. New approaches to the typing and identification of bacteria using the 16S-23S rDNA spacer region. *Microbiology* 142:3–16.

Hansen, B., Hendriksen, N. 2001. Detection of enterotoxic *Bacillus cereus* and *Bacillus thuringiensis* strains by PCR analysis. *Appl. Environ. Microbiol.* 67:185–189.

Hansen, M., Leser, T., Hendrikson, N. 2001. Polymerase chain reaction assay for the detection of *Bacillus cereus* group cells. *FEMS Microbiol. Lett.* 202:209–213.

Haque, A., Russell, N. 2005. Phenotypic and genotypic characterization of *Bacillus cereus* isolates from Bangladeshi rice. *Int. J. Food Microbiol.* 98:23–34.

Heinrichs, J., Beecher, D., Macmillan, J., Zilinskas, B. 1993. Molecular cloning and characterization of the *hbla* gene encoding the B component of hemolysin BL from *Bacillus cereus*. *J. Bacteriol.* 175:6760–6766.

Helgason, E., Caugant, D., Lecadet, M., Chen, Y., Mahillon, J., Lovgren, A., Hegna, I., Kvaloy, K., Kolsto, A. 1998. Genetic diversity *of Bacillus cereus/B. thuringiensis* isolates from natural sources. *Curr. Microbiol.* 37:80–87.

Helgason, E., Tourasse, N., Meisal, R., Caugant, D., Kolsto, A. 2004. Multilocus sequence typing scheme for bacteria of the *Bacillus cereus* group. *Appl. Environ. Microbiol.* 70:191–201.

Hoton, F., Andrup, L., Swiecicka, I., Mailon, J. 2005. The cereulide genetic determinants of emetic *Bacillus cereus* are plasmid-bone. *Microbiol. Comment.* 151:2121–2124.

Hsieh, Y., Sheu, J., Chen, Y., Tsenk, H. 1999. Enterotoxigenic profiles and polymerase chain reaction detection of *Bacillus cereus* group cells and *B. cereus* strains from foods and food-borne outbreaks. *J. Appl. Microbiol.* 87:481–490.

In't Veld, P., Ritmeester, W., Asch, E., Dufrenne, J., Wernars, K., Smit, E., van Leusden, F. 2001. Detection of genes encoding for enterotoxins and determination of the production of enterotoxins by HBL blood plates and immunoassays of psychrotrophic strains of *Bacillus cereus* isolated from pasteurized milk. *Int. J. Food Microbiol.* 64:63–70.

Jensen, G., Fisker, N., Sparso, T., Andrup, L. 2005. The possibility of discriminating within the *Bacillus cereus* group using *gyrB* sequencing and PCR-RFLP. *Int. J. Food Microbiol.* 104:113–120.

Jensen, M., Webster, J., Straus, N. 1993. Rapid identification of bacteria on the basis of polymerase chain reaction-amplified ribosomal DNA spacer polymorphisms. *Appl. Environ. Microbiol.* 59:945–952.

Kim, Y., Czajka, J., Batt, C. 2000. Development of a fluorogenic probe-based PCR assay for detection of *Bacillus cereus* in nonfat dry milk. *Appl. Environ. Microbiol.* 66:1453–1459.

Kuo, C., Sung, S., Kim, K., Lee, K., Heu, S., Ryuk, S. 2007. Prevalence and genetic diversity of *Bacillus cereus* in dried red pepper in Korea. *J. Food. Prot.* 70:917–922.

Kuo, W., Chak, K. 1996. Identification of novel *cry*-type genes from *Bacillus thuringiensis* strains on the basis of restriction fragment length polymorphism of the PCR-amplified DNA. *Appl. Environ. Microbiol.* 62:1369–1377.

Liu, Y., Elsholz, B., Enfors, S., Gabig-Ciminski, M. 2007. Confirmative electric DNA array-based test for food poisoning *Bacillus cereus*. *J. Microbiol. Methods* 70:55–64.

López, A., Alippi, A. 2007. Phenotypic and genotypic diversity of *Bacillus cereus* isolates recovered from honey. *Int. J. Food Microbiol.* 117:175–184.

Mäntynen, V., Lindström, K. 1998. A rapid PCR-based DNA test for enterotoxic *Bacillus cereus*. *Appl. Environ. Microbiol.* 64:1634–1639.

Manzano, M., Cocolin, L., Canatoni, C., Comi, G. 2003a. *Bacillus cereus Bacillus thuringiensis* and *Bacillus mycoides* differentiation using a PCRS-RE technique. *Int. J. Food Microbiol.* 81:249–254.

Manzano, M., Giusto, C., Iacumin, L., Cantoni, C., Comi, G. 2003b. A molecular method to detect *Bacillus cereus* from a coffee concentrate sample used in industrial preparations. *J. Appl. Microbiol.* 95:1361–1366.

McIntyre, L., Bernard, K., Beniac, D., Isaac-Renton, J., Naseby, D. 2008. Identification of *Bacillus cereus* group species associated with food poisoning outbreaks in British Columbia Canada. *Appl. Environ. Microbiol.* 74:7451–7453.

Nakano, S., Maeshima, H., Matsumura, A., Ohno, K., Ueda, S., Kuwabara, Y., Yamada, T. 2004. A PCR assay based on a sequence-characterized amplified region marker for detection of emetic *Bacillus cereus*. *J. Food Prot.* 67:1694–1701.

Ngamwongsatit, P., Buasri, W., Pianariyanon, P., Pulsrikam, C., Ohba, M., Assavanig, A., Panbangred, W. 2008. Broad distribution of enterotoxin genes (*hblCZDA, bheABC, cytdK,* and *entFM*). *Int. J. Food Microbiol.* 121:352–356.

Pirttijärvi, T., Andersson, M., Scoging, A., Salkinoja-Salonen, M. 1999. Evaluation of methods for recognizing strains of the *Bacillus cereus* group with food poisoning potential among industrial and environmental contaminants. *Syst. Appl. Microbiol.* 22:133–144.

Prüb, B., Dietrich, R., Nibler, B., Märtlbauer, E., Scherer, S. 1999. The hemolytic enterotoxin HBL is broadly distributed among species of the *Bacillus cereus* group. *Appl. Environ. Microbiol.* 65:5436–5442.

Raevuori, M., Kiutamo, T., Kiskanen, A. 1977. Comparative studies of *Bacillus cereus* strains isolated from various foods and food poisoning outbreaks. *Acta Vet. Scand.* 18:397–407.

Rupf, S., Merte, K., Eschrich, K. 1999. Quantification of bacteria in oral samples by competitive polymerase chain reaction. *J. Dent. Res.* 78:850–856.

Ryan, P., Macmillan, J., Zilinskas, B. 1997. Molecular cloning and characterization of the genes encoding the L_1 and L_2 components of hemolysin BL from *Bacillus cereus*. *J. Bacteriol.* 179:2551–2556.

Schraft, H., Griffiths, M. 1995. Specific oligonucleotide primers for detection of lecithinase positive *Bacillus* spp. by PCR. *Appl. Environ. Microbiol.* 61:98–102.

Sergeev, N., Distdler, M., Vargas, M., Chizhikov, V., Herold, K., Rasooly, A. 2006. Microarray analysis of *Bacillus cereus* group virulence factors. *J. Microbiol. Methods* 65:488–502.

Shinagawa, K. 1993. Serology and characterization of toxigenic *Bacillus cereus*. *Neth. Milk Dairy J.* 47:89–103.

Somerville, H., Jones, M. 1972. DNA competition studies within the *Bacillus cereus* group of Bacilli. *J. Gen. Microbiol.* 73:257–265.

Stackebrandt, E., Liesack, W. 1992. The potential of rDNA in identification and diagnostics. In: *Nonradioactive Labeling and Detection of Biomolecules.* C. Kessler, Ed., Springer: Berlin, pp. 232–239.

Stenfors, L., Granum, P. 2001. Psychrotolerant species from the *Bacillus cereus* group are not necessarily *Bacillus weihenstephanensi. FEMS Microbiol. Lett.* 197:223–228.

Stenfors, L., Mayr, R., Scherer, S., Granum, P. 2002. Pathogenic potential of fifty *Bacillus weihenstephanensis* strains. *FEMS Microbiol. Lett.* 215:47–51.

Svensson, B., Monthan, A., Guinebretiere, M., Nguyen-The, C., Christiansson, A. 2007. Toxin production potential and the detection of toxin genes among strains of the *Bacillus cereus* group isolated along the dairy production chain. *Int. Dairy J.* 17:1201–1208.

Svensson, B., Monthan, A., Shaheen, R., Andersson, M., Salkinoja-Salonen, M., Christianson, A. 2006. Occurrence of emetic toxin producing *Bacillus cereus* in the dairy production chain. *Int. Dairy J.* 16:740–749.

Swiecicka, I., Mahillon, J. 2006. Diversity of commensal *Bacillus cereus sensu lato* isolated from the common sow bug (*Porcellio scaber*, Iopoda). *FEMS Microbiol. Ecol.* 56:132–140.

Thaenthanee, S., Wong, A., Panbangred, W. 2005. Phenotypic and genotypic comparisons reveal a broad distribution and heterogeneity of hemolysin BL genes among *Bacillus cereus* isolates. *Int. J. Food Microbiol.* 105:203–212.

Toh, M., Moffitt, M., Henrichsen, L., Raftery, M., Barrow, K., Cox, J., Marquis, C., Neilan, B. 2004. Cereulide, the emetic toxin of *Bacillus cereus*, is putatively a product of nonribosomal peptide synthesis. *J. Appl. Microbiol.* 97:992–1000.

Van Netten, P., Moosdijk, A., Hoensel, P., Mossel, D., Perales, I. 1990. Psychrotrophic strains of *Bacillus cereus* producing enterotoxin. *J. Appl. Bacteriol.* 69:73–79.

Versalovic, J., Schneider, M., De Bruijn, F., Lupiki, J. 1994. Genomic fingerprinting of bacteria using repetitive sequence-based polymerase chain reaction. *Meth. Mol. Cell. Biol.* 5:25–40.

Yamada, S., Ohashi, E., Agata, N., Venkateswaran, K. 1999. Cloning and nucleotide sequence analysis of *gyrB* of *Bacillus cereus, B. thuringiensis, B. mycoides*, and *B. anthracis* and their application to the detection of *B. cereus* in rice. *Appl. Environ. Microbiol.* 65:1483–1490.

Yang, I., Shih, D., Huang, T., Huang, Y., Wang, J., Pan, T. 2005. Establishment of a novel multiplex PCR assay and detection of toxigenic strains of the species in the *Bacillus cereus* group. *J. Food Prot.* 68:2123–2130.

Zhou, G., Liu, H., He, J., Yan, Y., Yan, Z. 2008. The occurrence of *Bacillus cereus, B. thuringiensis* and *B. mycoides* in Chinese pasteurized full fat milk. *Int. J. Food Microbiol.* 121:195–200.

Index